PROCEEDINGS OF THE 5TH INTERNATIONAL CONFERENCE ON MACHINERY, MATERIALS SCIENCE AND ENGINEERING APPLICATIONS (MMSE 2015), WUHAN, HUBEI, CHINA, 27–28 JUNE 2015

T0173947

Advances in Engineering Materials and Applied Mechanics

Editors

Guangde Zhang & Quanjie Gao
Wuhan University of Science and Technology, Wuhan, China

Qiang Xu
School of Computing and Engineering, Huddersfield University, UK

CRC Press
Taylor & Francis Group
Boca Raton London New York

CRC Press is an imprint of the
Taylor & Francis Group, an **informa** business

A BALKEMA BOOK

Published by: CRC Press/Balkema
P.O. Box 11320, 2301 EH Leiden, The Netherlands
e-mail: Pub.NL@taylorandfrancis.com
www.crcpress.com – www.taylorandfrancis.com

First issued in paperback 2020

ISBN 13: 978-0-367-73772-6 (pbk)
ISBN 13: 978-1-138-02834-0 (hbk)

Visit the Taylor & Francis Web site at
http://www.taylorandfrancis.com

and the CRC Press Web site at
http://www.crcpress.com

Typeset by V Publishing Solutions Pvt Ltd., Chennai, India

Table of contents

Applied mechanics

Automation

Automation, control, system modeling and simulation

Dispose of communication signals

Electronic design engineering

Machinery

Preface

Nowadays, more and more experts and scholars are paying their attention on the topics of Advanced Engineering Materials, Advanced Manufacturing and Automation Technology, and Applied Mechanics and Aerospace Science and Technology. It is very important to combine these topics together from the angle of theory and practice in the field of industrial engineering. The 2015 5th International Conference on Machinery, Materials Science and Engineering Applications (MMSE 2015) provides a platform for scientists and engineering managers to exchange their thoughts and co-operation. The contributions discuss the latest achievements in Advanced Engineering Materials, Advanced Manufacturing and Automation Technology, Applied Mechanics and Aerospace Science and Technology and others related topics.

The 2015 5th International Conference on Machinery, Materials Science and Engineering Applications (MMSE 2015) is hold by Wuhan University of Science and Technology, co-Sponsored by Hubei University; University of Huddersfield, UK; University of Teesside, UK; University of Nottingham, UK; Worcester Polytechnic Institute, Dong-Eui University, Korea; China University of Geosciences; Hubei Mechanical Engineering Society; and Chinese Mechanical Engineering Society.

About 120 representatives from domestic and overseas countries attended this international conference. Here, I would like to express my sincere gratitude to those Leaders, Colleges and Universities, Research Institutes, Companies and all representatives, and appreciate their great support on MMSE 2015.

MMSE 2015
June 27, 2015

Advances in Engineering Materials and Applied Mechanics – Zhang, Gao & Xu (Eds)
© 2016 Taylor & Francis Group, London, ISBN 978-1-138-02834-0

Sponsors

SPONSORED BY

Wuhan University of Science and Technology, China

CO-SPONSORED BY

Hubei University, China
University of Tesside, UK
University of Huddersfield, UK
University of Nottingham, UK
Hubei Mechanical Engineering Society, China

Conference committees

GENERAL CHAIR

Dr. Guangde Zhang, *Wuhan University of Science and Technology, China*
Dr. Jianmin Xiong, *Hubei University, China*

CONFERENCE VICE CHAIR

Dr. Quanjie Gao, *Wuhan University of Science and Technology, China*

HONORARY CHAIR

Dr. Yimin Rong, *Worcester Polytechnic Institute, USA*

CONFERENCE PROCEEDING EDITOR

Dr. Q. Xu, *Huddersfield University, UK*
Dr. Guangde Zhang, *Wuhan University of Science and Technology, China*

PROGRAM CHAIR

Dr. Wancheng Chen, *Hubei Mechanical Engineering Society, China*
Dr. Anping Liu, *China University of Geosciences, China*

INTERNATIONAL SCIENTIFIC COMMITTEE

Dr. Y.M. Rong, *Worcester Polytechnic Institute, USA*
Dr. Q. Xu, *Huddersfield University, UK*
Dr. Zhongyu Lu, *School of Computing and Engineering, Huddersfield University, UK*
Dr. Zhiming Zhang, *Wayne State University, USA*
Dr. Urszula Forys, *University of Warsaw, Poland*
Dr. Aihua Xia, *The University of Melbourne, Australia*
Dr. Chih-Hsing Chu, *National Tsing Hua University, Taiwan*
Prof. Nisar Ahmed Memon, *Mehran University of Engineering and Technology, Pakistan*
Dr. Manzoor Iqbal Khatak, *University of Balochistan, Pakistan*
Prof. Gang Feng, *City University of Hong Kong, China*
Prof. Xiaoping Li, *National University of Singapore, Singapore*
Prof. George Q. Huang, *University of Hong Kong, China*
Dr. Wong Wai-on, *Hong Kong Polytechnic University, China*

LOCAL PROGRAMME COMMITTEE

Dr. Guangde Zhang, *Wuhan University of Science and Technology, China*
Dr. Jianmin Xiong, *Hubei University, China*
Dr. Quanjie Gao, *Wuhan University of Science and Technology, China*
Dr. Guangde Zhang, *Wuhan University of Science and Technology, China*
Dr. Wancheng Chen, *Hubei Mechanical Engineering Society, China*
Dr. Anping Liu, *China University of Geosciences, China*
Dr. Huamin Zhou, *Huazhong University of Science and Technology, China*
Dr. Huamin Jin, *Wuhan University, China*
Prof. Jixiang Li, *Wuhan Polytechnic University, China*
Dr. Mucai Ye, *China University of Geosciences, China*
Prof. Shuiming He, *China University of Geosciences, China*
Dr. Xiaojin Wang, *Wuhan University, China*
Prof. Hanbao Li, *Wuhan University, China*
Prof. Shengwu Xiong, *Wuhan University of Technology, China*
Prof. Jie Yang, *Wuhan University of Technology, China*

Advanced engineering materials

Research on precipitation of Ti-containing compound in ER 50-G wire production

T. Chen, H. Li & K. Wu
School of Metallurgical and Ecological Engineering, USTB, Beijing, China

Q. Tong, N.B. Lv & M. Yi
Research Institute of Technology, Shougang General Corporation, Beijing, China

ABSTRACT: Characters of various Ti-containing compounds precipitated in ER50-G wire were studied in this research by using TEM, SEM, etc. testing methods. The curves of precipitation amount against temperature were deduced basing on the results of thermo dynamical calculations and phase analysis. The measures to control the precipitation behavior were raised and applied to production. Finally a ER50-G wire with good drawing performance was obtained.

1 INTRODUCTION

Titanium micro alloying technology has been widely used combining TMCP technology in steel production in recent years. The mechanical properties of steel, particularly the strength, is improved by controlling the precipitation of Ti-containing compounds[1-3]. But for those steel type for making welding electrodes, suitable strength and excellent ductility are required to be fit for the large amount of deformation during cold drawing. So for ER50-G wire, a new type of welding steel in which the range of titanium content is 0.12–0.25%, it is necessary to study the precipitation behaviors of Ti-containing compounds during the production process and find ways and means to weaken the strengthening effect of titanium micro alloying.

2 EXPERIMENTAL METHOD

ER50-G wire samples used in this research were obtained randomly from Shougang high-speed wire rod plant. The specimens were sectioned, mounted and polished following standard metallographic procedures, and then the morphology of precipitates in the specimens were observed by Transmission Electron Microscopy (TEM) and Scanning Electron Microscope (SEM). The specimens used in TEM test were obtained by carbon extraction replica technique. During the examinations, the compositions of precipitates were analyzed by Energy Dispersive Spectroscopy (EDS) equipped on the facilities. The sizes and amounts of each kind of precipitates were also measured. The qualitative and quantitative detection of precipitates were obtained by phase analysis and X-ray diffraction test. The solidus, liquidus and other critical temperatures were measured with a simultaneous thermal analyzer. Thus the phase features of the matrix could be determined when the Ti-containing compounds were precipitating.

3 RESULTS AND DISCUSSION

3.1 *Precipitates of ER50-G wire*

Table 1 shows the statistical results of main chemical components of samples in this study.

Table 1.   Statistical results of main chemical components of ER50-G samples, wt%.

	C	Si	Mn	P	S	Ti	T.O	N
Range	0.042–0.053	0.76–0.84	1.47–1.50	0.012–0.013	0.012–0.015	0.16–0.19	0.0010–0.0016	0.0052–0.0080
Mean	0.048	0.81	1.48	0.012	0.014	0.18	0.0012	0.0061

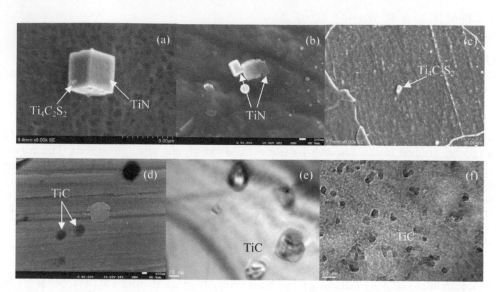

Figure 1.   Morphology of precipitates in ER50-G wire specimens observed by TEM and SEM.

The main precipitates observed in the specimens are Ti-containing compounds, like TiN, TiC etc., as shown in Figure 1, since there is such a high content of titanium in the steel.

TiN particles observed are rectangular, as shown in Figure 1a and b. Size of the most TiN particles is several microns, while a small amount of TiN are in nano-scale. The quantity of TiC particles is the largest, and the size ranges from several nanometers to hundreds of nanometers, as shown in Figure 1d, e and f. TiC particles have irregular shape, e.g. spherical or rod-like. The size of $Ti_4C_2S_2$ particles is between that of TiN and TiC, and most are hundreds of nanometers, while a few are about 1 μm. Some $Ti_4C_2S_2$ adhere to the TiN particles, as shown in Figure 1a.

The mass of the precipitates in ER50-G wire was tested by phase analysis method and the result was listed in Table 2.

The main components of the sample tested in the phase analysis experiment is: C 0.046, Si 0.77, Mn 1.48, P 0.012, S 0.012, Al 0.006, Ti 0.164, O 0.0016, N 0.0080. The results demonstrate that the major compounds precipitated in ER50-G wire are Ti-containing compounds, among which TiC has the largest amount and its proportion accounts for more than 70% of the total precipitates. The proportion of $Ti_4C_2S_2$, which ranks second, accounts for about 21%, and that of TiN is about 8%. However the total mass of the oxides is about 0.00167% and only accounts for about 1%. The content of Ti in the precipitates is about 0.142%. It means that more than 85% of the total Ti content is present as compounds, while less than 15% is present as solid solutions.

3.2   Solidus, liquidus and critical temperature of ER50-G steel

The solidus, liquidus and critical temperatures like Ar_1 and Ar_3 of ER50-G steel were detected by differential scanning calorimetry, in which the rate of changing temperature was 10 K/min. The results were listed in Table 3.

Table 2. Mass of precipitates in ER50-G wire, wt%.

Precipitate	TiC	TiN	$Ti_4C_2S_2$	Ti_3O_5	Other oxides (Al_2O_3, Cr_2O_3, etc.)
Mass	0.1298	0.0153	0.0384	0.00036	0.00131

Table 3. Solidus, liquidus and critical temperatures of ER50-G steel, K.

Phase transition	Liquidus	Solidus	$\delta \rightarrow \gamma$	Ar3	Ar1
Temperature	1795.1	1743.8	1573.3	1163.7	950.9

The results indicates that the liquidus, solidus, transformation finishing temperature from ferrite to austenite, Ar_3 and Ar_1 of ER50-G steel are approximate 1795.1 K, 1743.8 K, 1573.3 K, 1163.7 K and 950.9 K, respectively.

3.3   Thermodynamic analysis of the Ti-containing compounds' precipitation

Titanium is considerably stable at room temperature, however its activity rapidly increases at a high temperature. It is inclined to react with oxygen, nitrogen, carbon and sulfur. The main thermodynamical equations suitable for ER50-G steel are as follows[2,3,4,5].

$$[Ti] + [N] = TiN(s) \quad \lg K = \frac{17040}{T} - 6.4 \tag{1}$$

$$3[Ti] + 5[O] = Ti_3O_5(s) \quad \lg K = \frac{72813}{T} - 21.32 \tag{2}$$

$$[Ti] + [C] = TiC(s) \quad \lg K = \frac{7000}{T} - 2.75 \text{ (in austenite)} \tag{3}$$

$$[Ti] + [C] = TiC(s) \quad \lg K = \frac{9575}{T} - 4.40 \text{ (in ferrite)} \tag{4}$$

$$4[Ti] + 2[C] + 2[S] = Ti_4C_2S_2(s) \quad \lg K = \frac{17045}{T} - 7.9 \tag{5}$$

Once the product of titanium and carbon/nitrogen exceeds the critical value under a certain temperature, the solid TiN or TiC will form and precipitate. The equilibrium precipitation temperature (abbr. EPT) of TiN, TiC and other particles could be obtained by above five equations. For instance, the equilibrium constant of equation 1, which reflects the reaction of the formation of TiN in liquid steel, is as follows.

$$K = \frac{a_{TiN(s)}}{a_{[Ti]}a_{[N]}} = \frac{1}{f_{Ti}[\%Ti]f_N[\%N]} \tag{6}$$

wherein, K is the equilibrium constant. $a_{TiN(s)}$, $a_{[Ti]}$ and $a_{[N]}$ are the activities of solid TiN that generated in molten steel, the dissolved titanium and nitrogen, respectively. [%Ti] and [%N] are the corresponding concentration mass percentage. f_{Ti} and f_N is the corresponding activity coefficient, and they could be worked out by using the interaction coefficient provided from reference[6–8]. The following equation 7 could be deduced by equation 1 and 6.

$$\lg\{[\%Ti][\%N]\} = -\frac{17040}{T} + 6.456 \tag{7}$$

5

Thus EPT of TiN under the experimental conditions could be calculated with equation 7. Similarly, EPT of Ti_3O_5, $Ti_4C_2S_2$ and TiC might also be conducted by using equation 2, 3 and 5. It was assumed that the precipitation of TiC had started in austenite and then proved by the calculation. The results were shown in Table 4.

According to the thermodynamic calculation, Ti priors to react with nitrogen instead of oxygen in the steel. Furthermore there was about 0.006% Al in ER50-G, thus it's difficult for Ti to react with oxygen[9]. Actually the mass of titanium oxides was so little in ER50-G wire according to the phase analysis result that it could be neglected when studying precipitation. By comparing the critical temperatures of ER50-G in Table 3 with its EPT in Table 4, it could be found that EPT of all Ti-containing compounds in ER50-G were higher than Ar_3, and the precipitation sequence was successively TiN, $Ti_4C_2S_2$ and TiC. Among all precipitates, TiN particles formed first in liquid phase and grew up gradually in subsequent process, and thus the maximum size of TiN particles could be several micrometers. But for $Ti_4C_2S_2$ and TiC, both precipitated in solid phase. TiC is the smallest particle since its precipitation temperature is the lowest. Previous study[1] has found that TiC could precipitate in austenite, ferrite phase and the phase transition process. TiC particles that precipitated in austenite was relatively larger, while the other two types were mostly smaller than 10 nm.

The theoretical mass of precipitated TiN, $Ti_4C_2S_2$ and TiC in ER50-G at different temperature were calculated by the equation 1, 3 and 5. The calculation results were drawn as solid lines in Figure 2. While in practice, temperature and other conditions in the production process is changing constantly. When temperature decreases, the diffusion coefficient decreases, while the precipitation driving force is increased, and thus the actual amount of the precipitates, as shown in Table 2, can not reach the theoretical value. In combination with the values in Tables 2 and 4, the actual curves of precipitation were presumed and drawn as the dash lines in Figure 2.

According to the results of themodynamical calculation, the total amount of precipitated TiN in ER50-G should be more than 0.04% when reaching room temperature if the reaction between titanium and nitrogen took place under equilibrium state. In fact, the solid solubility of TiN in ferrite is much less than that in austenite or liquid steel[3,10]. This would further increase the

Table 4.   Equilibrium precipitation temperature of various Ti-containing compounds, K.

Ti-containing compound	TiN	Ti_3O_5	$Ti_4C_2S_2$	TiC
Equilibrium precipitation temperature	1824.8	1809.0	1647.4	1394.5

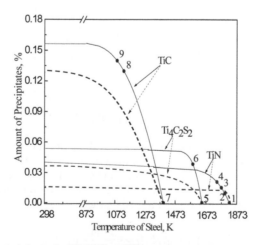

Figure 2.   Amount of precipitates in ER50-G at different temperature.

amount of precipitated TiN. Considering in the same way, the amount of precipitated $Ti_4C_2S_2$ would be up to 0.05% in theory, and for TiC it would be more than 0.15%. If it is assumed that the mass of precipitates listed in Table 2 were obtained under equilibrium state, the temperature that the precipitation behavior finishes also could be figured out with above equations. So does to the mass of precipitates corresponding to the critical temperatures of ER50-G. All of the results were marked as black solid dots in Figure 2, and the values were listed in Table 5.

The calculation results in Table 5 demonstrates that point 1 & 3 stand for the start and end of TiN precipitation, respectively. As to $Ti_4C_2S_2$ and TiC, the start and end of precipitation are point 5 & 6 and point 7 & 9, respectively. Point 2 represents the theoretical amount of TiN precipitation at liquidus line. And point 8 represents the theoretical amount of TiC precipitation at Ar_3. The amount of precipitated TiN would increase to 0.0153% at about 1772.1 K if under equilibrium state. In this way, about over 60% of TiN formed and precipitated in molten steel, while the remaining generated in the liquid-solid two phase region. However in reality, influenced by some factors such as steel cleanliness, super cooling and diffusion rate etc., the amount of precipitated TiN is over 90% in liquid steel and less than 10% in solid phase[10]. These two values deviated far from the theoretical value under equilibrium conditions. As to the precipitation of $Ti_4C_2S_2$, all should complete in austenite to obtain the actual amount under equilibrium conditions. While for TiC, it started precipitation in austenite and finished in two-phase zone of austenite and ferrite. In fact all of the precipitation behaviors extended until finished in ferrite for the influence of cooling conditions.

3.4 *Influence of precipitation behavior on strength of steel*

For titanium micro-alloyed steels, the strength is contributed by a variety of mechanisms, primarily by solid solution strengthening, grain refinement, precipitation hardening and dislocation strengthening etc.[1,3]. The effect of grain refinement on strengthening is inversely proportional to the square root of the grain size. The effect of precipitation hardening increases with the increase of precipitates volume fraction and the decrease of the particle size. Dislocation hardening effect is positively correlated to the steel dislocation density. If there is a large number of tiny particles precipitating in ER50-G wire, they could not only refine the grains and pinning dislocations, but also provide significant precipitation hardening effect. Thus the strength of ER50-G wire will be greatly improved. In order to restrain this phenomenon, the size and amount of the precipitated particles are required to be controlled effectively.

EPT of $Ti_4C_2S_2$ is about 1647.4 K in this study, and that of TiN is higher than liquidus. The higher the temperature is, the better the atom diffusion ability is. It's favorable for $Ti_4C_2S_2$ and TiN particles to grow up. Thus their precipitation hardening effect were weakened. EPT of TiC is the lowest, therefore the size of TiC is the smallest. Previous study indicated that the size of TiC precipitated in austenite was about 25 nm, but those precipitated in ferrite were smaller than 10 nm[1]. Furthermore the amount of TiC precipitated was the largest. Therefore, the effective control of the precipitation of TiC in ferrite could well optimize the drawing performance of ER50-G wire. In fact it was found that in those specimen which tensile strength were higher than 600 Mpa, the size of most precipitates was smaller than 10 nm, while in those specimen with a tensile strength about 500 MPa, most precipitates' size was between 10 to 30 nm. The statistical results of precipitates' size of specimens with different tensile strength but the same compositions was demonstrated in Figure 3.

Table 5. Temperature and mass of precipitates corresponding to the mark points in Figure 2.

Point	1	2	3	4	5	6	7	8	9
Temp./K	1824.8	1795.1	1772.1*	1743.8	1647.4	1588.2*	1667.5	1163.7	1131.9*
Precipitate	TiN	TiN	TiN	/	$Ti_4C_2S_2$	$Ti_4C_2S_2$	TiC	TiC	TiC
Mass/%	0	0.0094*	0.0153	/	0	0.0384	0	0.1207*	0.1298

*: All the values were calculated assuming precipitation took place under equilibrium state.

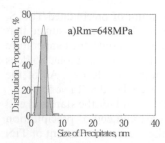 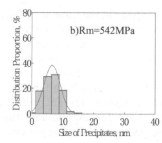 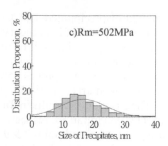

Figure 3. Distribution of precipitates size in ER50-G specimen with different tensile strength.

The results showed that the difference of tensile strength which was affected by precipitation behavior on ER50-G wire could be more than 100 MPa. Titanium alloying of ER50-G is conducted in RH refining process, and the titanium content is about 0.19–0.23%. Therefore TiN, $Ti_4C_2S_2$ and TiC particles come into being and precipitate in the continuous casting process of billet. Most of those precipitates, specially small TiC particles, dissolve in reheating process of rolling procedure, since the reheating temperature ultimately reached more than 1373 K. Dissolved titanium forms precipitates again during the subsequent rolling and cooling process. Thus to control the precipitation of small TiC particles in the two-phase region and ferrite, it requires to control the amount of dissolution and the secondary precipitation behaviors. Two primary measures were carried out based on above analysis. The first one is to increase the nitrogen and sulfur content and extend the period of high temperature zone during billets solidification. It is favor of the precipitation of TiN and $Ti_4C_2S_2$ with larger size, which could consume more titanium. Thereby the amount of the particles with small size is reduced, so does to the dissolution during reheating. The second one is to promote more TiC forms and precipitates in austenite by adjusting the rolling parameters, since the period of precipitation induced by deformation is shorter than that of equilibrium state[11,12]. This could also decrease the amount of TiC precipitation in ferrite. Before applying the above two measures, the mean tensile strength of ER50-G wire of Shougang was about 540 MPa, and the highest level was more than 600 MPa. While after that, the mechanical property of ER50-G wire was significantly improved to exhibit a good drawing ability. The mean tensile strength by testing 1008 samples has decreased to be about 486.9 MPa, and the fluctuation range was 432–596 MPa, and the proportion of samples with tensile strength less than 550 MPa was over 98%.

4   CONCLUSION

1. The primary precipitates in ER50-G wire were Ti-containing compounds in this study, like TiN, $Ti_4C_2S_2$ and TiC. Their equilibrium precipitation temperature decreases successively, and the value was 1824.8 K, 1647.5 K and 1394.5 K, respectively. The size of precipitation particles decreased gradually with the decreasing of precipitation temperature.
2. The amount of TiN, $Ti_4C_2S_2$ and TiC precipitated in ER50-G wire was 0.0153%, 0.0384% and 0.1298%, respectively in this study. The total titanium content in precipitates was about 0.142% which was more than 85% of the total titanium that existing in the steel. The rest titanium solutes in the matrix and the proportion was less than 15%.
3. To generate more large precipitate particles during continuous casting and rolling process could reduce the amount of TiC of small size which precipitated in ferrite. Thus a ER50-G wire with lower strength and better drawing performance could be obtained.

REFERENCES

[1] Mao, X.P. & Sun, X.J. & Kang, Y.L. et al. 2006. Physical metallurgy for the titanium microalloyed strip produced by thin slab casting and rolling process. *ACTA Metallurgica Sinic* 42(10):1091–1095.

[2] Cheng, G.G. & Zhu, X.X. & Peng, Y.F. et al. 2002. Foundation of solidification refinement technology with TiN in pure steel. *Journal of USTB*. 24(3): 273–275.

[3] Xie, L.Q. & Mao, X.P. & Huo, X.D. Effect of Ti on microstructure and property of steel. *Metallurgical Collections*. 1(2): 1–4, 8.

[4] Pak, J.J. & Jo, J.O. & Kim, S.I. et al. 2007. Thermodynamics of Titanium and Oxygen Dissolved in Liquid Iron Equilibrated with Titanium Oxides. *ISIJ International*. 47(1): 16–24.

[5] Yang, X.H. & Dirk, V. & Jozef, D. et al. 1996. Solubility Products of Titanium Sulfhpide and Carbosulphide in ultra-low Carbon Steels. *ISIJ International*. 36(10): 1286–1294.

[6] Huang, X.H. 1997. *Principle of ferrous metal/urgy:* 55. Beijng: The publishing company metallurgical industry.

[7] Chen, J.X. 2006. *Data book of major graph table of steelmaking*: 758–759. Beijng: The publishing company metallurgical industry.

[8] 19th Research Committee of Steelmaking of Japan Society for the Promotion of Science. Oct. 1984. *Commendatory equilibrium value of steelmaking reaction (Revised Edition)*: 255.

[9] Wang, M.L. & Cheng, G.G. & Zhao, P. et al. 2004. Formation Thermodynamics of Titanium Oxide during Solidification of Low Carbon Steel Containing Titanium. *Journal of Iron and Steel Research*: 16(3): 40–43.

[10] Chen, T & Li, H. & lv, N.B. et al. 2012. Thermodynamics Analysis on the Precipitation of Titanium Nitride in a welding Wire Steel. *Iron Steel Vanadium Titanium*. 33(6): 86–90.

[11] Lou, Y.Z. & Liu, D.L. 2009. Precipitation-Time-Temperature circle of Ti-microalloyed steel produced by compact strip process. *Modern Scientific Instruments*. 4(4): 112–114.

[12] Wang, Z.D. & Qu, J.B. & Liu, X.H. et al. 2000. Investigation of Strain-induced Precipitation Behavior in a Microalloying Steel by Stress Relaxation Method. *ACTA Metallurgica Sinica*. 36 (6): 618–621.

Advances in Engineering Materials and Applied Mechanics – Zhang, Gao & Xu (Eds)
© 2016 Taylor & Francis Group, London, ISBN 978-1-138-02834-0

Optimization of experimental parameters to the formation of ZrO_2 phase via hydrothermal method

H.Z. Yang, L.Z. Han & L.A. Zhao
China Medical University, Shenyang, P.R. China

Z.Z. Zhi
Shenyang Pharmaceutical University, Shenyang, P.R. China

Y. Ji
General Hospital of Shenyang Military Area Command, Shenyang, P.R. China

ABSTRACT: Hydrothermal method was adopted to prepare nano zirconia powders. Analytical reagent $Zr(NO_3)_4 \cdot 5H_2O$ and $Y(NO_3)_3 \cdot 6H_2O$ were adopted as reactors and the experimental parameters such as the pH, hydrothermal reaction time, temperature and doping contents were optimized through orthogonal experiment to prepare zirconia samples with high tetragonal phase purity and controllable grain size. X-ray diffraction was adopted to characterize the phase purity and crystallinity of samples derived. Results show that the four parameters have different influence weight on tetragonal structure and grain size of zirconia samples. It was demonstrated that the reaction time had the most remarkable influence among all of the experimental parameters. While for the influence of experimental parameters on grain size of samples, the pH value had the most remarkable influence.

1 INTRODUCTION

Zirconia (ZrO_2) is a kind of well-known structural and biomedical ceramic, which has attracted extensive attention in the past decades due to its excellent mechanical properties and biocompatibility. Especially, the superior mechanical properties (fracture toughness, strength and hardness) comparable to that of the metals enable ZrO_2 to be applied as all-ceramic crowns materials in the field of prosthodontics[1-2].

It has been well recognized that there are three different zirconia polymorphs depending on the temperature, i.e., monoclinic (*m*), tetragonal (*T*) and cubic (*c*) fluorite structures, and the yttria-stabilized tetragonal zirconia (Y-TZP) is considered as a desirable dental materials. Therefore, numerous of works have been done to obtain zirconia with high tetragonal phase purity. Zhang et al. investigated the influence of alumina content on low temperature degradation of Y-TZP systematically[3]. Hallmann et al. investigated the effect of dopants and sintering temperature on microstructure and low temperature degradation of Y-TZP[4]. Furthermore, Denry believed that the grain size of zirconia crystals also pose a remarkable influence on tetragonal phase stability[5-6], and a great number of investigations have been carried out to investigate the influence of experimental parameters on both the tetragonal phase stability phase stability and grain size of zirconia crystals. Ji et al. obtain zirconia nano crystals with high tetragonal phase purity via hydrothermal method, and found that experimental parameters such as the pH, hydrothermal reaction time, temperature and doping contents will have a remarkable influence on the phase stability and grain size[1]. However, to the best of our knowledge, there is still lack of systematic investigations comparing the influence weight of different parameters on the tetragonal phase and grain size.

In this paper, orthogonal experiment was devised and carried out to investigate the influence weight of different experimental parameters on the tetragonal phase and grain size. Furthermore, the optimal conditions to form desirable tetragonal phase and controllable grain size were also investigated. The research will be the catalyst to the further development of zirconia materials.

2 EXPERIMENTAL

Analytical reagent $Zr(NO_3)_4 \cdot 5H_2O$ and $Y(NO_3)_3 \cdot 6H_2O$ in different mole ratio (the mol% of Y^{3+} is from 2% to 5%) was respectively dissolved in distilled water to obtain two type of solution. Then NaOH or Ammonia water (mass fraction: from 25% to 28%) were added into the mixed the solution of $Zr(NO_3)_4$ and $Y(NO_3)_3$ to adjust pH of the solution (from pH = 9 to pH = 12). After forming the homogenous emulsion, it was pour into teflon lined autoclave for hydrothermal treatment. The reaction temperature was between 170°C and 200°C with interval of 10°C, and the reaction time was from 16 h to 28 h with interval of 4h. The detail of experimental parameters is set according to the orthogonal design, as listed in Table 1. Thereafter, the precursor powder was separated by centifugation and washed for 3 times with ethanol (99.9%). Finally, the powder was dried and grinded. The flow chart of experiment is shown in Figure 1. XRD (D/max 2000, Rigaku) was adopted to investigate the crystallinity of samples to provide the orthogonal design results according to the phase purity as well as the grain size.

Table 1. Orthogonal design of experimental parameters.

Samples	Temperature (°C)	Time (h)	pH	Y^{3+} content (mol/%)
1#	170	16	9	2
2#	170	20	10	3
3#	170	24	11	4
4#	170	28	12	5
5#	180	16	10	4
6#	180	20	9	5
7#	180	24	12	2
8#	180	28	11	3
9#	190	16	11	5
10#	190	20	12	2
11#	190	24	9	3
12#	190	28	10	2
13#	200	16	12	3
14#	200	20	11	2
15#	200	24	10	5
16#	200	28	9	4

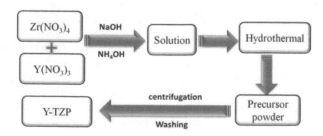

Figure 1. Flowchart of the experiment.

3 RESULTS AND DISCUSSION

XRD patterns of ZrO_2 samples derived from different experimental parameters are shown in Figure 2. Figure 2(a) shows samples derived at the same reaction temperatures. The major phase in the four samples was tetragonal phase, and there was a small amount of monoclinic phase in samples 4#, compared with the nearly pure tetragonal phase in other 3 samples. The relative purity of tetragonal phase of the samples could be calculated through calculating the strongest peak intensity ratio of tetragonal phase and monoclinic phase. Then, the value of purity was 0, 0, 0 and 0.27 respectively. In addition, the Full Width at Half Maximum (FWHM) of the strongest diffractions peaks (around 29.7°) in the four samples was 1.0467°, 1.1696°, 0.9418° and 0.7100°, respectively. Then, the grain size of samples can be calculated from the Scherrer formula

$$D = k \cdot \lambda / \beta \cdot \cos\theta$$

where k = 0.89, β is the FWHM of the strongest diffraction peak, λ is the X-ray wavelength and h is the diffraction angle. The grain size for samples 1#–4# was 7.76 nm, 6.95 nm, 8.63 nm and 11.61 nm, respectively. Therefore, grain size of samples can be controlled, and experimental parameters to prepare pure samples with lowest grain size was 20h in time, 10 in pH and 3% in Y^{3+} molar ratio.

Similarly, from Figure 2(b)–(d), we can also obtain the phase purity, FWHM and grain size for samples 5#–16#. Then, we can obtain the orthogonal design results according to the phase purity as well as the grain size, as listed in Table 2 and Table 3.

In orthogonal analysis, it is important to calculate the range (R) from factors levels through the following equation:

$$R = T_{max} - T_{min} \tag{1}$$

where T is the average value of factor levels (phase purity or grain size) provided that one of the experimental parameters is fixed while the other three parameters are different. Since there are four different factor levels for each group, we can calculate the range of each

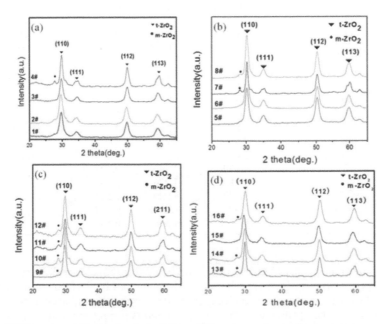

Figure 2. XRD patterns of samples derived from different experimental parameters.

Table 2. Orthogonal design results according to the phase purity.

Samples	Temperature (°C)	Time (h)	pH	Y^{3+} (mol%)	Purity
1#	170	16	9	2	0
2#	170	20	10	3	0
3#	170	24	11	4	0
4#	170	28	12	5	0.27
5#	180	16	10	4	0
6#	180	20	9	5	0
7#	180	24	12	2	0.41
8#	180	28	11	3	0.38
9#	190	16	11	5	0
10#	190	20	12	4	0.25
11#	190	24	9	3	0.28
12#	190	28	10	2	0.27
13#	200	16	12	3	0.23
14#	200	20	11	2	0.24
15#	200	24	10	5	0
16#	200	28	9	4	0.24
T_1	0.068	0.058	0.130	0.230	
T_2	0.198	0.122	0.068	0.223	
T_3	0.200	0.172	0.155	0.122	
T_4	0.177	0.290	0.290	0.068	
Range	0.132	0.232	0.222	0.162	

Table 3. Orthogonal design results according to the grain size.

Samples	Temperature (°C)	Time (h)	pH	Y^{3+} (mol%)	Grain size (nm)
1#	170	16	9	2	7.7648
2#	170	20	10	3	6.9475
3#	170	24	11	4	8.6306
4#	170	28	12	5	11.6079
5#	180	16	10	4	8.151
6#	180	20	9	5	8.1295
7#	180	24	12	2	13.3608
8#	180	28	11	3	9.9689
9#	190	16	11	5	8.8547
10#	190	20	12	4	13.3729
11#	190	24	9	3	7.7753
12#	190	28	10	2	8.2829
13#	200	16	12	3	11.5472
14#	200	20	11	2	8.3003
15#	200	24	10	5	8.3223
16#	200	28	9	4	8.1136
T_1	8.738	9.079	7.946	9.427	
T_2	9.903	9.188	7.926	9.060	
T_3	9.571	9.522	8.939	9.567	
T_4	9.071	9.493	12.472	9.229	
Range	1.165	0.443	4.546	0.507	

experimental parameter is 0.132, 0.232, 0.222 and 0.162 respectively for the results according to the phase purity. The optimal experimental parameter can be derived from the lowest of T value, and the influence weight of each experimental parameter can be decided by the value of range. Therefore, the optimal experimental parameters to prepare pure samples is 170°C in temperature, 16h in time, 10 in pH and 5% in Y^{3+} molar ratio. Besides, the influence weight for each parameter is time > pH > Y^{3+} molar ratio > temperature. Likewise, in Table 3, the experimental parameters to prepare samples with lowest grain size is 170°C in temperature, 16 h in time, 10 in pH and 3% in Y^{3+} molar ratio. Besides, the influence weight for each parameter is pH > temperature > Y^{3+} molar ratio > time.

4 CONCLUSIONS

Orthogonal design and analysis were carried out to investigate the influence of various experimental parameters on tetragonal phase purity and grain size of zirconia crystals. It is found that the optimal experimental parameters to prepare pure samples is 170°C in temperature, 16 h in time, 10 in pH and 5% in Y^{3+} molar ratio, and time is the most important parameter to improve the phase purity. While for the controlling of grain size, pH had the most important influence weight among all of the parameter.

ACKNOWLEDGEMENT

The work was supported by the project from the Foundation of the Education Department of Liaoning Province (Grant No. L2013285) and Science and Technology Planning Project of Shenyang City (Grant No. F11-262-9-16).
    Corresponding author: L.A. Zhao; E-mail: lazhao@mail.cmu.edu.cn.

REFERENCES

[1] Ji, Y. & Zhang, X.D. & Wang, X.C. & Che, Z.C. & Yu, X.M. & Yang, H.Z. 2013. Zirconia Bioceramics as All-Ceramic Crowns Materials: a Review. *Rev Adv Mater Sci* 34(1): 72–78.
[2] Pelaez, J. & Cogolludo, P.G. & Serrano, B. & Lozano, J.F.L. & Suarez, M.J. 2012. A Four-year prospective clinical evaluation of zirconia and metal-ceramic posterior fixed dental prostheses. *The International Journal of Prosthodontics* 25(5): 451–458.
[3] Zhang, F. & Vanmeensel, K. & Inokoshi, M. & Batuk, M. & Hadermann, J. & Van Meerbeek & Van Meerbeek, B. & Naert, I. & Vleugels, J. Critical influence of alumina content on the low temperature degradation of 2–3 mol% yttria-stabilized TZP for dental restorations. *Journal of the European Ceramic Society* 35(2): 741–750.
[4] Hallmann, L. & Ulmer, P. & Reusser, E. & Louvel, M. & Hammerie, C.H.F. Effect of dopants and sintering temperature on microstructure and low temperature degradation of dental Y-TZP-zirconia. *Journal of the European Ceramic Society* 32(16): 4091–4104.
[5] Denry, I. & Kelly, J.R. State of the art of zirconia for dental applications. *Dental Materials* 24(3): 299–307.
[6] Denry, I. How and when does fabrication damage adversely affect the clinical performance of ceramic restorations? *Dental Materials* 29(1): 85–96.

Advanced manufacturing and automation technology

Advances in Engineering Materials and Applied Mechanics – Zhang, Gao & Xu (Eds)
© 2016 Taylor & Francis Group, London, ISBN 978-1-138-02834-0

Conventional missile structural configuration design based on secondary development of UG

J. Li & X. Li

College of Aerospace Engineering, Beijing Institute of Technology, Beijing, China

ABSTRACT: In the missile general design, an aerodynamic characteristic is an important aspect that can affect a missile performance. Structural configuration is one of the most important aspects of the missile design. The paper combined Unigraphics NX (UG), Visual C++, C# and other software to make a missile design system, which used powerful 3D modeling capabilities and open secondary development function of UG. The system can change the parameters of missile without starting the UG software, and the convenient interface makes the system easy for missile general designer to use.

1 INTRODUCTION

The missile general design is in the frontier of the engineering application field and plays an important part of missile general design, and it should satisfy the tactical and technical index of missile and each subsystem work requirements, so it needs to seek a new method of optimum design and manufacturing process. Digital technology is a popular way to do produce design, produce processing and simulation by using computer software. UG, which use digital technology, is one of the most commonly used software in engineering.

There are many groups work together to finish the whole missile design, the structural group design the missile structural configuration, pneumatic group carry out analysis and calculate the aerodynamic date, etc. When some aspects of the missile performance cannot meet the requirements, an aerodynamic configuration need to be modified, while it is a huge work to change these structural parameters, this situation greatly reduces the working efficiency. So, it need to use digital technology to design a convenient system, and it can increase design efficiency[1].

This paper will detail the UG parameterized modeling method, secondary development method based on VC, introduce a method to get missile inertia and center of mass and other characteristics by UG/Open API, give a simple introduction how to create interactive platform based on Auto CAD and C#.

2 UG/OPEN

UG/Open module is the secondary developing tool sets of the UG software. With the UG/Open module, a user can develop a customizable interface. UG/Open module including the following several parts: UG/Open MenuScript, UG/Open UI Style, UG/Open API, and UG/Open GRIP[7].

This paper mainly gives an introduction about UG/Open API combined with VC, because the program is done in VC, so the UG/Open API can give full play to the powerful function of VC and abundant resources.

UG/Open API encapsulate much UG operation function, it allows C and C++ program to call all the library function in UG software, it can perform operations on UG graphics terminal, file management system and database, almost all operation can be realized with UG/Open API function.

3 ESTABLISH A GENERALIZED SYSTEM

3.1 *Modeling and assembly*

In the product design process, designer always meets the requirement of modifying the model size through analysis and calculation. At this time, in order to speed up the process of product design and reduce the repetitive work, designers often use the parameterization design method to reduce the workload. UG have the perfect function of system parameter auto-acquisition. The dimension constraints of input can be stored as characteristic parameters in the sketch design. The constraints associated with the named parameters, then UG achieves the goal of directly driving model with parameter[5]. Parts in the model update, assembly will be updated.

First, by stretching, rotating, Boolean and other operation commands, UG software can generate the model that satisfies the performance demand. With powerful function of modeling and parameterization, UG can performance every detail of missile body as much as possible and complete a precise missile configuration. When the parameterization is finished, the missile configuration in any position can be changed according to needs. Before the design, it is needed to analyze the parts or components carefully. Second, the general idea of modeling of this part is formed, then which features need to be created and the operations sequence should be clear. At the same time, it is also needed to pay attention to the internal relations of the various features and their respective characteristics. Finally, it is important to ensure the quantity of driven parameters. In the software, UG provides the function of creating relationship between the driving dimensions and the parts by using expression.

Before establishing table driven, the determined design variables are assigned to the corresponding size by renaming expression, the effect is shown in Figure 1.

It is important to note that if there are some features of the model is not generated from the sketch modeling, it is needed to choose the 'Associative' option in the dialog. One dialog is shown in Figure 2. And part of the operation cannot parameterize such as mirror, translation, rotation, etc.

3.2 *Secondary development of UG*

Based on difference in compiling environment, UG/Open API program can be divided into two mode, the Internal mode and the External mode[5,6].

Without starting UG software, external mode can be run directly under the operating system as an application, this mode can be combined with C# and other software to make a complex, beautiful and practical interface.

Figure 1.   The expression dialog of UG.

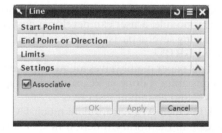

Figure 2.   The 'Associative' option in the dialog.

Internal type can be run through the UG software and VC environment. The internal mode can be started by calling the DLL (dynamic link library) through UG software, once it is resident in memory it will always exits until close of the UG.

It should be paid attention to when using the function that all the UG/Open API application must be initialized and termination in order to ensure access to or release of UG/Open API execute permissions.

UF_initialize is an initialization function. When secondary development began to call the function of UG/Open API, it should be first call the UF_initialize to obtain execute permissions.

UF_terminate is a termination function, when secondary development will not use UG/Open API function, it must first call the UF_terminate to release the execute permissions.

The program frame about UG secondary development is as follows:

```
if (!UF_CALL(UF_initialize()))
{
do_ugopen_api();
UF_CALL(UF_terminate());
}
```

The main functions about getting the expression of model are as follows:

```
UF_PART_load_status_t error_status;
UF_PART_open("XX\XX\XX\xx.prt",&part_tag,&error_status);
UF_PART_free_load_status(&error_status);
UF_MODL_ask_exps_of_part(part_tag,&number_of_exps,&exps);
UF_CALL(UF_PART_save());
UF_CALL(UF_PART_close(part_tag,scope,mode));
```

The main functions about model updating are as follows:

```
UF_PART_load_status_t error_status;
UF_PART_open("XX\XX\XX\xx.prt",&part_tag,&error_status);
UF_PART_free_load_status(&error_status);
UF_CALL( UF_MODL_import_exp( (char*)ep.c_str(), 0) );
UF_CALL( UF_MODL_update() );
UF_CALL( UF_PART_save() );
UF_CALL( UF_PART_close(prt_id,1,1) );
UF_CALL( UF_PART_free_load_status(&st) )
```

3.3 *Access to quality attributes*

Moment of inertia, mass and center position of missile can be used as a basis for the quality of aerodynamic performance calculation, ballistic calculation, load calculation, stability and maneuverability calculation, structure design, the design of transmission device and missile transportation loading equipment. It is convenient to use the UG built-in function to calculate the above content to be used[8].

API provides quality analysis interface function: UF_MODL_ask_mass_props_3d, the contents of solid analysis are solid surface area, volume, mass, center of mass, moment of inertia and product of inertia, spindle, torque and turning radius, etc[2].

The main functions about getting model properties are as follows.

```
UF_PART_load_status_t error_status;
UF_PART_open("XX\XX\xxx.prt",&part_tag,&error_status);
UF_PART_free_load_status(&error_status);
UF_OBJ_cycle_objs_in_part(part_tag, UF_solid_type, &solid);
do
```

```
{
UF_OBJ_ask_type_and_subtype(solid,&type,&sub_type);
if(sub_type = = UF_solid_body_subtype)
{
UF_MODL_ask_body_type(solid,&body_type);
if(body_type = = UF_MODL_SOLID_BODY)
tag_solid[body_count] = solid;
}
UF_OBJ_cycle_objs_in_part(part_tag,UF_solid_type,&solid);
}while(solid ! = NULL_TAG);
UF_MODL_set_body_density();
UF_CALL(UF_MODL_ask_mass_props_3d(solbody,
                                  count,
                                  type,
                                  unit,
                                  density,
                                  accuracy,
                                  acc_val,
                                  massprop,
                                  massprop_stat));
```
Through the above procedures, we can get the data in Table 1.

3.4 *Visualization*

The above content is just some procedure code in VC and those code can not achieve the objectives of visualization and easy to use. Then, the procedure code should be combined with Auto CAD and C# to design a User-Friendly Operator Interface with all the function. Auto CAD presents each parts' schematics of missile, C# makes input variable in VC code or output date displayed in the interface, and the effect is shown in Figure 3.

Table 1.   Some important properties of a solid part.

No.	Date	Note	No.	Date	Note
1	0.000000	Center of Mass	7	710.986400	Moments of Inertia
2	0.000000	(COFM), WCS	8	710.986400	(centroidal)
3	1.500000		9	1066.479600	
4	1244.226200	Moments of	10	2.794000	Density
5	1244.226200	Inertia, WCS	11	236.995467	Mass
6	1066.479600				

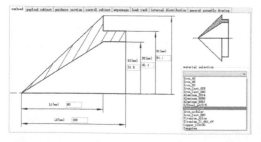 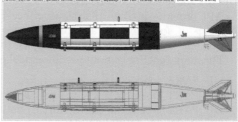

Figure 3.   A working sketch of warhead in system.   Figure 4.   General assembly drawing.

4 CONCLUSION

Based on the above method, the system through a combination of UG, VC, C# and Auto CAD can realize the external parameterization design and assembling, setting the density of missile, acquiring moment of inertia and outputting the total quality.

With the expansion of the application of UG software, secondary development tools, as a kind of indispensable auxiliary function, is applied more and more in the field of missile design. By using the UG OPEN API, it is completely feasible to realize 3D CAD model automatically update. An automatic update of model for designers not only eliminates the heavy manual operation, but also shortens the design cycle of the missile.

Combining with an instance, this paper expounds the external mode of secondary development of UG for missile design. Using external mode for missile aerodynamic configuration design, is a kind of new design method. This design method can be applied to missile general design, each subsystem design and design in any other field.

REFERENCES

[1] Y.L. Wang, 1993, A Graphic Generating Software for Design Cruise Configuration, *Tactical MIssile Technology*, vols (3), pp:57–60.
[2] L. Wang, Z.K. Li, Y. Tan, Z.X. Liu, 2006, The Study of Secondary Development of UG by The UG/Open API, *Development & Innovation of Machinery & Elactrical Products*, vols (19), pp:105–106.
[3] T. Zhang, B.J. Chang, Z.W. Dong, J.Y. Wang, 2005, Research on Parts-layout for Winged Missile, *Tactical MIssile Technology*, vols (1), pp:1–4.
[4] M.H. Yuan, Y. Ding, A.M. Ji, 2012, Design of the Gear Datebase Based on UG Secondary Development, *Advanced Technology for Manufacturing System and Industry*, vols (236–237), pp:1312–1315.
[5] X. Huang, Y.G. Li, 2005, *Application Development Tutorial & Examples for UG*, Beijing: Tsinghua University Press.
[6] R. Mo, Z.Y. Chang, H.J. Liu, 2008, *Chart Detailing About Second Development of UG*, Beijing: Electronic Industry Press.
[7] Z.W. Dong, T.L. Zhong, Y.L. Fu, 2002, *UG/OPEN API Program Foundation*, Beijing: Tsinghua University Press.
[8] J.Q. Yu, Z.H. Wen, Y.S. Mei, T. Teng, 2010, *Tactical Missile General Design*, Beijing: Beijing University of Aeronautics and Astronautics Press.
[9] Y. Huang, 2009, *UG/OPEN API, MFC and COM Development instance of the refined solution*, Beijing: National Defence Industry Press.

Advances in Engineering Materials and Applied Mechanics – Zhang, Gao & Xu (Eds)
© *2016 Taylor & Francis Group, London, ISBN 978-1-138-02834-0*

Research on data structure and data management of MBD for shipbuilding

M.Y. Li, W. Xu, W.J. Zhang & W.W. Wang
Wuhan Second Ship Design and Research Institute, Wuhan, China

ABSTRACT: The Model-Based Definition (MBD) modeling technology is a highlight in the shipbuilding digital design and manufacturing, and also is the inevitable direction of future development of the shipbuilding industry. In this paper, the MBD data structure and the MBD data management methods for shipbuilding are presented, and the data classification is researched. And then, the MBD data management is also established, the process for shipbuilding of management is presented, and divided the management level into design level, process level and manufacturing level. To explain the approach more clearly, a practical of a ship axis is used as an illustration in this paper.

1 INTRODUCTION

Shipbuilding is a typical large equipment manufacturing industry, which represents the comprehensive economic strength and technological strength of a country. With the rapid development of digital technology in the shipbuilding industry, the traditional design and manufacturing model based on two-dimensional drawings have been unable to adapt to the requirements of technology. The three-dimensional digital technology based on the three-dimensional model technology will become an important direction for the development of shipbuilding industry in the future.

The Model-Based Definition (MBD) is a kind of new method and technology for computer application in shipbuilding, which is aimed at adopting the integrated three-dimensional model to express the shipbuilding information. Thus, the digital assembly and shipbuilding under a single source of product data can be achieved.

The key point of MBD modeling is to redefine information of the traditional two-dimensional engineering drawings on the use of the three-dimensional digital. The information of design, process and assembly of shipbuilding can be completely expressed in the three-dimensional model design[1]. MBD is originated in aircraft manufacturing for shortening the development cycle and reducing the development costs with the development of digital technology[2]. As an information carrier for shipbuilding, the MBD modeling will play an important and supporting role in the design, process and assembly of shipbuilding in the future.

The application of MBD technology in foreign countries has been relatively years ago. The MBD modeling in design, process and manufacturing of products is used by the Airbus, Boeing, Lockheed Martin and many other manufacturers comprehensively, and the MBD as the only source of data in the process has been managed[3–5]. Through the comprehensive promotion and the use of MBD model method, the aircraft project development cycle reduced about 40% and rework reduced almost 50% by the B-787 program of the Boeing company in 2004[6].

At present, the modeling based on MBD in China is still in the research stage. There are a lot of problems as unified standards, level of data visualization, information of product structure should be researched and developed[7–10]. With the deep application of the three-dimensional digital technology in the shipbuilding industry, the existing three-dimensional model methods cannot be used to achieve real-time interactive information from assembly process to manufacturing process. The complete and consistent of the product in the process of shipbuilding is difficult to be ensured.

In this paper, the **MBD** data structure and the **MBD** data management methods for the information of design, process and assembly is proposed to solve the above problems, which can improve the efficiency and manufacturability in shipbuilding.

The **MBD** modeling regard as a unique source in shipbuilding, which supplies the main information resource, and this method not only saves development cycle, but also improve the efficiency. Therefore, in recent years, the **MBD** modeling is a highlight in modeling field.

The rest of this paper is organized as follows. In Section 2, the **MBD** data construction is presented, and data classification is described in detailed. In Section 3, based on the data construction, the data management is presented and meanwhile, the detailed of data management process is given, too. To explain the approach more clearly, a case study is used as an illustration in Section 4. Finally, Section 5 summarizes the conclusion.

2 MBD DATA CONSTRUCTION

The **MBD** data structure is consisted of geometry information and non-geometry information for shipbuilding, and including coordinate system, solid model, baseline, annotated information and process attribute, as shown in Figure 1.

2.1 Coordinate system information

Coordinate system information (*CS*) is the basis of the three-dimensional modeling, while the coordinate system can be divided into the world coordinate system CS_w and the local coordinate system CS_p.

$$CS = f(CS_w, CS_p) \tag{1}$$

In the process of defining shipbuilding information, the part coordinate system can be used as the baseline to design components in the three-dimensional modeling with interrelated components, which is used to reduce the repeat positioning in the design process.

2.2 Solid model information

Solid model information is the **MBD** carrier. The geometric characteristics of the solid model information (*SM*) can be described as geometric features SM_f, geometric surface SM_p and geometric area SM_a.

$$SM = f(SM_f, SM_p, SM_a) \tag{2}$$

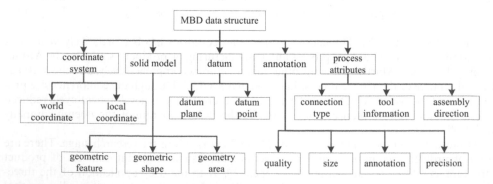

Figure 1.   MBD data structure for shipbuilding.

26

where SM_f is the geometric shape used in the description of geometric element in the process of modeling the three-dimensional model, such as external thread, grooves, draft angle. SM_p can be defined as the geometric shape elements in the constituting of the basic geometric shapes, such as flat, curved, rotating surface, cylindrical. SM_a is the geometric area, which is a region with specific requirements.

Solid model information is the basis for constructing MBD data in shipbuilding. And also is the carrier of other non-geometric information, such as datum, annotation and process attributes. Through the geometric shape, geometric shape and geometric area combined, the solid model information of MBD can be accurately described.

2.3 *Datum information*

The key point of MBD modeling is the datum information (BM). The geometric information are used to determine the datum information in the process of shipbuilding, and it includes datum point (BM_p) and datum plane BM_l.

$$BM = f(BM_p, BM_l) \qquad (3)$$

where BM_p can be divided into the design datum point, the process datum point and the manufacture datum point. The design datum point is widely used in the technology for MBD model shipbuilding, which can be used to define the location of the model and to establish other bases, such as the datum axis, the datum curve and so on.

BM_l is the datum plane for assembly, which is included of design datum plane and manufacture datum plane. And the design datum plane more often used in the design of components and parts, whereas the manufacture datum plane is oriented to the datum of process. In order to information exchange among design, process and manufacture, design datum plane and manufacture datum plane should be accurately distinguished in MBD modeling.

2.4 *Annotation*

Three-dimensional annotation is an important direction for the development of the ship-building, and it is also the basis of the three-dimensional digital technology. The three-dimensional annotation for shipbuilding is given the special advantage of visibility and simplicity compared with the two-dimensional drawings. According to labeled the process annotation and components annotation, the real-time interaction can be realized. Annotation (AC) of MBD for shipbuilding is focused on AC_q, AC_m, AC_n, and AC_p, and can be verified as follows:

$$AC = f(AC_q, AC_m, can, AC_p) \qquad (4)$$

where AC_q is the quality information, which is related to material and volume of components and parts.

AC_m is the size of components and parts, which included setting size, position size and overall size. The setting size is used to determine the basic dimensions of the desired size, such as pore size and side length. The position size is used to describe the relative position of the basic size between geometry. The overall size is used to describe the main shape information of assembly or components.

AC_n is the annotation information, which is used to instruct the process of design and manufacture. The annotation information is made up of standard instruction, component annotation, material description and label instructions, as shown in Table 1.

AC_p is the precision information, which is mainly used to describe the permitted variation and deviation of geometry and size. And AC_p is an important basis for ship design, assembly and manufacturing. Precision information can be divided into tolerance information and surface roughness information, as shown in Figure 2.

Table 1. Annotation information data structure.

Data structure	Description of annotation information
Standard instructions	Contains data of management and legal rights
Component annotation	Defined key features
Material description	The collection of raw materials characteristics
Annotation instruction	Describe specific process information

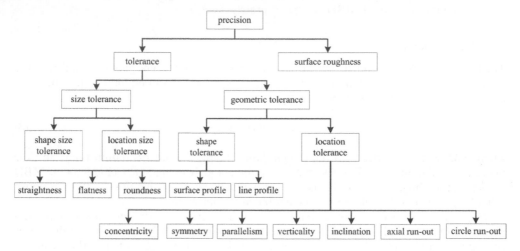

Figure 2. Precision information.

2.5 *Process attributes*

The process attributes information (*PI*) is an important carrier of manufacturing and assembly information, which included PI_c, PI_t and PI_{ac}, and can be verified as follows:

$$PI = f(PI_c, PI_t, PI_{ac}) \qquad (5)$$

where PI_c is the connection relation, which consisted of welding, riveting, screwing, bonding, sealing connection, and pin connection.

PI_t is the tool information, which is defined as the type and quantity of the tools that are used in the process of assembly and manufacturing.

PI_{ac} is the direction information, which is defined as a direction of the assembly information for parts in the coordinate system.

Finally, the MBD data structure can be defined as follows:

$$M = f(CS, SM, BM, AC, PI) \qquad (6)$$

3 MBD DATA MANAGEMENT

In the process of building MBD data for shipbuilding, different components are defined as different model information. Even though in the different stages of designs, process and manufacturing, the data information of three-dimensional is not the same with the same component. For example, in the design stage of three-dimensional components in shafts, whose coordinate axes, three-dimensional model, datum should be described in detail. In the stage of process design, the precision, tolerance and other attributes should be illustrated to assemble the shaft successfully. In order to facilitate manufacturing, after the process design is completed, flexural rigidity, materials of the cross section in the three-dimensional model are required detailed explanation.

For facilitating the management and interaction of information, different model attributes of **MBD** in different design and manufacture phases should be established. The original model data **MBD** should be managed separately in order to simplify the complexity of the **MBD** model and facilitate the exchange of information. Models of different components and model information have been distributed at different design stages, which can facilitate parts design, process design and manufacturing design.

After designing and releasing the **MBD** model in design department, the **MBD** is delivered to the process design department. Along with the process design and simulation, the **MBD** model is delivered to the manufacture department with the information of tool and process flow. Finally, the manufacturing department can achieve production and manufacturing by using the **MBD** information. The **MBD** data management process for shipbuilding is shown in Figure 3.

In order to explain the **MBD** data management model for shipbuilding completely and clearly, the data management model can divided into three levels, the design level, process level and manufacture level. The three levels are interrelated and interacted through the unified data exchange platform. The design level is the base level, the process level is the middle

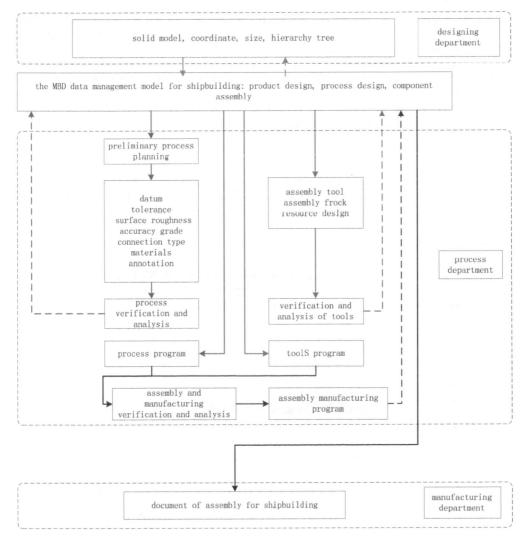

Figure 3.   The MBD data management process for shipbuilding.

level, and the manufacture level is the implement level. To a large extent, the levels can be satisfied with the requirements of three-dimensional model for the shipbuilding.

3.1 Design level

The design level is comprised of three-dimensional model information, coordinate information, size information, hierarchy tree and other three-dimensional information. The design level is the basis of **MBD** data construction. First, the purpose of design level is to establish and release the coordinate system and basic dimensions of components. Second, it is used to establish size information for process design. Finally, the hierarchy tree of **MBD** is the basis of the data management.

The design level is the core level of components design and it contains all the design information of components. The accurate definition and classification of the design level not only satisfy the demand of design department, but also can realize the interaction between components design and data management systems.

3.2 Process level

The process level is the core level of **MBD** data management, and it is the key level of **MBD** model construction for shipbuilding. The process level included all process information in the original two-dimensional drawings, such as machine baseline, tolerance, surface roughness, accuracy and other key process information of components. The purpose of establishing the process level is to explain the process information completely, and second, to establish the process planning, the third, to carry out the verification and analysis of manufacturing plan in virtual manufacturing, finally, the process level can achieve the information interactive between the design level and the manufacture level.

3.3 Manufacturing level

The manufacturing level is the top level of **MBD** data management. The model information in this level is simple and the purpose of this level is to realize the product manufacturing completely.

4 SAMPLE

The **MBD** model for a ship axis is established by **CATIA** V5R20. In order to realize the construction of **MBD** information data for a ship axis, the datum, standard information, attribute and other data information are marked, as shown in Figure 4.

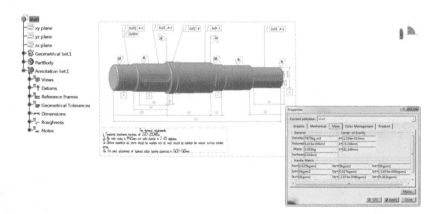

Figure 4. A MBD model for a ship axis.

5 CONCLUSION

The MBD data structure and the MBD data management methods for shipbuilding are presented in this paper, which is an inevitable trend for the digital development of shipbuilding in the future. The MBD data classification is researched and consisted of coordinate system, solid model, baseline, annotated information and process attribute. Then, MBD management process for shipbuilding is presented, and the MBD data management is divided into design level, process level, and manufacturing level.

Based on the sample, an MBD model for a ship axis is established, which included datum, standard information, attribute and other data information. In essential, this sample is necessary for design, assembly and manufacturing, and could be helpful to shipbuilding.

For future work, experiments to several shipbuilding businesses should be taken. Based on these tests, the MBD data structure and the MBD data management will be improved, so as to prove save development cycle and improve the business more efficiency.

FIRST AUTHOR AND CORRESPONDING AUTHOR

LI Ming-yu received his PhD from Huazhong University of science and technology. Now he is an engineer in Wuhan Second Ship Design and Research Institute. His main research interests include assembly sequence planning, green manufacturing, virtual maintenance, and maintenance process simulation.

E-mail: lmy1141@foxmail.com; **Mobile:** +86–18971698225.

Address: The 19th of Yangqiao Lake Road of Jiangxia District, Wuhan, Hubei Province.

REFERENCES

[1] Alemanni, M. Destefanis, F. & Vezzetti, E. 2011. Model-based definition design in the product lifecycle management scenario. *The International Journal of Advanced Manufacturing Technology* 52:1–14.

[2] Lu, Hu. Fan, Yu-qing. 2008. The study of the drawing functions in the digital aero product definition. *Journal of Engineering Graphics* 2:29–34.

[3] Quintana, V. Rivest, L. Pellerin, R. Venne, F. & Kheddouci, F. 2010. Will model-based definition replace engineering drawings throughout the product lifecycle? A global perspective from aerospace industry. *Computers in Industry* 61:497–508.

[4] Feng, Yan-yan. Jin, Xia. Wang, Min. 2010. Based model design for aircraft assembly process. *Aircraft Assembly Technologies* 24:95–98.

[5] An, Lu-ling. Jin, Xia. 2011. Application of model based definition in aircraft development. *Aeronautical Manufacturing Technology* 12:45–47.

[6] Quintana, V. Rivest, L. Pellerin, R. & Kheddouci, F. 2012. Re-engineering the engineering change management process for a drawing-less environment. *Computers in Industry* 63(1):79–90.

[7] Rivest, L. Kheddouci, F. Clement, F. 2011. Towards long term archiving of 3D annotated models: a sneak peek to a potential solution. *Integrated Design and Manufacturing in Mechanical Engineering-Virtual Concept*; *Proc. Bordeaux, France, 20–22 October 2010*.

[8] Zhen, Xi-jin. Yang, Run-dang. Wei, Nai-kun. Yang, An-hai. 2012. Research on assembly process simulation for shipbuilding. *Ship Science And Technology* 34(12):136–139,143.

[9] Chen, Xing-yu. Zhang, Hong-qi. Chen, Di-jiang. Zhang, Chong. Xiao, Cheng-xiang. 2010. Process design method vased on full three-Dimensional model for complex electromechanical product. *Radar Science and Technology* 5:474–479.

[10] Hao, Bo. Jiang, Bei-bei. 2011. Technology of 3D-CAPP based on MBD. *Journal of Sichuan Ordnance* 12:6–9.

Pressure sensor testing system based on PCI data acquisition card

C.L. Tan
School of Physic and Information Engineering, Jianghan University, Wuhan, Hubei Province, China

H.L. Liu & Y.B. Wang
Wuhan Shendong Automobile Electronics Co, Ltd., Wuhan, Hubei Province, China

ABSTRACT: In order to facilitate the pressure sensor manufacture to test a product quality, a kind of multi-channel pressure sensor testing system based on PCI8602 data acquisition card is designed. The system is composed of industrial computer, data acquisition card, standard pressure transmitter, signal processing circuit and multi-channel sensors. The VB developed application program controls the entire testing process. By placing the 4~20 mA standard pressure transmitter and multi-channel pressure sensors in the same testing environment, the VB application program can estimate the quality of the sensors by analyzing the real-time data acquisition of every channel. The working principle of the system is introduced; the key circuit diagram and software design method are given. The presented testing system can be widely applied in the pressure sensor testing to save manual work, reduce labor intensity and improve the testing quality.

1 INTRODUCTION

The sensor is a key component for measurement circuit, and its performance parameters determine the quality of the whole testing system. The pressure sensor is widely used in modern measurement and control field. In order to evaluate the pressure sensor's performance and its life cycle, to overcome the traditional artificial testing method of low efficiency, high labor intensity and low reliability so that the manufacturer can control the product's quality conveniently. This paper designs a set of quantitative analysis testing system based on PCI 8602 data acquisition card for multi-station pressure sensors. The system can be expanded easily with low cost (Cao, 2008)[2].

2 SYSTEM CONFIGURATION

In this system, the multi-channel pressure sensors and the standard pressure transmitter that are used to test the current atmospheric pressure are arranged in the same test environment. The industrial computer receives and analyses the output data of multi-channel sensors and standard pressure transmitter, so that the system can monitor the performance of each testing sensor, and the testing results are displayed on the computer screen, then the testing data can be stored in computer hard disk. The working principle of the whole system block diagram is shown in Figure 1 (Wei, 2012)[7].

The output signal of standard transmitter is 4~20 mA, this current signal should be converted to 1~5 V by current/voltage isolate conversion circuit unit so that the PCI data acquisition card can process the signal. Multi-station testing sensors can be a voltage-type sensor in which output signal is 0~5 V or a current-type sensor in which output signal should be converted to 0~5 V by the conversion circuit. In the voltage-type sensor, the output voltage has a linear relationship with pressure, whereas in the current-type sensor, the current that converted to voltage is approximately linear with pressure. The output signals of multi-channel

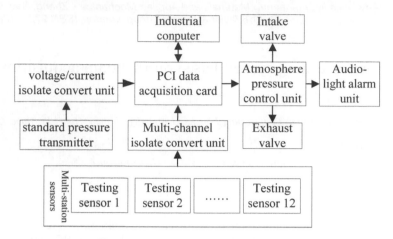

Figure 1. The system block diagram.

sensors are processed by the multi-channel isolate conversion circuit and then send to PCI data acquisition card. The application program on computer can read signals from the PCI data acquisition card to obtain the AD results of each channel, and can send command to PCI data acquisition card to control peripheral unit (Wu, 2014)[8]. The user can set the test parameters of sensors on the computer application program interface.

The testing gas resource is compressed atmosphere, the intake valve and exhaust valve that installed on the vent pipe are electromagnetic valves, and they are used to control the air pressure. Industrial computer automatically control the electromagnetic valves switching periodically by sending command to PCI data acquisition card. The pressure sensors and transmitter are arranged between intake and exhaust valves. Air pressure control cycle of impact fatigue test on sensors includes the process of inflation, holding and release. In inflation process, pressure in the pipe is increasing with intake valve turn on and exhaust valve turn off; in holding process, pressure in the pipe is stable with intake valve turn off and exhaust valve turn off; as to release process, pressure in the pipe is decreasing with intake valve turn off and exhaust valve turn on. The cycle times of the impact fatigue test that the sensor can bear indicate the life cycle of sensor (Huang, 2012)[5].

The VB application program controls the testing process automatically, periodic impact fatigue test is applying on sensors and transmitter continuously, output voltage on each channel can be read out and displayed on computer screen by application program. By comparing the output testing results of every sensor with the theoretical result that calculates according to the output voltage of standard transmitter, the program can estimate the quality of the sensors. The sensor is OK if the testing result is in the range of error, otherwise the sensor is failure. Real-time status of every sensor can be displayed on computer screen by application program, dynamic curve of output voltage can be showed and all the testing results can be saved on the hard disk automatically for review.

3   DESIGN OF KEY CIRCUIT

The PCI 8602 has 16-bit ADC of 32 channels and up to 8 digital I/O ports implemented[1]. The card is designed based on PCI bus and can be inserted in the PCI slot of IBM-PC/AT or any other PCI slot which is compatible with PC. It plays an important role in the field of data acquisition and industrial process monitoring system.

The range of standard pressure transmitter that measures the air pressure is 0~2.5 Mpa. The output signal of pressure transmitter is 4~20 mA, which must be converted to 1~5 V by isolation circuit and then sent to PCI 8602 data acquisition card. To avoid signal interfering, the author selected high-linearity analog optocoupler HCNR200 from USA Agilent

Company for isolation circuit. Figure 2 illustrates how the HCNR200 high-linearity opto-coupler is configured.

The basic optocoupler consist of an LED and two photodiodes. The LED and one of the photodiodes (PD1) are on the input leadframe, and the other photodiodes (PD2) are on the output leadframe. The package of the optocoupler is constructed so that each photodiode receives approximately the same amount of light from the LED[6]. The current in PD1 (I_{PD1}) and PD2 (I_{PD2}) is proportional to the LED current (I_F), the photodiode current ratio can be expressed as a constant K:

$$K = I_{PD2}/I_{PD1} \tag{1}$$

where the transfer gain of K is about $(100 \pm 15)\%$.

The Figure 3 shows the application isolation circuit that converts 4~20 mA to 1~5 V, the input signal I_{LOOP} is from the standard pressure transmitter, the output signal V_{OUT} is sent to PCI card from LMV358 output (Yang, 2003)[9].

As shown in the Figure 3, zener diode D1 on the input side of the circuit regulates the supply voltage for the input amplifier. As in the simpler circuits, the input amplifier adjusts the LED current so that both of its input terminals are at the same voltage. The loop current is then divided between R1 and R3. I_{PD1} is equal to the current in R1 and is given by the following equation:

$$I_{PD1} = I_{LOOP}*R3/(R1 + R3) \tag{2}$$

Amplifier U3A and resistor R5 on the output side of the circuit convert I_{PD2} back into voltage, V_{OUT}, where.

$$V_{OUT} = I_{PD2}*R5 \tag{3}$$

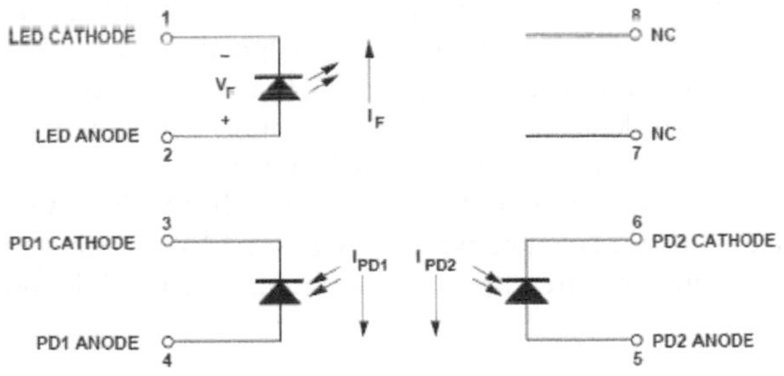

Figure 2.   HCNR200 schematic.

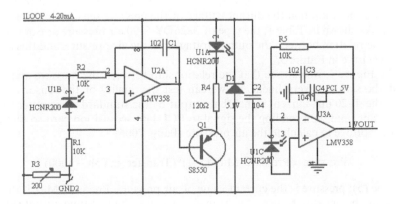

Figure 3.   4~20 mA isolation convert circuit.

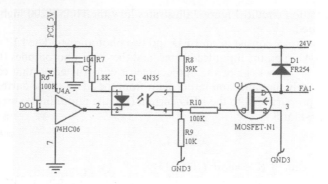

Figure 4.   Electromagnetic valve control circuit.

Combining the equation (2) and equation (3) with the equation (1) yields an overall expression relating the output voltage to the loop current,

$$V_{OUT}/I_{LOOP} = K*R3*R5/(R1 + R3) \qquad (4)$$

From the equation (4), we can see that the output voltage V_{OUT} is proportional to I_{LOOP}. By adjusting the adjustable resistor of R3, we can make the value of V_{OUT}/I_{LOOP} equal to number 250, the application circuit shown in Figure 3 can convert 4~20 mA to 1~5 V. Because the 4~20 mA current output from the standard pressure transmitter is proportional to air pressure 0~2.5 MPa, so that we can calculate the air pressure according to the output voltage of the circuit, which will be discussed detailed in the following part of the software design.

The test period of impact fatigue testing which includes three processes of charging, holding and releasing is controlled by a pair of electromagnetic valves whose supply voltage is DC 24 V. The driver circuit application is shown in Figure 4, the optocoupler 4N35 is used for isolated conversion. The electromagnetic valve is connected between 24 V and FA1-. DO1 stands for the digital I/O input of PCI8602 card, it receives command from computer, if DO1 receive digital "1" from computer, it will make the MOSFET of Q1 conduct so that the electromagnetic valve that connected between 24 V and FA1 will activate with power. If DO1 receive digital "zero" from computer, the electromagnetic valve will be power off.

4   THE SOFTWARE DESIGN

We studied all kinds of the pressure sensors for a certain automotive sensor manufacture, and draw the conclusion that the output voltage of the pressure sensor is proportional to the air pressure. As shown in Table 1, the type of 3826DY-5190 air pressure sensor is selected for explaining the relation between the output voltage and the air pressure, and this relationship is showed by curve in Figure 5.

From the Figure 5, we can see that the relationship between the pressure and the output voltage of the sensor is very linear. So we can test the quality of the sensor as following: converting the 4~20 mA current signal that output by the standard transmitter to 1~5 V by isolation circuit, and then sending the signal to PCI data acquisition card so that computer can read the voltage to calculate the air pressure (Fang, 2006)[3].

$$Pressure = Pressure\_Max/(5-1)*(Transfer\_mVolt - 1000) \qquad (5)$$

In the type (5): pressure is the current value of air pressure, Pressure_Max is the measurement range of the pressure transmitter, Transfer_mVolt is the real-time output voltage (unit: mV) of the sensor.

Table 1. Pressure sensor parameter.

3826DY-5190 pressure sensor					
Pressure (KPa)	0	500	590	1000	1470
Output voltage (V)	0.5	1.86	2.105	3.22	4.5
Permissible error (V)	±0.06				

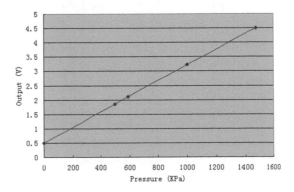

Figure 5. Output voltage of the pressure sensor.

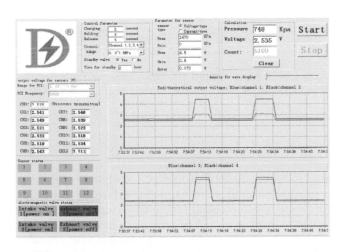

Figure 6. Software interface.
Figure 7. Flow chart. 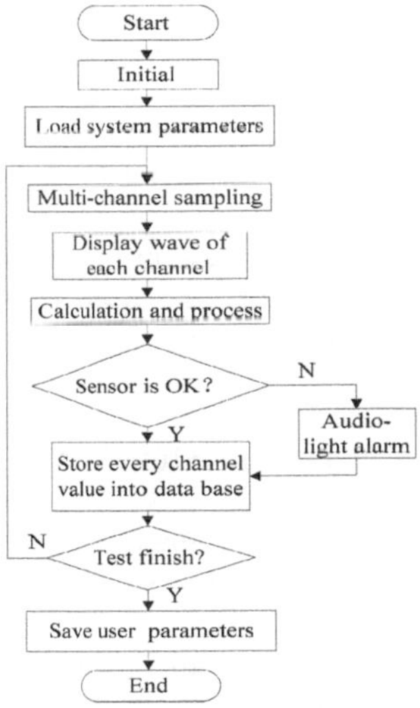

We denote the air pressure measurement range of the voltage-type sensor by P_{max} and P_{min}, and accordingly, the output voltage range is denoted by V_{max} and V_{min}, so we can calculate the theoretical output voltage of the sensor that denotes as $V_{calculate}$ according to the pressure value which calculates by equation (5).

$$V_{calculate} = V_{min} + (Pressure-P_{min})*(V_{max} - V_{min})/(P_{max} - P_{min}) \tag{6}$$

The computer will compare the experimental output voltage of each sensor with the theoretical result which calculates by equation (6), it will estimate the senor to be OK if the error is within the permissible range, otherwise the sensor is failure. The software interface that programmed with VB language is shown in Figure 6.

Table 2. Sampling data after 8 hours testing in −30°C ± 2°C condition.

Output (V)	Pressure (permissible error ± 0.072 V)					Quality
	50 KPa	100 KPa	200 KPa	300 KPa	400 KPa	
Theoretical value	0.5 V	1.071 V	2.214 V	3.357 V	4.500 V	
1#	0.516	1.089	2.232	3.372	4.517	OK
2#	0.510	1.081	2.227	3.378	4.515	OK
3#	0.508	1.083	2.230	3.374	4.510	OK
4#	0.507	1.087	2.232	3.383	4.522	OK
5#	0.494	1.076	2.232	3.389	4.545	OK
6#	0.517	1.092	2.238	3.387	4.526	OK
7#	0.512	1.082	2.231	3.368	4.512	OK
8#	0.499	1.083	2.244	3.398	4.558	OK
9#	0.510	1.083	2.225	3.377	4.510	OK
10#	0.491	1.071	2.231	3.385	4.538	OK
11#	0.511	1.085	2.226	3.374	4.512	OK
12#	0.506	1.084	2.231	3.375	4.513	OK

The application program can sample the multi-channel sensor's output voltage and display the wave on the screen, by analyzing the testing result of each channel, the software can estimate the performance of the sensors, and the testing parameters can be set on the software interface, all the test data can be stored in the hard disk (Ye, 2007)[10]. The Figure 7 shows the flow chart of the software (Hang, 2011)[4].

5 TEST RESULTS

The testing system for pressure sensor can improve the test efficiency greatly; Table 2 shows the experimental results for one kind of test, it lists out the output voltage of 12 tested sensors.

6 CONCLUSIONS

In this paper, pressure sensor testing system designed based on PCI 8602, with high accuracy standard pressure transmitter to measure air pressure, multi-channel isolate conversion circuit for AD sampling. The testing system with excellent function and simple structure is easy to be expanded according to different manufactures. The system can monitor the sensor quality and evaluate the life cycle of the pressure sensor. This testing system has been running in some automotive pressure sensor manufactures for more than two years, experimental results show that the testing system can improve the efficiency greatly and save cost for the manufacturer.

REFERENCES

[1] Art Beijing Science and Technology Development Co., Ltd. PCI8602 User's manual H[DB]. http://art-control.com/UploadFiles/PCI8602H(V6.406).pdf.

[2] Cao Lixue. 2008. Design of the further temperature system based on PCI data acquisition card. *Yiqi Yibiao Yonghu* 15(6):13–14.

[3] Fang Li-qian, Li Xiao-ming, Gao Hui, Xiong You-hong. 2006. A High Speed Travelling Wave Acquisition System based on PCI Bus. *DianWang Jishu* 30(3):80–84.

[4] Hang xiao-dong, Zhai Zheng-jun, Lu yan-hong. 2011. Multifunction and Customizable Data Gathering System Based on PCI Bus. *ce kong jishu* 30(10):15–18.

[5] Huang Lijuan, Cheng Zhixin, Yang Yunfei, Liao Xuebing. 2012. Research on Test and Control Technology of Self—propelled Artillery Servo System Based on PCI Bus. *Computer Measurement & Control.* 20(7):1898–1990.

[6] USA Agilent Technologies Co., Ltd. HCNR200 and HCNR201 High-Linearity Analog Optocouplers [DB]. http://www.waveshare.net/datasheet/AVAGO_PDF/HCNR200,HCNR201.PDF.

[7] Wei Jun-hui, Cai Jin-hui, Tang Jian-bin, Yao Yan, Jiang Qing. 2012. Comprehensive Performance Testing System for Pressure Transmitter. *Automation & Instrumentation.* 2012(2):16:19.

[8] Wu Zhengyang, Xu Huigang, Xie Qi, Dai Mei, Xu Wei. 2014. Design and Realization of Control and Proctive Switching Device Circuit Board Test System. *Computer Measurement & Control.* 22(3):674–676, 683.

[9] Yang Xiaochen, Wang Xin. 2003. High-Linearity Analog Optocoupler HCNR200/201 and its Application. *Yiqi Yibiao Yonghu* 10(5):41–42.

[10] Ye Xuesong, Gao Bo, Zhang Yingwei, Wang Bin, Wang Peng, Li Jiongxi. 2007. Research of a multi-channel neural signal acquisition system based on PCI and DSP. *Yiqi Yibiao Xuebao* 28(2):198–202.

Advances in Engineering Materials and Applied Mechanics – Zhang, Gao & Xu (Eds)
© 2016 Taylor & Francis Group, London, ISBN 978-1-138-02834-0

Research on fault online diagnosing method of motor based on fuzzy mathematics

Y. Wang
Zaozhuang University, Zaozhuang, China

ABSTRACT: It is of great importance for safe and economic running of motor to identify faults timely and accurately. On the basis of monitoring data system at the motor running workshop, this paper establishes fault online diagnosing system of motor, which includes several models and uses fuzzy fault diagnosis method. The fault diagnosis method was validated through experiment. Experiment results show that this method can get accurate diagnosis.

1   INTRODUCTION

Motor is the main motive force and driving gear in the modern industrial production, and its security and reliability is very important. Timely and accurately identifying the motor fault is one of the important measures to ensure the motor running. The motor fault generally has a variety of types and complex symptoms. As a result, it is of great difficulty to diagnose and maintain the motor fault. Contraposing fuzzification of the motor fault, diagnosing method based on fuzzy mathematics improves the accuracy of the fault diagnosis (Zhu, 2013) (Rao, 2006).

At present, the research on fuzzy diagnosis of motor is in the majority with fuzzy theory and method of fuzzy diagnosis, and less involves in field application (Xu, 2006). Besides, calculating fuzzy membership degree of the symptom only rely on the statistical data and expert experience (Xue, 2010), with no consideration of the characteristics of symptom parameters. This article applies fuzzy diagnosing method to running workshop of motor in order to realize the online real-time fault diagnosis. Considering variation characteristics of symptom parameter, this article extracts the fault symptoms through a variety of factors to describe fault symptom.

2   DIAGNOSING SYSTEM AND DIAGNOSING PROCESS

Based on Supervisory Information System (SIS) and Management Information System (MIS) of the motor running workshop, the fault symptom extraction model and the fault fuzzy recognition model are established in Integrated Modular Modeling Software (IMMS) (Gao, 2011). These models transmit data among SIS, MIS, and the client through the network server, as shown in Figure 1.

Fault symptom extraction model calculate the symptom value based on data from maintenance history and expert experience, the symptom value is sent to fault knowledge library establishing model to establish the motor fault knowledge library. In the running process of the motor, the running data from real-time monitoring data of SIS and MIS is sent to IMMS where the running data is transformed to symptom value. Using the symptom value, the fault diagnosis model judges whether the current running state is fault state and diagnoses what kind of fault the current state is. Fault diagnosis model diagnoses fault based on fault knowledge library. Data in fault knowledge library can also be reset or changed by the client through the network server.

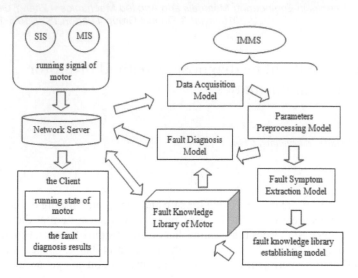

Figure 1. Fault fuzzy diagnosing system of motor.

3 FAULT SYMPTOM EXTRACTION MODEL

The steps of extracting the motor fault symptoms based on fuzzy method are as follows.

1. The emergence of motor fault symptoms is uncertain, the most important factor of describing symptoms is to get the probability that the symptom appear (Hu, 2009). According to the statistical data of on-site maintenance, when fault j occurs the probability y_{ij1} that the symptom appear can be calculated by the following formula:

$$y_{ij1} = \frac{q_{ij}}{p_j} \tag{1}$$

where p_j is the number that fault j occurs, q_{ij} is the number that symptom i appear when fault j occurs, $j = 1, 2 \ldots n$, n is the number of motor faults, $i = 1, 2 \ldots m$, m is the number of fault symptoms.

2. It is difficult to accurately describe the fault symptoms only depend on the type (1), so other factors must be taken into consideration. Consider the theoretical analysis result y_{ij2} that fault symptoms appear. For example, if fault "rotor imbalance" occurs, according to mechanism analysis of motor, the symptom "motor vibration" must appear, that is $y_2 = 1$.

3. Consider the experts' experience y_{ij3}, y_{ij3} is determined by the formula:

$$y_{ij3} = \frac{e_{ij}}{e} \tag{2}$$

where e_{ij} is the number of experts who think symptom i appears when fault j occurs, e is the total number of experts.

4. For fault symptom of phenotype and process-type, consider obvious degree y_{ij4}, This paper proposes the following two functions to describe the obvious degree of symptoms. For symptom that parameters increase (rise),

$$y_{ij4} = \begin{cases} 1 & x_i \geq a \\ \sin\left(\dfrac{\pi}{2} \cdot \dfrac{x_i - x_{ih}}{a - x_{ih}}\right) & x_{i0} < x_i < a \\ 0 & x_i \leq x_{ih} \end{cases} \tag{3}$$

where x_i is the actual value of characteristic parameters, x_{ih} is upper limit of normal value of characteristic parameters, a is threshold that characteristic parameters increase (rise) obviously.

For symptom that parameters reduce (fall),

$$
y_{ij4} = \begin{cases} 0 & x_i \geq x_{il} \\ \sin\left(\dfrac{\pi}{2} \cdot \dfrac{x_{il} - x_i}{x_{il} - b}\right) & b < x_i < x_{il} \\ 1 & x_i \leq b \end{cases}
\tag{4}
$$

where x_{il} is lower limit of normal value of characteristic parameters, b is threshold that characteristic parameters reduce (fall) obviously.

5. Give weights of y_{ij1}, y_{ij2}, y_{ij3}, y_{ij4} respectively as α, β, γ, δ, then calculate comprehensive value of symptom by the formula:

$$
y_{ij} = \alpha y_{ij1} + \beta y_{ij2} + \gamma y_{ij3} + \delta y_{ij4}
\tag{5}
$$

Among the formula

$$
\alpha + \beta + \gamma + \delta = 1
\tag{6}
$$

It is important to note that for different faults and different symptoms, weight values are different.

According to the symptoms value y_{ij} of fault j, the feature vector $Y_j = (y_{1j}, y_{2j}, \ldots y_{mj})$ of the fault can be got and fault knowledge library can be established.

4   FAULT DIAGNOSIS MODEL

Suppose the fault to be identified is labeled as u_0, characteristic vector of fault u_0 is $Y_0 = (y_1, y_2, \ldots y_m)$. The membership function which identify whether u_0 is fault j is as follow:

$$
\mu_j(u_0) = \sum_{i=1}^{m} y_i y_{ij}
\tag{7}
$$

According to the principle of maximum membership, u_0 is the fault corresponding to the maximum of $\mu_1(u_0)$, $\mu_2(u_0)$ … $\mu_n(u_0)$.

5   DIAGNOSTIC INSTANCE

Take a certain type of three-phase asynchronous motor as example to verify the fault identification method. 9 typical faults and 7 corresponding fault symptoms are selected, shown in Table 1 and Table 2 respectively.

By means of fault symptom extraction model, fault symptom value was calculated based on statistics data of motor workshop, and the fault knowledge library was established, as shown in Table 3.

Using fault diagnosis model, motor faults were diagnosed. Take 3 fault samples to be identified for example, the diagnosis results are shown in Table 4.

The Table 4 shows that the diagnosis result of 3 fault to be identified is respectively $u_1(\mu_1(u_0) = 1.85)$, $u_5(\mu_5(u_0) = 3.7)$, $u_7(\mu_7(u_0) = 2.48)$, which is accord with the actual fault. Therefore, using the diagnosis method can get accurate diagnosis result.

Table 1.   Set of faults of three-phase asynchronous motor.

Symbols	Typical faults
u_1	Stator core loosening
u_2	Stator winding interturn short circuit
u_3	Stator winding dislocation
u_4	Rotor imbalance
u_5	Rotor conducting bar fracture
u_6	The stator and rotor rubbing
u_7	Bearing wear
u_8	Low power supply voltage
u_9	Heavy load

Table 2.   Set of fault symptom of three-phase asynchronous motor.

Symbols	Fault symptom
x_1	Rotating speed decrease
x_2	Torque decrease
x_3	Vibration increase
x_4	Three-phase current imbalance
x_5	Noise increase
x_6	Temperature of stator winding rising
x_7	Current fluctuations

Table 3.   Fault knowledge library of three-phase asynchronous motor.

	u_1	u_2	u_3	u_4	u_5	u_6	u_7	u_8	u_9
x_1	0	0.71	0.57	0	0.81	0	0	0.96	0.87
x_2	0	0.89	0.71	0	0.93	0	0	0	0
x_3	0.96	0.67	0.82	1	0	0	0.84	0	0
x_4	0	0	0	0	0.61	0	0	0	0
x_5	0.89	0.73	0.71	0	0	0.95	0.92	0	0.88
x_6	0	0.69	0.59	0	0.62	0.9	0.72	0.89	0.96
x_7	0	0	0	0	0.73	0	0	0	0

Table 4.   Diagnosis results of three-phase asynchronous motor.

Fault samples	Actual fault	Diagnosis results								
		u_1	u_2	u_3	u_4	u_5	u_6	u_7	u_8	u_9
1	Stator core loosening	1.85	1.4	1.53	1	0	0.95	1.76	0	0.88
2	Rotor conducting bar fracture	0	2.29	1.87	0	3.7	0.9	0.72	1.85	1.83
3	Bearing wear	1.85	2.09	2.12	1	0.62	1.85	2.48	0.89	1.84

6   CONCLUSION

1. Fault diagnosis of motor has been achieved based on the fuzzy diagnosis method, and accurate diagnosis results are obtained.
2. The fuzzy diagnosis method is applied to the motor workshop, online real-time fault diagnosis is realized with the help of data monitoring system in factory.

3. A method to extract the fault symptoms is put forward, the method considers the characteristics of symptom parameters, and extract fault symptoms through a variety of factors.

REFERENCES

[1] Gao Jianqiang, Zhang Ying, Chen Hongwei. 2011, Development and Application of SIS Functional Models Based on the integrated modular modeling software. *Thermal Power Generation* 40(2): 64–68.
[2] Hu Wei, Hu Jingtao. 2009, Application of Weighted Fuzzy Relative Entropy to Fuzzy Recognition of Motor Rotor Fault. *Information and Control* 38(03): 326–330, 336.
[3] Rao Hong, Fu Mingfu, Jin Yuhua. 2006, Comparing and Analyzing the Membership Function in Fuzzy Fault Diagnosis. *Journal of Nanchang University (Natural Science)* 30(04): 383–385, 397.
[4] Xu Yuxiu, Xing Gang, Yuan Peixin. 2000. Research on Application of Fuzzy Mathematics to Fault Diagnosis System. *High Voltage Apparatus* (5): 19–21, 39.
[5] Xue Han, Xie Lili, Ye Liuyi. 2010, Study of Fuzzy Fault Diagnosis Expert System in Motor System. *Computer Measurement & Control* 18(1): 8–10.
[6] Zhu, Chaopeng. Hao, Wei. Hao, Wangshen. et al. 2013. Application of Fuzzy Fusion to Motor Fault Diagnosis. *Coal Mine Machinety* 34(09): 300–302.

Advances in Engineering Materials and Applied Mechanics – Zhang, Gao & Xu (Eds)
© 2016 Taylor & Francis Group, London, ISBN 978-1-138-02834-0

Mo coating on AISI 4340 steel deposited by supersonic plasma spraying and HVAS

Z.X. Yang, G.M. Liu & Y. Wang
Faculty of Remanufacturing Engineering, AAFE, Beijing, China

ABSTRACT: Two kinds of Mo coatings were sprayed by supersonic plasma spraying and HVAS on the AISI 4340 steel in order to improve the anti-wear performance, and the microstructure was also studied. The microstructure of super plasma sprayed Mo coating showed more compact and less pores than HVAS Mo coating, and the ball-on-disc sliding wear test with heavy loading and boundary lubrication indicated that the supersonic plasma Mo coating had a better anti-wear performance as compared with that of HVAS Mo coating which made nothing much difference to the anti-wear performance AISI 4340 steel. The worn surface revealed the abrasive wear mechanism accompanied with grooves and pits of supersonic plasma sprayed Mo coating, and the worn surface of HVAS Mo coating had a serious plastic deformation with oxidation, which belongs to oxidation and adhesive wear.

1   INTRODUCTION

AISI 4340 steel is widely applied in military equipment, aircraft components, drilling device and so on because of its excellent behavior in wear, corrosion, fatigue, high temperature and high speed operating conditions. However, with the rapid development of modern mechanical industry, severe environment resulted from the continuous working, which will bring about a steady performance decline of those components surface during their service life. Thus, many researchers have adopted lots of surface treatment to improve or enhance the surface properties for a longer service life[1–3].

Mo is a kind of low activity refractory metal with the 2630 °C melting point, as well as possess a comprehensive excellent physical and mechanical properties, such as high hardness, good wear resistance, high thermal conductivity, good adhesion and anti-oxidation, resistance to the erosion of molten copper or iron. Spraying Mo coating has been widely used in piston ring, gear wheels, synchronous ring and internal combustion engine cylinder parts, which have achieved good results[4,5]. Recently, Manjunath compared the plasma sprayed Al_2O_3 30%Mo and Mo coating on the dry sliding wear properties, which proved that the latter had a significantly better tribological performance[6].

However, the tribological performance with boundary lubrication of sprayed Mo coating is scarce in the literature, combined with the above-mentioned method, this paper was aimed to study the microstructure and tribological performance with boundary lubrication of Mo coatings prepared by supersonic plasma spraying and High Velocity Arc Spraying (HVAS) on AISI 4340 steel. Meanwhile, the difference of wear mechanism between Mo coatings and AISI 4340 steel was discussed through the worn surface analysis.

2   EXPERIMENTAL PROCEDURE

2.1   *Materials and spraying*

The spray materials are commercially available pure Mo powder (Sang Yao Development Co., LTD, Beijing) and Mo wire (Sunstone Tungsten Molybdenum Co., LTD, Beijing) for

Table 1. Process parameters for supersonic plasma spraying.

Spraying distance/mm	Voltage/V	Current/A	Air flow/ (L·min⁻¹)	H₂ flow/ (L·min⁻¹)	Powder feeding rate/(g·min⁻¹)
100	140	390	120	5.0	35

Table 2. Process parameters for HVAS.

Spraying distance/mm	Voltage/V	Current/A	Air pressure/ MPa	Wire feeding rate/(m·min⁻¹)
150	35	180	0.65	4

Table 3. The boundary lubrication wear test conditions.

Acquisition frequency/Hz	Loading/N	Time/min	Single stroke distance/mm
10	50	30	4

the supersonic plasma spraying and HVAS respectively. The particle size of pure Mo powder ranges from 45 um to 96 um and the diameter of Mo wire is 2 mm, the purity of both exceeds 99.9%.

Supersonic plasma sprayed Mo coating was prepared by the high-efficient plasma spraying system with the novel spraying gun (HEP-Jet), the excellent characteristics of this system have been described in[7], the process parameters are shown in the Table 1. As for the HVAS Mo coating, the QD10 HVAS system (Dahao Ruifa Thermal Spray Machinery Co., LTD, Shanghai) was chosen, and the process parameters are shown in Table 2. Thickness of all the coatings should be controlled within the range of 0.3 mm–0.4 mm.

2.2 *Experimentation*

After the Mo coating specimens were prepared, the microstructure of cross-section and surface was characterized by an FESEM (Nova NanoSEM50, FEI, USA) that accompanied with XRD (X-Max 80, OXFORD, UK). The wear test with boundary lubrication of two coatings was conducted by reciprocating ball-on-disc wear device (UMT-3, CETR, USA), the material of wear counterpart is zirconia ball of the hardness 1200 HV with a diameter of 4 mm, and the wear test conditions are shown in Table 3. The 50 N loading was selected to create a huge pressure on the surface of coating, which attempted to imitate the rigorous heavy load working condition. The anti-wear performance was evaluated by the wear volume measured by 3D laser high-precision topography tester (OLS40-SU, OLYMPUS, Japan). All of the specimens were cleaned with acetone in the ultrasonic cleaner before the SEM observation.

3 RESULTS AND DISCUSSION

3.1 *Microstructure*

The microstructure of the supersonic plasma sprayed and HVAS Mo coatings is shown in Figures 1 and 2 respectively, both of the two kinds Mo coating show a coarse surface pattern, but the supersonic sprayed Mo coating seems like more rough relatively, emerging many gibbous protuberances rather than many smooth local areas resulted from the spread of molten

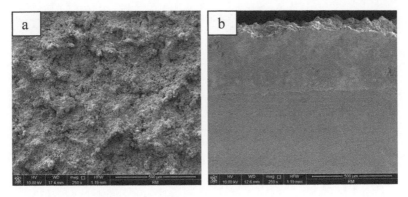

Figure 1. SEM micrograph showing of supersonic plasma sprayed Mo coating (a) surface and (b) cross-section.

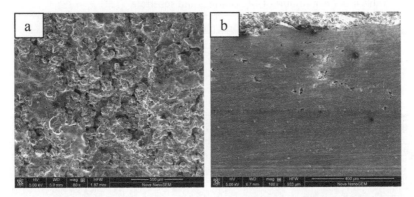

Figure 2. SEM micrograph showing of HVAS Mo coating (a) surface and (b) cross-section.

Mo particles in the HVAS Mo coating. However, what's interesting is that the cross-sectional micrograph revealed another feature, just on the contrary, the supersonic plasma sprayed Mo coating represented a more compact and homogeneous microstructure instead of many pores in the HVAS Mo coating. Generally speaking, the roughness of sprayed coating surface has little connection with the consistency, because the density of sprayed coating is decided by the molten degree and the speed of particles' impact[8]. The nonuniform molten degree and relatively low impact speed of HVAS Mo particles can explain this phenomenon, and other researchers also discovered a porous microstructure in the arc sprayed Mo coatings[9, 10].

Figure 3 is a magnified micrograph of the interface between Mo coating and substrate, from which we can find that the interfaces were compact and no detachment was observed. The supersonic plasma sprayed Mo coating was compact and almost no micro-pores could be found near the interface, while there exist some micro-pores between the coating and substrate in the HVAS Mo coating, even some of them are perforative, but beyond that, some pits can also be observed, which probably generated from the pullout of the half-molten particles during the polishing, this can also be inferred from the traces left on the coating and substrate.

3.2 *Anti-wear performance*

The wear volume of coatings and substrate are displayed in Table 4, from Table 4 it can be seen that the wear volume of supersonic plasma sprayed Mo coating is the least of 0.025 mm^3, which is just about 40% of the substrate, meanwhile, this number indicated that

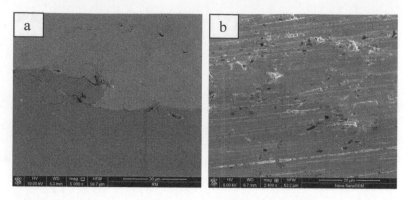

Figure 3. Magnified micrograph of the interface (a) supersonic plasma spraying and (b) HVAS.

Table 4. Wear volume of Mo coatings and substrate.

Item	Supersonic plasma sprayed Mo coating	HVAS Mo coating	Substrate
Wear volume/mm³	0.025	0.055	0.060

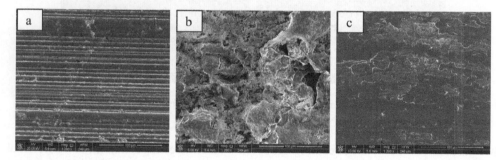

Figure 4. SEM micrographs of worn surface (a) supersonic plasma sprayed Mo coating, (b) HVAS Mo coating, and (c) substrate.

the supersonic plasma sprayed Mo coating on the surface of AISI 4340 steel improved the anti-wear performance. However, compared with the 0.06 mm³ of substrate, the wear volume of HVAS Mo coating is only less 0.005 than the substrate, which illustrated that the HVAS Mo coating makes no difference in the anti-wear performance of AISI 4340 steel when under the heavy load and boundary lubrication working condition.

SEM micrographs of the worn surface of Mo coatings and substrate are shown in the Figure 4, from which the three distinct wear failures can be seen. There is a obvious abrasive wear mechanism we can find in the worn surface of supersonic plasma sprayed Mo coating where many parallel grooves and micro-pits are left along the sliding direction, it can be deduced that the grooves was ploughed by the pullout of hard particles which belongs to three-body abrasive wear. As for HVAS Mo coating, the XRD results revealed that serious oxidation happened in the bright areas because of the contact with oxygen by the worn-through pores, and so much plastic deformation occurred that even have filled the pores, so the wear mechanism is the combination of adhesive wear and oxidation wear, similarly, the substrate worn surface also detected oxides, but the content is much lower than that of the HVAS Mo coating, at the same time, the failure mode is the delamination of materials, which can seen from the large area exfoliation around which many cracks caused by the repetitive cycle of compressive stress under subsurface of substrate. The reduplicative generation and wear-out of oxides may responsible for the big wear volume of HVAS Mo coating and substrate[11].

4 CONCLUSION

Two kinds of Mo coatings were deposited on the surface of AISI 4340 steel substrate by the supersonic plasma spraying and HVAS, the microstructure and anti-wear performance with boundary lubrication were studied in this paper.

The surface microstructure of two kinds Mo coatings are rough, but the cross-sectional microstructure of supersonic plasma sprayed Mo coating has a denser and lesser pores than that of HVAS Mo coating, the interface of both coatings were compact and no detachment was observed.

The wear test with boundary lubrication results indicate that the supersonic plasma sprayed Mo coating enforced the anti-performance of AISI 4340 steel substrate, while the HVAS Mo coating seems has nothing much advantage of AISI 4340 steel substrate.

The wear mechanism of supersonic plasma sprayed Mo coating is abrasive wear, and that of HVAS Mo coating belongs to oxidation wear and adhesive wear.

ACKNOWLEDGMENT

This work was partially supported by the Beijing Natural Science Foundation of China (2152031) and the project of State Key Laboratory of Remanufacturing of China (9140C8502010C85).

REFERENCES

[1] Marcelino, P.N. et al. 2001. Effects of tungsten carbide thermal spray coating by HP/HVOF and hard chromium electroplating on AISI 4340 high strength steel. *Surface and Coatings Technology* 138(2–3): 113–124.
[2] Voorwald, H.J.C. et al. 2005. Evaluation of WC-17Co and WC-10Co-4Cr thermal spray coating by HVOF on the fatigue and corrosion strength of AISI 4340 steel. *Surface and Coating Technology* 190(2–3): 155–164.
[3] Meng, F.J. et al. 2008. Microstructure and tribological performance of 45crnimova steel welded surfacing layer. *Chinese Journal of Mechanical Engineering* 44(4): 150–153.
[4] Patel, P.D. et al. 2013. Experimental investigation on life cycle analysis of the Moly(Mo) coated piston ring in C.I. Engine. *International Conference on Advances in Tribology and Engineering System(ICATES)*. 321–329. Ahmedabad, Springer India.
[5] Laribi, M. et al. 2007. Study of mechanical behavior of molybdenum coating using sliding wear and impact tests. *Wear* 262(11–12): 1330–1336.
[6] Manjunath, S.S. & Basavarajappa, S. 2014. Dry sliding wear behavior of plasma sprayed Al_2O_3 30-%Mo and Mo coating. *Procedia Materials Science* 5: 248–255.
[7] Piao, Z.Y. et al. 2013. Characterization of Fe-base alloy coating deposited by supersonic plasma spraying. *Fusion Engineering and Design* 88(11): 2933–2938.
[8] Sampath, S. et al. 2003. Development of process maps for plasma spray: case study for molybdenum. *Materials Science and Engineering A* 348(1–2): 54–66.
[9] Zajchowski, P. & Crapo, H.B. 1996. Evaluation of three dual-wire electric arc-sprayed coatings: industrial note. *Journal of Thermal Spray Technology* 5(4): 457–461.
[10] Naimi, A. et al. 2012. Microstructure and corrosion resistance of molybdenum and aluminum coatings thermally sprayed on 7075-T6 aluminum alloy. *Physicochemical Problems of Materials Protection* 48(5): 557–562.
[11] Stolarski, T.A. & Tobe, T. 2001. The effect of spraying distance on wear resistance of molybdenum coatings. *Wear* 249(12): 1096–1102.

Advances in Engineering Materials and Applied Mechanics – Zhang, Gao & Xu (Eds)
© 2016 Taylor & Francis Group, London, ISBN 978-1-138-02834-0

Research of the coating textiles' coating Gram weight measurement system based on infrared

A.H. Zhang & W.P. Du
School of Mechanical Engineering, Nantong University, Nantong, Jiangsu, China

S.Y. Ma
Quan Ji Textile Coating E&T Center, Tongzhou, Nantong, Jiangsu, China

P. Liao
Graduate School of Nantong University, Nantong, Jiangsu, China

ABSTRACT: Gram weight gauge can provide objective weight control guidance during the production of the coating textiles. Nowadays, the ray Gram weight gauge has been used in measuring the coating Gram weight of coating textiles, but it has the shortcomings: statistical error, radioactive hazard, expensiveness and others. To compensate for these shortcomings, the Gram weight measurement system based on infrared has been designed in general. Before designing the system, the physical structure of the coating textiles and the infrared spectra of the coating textiles have been studied in detail. The whole Gram weight measuring system includes two parts: mechanical system, and measuring and control system. Finally, through some preliminary experimental, infrared technology was initially validated to measure the coating Gram weight of the coating textiles.

1 INTRODUCTION

Coating textiles, a technical product, occupies an important position in the textile industry. It has the functions of waterproof and breathable, anti-static, moisture heat, radiation, pollution antibacterial, and so on, which is widely used in industrial, construction, garment, agriculture, military, medical and other industries. With development of economy and continuous improvement of living standards, more and more high quality coating textiles will be needed.

The coating textiles' Gram weight of the finished product is an important indicator of the physical, directly impacting on tear strength, fabric water pressure resistance, the costs of production and economic benefits. Therefore, On-line System for measuring the coating textiles' Gram weight of the production in the production process is the important question which is concerned by people, so measurement of the coating textiles' Gram weight is a base test case on manufacturing industry of coating textiles.

As the system of On-line measurement Gram weight, the ray Gram weight gauge is the most widely used, but it has the shortcomings: statistical error, radioactive hazard, expensiveness and others. Compared with the ray Gram weight gauge (the β-ray or γ-ray), the infrared thickness gauge has low price, high safeness and reliableness. So in recent years, infrared technology has been used to develop infrared Gram weight measurement for domestic and international people.

2 THE PRINCIPLE OF THE COATING TEXTILES' COATING GRAM WEIGHT MEASUREMENT SYSTEM

According to the principle that different molecular bonds (O-H, C-H, N-H etc) can absorb certain infrared wavelength strongly, the content of the number and composition is proportional to energy which is absorbed, so it is possible to calculate the content of the component by calculating the absorption of energy. Mostly, Polyurethane (-[-O-CONH-]n-) is used as a coating material in manufacturing industry for coating textiles. The structural model of typical coating textiles was obtained by means of micrography (Fig. 1). IR spectra of the base fabric and IR spectra of the coating textiles are shown in Figure 2 and Figure 3, according to the IR spectra of the base fabric and coating textiles, we can conclude that there is a new infrared absorption peak (the wave number is between 3400–3500 cm^{-1}) after coating, furthermore, we can get that N-H brings about this new absorption peak, so we can use the absorption properties of N-H to measure the coating textiles' coating Gram weight.

3 THE DESIGN OF OVERALL SCHEME

According to the requirements of manufacturing industry of coating textiles, based on the technology of coating Gram weight measurement by infrared, we design the coating textiles' coating Gram weight measurement system in general. Through this system, the problem of non-contact monitoring of the Gram weight of coating textiles, which occurs during the

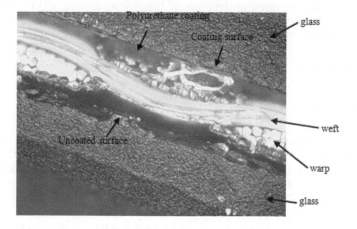

Figure 1. Picture of a typical coating textiles micrography.

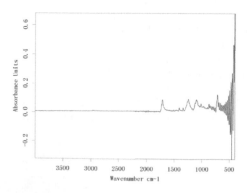

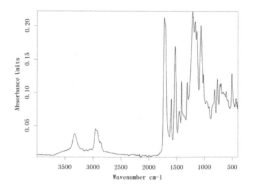

Figure 2. IR spectra of the base fabric.

Figure 3. IR spectra of the coating textiles.

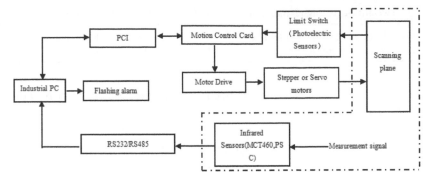

Figure 4. The hardware components of the measuring and control system.

process of production, can be solved. The whole Gram weight measuring system includes two parts: mechanical system, and measuring and control system.

3.1 Mechanical system

The mechanical system, in a word, is the scanning plane, which provides platform for the whole measuring system, and the measuring probe scanning, which was installed on the scanning plane, can scan in a specific movement path.

3.2 Measuring and control system

According to the composition, the measuring and control system can be divided into two parts: hardware and software.

The hardware, made of many different features modular components, is the foundation to realize various functions of the entire measurement and control system (shown in Fig. 4). The hardware of measuring and control system can be divided into two parts: the hardware of measuring system and the hardware of motion control system. Industrial PC is the key part of the hardware of measuring system and alarm system. The hardware of motion control system mainly includes the motion control cards. The industrial PC receives signals, completes the processing of data and sends out the motion commands. The motion control cards receive the commands and then execute the corresponding action.

Software is the core of measurement and control system. The measuring and control system software is made of data processing program, motion control program and alarm program, which is controlled by built-in program. The measurement and control system is developed by VC Programming. Motion control programs are aimed to achieve control of the measuring mechanism of action; data processing procedures are designed to achieve operational measurement data, display and storage; alarm program will control flash alarm to alert when weight does not comply with requirements.

4 PRELIMINARY TEST

Based on the N-H absorption properties of coating textiles, a proper Infrared Weight probe should be selected.

Infrared Engineering (NDC) Inc and Process Sensors Corporation (PSC) are the world's two best infrared sensor manufacturers. They can provide different infrared sensors, which can measure different material composition. The measurement principle of these probes is that particular wavelength emitted of infrared light will be absorbed by specific chemical bonds (e.g. NH, OH, CH). Taking applicability, price and performance and some others into consideration, MCT460 probe it stands out. The MCT460, provided by PSC, was used to measure the grams of coating textiles in a static environment. The relationship between the weight of coating textiles and the current output by MCT460 probe is given in Table 1.

Table 1. The weight of coating textiles and the current output by MCT460 probe.

The Gram weight of the coating textiles (x)/g	104	130	156	182	208
The output of current by sensor (y)/mA	6.30	6.68	7.08	7.47	7.85

The correlation coefficient can be calculated from the deflection of the strip by the following equation:

$$r = \frac{\sum\limits_{i=1}^{n}(x_i - \bar{x})(y_i - \bar{y})}{\sqrt{\sum\limits_{i=1}^{n}(x_i - \bar{x})^2 \cdot \sum\limits_{i=1}^{n}(y_i - \bar{y})^2}} \tag{1}$$

After calculating, $r = 0.999$, so there is a high degree of linear correlation between the weight of the coating textiles (x) and the output of current by sensor (y). So, we infer that accurately coating weight can be obtained after the calibration and error compensation of the sensor.

5 SUMMARY

The coating textiles' coating Gram weight measurement system, based on infrared, can be used to solve the traditional detection method by manual, enabling real-time detection of the coating textiles' Gram weigh, finally, acquiring lower defect rate and production cycle of the product.

ACKNOWLEDGEMENTS

This work was financially supported by Prospective Joint Research of Jiangsu Province (BY2013042-04), University Advantage Discipline Construction Engineering Projects of Jiangsu Province. Its corresponding author is Ping Liao (e-mail: liao.p@ntu.edu.cn Tel: 13815201558).

REFERENCES

Alan Owens, S.C. Bayliss, G.W. Fraser, S.J. Gurman. 1997. On the relationship between total electron photoyield and X-ray absorption coefficien. *Nuclear Instrument and Methods in Physics Research*, 385(6):557.

Fung, W. 2002. *Coated and laminated textiles*. 1. Woodhead Publishing Ltd.

Giessmann, A. 2012. *Coating Substrates and Textiles*. 2. Springer-Verlag Berlin Heidelberg.

H.X.F. 1989. RCT-β Thickness Gauge has been Developed. *Rare Metals*, 8(2):354–365.

Qiang, W. 2013. The Research Progress of Functional Coating Fabric Production Equipments. *Dyeing & Finishing Technology*, (7):87–90.

Tan, K.K., Huang, S.N., Lee, T.H. 2000. Development of a Gpc-based PID controller for unstable systems with deadtime. *ISA Transactions*, 39:57–70.

Xing, Z. 2011. Development of the Film Thickness Measuring Instrument Based on Infrared [D]. Harbin: *Harbin Institute of Technology*.

Yang Ni, Wang Yu-tian, Lv Jiang-tao, Li Huan-huan. 11–12 April 2009. "Research on Thickness Measurement of Transparent Object Based on CCD Vision System," *Measuring Technology and Mechatronics Automation, 2009. ICMTMA'09. International Conference on, vol. 1, no*, pp.113–116.

Yang Ni, Wang Yu-tian. 17–18 Oct. 2008. "Novel Float Glass Thickness Measurement based on Image Processing," *Computational Intelligence and Design, 2008. ISCID'08. International Symposium on, vol. 2, no*, pp. 94–97.

Advances in Engineering Materials and Applied Mechanics – Zhang, Gao & Xu (Eds)
© 2016 Taylor & Francis Group, London, ISBN 978-1-138-02834-0

Microstructure and properties of TiC-Ni(Mo) cermet composite coating prepared by Self-propagating High-temperature Synthesis

X.Y. Zhang & J. Tan
Science and Technology on Remanufacturing Laboratory, Academy of Armored Force Engineering, Beijing, China

W. Ren
Central Iron and Steel Research Institute, Beijing, China

M.L. Dong
Institute of Surface/Interface Science and Technology, Harbin Engineering University, Harbin, China

ABSTRACT: The TiC-Ni(Mo) cermet composite coating was prepared on 7 A52 aluminum alloy by the Self-propagating High-temperature Synthesis (SHS). The microstructure and properties of coating were investigated by means of SEM, EDS, XRD, micro-hardness test and bonding strength test. The results showed that: the coating consisted of TiC ceramic phase and Ni, Mo, Al metal phase. Each phase of coating combined very well. The surface of coating is the high strength metallurgical bonding. The micro-hardness of the coating is about 8 to 10 times of the substrate. The result of bonding strength test further proved the binding strength between the coating and the substrate was satisfactory. The experimental results showed that adding the metallic Mo and Ni, which has a good dissolvability in Al, to transition area of samples, and those alloying element improved the bonding strength of coating and substrate.

1   INTRODUCTION

SHS is a kind of highly effective new technical means of the preparation of coating material. This technology has the advantages of simple process, high product purity, reaction to the low cost and low environmental pollution[1]. The SHS combustion system is a high exothermic system that heat can be produced by itself, and the ways that coating bonding with substrate was high quality metallurgical bonding[2–4]. In order to improve the densification level of the product, experiment usually using the SHS method combined with hot-pressing and centrifugal technology[5–6].

7 A52 aluminum alloy has many advantages such as high specific strength, high specific rigidity and low density, so it widely used in aerospace manufacturing, armored equipment and machinery[7–9]. But the alloy material may appear as some coarse compounds, porosity in casting process, and it will appear as some practical matters such as deformation, wear, corrosion, etc[10]. Ceramic materials with high hardness, high melting point, good thermal stability, etc, but the cold and hot shock resistance of the ceramic material is poor[11]. Combining ceramic material with metal material can get high performance composite materials[12]. This paper prepared TiC-Ni(Mo) cermet composite coating on the surface of 7a52 aluminum alloy by using the SHS method combined with hot pressing technology, and analyzed the microstructure and properties of coating deeply.

Table 1. Properties of the raw material powders.

Powder	Size (μm)	Purity (%)
C	0.53	>99
Ti	48	>99
Ni	45	>99
Mo	45	>99
Al	380	>99

Table 2. Weight ratio of each chemical element.

Chemical element	Ti	C	Ni	Mo
Weight%	64	16	16	4

2 EXPERIMENTAL PROCEDURE

2.1 *Materials*

Five types of starting powders were prepared in advance. It is shown in Table 1. 7 A52 aluminum alloy was machined to the size of 128 mm*128 mm*17 mm and used as the substrate. The mass rate of elementary composition of 7 A52 aluminum alloy is 4.0~4.8 Zn, 2.0~2.8 Mg, 0.20~0.50 Mn, 0.30 Fe, 0.25 Si, and rest of Al.

2.2 *Coating procedure*

The preformed blank was made by four types of starting powders except Al powder. The specific test process was as follows: the mass ratio of those four powders is shown in Table 2, mixing the mixed powder by ball mill for 8~10 h, putting the mixed powder and Al powder to the 65°C drying box for 6~8 h. Double layers of reactant powders were placed on the substrate. The lower layer consisted of 4 g Al powder, and the upper layer consisted of 350 g mixed powder. The powder compacts were surrounded by asbestos mold. Under the pressure of 6.4 MPa, preloading was carried out on the sample. And make it as perform of relative density of 55%. Some igniting agent, which can be ignited by resistance wire, was placed on the top of mixed powder compact to trigger the reaction. Starting pressurized to mould when temperature of the substrate rising to 370°C. And when the pressure value reached 33 Mpa maintained the pressure 8 to 15 s. Finally, a coating with a thickness of 4.5 mm was produced on the aluminum alloy substrate.

2.3 *Characterizations*

After the preparing the TiC-Ni(Mo) coating, the microstructure of coating and substrate was characterized by Scanning Electron Microstructure (SEM); X-Ray Diffraction (XRD) and Energy Dispersive Spectroscopy (EDS) were used for chemical and phase composition analysis severally; Using the ΠMT-3 micro-hardness tester measured the micro-hardness in different areas of the substrate and the coating, the test load was 200 g; Using the stretching method tested the bonding strength between coating and substrate. Specimens were bonded with counterpart by E7 glue. Then the specimens were solidified for 3 hours under 120°C.

3 RESULTS AND DISCUSSION

3.1 *Macroscopic structure*

After experiment the average thickness of coating is about 4.5 mm, and the average thickness of substrate was about 16.3 mm. Through the result it can be seen that the thickness of substrate is lower than before. There are two reasons could explain this phenomenon. First, when the hydraulic press has pressed the mould, there are many molten aluminum would overflow from mould through the exhaust hole. Second, because the bottom layer of coating is Al powder, so some of molten aluminum would permeate to the coating layer with those molten Al powder. Macroscopic structure is shown in Figure 1, from Figure 1 it can be seen that the effect of layer binding between coating and substrate was beyond to compare. There is no obvious crack, hole and other flaws.

3.2 *Phase constitution*

The morphology of transition layer is shown in Figure 2. The coating area is on the left side and the substrate area is on the right side. It can be seen that the interface of transition layer is a discontinuous line, and some part of substrate has fused with the surface of coating. Figure 3 shows the XRD result of the coating area. It can be seen that the phase of coating layer consist of a large amount of ceramic phase TiC and a part of metal phase Ni, Al, Mo. The Al metal phase is shown in Figure 3 further explains that the coating is metallurgical bonded to the substrate.

3.3 *Microstructure*

The microstructure of the coating is shown in Figure 4. Figure 4-b and 4-c is the zoom for area B and C respectively. The result of EDS of Figure 4-b is shown in Table 3. From

Figure 1. Morphology of specimen.

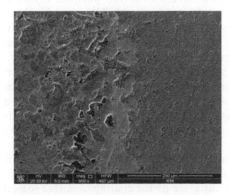

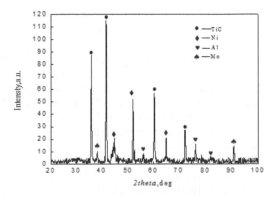

Figure 2. Interfacial microstructure of coating and substrate.

Figure 3. X-ray diffraction pattern of coating.

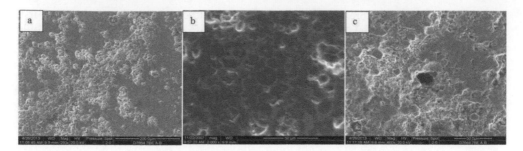

Figure 4. Microstructure of coating: a) microstructure and morphology; b) zoom for area B; c) zoom for area C.

Table 3. Result of spectrum analysis on b area.

Element	Weight%	Atomic%
C	16.88	53.15
Ti	60.79	47.85
Mo	5.99	2.36
Ni	16.34	10.64
Totals	100	

Table 4. Result of spectrum analysis on c area.

Element	Weight%	Atomic%
C	21.73	53.15
Ti	75.02	45.86
Mo	3.25	0.99
Totals	100	

Figure 4-b it can be seen that TiC particle is distributed intensively, the gray black sphere particle is TiC particle, and the light metallic binding phase surround the TiC particle. Through Table 3 it can be found that the atomic ratio between C atom and Ti atom is greater than one to one. It indicated that some of C atom does not dissolve and diffuse in the coating layer. So, this C atom exists in metallic binding phase in free form on high temperature condition. The result of EDS of TiC particle is shown in Table 4. From Table 4 it can be found that there is a little of Mo atom exists in TiC particle. It means that some part of Mo atom instead of Ti atom has been fused with TiC particle. From Tables 3 and 4 it can be ensured that metallic binding phase is consist of metal phase Ni, Mo, but it contains lower content of Mo atom. It is said that the wettability between Ni and TiC would be improved obviously when added trace amounts of Mo atom to Ni atom[13]. The metallic binding phase mainly has three functions in the coating layer. First, it makes the TiC particle uniting closely, and improves the density of coating layer. Second, it works as a kind of thinner in combustion system, metal phase will absorb a part of heat from system. Third, with the increase of the metallic binding phase, the difficulty of TiC particle continue to grow up is also increasing. So, it enhanced the stability of surface of the coating.

Some black holes can be found in Figure 4-c. The reason of forming this kind of defect is complicated. One reason is that some gas can be formed through the combustion reaction, and finally those various gas stay in the low-energy interface of the coating. Another reason is that the time of starting pressuring is legs behind the time of starting solidification, and

Table 5. Hardness of sample at different area.

Items	Hardness of substrate/HV	Hardness of transition area/HV	Hardness of cladding/HV
No. 1	110	280	810
No. 2	118	320	1020
No. 3	125	390	1230

Table 6. Results of bonding strength test.

Items	Measurement/mm	Bonding strength/MPa	Thickness of ceramic layer/mm	Average value/MPa
No. 1	φ25	32	2.3	31
No. 2	φ25	30	2.6	
No. 3	φ25	31	2.7	

under this situation the densification degree of the coating is lower then anticipation. The performance of coating would be restricted by free form C atom, micropore, porosity and other kind of defects.

3.4 *Micro-hardness and bonding strength*

The micro-hardness of the different area of specimens is measured, and the result is shown in Table 5. Each test area has three test points, and all of those test points are randomly selected. It can be seen that the value of micro-hardness is about 120 HV in substrate area. In transition area, it can be found that value of micro-hardness is higher than substrate area. The value of micro-hardness increased obviously in ceramic area, which is 8 to 10 times of substrate. So this process improved hardness of raw material obviously.

The result of bonding strength test is shown in Table 6. The processing size of each specimen is φ25. The total thickness of each specimen is the same but the thickness of coating layer shows difference in different specimens. From Table 6 it can be seen that the average bonding strength of each specimen is the same, and the average value of bonding strength is 31 Mpa. The thickest coating layer is No. 3 specimen, and the bonding strength of No. 3 specimen is also the biggest. It can be seen that the bonding strength of these three specimens is greater than 30 Mpa. So the result of measurement is satisfactory.

4 CONCLUSION

In this experiment, Ti-C-Ni(Mo) composite is prepared on the surface of aluminum alloy by using the SHS method. The bonding strength between coating and substrate is well; the surface of coating is flat; the TiC particle distributed equably in coating layer; the interior of coating layer does not have obvious defects such as cracks, porosity and loose.

The result of micro-hardness test shows that the hardness of different areas has improved obviously, which is about 8 to 10 times of substrate.

The coating is metallurgical bonded to the substrate. The Al powder plays an important role in enhancing the bonding strength between coating and substrate.

Ni and Mo are the major phase in the transition layer, both of these two atom have a good wettability with substrate. The wettability between Ni and TiC would be improved obviously when added trace amounts of Mo and Ni.

ACKNOWLEDGEMENT

The authors would like to express their gratitude for the financial support of Scientific Research Projects of the Army under grant No. 40401050201.

REFERENCES

[1] Subrahmanyam, J. Vijayakumar, M. 1992. Self-propagating high-temperature synthesis. *Journal of materials science* 27(23): 6249–6273.

[2] Mossino, P. 2004. Some aspects in self-propagating high-temperature synthesis. *Ceramics International* 30(3): 311–332.

[3] Sheng, L.Y. Yang, F. Guo, J.T. et al. 2013. Investigation on NiAl-TiC-Al$_2$O$_3$ composite prepared by self-propagation high temperature synthesis with hot extrusion. *Composites Part B: Engineering* 45(1): 785–791.

[4] Ando, R.K. Lee, L. 2008. Interative Residual Rescaling: An Analysis and Generalization of LSI. *Proceedings of the 24th SIGIR*: 154–162.

[5] Wang, W.M. Mei, B.C. Fu, Z.Y. Yuan, R.Z. 1996. Fabrication of Dense TiB$_2$/NiAl Intermetallic Compound-Matrix Composites by SHS+HP. *Acta Materiae Composite Sinica* 01: 30–34.

[6] Wei S.C., Zhang T.A., Yang H., et al. 2000. SHS-centrifugal Technique Research. *Materials Review* 09: 17–18.

[7] Zhang. Y. Qing, S.G. 2013. Performance Study of TiC Ceramic Coating on Aluminum Alloy Prepared by SHS + HP. *Journal of Academy of Armored Force Engineering* (3): 19.

[8] Yu, J. Wang, K. Xu, Y. et al. 2005. Microstructures and properties of 7 A52 aluminum alloy welded joint by twin wire welding. *Transactions-china Welding Institution* 26(10): 87.

[9] Zhang, Y.Y. Liu, H. Zhu, X.B. 2013. Microstructure and Property of Welded Joint of 7 A52 Al Alloy by MIG Welding Process. *Hot Working Technology* 42(19): 173–174.

[10] Teng, Z.G. Wang, L.J. Zhang, W.J. Zhang, J.L. 2009. Analysis of Compounds in 7 A52 Aluminum Alloy. *Light Alloy Fabrication Technology* (5): 13–14.

[11] Yang, M. Zou, Z.D. Liu, X.Z., Liu L. 2004. The Research and Prospect of Ceramic and Metal Bonding, *Shandong Metallurgy* 26(1): 37–40.

[12] Zhao, X. Liew, K.M. 2011. Free vibration analysis of functionally graded conical shell panels by a meshless method. *Composite Structures* 93(2): 649–664.

[13] Wang, L.S. 1994. *Special ceramics*. Hunan: Central south university of technology press.

Aerospace science and technology

Advances in Engineering Materials and Applied Mechanics – Zhang, Gao & Xu (Eds)
© 2016 Taylor & Francis Group, London, ISBN 978-1-138-02834-0

Measurement of the main performance of Model ZDB-1320 cryopump

R.P. Shao, W. Sun, Y.S. Zhao, L.C. Sun & Y. Wang
Beijing Institute of Spacecraft Environment Engineering, Beijing, China

ABSTRACT: Cryopumps have been widely used in large space environment simulators because of the large pumping speed (volume rate of flow), cleanness and the convenience to operate for the user. To meet the demand of the high vacuum system in a special thermal vacuum facility in which the large quantity of pure Oxygen gas was admitted, Model ZDB-1320 Refrigerator cooled cryopumps were designed in China Academy of Space Technology in 2007. The cryopump used one dual stage cold head with high second stage cooling capacity of 18 W at 20 K. The cryopanels were contacted directly with the second cold head stage, and were not covered with activated charcoal to pump the oxygen gas. The compressor unit supplied the cryopump with compressed helium gas to generate the low temperatures for the cold head and the cryopanels. The thermal radiation shield and the battle were not attached to the first cold head stage. They were cooled by the liquid nitrogen. In order to obtain the values of the main performance of the cryopump, some different measuring apparatus were established, thus the pumping speed, crossover, cooldown time and the maximum throughput were measured. For measuring the pumping speed, the standard conductance method was adopted. The test results indicated that the pumping speed (for N2) was up to 60,000l/s, and the cooldown time was about 190 min.

1 INTRODUCTION

The pumping speed, cooldown time and crossover value are the main performance of the refrigerator pump. There are two basic standard methods for measuring the pumping speed. The first method for measuring the pumping speed is the throughput method, in which a steady gas flow is injected into the test dome while the inlet pressure is measured. The second method for measuring the pumping speed is the orifice method, which is used when there is very small throughput occurring at very small inlet pressures. It is based on measuring the ratio of pressures in a two-chamber test dome where the two chambers are separated by a wall with a circular orifice. In practice, the measurement of small gas throughput may be complicated and the very small gas throughput may be not easy to be measured exactly. For this reason, the orifice method for the high and ultra-high vacuum pump is often adopted to measure the pumping speed, which can avoid the direct measurement of throughput. The measurement of the pumping speed of Model ZDB-1320 refrigerator cryopump with the orifice method is described below. Based on the pumping speed measurement apparatus, the cooldown time and crossover value are also measured.

2 STRUCTURE DESIGN AND FUNCTION

Model ZDB-1320 cryopump essentially consists of the cryopanel, cooled thermal radiation shield and the baffle, one two-stage cold head and the pump housing with flanges. The schematic drawing is shown in Figure 1. In the crypump, the flange of the second stage of the cold head is contacted to the cryopanels, and the flange of the first stage is empty and has no loads. The operating temperature of the baffle is sensed by a Pt 100 sensor. A silicon diode

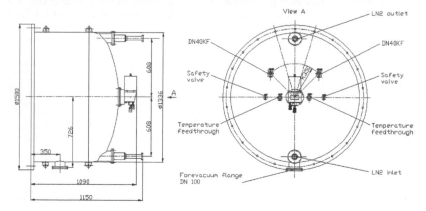

Figure 1.   Schematic drawing.

is installed at the cryopanel. The baffle and the thermal radiation shield are cooled by liquid nitrogen. The cryopanels are cooled by the second stage of the cold head to below 15 K, so that gases and vapors can be condensed and adsorbed on the very cold surface. The thermal radiation shield and the baffle can reflect the radiated heat from the outside and prevent the heat radiations onto the cryopanels. The cryopumps are designed to remove the gases in the environmental simulation facilities.

3   TEST METHODS

The orifice method is applicable to high vacuum pumps. Molecular flow conditions shall be present in the test dome. This method is recommended for low gas throughputs where no suitable gas flow meters are available. The orifice diameter in the test dome shall be adapted to the expected volume flow rate of the test pump in order to avoid excessively high pressures which would result in laminar flow conditions through the orifice.

A thin circular orifice plate divides the test dome into two volumes. The volume flow rate is given by

$$S = C\left(\frac{P_1 - P_{10}}{P_2 - P_{20}} - 1\right) \tag{1}$$

where C is the calculated conductance, taking into account the orifice size and the gas properties. The base pressures, P_{10} and P_{20}, in the upper and lower chamber of the test dome are measured after baking and before admission of the gas. The conductance of the orifice with diameter d, and thickness L may be calculated using the following equation:

$$C = \sqrt{\frac{\pi RT}{32M}}\left(\frac{1}{1 + L/d}\right)d^2 \tag{2}$$

where The term $(1 + L/d)$ is a correction factor (only valid for L << d) that can be defined as the average transition probability through the orifice.

Take care that the equation is used with consistent units. Specific values such as [R = 8.314 J/(mol · K)], [M_{air} = 28.96 × 10⁻³ kg/mol] and [T = 293 K (20 °C)] will give, in cubic meters per second,

$$C_{air} = 91d^2/(1 + L/d) \tag{3}$$

where L and d are measured in meters.

The arrangement of the measuring equipment is given in Figure 2.

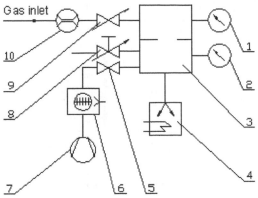

1,2—Vacuum gauge (P1, P2); 3—Test dome; 4—Model 1320 Cryopump; 5—gate valve; 6—Turbomolecular pump; 7—Roughing pump; 8—Adjustable valve; 9—Gas inlet valve; 10   Flowmeter.

Figure 2.   Arrangement for measuring the pumping speed with the orifice method.

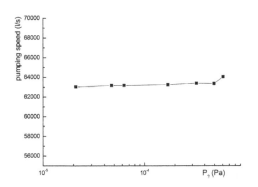

Figure 3.   Pumping speed curve of 1# cryopump.

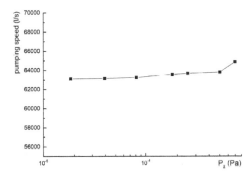

Figure 4.   Pumping speed curve of 2# cryopump.

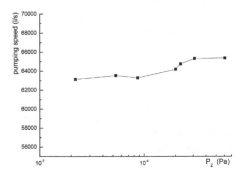

Figure 5.   Pumping speed curve of 3# cryopump.

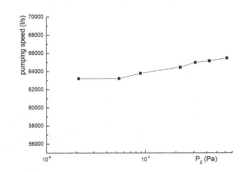

Figure 6.   Pumping speed curve of 4# cryopump.

4   TEST DATA AND ANALYSIS

During the test, the orifice with diameter 200 mm was adopted, and the ratio of the pressures P_1 and P_2 was about 18. The pumping speed curves of four cryopumps are given from Figure 3 to Figure 6. To ensure that the mean free path of the gas particles in the orifice is not smaller than twice the orifice diameter, the maximum value P1max when the gas was

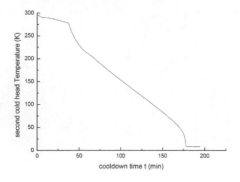

Figure 7. Cooldown time curve of 1# cryopump.

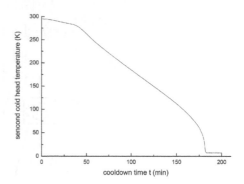

Figure 8. Cooldown time curve of 2# cryopump.

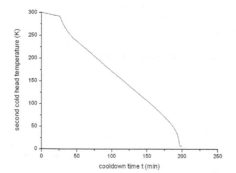

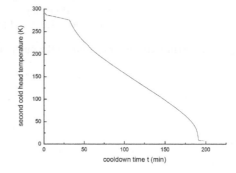

Figure 9. Cooldown time curve of 3# cryopump. Figure 10. Cooldown time curve of 4# cryopump.

introduced to the test dome must be not more than 1.47×10^{-2} Pa. The cooldown time curves of four pumps are shown in Figure 7 to Figure 10.

The crossover value was tested in a thermal vacuum facility. When the pump reached its operating temperature and the chamber of the facility was pumped down to Pcut-in 2.2 Pa by the forevacuum pumps, close the forevacuum valves and open the high vacuum gate valve between cryopump and vacuum chamber, the second head cold temperature rised to 20 K from 7.8 K, but for a while (not more than two minutes) the temperature went back to about 8 K, and the cryopump reached its normal operating mode again. And the crossover value can be obtained by calculating the multiplication of Pcut-in and volume of the vacuum chamber (V = 120 m³). Every test only one cryopump can be used to pump. The results showed the crossover value of four cryopumps tested was about 266000 Pa · L.

The test curves above show that pumping speed is constant from the magnitude of 10^{-6}Pa to 10^{-3} Pa for each cryopump, but during the magnitude of 10^{-2} Pa, the pumping speed increases. The value of the pumping speed for each cryopump is about 63000l/s to 64000l/s, which can be seen as a constant. The cooldown time is about 190 min.

5 CONCLUSIONS

Four cryopumps were firstly made, and from the measuring system these relative perform-ances were tested, such as the pumping speed, cooldown time and crossover value. The results here indicated that pumping speed for N_2 was not influenced if the cryopanel was not covered with activated charcoal, and these parameters all meet our needs. And the four cryopumps have been used successfully in a thermal vacuum facility.

Corresponding author: Rongping Shao (1972–), female, Senior Engineer, major in aerospace leak detection technology and vacuum system design, e-mail: shaovacuum@126.com; P.O.Box 5142-46, Beijing, 100094, China.

REFERENCES

Baechler, W.G. 1987. *Cryopumps for research and industry*. Vacuum 37(1–2): 21–29.

Daoan, Da. 2004. *Vacuum design handbook*. Beijing: National Defense Industry Press.

Gilankar, S.G. 2008. Experimental Verification of capture coefficients for a cylindrical cryopanel of closed cycle refrigerator cryopump. *Journal of Physics: Conference series*.

Hafner, H.U. & Klein, H.H. et al. 1990. New methods and investigations for regenerating refrigerator cryopumps. *Vacuum*. 41(7–9): 1840–1842.

Ikegami, K., S. Nakajima, et al. 1988. Design and performance characteristics of refrigerator-cooled cryopumps for the RIKEN ring cyclotron. *Vacuum* 38(2): 99–102.

James M. Lafferty. 1998. *Foundations of vacuum science and technology*. New York. John Wiley & Sons, Inc.

Juhnke, C. & Klein, H.H., et al. 1993. The crossover of refrigerator-cooled cryopumps. *Vacuum* 44(5–7): 717–719.

Kimo M. Welch, 1999, Recommended practices for measuring the performance and characteristics of closed-loop gaseous helium cryopumps. *Journal of Vacuum Science and Technology A*, 17(5): 3081–3095.

Walker, G. 1983. *Cryocoolers, Part I*. New York: Plenum Press.

Advances in Engineering Materials and Applied Mechanics – Zhang, Gao & Xu (Eds)
© 2016 Taylor & Francis Group, London, ISBN 978-1-138-02834-0

Evaluation of influence factors on the portable ultrasonic leak detection method for the space station on orbit

W. Sun, R.X. Yan, L.C. Sun, Y. Wang & W.D. Li
Beijing Institute of Spacecraft Environment Engineering, Beijing, China

ABSTRACT: It is very important to study the portable ultrasonic leak detection method for the on-orbit space station and astronaut safety. To insure the method availability, analyzing the influence factors is necessary. The method research was based on the aero acoustic theory, and the leak acoustic source production, propagation and attenuation were analyzed. The mathematic module of leak acoustic with some influence factors was established. Furthermore, the effects of the detection distance, angle and horizontal offset were analyzed by the experiment for different leaks. The research conclusions indicate that the main influence factors of the method include the leak size, differential pressure, detection distance, detection angle and horizontal offset and others. The leak ultrasonic signal is declining as the detection distance and the horizontal offset increasing, and the best detection angle range is ±30°. These conclusions provide some references and advices for improving the portable ultrasonic leak detection method availability.

1 INTRODUCTION

With the development of manned spacecraft technology, the sealing proper of the space station on orbit turns out to be very important for the safety of spacecraft and astronauts. Due to the influence of temperature mutation, micro-meteor, space debris and other space environment factors, the sealing cabin of space station on orbit for long time maybe leak, and the gas will lose largely to threaten the life of astronauts[1]. Hence, it is very important to study the on-orbit spacecraft leak detection method to find the leakage point location. The portable ultrasonic leak detection method is an effective way to locate the leak for the on-orbit space station at present. NASA developed the UL101 ultrasonic leak detector to find a leakage point in the international space station on orbit in 2001[2]. France Dynergys Company has developed a visual leak detector[3], and INFICON Company has a handheld ultrasonic leak detector named as 'whisper'[4]. And then Beijing University of Aeronautics and Astronautics, Tsinghua University, Xi'an Institute of Technology and other units also studied the acoustic detection theory, but there is little research about the spacecraft ultrasonic leak detection technology on orbit, Beijing Institute of Environmental Engineering has paid attention to studying the portable ultrasonic leak detection method for space station on orbit[5]. This paper focuses on analysis of influence factors on the portable ultrasonic leak detection method for the space station on orbit.

2 THEORY ANALYSIS

When there are some leaks in the cabin of space station on orbit, the gas flows from the atmosphere environment of the cabin inside to the vacuum environment outside. In the leak process, there are many burbles, which can produce some quadruple acoustical sources. The theory and the process of the leak acoustical source are very complex, but the gas leak process can be expressed by the continuity equation, momentum equation, flow equation and heat transfer equation[6]. Assuming gas density as ρ, flow velocity as v, time as t, pressure as p, specific heat as γ, dynamic viscosity as η, the analysis of leak process can be given by the flowing equations:

$$\frac{\partial \rho}{\partial t} + \nabla \rho v = 0 \tag{1}$$

$$\rho \frac{\partial v_i}{\partial t} + \rho v_i \frac{\partial v_i}{\partial x_i} = -\nabla \frac{\partial p}{\partial x_i} + \eta \frac{\partial}{\partial x_i}\left(\frac{\partial v_i}{\partial x_j} + \frac{\partial v_j}{\partial x_i} - \frac{2}{3}\delta_{ij}\frac{\partial v_k}{\partial x_k}\right) \tag{2}$$

$$p' = c^2 p' \tag{3}$$

where $i, j, k = 1, 2, 3$; $p' =$ acoustic pressure change; $\rho' =$ gas density change; $c =$ acoustic velocity.

According to Lighthill's tensor, the following equations can be given:

$$\frac{\partial^2 \rho}{\partial t^2} - c_0{}^2 \nabla^2 \rho = \frac{\partial^2 T_{ij}}{\partial x_i \partial x_j} \tag{4}$$

$$T_{ij} = \rho v_i v_j - \eta\left(\frac{\partial v_i}{\partial x_j} + \frac{\partial v_j}{\partial x_i} - \frac{2}{3}\delta_{ij}\frac{\partial v_k}{\partial x_k}\right) + \delta_{ij}(p - c_0{}^2\rho) \tag{5}$$

where $T_{ij} =$ Lighthill tensor; the $\delta_{ij} =$ Longneck symbol, while $i = j$, $\delta_{ij} = 1$ and then if $i \neq j$, $\delta_{ij} = 0$[7]. Because the leak is very small, the gas destiny change value and the gas viscosity tensor which are both little can be ignored. Hence, the Lighthill tensor can be given by the equation as $T_{ij} = \rho_0 v_i v_j$. And then the flow equation can be transformed to acoustic pressure wave equation. Through the Rayleigh sound analogy method and the dimension compare method, the acoustic pressure can be given by the following equation[8]:

$$p = N\frac{D\rho_0 V^4}{rc_0^2} \tag{6}$$

where $N =$ coefficient; $V =$ leak gas velocity; $r =$ sound field extent; $c_0 = 331.45$ m/s.

The leak acoustic power can be expressed in the form[8].

$$W = [p^2/(\rho_0 c_0^2)]c_0 S = KSD^2 \rho_0 V^8 / c_0^5 \tag{7}$$

where $S =$ accepting area; $K =$ Lighthill coefficient. The whole leak process can be analyzed as the adiabatic flow process. According to gas dynamics theory, the gas leak downstream velocity can be expressed in the following form[9].

$$V = \sqrt{\frac{2\gamma}{\gamma - 1}\frac{p(t)}{\rho_0}\left[1 - \left(\frac{p_{out}}{p(t)}\right)^{\frac{\gamma-1}{\gamma}}\right]} \tag{8}$$

where $\gamma =$ air specific heat coefficient; $P_{out} =$ pressure of the cabin outside; $p(t) =$ pressure of the cabin; $\rho_0 =$ gas density.

Because of the cabin leak, the cabin pressure will drop fast. Hence, the cabin pressure can be expressed in the form:

$$p(t) = (p_0 - p_{out})e^{-(c/v)t} + p_0 \tag{9}$$

where $v =$ volume of cabin; $p_0 =$ atmosphere; $p_{out} =$ pressure of the cabin outside; $c = 1.34 \times 10^3 \times D^4/l$; $D =$ leak diameter; $l =$ length of the leak.

However, the leak ultrasonic attenuation in the air propagation will affect the leak detection. Acoustic absorption and scattering attenuations obey the exponential rule. And then the sound pressure can be express in the form[10].

$$p = p_0 e^{-\alpha L} \tag{10}$$

where a = attenuation coefficient which is the sum of the absorption and scattering attenuation coefficients; L = distance; and acoustic diffusing attenuation is related with wave front area. Supposed leak ultrasonic as spherical wave, the acoustic diffusing attenuation obeys the rule as r^{-2}, r = spherical wave semi-diameter.

Hence, the mathematic module of leak acoustic power that is detected in the distance L can be established in the following form.

$$W = \frac{A}{L^2} \frac{KD^2}{c_0^5 \rho_0^3} S_c \cos\alpha \left(\frac{2\gamma}{\gamma-1}\right)^4 \left[\frac{p(t)}{\rho_0}\left(1 - \left(\frac{p_{out}}{p(t)}\right)^{\frac{\gamma-1}{\gamma}}\right)\right]^4 e^{-2aL} \tag{11}$$

where S_c = receiving area of ultrasonic sensor; α = angle between the accepting face of the sensor and leak normal direction.

Some conclusions can be given by the Equation (11). The leak acoustic influence factors mainly include the leak size, differential pressure, detection distance, detection angle, horizontal offset, acoustic attenuation and others.

3 EXPERIMENT

3.1 *Experiment method*

A vacuum environment is provided by the ϕ400 mm vacuum simulation system as space environment. And the laboratory environment simulates the inside environment of the space station sealing cabin. Then the gas can flow from the laboratory into the ϕ400 mm vacuum simulation system through the different diameter leaks. The portable ultrasonic on-orbit leak detector is fixed on a 3D coordinate 4 free degree equipment which makes the detector move along X, Y, Z axis and rotate around Z for the different distance, offset and angle experimentations. The experiment system is built as shown in the Figure 1.

There are three experiment steps in the whole process as follows:

1. Pressure environment. The vacuum environment of the vessel inside is gained by vacuum pump. At the same time, the valve I is opened and the valve II is closed.
2. Leak detection. When the pressure of the vessel is less than 100 Pa, the valve II is opened, and the air flows into the vessel through the leak. Then the portable ultrasonic on-orbit

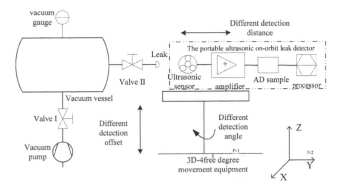

Figure 1. Experimental system.

leak detector tests the leak ultrasonic signals in different conditions by moving the 3D-4 free degree movement equipment to changing the detector position.

3. Analysis. The leak ultrasonic signal in one position can be tested thirty times and recorded. The results can be analysis by extracting character frequency spectrum including the sound pressure in 40 kHz frequency and averaging the thirty results for studying the influence factors.

3.2 Detection distance experiment result and analysis

The formula (11) shows that the leak sound power will attenuate as the detection distance change. Hence, the experiment is carried out for evaluating the detection distance influence. During the experiment, the leak detector is fixed on the 3D-4 free degree movement equipment and moving along the Y-axis from 20 mm to 500 mm as shown in Figure 1. Measure the sound pressure ($P = W^{1/2}/\rho_0 c_0$) in different distance for 3 different leaks. Under the same conditions, the leak sound pressure is only the function of distance L. Some experiment results are given by the curves as shown in Figure 2, in which the X-coordinate denotes the distance value (mm). And the Y-coordinate denotes the leak sound pressure value (mV). The result curves are gained by the third-order polynomial fitting to the data. The fitting formula is shown as the following form:

$$P(L) = AL^3 + BL^2 + CL + D \tag{12}$$

where A, B, C, D = coefficients.

According to the fitted curve and the coefficients of the polynomial equation as shown in Figure 2, it is clearly gain that the model of sound pressure change with distance is accord with the 3-order Taylor polynomial as the following form.

$$f(x) = A'(1+x)^{-1} \tag{13}$$

According to the Taylor expansion, the formula (11) will change as the following form.

$$P(L) = W_{(L)}^{\frac{1}{2}}/\rho_0 c_0^2 = A_1 \frac{1}{L} = F(L) = A'(1+(L-1))^{-1} = f(x) \tag{14}$$

And then the following equation can be gained by 3-order Taylor polynomial expansion.

$$P(L) = A' \frac{1}{1+L} = -aL^3 + bL^2 - cL + d \tag{15}$$

where A', a, b, c, d are the coefficients.

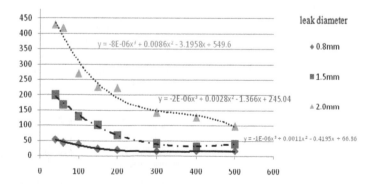

Figure 2.　Acoustic pressure curve with distance of different leaks.

Hence, the experiment result is accord with theory analysis, the detection distance can affect the intensity of the leak signal detected. When the detection distance is increased, there will be a sharp decline in the sound pressure detected.

3.3 Detection angle experiment and results

The formula (11) shows that the received leak ultrasonic sound pressure is related with the detection angle as cosine ruler. There are the detection angle experiments in a position. The leak detector is fixed on the 3D-4 free degree movement equipment in the point P which denotes the detection position (20 mm, −200 mm) and rotated around Z-axis as shown in Figure 1.

Figure 3 shows that R denotes the flat radius, c is defined as the rotation angle, l is the distance from detector central axis to the leak central axis, l1 denotes the distance between detector and leak plane, S is known as the receiving area of sensor, Sc denotes the efficiency receiving area, and a is the detection angle between sensor normal and the connect line leak to receiver center. So, the detection angle can be expressed in the following form.

$$a = \left| \arctan\left(\frac{l - R\sin c}{l1 + R - R\cos c} \right) - c \right| \tag{16}$$

The detection angle can be achieved by the rotating flat along Z-axis as the angle c including 0°, 30°, 60°, 90°. The experiment results are shown in Figure 4 in which the Y coordinate is sound pressure and the X coordinate is the detection angle between sound source and receiver sensor surface. The Y coordinate is the sound pressure mV.

As the shown in Figure 4, at the rotation angle of 30°, the receiving signal increases largely, and at the rotation angle of 60° the sound pressure gets the peak value, but at the rotation

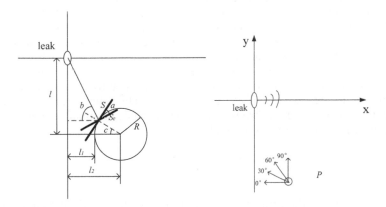

Figure 3. A schematic view of the sound pressure test on leak detection direction.

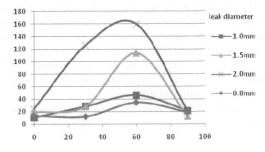

Figure 4. Sound pressure test results at receiving point P with angle changes.

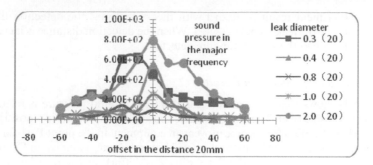

Figure 5. Different detection offset experiment result curve.

angle c of 90°, the signal decreases rapidly. The detection angle value a is 88.6°, 50°, 4°, and 47° when the rotation angle c is 0°, 30°, 60°, 90° as the formula (16). And the effective detection area is correlation with $cos\ a$, which is 0.024, 0.643, 0.99, 0.681, the biggest signal in the experiments acquired at the rotation angle c of 60° ($a = 4°$), which is the same with theoretical analysis value.

Furthermore, the sensor has effective receiving angle ±30°, which is related with the detection angle. Hence, the best detection angle is 0° that the detector is face to the leak center, but the angle is not more than ±30°.

3.4 Detection offset experiment and results

For analyzing the detection offset influence, the experiment can be carried out in the different detection offsets. During the experiment, the leak detector is fixed on the 3D-4 free degree movement equipment and moving on the Z- or X-axis at different offsets from 20 mm to 60 mm. Measure the sound pressure changing as the offset change for different leaks.

The experiment results are given by the curves of the different leaks as shown in Figure 5, in which the X-coordinate denotes the offset value (mm) and the Y-coordinate denotes the leak sound pressure value (mV) in the major frequency in the same detection distance 20 mm.

According to the experiment data, it is found that the leak ultrasonic sound pressure in the major frequency is larger as the offset value is smaller. The sound pressure is the largest in the leak center position. Hence, through detecting the value of sound pressure in the major frequency, the leak center position is found.

4 CONCLUSION

This paper has analyzed the principle of leak ultrasonic and established the theory module which has given some influence factors. At the same time, some experiments have carried out for validating the theory analysis. Some conclusions can be gained as the following:

1. The method is related with the detection distance, detection angle and detection offset and others. And the detection sensitivity is best when the detector is face to the leak.
2. When the leak location is unknown, the detector should be vertical with the surface tested and the detection angle is not more than ±30°.
3. The detection offset affects the detection, the leak ultrasonic sound pressure is largest in the leak center. Hence, the leak can be found easily by the moving detector along the surface to observe the sound pressure variety.
4. These conclusions will provide some references and guides for researching the leak ultrasonic detection method.

CORRESPONDING AUTHOR

Wei Sun (1984–), male, Engineer, major in aerospace leak detection technology; e-mail: sunwei84@163.com; P.O. Box 5142-46, Beijing, 100094, China.

REFERENCES

[1] Xinfa, Y. & Rongxin, Y. 2008. The study of space station leak detection technology on orbit. Spacecraft Environment Engineering, 25(2): 177~182.
[2] Hoover, A. 2002. *Maryland Company Expanding Technology in Space—NASA Won't Leave Earth Without the CTRL UL101*. New York: CTRL Systems, Inc.
[3] http://www.leakshooter.icoc.cc.
[4] http://www.inficon.com.
[5] Lei, Q. & Wei, S. et al. 2014. A method of non-contact ultrasonic leak detection. *Spacecraft Environment Engineering*, 31(2):212–216.
[6] Goldstein, M.E. 2014. *Aeroacoustics*. Beijing: National Defense Industry Press.
[7] Xiaofeng, S. & Sheng, Z. 1994. *Aeroacoustics*. Beijing: National Defense Industry Press.
[8] Dayou, Ma. 2005. *Foundations of modern acoustic theory*. Beijing: Science press.
[9] Zhuoru, Chen & Zhaoming, Jin. 2004. *Fluid Mechanics*. Beijing: Higher education press.
[10] Ruo, Feng. 1999. *Ultrasonics Handbook*. Nanjing: Nan Jing University press.

Advances in Engineering Materials and Applied Mechanics – Zhang, Gao & Xu (Eds)
© 2016 Taylor & Francis Group, London, ISBN 978-1-138-02834-0

Leak testing of Chinese Space Laboratory hatch in a thermal vacuum environment

L. Wang, E.J. Liu & L.C. Sun
Beijing Institute of Spacecraft Environment Engineering, Beijing, China

ABSTRACT: Chinese Space Laboratory (CSL) hatch successfully passed an extensive leak testing in a thermal vacuum environment. The testing was performed at China Academy of Space Technology (CAST) in the 2 meters diameter by 3 meters length "CMK2000" vacuum chamber. As the hatch was thermally cycled through its analytically predicted temperature extremes, a leak rate test was performed on all of the hatch seals using the helium mass spectrometer leak detectors. The test was part of the qualification program for the hatch used on the Manned Spaceships.

1   INTRODUCTION

Due to the expected longevity of the program and the limitations on resupply gas, the CLS leak rate requirements are stringent and testing on the ground is extensive. The leak rate requirement for the entire CLS assembly is less than $6.0E\text{-}3\ Pa \cdot m^3/s$. Potential leakage paths to space include feed through penetrations through which fluids and power pass to different areas on the laboratory, Common Berthing Mechanisms that mate the various modules together, and windows that allow crew observations of external activities and space. This paper focuses on another leakage path, the hatches that provide crew access to the CLS modules.

Throughout this paper, references will be made to certain sides of the Hatch assembly. The external side will sometimes be referred to as the Outboard or Extravehicular Activity (EVA) side, while the internal side will sometimes be referred to as the Inboard or Intravehicular Activity (IVA) side. The EVA side is the side that will face away from the modules toward space if so exposed and the IVA side will face toward the modules.

2   CHINESE SPACE LABORATORY HATCH

When two modules between laboratory and spaceship are connected, the hatches between them will generally be open, however, for an unused port the Hatch is exposed to the vacuum of space. CLS modules include one hatch. The leakage requirement for the Hatch seals is less than $1.0E\text{-}4\ Pa \cdot m^3/s$ helium and is verified by testing on the ground. The aluminum Hatch seals to the module via two Perimeter seals. The seal is installed on the Hatch, and seals the outer perimeter of the module when the Hatch is latched. (See Fig. 1 for an illustration of the Hatch. Fig. 2 shows perimeter seal cross section).

The Hatch has a 180-millimeter diameter double—paned window at its center. The window assemblies are sealed using two silicone O-rings. The Hatch has a Manual Pressure Equalization Valve (MPEV), used to equalize pressure across the Hatch prior to use. The MPEV has one silicone O-ring in its flange.

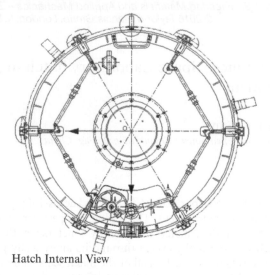

Hatch Internal View

Figure 1.   Chinese Space Station Hatch.

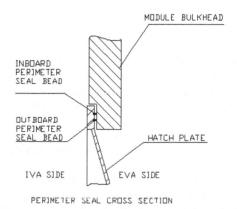

Figure 2.   Perimeter seal cross section of hatch and module.

3   THERMAL VACUUM TEST

Completion of the Thermal Vacuum test was a major milestone for the Qualification program of the CLS Hatch. The purpose of this test was to verify that the Hatch assembly maintained the structural integrity and performance characteristics after exposure to the thermal vacuum environment. The thermal vacuum test temperatures were defined as −45ºC to +45ºC for vacuum containment. The leak tests were performed at ambient temperature and at the different temperature extremes throughout 2.5 thermal cycles. (Fig. 3 shows the thermal vacuum cycles and leak tests). All the Hatch seals were tested during the thermal vacuum testing.

4   DESCRIPTION OF LEAK TEST FIXTURES AND PLUMBING

The Qualification Hatch was latched to an adapter plate during the thermal vacuum testing. The adapter plate simulated the module bulkhead and retained the perimeter seal. The test setup schematic is shown in Figure 4.

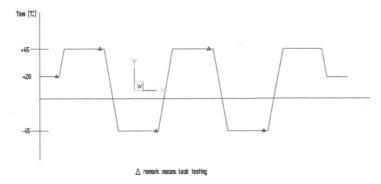

Figure 3.　Thermal vacuum cycles and leak tests.

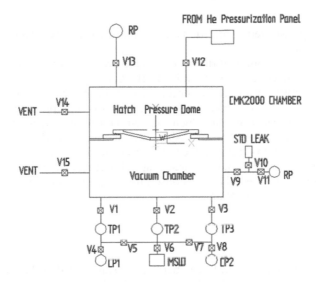

Figure 4.　Hatch thermal vacuum leak test setup.

All chamber plumbing was stainless steel tube or flex hoses, and metal gaskets were used in the joints wherever possible. Thorough helium leak tests were performed on all of the test plumbing as it was built up and installed into the chamber.

5　LEAK TESTING VIA THE CMK2000 VACUUM CHAMBER ENVIRONMENT

CAST's CMK2000 Space Simulation Chamber is a 2 meters diameter by 3 meters length horizontal vacuum chamber capable of a 1.0E-5 to 1.0E-6 Pascal vacuum. The chamber is pumped by 6 turbomolecular pumps (KYKY FF-250/2000) and by 4 dry compressing vacuum pumps (Leybold SP 250). The turbopump foreline was temporarily modified to allow the installation of a portable leak detector that was used to back the turbopump, thus sampling the chamber environment. The leak detector used on the chamber was a Leybold L300i, a counter-flow leak detector whose higher inlet pressure made it acceptable for the turbopump foreline pressure of 1.3 Pa. A three-valve manifold was installed on the top of the chamber and was used in conjunction with a dry pump to evacuate the helium leaks, which used to determine the chamber sensitivity. The manifold for the standard leaks was located as far as was practical from the MSLD and turbopump. The Helium leaks used for the test

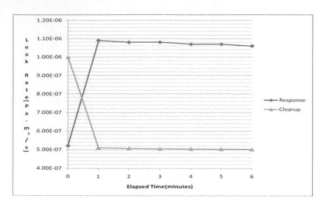

Figure 5. CMK2000 chamber calibration curve.

were ALCATEL's pressurized permeation and capillary leaks in sizes ranging from 1.0E-5 to 1.0E-6 Pa·m³/s helium.

Once the chamber was pumped down to a stable pressure, the leak detector was valved onto the turbopump foreline. Then the valve to the backing pump was slowly closed, while the foreline pressure was monitored to ensure that it did not rise above 1.3 Pa, the maximum desired foreline pressure. Once the leak detector was fully backing the turbopump, a system calibration was performed in which the helium leak was evacuated by a dry pump, then valved onto the chamber. This procedure was referred to as the "system calibration". Response and cleanup results from a typical system calibration with a 1.5E-6 Pa·m³/s helium leak are shown in Figure 5. The data showed that leak testing of the chamber volume in this configuration resulted in an excellent response time. Sensitivity of the chamber system was approximately 2.5E-8 Pa·m³/s helium.

6 TEST CONDUCT

The thermal vacuum test began on December 20, 2010 and ended on January 5, 2011. When the Hatch was assembled to the chamber, the chamber becomes two parts: the Hatch pressure dome (IVA side) and the vacuum vessel (EVA side). During the initial pumpdown of the chamber and during subsequent temperature shifts, the pressure in the Hatch pressure dome volume was equalized with the vacuum chamber pressure. Chamber pressure was maintained at 1.3E-3 Pa, or less during the tests, with an average pressure of approximately 5.0E-5 Pascal.

The total leak rate of all the Hatch seals was tested during every temperature extreme of the initial 2 cycles. The seal testing was chosen in an attempt to optimize chamber helium background, which would rise as testing continued, and to shorten test time.

The test of the total leak rate requires the MSLD to sample the chamber volume. The MSLD connected to the chamber turbopump foreline was started, calibrated, and brought on line to the turbopump foreline. After the initial pumpdown of the vacuum chamber, a leak test was performed on the Hatch itself: a helium tracer probe was used to spray helium on all leakage paths, such as the Shaft, the Window, the Manual Pressure Equalization Valve (MPEV), and the Perimeter seals. After the pumpdown of the total chamber, a leak test was performed on the chamber itself: a helium tracer probe was used to spary helium on all chamber feedthroughs and leakage paths. A chamber calibration was performed to obtain the chamber sensitivity; then the leak was closed and the background was allowed to return to its original value.

The Hatch pressure dome was evacuated with three turbopumps. Then the pumps were valved off. The MSLD background reading was recorded, and the Hatch pressure

Table 1. Leak rate test results.

Hatch cycle	Hatch temp (°C)	Total seals (Pa·m³/s he)	Remarks
Leak rate requirement (Pa·m³/s he)	All	1.0E-4	–
Baseline ambient	+20	3.4E-7	–
First cycle hot	+45	2.1E-7	–
First cycle cold	–45	2.0E-7	–
Second cycle hot	+45	1.8E-6	–
Second cycle cold	–45	1.1E-7	–

dome was backfilled with nitrogen to 50 kPa, then backfilled with helium to 100 kPa, and then the MSLD stably reading was recorded again. The leak rate was only observed for few minutes (a minimum of five minutes) to determine whether there was a leakage around the seals. The pressure shall be maintained until stabilization of the leak detector output is achieved. The Hatch pressure dome was evacuated again by using a dry compressing vacuum pump. When the MSLD background reading was recovered, another chamber calibration was performed to ensure that there had been no loss in chamber sensitivity. Finally, the MSLD connected to the chamber turbopump was valved off and shut down.

Following this leak test, the Hatch pressure dome was evacuated, then equalized with the vacuum vessel pressure in preparation for heating or cooling to the next temperature extreme.

7 TEST RESULTS

Leak rate data, consisting of the test requirements and test results, are summarized in Table 1. The data show that the leak rate of the Hatch was anomaly in the second +45°C cycle. Uncertainty of the leak test and stabilization of the leak detector output should be the mainly reasons.

8 CONCLUSIONS

Optimal test sensitivity was obtained on the test system by installing the MSLD on the turbopump forline. The leak rate tests were a successful example of complex quantitative leak rate testing in a middle vacuum chamber.

ACKNOWLEDGEMENTS

The authors would like to thank their team members, the CAST CSL leak test technicians: Qi Feifei, Li Wenbin, Zhang Haifeng, Shi Jijun, Sun Lizhi, and Zhao Jianchao, for their hard work and dedication in the performance of Chinese Space Laboratory leak testing.

REFERENCES

[1] A. Holt. S. Underwood. 1998. Leak Rate Testing of the International Space Station Hatch in a Thermal Vacuum Environment, *20th Space Simulation Conference*, NASA/CR-1998-208598.
[2] S. Underwood. A. Holt. 1998. *International Space Station Module Acceptance Leak Testing*.

[3] S. Underwood. A. Holt. 1998. Structural Weld Leak Testing in Support of the International Space Station Node1, Airlock, and Laboratory Flight Elements, *20th Space Simulation Conference*, NASA/CR-1998-208598.

[4] S. Underwood. A. Holt. 1998. Verification of the International Space Station Redundant Seals by Analysis of Helium Permeation Data, *18th Aerospace Testing Seminar*.

[5] S. Underwood. A. Holt. 1998. International Space Station Node1 Helium Accumulation Leak Rate Test, *20th Space Simulation Conference*, NASA/CR-1998-208598.

[6] Robert L. Malin. 1969. Leak Testing various seal configurations on a manned spacecraft, *AIAA Paper* No. 69–1030.

[7] Barry T. Neyer. John T. Adams. Terry S. Stoutenborough. 2000. Aerospace Leak Test Requirements, *AIAA* 2000–3735.

[8] Lvovsky O. Grayson C. 2005. *Aerospace Payloads Leak Test Methodology*.

[9] Jackson, Charles N. 1997. *Nondestructive Testing Handbook, Third edition: Volume 1*, Leak Testing, American Society for Nodestructive Testing.

[10] William J. Marr. 1967. *Leakage Testing Handbook*, New York, Report No. S-67-1014.

[11] ASTM432. 1997. *Stand Guide for selection of a Leak Testing Method*.

Advances in Engineering Materials and Applied Mechanics – Zhang, Gao & Xu (Eds)
© *2016 Taylor & Francis Group, London, ISBN 978-1-138-02834-0*

Coaxial compound helicopter transition balancing analysis

M.L.W. Wu & M. Chen
Beijing University of Aeronautics and Astronautics, School of Aeronautic Science and Engineering, Beijing, China

ABSTRACT: High speed compound configuration has become the development trend of modern helicopters, which needs deeper theoretical researches and experiments. In this article a new configuration of compound helicopter remodeled by a prototype with coaxial rotor, was applied a series of concrete aerodynamic analyses. The model was added wing and propeller parts to be compound. Its transition period was divided into three stages from hovering to the final autorotation status. Based on a new control scheme and a linear analysis method, the study obtains the balancing dynamic performance by simulating the model. The author suggests that the autorotation mode could be a proper way for high speed controlling and drag reducing. The study accumulates theoretical experience for later product development and optimization.

1 INTRODUCTION

Helicopter is a VTOL vehicle that can hover and perform excellently at low altitude. Normal configured helicopter has inherent defects because of the rotor motion mode. Since the mid-twentieth century, countries in the world began to try a variety of compound configurations to break through the speed and performance limitation of helicopters. In 2010, X2 coaxial compound helicopter from Sikorsky Aircraft Corporation broke the record of 'Lynx MK', 1986, by 259 miles per hour[1], sparking great concern, while domestic development still remains in theoretical area.

The compound model analyzed in this article is a new feasible configuration scheme, based on a prototype model 'FH1' unmanned helicopter of Beihang helicopter institute. The scheme chooses wing for unloading the rotor and adds propeller for motivation. According to preliminary estimations, this configuration can double the flying range of the original one, meanwhile increase the maximum flying speed by about 20% and the zoom altitude by about 40%, and adds an extra weight by 15%. So in generally it is a nice program.

The model is analyzed by designing an integrated control scheme and establishing the flight dynamics model of all parts. With the help of a double linear transition analysis method, the

Table 1. Overall parameters.

FH1 model overall parameters		Compound model size parameters	
Items	Value	Items	Value
Blade tip speed	150 m/s	Wing area	0.673 m^2
Power loading	51 N/kw	Half span	1 m
Rotor diameter	3 m	Propeller diameter	0.6 m
Disk loading	70 N/m	Propeller blade number	3
Rotor solidity	0.035	Tail moment arm	0.9 m
Hovering ceiling	1550 m	Tail support spacing	0.8 m
Engine	9 kw (double)	Engine	26 kw (single)
Empty weight	54 kg	Empty weight	<70 kg

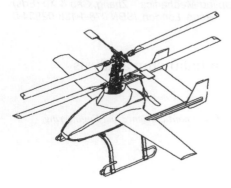

Figure 1. Compound model stereogram.

Figure 2. Compound model front view.

accurate dynamics performance data are collected through analyzing the transition period from take-off to autorotation. The overall parameter of the prototype model and the new configuration are laid below (see Table 1).

2 SCHEME DESCRIPTIONS

2.1 *Mission analyses*

The transition process can be divided into three stages. In the low speed stage, the wing is the main drag source, therefore it should be rotated to erect position for drag reducing. As speed goes up, to the full speed stage, the rotor speed limitation will be highlighted out. Without the ABC blade technique, the autorotation mode is selected to be the final status, retaining the fully developed coaxial rotor system. This scheme cuts the power input to the rotor in the final stage and will reduce the flight drag of the main rotor dramatically.

The process is divided as described below:

1. Low speed hovering stage: Power is fully input to the main rotor. The vehicle takes off in helicopter mode, controlled by adjusting collective pitch and cyclic pitch.
2. Middle speed transition stage: Power is input to both the main rotor and the propeller. The control unit of both helicopter and fixed wing aircraft work in coordinating.
3. Full speed stage: Power is fully input to the thrust propeller. The main rotor gets into the autorotation status. The vehicle is controlled by regulating ailerons, elevator, rudders and propeller pitch.

2.2 *Control mode analyses*

With the speed increases, the control mode transforms from helicopter to fixed wing type. The control variables change from $[\theta_0, A_{1c}, B_{1c}, \theta_c]$ (collective pitch, cyclic pitch and differential control) to $[\delta_a, \delta_e, \delta_r, \theta_{clj}]$ (aileron, elevator, rudder and propeller pitch) to adjust the aircraft attitude. Finally the aircraft reaches the steady state and completes the flight mission.

There are two types of transitions. One is the control mode transition and the other is the power distribution transition. At the beginning of the forward flight process, four control variables are working. As speed increases, eight variables work in coordinating and end in just four variables of fixed wing mode in the final full speed stage. Along with the control transition, the power transition moves the focus from the main rotor to the propeller. As trust from the propeller increases, the wing will unload the lift of rotor. Until the velocity increases to a certain level, the power distribution system cuts the power to the main rotor, letting it get into the autorotation status.

After repeatedly calculations and corrections, an optimal transition interval is decided.

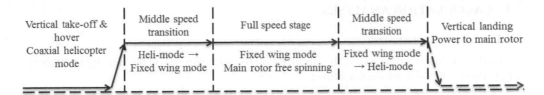

Figure 3.　Mission curve.

1. Control mode transition interval: forward speed ratio from 0.1 to 0.16.
2. Power distribution transition interval: forward speed ratio from 0.1 to 0.3.
3. Initial conditions: $[\theta_0, A_{1c}, B_{1c}, \theta_c] = [5^0, 0, 0, 0]$; $[\delta_a, \delta_e, \delta_r, \delta_{clj}] = [0, 0, 0, 14^0]$.

3　ESTABLISH PARTS MATHEMATICAL MODELS

Establish the dynamic model for main rotor, servo vane, body, tail, wing and propeller. Because the first four parts in the coaxial prototype have already been simulated[2], we just focus on the last two parts.

3.1　Wing model

The left wing marks 1, and the symmetry marks 2. Since the method between the left and right is the same, the left wing is shown as an example below.

Lift, drag & moment:

$$L_{w1} = K_{lw1}C_{Lw1}q_wS_{w1}; \quad D_{w1} = K_{dw1}C_{Dw1}q_wS_{w1}; \quad M_{w1} = C_{Mw1}q_wS_{w1}b_A \tag{1}$$

where b_A = aerodynamic chord; q_w = dynamic pressure; and C_{Lw1}, C_{Dw1}, C_{Mw1}: lift coefficient, drag coefficient and moment coefficient of left wing.

The force and moment generated by the wing:

$$
\begin{bmatrix} F_{xw1} \\ F_{yw1} \\ F_{zw1} \end{bmatrix} =
\begin{bmatrix} -\cos\alpha_{wx}\cos\beta_{w1} & \sin\alpha_{wx} & 0 \\ \sin\alpha_{wx}\cos\beta_{w1} & \cos\alpha_{wx} & 0 \\ -\sin\beta_{w1} & 0 & 1 \end{bmatrix}
\begin{bmatrix} D_{w1} \\ L_{w1} \\ 0 \end{bmatrix}
\tag{2}
$$

$$
\begin{bmatrix} M_{xw1} \\ M_{yw1} \\ M_{zw1} \end{bmatrix} =
\begin{bmatrix} 0 \\ 0 \\ M_{w1} \end{bmatrix} +
\begin{bmatrix} 0 & -z_{w1} & y_{w1} \\ z_{w1} & 0 & -x_{w1} \\ -y_{w1} & x_{w1} & 0 \end{bmatrix}
\begin{bmatrix} F_{xw1} \\ F_{yw1} \\ F_{zw1} \end{bmatrix}
\tag{3}
$$

3.2　Propeller model

The variable from the propeller part is the pitch, marked as θ_{clj}. Firstly the AOA of the airfoil is calculated by defining the induced velocity v_3 of the propeller. Then by carrying out the integral operation of blade sections, the force and moment of the propeller are finally gotten.

$$
\begin{bmatrix} F_{xlj} \\ F_{ylj} \\ F_{zlj} \end{bmatrix} =
\begin{bmatrix} 0 & 1 & 0 \\ -1 & 0 & 0 \\ 0 & 0 & 1 \end{bmatrix}
\begin{bmatrix} F_{xL} \\ F_{yL} \\ F_{zL} \end{bmatrix};\quad
\begin{bmatrix} M_{xlj} \\ M_{ylj} \\ M_{zlj} \end{bmatrix} =
\begin{bmatrix} 0 & 1 & 0 \\ -1 & 0 & 0 \\ 0 & 0 & 1 \end{bmatrix}
\begin{bmatrix} M_{xL} \\ M_{yL} \\ M_{zL} \end{bmatrix} +
\begin{bmatrix} 0 & -z_{lj} & y_{lj} \\ z_{lj} & 0 & -x_{lj} \\ -y_{lj} & x_{lj} & 0 \end{bmatrix}
\begin{bmatrix} F_{xlj} \\ F_{ylj} \\ F_{zlj} \end{bmatrix}
\tag{4}
$$

4   CALCULATION ANALYSES

4.1   *Autorotation analyses*

In order to study the final stage, the autorotation rotor is analyzed on the velocity of 40 meters per second. Results are shown below:

The rotor is regarded as steady when the rotor torque is judged to be zero by the program. As shown in the AOA distribution map (Fig. 4), the white hole marks the reversed flow region and the left direction means the forward way. By integrating the forces on the vertical force map (Fig. 5), it is concluded that the rotor rotate speed becomes steady at 350 revolutions per minute on 40 meters per second. The rotor generates lift about 550 N and reduces drag by 40% compared to the powered rotor. At this final stage, power is fully input to the propeller.

4.2   *Control variable curve analyses*

The control variable curves of both 'FH1' prototype and the compound model are compared as below.

Some conclusions are gained from the curves:

1. Notice that the normal helicopter collective pitch curve (Fig. 6) is a saddle shape, fitting the helicopter rule. But the curve of compound model gradually declines to the final site of transition, meaning that the main lift system transfers from the rotor to the wing part. At the final stage, the wing unloads a large part of lift from the rotor.
2. The cycle pitch curve of the 'FH1' prototype tend to be infinite, meaning that the horizontal and vertical balancing force increase as speed goes up (Fig. 7, 8). But the curve of

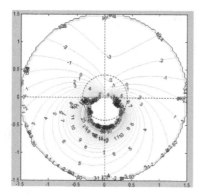

Figure 4.   AOA distribution map.

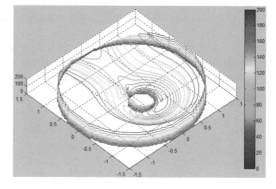

Figure 5.   Vertical force contour map.

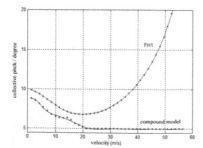

Figure 6.   Collective pitch.

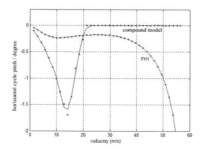

Figure 7.   Horizontal cycle pitch.

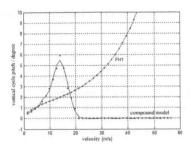

Figure 8.   Vertical cycle pitch.

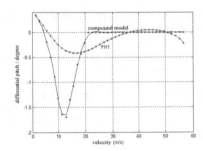

Figure 9.   Differential pitch.

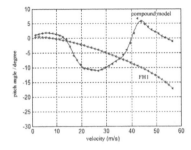

Figure 10.   Pitch angle.

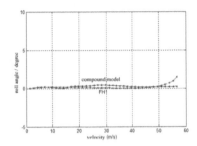

Figure 11.   Roll angle.

the compound model tends to be zero after the transition, meaning an extension of rotor limitation.

3. The differential pitch (Fig. 9) trend is similar between the two models, declining as speed goes up, meaning the reduction of interaction between the coaxial rotors.
4. The pitch angle of the prototype declines to be uncontrolled (Fig. 10), while the compound one rises up steady to about 5 degree, letting the rotor autorotate.
5. The roll angle changes around zero inconspicuously (Fig. 11).

5   CONCLUSIONS

1. By establishing dynamics models and designing the control scheme, the whole transition process is simulated and calculated precisely. It is discovered that the autorotation mode helps break through the speed limitation and reduce drag, showing that this configuration is valuable for compound helicopter product development.
2. The linear analysis method is an efficient way to handle transition problems. As the above example alludes, it solves this complex transition balancing problem very well and verifies the availability of autorotation status for broadening the speed range. But there are discontinuous points on some maps, which may be improved by later nonlinear analysis research.

REFERENCES

[1] Hua Tang. 2001. US Exploring A New Compound Helicopter Using VTDP Technology. *International Aviation*: 2001(07).
[2] Guoyi Zhou, Jizhong Zhou, Yihua Cao. 2003. Mathematical Model for Twin Rotor Loads of a Coaxial Helicopter. *Journal of Aerospace Power*. 18(3): 343–347.

[3] Kvokov, V.N. 1992. Factor Analysis of Coaxial Rotor Aerodynamics in Hover. *18th European Rotorcraft Forum*: 99–103.

[4] Padfield, G.D. 2007. *Helicopter Flight Dynamics*. Washington: Blackwell Publishing.

[5] Raptis, I.A., Valavanis, K.P. 2011. *Linear and Nonlinear Control of Small-Scale Unmanned Helicopters*. New York: Spinger.

[6] Leishman, J.G. 2006. *Principles of Helicopter Aerodynamics*. Cambridge University Press.

[7] Bramwell, A.R.S. 2001. *Bramwell's Helicopter Dynamics*. Butterworth-Heinemann.

[8] Chen, R.T.N. 1990. A Survey of Nonuniform Inflow Models for Rotorcraft. *Flight Dynamics and Control Applications*. Vertica, 2: 147–184.

[9] Bagai, A. 1996. Free-Wake Analysis of Tandem, Tilt-Rotor and Coaxial Rotor Configurations. *J. of AHS*. 41(3): 196–207.

[10] Coleman, C.P. 1993. A Survey of Theoretical and Experimental Coaxial Rotor Aerodynamic Research. *Proceedings of the 19th European Rotorcraft Forum*.

Advances in Engineering Materials and Applied Mechanics – Zhang, Gao & Xu (Eds)
© 2016 Taylor & Francis Group, London, ISBN 978-1-138-02834-0

Based on the orthogonal experiment optimization of UAV wing bionic airfoil design

J. Zhang & B.S. Chen
Department of Aircraft and Propulsion System, Aviation University of Air Force, Changchun, China

X. Hua
Department of Aviation Theory, Aviation University of Air Force, Changchun, China

ABSTRACT: Due to the problem in the Unmanned Aerial Vehicle (UAV) airfoil design, the method of bionics is used to get the interception of airfoil from biological prototype seagulls wings, and the orthogonal experiment is conducted to obtain the original airfoil optimization. Orthogonal experiment optimization results show that it is the largest influence of the airfoil tail thickness on aerodynamic parameters lift-to-drag ratio. The aerodynamic parameters of the optimized seagull airfoils in low-altitude UAV flight in the environment of air are higher than common airfoil FX60-126, NACA0015, NASA0417, maximum lift-to-drag ratio increased by 58.70%, 58.70% and 58.70%, respectively. Under the condition of large Angle of attack, it can effectively delay the separation of flow. At the same time, it has strong applicability during the Angle of attack changing.

1 INTRODUCTION

The outstanding performance of military UAVS in the gulf and Afghanistan war, brought the attention of more and more countries. Many countries put the development of military UAVS in priority. In recent years, large-load, long-endurance, low-detect will be the future trend for the development of the UAV, at the same time, it became a hot issue in the international aviation research[1]. To increase the load and distance, it needs to improve the design of UAV, in order to improve the aerodynamic parameters of UAV effectively. One of the key problem is the airfoil design, the stand or fall of UAV airfoil, determines the performance of UAV to a large extent.

At present, the international aerodynamic shape design on airfoil mainly adopts two ways that is the reverse design and optimization design. Zhu Xiongfeng, Guo Zheng, zhong-xi Hou[2][3] proposed the airfoil optimization design based on dynamic grid increased the performance of the airfoil greatly, omit a lot of repetitive operations. Zuo Lin-xuan, Wang Jin-jun[4] developed the perturbation function of airfoil, and discussed the multi-parameter and multi-objective in the process of airfoil optimization achieved good effect. These methods are used to obtain good-performance airfoil. However, those methods still cannot avoid a large amount of computation.

Biological after a long evolution in nature, their structure completely applicable to the harsh environmental requirements and has reached the optimal, so adopt the method of bionic to solve the problem of actual engineering is simple and effective. The Tianshu Liu, etc.[5] have extracted the airfoil of seagulls, merganser, teal, owl, which provided a feasible way to extract the bionic airfoil. Liu Yu-rong and others[6] of Jilin University respectively extracted the swallow wing's airfoils in glide and sweepback of two different flight, and in different cases to analyze the aerodynamic performance of airfoil, concluded that in 35% place of the sweepback airfoil have the optimal efficiency. Jing-fu Jin et al.[7][8] intercepted the long-eared owl wings airfoils of the aspect ratio in different parts, and analyzed the aerodynamic characteristics of the different airfoils, and obtained the conclusion that the airfoil aerodynamic efficiency is the highest at 50%. And applied it to the wind turbine that showed very good effect.

For UAV airfoil design, this article starts from the perspective of bionics, first of all to intercept the seagulls wings model for cross-section surface profile, and then analyze the cross section of airfoil to extract the characteristics of the control points. Adopt the method of orthogonal experiment to optimize wings model section airfoil, point out the control points, which is the biggest factors, affect the lift-to-drag ratio at lower surface of seagull airfoil. Finally, get bionic airfoils that are suitable for the environment of UAV flight, and refer to the aviation airfoil aerodynamic characteristics of the commonly used by contrast analysis.

2 THE ACQUISITION AND ANALYSIS OF SEAGULL WINGS CROSS SECTION MODEL

2.1 *The acquisition of seagull airfoil*

The wings of the seagulls is scanned by 3D scanner to get point cloud, and deal with all the point cloud by reverse engineering software to establish 3D model. As shown in Figure 1, the X direction is wings string direction, the Y direction is the wings of the exhibition, the Z direction is the vertical direction of X, Y. Wings from the wing root to the end for the direction of the intercept, capture cross-section profile in half a wingspan every 10% of the location, the selection of airfoil optimal airfoil section as basic bionic wing airfoils[9].

As shown in Figure 1, the airfoil is intercepted which is based on the seagulls wings scanning point cloud. The seagull airfoil has the following several characteristics: First, the upper and lower surface of profile are bent upward, belongs to the bending airfoil; Second, seagull airfoil in front is thicker and the behind is thinner; Third, the seagull airfoil on the wing oriented on a raised, excessive is relatively smooth, the above downward raised first, after the upward transition.

2.2 *The analysis of seagull airfoil feature point*

This paper selected the most close to the seagulls airfoils on the upper surface by conventional airfoil FX63-137 to form on the above of bionic configuration model. Using b-spline fits the lower surface of the seagull airfoil. The feature points of seagull airfoil (shown in Fig. 2 black spots) to control the shape of a b-spline curve which makes a curve and a seagull

(a) Wing model in the direction of the XYZ   (b) Airfoil

Figure 1.   The wing model and airfoil of seagull.

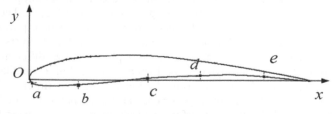

Figure 2.   The lower surface feature points of seagull airfoil.

92

airfoils lower curve is similar. The seagull airfoil front arc curvature is controlled by point a; Airfoil leading edge thickness is controlled by point b; Airfoil front Angle is controlled by point c; Seagulls airfoil surface curvature points is controlled by point d; Airfoil thickness of the tail is controlled by point e.

3 NUMERICAL MODELS

3.1 Computational domain and computing grid partition

In this paper, meshing model by ICEM-CFD before numerical simulation. With large scale, to simulate the flow field calculation domain set airfoil chord length as c. The front of the computational domain is semicircle region, the radius of half round is 12.5c, rear flow field using rectangular area, length of 20c, width of 12.5c. Because of the aerofoil structure is simple, so this paper adopts structure mesh with C type, compute grid as shown in Figure 3.

3.2 The boundary conditions and the turbulence model

According to the seagulls flight environment features we choose sea level atmospheric environment as a numerical computing environment, to flow rate $v = 40\,\text{m}/\text{s}$, the air velocity of the boundary conditions is less than 0.3M, so the air is incompressible flow and computational domain by velocity inlet and pressure outlet. Turbulence model selection Spalart–Allmaras model, Spalart–Allmaras model is designed to jet aerodynamic problems. It is good and effective for predicting the low Reynolds number model. In the process of fluid flow, can be a very good area to deal with the impact of viscosity in the boundary layer. So choosing this model can achieve requirements for accuracy.

4 THE RESULTS AND ANALYSIS OF THE CALCULATION

4.1 The results and analysis of orthogonal experiment

In this paper, selecting lift-to-drag ratio as indexes for seagull airfoil to carry out the orthogonal experiment, the Table 1 presents the lift-to-drag ratio which is experiment index of orthogonal experiment and calculation results of each scheme. Each of the factors to select four levels.

The lift-to-drag ratio is range analysis according to the results of the orthogonal experiment as Table 2 shown. For the y_i average income in the table, the greater its value, illustrates that the effects on the airfoil lift-to-drag ratio influence factors on the level is. Defining R_i as range. This is defined as the change of various factors affects the size of the test index variations, each feature points on the lift-to-drag ratio is related to the size of the range, the influence of the feature points according to the size of the range of a primary and secondary sort is eadcb. As we can see seagull airfoil tail thickness feature points is the main factors to affect t lift-to-drag ratio of seagull airfoil. According to the size of the average k

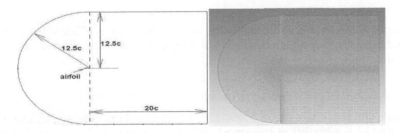

Figure 3.   Computational domain size figure and seagull airfoil computing grid.

Table 1. Orthogonal experiment scheme and results.

Experiment number	A	b	c	d	e	Lift-to-drag ratio
1	−1.5	−2	0	1.4	1.3	51.79
2	−1.5	−1.8	0.4	1.8	1.5	53.35
3	−1.5	−1.6	0.8	2.2	1.7	54.51
4	−1.5	−1.4	1.2	2.6	1.9	55.89
5	−1.3	−2	0.4	2.2	1.9	56.78
6	−1.3	−1.8	0	2.6	1.7	54.01
7	−1.3	−1.6	1.2	1.4	1.5	55.84
8	−1.3	−1.4	0.8	1.8	1.3	53.01
9	−1.1	−2	0.8	2.6	1.5	54.16
10	−1.1	−1.8	1.2	2.2	1.3	53.68
11	−1.1	−1.6	0	1.8	1.9	58.01
12	−1.1	−1.4	0.4	1.4	1.7	57.55
13	−0.9	−2	1.2	1.8	1.7	56.44
14	−0.9	−1.8	0.8	1.4	1.9	58.18
15	−0.9	−1.6	0.4	2.6	1.3	52.33
16	−0.9	−1.4	0	2.2	1.5	54.41

Table 2. The results analysis of bionic airfoil range test.

y1	215.54	219.17	218.22	223.38	210.80
y2	219.64	219.22	220.01	220.81	217.76
y3	223.40	220.70	219.86	219.37	222.51
y4	221.36	220.86	221.86	216.39	228.87
y1 on average	53.89	54.79	54.56	55.84	52.70
y2 on average	54.91	54.80	54.00	55.20	54.44
y3 on average	55.85	55.17	54.96	54.84	55.63
y4 on average	55.34	55.22	55.46	54.10	57.22
R_j	1.96	0.42	0.91	1.75	4.52
Optimal levels	In y_i on average, the greater number is priority level				
Primary and secondary factors	e > a > d > c > b				

Figure 4. The optimization results of seagull airfoil.

under various factors, the most combination of seagulls airfoil is a3b2c4d1e4, according to orthogonal experiment design optimization, it is concluded that the optimal combination of the data and obtained airfoil as shown.

4.2 *The optimized seagull airfoil aerodynamic characteristics analysis*

Low-altitude aircraft often chose FX60-126, FX63-137, NACA0015, NASA0417 for its wing. FX63-137 is a big lift at low speed airfoils, often used in human kind slow planes. FX60-126 better resistance to stall characteristics, usually used for aircraft winglet. Refer to the aviation airfoil aerodynamic characteristics of the commonly used by contrast analysis with the optimized airfoil.

In low altitude long-endurance UAV is usually cruising speed under 200 km/h, flying height under 5000 m. In the process of simulation, with 1000 m altitude atmospheric conditions. Control flow velocity $v = 60$ m/s, take numerical simulation analysis of aerodynamic characteristics for the selected airfoils.

4.2.1 The aerodynamic parameters change

Figure 5 (a) the curve of the lift coefficient along with the change of Angle of attack of the optimized bionic airfoil and 4 kinds of standards of airfoil, can be seen from the diagram, all airfoils lift coefficient change trend are presented along with the increase of Angle of attack, lift coefficient decreases after increasing first. Between 8° and 12° appeared airfoil maximum lift coefficient. The airfoil maximum lift coefficient of sorting as follows:

FX63-137 > seagull airfoil > NASA0417 > FX60-126 NACA0015

The maximum lift coefficient of optimized bionic airfoil is 1.4533. The maximum lift coefficient of optimized bionic airfoil than NACA0015, NACA0417, FX60-126 increased by 65.28%, 65.28% and 27.36%, respectively. Slightly lower than FX63-137. Figure 5 (b) the lift-to-drag ratio relationship along with the change of Angle of attack of the optimized bionic airfoil and 4 kinds of standards of airfoil. It can be seen that all airfoils lift-to-drag ratio varies with Angle of attack first increases and then decreases. At 5°, the optimized bionic airfoils have maximum lift-to-drag ratio, $K_{seagull} = 52.0901$, the maximum lift-to-drag ratio of optimized bionic airfoil than NACA0015, NACA0417, FX60-126 increased by 58.70%, 58.70% and 58.70%, respectively. Slightly lower than FX63-137 also.

4.2.2 The static pressure contours

Figure 6 (a), (b), (c), (d), (e) is the static pressure contours of NACA0015, FX63-137, FX60-126, NASA0417 and bionic airfoil respectively. It can be seen from the diagram that the negative pressure area are mainly distributed on the airfoil surface, positive pressure is mainly distributed in the airfoil surface, and the maximum stress was in streamline stagnation point place, near the airfoil leading edge, the profile of the above has obvious negative pressure gradient. For bionic airfoils, under the airfoil surface than NACA0015, NASA0417, FX60-126 have larger positive pressure zone that can provide airfoils with higher lift. On bionic airfoil surface pressure gradient is obvious, so on the above surface will produce greater suction, the airfoil can produce much more lift.

4.2.3 The flow chart

Figure 7, the static pressure contours and the flow chart of bionic airfoil and FX63-137 at 8°. Bionic airfoil is in a laminar flow state at 8°, the airfoil trailing edge of FX63-137 has started to separation. The viscous resistance on the surface of the airfoil boundary layer is the cause

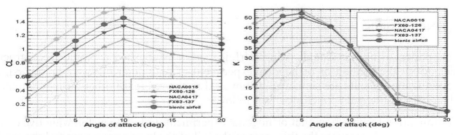

(a) Lift coefficient curve along with the change of Angle of attack
(b) Lift-to-drag ratio curve along with the change of Angle of attack

Figure 5.   Airfoil aerodynamic parameters change curve.

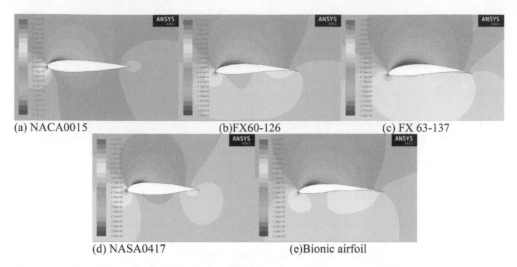

(a) NACA0015 (b)FX60-126 (c) FX 63-137

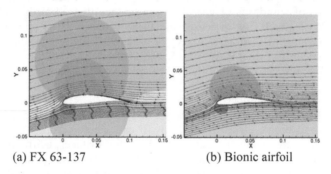

(d) NASA0417 (e)Bionic airfoil

Figure 6. The static pressure contours of seagull airfoils and the standard of airfoil.

(a) FX 63-137 (b) Bionic airfoil

Figure 7. The static pressure contours and the flow chart.

of the separation of the airfoil. Viscous resistance to air flow cannot flow on along the profile surface and make airfoil trailing edge of the air lift, produce low pressure area, further development will flow back to the vortex generated in the low pressure area airfoil stall. It can be seen from the flow chart, bionic airfoils in 8°, the above air separation has not occurred, which explain the bionic airfoil has a wider range of Angle of attack to use than FX63-137, to a greater degree of under the condition of complicated airflow to keep the stability of the airfoil aerodynamic parameters.

5 CONCLUSION

In this paper, using the seagulls wings airfoil section established a bionic reconstruction model, adopt the method of orthogonal experiment of seagull airfoils were optimized, and the optimization results are simulated. Compared to the reference other airlines commonly used airfoil, get the main conclusion is as follows:

1. Optimization results show that orthogonal experiment under the airfoil wing five control points of the plane curve, the thickness of the tail of the bionic wing type had the greatest influence.
2. In a simulated attack Angle range the aerodynamic parameters of optimized bionic airfoil is higher than aviation common seagull, NASA0417, NACA0015, FX60-126 and slightly

below FX63-137. Along with the change of attack Angle bionic airfoils is in separate later, has a strong ability of inhibition of separation. Bionic airfoil has a wider range of Angle of attack to use, to a greater degree of under the condition of complicated airflow to keep the stability of the airfoil aerodynamic parameters.

REFERENCES

[1] Zhu Ziqiang, Chen Yingchun, Wang Xiaolu. et. al. 2011. Modern aircraft aerodynamic design. *National defence industry press*, 78–79. Beijing.

[2] Zhu Xiongfeng Guo Zheng, Hou zhong-xi. et. al. 2013. Based on dynamic grid of airfoil optimization design. *Journal of national university of defense technology* 35 (2): 1–6.

[3] Zhu Xiongfeng Guo Zheng, Hou zhong-xi. 2011. High-altitude long-endurance airfoil optimization based on dynamic grid. *Journal of air dynamics* 32 (4).

[4] Zuo Lin-xian, Wang Jin-jun. 2009. The optimal design of the low Reynolds number airfoil. *Journal of armaments factories* 30 (8): 1073–1078.

[5] Tianshu Liu, Kuykendoll K, Rhew R, Jones S. Avian Wings. *AIAA* 2004–2186.

[6] Liu Yu-rong. 2012. *Barn swallow feather surface structure and wings airfoil aerodynamic characteristics research*. Jilin university.

[7] CongQian Liu Yu-rong, Ma Yi, Jin Jing-fu. 2011. Barn swallow wings show aerodynamic characteristics of airfoils. *Journal of jilin university (engineering science)* S2: 231–235.

[8] Ma Yi. 2012. *Based on the wings of birds bionic horizontal axis wind turbine blade optimization analysis*. Jilin university.

[9] Hua Xin. 2013. *Seagulls wings pneumatic performance research and its application in wind turbine bionic blade design*. Jilin university.

Applied mechanics

Advances in Engineering Materials and Applied Mechanics – Zhang, Gao & Xu (Eds)
© 2016 Taylor & Francis Group, London, ISBN 978-1-138-02834-0

Comparison of simulation on aerial conductor vibration damper power characteristics tested by two methods

L.Q. An, B. Yang & W.Q. Jiang
School of Energy, Power and Mechanical Engineering, North China Electric Power University, Baoding, Hebei, China

ABSTRACT: In this paper, a mathematical model of Stockbridge damper dynamic characteristics is established, theoretical derivation and calculation programs for the power characteristics, and simulation on power characteristics of FD-1 vibration damper tested by force-amplitude and force-velocity methods are completed. The results shows: vibration damper consumes more power in the case of force-amplitude throughout the whole high-frequency range, which easily leads to the response amplitude of vibration damper approaching or exceeding the elastic deformation limit of steel strand, and wears the steel strand excessively. By contrast, the force-velocity method makes less damage to the steel strand, by which the power calculated is closer to the actual operation of the vibration damper, proved to be an improvement. Moreover, the simulation program compiled in this paper can estimate the power characteristics curve in an absence of a specific vibration damper, which provides an effective analytical tool for the anti-vibration performance prediction and design optimization of Stockbridge damper, compared with the test methods.

1 INTRODUCTION

As an important fittings to inhibit vibration of overhead transmission conductors, vibration damper can limit aeolian vibration to a level of basic safety, if well-designed and appropriately-installed[1,2]. Generally, the vibration damper reduces the resonance of conductors caused by self-resonant and aeolian vibration etc., through absorbing and consuming vibration energy, to play a role in protecting the conductors. Therefore, projects always adopt power characteristics curve as an important basis for the anti-vibration performance evaluation and optimization of vibration damper.

Currently, the power characteristics of vibration damper have been measured by shaking table test[3,4]. Earlier, many countries adopt the force-amplitude method to test its power characteristics[5], and now always use the force-velocity method, which is recommended by latest industry standards in our country[6]. Typically, the test method is accurate and reliable, suitable for quality inspection, design verification and other occasions. While, it is difficult to be adapted to anti-vibration performance prediction or optimization of vibration damper, because that it cannot predict the changes of power characteristics when the vibration damper is no physical products, or occurs structure change. By contrast, the simulation analysis by establishing a physical model is an important research tool, which can predict power characteristics in an absence of physical products[7,8].

This paper will focus on common Stockbridge damper, establish its dynamic characteristics mathematical model, and complete the theoretical derivation and calculation programs of power characteristics. Then, chose FD-1 vibration damper as an example, calculate and analyze the vibration damper power characteristics under two testing methods of force-velocity and force-amplitude, to provide a reference for the anti-vibration performance prediction and design optimization of vibration damper.

2 POWER CHARACTERISTICS DERIVATION

Stockbridge damper will be simplified to two degrees of freedom system and divided into two parts from the middle, since that one side of the vibration damper is studied, the other side can be analyzed in the same theory[3,9,10]. Figure 1 shows a physical model of the right side of vibration damper, points O is the centroid of hammer, and O' is the connection point of hammer and steel strand.

Define $u(t) = 0$, the system has $y(t)$ and $\varphi(t)$ two degrees of freedom, so its Lagrange equation is:

$$\frac{d}{dt}\left(\frac{\partial L}{\partial \dot{q}_i}\right) - \frac{\partial L}{\partial q_i} = Q_i \quad (i = 1,2) \tag{1}$$

Assuming the damper do minor vibration, schematic diagram of vibration damper displacement is shown in Figure 2. Therefore, the vertical displacement y_O of hammer centroid, and the angular displacement φ_O of hammer around its centroid can be expressed as follows:

$$\begin{cases} y_O = y_{O'} - s\varphi_O \\ \varphi_O = \varphi_{O'} \end{cases} \tag{2}$$

Simplify all the forces to the point O', as a concentrated force $F_{O'}$ and a concentrated moment $M_{O'}$. Minor vibration meets the principle of superposition, so

$$\begin{cases} F_{O'} = \dfrac{6EI}{l^3}(2y_{O'} - \varphi_{O'}l) \\ M_{O'} = \dfrac{2EI}{l^3}(2\varphi_{O'}l^2 - 3y_{O'}l) \end{cases} \tag{3}$$

In addition, bending moment at any point on the steel strand is: $M(x) = F_{O'}x + M_{O'}$ So, the kinetic energy of the system is:

$$T = \frac{1}{2}m\dot{y}_O^2 + \frac{1}{2}J_O\dot{\varphi}_O^2 = \frac{1}{2}m\left(\dot{y}_{O'} - s\dot{\varphi}_{O'}\right)^2 + \frac{1}{2}J_O\dot{\varphi}_{O'}^2 \tag{4}$$

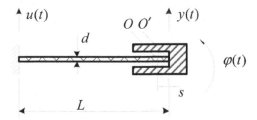

Figure 1.   Physical model of Stockbridge damper.

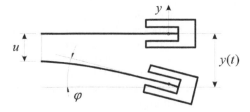

Figure 2.   Schematic diagram of vibration damper displacement.

And the potential of the system is:

$$U = \int_L \frac{M^2(x)}{2EI}\,dx = \frac{2EI}{L^3}\left(3y_{O'}^2 + L^2\varphi_{O'}^2 - 3Ly_{O'}\varphi_{O'}\right) \tag{5}$$

Combining Equations (1), (4) and (5)

$$\begin{cases} m\ddot{y} - ms\ddot{\varphi} + k(6y - 3L\varphi) = P_{O'} \\ -ms\ddot{y} + (J_O + ms^2)\ddot{\varphi} + k(-3Ly + 2L^2\varphi) = M_{O'} \end{cases} \tag{6}$$

Define $k = 2EI/L^3$, transform Equation (6) to matrix form, and obtain the motion differential equation of system

$$[M]\begin{Bmatrix} \ddot{y}_{O'} \\ \ddot{\varphi}_{O'} \end{Bmatrix} + [K]\begin{Bmatrix} y_{O'} \\ \varphi_{O'} \end{Bmatrix} = \begin{Bmatrix} P_{O'} \\ M_{O'} \end{Bmatrix} \tag{7}$$

where the mass matrix $[M] = \begin{bmatrix} m & -ms \\ -ms & J_O + ms^2 \end{bmatrix}$, stiffness matrix $[K] = k\begin{bmatrix} 6 & -3L \\ -3L & 2L^2 \end{bmatrix}$.

Considering the damping effects, assume that the system has a constant damping ratio D, so the control differential equation of system is:

$$[M]\ddot{X} + [C]\dot{X} + [K]X = F \tag{8}$$

where $C = \dfrac{2D}{\omega}K$, $X = \begin{Bmatrix} y \\ \varphi \end{Bmatrix}$, $F = \begin{Bmatrix} -m\ddot{u} \\ 0 \end{Bmatrix}$.

For a stable vibration system, the solution of Equation (11) is:

$$X = \left(-M\omega^2 + K + j\omega C\right)^{-1} F \tag{9}$$

① For the force-amplitude method, the amplitude of hammer is:

$$\begin{Bmatrix} y_0 \\ \varphi_0 \end{Bmatrix} = \left(-M\omega^2 + K + j\omega C\right)^{-1}\begin{Bmatrix} m\omega^2 u_0 \\ 0 \end{Bmatrix} \tag{10}$$

② For the force-velocity method, the amplitude of hammer is:

$$\begin{Bmatrix} y_0 \\ \varphi_0 \end{Bmatrix} = \left(-M\omega^2 + K + j\omega C\right)^{-1}\begin{Bmatrix} -mj\omega v_0 \\ 0 \end{Bmatrix} \tag{11}$$

Therefore, the support of hammer strand system is:

$$F_0 = K_{11}y_0 + K_{12}\varphi_0 \tag{12}$$

And the power consumption of vibration damper in this side is:

$$P_0 = \frac{1}{2}|F_0|\cdot v_0 \cdot \cos\alpha \tag{13}$$

if described by amplitude

$$P_0 = \frac{1}{2}\omega|F_0|\cdot|u_0|\cdot\cos\alpha \tag{14}$$

103

where the phase difference between the force and velocity is:

$$\alpha = arctg\left(\frac{I_m\left(\mathbf{F_0}/v_0\right)}{R_e\left(\left(\mathbf{F_0}/v_0\right)\right)}\right) \tag{15}$$

The above is one side power of vibration damper, and the total power consumption is a sum of both sides of vibration damper.

3   SIMULATION EXAMPLES

3.1   *Calculation parameters*

Ignoring the quality of steel strand, and referring the "electrical fittings Handbook" to get calculation parameters: hammer quality $m = 0.75$ kg, strand length $L = 120$ mm, stranding diameter $d_1 = 7.5$ mm, and hammer diameter $D_1 = 40$ mm. Then obtain elastic modulus $E = 6e10$ Pa, rotation inertia $J_o = 4.971e\text{-}3$ kg$\cdot$m^2, and static moment of inertia $I = 1.816e\text{-}10$ m^4.

3.2   *Main calculate ideas*

The main ideas of programming to solve power characteristics of vibration damper are as follows:

a. Calculate the mass matrix $[M]$, stiffness matrix $[K]$ and damping matrix $[C]$, based on the above calculation parameters.
b. Give the vibration amplitude ($u_0 = 0.5$ mm, $u_0 = 0.8$ mm) and the vibration velocity ($v_0 = 0.1$ m/s, $v_0 = 0.2$ m/s), set the frequency ω (in steps of 0.1, the range is 0~600 Hz), calculate the $\{y_0 \ \varphi_0\}^T$ in the case of force-velocity and force-amplitude respectively, according to formulas (11) and (12).
c. Calculate the support of the hammer strand system $\mathbf{F_0}$, via the stiffness matrix $[K]$ calculated in step a) and $\{y_0 \ \varphi_0\}^T$ calculated in step b) into formula (13).
d. Calculate the one-side power $\mathbf{P_0}$ in the case of force-velocity and force-amplitude respectively, via the support of the hammer strand system $\mathbf{F_0}$ calculated in step c) into Equations (14) and (15). So, the total power $P = 2\mathbf{P_0}$, while FD type vibration damper is symmetrical.
e. Change the vibration frequency ω, and repeat the above steps to calculate the corresponding total power.
f. Finally, draw the total power versus frequency curve, namely power characteristic curve of the Stockbridge vibration damper.

3.3   *Simulation results*

First, draw the vibration damper power characteristic curves in the case of force-amplitude ($u_0 = 0.5$ mm and $u_0 = 0.8$ mm), as shown in Figure 3.

Figure 3 shows: vibration damper power peaks at about 12 Hz and 40 Hz in the case of force-amplitude. The vibration damper consumes more power throughout the whole high frequency range, which may lead to the response amplitude of vibration damper approaching or exceeding the elastic deformation limit of steel strand[11], and wears the steel strand excessively, which coincides with the test conclusions in the paper[12].

Second, draw the vibration damper power characteristic curves in the case of force-velocity ($v_0 = 0.1$ m/s and $v_0 = 0.2$ m/s), as shown in Figure 4.

Figure 4 shows, the vibration damper power also peaks at about 12 Hz and 40 Hz in the case of force-amplitude. The power closes to zero in the whole range except for two peaks, which can reduce the damage from testing. In addition, maximum resonance power is about 2 times of the minimum one in the condition of force-velocity, while the ratio is more than

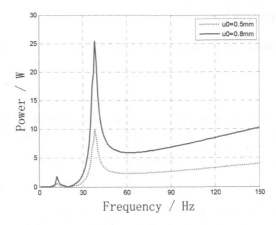

Figure 3.   Power characteristics curves of force-amplitude.

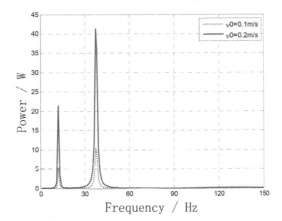

Figure 4.   Power characteristics curves of force-velocity.

10 of the simulation results by the force-amplitude method, and the simulation results are closer to the actual operation of the vibration damper, according to the conclusions (3) in the paper[12]. Therefore, the force-velocity method proves to be an improvement in comparison of the force-amplitude method.

4   CONCLUSION

In this paper, we established a mathematical model of Stockbridge damper dynamic characteristics, completed the theoretical derivation and calculation programs for power characteristics, then analyzed the power characteristics of FD-1 vibration damper simulated by the method of force-velocity and force-amplitude. The conclusions are as follows:

1. The force-velocity method can reduce the damage from testing, whose simulation results are closer to the actual operation of the vibration damper, proved to be an improvement in comparison with the force-amplitude method.
2. The simulation program compiled in this paper can estimate the power characteristics of vibration damper in an absence of a specific vibration damper, and has an ability to predict in comparison with the test experimental methods, which provides an effective analytical tool for the anti-vibration performance prediction and design optimization of Stockbridge damper.

ACKNOWLEDGEMENTS

This research was supported by Hebei Province Natural Science Foundation (E2013502291).

REFERENCES

[1] Ervik M., Berg A., Boelle A., et al. 1989. Report on aeolian vibration. *Electra*, 124: 40–77.
[2] Chen Jianhua, Chen Jingyan. 2003. Test and application of power properties of vibration dampers. *Power Construction*, 24(4): 42–44.
[3] Barbieri N., Barbieri R. 2012. Dynamic analysis of Stockbridge damper. *Advances in Acoustics and Vibration*.
[4] Barry O., Zu J.W., Oguamanam D.C.D. 2015. Nonlinear Dynamics of Stockbridge Dampers. *Journal of Dynamic Systems, Measurement, and Control*.
[5] IEEE Committee 664. 1993. Guide on the Measurement of the Performance of Aeolian Vibration Dampers for Single Conductors (IEEE Std.), pp. 664–1993.
[6] DL/T 1099–2009, Technical requirements and tests for damper. *China Electric Power Press*.
[7] Lu Mingliang. 1994. Computer Simulation on characteristics of dampers, *Northeast Electric Power Technology*, 02: 1–3.
[8] Wagner H., Ramamurti V., Sastry R.V.R., et al. 1973. Dynamics of Stockbridge dampers. *Journal of sound and Vibration*, 30(2): 207-IN2.
[9] Sauter D., Hagedorn P. 2002. On the hysteresis of wire cables in Stockbridge dampers. *International journal of non-linear Mechanics*, 37(8): 1453–1459.
[10] Verma H. 2002. The Stockbridge damper as a continuous hysteric system in single overhead transmission lines. *Indian Institute of Technology Bombay Mumbai*.
[11] Thomson W. 1996. *Theory of vibration with applications*. CRC Press.
[12] Sun Lin, He Xiaoxiong. 2003. Comparison between two methods for measuring power characteristics of damper. *East China Electric Power*, 37(5): 37–39.

Advances in Engineering Materials and Applied Mechanics – Zhang, Gao & Xu (Eds)
© 2016 Taylor & Francis Group, London, ISBN 978-1-138-02834-0

Heat propagation properties of short laser heating based on time fractional single phase lagging model

Z. Han & G. Yi Xu

Institute of Sea Transport and Port Architecture Engineering, Zhejiang Ocean University, Zhou-Shan, PC, P.R. China

ABSTRACT: In this paper, the laser short-pulse heating of a solid surface is considered. The time fractional heat conduction model is used as the constitutive heat diffusion model and the corresponding fractional heat conduction equation with a volumetric heat source is built. Inverse Laplace transform is applied to obtain the analytical solution for non Gaussian type laser. The numerical results are presented graphically for various values of model parameters. Fractional order parameter only influences the magnitude of temperature rising, having no effects on the speed of heat propagation. Higher the temperature rise, larger the fractional order parameter. This research provides some new points for further studying the laser heating or other non-Fourier heat conduction problems.

1   INTRODUCTION

Lasers are considered to be one of the effective tools for laser treatment of metallic surfaces. High power laser, when focused at the surface, generates excessive heat, which enables the surface to reach the melting temperature of the substrate material. In generally, Fourier law is applied to analyze the heat conduction during laser processing. But the known constitutive relation in Fourier model which is originally based on experimental observations. This model is pure diffusive in nature considers the instantaneous flow of heat in the medium of the presence of even a small temperature gradient. In other words, the velocity of thermal propagation is infinite according to the Fourier model, which is in conflict with physical laws. Although it works well in many physical and engineering applications, many experimental studies have shown the inadequacy of Fourier model in some situations of practical interest. Up to now, some non-Fourier constitutive models have been introduced among which Cattaneo and phase lagging models have found greater applications (Joseph & Preziosi 1989, Zhang & Liu 2000). On the other hand, in the past three decades fractional calculus has proved its efficiency in modeling the intermediate anomalous behaviors observed in different physical phenomena. Fractional calculus is the calculus of differentiation and integration of non-integer orders. It is as old as classical (integer-order) calculus. However, until the recent decades it had not found considerable applications in practical and engineering fields and had been studied only in pure mathematics.

Recently, considerable research studies have been carried out to investigate the laser pulse heating process based on the parabolic or hyperbolic heat conduction equation (Hilfer 2000, Podlubny 1999, Ready 1978, Blackwell 1990, Yilbas 2012). In this work, the laser short-pulse heating of a solid surface incorporating a volumetric heat source is considered. Here, based on the previous analysis, we use the time fractional single lagging model as the generalized heat conduction law. The corresponding fractional heat conduction equation is built and the volumetric heat source is considered to resemble the absorption of a laser irradiated pulse. The numerical results are presented graphically for various values of model parameters.

2 MATHEMATICAL MODEL

Considering the metal material irradiating by a short pulse laser, the nature of short pulse laser heating has two special characteristics, the one is short duration time whose scale is nano-second, pico-second or femto-second orders and the another is short depth of energy absorbed whose depth is nano-meters orders. On duration time of laser heating is equivalent to the relaxation time of material, the process of heat conduction is non-equilibrium process.

The well-known and extensively used non-Fourier model is based on phase-lagging or delayed constitutive models. The simplest model of this kind is *single-phase-lag* model that is based on the assumption of the existence of a delay between the heat flux and its temperature gradient in the following form.

$$q(r, t + \tau) = -k\nabla T \qquad (1)$$

where τ is the phase-lag in time. Expanding Equation (1) up to first order by Taylor series expansion (assuming τ is small compared with t) results to C-V model (Cattaneo, 1958)

$$-k\nabla T = q(r, t + \tau) \approx q(r, t) + \tau \frac{\partial q}{\partial t} \qquad (2)$$

Recently, Odibat et al. (Odibat & Shawagfeh 2007) introduced a fractional Taylor series expansion according to which we can expand single-phase lag constitutive model (Equation (1)) in the following form

$$-k\nabla T = q(r, t + \tau) \approx q(r, t) + \frac{\tau^p}{\Gamma(p+1)} D_t^p q, \ 0 < p < 1 \qquad (3)$$

where p is fractional order of differentiation.

From Equation (3), it is obvious that when no thermal lag $\tau = 0$, the fractional non-Fourier model reduces to Fourier model (pure diffusion) and for $\tau \neq 0 \ p = 1$. Finally for $\tau \neq 0 \ 0 < p < 1$, because of the non-local property of fractional derivative introduced through the memory kernel (memory integral) existing in the definition of fractional derivative, it models the intermediate processes between thermal wave and pure diffusion (parabolic) which cannot be captured through integer-order Taylor series expansion of single-phase lag model.

Since the generalized Taylor formula is based on Caputo definition in Equation (4), Caputo definition is following

$$D_t^\alpha f(t) \equiv \frac{1}{\Gamma(n-\alpha)} \int_0^t \frac{f^{(n)}(\tau)}{(t-\tau)^{\alpha+1-n}} d\tau, \ n-1 < \alpha < n \qquad (4)$$

Combining (3) with the conservation law (5) of energy

$$\rho c_p \frac{\partial T}{\partial t} = -\nabla \cdot q + Q \qquad (5)$$

and eliminating the thermal flux term results in the following generalized fractional heat conduction equation:

$$\frac{\partial T}{\partial t} + \frac{\tau^p}{\Gamma(p+14)} \frac{\partial^{p+1} T}{\partial t^{p+1}} = a\nabla^2 T + \frac{1}{\rho c_p}\left(1 + \frac{\tau^p}{\Gamma(p+1)} \frac{\partial^p}{\partial t^p}\right)Q \ \ 0 < p < 1 \qquad (6)$$

When $\tau \to 0$, Equation (6) is reduced to the classical heat conduction equation. Equation (6) with $p = 1$ reduces to the hyperbolic heat conduction equation.

In order to understand the effects of fractional order p on the thermal diffusion wave, we consider the laser short-pulse heating situation. That is, a volumetric source term should be incorporated into the Equation (6) to account for the absorption. The temporal profile of the laser pulse is considered as step pulse and given by the following:

$$I(t) = I_0 [H(t) - H(t - t_0)] \tag{7}$$

where $H(t)$ is the Heaviside function, I_0 id the intensity that is defined as the total energy carried by the pulse per unit cross-section of the beam. Therefore, the volumetric source term is expressed as follows:

$$Q = (1 - r_f) I(t) \delta \exp(-\delta x) \tag{8}$$

where r_f is the reflection coefficient, and δ is the absorption coefficient.

The initial and boundary conditions appropriate for the laser heating situation are given by:

$$T(x,t) = 0, \quad \frac{\partial}{\partial t} T(x,t) = 0, \quad x > 0, t = 0 \tag{9}$$

$$\frac{\partial}{\partial x} T(x,t) = 0, \quad x = 0, \ t > 0 \tag{10}$$

$$T(x,t) = 0, \quad x \to \infty, t = 0 \tag{11}$$

3 SOLUTION OF TEMPERATURE DISTRIBUTION

For convenience in the subsequent analysis, the following dimensionless quantities are introduced:

$$T^* = \frac{k\delta T}{(1 - r_f) I_0}, \quad t^* = a\delta^2 t, \, x^* = x\delta, \; \tau^* = a\delta^2 \tau, \, t_0^* = a\delta^2 t_0 \tag{12}$$

Then (6), (9), (10), (11) can be rewritten in dimensionless form (the superscript $*$ is omitted)

$$\frac{\partial T}{\partial t} + \frac{\tau^p}{\Gamma(p+1)} \frac{\partial^{p+1} T}{\partial t^{p+1}} = \frac{\partial^2 T}{\partial x^2} + \left(1 + \frac{\tau^p}{\Gamma(p+1)} \frac{\partial^p}{\partial t^p}\right) \exp(-x) f(t) \tag{13}$$

$$T(x,t) = 0, \quad \frac{\partial}{\partial t} T(x,t) = 0, \quad x > 0, t = 0 \tag{14}$$

$$\frac{\partial T}{\partial x} = 0, x = 0, t > 0 \tag{15}$$

$$T(x,t) = 0, x \to \infty, \ t = 0 \tag{16}$$

where $f(t) = H(t) - H(t - t_0)$.

Applying the Laplace transform method and the convolution theorem, analytical solution is derived

$$T(x,t) = T_g(t)\exp(-x) - \int_0^t T_g(\xi) T_2(x, t - \xi) d\xi \tag{17}$$

where

$$T_g(t) = \int_0^t T_1(\xi) f(t - \xi) d\xi \tag{18}$$

$$T_1(t) = \sum_{n=0}^{\infty} (-1)^{-n} \frac{\tau^{-p(n+1)} \left(\Gamma(p+1)\right)^{n+1} t^{p(n+1)}}{n!} E_{p+1,p+1-n}^{(n)} \left(\tau^{-p} \Gamma(p+!) t^{p+1}\right)$$

$$+ \sum_{n=0}^{\infty} (-1)^{-n} \frac{\tau^{-pn} \left(\Gamma(p+1)\right)^{n} t^{pn}}{n!} E_{p+1,1-n}^{(n)} \left(\tau^{-p} \Gamma(p+!) t^{p+1}\right) \qquad (19)$$

$$T_2(x,t) = \frac{1}{\sqrt{t\pi}} \sum_{n=1}^{\infty} \frac{(-1)^n}{n!} \frac{t^{pn+\frac{p}{2}}}{\tau^{pn+\frac{p}{2}}} H_{0,2}^{2,0} \left[\frac{x^2 \tau^p}{4t^{p+1}} \Big|_{(0,1),\left(n+\frac{1}{2},1\right)} \right] \text{(Hai-Tao et al. 2013)} \qquad (20)$$

4 RESULTS AND DISCUSSION

In the previous section, we have obtained the analytical solutions for the laser short-pulse heating. In order to display the relevant physical effects of the obtained results, we present the numerical results for the dimensionless temperature field when $t_0 = 4$. For convenience of analyses, we select the pure copper material, its properties of the material are as follows:

$$\rho = 8954 \text{ kg} \cdot \text{m}^{-3}, \ k = 386 \text{ W} \cdot (\text{m} \cdot \text{K})^{-1}, \ c_p = 383.1, \text{ J} \cdot (\text{kg} \cdot \text{K})^{-1}, \ \delta = 7.1 \times 10^7 \text{m}^{-1}, \ \tau_0 = 10^{-11}\text{s}.$$

4.1 *The variations of temperature on p = 0.5*

From the definition of dimensionless relaxation and copper' properties, $\tau = 5.6$ is calculated. Letting fractional order parameter $p = 0.5$, the dimensionless temperature of different locations $x = 0, 1, 3, 5$ is shown in Figure 1. What we are interested in are their different initiation times and different variation rate of temperature. The rate of temperature variations for inner zone apart from the surface are slower than that for surfaces. Surface's temperature rising time synchronizes with laser heating time. The temperature rising time within the material delays some certain times. The deeper places have longer delayed times. The numerical results show the wave's mechanism of heat propagation. From the Figure 2, the zones of the rising temperature at equal interval times are different. This phenomenon denotes that thermal wave is not constant.

4.2 *Effects of parameter P on the variations of temperature*

Figures 3 and 4(a) present the dimensionless temperature profiles at different depths for different values of fractional parameter p. At the surface ($x = 0$), the temperature rises sharply

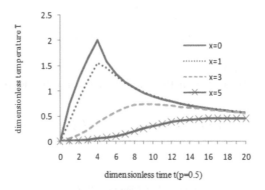

Figure 1. Dimensionless temperature variations ($p = 0.5$).

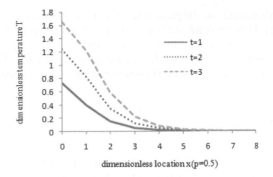

Figure 2. Dimensionless temperature distributions ($p = 0.5$).

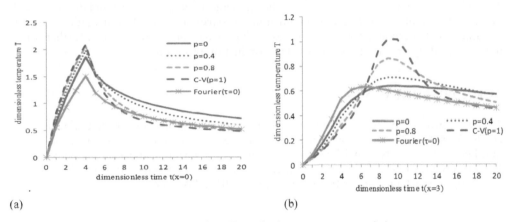

(a)   (b)

Figure 3. Effects of parameter p on surface dimensionless temperature variation versus t.

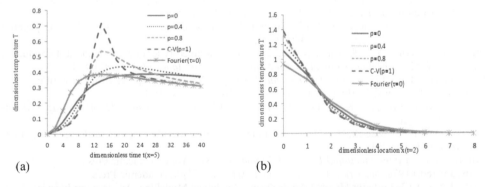

(a)   (b)

Figure 4. Effects of parameter p on dimensionless temperature variation and distribution.

to its maximum near the pulse ending. When the laser pulse heating ends, the temperature first decays sharply, and as time progresses, the temperature decay becomes gradual. Comparing the variable curve corresponding parameter $p = 0$, fastest rising temperature occurs on the surface and slowest rising temperature takes place in the inner zone. At the inner zones, cure of fastest and longest delayed temperature variation always is C-V ($p = 1$). Temperature results based on the single lagging model stand intermediate between the that of Cattaneo law ($0 < p < 1$) and that of Fourier law. The numerical results indicate the slower the thermal propagate, the bigger the parameter p is.

Figure 4(b) shows that effects of the parameter p on dimensionless temperature distribution. Generally, within certain zone from surface, the higher temperature is, the larger the fractional order index p. Thus from the Figures 3, 4, 5 and 6, the heat conduction depicted by the generalized single lagging model with fractional derivative cannot be viewed as an intermediate process ($0 < p < 1$), this conclusion conform with the conclusion from the literature (Mathai et al. 2010).

5 CONCLUSIONS

In summary, we studied the laser short-pulse heating of a solid surface. The generalized single lagging model with fractional derivative is used as the constitutive relation and the corresponding fractional heat conduction equation is built to describe the laser pulse heating. The Laplace transform solution is presented using the Laplace transformation technique. The effects of the model parameters p on the temperature field are shown and discussed with numerical simulations. The results obtained in this paper also provide a theoretical basis for the analysis of thermal stress. Thus, this research provides some new points for further studying laser heating or other non-Fourier heat conduction problems.

ACKNOWLEDGEMENTS

This work was supported by The National Sparking Plan Project under the project number 2013GA700254, Zhejiang Provincial Natural Science Foundation of China under the project number Y3110170, National Training Programs of Innovation and Entrepreneurship for Undergraduates under the project number 201310340007 and International S&T Cooperation Program of China under the project number 2012DFA30600.

REFERENCES

Blackwell, B.F. 1990. Temperature profile in semi-infinite body with exponential source and convective boundary condition, *Trans. ASME J. Heat Transfer* 112:567–571.
Cattaneo, C. 1958. Sur une forme de l'équation de la chaleur eliminant le paradoxe d'une propagation instantanée, *C.R. Acad. Sci. Paris* 247:431–433.
Hai-Tao & Qi, Huan-Ying & Xua, Xin-Wei. 2013. The Cattaneo-type time fractional heat conduction equation for laser heating. *Computers and Mathematics with Applications* 66:824–831.
Hilfer, R. (ed.)2000. *Applications of Fractional Calculus in Physics*. Singapore: World Scientific.
Joseph, D.D. & Preziosi, L. 1989. Heat waves. (*Rev.*) *Modern Phys* 61:41–73.
Mathai, A.M. & Saxena, R.K. & Haubold, H.J. (ed.) 2010. *The H-Function: Theory and Applications*. Berlin: Springer.
Odibat, Z.M. and Shawagfeh, N.T. 2007. Generalized Taylor's Formula. *Applied Mathematics and Computation* 186: 286–293.
Podlubny, I. (ed.) 1999. *Fractional Differential Equations*. New York: Academic Press.
Ready, J.F. (ed.) 1978. *Industrial Applications of Lasers*. New York: Academic Press.
Yilbas, B.S. (ed.) 2012. Laser Heating Applications: Analytical Modelling. Amsterdam: Elsevier.
Zhang, Z. & Liu, D.Y. 2000. Advances in the study of non-Fourier heat conduction. *Adv. Mech* 30: 446–456 (in Chinese).

Advances in Engineering Materials and Applied Mechanics – Zhang, Gao & Xu (Eds)
© *2016 Taylor & Francis Group, London, ISBN 978-1-138-02834-0*

The study of stress-strain reconstruction method based on EMD

D.L. Han
School of Reliability and Systems Engineering, Beihang University, Beijing, China

ABSTRACT: In the actual project, we need to get the stress of a point, but it cannot be measured directly by installing strain gauge. In this paper, the research of resistance strain gauge theory measuring strain will be carried, and the calculation formula is derived. Stress reconstruction method based on empirical mode decomposition will decompose the stress-strain into intrinsic mode function. Using the transformation equation, we can transforming it into the mode of unmeasured point, and then we get the stress-strain by modal superposition. Combined with the finite element model of aluminum alloy cantilever beam, the feasibility of the method is verified through numerical simulation.

1 PRINCIPLE OF ELECTRO MOTIVE STRAIN

Resistance strain measurement method is also called electro motive strain. The strain gauge will be pasted on the surface of a component to be measured, and then access the measurement circuit, the strain sensitive gate deformation along with the member force deformation will result in its resistance value to change. The change of resistance value is proportional to the surface strain. The output signal will be amplified by the amplifying circuit, then recorded by recording instrument. The data will be processed by the computer after analog digital conversion.

1.1 *The principle of strain gauge*

The wires in the deformation will cause the change of resistance, that's why metal wire can produce strain resistance effect. The relationship can be represented by the following formula.

$$\frac{dR}{R} = \frac{d\rho}{\rho} + \frac{dL}{L} - \frac{dA}{A} \tag{1}$$

Here, R, ρ, L, A refers to the lead resistance, resistivity, length and cross sectional area respectively. The Poisson's ratio of the wire can be written as μ. dL/L is the strain ε. The above formula can be rewritten as

$$\frac{dR}{R} = \frac{d\rho}{\rho} + (1 + 2\mu)\varepsilon \tag{2}$$

When

$$K_s = \frac{dR/R}{\varepsilon} = \frac{d\rho/\rho}{\varepsilon} + (1 + 2\mu) \tag{3}$$

the formula above can be written as

$$\frac{dR}{R} = K_s \varepsilon \tag{4}$$

1.2   *The measuring circuit*

The strain can be converted to the change of resistance by strain gauge. But this change of resistance is usually very small. In order to facilitate measurement, the change of resistance is converted to a voltage (or current) signal, then the signal amplified by the amplifier, and recorded by the recorder. This process is accomplished by the resistance strain gauge. Wheatstone bridge The change in resistance will be converted into a voltage change by Wheatstone bridge. Wheatstone bridge is as follows.

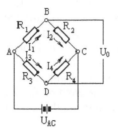

When the input voltage is U_{AC}, output voltage is U_0.

$$U_0 = \frac{R_1 R_4 - R_2 R_3}{(R_1 + R_2)(R_3 + R_4)} U_{AC} \tag{5}$$

In order to make the bridge balance, U_0 should be zero. The equation must be met.

$$R_1 R_4 = R_2 R_3 \tag{6}$$

If the bridge is balanced, which satisfies the equation (6), the bridge presents the output voltage when the bridge arm resistance changes. If changes in the bridges are $\Delta R_1, \Delta R_2, \Delta R_3, \Delta R_4$, respectively, the output voltage of the bridge by formula (5) is

$$U_0 = \frac{(R_1 + \Delta R_1)(R_4 + \Delta R_4) - (R_2 + \Delta R_2)(R_3 + \Delta R_3)}{(R_1 + \Delta R_1 + R_2 + \Delta R_2)(R_3 + \Delta R_3 + R_4 + \Delta R_4)} \tag{7}$$

Put the formula (6) into equation (7), we can get the following equation.

$$U_0 = \frac{R_1 R_2}{(R_1 + R_2)^2} \left(\frac{\Delta R_1}{R_1} - \frac{\Delta R_2}{R_2} - \frac{\Delta R_3}{R_3} + \frac{\Delta R_4}{R_4} \right) U_{AC} \tag{8}$$

It is called equal arm bridge when $R_1 = R_2 = R_3 = R_4 = R$. The relationship between the resistance and the output voltage of the bridge is

$$U_0 = \frac{U_{AC}}{4} \left(\frac{\Delta R_1}{R_1} - \frac{\Delta R_2}{R_2} - \frac{\Delta R_3}{R_3} + \frac{\Delta R_4}{R_4} \right) \tag{9}$$

According to the formula (4) the following formula can be obtained.

$$U_0 = \frac{U_{AC} K}{4} (\varepsilon_1 - \varepsilon_2 - \varepsilon_3 + \varepsilon_4) \tag{10}$$

114

If

$$\varepsilon_d = \varepsilon_1 - \varepsilon_2 - \varepsilon_3 + \varepsilon_4 \tag{11}$$

Then

$$U_0 = \frac{U_{AC}K}{4}\varepsilon_d \tag{12}$$

where the ε_d is called reading strain. According to (11) and (12), ε_d can be written as

$$\varepsilon_d = \frac{4U_0}{U_{AC}K} = \varepsilon_1 - \varepsilon_2 - \varepsilon_3 + \varepsilon_4 \tag{13}$$

To measure the bending strain of the cantilever beam, the full bridge strain gauge can be pasted on the cantilever beam. The strain each strain gauge feels can be written respectively as

$$\varepsilon_1 = \varepsilon_4 = \varepsilon_F + \varepsilon_t, \varepsilon_2 = \varepsilon_3 = \varepsilon_t$$

So the reading strain can be written as

$$\varepsilon_d = \varepsilon_1 - \varepsilon_2 - \varepsilon_3 + \varepsilon_4 = \varepsilon_F + \varepsilon_t - \varepsilon_t - +\varepsilon_t + \varepsilon_F + \varepsilon_t = 2\varepsilon_F$$

then the axial strain of cantilever beam can be expressed as

$$\varepsilon_F = \frac{1}{2}\varepsilon_d \tag{14}$$

2 RECONSTRUCTION PRINCIPLE

The axial strain could be got through the above analysis. Then we introduce the empirical mode decomposition method, and derive the mode conversion equation. So measured modal response can be reconstructed, and the strain can be calculated.

2.1 Obtain modal function using EMD

The instantaneous characteristics of nonlinear unstable signal can be effectively extracted by EMD. Its basic idea is to decompose the original signal into a set of intrinsic mode functions. Given strain $y(t)$ in time domain, EMD gets the intrinsic mode function empirical according to the following conversion process:

1. An approximate range for Natural frequency is estimated by Fourier transform.
2. Filter the wave according to that range.
3. Find local extreme value of the filtered signals, marked by y^+ and y^- respectively.
4. Interpolation on the y^+ and y^- through three times spline. Two envelope curves will be got which marked as $e^+(t)$ and $e^-(t)$. Its average value is $m1(t) = [e^+(t) + e^-(t)]/2$.
5. Calculate $h(t) = y(t) - m(t)$.
6. If $h(t)$ is not an IMF, continue the sifting process using $h(t)$ as the new signal data through Steps 4–6. The stopping criterion is

$$\sum_t \frac{[h_k(t) - h_{k-1}(t)]^2}{h_{k-1}^2(t)} \leq \varepsilon$$

where $h_k(t)$ is the sifting result in the kth iteration, and ε is a small value between 0.2 and 0.3. The resulting $h_k(t)$ is an IMF, denoted as $f_1(t)$.

The original signal can be expressed in the following form when we get all the modal functions.

$$y(t) \approx \sum_{i=1}^{m} x_i(t) + \sum_{i=1}^{n-m} f_i(t) + r(t) \tag{15}$$

where $x_i(t)$ is the modal response (that is also an IMF) for the ith mode. Terms $f_i(t)$ are other IMFs but not modal responses.

2.2 Derive the transformation equations to reconstruct strain

Consider a general FEM describing a structure under analysis, the system dynamics equation can be expressed as

$$M\ddot{X} + C\dot{X} + KX = F$$

where M, K and C are mass, stiffness, and damping matrices, respectively. X is the displacement vector and F is the load vector. For practical structures subject to stochastic excitations, F is unknown, and direct solving equation to obtain the dynamical responses of a sensor inaccessible location is not possible. But based on the finite element method, we can solve the problem by solving the follow equation.

$$[\Phi, \lambda] = eig(M^{-1}K)$$

where Φ and λ are the eigenvectors and eigenvalues, respectively. Φ is also referred to as the mode shape matrix. λ corresponds to the natural frequencies of the structure, where f is the vector of natural frequencies.

$$\Phi = \begin{bmatrix} \Phi_{11} & \cdots & \Phi_{1n} \\ \vdots & \ddots & \vdots \\ \Phi_{n1} & \cdots & \Phi_{nn} \end{bmatrix} \tag{16}$$

The physical meaning of Φ can be interpreted as follows: each column of Φ represents a mode and each component in the column represents the displacement contribution of a DOF in the structure. This characteristic indicates that the responses of one DOF under modal coordinates allows for the calculation of responses of another DOF under modal coordinates. The physical meaning of the modal response relationship between two DOFs can be expressed as

$$\frac{\Phi_{ie}}{\Phi_{iu}} = \frac{\delta_{ie}}{\delta_{iu}} \tag{17}$$

where the subscript e represents the DOF which physical responses can be measured by sensors and u represents the DOF that is inaccessible for sensor measurements. $\delta_{ij}(t)$ corresponds to the modal responses components for the overall physical displacement responses of $X_j(t)$ for DOF j at a time index t.

The above equation can be rewritten as

$$\Phi_i = \alpha\delta_i \tag{18}$$

where Φ_i is the ith column vector in the mode shape matrix Φ, δ_i is the ith modal responses for all DOFs, α is a scalar constant for a given time index t.

From the finite element method we can know the relationship between strain and displacement are as follows.

$$\mathcal{E}^{(k)} = B^{(k)} X^{(k)}$$

where $B^{(k)}$ is the strain-displacement matrix for element k. The expression of $B^{(k)}$ usually has the format of

$$B^{(k)} = L N^{(k)}$$

where L is the differential operator and $N^{(k)}$ is the matrix of shape functions for element k. Using Eq. (18), the following equation is obtained under modal coordinates.

$$B^{(k)} \Phi_i^k = \alpha B^{(k)} \delta_i^{(k)}$$

We obtained the following conversion equation.

$$\frac{B^{(e)} \Phi_i^{(e)}}{B^{(u)} \Phi_i^{(u)}} = \frac{\alpha B^{(e)} \delta_i^{(e)}}{\alpha B^{(u)} \delta_i^{(u)}} = \frac{\eta_i^{(e)}}{\eta_i^{(u)}} \tag{19}$$

So the strain of unmeasured point can be reconstructed using the following transformation equation.

$$\mathcal{E}^{(u)}(t) \approx \sum_{i-1\ldots m} \left[\eta_i^{(e)}(t) \left(\frac{B^{(e)} \Phi_i^{(e)}}{B^{(u)} \Phi_i^{(u)}} \right)^{-1} \right] \tag{20}$$

Review of the above discussion, we can sum up the whole process of strain reconstruction as follow.

1. Decompose the strain into modal response according to EMD.

$$\mathcal{E}^{(e)}(t) \approx \sum_{i-1\ldots m} \eta_i^{(e)}(t)$$

2. Get the modal response of the unmeasured point using the transformation equation.

$$\eta_i^{(u)}(t) = \eta_i^{(e)}(t) \left(\frac{B^{(e)} \Phi_i^{(e)}}{B^{(u)} \Phi_i^{(u)}} \right)^{-1}$$

3. Get the strain of the unmeasured point through the modal superposition.

$$\mathcal{E}^{(u)}(t) \approx \sum_{i=1\ldots m} \eta_i^{(u)}(t)$$

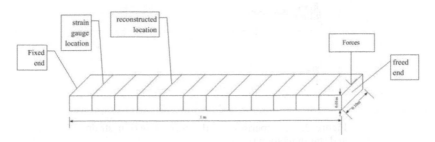

Figure 1.   The cantilever beam model.

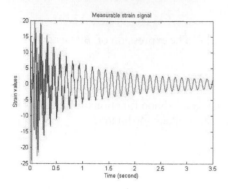

Figure 2.　Strain simulation signal.

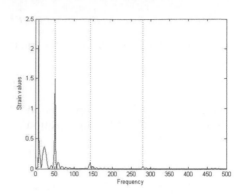

Figure 3.　The results of Fourier analysis.

Table 1.　Frequency ranges for each band-pass filters.

Model	1	2	3	4
Identified frequency	8.16	51.11	143.1	280.43
Passband corner frequency	[7 9]	[45 55]	[135 150]	[275 285]
Stopband corner frequency	[6 10]	[40 60]	[125 160]	[265 295]

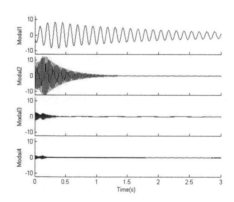

Figure 4.　Four modal responses of the strain.

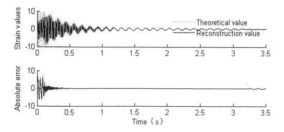

Figure 5.　Comparison of the reconstruction strain and the simulation strain.

3 THE SIMULATION TEST

Aluminum alloy cantilever beam is selected as the simulation model to verify the feasibility of the method. One end of the beam is fixed and the other end is free. The beam is 1m long, 0.1 m wide and 0.01 m thick. The Young's modulus is 69600 MPa and the density is 2730 kg/m^3. The beam structure is divided into 12 equal segments in the FE model, as shown in Figure 1. Measurement is the first node and unmeasured point is the fourth node.

Strain simulation signal is shown in Figure 2. The results of Fourier analysis of the signal is shown in Figure 3. The identified frequencies from the Fourier spectra are (8.16 Hz, 51.11 Hz, 143.1 Hz, 280.43 Hz). And they are used to design the band-pass filters. Frequency ranges for each band-pass filters are shown in Table 1.

Then four modal responses of the strain gauge measurement data obtained using EMD method are shown in Figure 4. According to the reconstruction process described earlier, reconstructed strain will be obtained and compared with the simulation strain in Figure 5.

The correlation coefficient between the simulation results and the value of the reconstruction is 0.916. The standard deviation of the error is 0.679. This shows that the reconstruction method is feasible, and has high accuracy.

REFERENCES

He, J., X. Guan, and Y. Liu. 2012. Structural response reconstruction based on empirical mode decomposition in time domain. *Mechanical Systems and Signal Processing*. 28: p. 348–366.

Kothamasu, R., S.H. Huang, and W.H. VerDuin. 2006. System health monitoring and prognostics—a review of current paradigms and practices. *The International Journal of Advanced Manufacturing Technology*. 28(9): p. 1012–1024.

Jacquelin, E., A. Bennani, and P. Hamelin. 2003. Force reconstruction: analysis and regularization of a deconvolution problem. *Journal of sound and vibration,* 265(1): p. 81–107.

Hurlebaus, S. and L. Gaul. 2006. Smart structure dynamics. *Mechanical Systems and Signal Processing*. 20(2): p. 255–281.

Kulkarni, S. and J. Achenbach. 2008. Structural health monitoring and damage prognosis in fatigue. *Structural Health Monitoring*. 7(1): p. 37–49.

Huang, N.E. and S.S. Shen. 2005. *Hilbert-Huang transform and its applications*. World Scientific Pub Co Inc.

Stephens, R.I. and H.O. Fuchs. 2001. *Metal Fatigue in Engineering*. Second Edition ed. New York: John Wiley & Sons, Inc.

Ribeiro, A., J. Silva, and N. Maia. 2000. On the generalization of the transmissibility concept. *Mechanical Systems and Signal Processing*. 14(1): p. 29–36.

Wanhill, M.J.J.Z.R.J.H., 2004. Fracture Mechanics. 1 ed. VSSD. 378.

Advances in Engineering Materials and Applied Mechanics – Zhang, Gao & Xu (Eds)
© 2016 Taylor & Francis Group, London, ISBN 978-1-138-02834-0

Energy method for wheelset hunting instability mechanism study

L. Wu, M.R. Chi & J.S. Qin
State Key Laboratory of Traction Power, Southwest Jiaotong University, Chengdu, China

ABSTRACT: In this paper, the object of the study is the wheelset that is restrained by primary suspension. The equation of wheelset motion has been decomposed into three parts: wheelset resilience section, suspension section and creep section. The wheelset hunting phenomenon was studied with the method of energy, from which the critical condition of hunting instability has been achieved. The results show that the hunting frequency calculated by energy method is just similar to the Kingel's hunting frequency.

1 INTRODUCTION

As early as the initial development of the railway transportation one hundred years ago, Stephenson from the Britain had noticed the lateral movement stability of the vehicle system and the phenomenon of hunting[4]. After 60 years, Kingel, the German scholar, illustrated the phenomenon officially from the kinematics view, who thought that the starting location of the vehicle during the travelling did not fit the centre line of the rail, because of the external excitation on the wheelset. And as the left and right wheels were fixed on one axle, these two wheels rolled at the same angular velocity, which caused the different rolling radius, then a force that would make the wheel set move to the centre line of the rail was produced. Kingel believed that the force produced was the main reason of motion instability. He also elicited the frequency and wavelength formula of the hunting movement. Along with the deep research on the creeping phenomenon, researchers found that the wheel set was affected by the excitation of creeping force while it was rolling on the straight rail. So dynamics was applied to study the hunting movement of the wheel set, which could reveal the essence, the characteristic and the rule of free wheel set hunting movement[6]. And this had brought a qualitative leap of the study on lateral movement stability of the vehicle system. DePater[1] was the first to consider the question of vehicle system lateral movement stability as the question of dynamic movement stability. Matsudaira[8] discussed the stability of the taper-tread wheel set and bogie by linear stability theory. After that, Wickens[2,3] discussed the stability question of worn wheel set and bogie with the consideration of gravity rigidity, and he thought that the creeping force and the taper of wheel set were the main reason of the system instability.

2 MATHEMATICAL MODEL

In this paper, the object of the study is the wheelset that is restrained by primary suspension. Lateral and yaw motion are considered in the model[5].

The differential equation of moving wheelset, as follows.

$$\mathbf{f}_G + \mathbf{f}_F + \mathbf{f}_S = \mathbf{M\ddot{u}} \tag{1}$$

where $\mathbf{f}_G$ is wheelset resilience section, $\mathbf{f}_F$ is suspension section, and $\mathbf{f}_S$ is creep section.

Figure 1. The single wheelsets model.

where the k_x is the longitudinal rigidity of the wheelset, k_y is the lateral rigidity of the wheelset, y is the lateral degree of freedom of the wheelset, ψ is the yaw degree of freedom of the wheelset, x is the direction of forward motion.

$$\mathbf{f}_G = -\begin{bmatrix} \dfrac{W\lambda}{b} & 0 \\ 0 & -W\lambda b \end{bmatrix}\begin{pmatrix} y \\ \psi \end{pmatrix}, \quad \mathbf{f}_F = -\begin{bmatrix} k_y & 0 \\ 0 & b^2 k_x \end{bmatrix}\begin{pmatrix} y \\ \psi \end{pmatrix}$$

$$\mathbf{f}_S = -\begin{bmatrix} 0 & -2f_{22} \\ \dfrac{2f_{11}\lambda b}{r_0} & 0 \end{bmatrix}\begin{pmatrix} y \\ \psi \end{pmatrix} - \dfrac{1}{v}\begin{bmatrix} 2f_{22} & 0 \\ 0 & 2f_{11}b^2 \end{bmatrix}\begin{pmatrix} \dot{y} \\ \dot{\psi} \end{pmatrix}$$

where W is axle load, λ is equivalent conicity, b is half of rolling cycle gauge, f_{11} is longitudinal creep coefficient, and f_{22} is lateral creep coefficient.

Also the equation (1) as follows,

$$\mathbf{M\ddot{u}} + \mathbf{C\dot{u}} + \mathbf{Ku} = 0 \tag{2}$$

$$\mathbf{M} = \begin{bmatrix} M_w & 0 \\ 0 & J_{wz} \end{bmatrix}, \quad \mathbf{C} = \dfrac{1}{v}\begin{bmatrix} 2f_{22} & 0 \\ 0 & 2f_{11}b^2 \end{bmatrix}$$

$$\mathbf{K} = \begin{bmatrix} \dfrac{W\lambda}{b} + k_y & -2f_{22} \\ \dfrac{2f_{11}\lambda b}{r_0} & -W\lambda b + k_x b^2 \end{bmatrix}$$

3 THE ENERGY METHOD OF HUNTING INSTABILITY

First, a differential equation constituted by a symmetric mass matrix and a symmetric stiffness matrix should be considered[7].

$$\mathbf{M\ddot{u}} + \mathbf{Ku} = 0 \tag{3}$$

In order to determine the power that various forces impose to the wheel-rail system, both sides of the Eq. 1 is multiplied by $\mathbf{du}^T$. Where $\mathbf{du}^T = \dot{\mathbf{u}}^T dt$.

$$\dot{\mathbf{u}}^T(\mathbf{M\ddot{u}} + \mathbf{Ku})\,dt = 0 \tag{4}$$

The mass matrix $\mathbf{M}$ and the stiffness matrix $\mathbf{K}$ are symmetric, so is the whole vehicle system, which including wheelsets, bogies and the carbody.

$$\dfrac{d}{dt}\left(\dfrac{1}{2}\dot{\mathbf{u}}^T\mathbf{M\dot{u}} + \dfrac{1}{2}\mathbf{u}^T\mathbf{Ku}\right)dt = 0 \tag{5}$$

In Eq. 5, the first item of the bracket is the kinetic energy T, and the second item is the potential energy of the primary spring V. Eq. 6 can be obtained by integrating the Eq. 5 from the time 0 to t.

$$\dfrac{1}{2}\dot{\mathbf{u}}^T\mathbf{M\dot{u}} + \dfrac{1}{2}\mathbf{u}^T\mathbf{Ku} = \dfrac{1}{2}\dot{\mathbf{u}}_0^T\mathbf{M\dot{u}}_0 + \dfrac{1}{2}\mathbf{u}_0^T\mathbf{Ku}_0 = T_0 + V_0 = \text{constant} \tag{6}$$

This is the energy conservation theorem of the conservative system. When the wheelset is running, potential energy and kinetic energy will convert to each other by the same amount. Then, the motion differential equation including damping item is considered as follows,

$$\mathbf{M\ddot{u}} + \mathbf{C\dot{u}} + \mathbf{Ku} = 0 \tag{7}$$

The matrix $\mathbf{M}$, $\mathbf{C}$ and $\mathbf{K}$ are symmetric. Eq. 8 can be obtained by the derivation as foreside.

$$\frac{d}{dt}\left(\frac{1}{2}\dot{\mathbf{u}}^T\mathbf{M}\dot{\mathbf{u}} + \frac{1}{2}\mathbf{u}^T\mathbf{Ku}\right)dt + \dot{\mathbf{u}}^T\mathbf{C}\dot{\mathbf{u}} = 0 \tag{8}$$

Eq. 9 can be obtained by integrating the Eq. 8 from the time 0 to t.

$$\frac{1}{2}\dot{\mathbf{u}}^T\mathbf{M}\dot{\mathbf{u}} + \frac{1}{2}\mathbf{u}^T\mathbf{Ku} = \frac{1}{2}\dot{\mathbf{u}}_0^T\mathbf{M}\dot{\mathbf{u}}_0 + \frac{1}{2}\mathbf{u}_0^T\mathbf{Ku}_0 - \int_0^t \dot{\mathbf{u}}^T\mathbf{C}\dot{\mathbf{u}}\,d\tau \tag{9}$$

or

$$T + V = T_0 + V_0 - \int_0^t \dot{\mathbf{u}}^T\mathbf{C}\dot{\mathbf{u}}\,d\tau \tag{10}$$

From Eq. 10, the system is called the non-conservative system, when the sum of the kinetic and the potential energy no longer remains constant. As part energy of the damper converts into heat energy, the sum of the kinetic and the potential energy must continue to decrease. The expression $\dot{\mathbf{u}}^T\mathbf{C}\dot{\mathbf{u}}$ is Rayleigh dissipation function, which is constantly positive or zero.

4   CALCULATION RESULTS

In summary, energy change is not affected by the symmetric mass matrix $\mathbf{M}$ and the stiffness matrix $\mathbf{K}$, and the symmetric damping matrix always output energy[9]. However, only input energy can stimulate wheelset hunting movement divergence, which attributes to integrated vector $\mathbf{f}_s$ of creeping force. Integrating for a period T:

$$\int_0^T \dot{\mathbf{u}}^T\mathbf{f}_s dt \geq 0 \tag{11}$$

The first integral equation of Eq. 11 is as follows:

$$\int_0^T \dot{\mathbf{u}}^T\mathbf{f}_s dt = \int_0^T \{\dot{y}\ \ \dot{\psi}\}\left\{-\begin{bmatrix} 0 & -2f_{22} \\ \dfrac{2f_{11}\lambda b}{r_0} & 0 \end{bmatrix}\begin{pmatrix} y \\ \psi \end{pmatrix} - \frac{1}{v}\begin{bmatrix} 2f_{22} & 0 \\ 0 & 2f_{11}b^2 \end{bmatrix}\begin{pmatrix} \dot{y} \\ \dot{\psi} \end{pmatrix}\right\}dt$$

$$= \int_0^T -\left[\frac{2f_{11}\lambda b}{r_0}y\dot{\psi} - 2f_{22}\dot{y}\psi\right] - \frac{1}{v}\left(2f_{22}\dot{y}^2 + 2f_{11}b^2\dot{\psi}^2\right)dt \tag{12}$$

The assumption of movement progress must be taken to further consider the energy. Same as the kinematic sinusoidal motion, it is verified by a non-damped movement, where there is a 90° phase between y and ψ.

$$y(t) = y_0 \sin \omega t \tag{13}$$

$$\psi(t) = \psi_0 \cos \omega t \tag{14}$$

Substituting these equations in Eq. 12:

$$\int_0^T \dot{\mathbf{u}}^T\mathbf{f}_s dt = 2\pi\left(f_{11}\frac{\lambda b}{r_0}y_0\psi_0 + f_{22}y_0\psi_0 - f_{22}\frac{\omega}{v}y_0^2 - f_{11}\frac{\omega}{v}b^2\psi_0^2\right) \tag{15}$$

123

Damping part and stiffness part are included in creeping force vector. From Eq. 15, the damping part energy is output and the stiffness part energy is input. When Eq. 15 satisfies the following relations:

$$f_{22} \frac{\omega}{v} y_0^2 - f_{22} y_0 \psi_0 = 0 \tag{16}$$

$$f_{22} \frac{\omega}{v} y_0^2 - f_{22} y_0 \psi_0 = 0 \tag{17}$$

Namely:

$$\frac{\psi_0}{y_0} = \frac{\omega}{v} \tag{18}$$

$$\omega^2 = v^2 \frac{\lambda}{br_0} \tag{19}$$

Extracting the root of Eq. 19:

$$\omega = v \sqrt{\frac{\lambda}{br_0}} \tag{20}$$

The ω shown in Eq. 20 is similar to the Kingel's hunting frequency.

ACKNOWLEDGEMENTS

This work has been supported by the State Key Program of National Natural Science of China (61134002), the National Key Basic Research Program of China (973 Program) (2011CB711100) and Innovation Group of Ministry of Education funded project (IRT1178).

REFERENCES

[1] A.D. DePater. 1960. The Approximate Determination of the Hunting Movement of a Railway Vehicle by aid of the Method of Krylov and Bogoljuow. Proceedings of the 10th Intenational Congress of Applied Mechanics. *Applied Scientific Research.* 10(1):205–228.
[2] A.H. Wickens. 1965. The dynamic stability of railway vehicle wheel-sets and bogies having profiled wheels. *International Journal of Solids and Structures.* 1(3):319–341.
[3] A.H. Wickens. 1965. The dynamic stability of a simplified four-wheeled railway vehicle having profiled wheels. *International Journal of Solids and Structures.* 1(4):385–406.
[4] J. Kingel. 1883. Uber den Lauf von Eisenbahnwagen auf gerader Bahn. Organ fur die Fortchrltte des Eisenbahnwesens in technischer Beziehung. NeueFolge. 20(4):113–123.
[5] O. Polach. 1999. A Fast Wheel-Rail Forces Calculation Computer Code [C]. Proc. of the 16th IAVSD Symposium, Pretoria, *Vehicle System Dynamics,* 33(Sup.):728–739.
[6] O. Polach. 2006. Comparability of The Non-linear and Linearized Stability Assessment during Railway Vehicle Design. *Vehicle System Dynamics,* VoL. 44, Supplement, 129–138.
[7] S.D. Iwnicki. A.H. Wickens. 1998. Validation of a MATLAB railway vehicle simulation using a scale roller rig. *Vehicle System Dynamics,* 30: 257–270.
[8] T. Matsudeira. 1960. The stability of complete vehicle with coned wheels. Papers awarded Prizes in the competition sponsored by office of Research and Experiment of the International Union of Railways. 5(3):13–23.
[9] Yang, Y.R. 1995. Limit cycle hunting of a bogie with flanged wheels. *Vehicle System Dynamics,* 24(3):185–196.

Advances in Engineering Materials and Applied Mechanics – Zhang, Gao & Xu (Eds)
© 2016 Taylor & Francis Group, London, ISBN 978-1-138-02834-0

Insertion loss calculation and parameter sensitivity analysis for noise barrier with T-shape top structure

Y.S. Liu & Y. Wang
Dalian Scientific Test and Control Technology Institute, Dalian, China

ABSTRACT: This paper mainly focuses on predicting and analyzing acoustical performance of noise barrier with T-shape top structure. A theoretical model describing the noise barrier under the influence of ground effect is investigated and built based on the Uniform Geometrical Theory of Diffraction (UTD) and the principle of image source, and corresponding method of Insertion Loss calculating is also given in detail. On this basis, an analysis method of parameter sensitivity is further proposed based on finite difference. The validity of the theoretical model and calculation analysis method are verified by the typical example simulation research, and the influence and rules of the main parameters on acoustical performance of the noise barrier are discussed including central structure height, top structure width and the incident sound frequency. This achievement can lay the foundation for optimization design of noise barrier with T-shape top structure, and also has a practical significance to guide the engineering application.

1   INTRODUCTION

Because of the advantages such as economy, simpleness and efficiency, the noise barrier structure or baffle is widely used in the noise pollution control of traffic, environmental and industrial noise. A large number of studies are shown that the top structure design can improve the noise reduction effect of sound barrier effectively, and the T-shape top structure is one of the most effective top types. Since 1980, May and Osman etc. (D.N. May, 1980) first obtained the T-shaped top sound barrier has a better acoustic performance than ordinary barrier, and there have been many scholars engaged in the noise barrier (Hothersall, 1991, J. Defrance, 2003, Ishizuka, 2004, D.H. Crombie, 1995, H. Shima, G.R. Watts, 1999, M.R. Monazzam, 2005, J.B. Keller, 1962). Hothersall, J. Defrance and P. Jean by numerical calculation and scale model test method have obtained an important conclusions that the insertion loss of T-shaped top noise barrier can improve 2~3 dB (Hothersall, 1991, J. Defrance, 2003) more than the height of the same general noise barrier. Moreover, Takashi Ishizuka (Ishizuka, 2004) investigated many different top noise barrier performances in detail by numerical and experimental analysis, and the results are shown that the T-shaped top structure can effectively improve the performance of traditional noise barrier. However, due to the effect of additional ground attenuation, including the terrain surface, and the economic cost constraints, this type of noise barrier is not adopted widely in practical engineering. In order to solve the problem, it is necessary to carry out a study in-depth in theoretical prediction, parameter sensitivity analysis and structural optimization design of noise barrier with T-shape top structure. And this paper is proposed under this background.

2   INSERTION LOSS CALCULATING METHOD

2.1   *Calculating multiple edge diffraction*

In this paper, the insertion loss calculation of noise barrier with T-shape top structure is carried out mainly based on the Uniform Geometrical Theory of Diffraction (UTD) and the principle of image source, which involves computing the single and multiple edge diffraction field. The theory of UTD is first proposed by J.B. Keller (J.B. Keller, 1962), and formed ultimately through the improvement and development by Pathak (P.H. Pathak, 1974). Pierce A.D. has succeeded to first apply the theory to calculating edge diffraction sound field originated from top structure of noise barrier (A.D. Pierce, 1974). Hyun-Sil Kim further validated the reliability of the theory and predicting method (Kim Hyun-Sil, 2005).

2.1.1 Single edge diffraction

Considering the double edge diffraction effect, that is the edges 1 and 2, and the two edges are coplanar, the edge diffraction sound pressure of noise barrier with T-shape top structure can be calculated as (A.D. Pierce, 1974):

$$
P_{ed}\left(f, n, r_s, \theta_s, r_r, \theta_r\right) = \frac{e^{-jkr_s}}{r_s} \cdot D_{ed}\left(f, n, r_s, \theta_s, r_r, \theta_r\right) \cdot \sqrt{\frac{r_s}{r_r\left(r_r + r_s\right)}} \cdot e^{-jkr_r} \tag{1}
$$

where θ_s is the angle between incident wave and edge structure, θ_r is the angle between edge diffraction of acoustic waves and edge structure; r_s is the distance between sound source and edge diffraction point, r_r is the distance between the field point and edge diffraction point; c is the acoustic velocity; n is a wedge exterior angle index; f is the incident sound wave frequency; $D_{ed}\left(f, n, r_s, \theta_s, r_r, \theta_r\right)$ is the edge diffraction coefficient, that is $D_{ed}\left(f, n, r_s, \theta_s, r_r, \theta_r\right) = \psi\left(f, n, \xi, \zeta, \theta_r - \theta_s\right) + \psi\left(f, n, \xi, \zeta, \theta_r + \theta_s\right)$, and the coefficient ζ is proposed for describing multiple edge diffraction, for example, single edge diffraction $\zeta = 1$. The expressions of the other parameters are as follows:

$$
\xi = \frac{r_s r_r}{r_s + r_r}, \psi\left(f, n, \xi, \zeta, \vartheta\right) = \frac{e^{-j\frac{\pi}{4}}}{\sqrt{2\pi k}} \frac{1}{2n} \cot\frac{\pi + \vartheta}{2n}\left\{F\left[\zeta \kappa^+\left(f, n, \vartheta\right)\right] + F\left[\zeta \kappa^-\left(f, n, \vartheta\right)\right]\right\} \tag{2}
$$

In above expressions, $F(x)$ are the modified Fresnel functions, and the expressions of the other parameters are as follows:

$$
\kappa^\pm\left(f, n, \varphi\right) = 2k\xi\cos^2\left(\frac{2v^\pm n\pi - \varphi}{2}\right), v^+ = \begin{cases} 0 & \varphi \le \pi(n-1) \\ 1 & \varphi > \pi(n-1) \end{cases},
$$

$$
v^- = \begin{cases} -1 & \varphi < \pi(1-n) \\ 0 & \pi(1-n) \le \varphi \le \pi(n+1) \\ 1 & \varphi > \pi(n+1) \end{cases}
$$

According to aforementioned mathematical expressions, the calculating formula of single edge diffraction sound pressure can be simplified as follows:

$$
P_{ed}\left(f, n, r_s, \theta_s, r_r, \theta_r\right) = \frac{e^{-jk\varsigma}}{\varsigma}\Gamma\left(f, n, \xi, \zeta, \theta_r, \theta_s\right) \tag{3}
$$

where $\varsigma = r_s + r_r$.

2.2 Multiple edge diffraction

The edge diffraction sound pressure of the noise barrier with top multiple edge structure, can be calculated based on the UTD theory and mathematical expression of single edge diffraction sound pressure as follows:

$$
P_{ed} = \left(\frac{1}{2}\right)^y \frac{e^{-jk\varsigma}}{\varsigma}\prod_{m=1}^{\upsilon}\Gamma_m\left(f, n_m, \xi_m, \zeta_m, \theta_{rm}, \theta_{sm}\right) \tag{4}
$$

where υ is the edge number, γ is the number of two adjacent coplanar edges, and ς is the sound distance of diffracted wave.

2.3 Calculating insertion loss with ground effect

Considering the ground additional attenuation effect with the noise barrier performance, we usually use the parameter of IL (Insertion Loss) to evaluate as following (HJ/T90-2004, 2004):

$$IL = 20 \cdot \lg \left| \frac{p_g}{p_{ted}} \right| (\text{dB}) \tag{5}$$

where p_{ted} is the total edge diffraction sound pressure of noise barrier, formed by the edge diffraction of sound barrier, including edge diffraction contribution produced by additional ground effect; p_g is the superposition of incident direct sound pressure and ground sound reflection, as shown in Figure 1, when the incident wave is a spherical wave, the normalized value can be calculated as follows:

$$p_g = \frac{e^{jkd_{sr}}}{d_{sr}} + \frac{e^{jkd_{s'r}}}{d_{s'r}} \cdot Q_{gs} \tag{6}$$

where d_{sr} is the straight line distance between sound source and received space position, $d_{s'r}$ is the straight distance between image of sound source and received space position, Q_{gs} is the spherical wave complex reflection coefficient of ground, and k is the wave number.

According to above-mentioned model and definition, we can see that, for the insertion loss calculation for noise barrier with T-shape top structure considering additional surface effect, the calculation of pg and pted are pivotal. In this paper, pg can be calculated by the principle of image source, and pted can be calculated based on the Uniform Geometrical Theory of Diffraction (UTD) and the principle of image source. As shown in Figure 2, the total dif-

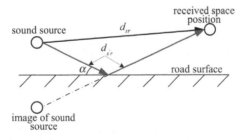

Figure 1. Sketch map of ground effect.

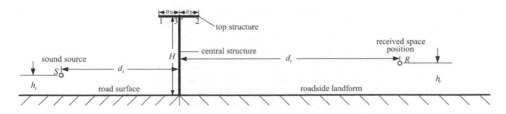

Figure 2. Sectional drawing of noise barrier with T-shape top structure on the horizontal landform.

fracted sound pressure of received space position of T-shape noise barrier is obtained by multiple diffraction pressure superposition, as shown in Table 1.

Unlike generic noise barrier, the noise barrier with T-shape top structure has the multiple edge top, which would be affected by higher order edge diffraction. Therefore, in order to assess the noise reduction effect of the noise barrier accurately, this paper considered double and triple edge diffraction effect top originated from T-shape top structure in the process of calculating the insertion loss. As shown in Figure 2, the single and multiple edge diffractions of sound pressure can be calculated as follows:

1. Considering the double edge diffraction effect, that is the edges 1 and 2, and the two edges are coplanar, the edge diffraction sound pressure of noise barrier with T-shape top structure can be calculated as (Pierce A.D., 1974):

$$p_{ed}\left(f,n_1,n_2,r_{s1},\theta_{s1},\theta_{r1},w_{12},\theta_{s2},r_{r2},\theta_{r2}\right)$$
$$=\frac{e^{-jk\varsigma}}{2\varsigma}\Gamma_1\left(f,n_1,\xi_1,\zeta_1,\theta_{r1},\theta_{s1}\right)\Gamma_2\left(f,n_2,\xi_2,\zeta_2,\theta_{r2},\theta_{s2}\right) \tag{7}$$

where

$$\varsigma = r_{s1}+w_{12}+r_{r2},\ n_1=2,\ n_2=2,$$

$$\Gamma_1\left(f,n_1,\xi_1,\zeta_1,\theta_{r1},\theta_{s1}\right)=\frac{1}{\sqrt{\xi_1\zeta_1}}\Big[\psi\left(f,n_1,\xi_1,\zeta_1,\theta_{r1}-\theta_{s1}\right)+\psi\left(f,n_1,\xi_1,\zeta_1,\theta_{r1}+\theta_{s1}\right)\Big]$$

$$\Gamma_2\left(f,n_2,\xi_2,\zeta_2,\theta_{r2},\theta_{s2}\right)=\frac{1}{\sqrt{\xi_2\zeta_2}}\Big[\psi\left(f,n_2,\xi_2,\zeta_2,\theta_{r2}-\theta_{s2}\right)+\psi\left(f,n_2,\xi_2,\zeta_2,\theta_{r2}+\theta_{s2}\right)\Big]$$

$$\xi_1=\frac{(w_{12}+r_{r2})r_{s1}}{\varsigma},\ \xi_2=\frac{(w_{12}+r_{s1})r_{r2}}{\varsigma}\quad \begin{cases}\zeta_1=\rho,\zeta_2=1 & \kappa^-\left(f,n_1,\theta_{r1}-\theta_{s1}\right)\le\kappa^-\left(f,n_2,\theta_{r2}-\theta_{s2}\right) \\ \zeta_1=1,\zeta_2=\rho & \kappa^-\left(f,n_1,\theta_{r1}-\theta_{s1}\right)>\kappa^-\left(f,n_2,\theta_{r2}-\theta_{s2}\right)\end{cases},$$

$$\rho=\frac{w_{12}\varsigma}{(w_{12}+r_{s1})(w_{12}+r_{r2})}.$$

2. The edge diffraction sound pressure of the noise barrier with top multiple edge structure, can be calculated based on the UTD theory and mathematical expression of single edge diffraction sound pressure as follows:

$$p_{ed}\left(f,n_1,n_2,n_3,r_{s1},\theta_{s1},\theta_{r1},w_{12},\theta_{s2},\theta_{r2},w_{23},\theta_{s3},r_{r3},\theta_{r3}\right)$$
$$=\frac{e^{-jk\varsigma}}{4\varsigma}\Gamma_1\left(f,n_1,\xi_1,\zeta_1,\theta_{r1},\theta_{s1}\right)\Gamma_2\left(f,n_2,\xi_2,\zeta_2,\theta_{r2},\theta_{s2}\right)\Gamma_3\left(f,n_3,\xi_3,\zeta_3,\theta_{r3},\theta_{s3}\right) \tag{8}$$

Table 1. Calculating edge diffraction sound pressure of noise barrier with T-shape top structure.

Top structure type of noise barrier	The path of edge diffraction	Total edge diffraction sound pressure
T-shape structure	Radiated sound source → edge of top structure → received space position	$p_{ted}=p_{tsr}+p_{ts'r}+p_{tsr'}+p_{ts'r'}$
	Radiated sound source → road surface → edge of top structure → received space position	
	Radiated sound source → edge of top structure → roadside ground surface → received space position	
	Radiated sound source → road surface → edge of top structure → roadside ground surface → received space position	

128

where

$$\varsigma = r_{s1} + w_{12} + w_{23} + r_{r3},\ n_1 = 1.5,\ n_2 = 2,\ n_3 = 2,$$

$$\Gamma_1\left(f,n_1,\xi_1,\zeta_1,\theta_{r1},\theta_{s1}\right) = \frac{1}{\sqrt{\xi_1\zeta_1}}\left[\psi\left(f,n_1,\xi_1,\zeta_1,\theta_{r1}-\theta_{s1}\right)+\psi\left(f,n_1,\xi_1,\zeta_1,\theta_{r1}+\theta_{s1}\right)\right]$$

$$\Gamma_2\left(f,n_2,\xi_2,\zeta_2,\theta_{r2},\theta_{s2}\right) = \frac{1}{\sqrt{\xi_2\zeta_2}}\left[\psi\left(f,n_2,\xi_2,\zeta_2,\theta_{r2}-\theta_{s2}\right)+\psi\left(f,n_2,\xi_2,\zeta_2,\theta_{r2}+\theta_{s2}\right)\right]$$

$$\Gamma_3\left(f,n_3,\xi_3,\zeta_3,\theta_{r3},\theta_{s3}\right) = \frac{1}{\sqrt{\xi_3\zeta_3}}\left[\psi\left(f,n_3,\xi_3,\zeta_3,\theta_{r3}-\theta_{s3}\right)+\psi\left(f,n_3,\xi_3,\zeta_3,\theta_{r3}+\theta_{s3}\right)\right]$$

$$\xi_1 = \frac{\left(w_{12}+w_{23}+r_{r3}\right)r_{s1}}{\varsigma},\ \xi_2 = \frac{\left(w_{12}+r_{s1}\right)\left(w_{23}+r_{r3}\right)}{\varsigma},\ \xi_3 = \frac{\left(w_{12}+w_{23}+r_{s1}\right)r_{r3}}{\varsigma},$$

$$\begin{cases} \zeta_1=\rho_1,\zeta_2=\rho_2,\zeta_3=1 & \left|\kappa^-\left(f,n_1,\theta_{r1}-\theta_{s1}\right)\right|\leq\left|\kappa^-\left(f,n_2,\theta_{r2}-\theta_{s2}\right)\right|\&\left|\kappa^-\left(f,n_2,\theta_{r2}-\theta_{s2}\right)\right|\leq\left|\kappa^-\left(f,n_3,\theta_{r3}-\theta_{s3}\right)\right| \\ \zeta_1=\rho_1,\zeta_2=1,\zeta_3=\rho_2 & \left|\kappa^-\left(f,n_1,\theta_{r1}-\theta_{s1}\right)\right|\leq\left|\kappa^-\left(f,n_2,\theta_{r2}-\theta_{s2}\right)\right|\&\left|\kappa^-\left(f,n_2,\theta_{r2}-\theta_{s2}\right)\right|>\left|\kappa^-\left(f,n_3,\theta_{r3}-\theta_{s3}\right)\right| \\ \zeta_1=1,\zeta_2=\rho_1\rho_2,\zeta_3=1 & \left|\kappa^-\left(f,n_1,\theta_{r1}-\theta_{s1}\right)\right|>\left|\kappa^-\left(f,n_2,\theta_{r2}-\theta_{s2}\right)\right|\&\left|\kappa^-\left(f,n_2,\theta_{r2}-\theta_{s2}\right)\right|\leq\left|\kappa^-\left(f,n_3,\theta_{r3}-\theta_{s3}\right)\right| \\ \zeta_1=1,\zeta_2=\rho_1,\zeta_3=\rho_2 & \left|\kappa^-\left(f,n_1,\theta_{r1}-\theta_{s1}\right)\right|>\left|\kappa^-\left(f,n_2,\theta_{r2}-\theta_{s2}\right)\right|\&\left|\kappa^-\left(f,n_2,\theta_{r2}-\theta_{s2}\right)\right|>\left|\kappa^-\left(f,n_3,\theta_{r3}-\theta_{s3}\right)\right| \end{cases},$$

$$\rho_1 = \frac{w_{12}\varsigma}{\left(w_{12}+r_{s1}\right)\left(w_{12}+w_{23}+r_{r3}\right)},\ \rho_2 = \frac{w_{23}\varsigma}{\left(w_{23}+r_{r3}\right)\left(w_{12}+w_{23}+r_{s1}\right)}.$$

3 SENSITIVITY ANALYSIS METHOD

According to the definition of sensitivity, it is derivative or partial derivative of the system state variables on the design variables, which can reflect the influence of design variables on the system state. In the practical engineering, parameter sensitivity analysis is the foundation of structure optimization design, reliability analysis and inverse problem research. The structure optimization design of noise barrier is usually based on parameter sensitivity analysis, and the results will directly affect the accuracy and reliability of the optimization design. In this paper, the target function expression of parameter sensitivity analysis is not specified clearly because of relative complexity of Insertion Loss calculating for noise barrier with T-shape top structure. Therefore, the parameter sensitivity analysis is carried out by the numerical analysis method of finite difference in the paper. As shown in Figure 2, the calculating expression of the noise barrier Insertion Loss is presumed as following:

$$IL = func\left(f,W_n,W_p,H,d_s,h_s,d_r,h_r\right) \tag{9}$$

where h_s is vertical distance between sound source and road surface, d_s is horizontal distance between sound source and noise barrier, h_r is vertical distance between received space position and ground surface, d_r is horizontal distance between received space position and noise barrier, W_n and W_p are the width of T-shape top structure on the side of sound source and received space position, respectively, H is height of central structure, and f is frequency of incident sound wave.

The reduced noise effect of noise barrier with T-shape top structure is mainly dependent on the structural parameters W_n, W_p and H, and so they are selected as primary object of sensitivity analysis. The corresponding mathematical expressions of parameter sensitivity analysis are given respectively as following:

$$\frac{\partial IL}{\partial H} = \frac{\partial func\left(f,W_n,W_p,H,d_s,h_s,d_r,h_r\right)}{\partial H},\ \frac{\partial IL}{\partial W_n} = \frac{\partial func\left(f,W_n,W_p,H,d_s,h_s,d_r,h_r\right)}{\partial W_n}$$

$$\frac{\partial IL}{\partial W_p} = \frac{\partial func\left(f, W_n, W_p, H, d_s, h_s, d_r, h_r\right)}{\partial W_p}, \ \frac{\partial IL}{\partial f} = \frac{\partial func\left(f, W_n, W_p, H, d_s, h_s, d_r, h_r\right)}{\partial f}, \quad (10)$$

where

$$f \in \left(f_{\min}, f_{\max}\right), \ W_n \in \left(W_{n\min}, W_{n\max}\right), \ W_p \in \left(W_{p\min}, W_{p\max}\right), \text{ and } H \in \left(H_{\min}, H_{\max}\right).$$

4 NUMERICAL SIMULATION ANALYSIS

In this section, the simulating validation researches are carried out based on above insertion loss calculation and parameter sensitivity analysis method of noise barrier with T-shape top structure. The influence and rules of acoustical and structure parameter on acoustical performance of the noise barrier are discussed in detail including the incident sound frequency, central structure height and top structure width. The spatial geometrical parameters such as sound source and received space position are specified respectively as following: $h_s = 0.5$ m, $d_s = 5$ m, $h_r = 1.5$ m and $d_r = 10$ m.

As shown in Figure 3(a), when the relative spatial geometrical configuration and structural parameters are specified respectively, the frequency parameter sensitivity of insertion loss almost fluctuates around zero in the noise frequency range from 50 Hz to 2.5 kHz, but the amplitude is relatively smaller. Moreover, it is worth to note that there is one break frequency point in the high frequency band, and further studies show that the number and distribution positions of the frequency point are closely related with the geometrical configuration and structural parameters. Accordingly, there is larger fluctuations in the frequency distribution of insertion loss of the noise barrier, but the insertion loss in the break frequency points of sensitivity value is minimum, as shown in the Figure 3(b). Therefore, we can effectively judge the frequency distribution characteristics and influence rules for insertion loss of noise barrier by frequency parameter sensitivity analysis, and it is helpful for guiding the acoustic design.

As shown in Figure 4(a), the effect of central structure height on insertion loss of noise barrier with T-shape top structure is closely related with the frequency. The parameter sensitivity values of central structure height will fluctuate gradually with frequency increasing, and the higher the frequency is, the more observably fluctuates. As shown in Figure 4(b), the sensitivity values are mostly positive in the low frequency range, that is, the insertion loss gradually increases with the height of central structure, but in the high frequency band, sensitivity values will start to fluctuate around zero. At this time, the insertion loss will increase or decrease with the central structure height increasing. Therefore, the results illuminate that the effect of central structure height is complicated on the acoustical performance of noise barrier.

As shown in Figure 5, when the width of T-shape top structure on the side of sound source is specified, the insertion loss and its changing trend with different frequency are

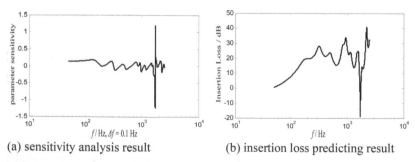

(a) sensitivity analysis result (b) insertion loss predicting result

Figure 3. The results of frequency parameter sensitivity analysis and corresponding theoretical predicting insertion loss.

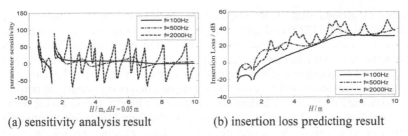

(a) sensitivity analysis result   (b) insertion loss predicting result

Figure 4. The results of central structure height parameter sensitivity analysis and corresponding theoretical predicting insertion loss.

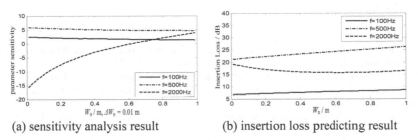

(a) sensitivity analysis result   (b) insertion loss predicting result

Figure 5. The results of parameter sensitivity analysis for width of T-shape top structure on the side of sound source and corresponding theoretical predicting insertion loss.

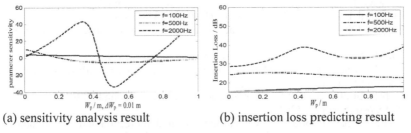

(a) sensitivity analysis result   (b) insertion loss predicting result

Figure 6. The results of parameter sensitivity analysis for width of T-shape top structure on the side of received space position and corresponding theoretical predicting insertion loss.

both distinct clearly. For low frequency, insertion loss value increases gradually with the width increasing, and for high frequency, insertion loss value will first decrease and then increase gradually. The critical value is associated with other acoustic parameters and structure parameters.

As shown in Figure 6, although depending on frequency similarly, but compared with the width of T-shape top structure on the side of sound source, the influences and trends of width on the side of received space position on the insertion loss of noise barrier have some difference in different frequency band, and it is more sensitive to incident sound frequency. Thus, it can be seen that there has obvious difference in the effect of the top width of both side on the acoustical performance of noise barrier with T-shape top structure, and this conclusion has an important guiding significance in the practical engineering application.

5 CONCLUSIONS

This paper mainly focuses on predicting and analyzing acoustical performance of noise barrier with T-shape top structure. A theoretical model describing the noise barrier under the influence of ground effect is investigated and built based on the Uniform Geometrical Theory of Diffraction (UTD) and the principle of image source, and corresponding method of Insertion Loss calculating is also given in detail. On this basis, an analysis method of parameter sensitivity is further proposed based on finite difference. The validity of the theoretical model and calculation analysis method are verified by the typical example simulation research, and the influence and rules of the main parameters on acoustical performance of the noise barrier are discussed including central structure height, top structure width and the incident sound frequency. This achievement can lay the foundation for optimization design of noise barrier with T-shape top structure, and also has a practical significance to guide the engineering application.

REFERENCES

Crombie D.H. etc. 1995. Multiple-edge noise barriers. *Applied Acoustics*, 44: 353~367.
Defrance J., P. Jean. 2003. Integration of the efficiency of noise barrier caps in a 3D ray tracing method. Case of a T-shape diffracting device. *Applied Acoustics*, 64: 765~778.
Keller J.B. 1962. Geometrical Theory of Diffraction. *J. Opt. Soc. Amer.* 52: 116~130.
Kim Hyun-Sil, Kim Jae-Sueng, Kang Hyun-Ju etc. 2005 Sound diffraction by multiple wedges and thin screens [J]. *Applied Acoustics*, 66: 1102~1119.
Hothersall etc. 1991. The Performance of profile and Associated Noise T-Barriers. *Applied Acoustics*, 32: 269~287.
May D.N. etc. 1980. The performance of sound absorptive. Reflective and T-profile noise barriers in Toronto. *Journal of sound and Vibration*, 72: 65~71.
Monazzam M.R., Y.W. 2005. Lam. Performance of profiled single noise barriers covered with quadratic residue diffusers. *Applied Acoustics*, 66: 709~730.
Pathak P.H., R.G. Kouyoumjian. 1974 An Analysis of Radiation from Apertures in curred Surface by the Geometrical Theory of Diffraction. *Proc. IEEE.* 62: 1438~1447.
Pierce A.D. 1974. Diffraction of sound around corners and over wide barriers. *J. Acoust. Soc. Am.*, 55: 941~955.
Shima H. etc. Noise reduction of a multiple noise barrier. *Inter-Noise* 96: 791~794.
Takashi Ishizuka, Kyoji Fujiwara. 2004 Performance of noise barriers with various edge shapes and acoustical conditions. *Applied Acoustics*, 65: 125~141.
Watts G.R., N.S. Godfrey, 1999. Effects on roadside noise levels of sound Absorptive materials in noise barriers. *Applied Acoustics*, 58: 385.
2004.10.1, The People's Republic of China environmental protection industry standards HJ/T90-2004: Specification for design and measurement of noise barrier. *State Environmental Protection Administration.*

Advances in Engineering Materials and Applied Mechanics – Zhang, Gao & Xu (Eds)
© 2016 Taylor & Francis Group, London, ISBN 978-1-138-02834-0

A study of normal stiffness model of joint interfaces based on fractal theory

W.W. Liu, J.J. Yang, N. Xi & L.L. Li
School of Mechanical Science and Engineering, Huazhong University of Science and Technology, Wuhan, Hubei, China

J.F. Shen
College of Sciences, Huazhong Agricultural University, Wuhan, Hubei, China

ABSTRACT: In order to obtain more accurate normal contact stiffness of joint interfaces theoretically, a contact stiffness model was developed and studied. The proposed model uses the fractal theory for characterizing surface topography and the contact mechanics theory for modeling contact stiffness of microcontacts, while the influence of the elastic, elastoplastic and fully plastic deformation regimes of contacting asperities are considered. Solutions for the force–displacement relationship in the elastoplastic regime are done by the well-supported assumptions: the microcontact area and microcontact normal load are enforcing continuity between the elastic and fully plastic regimes. Numerical calculation and simulation results reveal the effect of fractal parameters including fractal dimension and fractal roughness, and normal contact load on the normal stiffness. It is shown that normal contact stiffness increases with the increment of normal load and fractal dimension, and decreases with the increment of roughness parameter.

1 INTRODUCTION

The experimental results shown that about 60% of the total stiffness in assembled structures are derived from joint interfaces (Qu et al., 2013, Zhang et al., 2014). Because of the complication and nonlinear performances of the dynamic contact behavior, the mechanism of contact stiffness and influence of surface roughness are not fully understood for long periods of time. Jiang et al. (Jiang et al., 2010) proposed that contact stiffness model uses fractal geometry. Shi and Polycarpou (Polycarpou, 2005) measured the normal contact stiffness at the meso scale level based on the resonant frequency method. Asif et al. (Syed Asif et al., 2001) presented an imaging method to measure contact stiffness at the nano scale level. However, they did not consider elastic–plastic transition regime, and this may direct influence on the accuracy of constructing their models.

In this study, to accurately describe the contact interface and reveal the stiffness characteristic of joint interfaces, based on contact fractal theory and comprehensive contact mechanics theories as well as considering elastic, elastoplastic and fully plastic deformations of contacting asperities of joint interfaces, stiffness model of joint interfaces are obtained. Moreover, numerical simulation and calculation are employed and the results reveal the varying relations of the total stiffness versus corresponding parameters such as fractal dimension, fractal roughness parameters and dimensions normal load.

2 FRACTAL MODAL

2.1 *Characterization of surface topography*

A modified W-M function (Yan and Komvopoulos, 1998) is widely used for contact surface characteristic, and can be written as follows:

$$z(x) = L\left(\frac{G}{L}\right)^{D-1}(\ln\gamma)^{1/2}\sum_{n_{min}}^{n_{max}}\gamma^{(D-2)n}\left[\cos\phi_n - \cos\left(\frac{2\pi\gamma^n x}{L} - \phi_n\right)\right] \tag{1}$$

where $z(x)$ is the surface profile height, x is the lateral distance, L is the sample length, D is the fractal dimension, G is the fractal roughness parameter, γ is the scaling parameter, ϕ_n is a random phase, and n is a frequency index. The asperity interference δ and the size distribution function of microcontacts $n(a')$, can be expressed as, respectively,

$$\delta = 2G^{D-1}(\ln\gamma)^{1/2}(2r')^{2-D} \tag{2}$$

$$n(a') = \frac{D}{2}\psi^{(2-D)/2}a_l'^{D/2}a'^{-(D+2)/2} \tag{3}$$

where a' is the truncated area of microcontact with the radius r', a_l' is the truncated area of the largest microcontact, and ψ is the domain extension factor.

2.2 *Contact modeling*

Two rough surfaces in contact can be modeled as an equivalent rough surface in contact with a rigid smooth surface. In the elastic deformation regime, when a sphere with a radius of R comes into a smooth rigid flat with the interference δ, based on the Hertz contact theory, the microcontact area a_e and the elastic contact load p_e of one asperity can be expressed as (Liou and Lin, 2007)

$$a_e = \pi R\delta, \ p_e = 4ER^{1/2}\delta^{3/2}/3 \tag{4}$$

In the fully plastic deformation regime, the microcontact area a_p and the contact load p_p of one asperity can be expressed as (Liou and Lin, 2007)

$$a_p = 2\pi R\delta, \ p_p = Ha_p \tag{5}$$

In the first and second elastoplastic deformation regimes, the microcontact area a_{ep} and the microcontact load p_{ep} of one asperity are presented as (Liou and Lin, 2006),

$$a_{ep1} = \pi R\delta_c a_1\left(\delta/\delta_c\right)^{b_1}, \ p_{ep1} = 2KH\pi R\delta_c c_1\left(\delta/\delta_c\right)^{d_1}/3 \tag{6}$$

$$a_{ep2} = \pi R\delta_c a_2\left(\delta/\delta_c\right)^{b_2}, \ p_{ep2} = 2KH\pi R\delta_c c_2\left(\delta/\delta_c\right)^{d_2}/3 \tag{7}$$

where a_1, b_1, c_1, d_1, a_2, b_2, c_2 and d_2 are constants, Liou and Lin (Liou and Lin, 2006) summarized these constants through employing finite element method. If the normal microcontact load is supposed to vary monotonously and continuously during the approach of the two contact asperities, the real area should also be varying monotonously and continuously. However, the functions of area and load obtained from the finite element method did not follow this changing rule at the critical point. In other words, when the asperity deforms from the elastic regime to the elastoplastic regime, or from the elastoplastic regime to the plastic regime, the real area did not vary continuously. So to overcome this shortcoming, the microcontact area and microcontact normal load are supposed to vary continuously during the process of deformation, i.e.,

$$a_e(a_c') = a_{ep1}(a_c'), \ a_{ep1}\left((1/6)^{1/(D-1)}a_c'\right) = a_{ep2}\left((1/6)^{1/(D-1)}a_c'\right), \ a_{ep2}\left((1/110)^{1/(D-1)}a_c'\right) = a_p\left((110)^{1/(D-1)}a_c'\right)$$
$$p_e(a_c') = p_{ep1}(a_c'), \ p_{ep1}\left((1/6)^{1/(D-1)}a_c'\right) = p_{ep2}\left((1/6)^{1/(D-1)}a_c'\right), \ p_{ep2}\left((1/110)^{1/(D-1)}a_c'\right) = p_p\left((110)^{1/(D-1)}a_c'\right)$$

(8)

Thus, from Eqs. (4), (5), (6) and (7), it is found that

$$a_1 = 1, \ c_1 = 1, \ a_1/a_2 = (1/6)^{b_1-b_2}, \ c_1/c_2 = (1/6)^{d_1-d_2}, \ a_2(1/110)^{1-b_2} = 2, \ c_2(1/110)^{1-d_2} = 3/K \quad (9)$$

The differential of microcontact area and microcontact normal force is also assumed to vary continuously during the process of deformation, i.e.,

$$\mathrm{d}a_{ep1}\left((1/6)^{1/(D-1)}a_c'\right)\Big/\mathrm{d}a' = \mathrm{d}a_{ep2}\left((1/6)^{1/(D-1)}a_c'\right)\Big/\mathrm{d}a', \ \mathrm{d}p_{ep1}\left((1/6)^{1/(D-1)}a_c'\right)\Big/\mathrm{d}a'$$
$$= \mathrm{d}p_{ep2}\left((1/6)^{1/(D-1)}a_c'\right)\Big/\mathrm{d}a'$$

(10)

From Eqs. (6), (7), (9) and (10), we can obtain that

$$a_1 = a_2 = 1, \ b_1 = b_2 = 1 + \log_{110}(2), \ c_1 = c_2 = 1, \ d_1 = d_2 = 1 + \log_{110}(3/K) \quad (11)$$

2.3 The total normal contact load

Based on the contact mechanics model, the total contact regime of joint interface is represented in four kinds of forms, i.e., elastic, first elastoplastic, second elastoplastic and fully plastic regimes. The normal contact load P_n is then represented in the form of four kinds, i.e.

$$P_n = P_e + P_{ep1} + P_{ep2} + P_p$$
$$-\int_{a_c'}^{a_l'} p_e n(a')\mathrm{d}a' + \int_{(1/6)^{1/(D-1)}a_c'}^{a_c'} p_{ep1}n(a')\mathrm{d}a' + \int_{(1/110)^{1/(D-1)}a_c'}^{(1/6)^{1/(D-1)}a_c'} p_{ep2}n(a')\mathrm{d}a' + \int_0^{(1/110)^{1/(D-1)}a_c'} p_p n(a')\mathrm{d}a'$$

(12)

where P_e, P_{ep1}, P_{ep2} and P_p denote normal contact load of four regimes, respectively. Substituting Eqs. (3), (4), (5) (6), and (7) into Eq. (12) can yield expression of P_n

$$P_n = \begin{cases} f_1(D)EG^{(D-1)}a_l^{D/2}\left(a_l^{(3-2D)/2} - a_c^{(3-2D)/2}\right) + f_3(D)EG^{(D-1)}a_l^{D/2}a_c^{(3-2D)/2} & \text{for } D \neq 1.5 \ \& \ D \neq \\ & 2d/(2d-1) \\ f_2(D)EG^{(D-1)}a_l^{D/2}(\ln a_l - \ln a_c) + f_3(D)EG^{(D-1)}a_l^{D/2} & \text{for } D = 1.5 \\ f_1(D)EG^{(D-1)}a_l^{D/2}\left(a_l^{(3-2D)/2} - a_c^{(3-2D)/2}\right) + f_4(D)EG^{(D-1)}a_l^{D/2}a_c^{(3-2D)/2} & \text{for } D = 2d/(2d-1) \end{cases}$$

(13)

where $f_1(D)$, $f_2(D)$, $f_3(D)$, and $f_4(D)$ are the functions of fractal dimension D shown as follows:

$$f_1(D) = \frac{2^{(12-3D)/2}\pi^{(D-3)/2}D}{3(3-2D)}(\ln\gamma)^{1/2}\psi^{(2-D)/2}, \ f_2(D) = \frac{2^{(7-D)/2}\pi^{(D-3)/2}D}{3}(\ln\gamma)^{1/2}\psi^{(2-D)/2}$$

$$f_3(D) = 2^{(12-3D)/2}\pi^{(D-3)/2}D(\ln\gamma)^{1/2}\psi^{(2-D)/2}\left[\frac{(1-(1/110)^{((1+2\log_{110}(3/K))(1-D)+1)/(2(D-1))})}{3((1+2\log_{110}(3/K))(1-D)+1)} + \frac{(1/110)^{(2-D)/(2(D-1))}}{K(2-D)}\right]$$

$$f_4(D) = 2^{(12-3D)/2}\pi^{(D-3)/2}D(\ln\gamma)^{1/2}\psi^{(2-D)/2}\left[\frac{\ln(110)}{6(D-1)} + \frac{(1/110)^{(2-D)/(2(D-1))}}{K(2-D)}\right]$$

According to Eq. (13), the dimensionless total normal contact load P_n^* over the whole joint interfaces can be written as:

$$P_n^* = \begin{cases} f_1(D)G^{*(D-1)}a_l^{*D/2}\left(a_l^{*(3-2D)/2} - a_{yc}^{*(3-2D)/2}\right) + f_3(D)EG^{*(D-1)}a_l^{*D/2}a_{yc}^{*(3-2D)/2} & \text{for } D \neq 1.5 \& D \neq \\ & 2d/(2d-1) \\ f_2(D)G^{*(D-1)}a_l^{*D/2}\left(\ln a_l - \ln a_{yc}\right) + f_3(D)G^{(D-1)}a_l^{D/2} & \text{for } D = 1.5 \\ f_1(D)G^{*(D-1)}a_l^{*D/2}\left(a_l^{*(3-2D)/2} - a_{yc}^{*(3-2D)/2}\right) + f_4(D)G^{*(D-1)}a_l^{*D/2}a_{yc}^{*(3-2D)/2} & \text{for } D = 2d/(2d-1) \end{cases}$$

(14)

where a_c^* and G^* are dimensionless critical area and dimensionless fractal roughness parameter, respectively. These dimensionless parameters are written as:

$$P_n^* = P_n/EA_a, a_c^* = a_c/A_a, G^* = G/A_a^{1/2}$$

where A_a is total apparent contact area of the contact regime.

2.4 *Fractal modal of normal contact stiffness*

The contact asperity that deformed elastically or elastoplasticlly stored the elastic strain energy, so we can assume that stiffness characteristics of two contact surfaces are derived from the deformed asperities in the elastic and elastoplastic regime. The normal contact stiffness of a single microcontact under elastic regime, the first and the second elastoplastic regimes can be, respectively, written as:

$$k_{ne} = dp_e/d\delta, \; k_{nep1} = dp_{ep1}/d\delta, \; k_{nep2} = dp_{ep2}/d\delta$$

(15)

where k_{ne}, k_{ne1} and k_{ne2} are the contact stiffness of a single microcontact asperity of three regimes, respectively.

Introducing Eqs. (2), (4), (6) and (7) into Eq. (15), the normal contact stiffness of a single microcontact asperity are, respectively, represented as:

$$k_{ne} = \frac{4(3-D)E}{3(2-D)(2\pi)^{1/2}}a'^{1/2}, \; k_{nep1} = \frac{1}{3}KH\frac{2c_1[(d_1-1)(1-D)+1]}{(2-D)\left[2^{3-D}\pi^{(D-2)/2}G^{D-1}(\ln\gamma)^{1/2}\right]}a_c'^{(d_1-1)(D-1)}a'^{[(2d_1-3)(1-D)+1]/2}$$

$$k_{nep2} = \frac{1}{3}KH\frac{2c_2[(d_2-1)(1-D)+1]}{(2-D)\left[2^{3-D}\pi^{(D-2)/2}G^{D-1}(\ln\gamma)^{1/2}\right]}a_c'^{(d_2-1)(D-1)}a'^{[(2d_2-3)(1-D)+1]/2}$$

(16)

Therefore, the total normal stiffness K_n over the whole joint interfaces can be expressed as:

$$K_n = \int_{a_c'}^{a_l'}k_{ne}n(a')da' + \int_{(1/6)^{1/(D-1)}a_c'}^{a_c'}k_{nep1}n(a')da' + \int_{(1/110)^{1/(D-1)}a_c'}^{(1/6)^{1/(D-1)}a_c'}k_{nep2}n(a')da'$$

(17)

Substituting Eqs. (3) and (16) into Eq. (17) can yield

$$K_n = f_5(D)Ea_l^{1/2} + f_6(D)Ea_l^{D/2}a_c^{(1-D)/2}$$

(18)

where $f_7(D)$ and $f_8(D)$ are the functions of fractal dimension D shown as follows:

$$f_5(D) = \frac{4D(3-D)}{3\pi^{1/2}(2-D)(1-D)}\psi^{(2-D)/2}$$

$$f_6(D) = \frac{4D}{3\pi^{1/2}(2-D)(1-D)}\psi^{(2-D)/2}\left[(D-3) + \frac{(1-3/K)[1+\log_{110}(3/K)(1-D)]}{\log_{110}(3/K)}\right]$$

136

The dimensionless total normal stiffness K_n^* over the whole joint interfaces can be written as:

$$K_n^* = f_5(D)a_l^{*1/2} + f_6(D)a_l^{*D/2}a_c^{*(1-D)/2} \qquad (19)$$

where K_n^* is dimensionless total normal stiffness, and it can be written as:

$$K_n^* = K_n/EA_a^{1/2}$$

3 NUMERICAL SIMULATIONS AND DISCUSSION

Dimensionless total normal stiffness K_n^* can be figure out by Eq. (19). Simulations are given for 0.1 m × 0.1 m isotropic surface, fractal dimension $D = 1.2 \sim 1.9$, dimensionless fractal roughness parameter $G^* = 1 \times 10^{-11}$, 1×10^{-12}, 1×10^{-13} respectively, the Hardness $H = 9 \times 10^9$ Pa, the equivalent elastic moduli $E = 113 \times 10^9$ Pa, and the equivalent Poisson's ratios $v = 0.3$.

Figure 1 presents the effect of fractal dimension D and fractal roughness G^* on dimensionless total normal stiffness K_n^*. From Figure 1, for a fixed dimensionless normal load ($P_n^* = 1 \times 10^{-7}$), it can be seen that increasing the fractal dimension D increases the dimensionless total normal stiffness K_n^*. When the value of D is less than 1.6, K_n^* increases at a faster rate, and when the value of D is more than 1.6, its growth decreases gradually. Figure 2(a) and (b) shows the effect of the dimensionless total normal load P_n^* on the K_n^*. From Figure 2, it can be

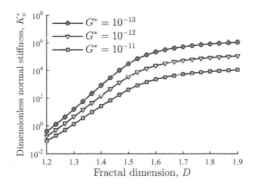

Figure 1. Variation relationship of parameters K_n^*, D, and G^*.

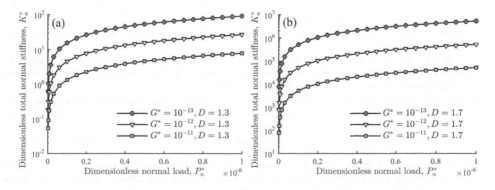

Figure 2. Variation relationship of parameters K_n^*, P_n^*, and G^* (a) as $D = 1.3$; (b) as $D = 1.7$.

seen that there is a nonlinear relation between K_n^* and P_n^*, the relationship between the stiffness and load approximately follows a power law. K_n^* increases along with the increment of P_n^* according to the two representative values of D ($D = 1.3$ and $D = 1.7$). From Figures 1 and 2, it can be also concluded that, for a fixed P_n^* or a fixed fractal dimension D, K_n^* decreases along with the increment of G^*, i.e., the coarser the surface is, the smaller the K_n^* is. Thus, decreasing the surface roughness is helpful to improve the stiffness characteristic of joint interface.

4 CONCLUSIONS

1. A normal contact stiffness model of joint interfaces has been developed based on the fractal theory and contact mechanics theories. The analysis is based on a fractal description of the equivalent interfaces and constitutive contact mechanics for elastic, elastoplastic and fully plastic deformations of contacting asperities. Solutions for the force–displacement relationship in the elastoplastic regime are done by the well-supported assumptions: the microcontact area and microcontact normal load are enforcing continuity between the elastic and fully plastic regimes.
2. For a fixed load, increasing the fractal dimension D increases the dimensionless total normal stiffness and dimensionless total normal stiffness decreases along with the increment of fractal roughness. The coarser the surface, the smaller the dimensionless total normal stiffness. Meanwhile, normal contact stiffness increases with the increment of the dimensionless total normal load, and the normal contact stiffness and load approximately follows a power law.
3. It is helpful to better understand the interfacial stiffness of mechanical structures and the influence of fractal parameters to the interfacial stiffness.

ACKNOWLEDGMENT

This research was supported by the National Science and Technology Major Project of China under grant no. 2013ZX04008-031.

REFERENCES

Jiang, S., Zheng, Y. & Zhu, H. (2010), "A contact stiffness model of machined plane joint based on fractal theory", *Journal of Tribology-Transactions of the Asme,* Vol. 132 No. 1, pp. 011401–7.

Liou, J.L. & Lin, J.F. (2006), "A new method developed for fractal dimension and topothesy varying with the mean separation of two contact surfaces", *Journal of Tribology-Transactions of the ASME,* Vol. 128 No. 3, pp. 515–524.

Liou, J.L. & Lin, J.F. (2007), "A new microcontact model developed for variable fractal dimension, topothesy, density of asperity, and probability density function of asperity heights", *Journal of Tribology-Transactions of the ASME,* Vol. 74 No. 4, pp. 603–613.

Polycarpou, A.A. (2005), "Measurement and modeling of normal contact stiffness and contact damping at the meso scale".

Qu, C., Wu, L., Ma, J., Xia, Q. & Ma, S. (2013), "A fractal model of normal dynamic parameters for fixed oily porous media joint interface in machine tools", *The International Journal of Advanced Manufacturing Technology,* Vol. 68 No. 9, pp. 2159–2167.

Syed Asif, S.A., Wahl, K.J., Colton, R.J. & Warren O.L. (2001), "Quantitative imaging of nanoscale mechanical properties using hybrid nanoindentation and force modulation", *Journal of Applied Physics,* Vol. 90 No. 3, pp. 1192–1200.

Yan, W. & Komvopoulos, K. (1998), "Contact analysis of elastic-plastic fractal surfaces", *Journal of Applied Physics,* Vol. 84 No. 7, pp. 3617–3624.

Zhang, X., Wang, N., Lan, G., Wen, S. & Chen, Y. (2014), "Tangential Damping and its Dissipation Factor Models of Joint Interfaces Based on Fractal Theory with Simulations", *Journal of Tribology-Transactions of the Asme,* Vol. 136 No. 1, pp. 1–10.

Advances in Engineering Materials and Applied Mechanics – Zhang, Gao & Xu (Eds)
© 2016 Taylor & Francis Group, London, ISBN 978-1-138-02834-0

Research on the design method of combat resilience based on conflict matrix

S.S. Ma
Institute of Ordnance Technology, Shijiazhuang, China

S.Y. Yang & Y.X. Xu
Huang pi NCO School, AFEWA, Wuhan, China

ABSTRACT: In allusion to the technical problem in innovative design of combat resilience, the method that utilizes the conflict matrix in the TRIZ is put forward in this paper. First, the specific existing problems are transformed into 39 general engineering parameters, then looked up for the corresponding innovative principle using the conflict matrix, and the best scheme design is confirmed by combining with the actual equipment. This method is used to analyze and improve the design problem in phased array antenna. The research results indicate that using the TRIZ conflict matrix can effectively solve the technical problem in innovative design of combat resilience, and improve the equipment innovation design efficiency.

1   INTRODUCTION

Combat resilience is its own supportability design characteristics. It refers a required function or the self-saving ability to make the damaged equipment recovery to complete certain tasks by emergency maintenance means and methods in the actual combat and the prescribed period[9]. Combat resilience is one of the main technical indexes of performance evaluation, and has a very important position. In the formulation of design scheme, combat resilience index need to be considered with other important technology indexes according to combat resilience design requirements[5]. And the combat resilience index should be given priority support for entirely resilience control. Combat resilience design takes the improvement of battlefield damage assessment and maintenance capability as the goal, and it is an important index for performance design.

Since America B.J. Stoll brigadier Karp first proposed the concept of resilience in 1986, resilience as a kind of equipment design characteristic that becomes the key research contents of combat resilience, and combat resilience design has been incorporated into the standard requirements document of new equipment development. At present, a series of research has been carried out on the combat resilience design in our army, which is mainly focused on the development and design of emergency resilience tools and technologies. But the basic theory is not enough. Because of many conditions, like the tactics and technology index, it is difficult for combat resilience design to make a breakthrough. Combat resilience design belongs to innovation design in essence, so we put forward conflict Matrix in TRIZ theory to analyze the technical problems encountered in resilience design, and then determine the best scheme for combat resilience using innovation principles of TRIZ theory.

2   CONFLICT MATRIX IN TRIZ THEORY

2.1   *Technical conflict*

The concept of conflict matrix[2,7] in TRIZ theory (the theory to solve invention problem) originates from Dialectics. Dialectics show that conflict is the basic power of the development, and exists in everything. The core problem of innovation is to solve the conflict. Technical conflict refers that an operation results in both useful and harmful results at the same time, or refers that introducing the beneficial effect or eliminating the harmful effect leads to the deterioration of one or several subsystem or the whole system. Technical conflict is often shown as the conflict between the two subsystems, and can be divided into several situations: ① introducing a useful function in a subsystem will lead to another sub system to produce a harmful function, or strengthen a harmful function; ② eliminating a harmful function leads to the weakening or deterioration of another subsystem; ③ the strengthen of the useful function or the reducing of the harmful function makes another subsystem or system complex[8]. The technical conflict generally involves two parameters A and B. when A is improved, B is becoming worse and worse. There are many examples of technical conflict. For example, increasing the system weight to improve the stability of the system, then the system movement speed is reduced.

To solve the contradiction, first the existing problems should be transformed to technical contradiction which can be expressed by 39 general engineering parameters, then querying innovation principles corresponding to the beneficial and harmful parameters by the contradiction matrix table, and finally, establishing the final solution to solve the specific problem according to the hints given by innovation principle[4]. The basic process to solve the technical conflict problem is shown in Figure 1.

2.2 Conflict parameters

In the TRIZ theory, Archie Schuler put forward that the specific conflict can be described by 39 general engineering technical parameters[1,3,6], the 39 engineering technical parameters is: 1. Weight of moving object; 2. Weight of static object; 3. Length of moving object; 4. Length of static object; 5. Area of moving object; 6. Area of static object; 7. Volume of moving object; 8. Volume of static object; 9. Speed; 10. Force; 11. Stress or pressure; 12 Shape; 13. Stable structure; 14. Strength; 15. Action time of moving object; 16. Action time of static object; 17. Temperature; 18. The illumination; 19. Energy of moving object; 20. Energy of static object; 21. Power; 22. Power, means the work in a unit time, namely the energy consumption rate; 23. The material loss; 24. The loss of information; 25. Loss of time; 26. The amount of substance; 27. Reliability; 28. The measurement accuracy; 29. The manufacturing accuracy; 30. The harmful factors acting on the object; 31. The harmful factors produced by the object; 32. Manufacturing convenience; 33. The operation convenience; 34. Maintenance convenience; 35. Practicality, versatility; 36. The complexity of equipments; 37. The detection complexity; 38. The degree of automation; 39. Productivity.

Finding out the parameters corresponding to conflict in 39 engineering technical parameters, transferring the specific problem into general parameters, can lay the foundation for the analysis of conflict, and make it easy to solve the problem by conflict matrix.

2.3 Conflict matrix

Conflict matrix is an important tool to solve problem in TRIZ theory[6], the horizontal and vertical coordinates of conflict matrix are all natural serial numbers that range from 1 to 39,

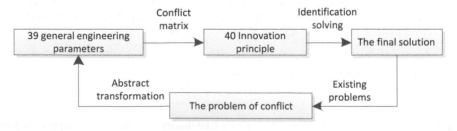

Figure 1.   The process to solve technical conflict problem.

Table 1. Conflict matrix.

		Technical characteristics need to be improved						
		No. 1	No. 2	No. 3	No. 4	No. 5	...	No. 39
The deterioration of technical characteristics	No. 1			15, 8, 29, 34		29, 17, 38, 34		35, 3, 24, 37
	No. 2				10, 1, 29, 35			1, 28, 15, 35
	No. 3	8, 15, 29, 34			15, 17, 4			14, 4, 28, 29
	No. 4		35, 28, 40, 29					30, 14, 7, 26
	No. 5	2, 17, 29, 4		14, 15, 6, 4				10, 26, 34, 2
	...							
	No. 39	35, 26, 24, 37	28, 27, 15, 3	18, 4, 28, 38	30, 7, 17, 26	10, 26, 34, 3		

each serial number corresponding to a general engineering parameter[7]. The first column of the matrix displays the 39 general engineering parameters causing deterioration; the first row of the matrix displays the 39 general engineering parameters need to be improved. The cross of each column and row formulates the conflict of the system, at each intersection squares, the proposed priority use of invention and creation principle serial numbers are displayed. For example, the cross of the first column of deterioration parameter No. 1 and the first row of parameter No. 3 need to be improved display "15, 8, 29, 34". These four natural numbers refer to the 15th, 8th, 29th, 34th innovation principles (the 40 invention principles can be found in the literature[4]). A corresponding relationship is established between 40 inventive principles and 39 general engineering parameters by conflict matrix. As shown in Table 1.

3 IMPROVED DESIGN METHOD OF COMBAT RESILIENCE BASED ON CONFLICT MATRIX

3.1 *The basic process*

Converting combat resilience design problem into conflict matrix, the recommended principles of invention and innovation can be obtained by conflict matrix. With these innovative principles as inspiration, you can find some practical solutions for real problems. According to the TRIZ technical conflict matrix theory, combined with the actual situation of combat resilience, the combat resilience design process is shown in Figure 2.

As shown in the above flow chart, the technical conflict principle is divided into four major steps. The combat resilience design process based on technical conflict is:

1. Describing problems. Describing concrete resilience problems to be solved in popular language.
2. Conflict conversion. According to 39 general engineering parameters, converting concrete resilience problems to be solved into technical conflict.
3. Determining innovation principle. Finding out innovation principles for resilience problems, according to conflict matrix provided by TRIZ.
4. Determining the final solution. According to the innovation principles for concrete resilience problems, inspiring thinking, after deduction and specific, finally finding a solution to the problem.

The following example of a radar antenna design was analyzed individually.

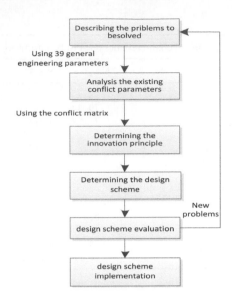

Figure 2.　The combat resilience design process based on the analysis of the technical conflict matrix.

3.2　*Specific analysis*

3.2.1　*Problem description*

The ideal solution for combat resilience can be summarized into three basic requirements: ① Requiring shorter recovery time; ② Requiring repair conditions as simple as possible; ③ Requiring good state after the restoration.

Under the combat environment, the radar antenna in the whole radar equipment is the most vulnerable part. Because of unprotected exposure to cause a high probability of damage, the resilience issues should be considered in the design of the radar antenna to achieve that a partial area of the radar antenna suffered damage can quickly return to work.

The best way for radar antenna to recover from the local destroyed to working state, and meet the three basic requirements, is to replace a spare antenna and switch directly restored to working condition. However, this method costs high and spare parts requirements cannot be met. The key of the problem is to design antenna to save the cost and repair quickly.

3.2.2　*Problem analysis*

First, extracting technical conflict of the problem to be solved.

For example, analyzing the structure design method of phased array antenna for a radar. The phased array antenna is mainly consists of antenna array and T/R component.

The radar antenna array and T/R assembly structure design together, located on the antenna base, its structure layout as shown in Figure 3. Antenna array and T/R assembly of the radar are designed together in structure, located in the antenna seat, and its structure layout is shown in Figure 3.

The remarkable characteristics of the phased array antenna are that the number of s is huge, each array element in a line in horizontal direction is printed at the same substrate block. Then array element assembly is formulated through adhesive by the thin film circuit hot pressing bonding technique. Thus the strength, stiffness of array element antenna is increased greatly, and the array element spacing and the consistency of each radiating element are guaranteed. Array element assembly structure is shown in Figure 4.

Analysis of resilience for the phased array antenna:

Question 1: Considering from the battlefield resilience, whether antenna arrays can be designed separately with T/R component. When the antenna array is suffered hard destruction, T/R component that is the most expensive component can keep safe. For this design idea, the

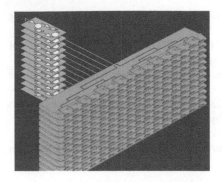

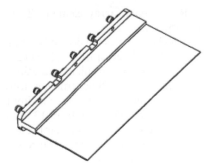

Figure 3.   Structure layout of phased array antenna.

Figure 4.   Array element assembly.

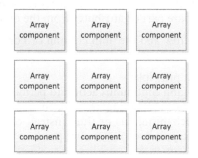

Array component	Array component	Array component
Array component	Array component	Array component
Array component	Array component	Array component

Figure 5.   The ideal array antenna modular.

array elements number and RF ports number are huge, a large number of RF cable is required for connection, and there are dozens of radio frequency port. In addition, there are a total of more than 100 RF port for switch array and power component, T/R component also contains dozens of RF port, thus the number of RF port is huge, RF cables are so many. On the other hand, RF interface of phased array antenna is intensive, and the phase for connectors must be consisted. Therefore, if the antenna array is far away from the T/R component, we must think of a way to solve the problem of transmission loss, complexity and manufacturing process. So the design protected T/R components, but also increased the complexity of the system, increased the energy loss of signal transmission system.

Question 2: For antenna array design, whether antenna array element and its corresponding T/R component can be combined, and make the whole antenna system consisted of a plurality of square array. The element component module contains a number of array elements in both level and vertical direction, as shown in Figure 5. This design style is more in line with the actual situation of war wounds than long strip array element component, can ensure that the array element needed to be replaced is few. For the convenience of resilience, each array element components should be designed to direct plug replacement.

For the design idea, the convenience of operation (direct swap) need to be resolved, and it brings a problem that the physical structure of the antenna is firmness.

In addition, the arrangement of multi array element components also has strict spacing requirements. Similarly, dielectric material between array components can only be with small dielectric constant, a certain strength and temperature stability, to ensure the array element spacing and improve the strength and rigidity of antenna array, and make the antenna stable and reliable.

3.2.3   *Conflict analysis*

Conflict 1: Reducing the damage to the T/R component, the energy loss caused by longer signal transmission path, and the system complexity caused by information exchange between

T/R and each array element. The conflict can be transferred into 39 general engineering parameters.

Characteristics needed to be improved: 30 harmful factors to the object. Characteristics of the system deterioration: 22 energy loss, and 36 system complexity.

Conflict 1: Convenient operation process and robust physical structure. The conflict can be transferred into 39 general engineering parameters.

Characteristics needed to be improved: 33 convenient operation processes. Characteristics of the system deterioration: 13 the stability of the object, and 14 robust.

(4) The innovation principle based on conflict matrix.

Corresponding to the 30th row and the 22th column of conflict matrix, we can get four innovation principles:

21# Reduce harmful effect time, and make dangerous or harmful work with high speed;
22# Variable harm into benefit principle;
35# A physical or chemical change in state;
02# Extraction principle, extracting spare part or property cause negative effects from objects, and extracting necessary spare part or attributes from the object.

Corresponding to the 30th row and the 36th column of conflict matrix, we can get four innovation principles:

22# Change harm into benefit principle;
19# Periodic action: (1) Changing continuous action to periodic action; (2) For the periodic motion, changing its frequency; (3) adding periodic in pulse;
29# Pneumatic, hydraulic mechanism: gas or fluid instead of fixed portion of the object, using air pressure, hydraulic pressure, oil pressure produce the buffer function;
40# Compound material.

Corresponding to the 33th row and the 13th column of conflict matrix, we can get three innovation principles:

32# Change the color;
35# Change physical or chemical state;
30# Flexible membrane or film.

Corresponding to the 33th row and the 14th column of conflict matrix, we can get four innovation principles:

32# Change the color;
40# Compound material;
03# Local change principle;
28# Mechanical system instead.

(5) The concrete solution based on the innovation principle.

According to the innovation principle prompt corresponding to two technical conflict parameters, combining specific design problem, inspiring thinking, after interpretation, we can get some feasible design schemes to solve resilience problem.

Aiming at the resilience design problem 2, we can get 6 innovation principles. Analyzing the feasibility and practicability of this kind of innovation theory in solving design problems, we can preliminarily exclude: 32# Change the color, 35# Change physical or chemical state, 30# Flexible membrane or film. The available innovation principle includes: 40# Compound material, 03# Local change principle, 28# Mechanical system instead. Considering from the angle of antenna design, 03# Local change principle and 28# Mechanical system instead are better design direction. Therefore, improving the design of the antenna installed on a certain type of radar, and replacing the material of scalable array (SAM) based on the innovation theory. According to the bearing mechanics analysis, taking different alloy material respectively for the antenna frame U shaped materials, web, end shaft and a baffle with, and increasing reinforcement and transition pieces. At the same time the positioning hole is replaced by embedded steel sleeve, and all connectors are using blind plug design. The mounting plate of array power supply is changed into a bearing plate, in order to improve the strength of antenna design. After that, the performance indexes of the antenna system are significantly improved to meet the index requirements of ordering sector.

4 CONCLUSION

As the tool to solve technical conflict, conflict matrix explains the concrete circumstances for using what innovation principle to solve specific problems, and the result is that both conflict sides can achieve "win-win", but also improve the efficiency of solving technical conflict. At the same time, conflict matrix also has some shortcomings, even if the system has a problem, but when the "parameter properties" is not obvious, it is difficult to find the conflict for specific parameters describing, not easy to use the conflict matrix. Therefore, how to make full use of TRIZ theory to extract the most important key point in perplexing paradox, how to clear corresponding relationship between technical conflict and innovation principle by conflict matrix, how to simplify the conflict and improve design which is the key for applying innovation theory to practice. They all need further research and more practical case to proof. The application of TRIZ theory in improvement design for equipment can improve the "five characters" index, and has certain innovation and application prospect, but the practice verification is a lengthy process.

REFERENCES

[1] Du Cun-chen, Yan Hui-geng, Yuan Qiang. 2011. Study on Problem Solving Methods by Contradiction Analysis Based on TRIZ. *Journal of Huaihai Institute of Technology (Natural Science Edition)*.

[2] Jiang Fan, Wang Yi-jun, Hu Yi-dan. 2013. The Mechanism Innovation Design Example Analysis based on TRIZ Theory. *Journal of Guangzhou University (Natural Science Edition)*. 2(1):76–79.

[3] Liu Liang, Chen Tao, 2012. In Product Design Theory of TRIZ Application. *Hunan Agricutural Machinery*.

[4] Lu Xi-mei, Zhang Fu-ying, Zhang Qing-qing. 2010. Product Innovation Design based on the Theory of TRIZ and Functional Analysis. *Machinery Design & Manufacture*. 12(12):255–257.

[5] Qiu Cheng. 2010. Analysis of the Evolution of Enterprise Management Information System Based on TRIZ. *Manufacture Information Engineering of China*. 1(6):23–26.

[6] Shi Xiao-ling, Xu Dong-shuang, Fan Yan-feng. 2009. TRIZ Simple Tutorial. *Beijing Yiweixun Science and Technology Co., Ltd*. 76–79.

[7] Yang Qing-liang. 2008. TRIZ Theory Full Contact *Beijing Machinery Industry Press*. 69–72.

[8] Yu Jiang-hong. 2012. A Creative Design of Elastic Composite Cylindrical Roller Bearing Based on TRIZ Contradiction Matrix. *Design & Research*. 2(2):61–63.

[9] Zhu Li. 2008. Solving the Thin Plate Glass Processing Problem by Contradiction Matrix of TRIZ Theory. *New Technology & New Process*. 3(4):87–89.

Advances in Engineering Materials and Applied Mechanics – Zhang, Gao & Xu (Eds)

Performance analysis of a lattice material shock isolator

H.B. Mao & M.J. Wang
State Key Laboratory of Digital Manufacturing Equipment and Technology,
Huazhong University of Science and Technology, Wuhan, P.R. China

Y.Z. Huang
College of Mechanical Engineering, Guangxi University, Nanning, P.R. China

ABSTRACT: A new shock isolation system with lattice material is proposed. The stiffness of the system is based on the stress–strain characteristic of the lattice material. The Runge–Kutta method is employed to analyze the performance of this shock isolation system when it suffered a shock excitation, and the maximum displacement and the maximum acceleration of the object are the main indexes of the performance. In the meantime, the energy transmissibility characteristic is calculated. The results show that stress plateau of the lattice material is beneficial to reduce the shock excitation.

1   INTRODUCTION

The performance of the conventional linear isolator cannot satisfy with the requirement of the designer (Ibrahim 2008). Snowdon (Snowdon 1970) pointed out that it is possible to simultaneously reduce both the maximum acceleration and the maximum displacement with linear shock mounts. The performance of different nonlinear shock isolators was discussed and compared (Shekhar et al. 1999). The results showed that the appropriate structure of the isolator could improve the performance under a given excitation. In the recent years, an isolator with switchable stiffness under appropriate control strategy was proposed as it proved that the response to a shock excitation could be effectively reduced (Ledezma-Ramirez et al. 2011, Ledezma-Ramirez et al. 2012).

In the meantime, cellular materials and structures, such as honeycombs and lattice material aroused the interest of researchers, because of the long-plateau range and controllable plateau stress (Gao & Yu 2006). Shim and Yap (Shim & Yap, 1997, Shim & Yap, 1997) tested the force-deformation response of foam plate systems for four geometries. Based on the test results, various foam-plate systems were represented by a one-dimensional mass-spring chain model to predict the transient impact force responses. The theoretical simulations provided useful insights into the complex interaction of strain rate effects.

In this paper, on the basis of the stress–stain characteristics of diamond lattice material, a new shock isolator is proposed. The performances of the isolator are evaluated, and the maximum responses and the energy transmissibility are analyzed.

2   THE MECHANICAL PROPERTY OF DIAMOND LATTICE MATERIAL

As the research of Gibson (Gibson & Ashby 1997), the stress–strain curve of elastic lattice material under axial compression exhibits three deformation phases (shown in Fig. 1): linear-elastic phase, elastic buckling, and densification phase. In the linear-elastic phase, cells throughout the structure will deform uniformly under pressure; the elastic buckling phase

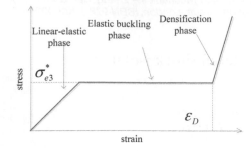

Figure 1.   Stress–strain curve of lattice material.   Figure 2.   Unit structure of diamond lattice.

is the dominant characteristic of crushing in cellular materials, and the stress–strain curve presents a long plateau; and in the densification phase, the cells deform to the extent that voids within them have been completely eliminated.

For the diamond lattice structure, as shown in Figure 2, the equivalent density of the diamond lattice material can be expressed as follows:

$$\bar{\rho} = \frac{\rho^*}{\rho} = \frac{2 + \sin\frac{\theta}{2}}{\sin\theta}\frac{t}{l} \tag{1}$$

where l is the length of the diamond sides, t is the thickness of the cell, θ is the included angle of the sides, and ρ^* and ρ are the density of the lattice materials and cell wall materials, respectively.

It is assumption that, the compressive force is applied in the out-plane direction. In the linear-elastic phase, the elasticity modulus of the lattice structure can be expressed as:

$$E_{(1)}^* = \bar{\rho}E_s \tag{2}$$

where E_s is the elasticity modulus of the parent material.

In the elastic buckling phase, the plateau stress can be expressed as:

$$\sigma_{e3}^* = 4\bar{\rho}\frac{E_s}{(1-v_s^2)}\left(\frac{t}{l}\right)^2 \tag{3}$$

In the densification phase, the theoretical strain is:

$$\varepsilon_D = 1 - \bar{\rho} \tag{4}$$

And the elasticity modulus of the lattice material is equal to the parent material, i.e.

$$E_{(3)}^* = E_s \tag{5}$$

3   SHOCK RESPONSE ANALYSIS OF SHOCK ISOLATION SYSTEM WITH ATTICE MATERIAL ELEMENT

The schematic of a one-degree-of freedom shock isolation system is shown in Figure 3. The system consists of a base, isolated object and the elastic element is made of lattice material. The role of the lattice material isolator is to insulate the shock from the base excitation.

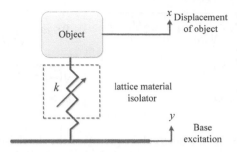

Figure 3. Mechanical model of the shock isolation system with lattice material isolator.

The stiffness of the system is piecewise, according to the above analysis, and it can be expressed as follows:

$$k(x-y)=\begin{cases} k_3(x-y+l_2)-k_1l_1-k_2(l_2-l_1) & x-y<-l_2 \\ k_2(x-y+l_1)-k_1l_1 & -l_2 \le x-y<-l_1 \\ k_1(x-y) & |x-y|\le l_1 \\ k_2(x-y-l_1)+k_1l_1 & l_1<x-y\le l_2 \\ k_3(x-y-l_2)+k_1l_1+k_2(l_2-l_1) & x-y>l_2 \end{cases} \quad (6)$$

where

$$k_1 = AE^*_{(1)}, \qquad k_2 = 0, \qquad k_3 = AE_s$$
$$l_1 = \frac{\sigma^*_{e3}}{E^*_{(1)}}d, \qquad\qquad l_2 = \varepsilon_D d - l_1 \qquad (7)$$

A and d are the area and thickness of the lattice material element, respectively.

Based on the research of Balandin (Balandin et al. 2001), the maximum absolute acceleration responses and the relative displacement between the base and the object are regarded as the indexes of the shock isolation performance. The former reflects the force acting on the object, the latter influences the designing size of the absorber.

The motion equation of the system suffering shock excitation can be expressed as:

$$\ddot{z} + k(z) = -\ddot{y}_2 \qquad (8)$$

where $z = x - y$, is the relative displacement between the object and base; $\ddot{y}_2$ is the shock excitation, and can be regarded as an acceleration impulse:

$$\ddot{y}_2 = v_0 \delta(t) \qquad (9)$$

where $\delta(t)$ is the Dirac's delta-function. The initial conditions of the system are $z(0) = 0$, $\dot{z}(0) = v_0$.

The solutions of the Equation (8) can be expressed as follows:

$$|z|_m = \begin{cases} \sqrt{\bar{\rho}\dfrac{v_0^2}{\omega_1^2}+\bar{\rho}l_1^2-2\bar{\rho}l_1l_2+\bar{\rho}^2l_1^2}+l_2-l_1\bar{\rho} & 0\le \omega_1 < \dfrac{v_0}{\sqrt{(2l_2-l_1)l_1}} \\[3mm] \dfrac{v_0^2}{2l_1\omega_1^2}+\dfrac{l_1}{2} & \dfrac{v_0}{\sqrt{(2l_2-l_1)l_1}} \le \omega_1 < \dfrac{v_0}{l_1} \\[3mm] \dfrac{v_0}{\omega_1} & \omega_1 \ge \dfrac{v_0}{l_1} \end{cases} \qquad (10)$$

$$|\ddot{x}|_m = \begin{cases} \omega_1^2 \sqrt{\dfrac{v_0^2}{\bar{\rho}\omega_1^2} + \dfrac{l_1^2}{\bar{\rho}} - 2\dfrac{l_1 l_2}{\bar{\rho}} + l_1^2} & 0 \le \omega_1 < \dfrac{v_0}{\sqrt{(2l_2 - l_1)l_1}} \\ \omega_1^2 l_1 & \dfrac{v_0}{\sqrt{(2l_2 - l_1)l_1}} \le \omega_1 < \dfrac{v_0}{l_1} \\ v_0 \omega_1 & \omega_1 \ge \dfrac{v_0}{l_1} \end{cases} \tag{11}$$

where $\omega_1 = \sqrt{k_1/m}$, represents the inherent frequency of the shock isolation system.

The displacement shock spectrum is plotted in Figure 4(a) with a given parameters $v_0 = 10$, $l_1 = 0.1$, $l_2 = 0.3$, $\bar{\rho} = 0.05$. Subsequently, the acceleration shock spectrum is plotted in Figure 4(b). The displacement monotonically decreases from infinite to zero as frequency ω_1 changing from zero to infinite. The acceleration presents opposite tendency with the displacement generally. It means that, the large stiffness of the absorber is beneficial for decreasing the displacement of the object, but it will increase the acceleration response. What should be noticed that, the acceleration does not monotonically increase with frequency. In the range of $\omega = 0 \sim 44.7$, the curve of maximum acceleration presents parabolic.

To evaluate the performance of the lattice material elastic element more effectively, a dimensionless parameter η represents the energy transmission rate is introduced as:

$$\eta = \frac{|\ddot{x}|_m |z|_m}{v_0^2} \tag{12}$$

The similar definition of the energy transmissibility was introduced by (Ibrahim, 2008) to study the transmissibility of nonlinear system.

The energy transmissibility η can be regarded as a function of frequency ω_1. Figure 5 shows a representative example of the transmissibility η. The value of η is obviously equal to one in the region $\omega_1 \ge 100$. In this region, the lattice material is in the linear-elastic phase and the system presents the linear characteristic. Obviously, shock energy is not attenuated.

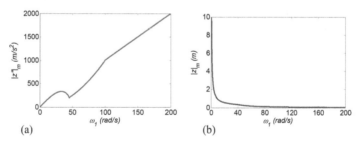

(a) (b)

Figure 4. Shock response spectrum of the shock isolation system excited by acceleration impulse, for (a) relative displacement, (b) absolute acceleration.

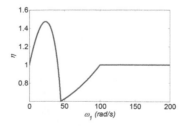

Figure 5. Energy transmissibility of the shock isolation system.

150

In the region $44.7 \leq \omega_1 < 100$, the transmissibility η is less than one, as shown in Figure 5. The value of transmissibility will trend to 0.5, when ω_1 trends to 44.7. The shock isolation performance is improved because of the introduction of the zero-stiffness spring into the system, as the lattice material is in the elastic buckling phase. It can be seen that, the stress plateau could effectively reduce the shock energy. In the region $0 \leq \omega_1 < 44.7$, the curve presents parabolic trend with the increase of ω_1. In the range of $0 \leq \omega_1 < 40.4$, the transmissibility η is larger than one as the lattice material is in the densification phase. It is harmful to the object as the isolator magnifies the shock energy.

4 CONCLUSION

This paper has illuminated the isolation mechanism as the lattice material used to be an isolator, and studies the shock isolation performance when the lattice material is in the different stress–strain phases. The results indicate that the stress plateau of the lattice material can reduce the shock energy effectively when it used as a shock-isolation element.

ACKNOWLEDGMENT

This work was supported by the National Natural Science Foundation of China (No. 51175195).

REFERENCES

Balandin D.V, Bolotnik N.N. & Pilkey W.D. 2001. Optimal protection from impact, shock and vibration. CRC Press.

Gao Z.Y. & Yu T.X. 2006. One-dimensional analysis on the dynamic response of cellular chains to pulse loading. Proceedings of the Institution of Mechanical Engineers, Part C: *Journal of Mechanical Engineering Science*, 220(5): 679–689.

Gibson L.J. & Ashby M.F. 1997. *Cellular solids: structure and properties*. Cambridge university press.

Ibrahim R.A. 2008. Recent advances in nonlinear passive vibration isolators, *Journal of sound and vibration*, 314(3): 371–452.

Ledezma-Ramirez D.F, Ferguson N.S. & Brennan M.J. 2011. Shock isolation using an isolator with switchable stiffness. *Journal of Sound and Vibration*, 330(5): 868–882.

Ledezma-Ramirez D.F, Ferguson N.S. & Brennan M.J. 2012. An experimental switchable stiffness device for shock isolation. *Journal of Sound and Vibration*, 331(23): 4987–5001.

Shekhar N.C, Hatwal H. & Mallik A.K. 1999. Performance of non-linear isolators and absorbers to shock excitations. *Journal of Sound and Vibration*, 227(2): 293–307.

Shim V.P.W. & Yap K.Y. 1997. Modelling impact deformation of foam-plate sandwich systems. *International journal of impact engineering*, 19(7): 615–636.

Shim V.P.W. & Yap K.Y. 1997. Static and impact crushing of layered foam-plate systems. International *Journal of Mechanical Sciences*, 39: 69–86.

Snowdon J.C. 1970. Isolation from mechanical shock with a mounting system having nonlinear dual-phase damping. *The Shock and Vibration Bulletin*, 41: 21–45.

Study the effect of long-period waves on mooring large vessel

W.J. Shen & F. Gao
Key Laboratory of Engineering Sediment of Ministry Communications, Tianjin Research Institute for Water Transport Engineering, M.O.T., Tianjin, China

Q. Chen
Zhejiang Ocean University, Tianjin, China

C.L. Mao
Offshore Oil Engineering Co. Ltd., Tianjin, China

ABSTRACT: In recent years, more and more ports are built away from the coastline, located in the open deep waters. These offshore deep-water ports face more hostile natural conditions than coastal ports, especially the long-period waves. It has a strong penetrating power and considerable energy that can easily cause the large amplitude motion of mooring ship, and thus leading to accidents. The effect of wave periods and wave heights on the mooring vessel is studied numerically in this paper, using the software Moses. The change of movement with wave period is analyzed in detail, and a forecast is made for the dangerous wave period. For the wave height, the change of movement with wave height is also analyzed in detail. The results can provide technical reserves and scientific basis for further research and exploration of mooring safety problems.

1 INTRODUCTION

Long-period wave refers to the period between tides and surge waves. The wave length is longer, which is different from the regular wave. Although the wave height is limited (usually a few centimeters), it has a strong ability of penetrate. Besides that, its propagation velocity is large, with a high energy, which is one of the important factors affecting a port structure design. In the harbor, it is easy to cause the resonance motion of a mooring vessel, so the values of vessel motion mooring tension and fender force will be increased largely, which will bring huge security for the terminal operation and the safety of mooring vessel.

However, research on the influence of long-period wave on a moored vessel is relatively less. Thus, the rule of motion response of a mooring large vessel under the long period wave is studied in this paper, and the effect of wave period and height on vessel motion is discussed.

2 THE ESTABLISHMENT OF MATHEMATICAL MODEL

A moored 300,000 DWT bulk cargo ship under the action of long-period wave is calculated, and 20 nylon cables with the diameter is 75 mm are used in the model, the mooring pattern is 4:4:2, which means there are 4 head lines, 4 stern lines, 4 breast forward lines, 4 breast aft lines, 2 spring forward lines and 2 spring aft lines. And the main parameter of the cargo is shown in Table 1.

Four fenders are installed on the side of breasting dolphins, the fender type is SUC2250H, and the stiffness curve simulated in the mathematical model is shown in Figure 1.

The stiffness curves of mooring lines are obtained according to the Wilson formula, which is shown as follows:

Table 1. Main parameter of 300,000 DWT bulk cargo ship (full load).

Length (m)	Beam (m)	Depth (m)	Draft (m)	Displacement (t)	Roll Period (s)	Pitch Period (s)
332	58	30.4	23.0	357479.98	17.60	15.4

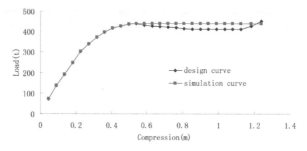

Figure 1. Simulated result for SUC2250H fender.

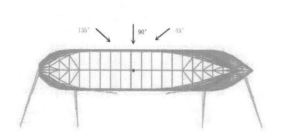

Figure 2. Side view of the model with mooring lines.   Figure 3. Comparison of the results.

$$T_m = \frac{C_p d_p^2 (\Delta S / S)^n}{\lambda^3}$$

where T_m (N) is the mooring tension in the scale model, C_p is the elastic coefficient of mooring line in prototype, and when is nylon cable $C_p = 1.540 \times 10^4$ MP$_a$; d_p (m) is the diameter of mooring line in prototype; $\Delta S / S$ is the relative elongation in prototype; n is index number, and it is equal to 3 when the line is nylon. λ is the length scale.

So the stiffness curves of mooring lines in prototype model can be obtained by a change form of the above formula, $T_p = C_p d_p^2 (\Delta S / S)^n$.

The hull model is established using the strip theory according to the offset table of this vessel, and the model of fenders and mooring lines are also considered, then the whole mathematical model is shown in Figure 2. And the wave direction in the mathematical model also can be seen in Figure 2.

In order to verify the feasibility of the numerical simulation method and the credibility of the results, the sway motion is compared with an empirical formula, as shown in Figure 3. It can be seen that, the results of this method and the formula values agree well on the whole, except some individual values.

3 ANALYSIS OF THE CALCULATION RESULTS

The motion values of 6 freedom degrees are obtained using the Moses software, including surge, sway, heave, pitch, roll and yaw, with the wave period from 6 second to 36 second.

JONSWAP is used to simulate the wave, and the spectral peak factor is 3.3. The wave direction is perpendicular to the dock. The water depth is 28 meter, and the mooring arrangement is 4:4:2, as mentioned above.

3.1   The influence of wave period on motion response

The wave height is 0.8 meter in this condition, and the calculation result is shown in Figure 4.

It can be obtained that, motion of surge, heave and pitch degrees have a strong cyclical. And the values increase gradually form 6 second, and reach the maximum when the wave period is 10 second. Then with the increase of wave period, these three degrees have a decreasing trend, and the surge motion remains basically stable after the wave period exceeds 15 second. And as the wave direction is 90°, so the motions of surge and pitch are relative small.

The sway motion increases with the wave period, and when the wave period reaches 18 second, it will not increase, and keeps a value of about 3 meter. The roll motion also has a strong periodicity, reaches peak value when wave period is about 15 second, and the maximum value is around 2.7°. On the other hand, the yaw motion increases with the increasing of period, and when the period reaches 20 second, it keeps a value of about 0.7°.

As mentioned above, the motion values of 6 freedom degrees have different response characteristics. When the wave period is near to the natural period of vessel, the motion value becomes larger, which is the resonance phenomenon. When the incident wave is perpendicular to the vessel, the motion is dominated by roll and sway, and heave motion comes second. And the long period wave plays an important role in the sway and roll degrees of vessel motion, the maximum values of the two degrees are increased nearly six times more than the wave period is 8 second. And the sway and roll motion are both large at the time when the wave period is 18 second, which means that this situation is very bad. In the actual situation, safeguard procedures must be arranged and avoid vessel mooring at the dock.

3.2   The influence of wave height on motion response

In this section, the main object is to study the effect of wave height on the calculation results. On the basis of the above calculation, two heights (0.3 meter and 0.5 meter) are added with other conditions unchanged. The calculation results are shown in Figures 5~10.

A conclusion can be drawn from the Figures 4 to 9, all the vessel motion have a good trend with wave height except the yaw degree, and the closer it gets to the peak the greater in the increment. The periodicity of surge, heave, pitch, sway and roll motion of a moored vessel is not affected by wave height. The trends are similar under different wave heights. And surge,

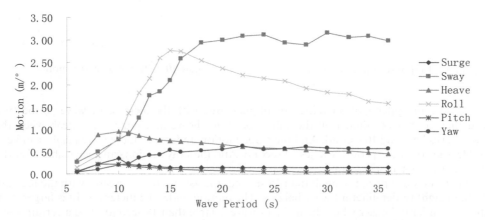

Figure 4.   Ship motion-wave period response curves of surge, sway, heave, roll, pitch and yaw.

155

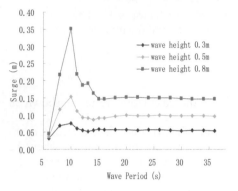

Figure 5.   Response of surge motion.

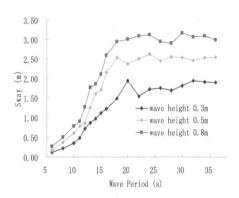

Figure 6.   Response of sway motion.

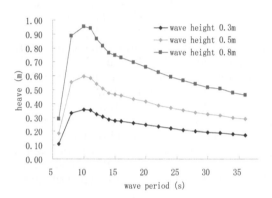

Figure 7.   Response of heave motion.

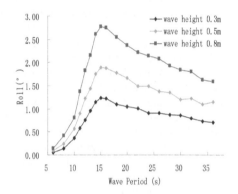

Figure 8.   Response of roll motion.

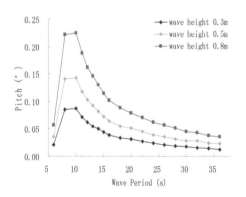

Figure 9.   Response of pitch motion.

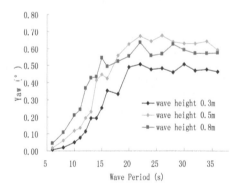

Figure 10.   Response of yaw motion.

heave, roll and pitch motion values are all increased with the increase of wave period first and then decreases. Almost all the values of surge, heave and pitch reach peak when the wave period around 10 second. And the surge motion under the all the wave height can keep stable after the wave period reaches 15 second. And the sway motion can keep stable after the wave period reaches 20 second.

When the wave period is small than 18 second, the yaw motion increases with the increase of the height under different wave heights. The motion under 0.5 meter height is larger than the one under 0.8 meter when the incident wave is large than 18 second. And a certain similarity also exists in the result.

4 CONCLUSION

Based on the numerical study conducted on the Truss Spar platform, the following conclusions can be drawn:

1. The long period wave plays an important role in the sway and roll degrees of vessel motion, the maximum values of sway and roll degrees are increased nearly six times more than the wave period of 8 second.
2. All the vessel motions have a good trend with wave height except the yaw degree, and the closer it gets to the peak the greater in the increment.
3. The change of wave height will not affect the trend of six freedom degrees motion of a mooring vessel along with wave period.

ACKNOWLEDGEMENTS

The project was supported by Central Commonweal Research Institute Basic R&D Special Foundation of TIWTE (Grant No. TKS130203).

REFERENCES

Gao Feng, Zheng Bao-you, Chen Han-bao et al. 2007. Analysis of mooring conditions under different mooring line arrangements. *Port & Waterway Engineering* (3):48–52.

Lee, S.K., H. Choi, Surendran S. 2006. Experimental Studies on the Slowly Varying Drift Motion of a Berthed Container Ship Model. *Ocean Engineering* (33):2454–2465.

Li Yue. 2013. Study on the influence of long-period waves on vessel mooring stability. Master Dissertation, Dalian, Dalian University of Technology.

Meng Xiang-wei, Gao Feng, Li Yan. 2011. Study mathematical model of the effect of wave period on the mooring conditions. *Proceedings of the 15th International Conference on Ocean and Offshore Engineering, 2011*. China: Taiyuan.

MA Xiao-jian, SUN Zhao-chen, Zhang Zhi-ming, et al. 2011. The Effect of Wave Groupiness on a Moored Ship Studied by Numerical Simulations. *Journal of Hydrodynamics* (2):145–153.

Sakakibara Shigeki, Takeda Shoji, Iwamoto Yuji, Kubo Masayoshi. 2010. A Hybrid PotentialTheory for Predicting the Motions of a Moored Ship Induced by Large-Scaledtsunami. *Ocean Engineering* (37):1564–1575.

Yang Xian-zhang.1989. Characteristics of long period wave and the influence on the mooring ship. *China Harbour Engineering* (6):37–42.

Yu-Cheng Li. 1982. Impact Energy of a Moored Tanker Under the Action of Regular Waves. *Journal of Energy Resources Technology* 104(3):235–240.

Zhang Zhi-ming, Zhou Feng, Yang Guo-ping et al. 2010. Research for the moored ship of offshore deepwater terminal. *China Harbour Engineering* (S1):49–52.

Zhili Zou, E C Bowers. 1995. A mathematical model simulation of a ship moored against a quay in harbour. *The Ocean Engineering*: 13(3):25–36.

Advances in Engineering Materials and Applied Mechanics – Zhang, Gao & Xu (Eds)
© 2016 Taylor & Francis Group, London, ISBN 978-1-138-02834-0

Sensor placement method for structural response reconstruction

Z. Wan
State Key Laboratory of Digital Manufacturing Equipment and Technology,
Huazhong University of Science and Technology, Wuhan, P.R. China

Y.Z. Huang
School of Mechanical Engineering, Guangxi University, Nanning, P.R. China

T. Wang & H.B. Mao
State Key Laboratory of Digital Manufacturing Equipment and Technology,
Huazhong University of Science and Technology, Wuhan, P.R. China

ABSTRACT: This paper proposes an uncertain method for structural response reconstruction based on a joint force-state estimation algorithm which was originally presented for optimal control applications. The modally reduced order model is used for structural dynamics. In addition, optimal sensor placement is performed by taking the minimum average variance of the reconstruction errors as the optimal objective. A simply but effectively heuristic algorithm is adopted for optimizing the sensor placement. Finally, a numerical study on a cantilever beam is conducted to demonstrate the effectiveness of the proposed method.

1 INTRODUCTION

Structural dynamic response reconstruction has received increased attention in the past decades. Accurate knowledge of response produced at discrete locations during its operation is vital to many engineering applications such as structural control, monitoring, parameter identification, damage detection and loads analysis. However, in some instances, it is not always possible to measure the responses at the desired locations in operational conditions (Limongelli 2003). For example, the locations of interest are at interface between substructures in a large structural system (Kammer 1997). In addition, in real life, it is impossible to arrange sensors at every location due to economic reasons.

Literature review on recent studies reveals several types of approaches for structural dynamic response reconstruction, including the transmissibility based methods (Ribeiro et al. 2000, Law et al. 2011), the modal methods (Wan et al. 2014, He et al. 2012, Wan et al. 2014), the Markov parameter based methods (Kammer 1997, Wang et al. 2014), and some other methods (Setola 1998, Djamaa et al. 2007). However, these methods are all the certain methods considering no model errors in structures. In 2007, Gillijns and DeMoor developed a joint input-state estimation algorithm which was originally proposed for optimal control applications (Gillijns & De Moor 2007). The algorithm can perform the unbiased minimum-variance input and state estimation for linear systems, which considers both of the process noise and measurement noise. In this paper, this algorithm is adopted for structural response reconstruction involving unknown forces, and extends for a modally reduced order model of structural dynamics. Moreover, a good sensor placement is beneficial for accurate reconstruction responses (Zhang et al. 2014). In this paper, optimal sensor placement is performed by tanking the minimum average variance of the reconstruction errors as the optimal objective.

2 STRUCTURAL RESPONSE RECONSTRUCTION

In structural dynamics, modally reduced order models are applied in many cases. When proportional damping is assumed, the continuous time decoupled equations of motion for modally reduced order models are given by

$$\ddot{\mathbf{q}}(t) + \mathbf{\Gamma}\dot{\mathbf{q}}(t) + \mathbf{\Omega}^2 \mathbf{q}(t) = \mathbf{\Phi}^T \mathbf{B}_u \mathbf{u}(t) \tag{1}$$

where $\mathbf{q}(t) \in \mathbb{R}^{n_r}$ is the vector of modal coordinates with n_r being the number of modes taken into account in the model. $\mathbf{\Phi} \in \mathbb{R}^{ndof \times n_r}$ is the mass normalized mode shapes matrix. $\mathbf{\Gamma} \in \mathbb{R}^{n_r \times n_r}$ is a diagonal matrix containing the terms $2\xi_j \omega_j$ on its diagonal, where ω_j and ξ_j are the natural frequency and modal damping ratio corresponding mode j respectively. $\mathbf{\Omega} \in \mathbb{R}^{n_r \times n_r}$ is a diagonal matrix as well, containing the natural frequencies ω_j on its diagonal. A time history vector $\mathbf{u}(t) \in \mathbb{R}^{n_u}$, with n_u being the number of forces and $\mathbf{B}_u$ represents the force locations.

The discrete dynamic equations in state space containing the state transition and observation equations are written as follows

$$\mathbf{x}_{k+1} = \mathbf{A}\mathbf{x}_k + \mathbf{B}\mathbf{u}_k + \mathbf{w}_k \tag{2}$$

$$\mathbf{y}_k = \mathbf{H}\mathbf{x}_k + \mathbf{D}\mathbf{u}_k + \mathbf{v}_k \tag{3}$$

where $\mathbf{x}$ is the modal state vector. The matrices $\mathbf{A}$, $\mathbf{B}$ depends on the time discretization scheme, and the reader is referred to (Franklin et al. 1998) for a detailed overview of common time discretization schemes. Assuming the measurements are the acceleration signals, then the matrices $\mathbf{H}$ and $\mathbf{D}$ are respectively written as

$$\mathbf{H} = \begin{bmatrix} -\mathbf{H}_0 \mathbf{\Phi}\mathbf{\Omega} & -\mathbf{H}_0 \mathbf{\Phi}\mathbf{\Gamma} \end{bmatrix}, \quad \mathbf{D} = \begin{bmatrix} \mathbf{H}_0 \mathbf{\Phi}\mathbf{\Phi}^T \mathbf{B}_u \end{bmatrix} \tag{4}$$

The deterministic system is described by the system matrices $\mathbf{A}$, $\mathbf{B}$, $\mathbf{H}$ and $\mathbf{D}$, and the process noise $\mathbf{w}_k$ and measurement noise $\mathbf{v}_k$ are not considered (the corresponding variance matrix are $\mathbf{G}_k$ and $\mathbf{R}_k$). In practice, these two types of noise are unavoidable. Additionally, the forces are also unknown due to the difficulty of force measurement. In this paper, we consider not only the noise but also the unknown forces for structural response reconstruction. Gillijns and De Moor proposed a filter to estimate input-state for linear systems (Gillijns & De Moor 2007), and this algorithm is adopted in this paper for structural response reconstruction. The filter is initialized using the initial state and its variance $\hat{\mathbf{x}}_{0|-1}$ and $\mathbf{P}_{0|-1}$, hereafter it identifies the unknown forces and states recursively in three steps: the force estimation, the measurement update and time update:

Force estimation:

$$\tilde{\mathbf{R}}_k = \mathbf{H}\mathbf{P}^x_{k|k-1}\mathbf{H}^T + \mathbf{R}_k \tag{5}$$

$$\mathbf{J}_k = (\mathbf{D}^T \tilde{\mathbf{R}}_k^{-1}\mathbf{D})^{-1}\mathbf{D}^T \tilde{\mathbf{R}}_k^{-1} \tag{6}$$

$$\hat{\mathbf{u}}_k = \mathbf{J}_k (\mathbf{y}_k - \mathbf{H}\mathbf{x}_{k|k-1}) \tag{7}$$

$$\mathbf{P}^u_k = (\mathbf{D}^T \tilde{\mathbf{R}}_k^{-1}\mathbf{D})^{-1} \tag{8}$$

Measurement update:

$$\mathbf{K}_k = \mathbf{P}^x_{k|k-1}\mathbf{H}^T \tilde{\mathbf{R}}_k^{-1} \tag{9}$$

$$\mathbf{x}_{k|k} = \mathbf{x}_{k|k-1} + \mathbf{K}_k(\mathbf{y}_k - \mathbf{H}\mathbf{x}_{k|k-1} - \mathbf{D}\hat{\mathbf{u}}_k) \tag{10}$$

$$\mathbf{P}^x_{k|k} = \mathbf{P}^x_{k|k-1} - \mathbf{K}_k(\tilde{\mathbf{R}}_k - \mathbf{D}\mathbf{P}^u_k\mathbf{D}^T)\mathbf{K}_k^T \tag{11}$$

$$\mathbf{P}^{xu}_k = (\mathbf{P}^{ux}_k)^T = -\mathbf{K}_k\mathbf{D}\mathbf{P}^u_k \tag{12}$$

Time update:

$$\mathbf{x}_{k+1|k} = \mathbf{A}\mathbf{x}_{k|k} + \mathbf{B}\hat{\mathbf{u}}_k \tag{13}$$

$$\mathbf{P}^x_{k+1|k} = \begin{bmatrix} \mathbf{A} & \mathbf{B} \end{bmatrix} \begin{bmatrix} \mathbf{P}^x_{k|k} & \mathbf{P}^{xu}_k \\ \mathbf{P}^{ux}_k & \mathbf{P}^u_k \end{bmatrix} \begin{bmatrix} \mathbf{A}^T \\ \mathbf{B}^T \end{bmatrix} + \mathbf{G}_k \tag{14}$$

According to the above filter, the state vector $\mathbf{x}_{k|k}$ and the force vector $\hat{\mathbf{u}}_k$ can be obtained. Therefore, the reconstructed responses can be computed by the estimated $\mathbf{x}_{k|k}$ and $\hat{\mathbf{u}}_k$ as

$$\mathbf{y}^r_k = \mathbf{H}^r \mathbf{x}_{k|k} + \mathbf{D}^r \hat{\mathbf{u}}_k \tag{15}$$

where r represents the information involving reconstruction. Here, $\mathbf{y}_k$ represents the actual response vector, then the reconstruction error vector of the corresponding responses is written as

$$\delta_k = \mathbf{y}_k - \mathbf{y}^r_k = \mathbf{H}^r(\mathbf{x}_k - \mathbf{x}_{k|k}) + \mathbf{D}^r(\mathbf{u}_k - \hat{\mathbf{u}}_k) = \begin{bmatrix} \mathbf{H}^r & \mathbf{D}^r \end{bmatrix}\begin{bmatrix} \mathbf{x}_k - \mathbf{x}_{k|k} & \mathbf{u}_k - \hat{\mathbf{u}}_k \end{bmatrix}^T \tag{16}$$

Therefore, the error variance matrix of the reconstructed responses is obtained as

$$\Delta = \mathrm{cov}(\delta_k) = \begin{bmatrix} \mathbf{H}^r & \mathbf{D}^r \end{bmatrix} \mathrm{cov}\left(\begin{bmatrix} \mathbf{x}_k - \mathbf{x}_{k|k} \\ \mathbf{u}_k - \hat{\mathbf{u}}_k \end{bmatrix} \right) \begin{bmatrix} (\mathbf{H}^r)^T \\ (\mathbf{D}^r)^T \end{bmatrix} = \begin{bmatrix} \mathbf{H}^r & \mathbf{D}^r \end{bmatrix} \begin{bmatrix} \mathbf{P}^x_{k|k} & \mathbf{P}^{xu}_k \\ \mathbf{P}^{ux}_k & \mathbf{P}^u_k \end{bmatrix} \begin{bmatrix} (\mathbf{H}^r)^T \\ (\mathbf{D}^r)^T \end{bmatrix} \tag{17}$$

3 SENSOR PLACEMENT

Each diagonal element of matrix Δ represents the variance of the reconstruction error for the corresponding response. Therefore, the trace of the matrix Δ represents the sum of the reconstruction errors at all locations of interest. The optimal sensor placement can be performed with the objective to minimize the average of the reconstruction error. The average reconstruction errors at all interesting locations yield

$$\sigma_{avg}^2 = tr(\Delta)/N \tag{18}$$

The objective function of the sensor location selection can be expressed as

$$\min \quad \sigma_{avg}^2 \tag{19}$$

Assuming the number of measurement sensors is *num*, a simply but effectively heuristic algorithm is adopted to optimize the sensor placement. The optimal algorithm is as follows:

Step 1: Take all the key locations as the candidates based on the consideration of some practical issues. For example, some locations which cannot be assessed by workers are not considered as the candidates. The number of the candidates is assumed as n, then we have $n < num$.

Step 2: Calculate ith average reconstruction error variance successively by deleting the ith element (location) only. The location with the minimum average reconstruction error variance is then removed.

Step 3: If the number of the remaining candidates is equal to the sensor number *num*, the iteration is stop; if it is not, then turn to Step 2 going on iteration.

It is noted that this optimal algorithm cannot be assured that the final sensor configuration is global optimal but it is sub-optimal at least.

The method for structural response reconstruction with unknown forces is summarized as:

Step 1: Determine the number of the target modes by the measured responses using Fourier spectrum and form the system matrices **A** and **B**.

Step 2: Form the observation matrices $\mathbf{H}^m$ and $\mathbf{D}^m$ according to the optimal sensor placement. The superscript m represents the information of measurement.

Step 3: Estimate the state vector $\mathbf{x}_{k|k}$ and the force vector $\hat{\mathbf{u}}_k$ using the filter of joint input-state estimation.

Step 4: Reconstruct the responses according to Eq. (15), and compare the reconstructed results with the actual values.

4 NUMERICAL EXAMPLE

To illustrate the validation of the proposed approach, a numerical study of a cantilever steel beam, 20 2D beam elements of length 1000 mm, breadth 50.8 mm and thickness 25.4 mm, is considered as shown in Figure 1. Each node has two DOFs (translational and rotational). The modulus of elasticity and density are taken as 2.1e+11 N/m^2 and 7750 kg/m^3 respectively. The first three natural frequencies are found to be 21.35 Hz, 134.1 Hz, and 377.6 Hz.

The applied force is a double-sine function at node 18 in vertical direction, which is given as

$$u = 100 \sin (80\pi t) + 100 \sin (160\pi t) \tag{20}$$

The force is assumed unknown, which will be estimated. The first four modes are utilized to reduce the structural system according to this example. 5% noise are added into the measured signals, and $\mathbf{x}_{0|-1}$ is assumed as zero vector, and the matrices **G**, **R** and $\mathbf{P}_{0|-1}$ are diagonal matrix containing the terms 1e^{-10}, 1e^{-3} and 1e^{-10} on its diagonal, respectively. In practice, acceleration sensors are used widely, so only acceleration sensors are used for measurements in the example. The DOFs in vertical direction are all considered as the candidates of sensor optimal placement due to the force applied in vertical direction. The number of sensors is chosen as 4, 5, 6 and 7 respectively. The optimal sensor placement is listed in Table 1. From it, it is shown that the sensor placement of best combinations has much smaller reconstruction errors, and is more symmetrical than that of worst combinations. According to the cases of best combinations, it is found that the reconstruction error becomes smaller along with the sensor number becoming more, and all of the reconstruction errors are very small except for the sensor number being 4.

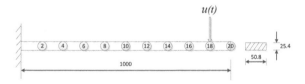

Figure 1.   The cantilever beam model.

Table 1.   Different number of sensor and sensor placement.

Type	Number of sensor	Location	σ_{avg}^2
Best combinations	4	9, 19, 27, 35	0.0011
	5	9, 19, 27, 35, 39	6.827e-4
	6	9, 11, 19, 27, 35, 39	5.635e-4
	7	9, 11, 19, 27, 33, 35, 39	4.882e-4
Worst combinations	4	1, 3, 5, 35	178.214
	5	1, 3, 5, 7, 35	65.105
	6	1, 3, 5, 7, 9, 35	15.038
	7	1, 3, 5, 7, 9, 11, 35	3.352

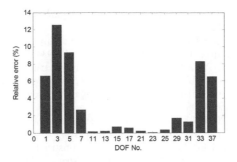

Figure 2. Relative erros of the reconstructed acceleration responses.

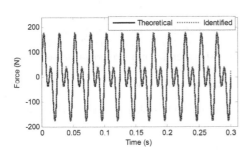

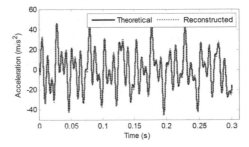

Figure 3. The actual and identified force.   Figure 4. The actual and reconstructed acceleration.

Sensor placement of sensor number being 5 is used to validate the reconstruction method in detail. The reconstruction errors of the responses at the remaining DOFs are shown in Figure 2. As seen from it, all relative errors are smaller than 10% except for one reconstruction response at DOF-1. The identified force is plotted in Figure 3, and the reconstruction acceleration at DOF-15 is plotted in Figure 4. From the two figures, we can see good results compared with the theoretical values, which indicates the effectiveness of the reconstruction method including the sensor optimal placement.

5 CONCLUSIONS

This paper deals with the structural response reconstruction with unknown forces. The reconstruction method is proposed based on the Gillijns-DeMoor algorithm of joint input-state estimation which is originally proposed for optimal control application. The response is reconstructed with the identified force and state (displacement and velocity). And also, optimal sensor placement is performed by considering the average reconstruction variance error minimum. A numerical example of a cantilever beam demonstrates the effectiveness of the proposed method.

ACKNOWLEDGMENTS

This work was supported by the National Natural Science Foundation of China (No. 51175195).

REFERENCES

Djamaa, M.C., Ouelaa, N. & Pezerat, C., et al. 2007 Reconstruction of a distributed force applied on a thin cylindrical shell by an inverse method and spatial filtering. *Journal of Sound and Vibration* 301: 560–575.

Franklin, G.F., Powell, J.D. & Workman, M.L. 1998 *Digital control of dynamic systems*, CA: Addison-wesley Menlo Park.

Gillijns, S. & De Moor, B. 2007 Unbiased minimum-variance input and state estimation for linear discrete-time systems with direct feedthrough. *Automatica* 43: 934–937.

He, J., Guan, X. & Liu, Y. 2012 Structural response reconstruction based on empirical mode decomposition in time domain. *Mechanical Systems and Signal Processing* 28: 348–366.

Kammer, D.C. 1997 Estimation of structural response using remote senor locations. *Journal of Guidance Control and Dynamics* 20: 501–508.

Law, S.S., Li, J. & Ding, Y. 2011 Structural response reconstruction with transmissibility concept in frequency domain. *Mechanical Systems and Signal Processing* 25: 952–968.

Limongelli, M.P. 2003 Optimal location of sensors for reconstruction of seismic responses through spline function interpolation. *Earthquake Engineering & Structural Dynamics* 32: 1055–1074.

Ribeiro, A., Silva, J. & Maia, N. 2000 On the generalisation of the transmissibility concept. *Mechanical Systems and Signal Processing* 14: 29–35.

Setola, R. 1998 A spline-based state reconstruction for active vibration control of a flexible beam. *Journal of Sound and Vibration* 213: 777–790.

Wan, Z. & Li, S., et al. 2014 Structural response reconstruction based on the modal superposition method in the presence of closely spaced modes. *Mechanical Systems and Signal Processing* 42: 14–30.

Wan, Z. & Wang, T., et al. 2014 Structural response reconstruction for non-proportionally damped systems in the presence of closely spaced modes. *Journal of Vibroengineering* 16: 3740–3758.

Wang, J., Law, S.S. & Yang, Q.S. 2014 Sensor placement method for dynamic response reconstruction. *Journal of Sound and Vibration* 333: 2469–2482.

Zhang, X. & Xu, Y., et al. 2014 Dual-type sensor placement for multi-scale response reconstruction. *Mechatronics* 24: 376–384.

Advances in Engineering Materials and Applied Mechanics – Zhang, Gao & Xu (Eds)
© *2016 Taylor & Francis Group, London, ISBN 978-1-138-02834-0*

Optimized implicit DRP scheme using exponential objective function

J.L. Wang, Q.B. Huang, H.B. Mao & T.L. Yang
State Key Laboratory of Digital Manufacturing Equipment and Technology,
Huazhong University of Science and Technology, Wuhan, P.R. China

ABSTRACT: Computational aeroacoustics needs to use finite difference schemes which have high accuracy and high resolution characteristics. In this article, a penta-diagonal compact implicit finite difference scheme is optimized using exponential objective function to obtain maximum resolution characteristics. Through Fourier analysis, the optimization is reduced to the problem of finding the minimum of a multivariable nonlinear function with multiple constraints. A uncomplicated algorithm was proposed to find the best performance scheme under the exponential objective function. Actual performance of the optimized compact scheme is applied to numerical simulations of a simple wave convection benchmark problem, the effective accuracy is then discussed.

1 INTRODUCTION

Compact spectral-like finite difference schemes are widely used in computational acoustics (CAA) because their robustness and the high accuracy (Ekaterinaris 1999, Meitz et al. 2000). In recent years, many optimized compact finite difference schemes have been proposed (Appadu et al. 2009, Appadu et al. 2011, Lele 1992, Kim et al. 1997, Tam et al. 1993a, Tam et al. 1993b). Lele (Lele 1992) presents the well-known improved Padé schemes with good resolution. Kim and Lee (Kim et al. 1997) optimized the tri-diagonal and penta-diagonal compact finite difference schemes with a variety of truncation orders, they found the optimized sixth-order tri-diagonal and fourth-order penta-diagonal schemes have better performance. During the optimization, they use a Minimized Integrated Exponential Error for Low Dispersion and Low Dissipation (MIELDLD) technique to get the optimized coefficients, the best integral interval was gotten by a trial-and-error method. In this article, we will use the Minimized Integrated Exponential Error for Low Dispersion and Low Dissipation (MIEELDLD) technique to get the optimized coefficients of the penta-diagonal compact scheme, the best integral interval will be gotten by using a simple algorithm automatically.

2 PENTA-DIAGONAL COMPACT FINITE DIFFERENCE SCHEMES

The penta-diagonal compact finite difference scheme (Lele 1992) is examined in this article. It can be written as

$$\beta f'_{j-2} + \alpha f'_{j-1} + f'_j + \alpha f'_{j+1} + \beta f'_{j+2} = a \frac{f_{j+1} - f_{j-1}}{2\Delta x} + b \frac{f_{j+2} - f_{j-2}}{4\Delta x} + c \frac{f_{j+3} - f_{j-3}}{6\Delta x} \tag{1}$$

where f_j and f'_j are the approximation of the spatial function value and its first order derivative $\partial f/\partial x$ at the jth node of a uniform grid, respectively, $\Delta x = x_{j+1} - x_j$ is the spatial interval which is assumed as a constant. A penta-diagonal scheme can be obtained when β

is nonzero otherwise when β is zero we get a tri-diagonal scheme. The relations between the coefficients a, b, c and α, β are determined from matching the Taylor series coefficients of various orders (Lele 1992).

3 OPTIMIZATION PROCEDURE

Take the Fourier transform of Eq. (1) one can get

$$ik(\beta e^{-2ik\Delta x} + \alpha e^{-ik\Delta x} + 1 + ae^{ik\Delta x} + \beta e^{2ik\Delta x})\tilde{f}$$
$$= \left(a\frac{e^{ik\Delta x} - e^{-ik\Delta x}}{2\Delta x} + b\frac{e^{2ik\Delta x} - e^{-2ik\Delta x}}{4\Delta x} + c\frac{e^{3ik\Delta x} - e^{-3ik\Delta x}}{6\Delta x} + \right)\tilde{f} \tag{2}$$

where $i = \sqrt{-1}$, k is the wavenumber, and $\tilde{f}$ represents the Fourier transform of the function f. By using Euler's formula with Eq. (3), one can get

$$\bar{k}\Delta x = \frac{a\sin(k\Delta x) + (b/2)\sin(2k\Delta x) + (c/3)\sin(3k\Delta x)}{1 + 2\alpha\sin(k\Delta x) + 2\beta\sin(2k\Delta x)} \tag{3}$$

where $\bar{k}$ is the modified (effective) wave number and $k\Delta x$ represents the scaled-wave number. The numerical dispersive errors come from the deviation between the effective and exact wavenumber.

The following two quantities are essential for the two techniques of optimization namely, MIELDLD and MIEELDLD. These are given as follows

$$\text{IELDLD} = \int_{(k\Delta x)_l}^{(k\Delta x)_h} \text{eldld } d(k\Delta x) \tag{4}$$

$$\text{IELDLD} = \int_{(k\Delta x)_l}^{(k\Delta x)_h} \text{eeldld } d(k\Delta x) \tag{5}$$

where

$$\text{eldld} = (k\Delta x - \bar{k}\Delta x)^2 \tag{6}$$
$$\text{eeldld} = \exp((k\Delta x - \bar{k}\Delta x)^2) - 1 \tag{7}$$

The quantities eldld (Appadu et al. 2009) and eeldld denote the square difference error for low dispersion and low dissipation and exponential error for low dispersion and low dissipation, respectively.

Using the objective function in Eq. (5), we can use the following optimization algorithm to get the best integral optimization integral interval and corresponding coefficients for the penta-diagonal compact scheme. A simple algorithm is summarized to implement the proposed method. In the algorithm, $r = k\Delta x$, $r_{h*} = k\Delta x$.

1. Setting $r = r_0$, $r_1 = r_0 + \delta r$, optimization the problem using the Taylor series matching conditions (Kim et al. 1997) and the corresponding Objective function of Eq. (7) (Kim et al. 1997), get an initial range r_1 and the corresponding coefficient vector $X_1 = [\,r_1, a_{11}, b_{12}, c_{13}, \alpha_{14}, \beta_{15}\,]$.
2. Loop in ranges $0 < r < \pi$, get a matrix of ranges and coefficients

$$X_n = \begin{bmatrix} r_1 & a_{1,1} & b_{1,2} & c_{1,3} & \alpha_{1,4} & \beta_{1,5} \\ r_2 & a_{2,1} & b_{2,2} & c_{2,3} & \alpha_{2,4} & \beta_{2,5} \\ \vdots & \vdots & \vdots & \vdots & \vdots & \vdots \\ r_n & a_{n,1} & b_{n,2} & c_{n,3} & \alpha_{n,4} & \beta_{n,5} \end{bmatrix}$$

166

where n is the number of r_i in ranges $[0, \pi]$, each row represents a integral upper limit value and the corresponding coefficients at that integral upper limit.

3. From the matrix, X_n find r_{h*i} with $|k\Delta x - \overline{k}\Delta x| < \varepsilon_{ph}$, for every row, the phase velocity ε_{ph} is specified to 1×10^{-4} in this work, which is the phase velocity error tolerance. Add r_{h*i} to the list of r_{h*} obtained in previous steps, at last we get a list of r_{h*i}.

4. Find the best $r_{h*i} = \max(r_{h*})$ and the corresponds $X_i = [\, r_i, a_{i1}, b_{i2}, c_{i3}, \alpha_{i4}, \beta_{i5}]$ in the matrix X_n, the obtained r_i is the best integral upper limit and best optimization coefficients of the finite difference scheme are $[a_{i1}, b_{i2}, c_{i3}, \alpha_{i4}, \beta_{i5}]$.

Follow the algorithm above, we can easily find $r_{h*i} = \max(r_{h*}) = 0.7650\ \pi$, and the coefficients is listed in Table 1 within the first row, the Kim's scheme's coefficients is also being listed in the second row for comparison.

Table 1. Optimized coefficients and maximum r of the penta-diagonal compact finite difference scheme.

rmax		a	b	c	α	β
Opt-Exp	0.7650	1.317737779	0.955441097	0.032000321	0.568591406	0.083998192
Kim's	0.3923	1.285617622	1.04930907	0.04446583	0.58959552	0.09751235

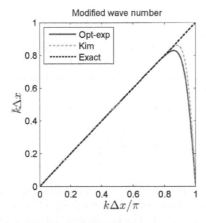

Figure 1. Modified wave number of the Opt-Exp and Kim schemes.

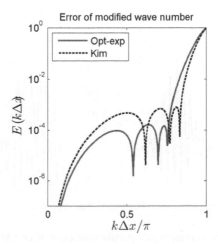

Figure 2. Error of the modified wave number of Opt-Exp schemes.

4   ILLUSTRATION AND DISCUSSION

In this section, the scalar wave convection problem is considered to investigate the performances of the proposed various truncation error orders finite difference scheme. During the computation, various truncation error order schemes are employed to spatial discretization and 6-stage Runge-Kutta optimized scheme of Hu (Hu et al. 1996) is applied to temporal integration.

4.1   *Convection of a harmonic signal*

To illustration the strengths of the large stencil with a relatively high truncation order, we consider the one-dimensional convection equation, which the solution consists of an initial disturbance that travels along the x-axis at a unitary dimensionless speed.

$$\frac{\partial u}{\partial t} + \frac{\partial u}{\partial x} = 0 \tag{8}$$

With initial condition in the form of

$$u(x,t=0) = \left(a + b\sin\left(\frac{2\pi(x-x_0)}{c\Delta x}\right)\right)\exp\left(-\ln(2)\left(\frac{x-x_0}{d\Delta x}\right)^2\right) \tag{9}$$

$$u(x=0,t) = 0$$

where $a = 1$, $b = 5$, $c = 2.86$, $d = 9.5$. The calculations are performed over the domain $x \in [0, N\Delta x]$. The time integration is numerically handled by 6 stage Runge-Kutta with a time step of $\Delta t = 0.1$, such the $CFL = c\Delta t/\Delta x = 0.1$. With a low CFL value, the temporal approximation induces only negligible errors. The total errors are almost equal to the discrete spatial differential operators.

For quantitative comparison, an L^2 norm error of solution is defined as

$$err = \sqrt{\frac{\sum(u_{num} - u_{exact})^2}{\sum u_{exact}^2}} \tag{10}$$

where u_{num} is the numerical solution, u_{exact} is the exact solution of Eq. (8).

Figure 3 shows the L^2 norm error of solution of Eq. (11), the absolute error of Opt-Exp scheme is less than Kim's scheme, it can be concluded that the Opt-Exp scheme is more accurate than Kim's scheme. Compared with Kim's scheme, the Opt-Exp scheme performs better for short waves.

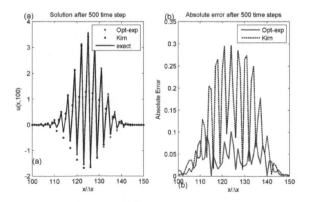

Figure 3.   Solution after 500 time steps (a) and absolute error after 500 time steps (b) with Opt-Exp and Kim's schemes.

5 CONCLUSION

A penta-diagonal compact finite difference scheme with high accuracy and maximum resolution has been suggested in this article. During the optimization, the following strategies have been applied: (a) an MIEELDLD error criterion has been used; (b) a simple algorithm to determine the best integral interval. The performance of the scheme has been demonstrated by applying it to a convective wave test case. The Opt-Exp scheme is more accurate than Kim's scheme in short wave ranges.

ACKNOWLEDGMENTS

This work received the support of the National Natural Science Foundation of China (No. 51175195).

REFERENCES

Appadu, A.R. & M.Z. Dauhoo, 2009. An Overview of Some High Order and Multi-Level Finite Difference Schemes in Computational Aeroacoustics. *Proceedings of World Academy of Science: Engineering & Technology*, 50.

Appadu, A.R. & M.Z. Dauhoo, 2011. The concept of minimized integrated exponential error for low dispersion and low dissipation schemes. *International Journal for Numerical Methods in Fluids*, 65, 578–601.

Ekaterinaris, J.A., 1999. Implicit, High-Resolution, Compact Schemes for Gas Dynamics and Aeroacoustics. *Journal of Computational Physics*, 156, 272–299.

Hu, F.Q. & M.Y. Hussaini & J.L. Manthey, 1996. Low-Dissipation and Low-Dispersion Runge–Kutta Schemes for Computational Acoustics. *Journal of Computational Physics*, 124, 177–191.

Kim, J.W., 2007. Optimised boundary compact finite difference schemes for computational aeroacoustics. *Journal of Computational Physics*, 225, 995–1019.

Kim, J.W. & D.J. Lee, 1996. Optimized compact finite difference schemes with maximum resolution. *AIAA Journal*, 34, 887–893.

Lele, S.K., 1992. Compact finite difference schemes with spectral-like resolution. *Journal of Computational Physics*, 103, 16–42.

Meitz, H.L. & H.F. Fasel, 2000. A Compact-Difference Scheme for the Navier–Stokes Equations in Vorticity–Velocity Formulation. *Journal of Computational Physics*, 157, 371–403.

Tam, C.K. & J.C. Webb & Z. Dong, 1993a. A study of the short wave components in computational acoustics. *Journal of Computational Acoustics*, 1, 1–30.

Tam, C.K.W. & J.C. Webb, 1993b. Dispersion-Relation-Preserving Finite Difference Schemes for Computational Acoustics. *Journal of Computational Physics*, 107, 262–281.

Advances in Engineering Materials and Applied Mechanics – Zhang, Gao & Xu (Eds)
© 2016 Taylor & Francis Group, London, ISBN 978-1-138-02834-0

Numerical simulation of flow field of a novel helical static mixer for polymer flooding

W.P. Song, Y.Q. Li, L.Q. Li, H.Q. Zhang & B. Guo
Harbin Institute of Technology, Nangang District, Heilongjiang Province, China

ABSTRACT: For polymer (PAM) flooding, the viscosity of the PAM solution injected into the bottom is a key factor for affecting production costs and oil recovery. In this paper, a three dimensional model is developed to study the mixing process of Polyacrylamide (PAM) and water through a helical static mixer. The viscosity of the PAM solution at the outlet was calculated using the Computational Fluid Dynamics (CFD) method. The effect of helical pitch and helical unit on viscosity of PAM solution are investigated and discussed. As the helical pitch increases, the PAM solution viscosity at the outlet of the static mixer also increased, the helical units spacing has the same influences on the outlet PAM solution viscosity with the helical pitch, but the impact is not so obvious. It means that we can reduce the polymer solution viscosity loss by increasing these two parameters appropriately.

1 INTRODUCTION

Polymer flooding technology has been widely used to increase the oil production in oilfield, especially for Daqing Oilfield in China. In the Polymer flooding technology, mixer are needed to mix the concentrated PAM solution and water. After that, the mixed PAM solution is injected to specific sections of wells. Because of its low energy consumption, high efficiency and other advantages, static mixers have been widely used to mix the concentrated PAM solution and water. Different to the chemical industry, the static mixers used in the polymer flooding technology are expected to induce low shear stress for the PAM solution. This is because the shear stress can decrease viscosity of the PAM solution, which determines the efficiency of polymer flooding[1]. However, the static mixers currently used in polymer flooding are most come from chemical industry. When the PAM solution flow through these static mixers, high shear stress is induced and the viscosity of the PAM solution is decreased obviously.

The static mixers most commonly used are the United States Kenics static mixer, Ross-ISG, Switzerland (Sulzer) SMV, SMX, SMXL and so on[2]. In recent years, phase Doppler (PDA) and Particle Image Velocimetry (PIV) are applied to create advanced instruments measuring droplet[3], but it has not been used in liquid–liquid dispersion measurements in a static mixer. With the development of computational fluid dynamics technology, numerical simulation has been an important and efficient way of fluid dynamics research[4–5]. Hobbs and Muzzio[6] have studied fluid's tracer particle trajectory at the cross-section of the mixing unit inside the mixer. The mixing process and mixing characteristics of the fluid qualitatively were also investigated in Ref.[6]. Hester[7] improved that the viscosity of the polymer solution is determined by the molecular contour length of the polymer and polymer molecular coil size. Zhang[8] studied the effects of temperature, concentration and shear rate and metal ion concentration on the rheological behavior of PAM solution.

At present, most studies have focused on the effect of temperature, stirring speed, concentration, solution composition and other factors on the viscosity of polymer. Research methods used are mostly the theoretical derivation and experimental verification. There has been no mature theory to guide mixed unit structure optimization and design, to minimize

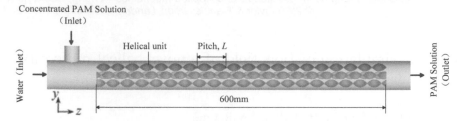

Figure 1.   Geometrical model of helical static mixer.

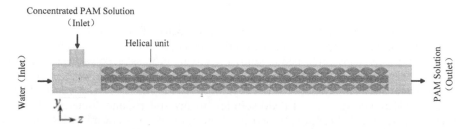

Figure 2.   Finite element model of helical static mixer for mixing PAM solution and water.

damage viscous[9–10]. In addition, only a few researchers focus on the static mixer for polymer flooding. In this study, a novel helical static mixer is proposed and the flow field of mixing concentrated PAM solution and water are simulated. The effect of geometrical parameters are also investigated.

2   MODELING

Figure 1 shows the geometrical model of the novel helical static mixer for polymer flooding. The helical static mixer consists of seven helical units, i.e., one helical unit located at the middle and other six helical units located around. The pitch of the helical unite is defined as L. The length of the helical unit is chosen to be 800 mm. The internal diameter of the helical static mixer is 50 mm in this study.

Figure 2 shows the finite element model of the helical static mixer. Fine meshes are used for the helical units. The elements in this model are approximately 160,000. The flow field for mixing the PAM solution and water is calculated by a commercially available finite element solver, ANSYS/CFX. Typical calculation time for one case is around 24 hours on a HP computer with four processing units of 3.3 GHz and 8 Gb RAM.

3   RESULTS AND DISCUSSION

In this study, the following parameters are used: flow rate of concentrated PAM solution $Q_p = 25$ m³/d (i.e., cubic meter per day), the viscosity of concentrated PAM solution $\eta = 200$ mPa·s, flow rate of water $Q_w = 50$ m³/d, and pressure at outlet $P_o = 12$ MPa.

3.1   *Simulation research on the impact of helical unit pitch on the PAM viscosity*

Figure 3 shows the viscosity distribution of the PAM solution at outlet for helical static mixer with different values of helical pitch (L). We observe that the viscosity of the PAM solution at outlet increases with an increase in the helical pitch. This is because that the helical with smaller pitch induce larger shear stress, and finally cause larger decrease of PAM viscosity.

In addition, we observe that the distribution of the PAM viscosity is more uniform for the helical static mixer with smaller helical pitch. It indicates that the helical static mixer with L = 30 mm is able to mix the concentrated PAM solution and water well.

Figure 4 shows the average viscosity of PAM solution at outlet as a function of the helical pitch. We observe that the viscosity of PAM solution decrease to be about 40 mPa·s if the helical pitch L = 30 mm. In addition, PAM solution with the viscosity of 40 mPa·s is good for the polymer flooding.

3.2 Simulation research on the impact of helical units spacing on the PAM viscosity

Using the same method, the viscosity changes in different helical units spacing (S) static mixers were simulated respectively, simulation models were spacing 1 mm, 1.5 mm, 2 mm, 2.5 mm, 3 mm. PAM solution viscosity at the outlet of the mixer clouds are shown in Figure 5.

It can be seen from Figure 5, with the helical unit spacing increases, PAM viscosity at the outlet of static mixer also tended to increase, but the effect is not very obvious. The average PAM viscosity value at the outlet was calculated, the trend curve is shown in Figure 6.

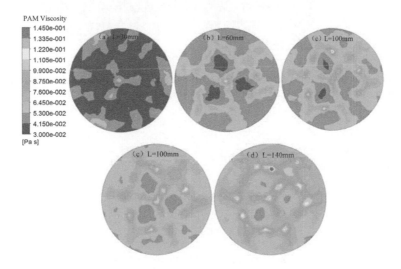

Figure 3. Viscosity distribution of PAM solution at outlet of helical static mixer for (a) L = 30 mm, (b) L = 60 mm, (c) L = 80 mm, (d) L = 100 mm and (e) L = 140 mm.

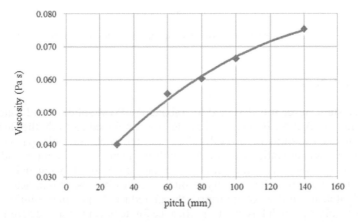

Figure 4. Average viscosity of PAM solution at outlet as a function of helical pitch.

173

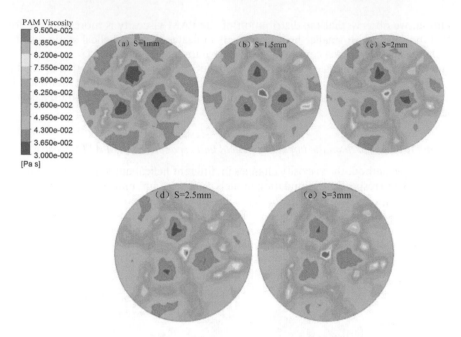

Figure 5. Viscosity distribution of PAM solution at outlet of helical static mixer for (a) S = 140 mm, (b) S = 100 mm, (c) S = 80 mm, (d) S = 60 mm and (e) S = 30 mm.

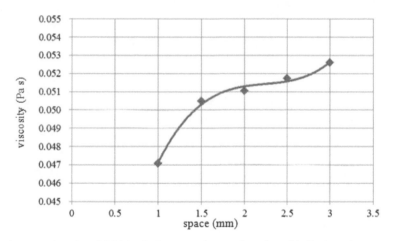

Figure 6. Average viscosity of PAM solution at outlet as a function of helical spacing.

4 CONCLUSION

A novel helical static mixer was designed for mixing the PAM solution and water in polymer flooding technology. A three dimensional model was developed to study the flow field of the helical static mixer. The simulation results show that the pitch and unit spacing will have an impact on the viscosity at static mixer outlet. With the increase of the pitch and unit spacing values, PAM solution viscosity at the outlet of the static mixer are showing an increasing trend, and the impact of changes in the viscosity of pitch is more obvious. This indicates that increasing the helical pitch and unit space appropriately can reduce the polymer solution viscosity loss caused by static mixers, which has great significance for achieving cost-effective EOR.

REFERENCES

[1] Zhang Chunmei, Wu Jianhua, Gong Bing. 2006. Turbulent drag research of SK static mixer. *Chemical Engineering*, 34(10): 27–30.

[2] Liu Renheng, Wang Jinhua, Xie Zhijun, etc. 2014. New low shear static mixer structure optimization. *Petrochemical Equipment*, 43(4): 7–10.

[3] D.M. Hobbs, F.J. Muzzio. 1997. Effects of Injection Location, Flow Ratio and Geometry on Kenics Mixer Performance. *AICHE Journal*, 12(43): 3121–3132.

[4] Gandhi M.S., Sathe M.J., Joshi J.B., Vijayan P.K. 2011. Two-phase natural convection: CFD simulation and PIV measurement. *Chemical Engineering Science*, 66(14): 3152–3171.

[5] Pianko-Oprych P., Jaworski Z. 2010. Prediction of liquid-liquid flow in an SMX static mixer using large eddy simulations. *Chem. Pap.*, 64(2): 203–212.

[6] Maa S., Metz F., Rehm T., Kraume M. Prediction of drop sizes for liquid-liquid systems in stirred slim reactors (I): Single stage impellers. *Chemical Engineering Journal*, 2010, 162(2): 792–801.

[7] R.D., L.M. Flesher, and C.L. McCormic. 1994. Polymer Solution Extension Viscosity Effects During Reservoir Flooding. *SPE/DOE Ninth Symposium*, 447–456.

[8] Luhong Zhang, Dan Zhang, and Bin Jiang. 2006. The Rheological Behavior of Salt Tolerant Polyacrylamide Solutions. Chem. *Eng. Technol,* 29(3): 395–400.

[9] Paricha Y.K.D., Legrand J., Moranca I.S.P., etc 2005. Drop breakage model in static mixers at low and intermediate Reynolds number. *Chemical Engineering Science*, 60(1): 231–238.

[10] Yu Yanfang, Wu Jianhua, Meng Huibo. 2007. The micro research and industrial application advances of static mixer. *Mechanical Design and Manufacturing*, 5: 211–212.

Advances in Engineering Materials and Applied Mechanics – Zhang, Gao & Xu (Eds)
© 2016 Taylor & Francis Group, London, ISBN 978-1-138-02834-0

Research and experiment analysis on the bending stress of the collet connector

D.M. Wang, C.P. Yang & X. He
Offshore Oil Engineering Co. Ltd., Tanggu, China

J. Liu, F.H. Yun, T. Liu, R.H. Wang & L.Q. Wang
Harbin Engineering University, Harbin, China

ABSTRACT: The bending load on the collet connector could lead to serious accidents. So this paper is concerned with analyzing the bending stress of the collet connector, which is properly sealed. The principle of the collet connector was described based on its mechanical structure. Bending loads on the connector came mainly from the waves and currents which acted on the jumper. The bending property of the collet connector was analyzed based on the load capacity of the jumper. Therefore, we could obtain the maximal moment, the tensile stress on dangerous cross-section, the extension value and the elongation. The bending experiments were carried out on the subsea connector load testing-device, the seal performance of the collet connector was perfect, the deviation of the extension value in the experiment compared with the theoretical calculation was small. These experimental results verified the validity of the theoretical calculation of the collet connector bending stress.

1 INTRODUCTION

A collet connector is used to connect pipeline ends in a sub-sea production system, which has been widely used in large subsea oil-gas field production systems from different areas. The collet connector which can be installed in more than 3000 meters deep works under extremely severe operating conditions, such as internal high temperature and high pressure, external low temperature and high pressure, and high tensile force from the jumper. All these factors lead to a large quantity of loads, of which bending load is the most principal. In fact, it will result in seal leakage and fatigue damage of the connector. In order to avoid it, it is very significant to examine thoroughly and to acquire knowledge about bending stress of collet connector.

The research of related technology on the connectors has been carried out for more than 50 years in abroad. Some scholars have studied the bending load of different connectors. Sato, T. (1988) conducted finite element analyses on the threaded marine riser connector which was under bending load, and compared the results of finite element analyses with the experimental results. The algorithm about axial and bending stress concentration factors in screw connectors was described by Bahai, H. & Esat, I.I. (1994). Burguete, R.L. et al. (1994) carried out a study of effect of bending on the normalized stress at roots of threaded connectors. It was found that bending load on connectors will affect the normalized stress and that it is possible to determine this effect in a similar way to the method used for axially loaded connections. Although a number of papers have been published in the general area of collet connectors, little work has been carried out for bending stress analysis of the collet connector at home and abroad. In this paper, we intend to focus on the bending load capacity analysis and the bending experiment of a collet connector.

2 THE PRINCIPLE OF THE COLLET CONNECTOR

The collet connector efficiently transfers the thrust of the hydraulic cylinder to the axial of the hub with the help of contacts of the actuator ring, finger and hub. The two hubs are pressed to produce a sufficient sealing force to ensure the sealing of the connector fitting. Figure 1 is a schematic diagram that shows the force conduction of the connector fitting. A larger radial force which applied to fingers is generated by the actuator ring under the loading force. The contact between the finger and the actuator ring is an inclined plane so that the radial force is transformed into an axial force which causes the hub locking. The seal is deformed by axial force. The axial force makes the seal sphere and hubs contact closely and achieve seal. The internal oil that with high-speed production outward pressure on the finger when the actuator ring loading is complete. The pressure has a component which makes the actuator ring upward. The angle between the rear surface of the finger and the inner surface of the actuator ring is less than a certain value. Therefore the self-locking of the connector is achieved and the connector seal's reliability is ensured.

3 BENDING STRESS ANALYSIS OF THE COLLET CONNECTOR

3.1 *Bending load origin*

The collet connector works in deepwater and will be influenced by environmental factors. Environmental factors including water, shoal disturbance, subsea earthquakes, volcanic eruptions, tsunamis and so on. The bending load is an important factor affects the reliability of the connector seal. Bending load is mainly caused by waves and currents. The dragging force which applied on the jumper is generated by ocean currents. A larger bending load is generated by the dragging force act on connector. Figure 2 is the schematic of a bending load.

3.2 *The calculation of bending load capacity of the connector*

The pipeline materials for the collet connector connect select X65. The minimum yield strength is 448 MPa. That is to say $\sigma_s = 448$ MPa. Take 80% of the minimum yield strength as the allowable stress σ.

$$\sigma = 80\%\sigma_s = 358.4 \text{ MPa} \tag{1}$$

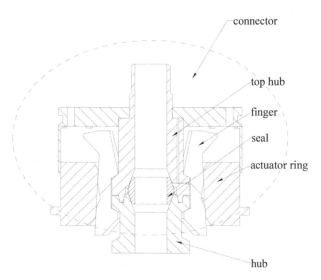

Figure 1.   The structure of the collet connector.

Figure 2.   The schematic of a bending load.

Ensure the connector internal pressure is 34.5 MPa and the preload is 120t. Based on this, the bending moment is calculated.

3.2.1   *The maximum tension of the connector*

In order to calculate the maximum moment that the pipeline can afford, we need to calculate the maximum tension of the connector first. The maximum tension of the connector F is the maximum tension of the pipeline when the connector is unsealed.

$$F = A \cdot \sigma = \frac{1}{4}\pi \cdot (D^2 - d^2) \cdot \sigma = \frac{1}{4}\pi \times (170^2 - 140^2) \times 358.4 = 261.8\text{t} \tag{2}$$

where A is the pipeline sectional area, m²; D is the outer diameter of the pipeline size, 170 mm; d is the inner diameter size, 140 mm.

The 120t axial force is required when the connector seal pressure is 34.5 MPa. We can obtain the maximum tension of connector F' as:

$$F' = F - 120 = 261.8 - 120 = 141.8\text{t} \tag{3}$$

3.3   *The maximum moment of the connector*

12 fingers are distributed around the hubs. When they work, the seal of the connector is realized, which equals to bolted connections tension the hubs. So we can check the fingers according to the overturning moment. The maximum moment of the collet connector is obtained through the collet connector's maximal tension multiplied by the contact radius of the hub and the seal.

The contact radius r of the hub and the seal was obtained by the practical measure. The contact radius was shown in Figure 3, $r = 86.7$ mm.

The maximum moment W of the connector is:

$$W = F'r' = 141.8 \times 86.7 = 122.94 \text{ kN} \cdot \text{m} \tag{4}$$

3.3.1   *The check of the dangerous section*

Check the dangerous section according to preset parameters of the applied load. Figure 4 shows a cyclic evident geometrical symmetry of the distance between finger and central line. Calculate the maximum force that the moment applied to a finger, F_{max} is:

$$F_{max} = \frac{WL_{max}}{4 \times \sum_{i=1}^{3} L_i^2} = 7.94 \times 10^4 \text{ N} \tag{5}$$

179

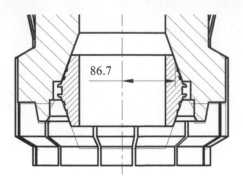

Figure 3. The contact radius.

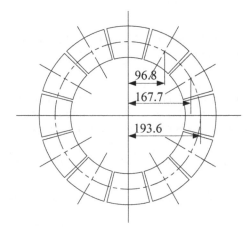

Figure 4. The distance between the finger and axis of hub.

where W is the maximum moment, 122940 N·m; L_{max} is maximum distance between central line of finger and axis of hubs, m; L_i is the distance between each finger and the central line, m.

Calculate the sum of all the fingers' minimum section, S_{12} is:

$$S_{12} = \frac{1}{4}\pi(d_1^2 - d_2^2)\frac{\alpha n}{360} \tag{6}$$

where d_1 is the outer diameter of finger's minimum section, 0.438 m; d_2 is the inner diameter of finger's minimum section, 0.346 m; α is the sector angle between the fingers, 28°; n is the number of fingers, 12.

The calculation result is $S_{12} = 0.0528$ m².

The minimum section of one finger is:

$$S_d = \frac{S_{12}}{12} = 0.0044 \text{ m}^2 \tag{7}$$

When the moment is 122.94 kN·m, the maximal stretched stress of one finger's dangerous section is:

$$\sigma_{max} = \frac{F_{max}}{S_d} = 18.05 \text{ MPa} \tag{8}$$

180

Then we can learn about that the fingers have strong moment loading capability and meet preset requirement. Elongation of the fingers caused by the applied moment may result in the failure of the seal. Therefore, extension value of the fingers is:

$$\Delta S_d = \frac{F_{max} \cdot L}{E \cdot S_d} = 0.0109 \text{ mm} \tag{9}$$

where L is the length of stretching part of the finger, 0.1275 m; E is the elastic modulus, 2.11×10^5 MPa.

The finger is the furthest stressed member from the center. We can obtain that the extension value is 0.0109 mm, and the elongation is 0.0085%, which can be ignored. As noted previously, the seal was lower failure when afford this moment.

4 EXPERIMENTAL INVESTIGATION

4.1 *Experimental equipment*

The connector bending experiment was performed by means of the subsea connector load testing-device. In detail, as is shown in Figure 5 and Figure 6, there are two hydraulic cylinders for supporting at the bottom of the testing-device and two hydraulic cylinders for bending on the sidewall of the testing-device. The bending hydraulic cylinders, supporting hydraulic cylinders and the collet connector are linked together by the link plate. With the aim to evaluate the moment on the connector, the control system converts the bending hydraulic cylinders force value, which is measured by force sensors installed on the bending hydraulic cylinders,

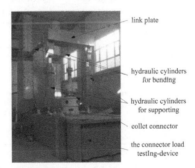

Figure 5.   Subsea connector load testing-device.

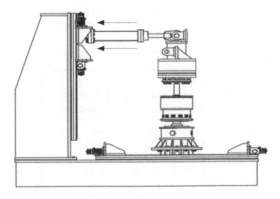

Figure 6.   Schematic diagram of bending.

to the moment value. In the same way, we can evaluate the extension value of the hub with the help of strain gages.

4.2 Experimental procedures and results

The testing program of the bending experiment is carried out by the following standards, such as **API 6 A, API 17D** and **ISO 13628**. The experimental procedures are formulated as follows:

1. The hydrostatic test of the collet connector is carried out. Make sure that the internal pressure is 34.5 MPa. Observe the leakage of the collet connector and record the data of pressure while holding the pressure for 15 min.
2. The first moment applied to the collet connector is considered equal to 50% of the maximum moment of the connector. Record the pressure and extension value of the finger.
3. The hydrostatic test of the collet connector with 51.75 MPa (1.5 times of 34.5 MPa) internal pressure is carried out. Observe the leakage of the collet connector and record the data of pressure while holding the pressure for 15 min.
4. The moment that applies to the collet connector is increased by 10% of the maximum moment until up to the maximum moment. Carry out the hydrostatic test after the increase has finished. When the leakage is detected, check the connector after pressure relief.
5. Repeat the procedure (2), (3) and (4) for 3 times.

The measurement data of the 3 experiments are given in Table 1, 2, 3 and 4. We can see that there is no leakage during the holding time both in the hydrostatic test and the

Table 1. The hydrostatic test of the collet connector.

Theoretical pressure (MPa)	Actual pressure (MPa)	Holding onset	Holding termination	Holding termination pressure (MPa)
34.5	34.5	15:27	15:42	34.5

Table 2. The 1st bending experiment.

Percentage of the moment	Theoretical moment (kN·m)	Actual moment (kN·m)	Actual pressure (MPa)	Holding onset	Holding termination	Holding termination pressure (MPa)	Displacement of the hub (mm)
50%	61.45	62	52	15:45	16:00	52	0.0035
60%	73.74	75	52	16:02	16:17	52	0.0043
70%	86.03	86	52	16:18	16:23	52	0.0058
80%	98.32	100	52	16:24	16:39	52	0.0085
90%	110.61	112	52	16:40	16:55	52	0.0102
100%	122.9	123	52	16:57	17:12	52	0.0113

Table 3. The 2nd bending experiment.

Percentage of the moment	Theoretical moment (kN·m)	Actual moment (kN·m)	Actual pressure (MPa)	Holding onset	Holding termination	Holding termination pressure (MPa)	Displacement of the hub (mm)
50%	61.45	62	52	17:50	18:05	52	0.0034
60%	73.74	74	52	18:07	18:22	52	0.0042
70%	86.03	86	52	18:23	18:38	52	0.0060
80%	98.32	99	52	18:39	18:54	52	0.0083
90%	110.61	113	52	18:55	19:10	52	0.0103
100%	122.9	125	52	19:11	19:26	52	0.0114

Table 4. The 3rd bending experiment.

Percentage of the moment	Theoretical moment (kN·m)	Actual moment (kN·m)	Actual pressure (MPa)	Holding onset	Holding termination	Holding termination pressure (MPa)	Displacement of the hub (mm)
50%	61.45	62	52	19:43	19:58	52	0.0034
60%	73.74	73	52	19:59	20:14	52	0.0041
70%	86.03	87	52	20:15	20:30	52	0.0063
80%	98.32	100	52	20:31	20:46	52	0.0084
90%	110.61	112	52	20:48	21:03	52	0.0101
100%	122.9	124	52	21:04	21:19	52	0.0114

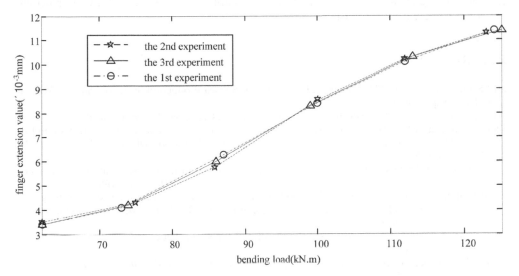

Figure 7. The finger extension value-bending load curve.

bending experiment. In theory, the extension value data in Table 2, 3 and 4 must be the same. However, each equipment used in the experiment has its own accuracy and errors, so the finger extension values are different in the three experiments.

As we see in Figure 7, the finger extension values of the three experiments have the same trends, and the biggest deviation value is less than 0.0002, which is negligible in the whole system. The biggest extension value of the finger under the 122.9 kN·m bending load is less than 0.0123 mm, which means that the elongation is less than 0.0096%. The extension value is 3.67% larger than 0.0109 mm compared with the result of the theoretical calculation, which is acceptable. The cause of the deviation is that we have simplified the model during the theoretical calculation.

5 CONCLUSION

In this paper, the bending stress of collet connector has been considered. Based on the load capacity of the jumper, the bending property of the collet connector was analyzed. The maximal moment of the connector was 122.94 kN·m; the tensile stress on dangerous cross-section was 18.05 MPa; the extension value is 0.0109 mm; the elongation was 0.0085%, which fits the equipment operational requirement.

The bending experiments were conducted on the subsea connector load testing-device. The experimental results showed that the collet connector could maintain its internal pressure perfectly, which proved that the collet connector could overcome large moment to keep a nice seal performance. The deviation of the extension value in the experiment and in the theoretical calculation is less than 3.67%, which verifies the correctness of the calculation. The research on the bending stress of the collet connector provides a new method for subsea connector stress analysis, and provides an effective new idea for other subsea structures' experimental study on mechanical.

ACKNOWLEDGEMENTS

This work was supported by a grant from the National High-Tech Research and Development Program of China (863 Program) (No. 2013AA09A217). The authors also would like to express our gratitude to all the friends in our scientific team who have always been helping and supporting without a word of complaint.

REFERENCES

API Specification 17D. 2011. Design and Operation of Subsea Production Systems—Subsea Wellhead and Tree Equipment.

API Specification 6 A. 2004. Specification for Wellhead and Christmas Tree Equipment.

Bahai, H. & Esat, I.I. 1994. Study of axial and bending stress concentration factors in screw connectors. In ASME, *American Society of Mechanical Engineers, Petroleum Division (Publication) PD*: 753–761. New York.

Burguete, R.L. & Patterson, E.A. 1994. Effect of bending on the normalized stress at roots of threaded connectors. *Journal of Offshore Mechanics and Arctic Engineering* 116(3): 163–166.

Di, P.A. et al. 2014. Reliability/availability methods for subsea risers and deepwater systems design and optimization. In Techno-Info Comprehensive Solutions (TICS), *PSAM 2014—Probabilistic Safety Assessment and Management.*

Fassina, P. et al. 2005. Failure analysis of a non-integral pipeline collet connector. *Engineering Failure Analysis*, 12(5): 711–719.

Hsu, J.C. 1999. Deepwater high-capacity collet connector. *Proceedings of the Annual Offshore Technology Conference* 3: 501–504.

ISO 13628-1. 2005. Petroleum and Natural Gas Industries Design and Operation of Subsea Production Systems-Part 1: General Requirements and Recommendations.

Junpeng, L. et al. 2012. An optimal design for locking mechanism of mechanical collet connector. In International Society of Offshore and Polar Engineers, *Proceedings of the International Offshore and Polar Engineering Conference*: 206–212.

Liquan, W. et al. 2011. Study and design of a deepwater sub-sea pipeline collet connector. *Journal of Harbin Engineering University* 32(9): 1103–1107.

Sato, T. et al. 1988. Finite element analysis and fatigue evaluation of the threaded marine riser connector. *Journal of Energy Resources Technology, Transactions of the ASME* 110(2): 85–92.

Yanshun, Z. et al. 2012. Installation technology of deepwater vertical collet connector. In Trans Tech Publications, *Advanced Materials Research* 524–527: 1465–1470. Clausthal-Zellerfeld.

Advances in Engineering Materials and Applied Mechanics – Zhang, Gao & Xu (Eds)
© 2016 Taylor & Francis Group, London, ISBN 978-1-138-02834-0

In-plane vibration analysis of moderately thick rectangular plates

J.F. Wang, S.W. Wang, W. Mu & P. Gao
R&D Institute, CSIC, Yichang, P.R. China

ABSTRACT: An improved Fourier series method is presented for the in-plane vibration analysis of rectangular plates. The in-plane displacement fields are expressed by the linear combination of a double Fourier series and auxiliary series functions. The general boundary conditions can be represented by two sets of linear springs along each edge, and the matrix eigenvalue equation of the plate with arbitrary boundary conditions can be derived using the Rayleigh–Ritz method. Modal analysis and harmonic response analysis are carried out to validate the accuracy of the current approach.

1 INTRODUCTION

Rectangular plates are the important structural elements and the analysis of the vibration is very important for the design of plate-type structures in aerospace, electronic, mechanical, marine, nuclear and structural engineering. Many researchers have researched the tranverse bending vibration of plates, and a lot of results have been published. Due to the couple effect between the longitudinal displacement and the shear displacement, the traditional series method cannot solve the in-plane problem effectively, so the in-plane vibration of plates is far less studied by few publications available in the literature.

Bardell[1] studied the in-plane vibration of isotropic rectangular plates have either free or fully constrained boundaries, and the results were presented for the lower six modes at two particular aspect rations. The work of Xing and Liu[2] was another significant contribution to solve free in-plane vibrations of rectangular plates, they used the direct separation of variables method to obtain exact solutions of the natural frequencies and the mode shapes for free in-plane vibrations of the rectangular plates, where at least two opposite edges had either type of simply supported conditions, and there were no problems in the interpretation of the computed mode shapes. Other researchers have used the superposition method[3–4], the Hamilton principle[5], the variational method[6–8], and the Rayleigh–Ritz method[9–10] to research the in-plane vibration of rectangular plates.

Most existing studies on the free in-plane vibrations of rectangular plates are limited to the classical homogeneous boundary conditions. To solve the limitations of the analysis method to the in-plane vibration of rectangular plates with boundary supports, an improved Fourier series method is presented for the in-plane vibration analysis of rectangular plates. The in-plane displacement fields are expressed by the linear combination of a double Fourier series and auxiliary series functions. The general boundary conditions can be represented by two sets of linear springs along each edge, and the matrix eigenvalue equation of the plate with arbitrary boundary conditions can be derived using the Rayleigh–Ritz method. Finally, the numerical results are presented to validate the correct of the method.

2 STRUCTURAL MODEL OF COUPLED PLATES

In order to establish a general model, the structural model of rectangular plate is shown in Figure 1. The length and width of rectangular plate are *a*, *b*, and the thickness of the plate is denoted by *h*. The boundary conditions of in-plane vibration are simulated by setting

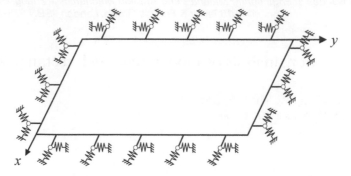

Figure 1. The in-plane vibration of rectangular with general elastic boundary support.

restraining springs along edges. Two types of linear spring (normal and tangential springs) are needed to simulate the general boundary conditions, and different boundary conditions can be directly obtained by changing the stiffness of springs. All the classical homogeneous boundary conditions can be easily derived by simply setting each of the spring constants to be infinite or zero.

3   THE DISPLACEMENT FUNCTIONS OF COUPLED PLATES

According to the in-plane vibration theory, the in-plane longitudinal and shear displacements are utilized. In this study, these quantities are expressed in form of improved Fourier series expansions:

$$
u(x,y) = \sum_{m=0}^{\infty}\sum_{n=0}^{\infty} D_{mn}\cos(\lambda_m x)\cos(\lambda_n y) + \sum_{m=0}^{\infty} d_{1m}^4 \xi_{1b}(y)\cos(\lambda_m x) + \sum_{n=0}^{\infty} f_{1n}^4 \xi_{1a}(x)\cos(\lambda_n y)
$$
$$
+ \sum_{m=0}^{\infty} d_{2m}^4 \xi_{2b}(y)\cos(\lambda_m x) + \sum_{n=0}^{\infty} f_{2n}^4 \xi_{2a}(x)\cos(\lambda_n y) \tag{1}
$$

$$
v(x,y) = \sum_{m=0}^{\infty}\sum_{n=0}^{\infty} E_{mn}\cos(\lambda_m x)\cos(\lambda_n y) + \sum_{m=0}^{\infty} d_{1m}^5 \xi_{1b}(y)\cos(\lambda_m x) + \sum_{n=0}^{\infty} f_{1n}^5 \xi_{1a}(x)\cos(\lambda_n y)
$$
$$
+ \sum_{m=0}^{\infty} d_{2m}^5 \xi_{2b}(y)\cos(\lambda_m x) + \sum_{n=0}^{\infty} f_{2n}^5 \xi_{2a}(x)\cos(\lambda_n y) \tag{2}
$$

where $l = 1, 2$, $\lambda_m = m\pi/a$, $\lambda_n = n\pi/b$, D_{imn} and E_{imn} are the expansion coefficients, and

$$
\xi_{1a}(x) = \frac{a}{2\pi}\sin\frac{\pi x}{2a} + \frac{a}{2\pi}\sin\frac{3\pi x}{2a} \tag{3}
$$

$$
\xi_{2a}(x) = -\frac{a}{2\pi}\cos\frac{\pi x}{2a} + \frac{a}{2\pi}\cos\frac{3\pi x}{2a} \tag{4}
$$

$$
\xi_{1b}(y) = \frac{b}{2\pi}\sin\frac{\pi y}{2b} + \frac{b}{2\pi}\sin\frac{3\pi y}{2b} \tag{5}
$$

$$
\xi_{2b}(y) = -\frac{b}{2\pi}\cos\frac{\pi y}{2b} + \frac{b}{2\pi}\cos\frac{3\pi y}{2b} \tag{6}
$$

Theoretically, there are an infinite number of these supplementary functions. However, one needs to ensure that the selected functions will not nullify any of the boundary conditions. It is easy to verify that $\xi_{1a}(0) = \xi_{1a}(0) = \xi_{1a}'(a) = 0$, $\xi_{1a}'(0) = 1$, $\xi_{2a}(0) = \xi_{2a}(0) = \xi_{2a}'(0) = 0$, $\xi_{2a}'(a) = 1$, similar conditions exist for the supplementary function in y-direction. Although these conditions are not necessary, they can simplify the subsequent mathematical expressions and the corresponding solution procedures.

One shall notice from Equations (1) and (2) that beside the standard double Fourier series, four single Fourier series are also included. The potential discontinuity associated with the x-derivative and y-derivative of the original function along the four edges can be transferred onto these auxiliary series functions. Then, the Fourier series would be smooth enough in the whole solving domain. Therefore, not only this Fourier series representation of solution is applicable to any boundary conditions, but also the convergence of the series expansion can be improved.

4 ENERGY MODEL OF COUPLED PLATES

The Rayleigh–Ritz method will be used to find the solution, specifically the Fourier expansion coefficients in Equations (1) and (2). The Lagrangian's function L for the coupled plate system can be generally defined as:

$$L = V_{spring} + V_{in} - T \tag{7}$$

In the above equation, the strain potential energy of in-plane vibration V can be expressed as:

$$V_{in} = \iiint_V \sigma_x \varepsilon_x + \sigma_y \varepsilon_y + \tau_{xy} \gamma_{xy} dxdydz$$
$$= \frac{G}{2} \int_0^a \int_0^b \left\{ \left(\frac{\partial u}{\partial x} + \frac{\partial v}{\partial y} \right)^2 - 2(1-\mu) \frac{\partial u}{\partial x} \frac{\partial v}{\partial y} + \frac{(1-\mu)}{2} \left(\frac{\partial v}{\partial x} + \frac{\partial u}{\partial y} \right)^2 \right\} dxdy \tag{8}$$

The spring potential energy of in-plane vibration V can be expressed as:

$$V_{spring} = \frac{1}{2} \int_0^b \left\{ \left[k_{nx0} u^2 + k_{px0} v^2 \right]_{x=0} + \left[k_{nxa} u^2 + k_{pxa} v^2 \right]_{x=a} \right\} dy$$
$$+ \frac{1}{2} \int_0^a \left\{ \left[k_{py0} u^2 + k_{ny0} v^2 \right]_{y=0} + \left[k_{pyb} u^2 + k_{nyb} v^2 \right]_{y=b} \right\} dx \tag{9}$$

And the total kinetic energy T is as follows:

$$T_{in} = \frac{1}{2} \rho h \int_0^a \int_0^b \left\{ \left(\frac{\partial u}{\partial t} \right)^2 + \left(\frac{\partial v}{\partial t} \right)^2 \right\} dxdy$$
$$= \frac{1}{2} \rho h \omega^2 \int_0^a \int_0^b \left\{ u^2 + v^2 \right\} dxdy \tag{10}$$

By substituting the displacement functions (1) and (2) into the Lagrangian (8) and minimizing the result against all unknown Fourier coefficients, a final system of linear equations is obtained as follows:

$$(\mathbf{K} - \rho h \omega^2 \mathbf{M}) \mathbf{A} = \mathbf{0} \tag{11}$$

where $\mathbf{K}$ and $\mathbf{M}$ are the stiffness and mass matrices, and $\mathbf{A}$ is a vector of all the unknown Fourier expansion coefficient. The natural frequencies and eigenvectors of coupled plates can be obtained through solving Equation (11).

5 RESULT AND DISCUSSION

The numerical examples will be discussed in this section. To avoid any comparison of the round off results that might be unrealistic, the non-dimensional frequency is used. For the analysis, the Poisson's ratio $\mu = 0.3$ and the shear correction factor $k = 5/6$ are used. The thickness of plate is $h = 1.0$ m. In identifying the boundary conditions, letters F, E and C have been used to indicate the free, elastic and clamped boundary conditions along an edge, respectively.

Table 1 gives the first sixth frequency parameters of the in-plane vibration for the ECEF rectangular plates. The elastic boundary at $x = 0$ and $x = a$ can be obtained by setting the spring constants to be $K_{x0} = 40 \times G$, $K_{x0} = 0$, $K_{xa} = 4000 \times G$ and $K_{xa} = 0$ ($G = Eh/(1-\mu^2)$). The free and clamped edge can be obtained when all the spring constants are zero and infinitely, respectively. The frequency parameters solved by FEA method are also given as a comparison. The first fourth mode shapes are plotted in Figure 2 for the square plate ($a/b = 1$) with this method, and the solutions of FEA method are plotted in Figure 3. Through the examples, the good agreement is also observed among these solutions.

Table 1. The frequency parameters $\Omega = (\omega b^2/\pi^2)(\rho h/G)^{1/2}$ for ECEF rectangular plates.

a/b	Method	1	2	3	4	5	6	The error
1	IFSM	1.5580	1.9143	2.6048	3.2849	3.8469	4.3127	0.48%
	FEA	1.5581	1.9188	2.6106	3.3007	3.8528	4.3264	
1.5	IFSM	2.3412	2.4779	2.9484	3.5199	4.9178	5.1679	–
2	IFSM	3.0920	3.1488	3.3213	3.8959	5.1868	5.6371	0.38%
	FEA	3.0948	3.1507	3.3238	3.9024	5.2070	5.6469	
2.5	IFSM	3.5476	3.9183	3.9522	4.4037	5.4720	5.9302	–
3	IFSM	3.9100	4.6926	4.7092	5.0037	5.8607	6.2119	–

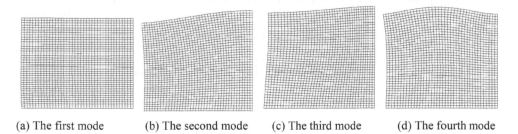

(a) The first mode   (b) The second mode   (c) The third mode   (d) The fourth mode

Figure 2. The first eight mode shapes for the ECEF rectangular plate.

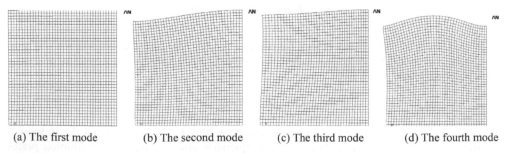

(a) The first mode   (b) The second mode   (c) The third mode   (d) The fourth mode

Figure 3. The first eight mode shapes for the ECEF rectangular plate (FEA).

6 CONCLUSIONS

Due to the couple effect between the longitudinal displacement and the shear displacement, the traditional series method cannot solve the in-plane problem effectively. To solve this problem, an improved Fourier series method is presented for the in-plane vibration analysis of rectangular plates. The in-plane displacement fields are expressed the linear combination of a double Fourier series and auxiliary series functions. The general boundary conditions can be represented by two sets of linear springs along each edge, and the matrix eigenvalue equation of the plate with arbitrary boundary conditions can be derived using the Rayleigh–Ritz method. The modal analysis is carried out to validate the accuracy and convergence characteristic of the current approach.

REFERENCES

[1] N.S. Bardell, R.S. Langley, J.M. Dunsdon. 1996. On the free in-plane vibration of isotropic rectangular plates. *Journal of Sound and Vibration* 191(3): 459–467.
[2] B. Liu, Y.F. Xing. 2011. Exact solutions for free in-plane vibrations of rectangular plates. *Acta Mechanica Sinica* 24(6): 556–567.
[3] D.J. Gorman. 2005. Free in-plane vibration analysis of rectangular plates with elastic support normal to the boundaries. *Journal of Sound and Vibration* 285(4–5): 941–966.
[4] D.J. Gorman. 2009. Accurate in-plane vibration analysis of rectangular orthotropic plates. *Journal of Sound and Vibration* 323(1–2): 426–443.
[5] K. Hyde, J.Y. Chang, C. Bacca, et al. 2001. Parameter studies for plane stress in-plane vibration of rectangular plates. *Journal of Sound and Vibration* 247(3): 471–487.
[6] J. Seok, H.F. Tiersten, H.A. Scarton. 2004. Free vibrations of rectangular cantilever plates. Part 2: in-plane motion. *Journal of Sound and Vibration* 271(1–2): 147–158.
[7] Y.F. Xing, B. Liu. 2009. Exact solutions for the free in-plane vibrations of rectangular plates. *International Journal of Mechanical Sciences* 51(3): 246–255.
[8] B. Liu, Y.F. Xing. 2011. Exact solutions for free in-plane vibrations of rectangular plates. *Acta Mechanica Sinica* 24(6): 556–567.
[9] Lorenzo Dozio. 2010. Free in-plane vibration analysis of rectangular plates with arbitrary elastic boundaries. *Mechanics Research Communications* 37: 627–635.
[10] Lorenzo Dozio. 2011. In-plane free vibrations of single-layer and symmetrically laminated rectangular composite plates. *Composite Structures* 93(7): 1787–1800.

Advances in Engineering Materials and Applied Mechanics – Zhang, Gao & Xu (Eds)
© *2016 Taylor & Francis Group, London, ISBN 978-1-138-02834-0*

Isogeometric Analysis for nonlocal elasticity problems

T. Wang, X. Wang, Y. Hu & Z. Wan
School of Mechanical Science and Engineering, Huangzhong University of Science and Technology, Wuhan, P.R. China

ABSTRACT: Based on the concept of the Isogeometric Analysis (IGA), the numerical solution of the 2D nonlocal elasticity problems is presented in this paper. In the numerical solution of the nonlocal material behavior described by the Eringen-type nonlocal model, a fine discretization is required when the stain field solution exhibits high gradients at boundary layers. The proposed nonlocal IGA approach provides an exact representation of the geometry and an efficient local improvement process by the reparameterization and refinement operations. Due to employing the higher order continuity of the NURBS-based shape functions and the superior properties during mesh refinement, the nonlocal IGA approach can improve the solution accuracy of the boundary layers.

1 INTRODUCTION

The nonlocal elasticity theory has received a considerable attention by the researchers intending to analyze the size-dependent structural materials. The key idea of the nonlocal theory is to use a continuum approach endowed with information regarding to the behavior of the material microstructure. Kröner (Kröner 1967) formulated a continuum theory for elastic materials with long range cohesive forces. A nonlocal elasticity theory for linear homogeneous isotropic continua was developed by Eringen and co-workers (Eringen et al. 1977, Eringen 1978, Eringen & Kim 1974). Polizzetto (Polizzotto 2001) proposed a FEM-based technique for the nonlocal elasticity of integral-type. Pisano et al. (Pisano et al. 2009, Pisano et al. 2009) developed a nonlocal FE model to study the homogeneous and non-homogeneous two-dimensional nonlocal elasticity problems. However, in the nonlocal integral elasticity the strain field solution exhibits high gradients in the thin boundary layer, these traditional FE approaches were usually linear approximation of the strain, which can lead to inaccuracies in the region close to the boundaries.

In this paper, a new numerical approach is employed for solving the nonlocal elastic mechanical problems by means of the isogeometric analysis (IGA) method. The conception of IGA has been introduced by Hughes et al. (Hughes et al. 2005). More details can be referred to references (Bazilevs et al. 2012, Bazilevs et al. 2006, Piegl & Tiller 1996). Main features of the IGA approach are the geometry generation step and the *k*-refinement process in which the polynomial order and the continuity of the basis functions are likewise increased. With these features, IGA has several advantages over the traditional FE analysis such as geometric exactness and simple refinements due to the use of NURBS. Using the *k*-refinement method of NURBS mesh, approximation of the geometry and analysis field can be achieved higher order and higher continuity. In the numerical implement of the nonlocal material behavior described by the Eringen-type nonlocal model, the solution of boundary value problems is intrinsically prone to exhibit boundary layers in the strain field (Pisano et al. 2009). A fine discretization is required in a region close to the boundaries. By employing the NURBS-based shape functions owning higher order continuity, the accuracy of the solutions at capturing of the stain profile at the boundary layers can be improved.

2 IGA METHOD FOR NONLOCAL ELASTICITY

The domain V is occupied by the material of a continuous and homogeneous, nonlocal elastic solid body in its undeformed state. The body is subjected to external actions as volume forces $\overline{b}(\mathbf{x})$ in V, surface forces $\overline{t}(\mathbf{x})$ on the portion S_f of the boundary surface ∂V, and imposed displacements $\overline{u}(\mathbf{x})$ on the constrained portion S_c of ∂V. The above loads act in a quasi-static manner and infinitesimal displacement are assumed. The Eringen-type nonlocal boundary value problem is set up in the following forms:

$$div\,\sigma(\mathbf{x}) + \overline{b}(\mathbf{x}) = 0 \quad in\,V \tag{1}$$

$$\varepsilon(\mathbf{x}) = \nabla^s u(\mathbf{x}) \quad in\,V \tag{2}$$

$$\sigma(\mathbf{x}) = \mathbf{D} : \hat{\varepsilon}(\mathbf{x}) = \alpha \mathbf{D} : \varepsilon(\mathbf{x}) + \beta \int_V A(\mathbf{x},\mathbf{x}')\mathbf{D} : \varepsilon(\mathbf{x}')dV' \quad in\,V \tag{3}$$

$$u(\mathbf{x}) = \overline{u}(\mathbf{x}) \quad on\,S_c \tag{4}$$

$$\sigma(\mathbf{x}) \cdot n(\mathbf{x}) = \overline{t}(\mathbf{x}) \quad on\,S_f \tag{5}$$

where, the Equation (1) expresses the field equilibrium; Equation (2) is the field compatibility equation; Equation (3) is the nonlocal elasticity constitutive law. Equations (4) (5) are the boundary equilibrium conditions and the boundary compatibility conditions, respectively. And $n(\mathbf{x})$ is the unit normal external vector to ∂V; $\partial V = S_c \bigcup S_f$ and $S_c \bigcup S_f = 0$. No imposed thermal-like strains are considered for simplicity.

Then, the nonlocal total potential energy functional can be given in the following form:

$$\Pi(\mathbf{u}(\mathbf{x})) = \frac{1}{2}\alpha \int_V \nabla^s \mathbf{u}(\mathbf{x}) : \mathbf{D} :\nabla^s \mathbf{u}(\mathbf{x})dV$$

$$+ \frac{1}{2}\beta \int_V \int_V A(\mathbf{x},\mathbf{x}')\nabla^s \mathbf{u}(\mathbf{x}) : \mathbf{D} : \nabla^s \mathbf{u}(\mathbf{x}')dV'dV$$

$$- \int_V \overline{b}(\mathbf{x}) \cdot \mathbf{u}(\mathbf{x})dV - \int_{S_f} \overline{t}(\mathbf{x}) \cdot \mathbf{u}(\mathbf{x})dS \tag{6}$$

The solution of the nonlocal elastic problem can be obtained by corresponding to a minimum of Equation (6). Similar to other numerical methods, the IGA method also requires the approximation of the unknown functions, as well as discretization of the domain of interest. Then, the volume V is discretized into n elements by the instinctive mesh of the NURBS. Hence, the discretized unknown displacement field $\mathbf{u}_i(\mathbf{x})$ of the i-th element is now given by:

$$\mathbf{u}_i(\mathbf{x}) = \mathbf{N}_i(\mathbf{x})\mathbf{d}_i, \quad i = 1, 2, ..., n \tag{7}$$

where $\mathbf{N}_i(\mathbf{x})$ is a matrix collecting the rational terms in Equation (7) and $\mathbf{d}_i$ is the vector of the unknown control point displacement. Also the related strain field $\varepsilon_i(\mathbf{x})$ is represented as:

$$\varepsilon_i(\mathbf{x}) = \nabla^s \mathbf{u}_i(\mathbf{x}) = \nabla^s \mathbf{N}_i \mathbf{d}_i = \mathbf{B}_i(\mathbf{x})\mathbf{d}_i \tag{8}$$

where $\mathbf{B}_i(\mathbf{x})$ is a matrix containing the shape functions partial derivatives.

Substituting Equations (7) and (8) into Equation (6) gets:

$$\Pi = \frac{1}{2}\alpha \sum_{i=1}^{n}\mathbf{d}_i^T \left(\int_{V_i} \mathbf{B}_i^T(\mathbf{x})D\mathbf{B}_i dV\right)\mathbf{d}_i + \frac{1}{2}\beta\sum_{i=1}^{n}\sum_{j=1}^{n}\mathbf{d}_i^T \left(\int_{V_i}\int_{V_j} A(x,x')\mathbf{B}_i^T(\mathbf{x})D\mathbf{B}_j(\mathbf{x}')dV'dV\right)\mathbf{d}_j$$

$$-\sum_{i=1}^{n}\mathbf{d}_i^T \left(\int_{V_i} \mathbf{N}_i^T(\mathbf{x})\overline{b}(\mathbf{x})dV + \int_{S_{f(i)}} \mathbf{N}_i^T(\mathbf{x})\overline{t}(\mathbf{x})dS\right) \tag{9}$$

where $S_{f(i)} := S_f \bigcap \partial V_i$. Then, the total potential energy associated with the element V_i can be

192

$$\Pi_i = \frac{1}{2}(\alpha \mathbf{d}_i^T \mathbf{k}_i^{loc} \mathbf{d}_i + \beta \mathbf{d}_i^T \mathbf{k}_{ij}^{nonloc} \mathbf{d}_j) - \mathbf{d}_i^T \mathbf{f}_i \qquad (10)$$

with

$$\mathbf{k}_i^{loc} := \int_{V_i} \mathbf{B}_i^T(\mathbf{x}) D \mathbf{B}_i(\mathbf{x}) dV \qquad (11)$$

$$\mathbf{k}_{ij}^{nonloc} := \int_{V_i} \int_{V_j} A(\mathbf{x}, \mathbf{x}') \mathbf{B}_i^T(x) D \mathbf{B}_j(\mathbf{x}') dV' dV \qquad (12)$$

$$\mathbf{f}_i := \int_{V_i} \mathbf{N}_i^T(\mathbf{x}) \overline{\mathbf{b}}(\mathbf{x}) dV + \int_{S_{f(i)}} \mathbf{N}_i^T(\mathbf{x}) \overline{\mathbf{t}}(\mathbf{x}) dS \qquad (13)$$

where $\mathbf{k}_i^{loc}$ is the classical local stiffness matrix. $\mathbf{k}_{ij}^{nonloc}$ is the non-local stiffness matrix. $\mathbf{f}_i$ is the external load vector. Then, the elastic equilibrium problem involving the solution of the control nodal displacement vector $\mathbf{d}$ of the NURBS mesh grid is obtained by the following condition:

$$\delta \Pi = \frac{\delta \Pi}{\partial \mathbf{d}} \delta \mathbf{d} \quad \forall \delta \mathbf{d} \Rightarrow \mathbf{Kd} - \mathbf{f} = 0 \qquad (14)$$

where the $\mathbf{K}$ represents the total stiffness matrix of the considered elastic solid as:

$$\mathbf{K} = \alpha \mathbf{K}^{loc} + \beta \mathbf{K}^{nonloc} \qquad (15)$$

where

$$\mathbf{K}^{loc} = \sum_{i-1}^{n} k_i^{loc}, \quad \mathbf{K}^{nonloc} = \sum_{i=1}^{n} \sum_{j=1}^{n} k_{ij}^{nonloc} \qquad (16)$$

It is worthy noticing that there are two contributions to the total stiffness matrix, the local terms $\mathbf{K}^{loc}$ and the nonlocal terms $\mathbf{K}^{nonloc}$, so that the band of the $\mathbf{K}$ is much larger than the traditional stiffness matrix only contains the local terms. And in this paper, the long-range influence is considered just when the element centroid locates in the influence circle for higher computational efficiency (Farin 1995). In the sake of increasing the accuracy of the solution at the boundary layers, the k-refinement method of the IGA is employed.

3 NUMERICAL EXAMPLE

An elastic bar under uniform tension is to validate the numerical solutions by comparing with the closed solution available in Ref. (Abdollahi & Boroomand 2013). The Eringen-type model is adopted for the material and with a bi-exponential attenuation function of the form $A(\mathbf{x}, \mathbf{x}') := \lambda_0 e^{-|x'-x|/l}$ (with $\lambda_0 = 1/(2\pi l^2 t)$ and $L_R = 6l$). The positive function $A(\mathbf{x}, \mathbf{x}')$ is the attenuation function related to the nonlocality effects of the (local) strain, it decays more or less rapidly with increasing distance $r = |\mathbf{x} - \mathbf{x}'|$. In practice $A(\mathbf{x}, \mathbf{x}') \approx 0$ for $r \geq L_R$, where L_R is the influence distance. NURBS basis function is used in the IGA method. Consider an elastic bar of length $L = 50$ cm, with uniform cross-section A = 0.1 cm^2 (width $h = 1$ cm and thickness $t = 0.1$ cm) and Young's modulus $E = 2.1 \times 10^6$ $daNcm^{-2}$, Poisson ratio $\upsilon = 0.3$, the internal length $l = 0.1$ cm and $\alpha = \beta = 0.5$. The bar is loaded with a self-equilibrated couple of external loads $F = 210 N$ and depicted in Figure 1. The quadratic, cubic and quartic NURBS basis functions are used to analyze the problem respectively for comparison. In this paper, we solve the problem by IGA as a 2D problem and set to $l = 0.1$ cm and $\alpha = \beta = 0.5$.

Figure 2 shows the strain distribution $\varepsilon_x(x, y)$ computed for the spline's order p = 3 with the continuity $C^2(c = 2)$, at $y = 0$, 2.5 and 5 mm. The solutions at $y = 0$ mm are most close to the exact solutions.

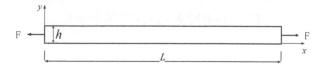

Figure 1. Geometry, loading and boundary conditions, and material data for nonlocal elastic bar.

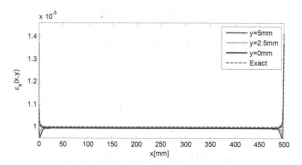

Figure 2. The comparison of the strain distributions $\varepsilon_x(x, y)$.

Table 1. The relative errors related to the boundary layers and the core domain.

$y = 0$ mm		Errors	$y = 0$ mm		Errors
$x = 0$ mm	$p = 2, c = 1$	2.65%	$x = 100$ mm	$p = 2, c = 1$	0.53%
	$p = 3, c = 2$	1.86%		$p = 3, c = 2$	0.46%
	$p = 4, c = 3$	1.21%		$p = 4, c = 3$	0.31%
$x = 1$ mm	$p = 2, c = 1$	2.37%	$x = 250$ mm	$p = 2, c = 1$	0.53%
	$p = 3, c = 2$	1.97%		$p = 3, c = 2$	0.46%
	$p = 4, c = 3$	1.13%		$p = 4, c = 3$	0.31%
$x = 2$ mm	$p = 2, c = 1$	3.24%			
	$p = 3, c = 2$	2.38%			
	$p = 4, c = 3$	1.64%			

Table 1 shows the relative errors of several stain solutions of the boundary layers and the middle domain along $y = 0$ mm. Compared with the closed solution, the relative errors are very small in the whole bar, and it almost close to zero in the middle domain. Both at the middle domain and each end portions, the error is reduced as the order and continuity of the NURBS basis function is increased. It is verified that the proposed method can provide an accurate solution for the nonlocal elasticity problem.

We further perform a convergence study and solve the problem with different polynomial orders of the shape function with various continuities. Figure 3 shows the solutions of the strain distribution ε_x with the ordinate $y = 5$ mm. These results are calculated with C^1-continuous shape functions of polynomial degree p = 2, C^2-continuous shape functions of polynomial degree p = 3, C^3-contious shape functions of polynomial degree p = 4, respectively. By observing the results plotted in Figure 3, the computed strain distributions with various polynomial degrees have little difference at the core domain. At the end portions with the width L_R of the bar, as shown in Figure 3(b), the computed strain is increased with elevation of polynomial degrees as expected. With the continuity increasing, the stain value derived at the boundary layer is also increased and the error value is lower for a higher polynomial degree p. The IGA approach provide an efficient local improvement process by the reparameterization and refinement operations to overcome the difficulties in capturing the high stain gradients at the boundary layers. Therefore, the solution accuracy of the boundary layers is improved well due to the higher order continuity of the NURBS-based shape functions.

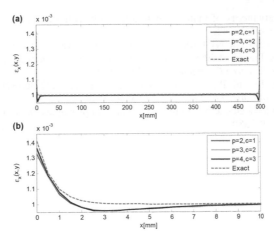

Figure 3. The comparison of the strain distribution $\varepsilon_x (x,y)$ at $y = 5$ mm: (a) the stain distribution along the bar length; (b) the details of the strain profiles in the end portion.

4 CONCLUSIONS

A NURBS-based IGA method for the nonlocal elasticity problems is implemented in this paper. In the proposed method, the Eringen-type model of nonlocal elastic theory is adopted, in which the stress-strain constitutive relation is characterized by a convolution integral form. By employing the NURBS-based shape function with higher order continuity, the accuracy of the solutions at the boundary zones of the nonlocal elasticity model improved. An example proves that the proposed method is an effective tool for numerical solution of the nonlocal elasticity problems.

ACKNOWLEDGMENTS

This research was supported by the National Natural Science Foundation of China (Grant nos. 51375184 and 51375182).

REFERENCES

Abdollahi, R. & Boroomand, B. 2013 Benchmarks in nonlocal elasticity defined by Eringen's integral model. *International Journal of Solids and Structures* 50: 2758–2771.

Bazilevs, Y., Beirao Da Veiga, L. & Cottrell, J.A., et al. 2006 Isogeometric analysis: approximation, stability and error estimates for h-refined meshes. *Mathematical Models and Methods in Applied Sciences* 16: 1031–1090.

Bazilevs, Y., Hsu, M. & Scott, M.A. 2012 Isogeometric fluid–structure interaction analysis with emphasis on non-matching discretizations, and with application to wind turbines. *Computer Methods in Applied Mechanics and Engineering* 249: 28–41.

Eringen, A.C. 1978 Line crack subject to shear. *International Journal of Fracture* 14: 367–379.

Eringen, A.C. & Kim, B.S. 1974 Stress concentration at the tip of crack. *Mechanics Research Communications* 1: 233–237.

Eringen, A.C., Speziale, C.G. & Kim, B.S. 1977 Crack-tip problem in non-local elasticity. *Journal of the Mechanics and Physics of Solids* 25: 339–355.

Farin, G.E. 1995 *NURB curves and surfaces: from projective geometry to practical use*: AK Peters, Ltd.

Hughes, T.J., Cottrell, J.A. & Bazilevs, Y. 2005 Isogeometric analysis: CAD, finite elements, NURBS, exact geometry and mesh refinement. *Computer Methods in Applied Mechanics and Engineering* 194: 4135–4195.

Kröner, E. 1967 Elasticity theory of materials with long range cohesive forces. *International Journal of Solids and Structures* 3: 731–742.

Piegl, L.A. & Tiller, W. 1996 The NURBS book (Monographs in visual communication).

Pisano, A.A., Sofi, A. & Fuschi, P. 2009 Finite element solutions for nonhomogeneous nonlocal elastic problems. *Mechanics Research Communications* 36: 755–761.

Pisano, A.A., Sofi, A. & Fuschi, P. 2009 Nonlocal integral elasticity: 2D finite element based solutions. *International Journal of Solids and Structures* 46: 3836–3849.

Polizzotto, C. 2001 Nonlocal elasticity and related variational principles. *International Journal of Solids and Structures* 38: 7359–7380.

Advances in Engineering Materials and Applied Mechanics – Zhang, Gao & Xu (Eds)
© 2016 Taylor & Francis Group, London, ISBN 978-1-138-02834-0

Filtering and de-noising of the rolling bearing vibration signal with EMD

X.C. Wang & Y. Xuan

College of Information Engineering, Inner Mongolia University of Science and Technology, Baotou, China

ABSTRACT: The principle and process of de-noising the rolling bearing vibration signal filtering by the algorithm of empirical mode (EMD) was studied in this paper. In this study, the acceleration sensor by the combination of hardware and software and the core of TMS320F28335 was used to acquire the vibration signal and then to transmit it to PC through a serial port communication interface for the display and analysis of result. The study results show that, if the method of de-noising signal filtering based on EMD is eliminating the noise selectively according to the spectrum character of the results of the signal decomposition, it will guarantee the inherent nature of the signal to show good adaptability, flexibility and feasibility.

1 INTRODUCTION

The rolling bearing vibration signal de-noising in signal processing field has been a concerned research. In the process of signal acquisition and transmission, there is no way to avoid the interference noise, for subsequent analysis process, which can have a big impact on getting useful information, hence, how to remove the noise, is a very important task.

2 A STATISTICAL MODEL OF NOISE

Random signal $x(t) = s(t) + n(t)(0 \leq t \leq T)$'s statistical characteristic depends largely on the interference noise $n(t)$'s nature. If in the process of the noise $n(t)$, for any $N \geq 1$ and all the moment $t_k(k = 1, 2, \ldots N)$, random variable $n(n_k, t_k)$ obeys the Gauss distribution. Hence, the random variable $n(n_k, t_k)$ is a Gauss noise, and the probability density function of Gauss, which can be written as:

$$p(n_k, t_k) = \left(\frac{1}{2\pi\delta_{n_k}^2} \right)^{1/2} \exp\left[-\frac{(n_k - \mu_{n_k})^2}{2\delta_{n_k}^2} \right] \tag{1}$$

In the formula, μ_{n_k}, $\mu_n(t_k)$ and $\delta_{n_k}^2$, $\delta_{n_k}^2(t_k)$ respectively represent the random variable $n(t_k)$'s mean value and variance value, in other words, is the time domain of white noise, and its frequency domain representation is evenly distributed in all the frequency axis with the power spectral density, can be written as:

$$P_n(\omega) = \frac{N_0}{2}(-\infty < \omega < +\infty) \tag{2}$$

We can conclude that white noise power spectral density in frequency domain is in the positive and negative half axis [1].

3 THE EMD FILTER CHARACTERISTICS

The statistical characteristic of white noise in the EMD algorithm is decomposed by EMD to get the IMF component can meet the normal distribution under any of the white noise, and the IMF energy density with its corresponding average cycle is a constant, which can be written as:

$$E_i \overline{T}_i = const \tag{3}$$

The average cycle of IMF is about two times the previous cycle of IMF component (formula 3.2), which can be written as:

$$\overline{T}_i = 2\overline{T}_{i-1} \tag{4}$$

where $E_i = 1/N \sum_{j=1}^{N} [imf_i(j)]$ is the first few IMF energy density, N is the data length, $\overline{T}_i$ is the first few IMF average cycle, $\overline{T}_i = N/N_{max}$. N_{max} is the first few IMF component's maximum numbers[2, 3].

EMD algorithm is from high to low frequency vibration signal decomposed into several IMF components, according to this characteristic; we can put the IMF component integrate to build a new filter. (1) If remove a number of low frequency IMF component, the rest of the IMF component reconstruction, is the equivalent of a high-pass filter; (2) If remove a number of high frequency IMF components, the rest of the IMF component reconstruction, is the equivalent of a low-pass filter; (3) If remove a number of low frequency and high frequency IMF components at the same time, the rest of the IMF component reconstruction, is the equivalent of a band-pass filter; (4) If remove the middle of several IMF components, the rest of the IMF component reconstruction, is the equivalent of a band-stop filter.

The EMD based de-noising has mainly the vibration signal affected by noise, its energy mainly concentrates on the low frequency band, hence, any noise of IMF decomposition components, there must be some IMF_k that can regard as the leading status for the subsequence IMF_{k+1} component, while the IMF_k before the component is the noise which occupying the leading status, so, the purpose of EMD de-noising is to look for IMF_k.[4,5,6]

4 VIBRATION SIGNAL ACQUISITION SYSTEMS

Acquisition system uses C2000 series TMS320F28335 of TI Company as the main control chip, and the sensor selected for acceleration sensor is LIS344ALH. The TMS320F28335 can be carried out simultaneously in 16 channels to data acquisition and signal processing, the vibration signals are transmitted to PC by serial bus, at the same time, and the vibration data can be stored. Figure 1 express the flowchart of signal acquisition system.

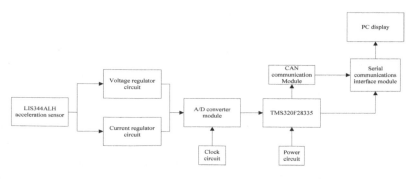

Figure 1.   The hardware block diagram of signal acquisition system.

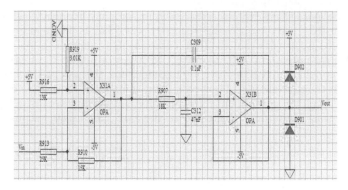

Figure 2. The voltage regulator circuit.

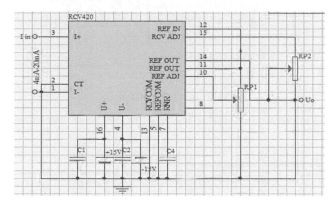

Figure 3. The current regulator circuit.

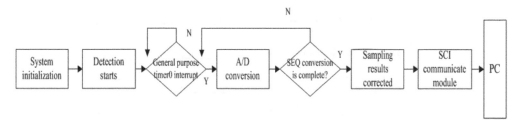

Figure 4. The flow chart of the software.

Because A/D input voltage from 0 to 3V, the output of the acceleration sensor is both voltage and current signals, so, a circuit is needed to put the atypical quantity into the amount received by the DSP. As shown in Figures 2 and 3.

The software parts, using the timer 1 interrupt to trigger the A/D conversion, when the conversion is completed, the vibration data by SCI module pass to the PC. As shown in Figure 4.

5 ROLLING BEARING VIBRATION SIGNAL DE-NOISING EXPERIMENT ANALYSIS

In order to analyze the EMD can effectively eliminate the noise of mechanical fault vibration signals, we use the signal acquisition system to obtain the vibration signal of rolling bearing inner ring, its basic parameters express: speed 600 rpm, sampling frequency 1 kHz, the length of intercepted signal 1024, and the operating frequency is 24.5 Hz, as shown in Figures 5 and 6.

From the results we can see that the IMF components from high to low frequency array reflect the multi resolution and self adaptive EMD. According to the EMD filter characteristic, we can conclude that if remove *imf*1 component, the rest of the IMF component reconstruction, is the equivalent of a low-pass filter; similarly, if remove *imf*6 ~ *imf*7 components, the rest of the IMF component reconstruction, is the equivalent of a high-pass filter; if reserve *imf*4 ~ *imf*5 components, the rest of the IMF component reconstruction, is the

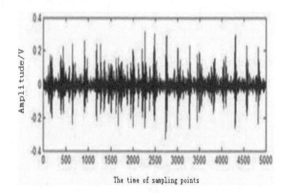

Figure 5.　The vibration signal of rolling bearing inner ring.

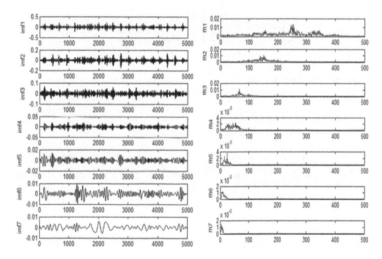

Figure 6.　EMD decomposition and its spectrum.

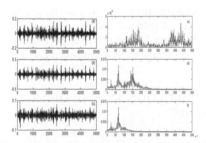

Figure 7.　EMD filtering and its spectrum.

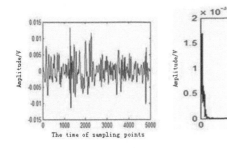

Figure 8.   EMD de-noising and its spectrum.

equivalent of a band-pass filter. As shown in Figure 7 (a)~(c), spectrum corresponding as shown in Figure 7 (d)~(f). If you want to analyze the natural vibration frequency of structure, just need *imf* 6 ~ *imf* 7 components for construction, can be obtained the results which after the de-noising, as shown in Figure 8.

6   CONCLUSION

Rolling bearing is the key part of the rotating machinery, is one of the highest parts of equipment failure. In the operation to acquire bearing signal with noise, put EMD filtering, denosing technique into the vibration of bearing processing can retain the inherent nonlinear signal and non-stationary characteristics, and achieve filtering, de-noising effect.

REFERENCES

Cohen, L. 1989, Time-frequency distributions—a review. *Proceedings of the IEEE*, 77(7):941–981.
Harris, S.L. 1958, Dynamic loads on the teeth of spur gears. *Proceedings of the Institution of Mechanical Engineers*, 172:87–112.
Lin, J., Zuo, M.J, 2003, Gearbox fault diagnosis using adaptive wavelet filter. *Mechanical Systems and Signal Processing*, 17(6):1259–1269.
Loutridis, S.J. 2004, Damage detection in gear systems using empirical mode decomposition. *Engineering Structure*, 26:1833–1841.
Mcfadden, P.D., Toozhy, M.M. 2000, Application of synchronous averaging to vibration monitoring of rolling element bearings. *Mechanical Systems and Signal Processing*, 14(6):891–906.
Yen, G.G., Lin, K.C. 2000, Wavelet packet feature extraction for vibration monitoring. *IEEE Transactions on Electronics*, 47(3):650–667.

Advances in Engineering Materials and Applied Mechanics – Zhang, Gao & Xu (Eds)
© 2016 Taylor & Francis Group, London, ISBN 978-1-138-02834-0

Effects of rotor–rotor interactions in a counter-rotating fan

H.M. Zhang, X.Q. Huang & X. Zhang
School of Power and Energy, Northwestern Polytechnical University, Xi'an, China

ABSTRACT: Taking a two-stage counter-rotating fan as the test case, the rotor–rotor interactions have been studied. A steady simulation and unsteady numerical simulation based on the non-linear harmonic method have been presented and discussed. The comparison between the steady and unsteady flow field shows that upstream wake is the main influencing factor in the rotor–rotor interactions effects. Specifically, the unsteady effects reduce the efficiency of the 2nd blade row above midspan, which is attributed to the interactions of wake-shock, wake-boundary layer, and wake-passage vortex. In addition, the efficiency of the 2nd row below midspan is increased by the wake–boundary interaction which could enhance the exchange of energy between boundary layer and mainstream, add energy to the boundary layer and delay the hub corner separation.

1 INTRODUCTION

The higher and higher demand for the performance of aircraft have brought great challenges to the design of aeroengine, which is required to be more and more compact. One of the feasibility and valid solutions is the counter-rotating technology, which simplifies the construction by canceling the stators and improves the thrust-weight ratio. Recently, more and more researchers have concentrated their attention to the counter-rotating compressor/fan. For instance, Freedman designed a three-stage counter-rotating compressor[1]. The total pressure ratio of this compressor has reached to 27, which is far higher than conventional axial compressor. Funded by European Union, VITAL Project was kicked off in 2005. As the work package leader, SNECMA developed a counter rotating low-speed fan-concept for a high bypass ratio engine[2]. And by taking this Counter-Rotating Turbo Fan (CRTF) as the test case, Eberhard[3] compared the measured data with high resolution CFD results. Also, the design and optimization of CRTF has been discussed. The results showed that the steep design goals have been meet. As a new revolutionary engine technology, CRTF has a lot advantages for economy and aerodynamics. Also, Liu Bo[4] and Ji Lu-Cheng[5] carried out some researches on the CRTF in aerodynamic design and optimization. These studies show the huge potential for the high performance and low cost requirements of the aviation industry in future. However, little research conducted on the unsteady effects of counter-rotating compressor/fan, which have a marked impact on the secondary flow, tip clearance flow, flow separation, etc. And, due to the absence of the stators, the relative rotation speed between two counter rotating rows is much higher than the conventional compressor/fan, which leads to the stronger interactions between blade rows. So, the investigation on unsteady interactions among rows is of great engineering significance.

This paper is focused on the unsteady effects in a two-stage counter-rotating fan by means of numerical simulation based on the Non-Linear Harmonic method (NLH). And the steady and NLH results have been compared and analyzed. The effects of upstream wake on downstream blade along span are discussed in detail.

2 TEST CASE AND SOLUTION METHOD

2.1 *Non-linear harmonic method*

The model equation of NLH is derived from the unsteady Reynolds Averaged Navier-Stokes equations (U-RANS). The integrated U-RANS equations could be expressed as follows:

$$\frac{\partial}{\partial t}\iiint U dV + \oiint (F - F_v)\cdot \vec{n}ds = \iiint SdV \tag{1}$$

where, U are the conservative variables, F and F_v are respectively the convective terms and viscous terms, and S are the source term. Non-Linear Harmonic (NLH) method is a reduced order model, which decomposes the conservative variables into a time-averaged value and a sum of periodic perturbations:

$$U = \overline{U} + U' \tag{2}$$

Substituting Eq. (2) into Eq. (1) and time averaging them. During the process of derivation, the laminar and turbulence viscosity coefficients are assumed to be frozen. The resultant equations are given as follows:

$$\oiint (\overline{F} - \overline{F_v})\cdot \vec{n}dA = \iiint \overline{S}dV \tag{3}$$

Eq. (3) contains the extra unsteady stress terms that are similar to the turbulence stress terms. So for the closure of equations, other equations should be adopted in this system of equations. The unsteady perturbations equations can be obtained by the difference between Eq. (3) and Eq. (1) under the assumption that the unsteady stress terms are dominated by the first-order terms, which gives:

$$\frac{\partial}{\partial t}\iiint U'dV + \oiint (F' - F_v')\cdot \vec{n}dA = \iiint S'dV \tag{4}$$

where U' are the perturbation terms, which can be expressed in the form of harmonic due to the periodicity, as follows:

$$U' = \tilde{U}e^{-i\omega t} \tag{5}$$

Applying Eq. (5) in Eq. (4), the harmonic perturbation equations can be given:

$$\iiint -i\omega\tilde{U}dV + \oiint (\tilde{F} - \tilde{F}_v)\cdot \vec{n}dA = \iiint \tilde{S}dV \tag{6}$$

Eqs. (3) and (6) constitute the physical model of NLH. Introducing the pseudo-time derivative term, this system of equations can be solved by finite volume and time-marching method. The simultaneous equations of NLH method consists of 2*N (perturbation equations) and one (time-averaging equation). By employing the phase-lag boundary condition, the NLH unsteady calculation can be conducted on a single-passage mesh. As a result, the amount of unsteady computation is as 2 N+1 as that of steady.

The accuracy and stability of NLH method have been validated by Knapke[6]. By adopting the commercial software NUMECA, Knapke simulated the flow field of a counter-rotating aspirated compressor by NLH and conventional unsteady method. The comparison among the flow fields of NLH, unsteady and experiment shows that NLH can capture the unsteady rows interactions accurately. Based on the validation of Knapke, this paper applies the NLH method to simulate the unsteady flow field by the NUMECA software.

2.2 Numerical method

The central difference method with second-order accuracy and four-stage Runge-Kutta method has been applied in this paper. In order to reduce time cost, local time stepping, implicit residual smoothing and multi-grid techniques are employed to accelerate convergence. The one-equation S-A model is chosen as turbulence model to calculate turbulence viscosity coefficient, which is widely used in turbomachinery numerical simulation.

2.3 Test case

The counter-rotating fan is very small with a hub and casing radii of 20 mm and 75 mm respectively. The rate of flow and total pressure ratio at design point is 3.4 kg/s and 1.4. As for the boundary condition, the total pressure (101325 pa), total temperature (288.15 K) and flow angle (axial direction) were imposed at the inlet, while the static pressure was specified at the outlet.

Before simulating this fan with NLH, we should know the frequencies of unsteady disturbance. In multi-stage environment, the kth frequencies of jth row could be given by:

$$\omega_k^j = \sum_{i=1}^{NR} n_{k,i} B_i (BPF_i - BPF_j) \tag{7}$$

where BPF means the blade passing frequency and NR is the blade count. The $n_{k,i}$ represents the k sets of user-defined integers that drive the frequency combinations. The parameters about the frequencies could be found in Table 1.

The number of harmonic wave should be large enough to make sure sufficient accuracy. According to Hembera[7], the three-harmonic-wave simulation can capture the unsteady effects between rows well. In this paper, upstream and downstream unsteady disturbances have been represented by five harmonic wave frequencies each other, which guarantee the reliability of numerical simulation.

Because of the adoption of phase-lag condition, the single passage grid could be used in the numerical simulation. The O-4H mesh topology has been applied to generation of grid, with an O-type mesh around blade surface. The grid number for row 1 and row 2 are

Table 1. Parameters about the unsteady frequency.

	Rotation speed (RPM)	Blade count	BPF (Hz)
Row 1	−31831	12	530.5
Row 2	22000	16	366.7

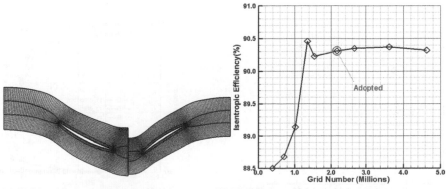

(a) Grid at blade-to-blade surface (b) Validation of mesh independence

Figure 1. Computational grid.

1,088,595 and 1,068,807 respectively with 97 grid points at radial direction, which is more enough to meet the requirement of grid independence. The blade-to-blade grid and validation of mesh independence are shown in Figure 1.

3 RESULTS AND ANALYSIS

By adopting the above-described mesh, the unsteady simulation based on non-linear harmonic method and steady simulation has been presented. Figure 2 compares the lift fan performances from unsteady and steady simulations at designed rotational speed. It is clear that, at choke operating point, the characteristic curve of NLH agrees with the steady one very well. However, near stall operating point, the total pressure ratio and efficiency from the NLH method are lower than that from steady computation. And be closer stall point, the difference between two methods becomes larger, which indicates that the unsteady effects and the circumferential non uniformity are more and more obvious with the operating point toward stall condition. Also, it is apparent that the stall margin of NLH (12.17%) is smaller than that of steady (13.18%). Above all, the unsteady effects have a negative impact on the performance of the lift fan.

The detailed and further analysis of this phenomenon could be conducted on the comparison of flow field between the steady and NLH at near stall point denoted by NS (as shown in Fig. 2), which are both given the same imposed static pressure at outlet boundary respectively. Figure 3 shows the distribution of efficiency of row 1 and row 2 along blade span at NS point.

The efficiency distributions of row 1 along span computed by NLH and steady are nearly uniform, while that of the second row (Row 2) have the apparent differences between two methods. This could be explained by the propagation characteristic of unsteady disturbance. According to the fluid mechanics, only one acoustic disturbance spreads upstream, while another acoustic disturbances, the entropy disturbance and vorticity disturbance convect downstream. As a result, the downstream of entropy disturbance of Row 2 changes more in efficiency.

A closer observation reveals that the unsteady effects make the efficiency at 55%~95% span reduced and the efficiency at 15%~55% span increased. Considering the different flow phenomenon along blade span, the different flow mechanisms may exist in different flow field.

For further analysis, normalized viscous stress at walls is defined as below:

$$Cf = \frac{\tau}{\rho V^2/2} \tag{8}$$

where, τ is the viscous stress at the blade wall, ρ and V is the reference density and velocity. Figure 4 shows the Cf distribution on R2 suction surface at NS condition. The separation bubble caused by the interaction between shock wave and boundary layer can be easily found in these pictures (red dotted line). And the Cf value of NLH in this area is lower and the separation bubble area of NLH is larger, compared with that of steady. This phenomenon indicates that the

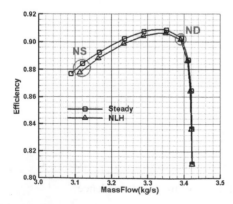

Figure 2. Lift fan characteristic.

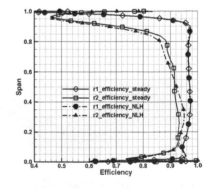

Figure 3. Distribution of efficiency along span at NS point.

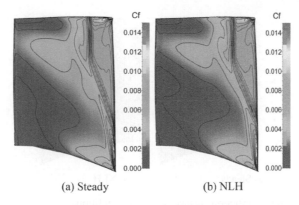

(a) Steady   (b) NLH

Figure 4.   *Cf* contour of R2 suction surface at NS.

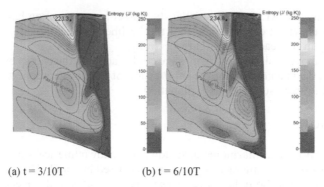

(a) t = 3/10T   (b) t = 6/10T

Figure 5.   Entropy instantaneous distribution on the suction surface R2 at NS condition.

separation bubble caused by the shock–boundary layer interaction is strengthened by upstream disturbances, which leads to the augment of loss. At the same time, two separation regions of NLH are larger than that of steady, namely, the unsteady disturbances widen the range of separation, which also results in efficiency drop. These increased loss are distributed on the blade surface. According to Montomoli[8], the boundary-layer perturbation due to the upstream wake should be the main reason. The upstream wake would cause the vorticity perturbation near blade surface, which promotes the energy exchange and also give rise to entropy production.

Figure 5 presents the instantaneous entropy distribution on the suction surface at t = 3/10 T and t = 6/10 T, where T is the time period. The upstream wakes have been marked in these pictures. Also, the passage vortexes could be found, which are surrounded by the red dotted lines in these pictures. From these pictures, the periodical interactions between upstream wakes and passage vortex are described. These intensive non-linear interactions lead to periodical formulation of circular high entropy regions, which indicate the loss increasing. And the centrifugal force caused by the rotation of blade makes the high entropy regions move toward the blade tip, which results in the reducing efficiency at the 55%~80% span. In addition, it is easy to find the interaction between wake and shock wave, and the entropy value after shock wave changes periodically. The entropy value (223.3) at 3/10T when the upstream wake is in front of the shock wave is lower than that (234.8) at 6/10T when the upstream wake acts on the shock wave, which manifests that the wake–shock interaction lead to the loss augment.

In addition, the unsteady perturbation plays a positive role in controlling the hub corner separation. Figure 6 illustrates the limit streamline on the suction surface of R2 at NS point. Three separation lines are marked by L1, L2 and L3 respectively, as shown in Figure 6. Evidently, the L1 line is caused by the shock–wave boundary interaction, which is corresponding to Figure 5. And the radial vortex results in separation line L2. Comparing with L2 of steady, L2 of NLH has a feeble displacement towards the blade trailing edge. As for separation line L3, it is leaded by the hub corner vortex. The comparison of separation line

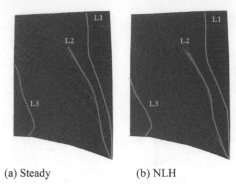

(a) Steady   (b) NLH

Figure 6.   Limit streamline on suction surface R2 at NS condition.

L3 between steady and NLH manifests that the hub corner separation has been suppressed by the unsteady effects. It can be found that the L3 line at radial direction expands to 55% span and 45% span by steady and NLH respectively.

Based on the above analysis, upstream wake could facilitate the energy exchange, which is considered to be the main reason for the suppression of hub corner separation. The research completed by Halstead[9] support this explanation.

4   CONCLUSIONS

In this paper, the non-linear harmonic unsteady and steady numerical simulations have been presented. The rotor–rotor interaction has been discussed in detail. The analysis shows that the interactions of wake–boundary layer, wake-passage vortex and wake-shock wave could increase entropy production and result in efficiency droop. Also, the wake–boundary layer interaction would intensify the energy exchange between mainstream and boundary layer, which supports to suppress the flow separation that is attributed to the lack of momentum and adverse pressure gradient. As a result, the hub corner separation is controlled. Therefore, it is possible and meaningful to study how to make use of upstream wakes.

REFERENCES

[1] Freedman J.H. 2000. Design of a Multi-Spool High-Speed Counter-Rotating Aspirated Compressor. *Massachusetts Institute of Technology*.
[2] Talbotec J. & Vernet M. 2010. Counter Rotating Fan Aerodynamic Design Logic & Tests Results. *ICAS 2010*. Nice, France.
[3] Eberhard N., Robert M., Bischoff A., Timea L.K. 2012. Design of an Economical Counter Rotating FanComparison of the Calculated and Measured Steady and Unsteady Results. *Proceedings of ASME Turbo Expo 2012, June 11–15, 2012*. Copenhagen, Denmark.
[4] Liu B., Chen Y.Y., Xiang X.R. Hou W.M. 2008. Experimental and Numerical Investigation of Dual Stage Counter-Rotating Compressor. *Journal of Propulsion Technology*, 29(4): 454–457.
[5] Ji L.C. 2007. Analysis for Technique Challenges on Counter-rotating Turbomachinery. *Journal of Propulsion Technology*, 28(1): 40–44.
[6] Knapke R.D. & Turner M.G. 2013. Unsteady Simulations of a Counter-rotating Aspirated Compressor. *Proceedings of ASME Turbo Expo 2013: Turbine Technical Conference and Exposition, June 3–7, 2013*. San Antonio, Texas, USA.
[7] Hembera M., Loos A., et al. 2009. Validation of the Non-linear Harmonic Approach for Quasi—unsteady Simulations in Turbomachinery. *Proceedings of ASME Turbo Expo 2009: Power for Land, Sea and Air, June 8–12, 2009*. Orlando, Florida, USA.
[8] Montomoli F., Naylor E., Hodson H.P. 2014. Unsteady Effects in Axial Compressors: A Multistage Simulation. *Journal of Propulsion and Power*, 29(5): 1001–1008.
[9] Halstead D.E., Wisier D.C., Okiishi T.H., et al. 1997. Boundary Layer Development in Axial Compressors and Turbines: Part 1 of 4-Composite Picture. *Journal of Turbomachinery*, 119(1): 114–127.

Advances in Engineering Materials and Applied Mechanics – Zhang, Gao & Xu (Eds)
© 2016 Taylor & Francis Group, London, ISBN 978-1-138-02834-0

Research on parameters influencing flutter characteristics of transonic fan blade

X. Zhang, X.Q. Huang, H.M. Zhang & Y.C. Chen
School of Power and Energy, Northwestern Polytechnical University, Xi'an, China

ABSTRACT: Three-dimensional Reynolds averaged Navier–Stokes equations are solved based on the harmonic balance method in frequency domain in order to numerically simulate unsteady flow around an oscillating blade, and the energy method is applied to calculate the aerodynamic damping coefficients in the one-way fluid-structure manner. The flutter characteristics of a transonic fan are predicted at design operation and near stall working point for 100% rotation speed line respectively. The flutter mechanisms is revealed by investigating the influence of Inter-Blade Phase Angle (IBPA), blade modes, shock wave and incidence flow angle on aeroelastic stability. The numerical results indicate that the unsteady shock wave disturbance on the pressure surface is a main inducement of aeroelastic instability and the other physical parameters have also varying degree of impact on the flutter characteristic.

1 INTRODUCTION

Flutter is a kind of aeroelastic instability phenomenon, which maybe causing blade failure[1]. The trend of pursuing high pressure ratio, low weight in aircraft propulsion systems needs a blade withstanding higher aerodynamic load and lower stiffness, so that this inevitably will lead to augmenting the probability of flutter. It should be noted that ensuring an aerodynamic performance while guaranteeing structural reliability by avoiding flutter is very urgent for designers. Therefore, more efforts are needed to gain an in-depth understanding into the mechanism of flutter.

The flutter involves two domains of structure and fluid around the blades in turbomachinery and its complexity leads to many inducements for instability in the aeroelastic field. Many researchers have applied a parametric method to investigate flutter mechanism. The effects of inter-blade phase angle were investigated by Xiaowei Zhang, Yanrong Wang and Kening Xu[2]. The study focusing on the influence of shock wave on aerodynamic damping was carried out by Isomura, Giles[3] and Srivastava[4]. The other studies have showed that tip clearance[5], flow separation[6] and blisks[7] having an impact on flutter characteristics.

The present work focuses on prediction of a transonic fan by considering the influence of different physical parameters including inter-blade phase angle, blade mode, shock wave and incidence angle. The variation of aerodynamic damping due to different values chosen of the key parameters is analyzed in detail to find the flutter mechanism in turbomachinery.

2 NUMERICAL APPROACH

2.1 Time domain harmonic balance technique

Integrating the Unsteady Reynolds-Averaged Navier-Stokes (U-RANS) equations over a control volume, we could obtain the following semi-discrete finite-volume form equations:

$$V\frac{\partial W}{\partial t} + R(W,s) = 0 \qquad (1)$$

where W is the conserved flow variables; V represents the volume of the cell; R is the residual and s is the velocity of the moving grid.

Since the solution W is periodic in time, we can represent it using a Fourier series approximately. Then the residual term R can be expressed in the similar form:

$$W(t) \approx \sum_{k=-N}^{N} \widehat{W}_k \, e^{i\omega_k t} \tag{2}$$

$$R(t) \approx \sum_{k=-N}^{N} \widehat{R}_k \, e^{i\omega_k t} \tag{3}$$

where $\widehat{W}_k$ and $\widehat{R}_k$ are the coefficients of the Fourier series for the disturbance angular frequency ω_k, N is the number of harmonics retained in the solution. Inserting the Eq. (2) and the Eq. (3) into Eq. (1) yields:

$$\sum_{k=-N}^{N} (i\omega_k V \widehat{W}_k + \widehat{R}_k) \, e^{i\omega_k t} = 0 \tag{4}$$

The formula (4) could be changed into 2N+1 equations in frequency domain by a harmonic balance. However, it is very difficult to solve it directly. By inverse discrete Fourier transformation, we could cast these equations back to a set of 2N+1 sub time levels in the time domain:

$$[F]^{-1}(iV[P]\widehat{W^*} + \widehat{R^*}) = 0 \tag{5}$$

$[F]^{-1}$ Inverse Discrete Fourier matrix (IDFT) read in the following form:

$$[F]^{-1} = \begin{bmatrix} e^{i\omega_{-N} t_0} & \cdots & e^{i\omega_0 t_0} & \cdots & e^{i\omega_N t_0} \\ \vdots & & \vdots & & \vdots \\ e^{i\omega_{-N} t_k} & \cdots & e^{i\omega_0 t_k} & \cdots & e^{i\omega_N t_k} \\ \vdots & & \vdots & & \vdots \\ e^{i\omega_{-N} t_{2N}} & \cdots & e^{i\omega_0 t_{2N}} & \cdots & e^{i\omega_N t_{2N}} \end{bmatrix} \tag{6}$$

So, the Fourier coefficients can therefore be computed by:

$$\widehat{W^*} = [F] W^* \tag{7}$$

$$\widehat{R^*} = [F] R^* \tag{8}$$

with

$$W^* = [W(t_0), \ldots, W(t_i), \ldots, W(t_{2N})]^T; \quad R^* = [R(t_0), \ldots, R(t_i), \ldots, R(t_{2N})]^T$$

Applying these above relationships, and finally Eq. (4) becomes:

$$iV[F]^{-1}[P][F]W^* + R^* = 0 \tag{9}$$

By introducing the pseudo-time derivative term to the above equations, we can solve these equations by a time marching method.

As for numerical calculation of inviscid flux, the central difference method with a second order and fourth order blended artificial viscosity is adopted. The hybrid 5–3 steps Runge–Kutta time marching method is applied for time integration. In addition, local time step, implicit residual smoothing and multi-grid techniques are adopted to accelerate convergence.

The Spalart–Allmaras turbulence model is employed to simulate turbulent flow in the strong coupled manner.

There are four types of boundary conditions applied into the numerical simulation. At the inflow boundary, total temperature, total pressure and incidence angle are used to specify the incoming flow values. For outgoing flow boundary, the back pressure needs to be specified. The non-reflecting boundary[8] has been applied at both inlet and outlet boundary. For viscous flow, the log-law is used to calculate the wall shear stress. In a steady flow simulation, a single-passage computational domain is typically employed with repeating boundary conditions. The phase lag periodic conditions are used at periodic boundary of single-passage based on the time and circumferential periodic assumption. Applying this condition in Eq. (2), we can obtain that:

$$\widehat{W}_k(\theta + \Delta\theta, t) = \widehat{W}_k(\theta, t) \cdot exp(i\beta_k) \tag{10}$$

where, β_k is the Inner-Blade Phase Angle (IBPA). Combining the Fourier transform Eq. (7), the periodic boundary condition in this paper could be given by:

$$W^*(\theta + \Delta\theta) = [F]^{-1}[M][F]W^*(\theta) \tag{11}$$

where

$$[M] = diag(-\beta_N, ..., \beta_0, ..., \beta_N).$$

2.2 Energy method

The energy method[9] is the most commonly used technique to evaluate the aeroelastic stability of blades by introducing the aerodynamic damping coefficient, which can be expressed as a non-dimensional parameter according to the concept of logarithmic decrement

$$LogDec = \frac{-W_{sum}}{2 \cdot E_{stain}} \tag{12}$$

where W_{sum} is the integration of aerodynamic work per cycle and E_{stain} represents the maximum kinetic energy. The positive damping coefficient indicates that flow damps the blade vibration, otherwise the flow amplifies the blade vibration.

3 RESULTS AND DISCUSSIONS

3.1 Steady flow simulation

The first row rotor of a two-stage fan is studied in the present work. The 3D solid modeling drawing of the objective row possessing 28 blades is shown in Figure 1. The designed rotation

 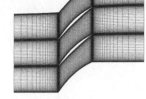 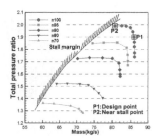

Figure 1.   Test fan overview.   Figure 2.   Numerical grid.   Figure 3.   Map of fan characteristics.

speed is 10720.0 r/min. The single-block H type grid used in the calculations of the flow field is shown in Figure 2 and the overall mesh dimensions are $56 \times 144 \times 64$ (pitchwise × axial × spanwise), with 2 cells used in the spanwise direction to resolve the 0.5 mm tip clearance.

The steady flow analysis was performed at the different rotation speed with the same inlet condition. Figure 3 shows the fan characteristic map from the CFD results. The red line represents 100% design speed line and only P1 and P2 were chosen for numerical flutter prediction among working points on this line. At the P1 condition the fan operates on the design point, and at P2 operation the total pressure ration is the highest, mass is the lowest, adjacent to the stall margin.

3.2 Structural dynamic analysis

Before processing flutter prediction, the structural dynamic analysis should be performed on the fan blade. The material of the blade is titanium alloy and the characteristic parameters are shown in Table 1. The modes of the rotor at 100% rotation speed have been determined via software ANSYS, which are illustrated in Figure 4. The first two modes represent the first bending (1F) and second bending (2F), respectively. The third mode represents the first torsion (1T).

3.3 Flutter characteristic analysis

Aerodynamic damping coefficients versus different nodal diameters corresponding to P1 and P2 operation are shown in Figure 5 and Figure 6, respectively. No matter what kind of working conditions and blade mode shapes, the aerodynamic damping coefficients present an approximate harmonic distribution and minimum value of these appears near the zero nodal diameter. However, what is different from the common characteristics described above is that the minimum damping state changes with vibration mode and working point. When the fan works at P1, the lowest value of damping is negative locating at $ND = 1$ of 2F mode, while vibration of blade could be damped at arbitrary ND for the other two modes. The similar

Table 1.   Material characteristics.

Density (kg/m³)	Poisson ratio	Modulus of elasticity (Gpa)
4440.00	0.34	109.00

Table 2.   The nature frequency corresponding to the first three modes.

Mode	1F	2F	1T
Nature frequency (Hz)	336.47	704.57	903.48

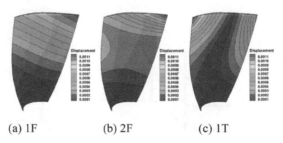

(a) 1F   (b) 2F   (c) 1T

Figure 4.   The first three modes of the fan blade.

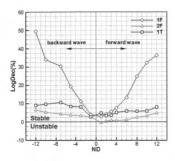

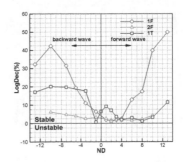

Figure 5. Aerodynamic damping coefficients at P1.

Figure 6. Aerodynamic damping coefficients at P2.

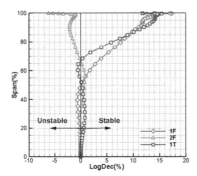

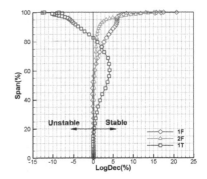

Figure 7. Aerodynamic damping coefficients versus different spans of the blade at P1.

Figure 8. Aerodynamic damping coefficients versus different spans of the blade at P2.

phenomenon can be found for the near stall operation P2. It is clear that **IBPA** is a key parameter to reflect the effect of coupled aerodynamics in circumferential direction influencing the flutter characteristics significantly.

Figure 7 gives span-wise distribution of aerodynamic damping coefficients for minimum damping state at the P1 working point. The same type distribution at the P2 operation is given in Figure 8. At the design point, there is a big damping effect in the upper half of the blade for 1F and 1T mode, aerodynamic excitation plays a major role for 2F mode at the similar region of the span on the contrary which produce the destabilizing force leading to a blade flutter risk. According to Figure 8, there is a similar distribution of damping in radial direction for 1F and 2F mode at near stall point, however the third mode is very different from the first two modes. The aerodynamic damping coefficients is positive near blade tip for 1F and 2F mode, otherwise, negative damping drive blade to vibrate in this region but larger positive damping is sufficient to offset destabilizing effect on the upper span.

In order to get insight on the influence of shock wave on blade flutter, relative Mach number distribution on the blade surface is given first. There is a passage shock wave from 40% span to blade tip on the suction surface while this shock is at the position closer to leading edge and extends from 50% span to tip with smaller strength due to lower Mach number in front of the wave on the pressure surface. Comparing the local aerodynamic work on the blade surface in Figure 10 with the relative Mach number contour shown in the Figure 9, it is clear that positive work always appears on the downstream of shock wave adjacent to this wave in both upper and lower blade surface. This region as an aerodynamic work concentration area always plays an important role into aerodynamic stability. This phenomenon is attributed to the unsteady pressure pulse resulting from shock wave oscillating in chordwise direction along with the vibration of blades. In addition, it is obvious that the positive work induced by shock wave on the pressure surface is more decisive accordingly to the strength of aerodynamic work.

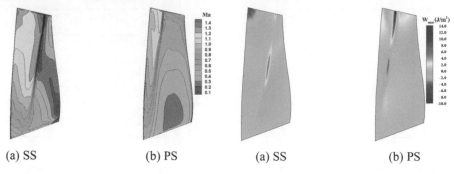

(a) SS   (b) PS   (a) SS   (b) PS

Figure 9. Local relative Mach number on the blade surface.

Figure 10. Local aerodynamic work on the blade surface.

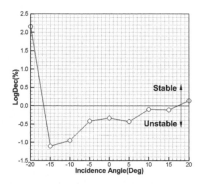

Figure 11. Relationship of aerodynamic damping with incidence angle.

Figure 11 shows the relationship of aerodynamic damping coefficient and incidence angle for $ND = 1$ mode at the P1 operation. The negative aerodynamic damping increase slowly with a variation from 0° to 15° for incoming flow angle and even change to positive damping at 20°. The trend is attributed to the decline of flow area, which results from the flow separation.

However, there is an opposite trend from 0° to −15° corresponding to the positive angles, meaning that a bigger risk of flutter. Then, it is found that a sharp change of aerodynamic damping exists from −15° to −20°. This phenomenon can be explained by the location change of shock wave.

4 CONCLUSIONS

The aeroelastic stability of a transonic fan stage has been investigated by computing aerodynamic damping of the blade using the energy method. The unsteady flow simulation around oscillating blade were based on different parameters influencing flutter characteristics, which can be concluded as follows:

1. The inter-blade phase angle is a key parameter that decides the aeroelastic stability. The dangerous state always occurs at low ND mode.
2. The unsteady pressure response is closely related to local motion of blade surface, so different modes will lead to an extreme variation of local aerodynamic work.
3. The vibrating shock wave induces aerodynamic work to concentrate and the shock wave on the pressure surface plays a crucial role in the flutter characteristics.
4. The variation of incidence angle has a significant impact on aeroelastic stability along with the characteristic change of shock wave and flow separation.

REFERENCES

[1] Arnold, S. 2005. Lessons Learned to Improve HCF Demonstration Tests. RTO-MP-AVT-121.

[2] Zhang X.W., Wang Y.R., Xu K.N. 2013. Mechanisms and Key Parameters for compressor Blade Stall Flutter. *ASME Journal of Turbomachinery*, 135(2): 024501.

[3] Isomura K. & Giles M.B. 1998. A Numerical Study of flutter in a Transonic Fan. *ASME Journal of Turbomachinery*, 120(3): 500–507.

[4] Srivastava R., Bakhle M.A., Keith Jr. TG. 2003. Numerical Simulation of Aerodynamic Damping for Flutter Analysis of Turbomachinery Blade Rows. *Journal of Propulsion and Power*, 19(2): 260–267.

[5] Huang X.Q., He L., Bell D.L. 2006. An Experimental Investigation into Turbine Flutter Characteristic at different tip clearances. *ASME 2006 Turbo Expo: Power for Land, Sea, and Air. Barcelona, 8–11 May 2006.*

[6] Vahdati M., Sayma A.I., Marshall J.G., Imregun M. 2001. Mechanisms and Prediction Methods for Fan Blade Stall Flutter. *Journal of Propulsion and Power*, 17(5): 1100–1108.

[7] Parthasarathy Vasanthakumar. Computation of Aerodynamic Damping for Flutter Analysis of a Transonic Fan. *ASME 2011 Turbo Expo: Turbine Technical Conference and Exposition. Vancouver, 6–10 June, 2011.*

[8] Saxer A.P. & Giles M.B. 1993. Quasi-Three-Dimensional Nonreflecting Boundary Conditions for Euler Equations Calculations. *Journal of Propulsion and Power*, 9(2): 263–271.

[9] Carta F.O. 1967. Coupled Blade-Disk-Shroud Flutter Instabilities in Turbojet Engine Rotors. *Journal of Engineering for Power*, 89(3): 419–426.

Advances in Engineering Materials and Applied Mechanics – Zhang, Gao & Xu (Eds)
© 2016 Taylor & Francis Group, London, ISBN 978-1-138-02834-0

A study on the self-excited vibration of driving system during locomotive slippage

L. Zhiyuan & D. Huanyun
State Key Laboratory of Traction Power, Southwest Jiaotong University, Chengdu, China

ABSTRACT: When the high-speed train is skidding, the wheel-rail adhesion force decreases with the creep velocity going up. That is a kind of self-excited vibration. This article proposes a dynamical model of the single frame driving system and the wheel-rail longitudinal coupling vibration when the high-speed train skids. By calculating the characteristic roots of the homogeneous equations that the model corresponds to, the effects of different rigidity and damping on the system characteristic roots are analyzed, thus providing a theoretical basis for parameter optimization. The theoretical analysis indicates that the parameters of the driving system should be appropriately set because of the inter-coupling between the torsional vibration from the driving system and the longitudinal vibration from the wheel set. Otherwise, the stability of the frame longitudinal movement will be affected.

1 INTRODUCTION

With the increasing speed of high-speed trains, the wheel-rail interaction forces of the high-speed trains become increasingly complicated. The wheel set, whose dynamic characteristics have a direct effect on the dynamic behavior of the driving system, works as an output terminal of the driving system. Therefore, it is attracting more and more attention from researchers to study the dynamic characteristics of the wheel set and the driving system as a whole[1][7][8].

It is known that the wheel-rail adhesion comes from the wheel-rail creep, but meanwhile the creep limits the maximum adhesion delivered by the wheel and rail. When the creep speed is small, the wheel-rail tangential force (F) is approximately proportional to the creep velocity (v). With the motor drive torque increasing, the creep speed gradually increases, and F tends to be saturated. Then v increases and F basically remains unchanged. At the moment, if the motor drive torque continues to increase, the wheel-rail tangential force cannot continue to maintain the maximum value so that the wheel-rail adhesion will go into the full slide area. After the wheel-rail adhesion enters the complete slide area, with v increasing, F gradually decreases, resulting in a negative slope section of the adhesion characteristic curve, which is different from Kallker's theory[5][6].

In terms of the research on the dynamic characteristics of the driving system during the wheel-rail stick-slip vibrations, Hirotsu and Ishida established a single-wheel driving system model to analyze the effects of the suspension parameters on the characteristic roots of the equation, and the results were compared to verify the correctness of the model[2]. Also, Leva, etc. established a mechatronic driving system model, offering further analyses of the effects of the electrical parameters on the driving system vibration characteristics[3]. Besides, Muller and Kogel utilized Adams and Matlab co-simulation model to study the influence of the driving system parameters on the wheel-rail stick-slip vibrations[4].

For a comprehensive analysis of the dynamic characteristics of the driving system and the effects of the suspension parameters on the system dynamic characteristics during the wheel-rail slip, this article proposes a dynamical model of the single frame driving system and the wheel-rail longitudinal coupling vibration. After linearizing the equation, the effects of different parameters on the system vibration characteristics are comprehensively analyzed.

Besides, this model can also be used to analyze the non-linear wheel-rail stick-slip vibration stability.

2 MATHEMATICAL MODEL

2.1 *The single frame driving system, the wheel-set longitudinal coupling vibration model*

$$I_m\ddot{\theta}_{mi} + C_m(\dot{\theta}_{mi} - \dot{\theta}_c) + k_m(\theta_{mi} - \theta_c) = \tau_{mi} \tag{1}$$

$$I_c\ddot{\theta}_{ci} - C_m(\dot{\theta}_{mi} - \dot{\theta}_{ci}) - k_m(\theta_{mi} - \theta_{ci}) + k_p(\theta_{ci} - \theta_{pi}) = 0 \tag{2}$$

$$I_p\ddot{\theta}_{pi} - k_p(\theta_{ci} - \theta_{pi}) - r_pP_i = 0 \tag{3}$$

$$I_g\ddot{\theta}_{gi} + r_gP_i + k_\theta(\theta_{gi} - \theta_{wi}) = 0 \tag{4}$$

$$I_s\ddot{\theta}_{si} + l_1P_i + l_2^2C_s\dot{\theta}_{si} + l_2^2k_s\theta_{si} = 0 \tag{5}$$

$$I_w\ddot{\theta}_{wi} + k_\theta(\theta_{wi} - \theta_{gi}) = -\mu_{si}Wr_w \tag{6}$$

$$M_w\ddot{X}_{wi} + C_w(\dot{X}_{wi} - \dot{X}_B - h_2\dot{\theta}_B) + k_w(X_{wi} - X_B - h_2\theta_B) = \mu_{si}W \tag{7}$$

$$M_B\ddot{X}_B + C_w \cdot (2\dot{X}_B + 2h_2\dot{\theta}_B - \dot{X}_{w1} - \dot{X}_{w2}) + k_w \cdot (2X_B + 2h_2\theta_B - X_{w1} - X_{w2}) = 0 \tag{8}$$

$$I_B\ddot{\theta}_B + C_wh_2 \cdot (2\dot{X}_B + 2h_2\dot{\theta}_B - \dot{X}_{w1} - \dot{X}_{w2}) + k_wh_2 \cdot (2X_B + 2h_2\theta_B - X_{w1} - X_{w2})$$
$$+ 2l^2C_B\dot{\theta}_B + 2l^2k_B\theta_B = 0 \tag{9}$$

Among them,

$$\tau_m = \frac{2\mu_mWr_w}{S} - \rho\dot{\theta}_m \tag{10}$$

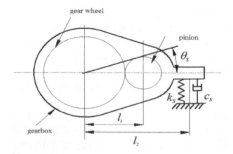

Figure 1.   Diagram of single frame nod and longitudinal movement.

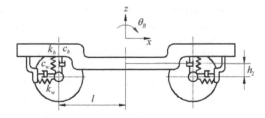

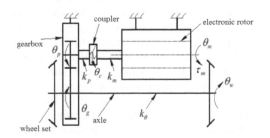

Figure 2.   Diagram of gearbox structure and movement.

Figure 3.   Diagram of the driving system structure.

Table 1. The meaning of symbols.

Symbols	Meaning
$I_m, I_c, I_p, I_g, I_s,$ I_w, I_B	Electronic rotor, coupler, pinion, gear wheel, gearbox, wheel set, rotational inertia of the frame
M_w, M_B	Wheel set, quality of the frame
C_m, C_s, C_w, C_B	Torsional damping of the rotor output shaft, gearbox hanging vertical damping, a longitudinal damping, a vertical damping
$k_m, k_p, k_\theta, k_w, k_B$	Rotor output shaft, coupling output shaft, axle torsional rigidity, a longitudinal rigidity, a vertical rigidity
r_p, r_g, r_w, S, ρ	Pinion, gear wheel, wheel rolling radius, gear ratio, suppression ratio of the motor torque speed curve
$W, \mu_{si}, k_{\mu i}, v, v_{mi}$	1/2 axle load, $(i = 1, 2)$ the adhesion coefficient of the front and the rear $(i = 1, 2)$ wheels, slope of the front and the rear wheels on adhesion characteristic curve sliding section, creep velocity, creep velocity under the maximum adhesion coefficient of the front and the rear wheels
$\theta_{mi}, \theta_{ci}, \theta_{pi}, \theta_{gi}, \theta_{si},$ $\theta_{wi}, \theta_B, X_{wi}, X_B$	Front and rear $(i = 1, 2)$ rotor, coupler, pinion, gear wheel, gearbox, nod displacement of the wheel set, nod displacement of the frame, the longitudinal displacement of the front and rear $(i = 1, 2)$ wheels, vertical displacement of the frame

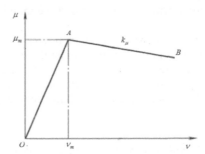

Figure 4. Diagram of linearization of adhesion characteristics curve.

$$\mu_{si} = \mu_{mi} + k_{\mu i}(v - v_{mi}), (i = 1, 2) \tag{11}$$

In the above equation, $\theta_p = S \cdot (\theta_g + \theta_s)$ (12)

$$r_p = \frac{r_g}{S} \tag{13}$$

2.2 Linearization of wheel-rail creep rate and adhesion characteristic curve

Compared with the vehicle system dynamics, the driving system dynamics defines the creep rate differently. This article adopts FREDERICH creep rate formula[9].

$$v = \frac{\omega r - V}{\omega r} = 1 - \frac{V}{\omega r} = 1 - \frac{V_0 + \dot{x}}{(\omega_0 + \dot{\theta})r} \tag{14}$$

In the formula, V is the wheel instantaneous speed, V_0 is the wheel average speed, ω is the wheel instantaneous angular velocity, ω_0 is the wheel average angular velocity and r is the wheel rolling radius.

If $\dot{x} = 0, \dot{\theta} = 0$, there is

$$v_0 = 1 - \frac{V_0}{\omega_0 r} \tag{15}$$

219

After linearizing the creep rate in Formula (14),

$$v = v_0 + \frac{V_0}{\omega_0^2 r}\dot{\theta} - \frac{1}{\omega_0 r}\dot{x} \qquad (16)$$

Substitute formulas (10), (11), (15), (16) into formulas (1), (2), (6), (7), (8), (9), and the linearized equations are obtained. Then the vibration characteristics of the system can be obtained by analyzing the characteristic roots and eigenvectors of homogeneous equation corresponding to the above equations. The homogeneous equation can be expressed as:

$$[\mathbf{M}]\{\ddot{\mathbf{X}}\} + [\mathbf{C}]\{\dot{\mathbf{X}}\} + [\mathbf{K}]\{\mathbf{X}\} = \{\mathbf{0}\} \qquad (17)$$

By solving the characteristic roots of the equation (17), the eigenvalues and the corresponding eigenvectors of the system can be obtained.

3 EIGENVALUE ANALYSIS

3.1 *Descending rate ρ of motor torque speed characteristic curve*

The descending rate of motor torque characteristic curve mainly influences the wheel nod vibration.

The descending rate (ρ) of motor torque speed characteristic curve almost has no influence on the frequency of the wheel nod vibration (The imaginary part has divided by 2π), but it has a greater effect on the real component of conjugate characteristic root. That is, when ρ is small, the real component of conjugate characteristic root is positive, and with ρ increasing gradually, the real component decreases and becomes a negative real number till ρ equals 9.

3.2 *Linearized slope k_u of the adhesion characteristic curve in slide area*

The linearized slope of the adhesion characteristic curve in slide area mainly influences the frame longitudinal vibration.

Figure 7 demonstrates that when k_u is negative, the real characteristic root 1 is over 0, and when k_u is positive, the real characteristic root 1 is less than 0. As stated above, the negative slope of the adhesion characteristic curve in slide area corresponds to the negative damping item, and constant energy is input into the system. Thus, when k_u is negative, the eigenvalue of the frame longitudinal vibration is a positive real number. When k_u is positive (It is then equal to the normal damping system.), the eigenvalue will be a negative real number within the normal scope of the system parameters.

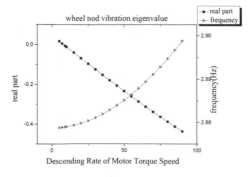

Figure 5.   Effects of the descending rate ρ of motor torque speed characteristic curve.

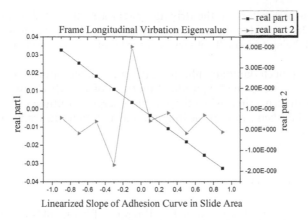

Figure 6. Effects of linearized slope k_u in slide area on the eigenvalue of the frame longitudinal vibration.

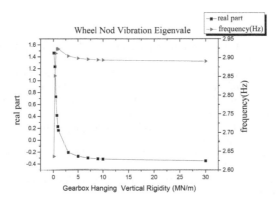

Figure 7. Effects of gearbox hanging vertical rigidity k_s on the eigenvalue of the wheel nod vibration.

3.3 *Gearbox hanging vertical rigidity k_s*

The gearbox hanging vertical rigidity k_s can affect whether the real component of wheel characteristic root is positive or negative.

When k_s is small, the real component of wheel characteristic root is positive and it gradually increases as k_s goes up. When k_s is around 3 MN/m, the real component begins to be negative. Besides, the vibration frequency increases rapidly at first, and after that it remains essentially unchanged.

4 CONCLUSIONS

This paper embarks on the analyses of the single frame driving System and the wheel-set longitudinal coupling vibration in terms of the time domain and characteristic roots. Here are the conclusions drawn from the analyses.

1. For a single frame, when the wheel-rail adhesion curve is in the negative slope section, the entire system is unstable and it is impossible to make it stable by adjusting the parameter matching of the system or changing the motor handle levels. Therefore, high-speed trains should remain travelling in the adhesion area by adhesion controlling.

2. The suspension parameters of the driving system are influential to the stability of the whole system. In particular, the gearbox hanging vertical rigidity must be chosen reasonably instead of being too small, otherwise the stability of the system movement will be affected.

3. Since there is not much damping inhibiting the frame longitudinal and nod movement in the entire vehicle system, the longitudinal motions of the wheel sets and the frame will lose stability at the same time, when the external energy is continuously transmitted to the wheel sets and they then transmit the longitudinal motion to the frame through a series of locating devices.

ACKNOWLEDGEMENTS

This work has been supported by the State Key Program of National Natural Science of China (61134002), the National Key Basic Research Program of China (973 Program) (2011CB711100) and Innovation Group of Ministry of Education funded project (IRT1178).

REFERENCES

[1] Bao Weiqian. 1999. Some Concepts and Improvement of Locomotive Adhesion. *Diesel Locomotives*, (1): 8–14.

[2] Hirotsu T., Ishida S. 1975. Adhesion Performance and Dynamic Phenomenon in Slipping of Electric Rolling Stock. *Transaction of the Japan Society of Mechanical Engineers*, 41(343): 784.

[3] Leva S., Morando A.P., Colombaioni P. 2008. Dynamic analysis of a high-speed train. *Vehicular Technology, IEEE Transactions on*, 57(1): 107–119.

[4] Muller S., Kogel R. 2000. The influence of mechanical design parameters on roll-slip oscillations in locomotive drives. *International ADAMS User Conference*.

[5] Polach O. 2005. Creep forces in simulations of traction vehicles running on adhesion limit. *Wear*, 258(7): 992–1000.

[6] Polach O. 2001. Influence of locomotive tractive effort on the forces between wheel and rail. *Vehicle System Dynamics*, 35(1): 7–22.

[7] Sun Xiang. 1994. System Design for Locomotives with High Adhesion. *Journal of Southwest Jiaotong University*, 29(3): 235–248.

[8] Sun Xiang. 1994. Relation of Adhesion to Power Transmission, Driving and System Control for Locomotives. *Journal of the China Railway Society*, (A06): 8–16.

[9] Zheng Lian-zhu, Zhang You-kun. 1997. Reascarch on the Self-Excited Vibration in Wheeled Vehicle Powertrain (Part III)—Stability Analysis of the Self-Excited Vibration System. *Transactions of the Chinese Society of Agricultural Engineering*, 13(1): 39–44.

Advances in Engineering Materials and Applied Mechanics – Zhang, Gao & Xu (Eds)
© *2016 Taylor & Francis Group, London, ISBN 978-1-138-02834-0*

Heat transfer analysis with transverse effect of a curved corrugated-core sandwich panel

W.C. Qi
Institute of Solid Mechanics, Beihang University, Beijing, China
Faculty of Aerospace Engineering, Shenyang Aerospace University, Shenyang, China

S.M. Tian
Faculty of Aerospace Engineering, Shenyang Aerospace University, Shenyang, China

ABSTRACT: A corrugated-core sandwich panel can be used on the integrated thermal protection system of a hypersonic space shuttle. The outside walls of wings are designed as curved shapes to maintain a good aerodynamic performance. This paper proposes an effective laying method that lays the integrated thermal protection system on a curved wall based on conformal transformation. Then a heat transfer analysis is carried out for different cells that have corresponding heat flux curves. A transverse flow of heat energies is observed. The results show that the transverse heat transfer between the different cells may have a significant impact on temperature distribution. The safety of hypersonic vehicles can be improved owing to the temperature of the inner panel being a key failure index.

1   INTRODUCTION

The Thermal Protection System (TPS), which can ensure the safety in the extreme thermal environment, is one of the key subsystems of hypersonic vehicles. To transfer the aerodynamic loads and maintain an aerodynamic shape, the traditional tiles require a relatively high density in order to ensure sufficient rigidity. However, this method makes a high thermal conductivity and thus the thickness of the TPS should be increased. Bapanapalli (2006) first proposed the concept of Integrated Thermal Protection System (ITPS) based on corrugated-core sandwich panels. The ITPS has the characteristics of lightweight, reusable and high structural efficiency compared with the traditional TPS.

At present, people mainly concentrate on the preparation and selection of insulation and bearing materials, the methods for fasten coupled thermal-mechanical response analysis and structural optimization design according to the service environment. Martinez (2007, 2010 & 2012) presented a rapid method for structural stress analysis to improve the efficiency of structural design. The main method is using a two-step analysis process named 'equivalent & anti-equivalent' according to the periodically among the cells. Gogu (2009) seeks the optimal dimensions and the best materials and solved the optimization problem in two steps. In the first step, good candidate materials are selected based on a spline interpolation of the maximum bottom face sheet temperature. Then the geometry of the integrated thermal protection system panel is optimized for different combinations of the materials. Kopp (2014) summarized the current activities on the preliminary structural analysis for the SpaceLiner. He focused on different materials and design options as well as the integration of structure and thermal protection system. Xie (2013) established an optimization procedure aiming to design an ITPS with minimum weight by developing a finite element simulation. The ITPS geometric dimensions are considered as the design variables. The objective function is the mass per unit area and the constraint conditions are the inner temperature and local stress. The optimization procedure might be useful in the future work.

Till now, one can find that only the thermo-mechanical behaviors of a cell are studied in almost all the literatures. However, an ITPS can be made in a large size to replace the traditional skin, thermal protection and corresponding support system. In this case, it's necessary to processing the outer panel in a curved way to maintain a good aerodynamic shape when lays ITPS on a wing. So, the preparation method of a whole ITPS panel with curved shapes should be considered first. In addition, different cells have different heat flux density inputs according to the service environment locations on the wing. Therefore, the transverse heat transfer between different cells also should be considered with great care.

2 PROBLEM STATEMENT

Considering the heat transfer analysis of a wing with ITPS as shown in Figure 1, the objective is to determine the temperature field of the corrugated sandwich structure that is mainly composed of the outer panel, the web, and the inner panel. And we used Saffil alumina fiber as the filling insulation material. Owing to the location-dependence of the heat flux density curve, a transient heat transfer analysis can be defined as follows:

$$H(x,t) \to S(x,t) \to T(x,t) \tag{1}$$

where H, S and T are the heat flow, ITPS, and the temperature respectively, and t is the time, and x is the position vector.

Our aim is to find the solution $T(x,t)$ satisfying Eq. (1). However, one cannot obtain it exactly by analyzing only a single cell because the wing structure is curved and the heat flux density curves are different with different locations. So, we should first develop a plan to lay the ITPS on a curved wing. Then the heat transfer should be observed not only in the direction of longitudinal but also in the transverse to obtain more exactly temperature distribution.

3 CONFORMAL TRANSFORMATION AND SERVICE ENVIRONMENT

A piece of ITPS panel with three cells is considered. What we know is only the curve of the outer wall. And what we should do is to design the inner panel and the webs. For a good topology performance, a conformal transformation method (Nehari, 1952 & Delillo, 2006) is introduced to design cells on a curved wing. Thus, we can determine the sizes and shapes of the curved ITPS by the following steps:

1. Draw three standard size cells below the curve. The demarcation points on the shape curve can be obtained from the intersections of the boundary lines of the standard cells with the shape curve.

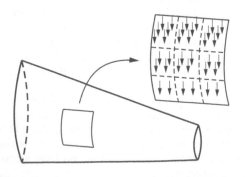

Figure 1.   A curved ITPS panel on a wing.

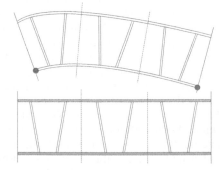

Figure 2.   The process of conformal transformation.

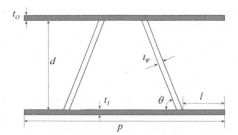

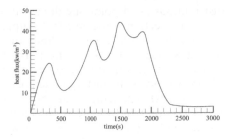

Figure 3.   A standard corrugated sandwich structure.   Figure 4.   The used heat flux curve.

Table 1.   The sizes of a cell in a corrugated sandwich structure.

Variables	t_o/mm	t_l/mm	t_w/mm	d/mm	p/mm	l/mm	$\theta(°)$
Size value	2	3	1.9	78.4	160	22	74.3

Table 2.   The thermal properties of used materials.

Material	Inconel718	Ti	Al2024	Saffil
Density (kg/m³)	8150–8300	4407–4451	2750	50
Conductivity (W/(m*k))	10.5–12.5	7.3–7.9	80–220	0.09
Specific heat (J/kg)	410–455	553–575.6	944–982	1260–1340

2. Draw normal curves of the shape curve at the points of intersection. This process ensures that the boundaries of cells are conformal.

3. According to the desired thicknesses of the cells, the point locations on the inner panel can be determined in the normal lines. The shape of inner panel can be obtained from connecting these points by a spline curve.

4. The positions of webs on the outer panel can be determined by the lateral coordinates, which can be measured from the standard cells. And the directions of webs should be kept the same angles with the normal curves.

In the study, we use a standard corrugated sandwich structure as shown in Figure 3. And the corresponding sizes are listed in Table 1. The used heat flux density curve (David, 2000) is shown in Figure 4. Considering the similarity of experienced time history, $H(x,t)$ on different cells are regarded as having the same shapes but only with different weight coefficients. Without loss of generality, these weight coefficients can be assumed as 1.0, 0.8 and 0.6 for different rows for convenience. The material of outer panel is Incorel718 that has a good impact performance. The web's material is Ti that can bear at a high temperature. The material of inner panel is Al to have a large heat sink. And the thermal insulation material is Saffil. All the material properties are list in Table 2.

4   NUMERICAL EXAMPLE

According to the proposed conformal transformation method, we construct a geometry model of a curved panel shown in Figure 5 based on the data of NACA0012 airfoil. The corresponding finite element model is built using ABAQUS 6.11, which contains 59475 nodes and 53760 elements as shown in Figure 6. We set four different feature points as shown in Figure 7 in all webs to observe temperature changes in cross section. So, temperature distributions at 24 points are discussed. Figure 8 shows the temperature distribution in the transverse section at the time 1500s (up) and 2500s (below).

For comparison, Figure 9 shows the time history of temperature without considering the transverse heat transfer effect. The heat flux load factor is set as 1.0. The four curves represent the temperature at different points (from top to bottom in order A, B, C & D). Figure 10 shows the history of temperature of the four group points at webs (each group has 6 points). The temperature of the inner panel is important to the safety of the ITPS. From Figure 9 and Figure 10, it can be observed that the maximum temperature reduces about 30°C in the first cell. In other words, owing to the non-uniform temperature distribution in the transverse direction, the energy flows from the high temperature zone to the low temperature region. It eases the requirements of the material performance in the high temperature zone. However, it is dangerous for the low temperature zone. Therefore, it is necessary to consider the transverse heat transfer effect in the ITPS design to ensure the safety of the structures.

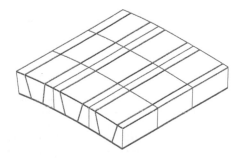

Figure 5.   The geometry model of a curved panel.

Figure 6.   The finite element model.

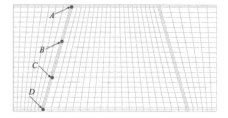

Figure 7.   The layout of the observation points.

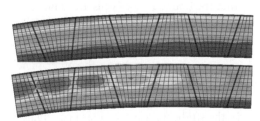

Figure 8.   Temperature contours at different time.

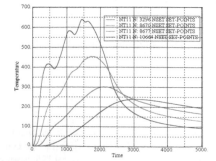

Figure 9.   Time histories of temperature without transverse effect at points A, B, C & D.

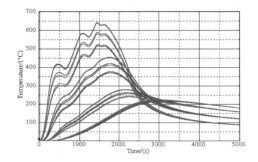

Figure 10.   Time histories of temperature at 24 points.

226

5 CONCLUSIONS

The paper calculates temperature response of a curved ITPS panel with corrugated sandwich structure under non-uniform transient temperature heat flux loads. The conformal transformation method ensures a good topology shape of cells when lays ITPS panel on an airfoil. More exactly temperature distribution can be obtained by considering the heat transfer in the transverse direction. The results show that the change of the temperature of the inner panel is obvious. The safety of ITPS structure can be improved when considering the transverse heat transfer effect.

ACKNOWLEDGEMENTS

This work was supported by the National Nature Science Foundation of the P. R. China Program (No. 11372025, No. XY201405) and the Defense Industrial Technology Development Program (No. JCKY2013601B, No. XY201462).

REFERENCES

[1] Bapanapalli, S.K., Martinez, O.M. & Gogu C. 2006. Analysis and design of corrugated-core sandwich panels for Thermal Protection Systems of space vehicles. *AIAA* 2006–1942.
[2] Martinez, O.A., Sankar, B.V., Haftka, R.T. & Bapanapalli S.K. 2007. Micromechanical analysis of composite corrugated core sandwich panels for integral thermal protection systems. *AIAA Journal*, 45(9): 2323–2336.
[3] Martinez, O.A., Sharma, A., Sankar, B.V. & Haftka, R.T. 2010. Thermal force and moment determination of an integrated thermal protection system. *AIAA Journal*, 8(1): 119–128.
[4] Martinez, O.A., Sankar, B.V., Haftka, R.T. & Blosser M.L. 2012. Two dimensional orthotropic plate analysis for an integral thermal protection system. *AIAA Journal*, 50(2): 287–398.
[5] Gogu, C., Bapanapalli, S.K., Haftka, R.T. & Sankar, B.V. 2009. Comparison of materials for an integrated thermal protection system for spacecraft reentry. *Journal of Spacecraft and Rockets*, 46(3): 501–513.
[6] Kopp, A., & Garbers, N. 2014. Investigation of structure, thermal protection system, and passenger stage integration for the hypersonic transport system spaceliner. *AIAA* 2014–2531.
[7] Xie, G.N., Wang, Q., Sunden, B. & Zhang W.H. 2013. Thermomechanical optimization of lightweight thermal protection system under aerodynamic heating. *Applied Thermal Engineering*, 59(2013): 425–434.
[8] Nehari, Z. 1952. *Conformal mapping*. McGraw-Hill, New York.
[9] Delillo, T.K. 2006. Schwarz-Christoffel mapping of bounded, multiply connected domains. *Comput. Methods and Function Theory*, 6(2): 275–300.
[10] David, E., Carl, J., Max, L. 2000. Parametric Weight Comparison of Advanced Metallic Ceramic Tile and Ceramic Blanket Thermal Protection System, NASA/TM-2000-210289.

Advances in Engineering Materials and Applied Mechanics – Zhang, Gao & Xu (Eds)
© 2016 Taylor & Francis Group, London, ISBN 978-1-138-02834-0

The simulation technique of spacecraft taking off on lunar based on cable-driven parallel mechanism

W.M. Yi & W. Feng

Beijing Institute of Spacecraft Environment Engineering, Beijing, China

ABSTRACT: In this paper, an attitude feedback force control simulation method based on the cable-driven parallel mechanism is put forward, to simulate the stress state at the moment of lunar spacecraft taking off. Eight cables-parallel mechanism of redundant constraints is adopted, and designed by the minimum cable force. The theoretical cable force of the eight cables is worked out reversely by measuring and calculating the spacecraft posture in real time. In different external disturbance force, spacecraft coupling force is analyzed and verified by Adams and Simulink. A new engineering solution is provided for the taking off simulation of lunar exploration engineering ascender. According to the simulation analysis, the force vector simulation error of the system is 3.7N when the disturbing force is 100N.

1 INTRODUCTION

The technology of taking off lunar is the key for lunar exploration project, which is of great importance for the returning of spacecraft from lunar. To verify the stability of spacecraft when taking off lunar, the force state of the spacecraft needs to be simulated accurately, which is the key of the take-off from lunar. Computer simulation technology can not fully reflect the complex stress state on take-off stage and separation, so need to carry out the physical simulation experiment to simulate the stress state of riser separation instant on the ground (including the engine thrust and low gravity environment and external disturbing force), verify the separation process of off-stability requirements and determine boundary conditions. The simulation of separation process shall be verified by the igniting flight of the thruster of ascender, which is characterized by accurate simulation and high validity, but it has high cost, great environment pollution and long preparation cycle, and it is also difficult to conduct repeated verification. However, the simulation system of the planar cable-driven parallel mechanism is low in cost, pollution free and short in preparation cycle, and it can conduct the verification of stability in the taking off of ascender repeatedly.

In recent years, planar cable-driven parallel mechanism receives extensive attention from domestic and foreign scholars, and there are numerous studies on the configuration design, space optimization, stiffness analysis and motion control, achieving substantial theoretical and experimental results. In 1987, Landsberger proposed the design of cable-driven parallel robot[1], and conducted positioning analysis, stiffness analysis and work space analysis, etc. In 1989, NIST (National Institute of Standards and Technology) regarded the gravity as an invisible cable, realizing the six-degree-of-freedom motion controlled by six cables[2]. Kawamura et al. from Japan have designed a seven-cable-driven parallel mechanism with large work space FALCON-7, and accomplished the high-velocity control analysis, stiffness analysis and work space analysis of the mechanism[3]. Ming and Higuchi conducted the substantial studies on the conformational analysis of the cable-driven parallel mechanism, and proposed a basic sorting technique that can restrain the locating mechanism completely[4, 5]. Since the 1990s, related studies have been conducted for the cable-driven parallel mechanism. Zheng Yaqing from Huaqiao University conducted a lot of studies on the application of cable-driven parallel mechanism in the wind tunnel experiment[6], as well as the design and control of cable-driven parallel mechanism, including the work space analysis of cable-driven parallel mechanism, motion planning, static rigidity, kinematics analysis, cable force optimization of redundant cable-driven parallel mechanism, etc. and proposed the work space quality as an indicator of mechanism optimization design. Duan Boayan[7–8] et al. from Xidian University have conducted in-depth study on the multi-axis servo control problem and chattering problem of the cable-driven parallel mechanism, including the analysis of chattering features, modeling and restraining control.

In conclusion, there are certain achievements in the configuration design, space optimization, stiffness analysis, and motion control of cable-driven parallel mechanism, which are launched by aiming at the control of pure location, with little demands on the control precision and response time. These research achievements have certain reference significance and theoretical foundation for the study on the force vector simulation based on cable-driven parallel mechanism and posture feedback. In this paper, a new force vector simulation method based on the planar cable-driven parallel mechanism is proposed, and it is characterized by such features as high precision, large-interaction space, high load capacity, and multi-degree of freedom simulation, etc.

2 LOW GRAVITY MODELING SCHEME OF PLANAR CABLE-DRIVEN PARALLEL MECHANISM

2.1 Overall scheme

According to the demands of stability test of lunar taking off, the ascender gravity, 3000N thrust, and the interactions between the ascender and lander, as well as the plume power shall be simulated. Consequently, the low-gravity simulation system of cable-driven parallel mechanism contains the support subsystem, cable-driven parallel mechanism subsystem, control subsystem and measurement subsystem, as shown in Figure 1.

In which, the supporting subsystem mainly provides high-rigidity support for the low-gravity simulation system, for fixing the cable-releasing structure of the cable-driven parallel mechanism. The cable-driven parallel mechanism subsystem consists of several cables, and the resultant force and resultant moment of all cables satisfy the combined resultant force and resultant moment of the test object. The measurement subsystem employs wire-drawing encoder to measure the length of tight-wire, and the wire-drawing encoder is high in precision (0.004 mm) and rapid in feedback frequency (1 Khz). Kinematics positive solution is calculated for the posture of ascender according to the length of rope reflected by the wire-drawing encoder, and the spatial position and posture of the ascender is achieved. The control subsystem takes the spatial position and posture of the ascender as the input, for achieving the theoretical cable force in this posture through reverse solution, so as to conduct force servo feedback control for each soft cable.

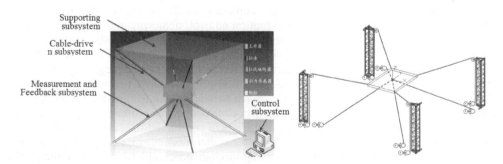

Figure 1.   Force vector simulation system of planar cable-driven parallel mechanism.

230

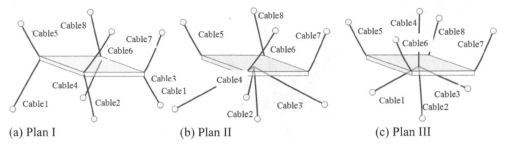

(a) Plan I   (b) Plan II   (c) Plan III

Figure 2.   Designing scheme of planar cable-driven parallel mechanism.

2.2 *Optimized design of cable force*

According to the relationship between the number of freedom degree and cables, the cable parallel mechanism is divided as IRPMs (Incompletely Restrained Positioning Mechanisms, m ≤ n), CPRMs (Completely Restrained Positioning Mechanisms, m = n+1), and PPRMs (Redundantly Restrained Positioning Mechanisms, m > n+1), in which m is the number of cables, while n is the number of freedom degree of the end of cable-driven parallel mechanism.

In order to simulate the six-degree-of-freedom of the ascender, the completely restrained parallel mechanism shall require seven cables. Meanwhile, considering that it shall carry out the work space and cable force optimization for cable parallel mechanism, for improving the controllability of the parallel mechanism, eight-cable restrain parallel mechanism is employed, and the redundancy in motion makes it possible for the control optimization. The determined original configuration consists of three types, as shown in Figure 2.

In order to guarantee the better accelerated velocity response characteristics, the minimum sum of cable force is the optimization condition, and it can be discovered from the simulation study on cable-drive parallel mechanism that the cable force distribution of plan II is relatively even, and the initial tensile force is relatively small, which can be selected and used in the first place.

3 SIMULATION ANALYSIS OF SYSTEM ERROR

3.1 *Simulation scheme*

The built physical simulation model shall be defined according to the required output and input interface with the interface of Adams and Simulink simulation analysis software, so as to achieve the system black box that can be applied for the direct Simulink simulation. And then, the simulation of control algorithm shall be accomplished with the strong control simulation of Simulink. Eventually, the process data achieved from simulation shall be led to Adams, and the motor process of model in the entire process can be observed intuitively through animation. The physical simulation model of eight-cable parallel mechanism established with Adams software is shown in Figure 3.

The physical simulation model obtained from the Adams software is applied for the follow-up Simulink joint simulation. The structure of physical model simulation model is shown in Figure 4.

In which, the nine inputs are on the left. External force is the external winding power, force 1 to force 8 are the cable forces of the eight cables, which are input into the MSC software for simulation calculation. The inputs of physical simulation model are on the right, and L1 to L8 are the lengths of eight cables.

3.2 *Simulation analysis result*

Considering the pulsation effect of 3000N engine and inter-engine coupling stress, the simulation analysis is conducted for the external force 10N and 100N respectively.

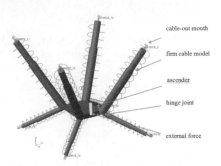

Figure 3. Adams motion simulation physical model.

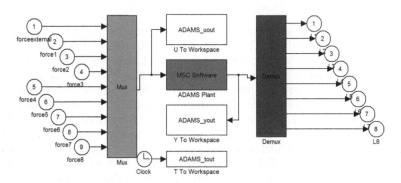

Figure 4. Simulation model of the physical model derived by Adams.

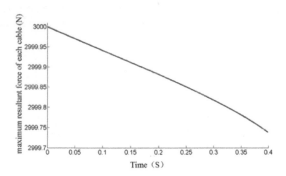

Figure 5. Resultant force simulation error of the system when the external disturbing force is 10N.

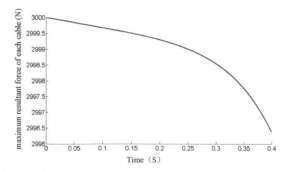

Figure 6 Resultant force simulation error of the system when the external disturbing force is 100N.

1. When the external disturbing force is 10N, the resultant force simulation analysis result of the system simulation detector is shown in Figure 5, and the maximum error in simulated resultant force is 0.23N.
2. When the external disturbing force is 100N, the resultant force simulation analysis result of the system simulation detector is shown in Figure 6, and the maximum error in simulated resultant force is 3.7N.

4 CONCLUSION

In this paper, the force vector simulation method based on cable-driven parallel mechanism and force control is proposed according to the technical indicator of stability test of lunar taking off, and the Adams simulation analysis is conducted, obtaining the following results:

1. Overall plan for the force vector simulation system based on the cable-driven parallel mechanism is proposed.
2. The number and configuration of cable force is designed and optimized, and a feasible cable force optimization scheme is proposed.
3. The Adams simulation analysis is conducted, and the simulation error of the system with 100N disturbing force is 3.7N.

ACKNOWLEDGEMENT

This study was supported by national natural science foundation of China (No. 51405024).

REFERENCES

[1] Landsberger, S.E. & Sheridan, T.B. 1993. A minimal, minimal linkage: the tension-compression parallel link manipulator. Robotics, *Mechatronics and Manufacturing Systems* (2):81–88.
[2] Albus, J. & Bostelman, R. et al. 1993. The NIST robocrane. *Journal of Robotic Systems*10 (5):709–724.
[3] Kawamura, S. & Choe, W. et al. 1995. Development of an ultrahigh speed robot FALCON using wire drive system. *Proceeding of the 1995 IEEE International conference on Robotics and Automation the Japan Society for Precision Engineering* 215–220.
[4] Ming, A. & Higuchi, T. 1994. Study on multiple degree-of-freedom positioning mechanism using wires. I: Concept, design and control. *International Journal of the Japan Society for Precision Engineering*28(2):131–138.
[5] Ming, A. & Kajitani, M. et al. 1995. On the design of wire parallel mechanism. *International Journal of the Japan Society for Precision Engineering*29(4):337–342.
[6] Hu, L. & Zheng, Y.Q. et al. 2008. Dynamic Analysis of a Wire-Driven parallel Manipulator for Low-Speed Wind Tunnels. *Journal of HuaQiao University(Natural Science)* 102(02):184–189.
[7] Su, Y.X. & Duan, B.Y. et al. 2004. Auto-disturbances rejection control of the feed-supporting system for the square kilometre array. *Control Theory & Applications*(6):956–960.
[8] Su, Y.X. & Duan, B.Y. et al. 2001. Wind Vibration Damping Control of Radio-telescope Feed Supporting System based on Electro-rheological Fluid(3): 46–50.

Automation

Advances in Engineering Materials and Applied Mechanics – Zhang, Gao & Xu (Eds)
© 2016 Taylor & Francis Group, London, ISBN 978-1-138-02834-0

Application of computer image processing technology in fuse safety detector

D.B. Zhang, W.Q. Huang, Z.W. Li & X.Y. Li
Wuhan Mechanical College, Wuhan, Hubei, China

ABSTRACT: In the process of assembling and transporting the ammunition, the fuse is possible to be armed when the case is dropped or tossed, but the fuse arming sometimes can not be found in time by the traditional detection methods with its poor effects, which easily causes the accident of ammunition explosion and even injuries to the on-site inspection personnel. Such computer image processing technologies as image de-noising, image edge detection, image enhancement and compression, are applied in the detection of the ammunition fuse safety, realizing the visualization of fuse safety detection, the remote real-time detection and technical support through the local area network, which improves the precision and efficiency of the detection and ensures its efficiency and safety.

1 INTRODUCTION

The computer image processing technology is an emerging discipline developed rapidly in the past thirty years. With the development of computer technology, computer image processing technology is now mature, and more and more widely applied in space exploration, remote sensing, biomedicine, artificial intelligence, military detection and many other fields. Computer image processing is such a technology that firstly turns the image analog signal into digital signal and then takes advantage of the computer's fast processing to translate, revolve, de-noise, smooth, enhance, segment and compress the digital image, so it is also named as digital image processing. Computer digital image processing technology is applied to process the digital images x-rayed by fuse safety detector of the ammunition through computer and the image processing software, after completing the digital processing of the image acquisition card. The processes include the pre-processing, mid-processing, post-processing of the images, as well as the display and print of the detecting results, on the basis of which the clear image of the ammunition fuse mechanism is obtained to judge whether the fuse is armed or not, ensuring the safety of ammunition in the process of assembling and transportation.

2 FUNCTIONS OF FUSE SAFETY DETECTOR

The ammunition fuse safety detector is mainly to detect whether the ammunition fuse is armed when the case is dropped or tossed in the process of its assembling and transportation, in order to ensure the safety of the ammunition. The detection needs to be timely, accurate, safe and effective: on the one hand, make a nondestructive detection of the fuse experienced by accidents in the process of storage, transportation and operation, and a real-time judge of its situation. On the other hand, make an auxiliary detection of the fuse happened on explosion in tube, pre-explosion and explosion during transportation, to provide the real-time scientific evidence for analyzing the causes of the accident. The fuse safety mechanism is located inside the metal structure of the bullet with its thickness of about 30 mm~80 mm, so the x-ray detection method is usually adopted. The fuse arming sometimes can not be found by the traditional methods of detection with poor effects, which easily causes the accidents of ammunition explosion, and even injuries to the on-site inspection personnel. With the development of the computer technology, the computer image processing technology is applied with x-ray imaging technology to develop the ammunition fuse safety detector that is able to meet the requirement of the above ammunition fuse safety detection with higher quality of the images, more advantages than that of other detecting devices. All the detecting processes, including video save and replay, image storage, consultation, printing and parameter setting are completed under the control of the computer without any operation of the detecting personnel, which largely improves the efficiency of the fuse safety detection, and realizes the visualization of the detection, as well as the remote real-time share and technical support of the detecting results through local area network, ensuring the efficiency and safety of the detection.

3   COMPUTER IMAGE PROCESSING TECHNOLOGY

Computer image processing generally includes the series of processes from the collection to output of the images. However, in the professional field of computer image processing, the image processing technology mainly refers to these technologies of processing the collected images, including the pre-processing technology such as translation, revolving, segmentation and de-noising, mid-processing technology such as enhancement, restoration, segmentation and reconstruction, and the post-processing technology such as identification, analysis, encoding and compressing. Based on the above description, the computer image processing technology is a comparatively complex discipline [1]. Here we will mainly introduce these technologies used to process the images of the fuse safety mechanism in the process of ammunition fuse detection, including image de-noising, edge detection, enhancement and compression.

3.1   *Image de-noising*

When processing the images of the fuse safety mechanism, it is especially important to filter out the noise component and keep the geometric structure of the images. Due to the influence of the physical property of hardware system, the images are unavoidably with some irregular noise, such as the in-born noise from fuse safety mechanism of the ammunition and the noise of the photons. It will greatly affect the quality of the images of the fuse safety mechanism, causing more difficulties on the identification and post-processing of the detected images, so it is very necessary to make a de-noising processing to all the images of fuse safety mechanism. Its main purpose is to eliminate the interference in the process of collecting images and to keep their high fidelity[4]. In the image signal, the edge carries the largest quantity of information, which requires keeping the edge of image as complete as possible, while de-noising the image.

Provided that the samples {X1, X2, ... Xn} are the random variables of the distribution function F(x), the estimated values are defined as Tn (X1, X2, ... Xn), so the estimator output value $T = Tn \cdot (X1, X2, ... Xn)$, the formula below is true:

$$\sum_{i=1}^{n} p(x_i - T) \rightarrow \text{minimum} \tag{1.1}$$

Provided that $P'(t) = \phi(t)$, so the formula (1.1) can be written as

$$\sum_{i=1}^{n} \phi(x_i - T) = 0$$

After the median filtering,
Provided that $p(t) = |t|$, $-\infty < t < \infty$, thus

238

$$f(t) = \begin{cases} 0 & t > 0 \\ 0 & 0 \\ -1 & t < 0 \end{cases}$$

When de-noising the image, the median filtering is adopted to eliminate the impulse inter-ference in image and keep its edge.

3.2 *Image edge detection*

Image edge detection is an important content of the computer image processing. The edge refers to the collection of these gray pixels with the stepping or roofing change around the images. The image is fundamentally characterized by the edge, which plays an important role in border detection, image segmentation, mode identification and computer visualization. Its detection is mainly to measure, detect and position the gray changes of the images.

The framed edge detection based on the histogram of gray scale is one of the most com-monly used approaches of edge detection, which is very effective in detecting the edge of the target in the background image. The gray scale histogram of the images appears to be twin-peaked. Provided that the gray scale histogram of the image $f(I, j)$ is shown in Figure 1, from which it can be found that in this image exist many pixels in the background zone, but the gray scale of other pixels is very plain, so the histogram is framed into two parts on the basis of the value T at the lowest point of the histogram to detect the edge of the object in the image. And then take the following steps: firstly, scan every line $(i = 0, 1, 2, ..., N-1)$ of the image $f(I, j)$[2], and compare the gray scale of the scanned pixel with the value T, we can make the conclusion

$$g_1(I, j) = \begin{cases} A \text{ while } f(I, j) \text{ and } f(I, j-1) \text{ are in different zones } (0 \le A \le 255) \\ B \text{ other } (0 \le B \le 255) \end{cases}$$

Secondly, scan every column $(j = 0, 1, 2, ..., N-1)$ of the image $f(I, j)$, and compare the gray scale of the scanned pixel with the value T, we can make the conclusion

$$g_2(I, j) = \begin{cases} A \text{ while } f(I, j) \text{ and } f(I, j-1) \text{ are in different zones } (0 \le A \le 255) \\ B \text{ other } (0 \le B \le 255) \end{cases}$$

Thirdly, combined $g_1(I, j)$ with $g_2(I, j)$, the boundary image of the object can be obtained

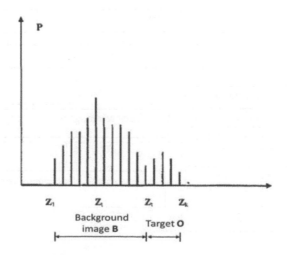

Figure 1.   Histogram of gray image.

$$g(I, j) = \begin{cases} A \text{ while } g_1(I, j) = A \text{ and } g_2(I, j) = A \\ B \text{ other} \end{cases}$$

3.3 *Image enhancement*

Image enhancement, mainly used to highlight the important information of the images and weaken the unnecessary information for the sake of the further analysis later, is an important approach to improve the quality of the images. This technology, mainly including enhancement of contrast ratio, edge, as well as the enhancement and sharpening of pseudo-color, adds no information to the original image, but highlight the property of some part of the image. Its aim is to display the difference more obviously between some information and other in the whole image and some part for the sake of post-processing. The image enhancement mainly refers to the edge enhancement which is a technology focused on the subtle changes of the gray scale, with its aim to highlight the edge of the image and improve the quality of the detected image pre-processing.

3.4 *Image compressing*

Image compressing is aiming at the data of the image. The data of some detected image is rated to more than mega bytes per second, and in the meantime the ammunition fuse safety detection demands remote real-time detection and technical support through the local area network. The remote detection demands to transmit the image data from one place to another place as soon as possible. To meet the requirements of the size and clearness of the image, an effective approach of image compressing is essential to realize the purposes of largely compressing the data of the image and reducing its infidelity as possible as it can[3].

4 APPLICATION OF COMPUTER IMAGE PROCESSING TECHNOLOGY IN FUSE SAFETY DETECTION

The ammunition fuse safety detector is composed of x-ray generator, control system, image enhancement system, computer image processing system, fixing and protection system, as shown in Figure 2.

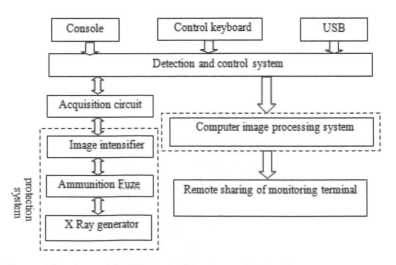

Figure 2. Composition diagram for the ammunition fuse safety functions.

4.1 X-ray generator

X-ray generator adopts mobile metal ceramic tube x-ray machine mainly launching x-ray with its penetrating ability to meet the requirements of fuse safety detection for all types of ammunition. X-ray generator is an assembled device composed of x-ray tube, high-voltage transformer, temperature sensor, air pressure sensor, cable-connected socket, warning light, x-ray tube cooling system, air valve and so on. The cooling and ventilation system are equipped at one end of the x-ray generator. To meet the needs of fuse safety detection for all types of ammunition, the technical parameters for the x-ray generator must be: Tube voltage: 20~160 KV, continuously adjustable, with the precision as ±1% and step length as 1 KV. Tube current: 0~4 mA (big focus), 0~4 mA (small focus), adjustable continuously, with the precision as ±1% and step length as 0.1 mA. Size of the focus: d = 1.0 mm (small focus) (according to the standard of EN12543), 0.4 × 0.4 (small focus) (according to the standard of IEC336). Way of cooling: anode-forced cooling. X-ray radiation angle: 40°. Maximum penetrating ability: 110 mm.

4.2 Detection and control system

The detection and control system, adopting computer control technology, is composed of computer and control software. The first is to control the beam radiation angle, maximum penetration power, focus, exposure time of x-ray generator. The second is to control the adjustment of the raster, camera zoom, camera aperture, and vision of the image enhancement system. The third is to control the operation of the detection mechanism, realizing the adjustment of parameters on the ammunition fuse safety and the real-time control of the inspection.

4.3 Image enhancement system

Image enhancement system combines x-ray imaging technology with low-light-level imaging technology by using image enhancement technology, realizing image acquisition and conversion of the ammunition fuse safety detection. Image enhancement system is mainly composed of the high-resolution CCD camera, the high-resolution main monitor with three variable electro-optical lens, optical connection system, monitoring system, collimator and so on. Through the control system of the detector, the image formed by x ray penetrating the ammunition fuse internal mechanism, is launched on the front-end fluorescence plate of the image intensifier. After the photoelectric conversion, optoelectronics is focalized by high voltage electric field in the closed vacuum cavity. The simulated image in the ammunition fuse internal mechanism is focalized on the CCD chip through the camera lens. The CCD accumulates the corresponding charge according to the intensity of light. The accumulated charge in each pixel under the control of the video sequence, moves out, forms the analog signal to output after filtering and amplification, then inputs the computer for the analog and digital conversion, and connects to the on-site monitoring system, realizing the on-site detection to the ammunition fuse safety.

4.4 Computer image processing system

The computer image processing system adopts computer image processing technology to process the image of ammunition fuse internal mechanism, mainly including hardware and software system. With such functions as multi-image storage and output, image saving, pseudo-color output, image printing, image amplification and reduction, image superimposition of the characters and Chinese characters, gray-scale display[5], geometrical measurement of the ammunition fuse safety mechanism, edge enhancement and average de-noising of multi-frame negative image, as well as window and port technology, it is able to realize the detection, printing and statistics of the ammunition fuse safety mechanism in the static single-image condition.

4.4.1 *Computer image processing system hardware*

The hardware of computer image processing system of the ammunition fuse safety detector is composed of host computer, video monitor, image acquisition card, recorder and printer. The host computer is equipped by the industrial control machine, the high-resolution graphics card, 160 G hard disk, 1 G memory CPU, IntelPentium4 processor, 19-inch ViewSonic LCD, the image acquisition card imported from Canada, the Asus read-write recorder, the Canon laser printer. It is able to save, print and burn the images of test results, realizing the remote real-time sharing and technical support of the detection results through LAN.

4.4.2 *Computer image processing system software*

Computer image processing system software is Windows XP. With the functions of static/dynamic, electronic filming, film selection, the positive/negative, size measurement, video storage, video playback, image storage, image query, image zoom, pseudo color, gray-level transformation, image printing, user management, parameter setting, exit the system, it meets the operational needs of the various types of ammunition fuse detection.

4.5 *The fixed and protection system*

The fixed system, with its design of portable and adjustable structure, can meet the fixing requirements of detecting all the ammunition fuse safety. The x-ray protection system is designed to send a red warning signal when the x-ray machine opens at a high voltage. The interlock protection device is equipped in the lead door and x-ray machine. As long as the lead door opens, the system immediately stops the occurrence of ray, which ensures the security of environment detection personnel, and meets the safety requirements of different types of Ammunition Fuse Safety detection.

5 DETECTING RESULTS

All the performances of the fuse safety detector developed with the computer image processing technology are tested, the result of which shows that all indexes of detecting instrument can meet the requirements of all types of ammunition fuse safety detection. For the sake of confidentiality, only parts of fuse detection results are described in this paper. The self-contained ammunition fuse internal structure is shown in Figure 3, in which it can be seen that the image of internal safety components for a certain type of self-contained ammunition fuse at its safety state is very clear and available to judge the condition of the fuse safety accurately.

The detector is applied to complete a nondestructive testing of a certain type of accidental ammunition fuse in an ammunition warehouse. It identifies accurately whether the accident ammunition fuse is armed or not, ensuring the normal and safe use of ammunition, and

Figure 3.   Internal structure of x-type self-contained ammunition fuse.

at the same time realizing the remote real-time detection and sharing of resources through LAN, as well as testing personnel's safety.

6 CONCLUSION

The fuse safety detector based on computer image processing technology, is a new type of ammunition fuse detecting instrument, which combines the computer technology with the x-ray generator, realizing the visualization of fuse safety detection, as well as the remote detection and the sharing of resources through the network. It improves the detection accuracy and effect, and is a major innovation in technology. It creates a new mode of detecting the ammunition fuse safety, solves the problem of the army that ammunition fuse safety is unable to be detected in real time, makes the technical approaches of ammunition security more abundant, and improves the security and efficiency of the ammunition support.

REFERENCES

[1] Guo Zhanhai, Jiao Hong. 2010. Research on the Fuse Electronic Safety and Arming Device and Its Design Criteria. *Foundation of National Defence Technology*. 26(4):6–10.
[2] Li Hongjun, Han Jiwan. 2002. Digital Image Processing and Its Application. *Computer Measurement & Control*. 12(9).
[3] Liu Zhonghe, Wang Ruixue, Wang Fengde. 2005. Status and Prospect of Digital Image Processing Technology, *Computer Era*. 24(9).
[4] Shang Yaling, Li Guangchao, Dan Bo. 2010. General Environmental Information of Fuse Safety System for Torpedo. *Journal of Exploration & Automation*. 32(4):30–33.
[5] Zhang Debao. 2009. Design of the Online Detector for Some Type Self-propelled Gun Communication System. *Control and Automation*. 20(9):6–16.

Advances in Engineering Materials and Applied Mechanics – Zhang, Gao & Xu (Eds)
© 2016 Taylor & Francis Group, London, ISBN 978-1-138-02834-0

Design of the detection and maintenance equipment for X-type fire control system

D.B. Zhang, X.Y. Li, Z.W. Li & C.A. Qiao
Wuhan Mechanical College, Wuhan, Hubei, China

ABSTRACT: Aiming at such features as complex structure, high-tech content, detection and maintenance difficulties of x-type fire control system, the high-performance detecting and maintenance equipment suitable to the army, is designed according to the current situation of the fire control system detection and maintenance. The equipment developed by applying the virtual instrument technology and fault diagnosis expert system, has good performance and strong compatibility and expansibility. It successfully solves the problems that affect the quick maintenance and the timely restoration of tactical and technical capability due to the faults of the fire control system, shortening the time of fault location, and greatly improving the capability to maintenance and support of x-type fire control system through the simplification of detection and maintenance.

1 INTRODUCTION

With the continuous application of high technology taking information technology as its core, great changes have taken place in the forms of war, as well as the style of combat and technical support. In the future high-tech local wars, there exist harder equipment support environment and exceptionally heavier support tasks, which requires that different ways of support are taken at each stage of the war, for the traditional maintenance approaches can not meet the need of equipment support. With the development of science and technology, x-type equipment with advanced fire control system is gradually equipped in the army. The fire control system which is composed of closely interrelated computer and optical, mechanical and electric system with complex structure, can comprehensively handle the information about the location, distance of the target and the working status of fire control system for x-type equipment, and send the results to the monitoring system, which is more convenient to the operators, but at the same time demands more reliability to the fire control system. Once the fire control system has something wrong, its observation to the target and real-time monitoring to x-type equipment are bound to be affected. In view of the above questions, the detecting and maintenance equipment which is developed with the latest achievement of virtual instrument technology, realizes the on-line and off-line detection and maintenance of the fire control system, and shortens the time of troubleshooting. Fire control system detector is one maneuverable detection and maintenance equipment which can realize the on-site detection and maintenance to the fire control system by using all the resources, and improve the maintenance and support capability of x-type fire control system when necessary.

2 KEY TECHNOLOGY

2.1 Virtual instrument technology

Virtual instrument, as the combination of instrument technology and computer technology, is one computer instrument system with such functions as instrument panel simulation and software testing, statistics and analysis, and can be designed and customized according to the user's requirements, on the hardware platform with the general-purpose computer as the core. The I/O interface equipment is used to fulfill the acquisition, measurement, amplification and modulation of the signals. The powerful functions of the microcomputer software are applied to realize the calculation, analysis and processing of the signals, the display function of the microcomputer monitor to simulate the control panel of the traditional instrument, to output the results of detection in various ways of expression[1]. As an innovation in the field of traditional instruments, it makes the measuring instrument and the computer effectively combined, and is a direction of equipment detection and maintenance research.

1. The detection accuracy is greatly improved. The I/O interface equipment mainly consists of signal conditioning and A/D circuit conversion, together with the digital processing function of the computer itself, so no mechanical and visual error of traditional instrument exists, and the analog signal errors is reduced, which easily ensures the high precision of the measurement.
2. The cost of the equipment is greatly reduced. Without the separate signal processing like the traditional instrument and the comprehensive measurement to mechanical part and an instrument panel, especially to the data, many functions separately measured by many instruments are focused and realized on one virtual instrument, so the cost is obviously lower.
3. The equipment customization is more convenient. The traditional instrument is difficult to meet the individual need of maintenance from the comprehensive equipment. The universal I/O interface equipment of the virtual instrument can fulfill most of its functions through software, so different functions of measurement can be realized through the design of the electric circuit and combination of different software modules according to the requirement of maintenance tasks. Additionally, the performance and functions of the instrument can be improved only by updating the related software according to the need of equipment development.
4. The equipment and computer are developed synchronously. The virtual instrument, with its open, flexible characteristics, can be synchronized with the development of computer and interconnected with the network and other equipment, to realize remote technical support to the equipment.

2.2　*Fault diagnosis expert system*

When something wrong exists in the fire control system, the detecting and maintenance equipment must analyze the fault phenomenon and BIT alarm information. Moreover, due to the growing complexion of the fire control system, one fault often causes working improperly in many places of the system and multiple BIT alarms. The fault diagnosis expert system need to be installed in the detection and maintenance equipment, to handle the fault phenomenon of the fire control system[2] and BIT alarm information and finally find the fault unit through eliminating the false alarm information and reducing the scope of detection as soon as possible.

1. Construction of RBD diagram of fire control system. Detection and repair equipment must understand the working state of the subsystems of the fire control system, as well as their mutual influence and interdependence, in order to analyze the influence of the faults from layer to layer, and construct reliable RBD diagram showing the relationship of reliability among the subsystems. On the basis of the diagram, the fault analysis of each subsystem forming the communication system is fulfilled, and the influence of the fault on the performance of higher layer system is made clear.
2. Design of FMECA system. FMECA system refers to using the diagnosis expert system of the software to make the analysis of reliability prediction, the fault mode effects and criticality to the detected, and then combining the original reliability of fire control system and fault analysis organically, to improve the accuracy and efficiency of fault analysis and ensure reliable operation of fire control system. FMECA is a bottom-up approach, through which the criticality is analyzed quantitatively, the known and potential fault

mode, as well as its causes and consequences are found out, the damage degree is estimated, and the measures and suggestions of improvement are provided. The detection and maintenance equipment, based on the input of fire control system RBD diagram, is required to conduct FMECA analysis of the communication system, and to draw the conclusion of online detection, and fault rate of various components and their typical fault modes and corresponding fault mode frequency ratio are also input as a diagnostic reference[5]. The fault ratio λ P of various electronic components can be obtained by prediction of the reliability, and be modified according to the actual situation. With the fault rate of the components, the probability of fault effect and the corresponding fault mode frequency ratio, the diagnosis expert system of the detection and maintenance equipment can preset a priority to criticality for the fire control system, and make the output of the system associated with various design parameters in the program, so that each component can send the report of fault in excess of design parameters range, and make a weighted calculation of harm degree of each mode of the fire control system. The detection and maintenance equipment chooses such parameters as the name of the related component, fault mode, fault rate and severity in the process of detection according to the suggestions from the expert system. When the detection is finished, a large amount of data is accumulated. If the rate of fault and severity is found to be unfit for the value preset by the expert system, the modification should be done according to the actual situation.

3 HARDWARE DESIGN

3.1 *Composition*

As for the hardware, the powerful processing function of computer and such advanced technology as the USB interface cable and power module are applied, so as to ensure the high quality of test function realization and make the whole system smaller in volume, lighter in weight, and a good flexibility, suitable for motor operation. The hardware of detection and maintenance equipment includes: computer system, signal conversion board, power supply module and testing cable. The diagram of hardware composition is in Figure 1.

The computer system, as the main equipment of detection and maintenance, is composed of the computer and the USB2008 A/D acquisition card. With the support of the computer hardware and the USB2008 A/D acquisition card, the command to detect and control is issued by the test system software. The signal stimulation, signal acquisition, data processing are completed through signal conversion board, and single-step test, continuous test, data browse, result print, data storage and maintenance, system reset, fault diagnosis and other functions are provided for the gun-controlled system.

In order to solve the contradiction between measured signal (analog signal) and computer-measured signal (digital signal), the signal transform circuit is added, to put the measured signal into digital signal at the TTL level, and send it to the A/D port of USB2008 A/D acquisition card for measurement via USB2008 A/D analog switch.

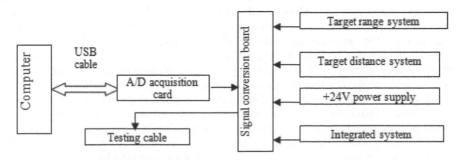

Figure 1.   Diagram of hardware composition.

Special cable for test includes 1 USB interface cable and 3 special gun-controlling system test cables, respectively, which separately realizes the connection of the signals from the computer and test case, and the signals between the detecting system and the tested object.

3.2 *Debugging*

Debugging the detecting and maintenance equipment for fire control system is one technical difficulty, and is divided into subsystem and the overall machine debugging after the prototype design is completed.

3.2.1 *Subsystem debugging*

1. Debug the target range system. After installing the purchased power module, debug the detection and maintenance function of target azimuth system, check whether output voltage of the purchased power is normal or not, insert the test circuit board No.1 into the test circuit board No.4, and check whether each working voltage of circuit board is normal or not[5]. Without anything abnormal, then debug the corresponding components.
2. Debug the target distance system. Insert the test circuit board No.2 into the test circuit board No.4, and check whether the working voltage of the circuit board is normal or not when the electricity is on. If it's normal, remove the 2D2 integration block, and use the signal generator to measure the AC signal. If it's abnormal, then debug the corresponding components.
3. Debug the integrated system. Insert the test circuit board No.3 of the integrated system into the test circuit board No.4, and check whether the working voltage of the electric circuit is normal or not. Without anything abnormal, then remove the 2D3 integration block, and use signal generator to measure AC signal, such as abnormal, then debug the corresponding components.

3.2.2 *The whole machine debugging.*

1. Install the **WINDOWSXP** operating system, LabWindows/CVI5.0 and C++Builder5.0 for the computer.
2. Connect the interface cable to the computer after debugging the subsystem.
3. Start the computer, enter the main menu, select the self-inspection, and start the test of gun-control system after passing the self-inspection.
4. The process of the test is followed according to the instructions of the Chinese interface, and its content should conform to the performance requirements of the technical specifications.

4 SOFTWARE DESIGN

Equipment detecting software shall c the principles of universal and combination design, and satisfies such requirements as the reusability of software, instrument independence, expansibility and upgrade, cross-platform operation. In order to meet the modular and general requirements, the system is developed under the control of the general test platform of the database. The selection of testing project and the control of test process are realized through the operation on the database, and then the effective development, application and maintenance of software are realized. System software consists of main program module, test module libraries, interactive capability module libraries, document library, function module library and online help. The diagram of software composition is shown in Figure 2.

The software is mainly functioned to display the digital signal after data collection through micro computer, and at the same time, compare the tested signal with the built-in circuit simulation software of the micro computer, so that the feasibility of simulation software is determined. The modular structure is adopted by the software of the equipment, and the LabVIEW virtual instrument software with its unique functions in the instrument control,

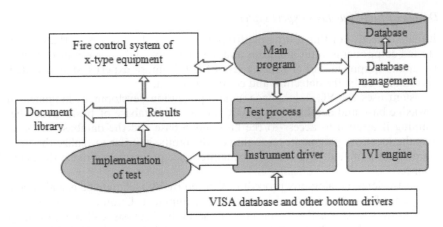

Figure 2.   Diagram of software composition.

virtual panel design and hardware access, and the IVI program testing the function module are chosen to develop the detecting function of the object to be tested.

4.1   *LabVIEW virtual instrument software*

The LabVIEW virtual instrument software of NI Company is applied in the detecting and maintenance equipment for fire control system, and its development environment is mainly oriented to test engineers and non-professional programmers. LabVIEW, with the powerful ability of man-machine interface design, is easy to realize all kinds of complicated instrument panel. Compared with traditional programming languages, the main characteristics of LabVIEW graphical programming language include:

1. The icons of a variety of test, control and data analysis function module are provided systematically, and customized or set by the detecting and maintenance equipment according to the actual need.
2. Programming is the process to design and define the flow chart, and establish the application program for the detection and maintenance by connecting the icons which represent the various functional modules.
3. The structured and modular advantages in the traditional programming language are inherited, which makes the detecting and maintenance equipment program more conveniently.
4. A number of special tools and objects, such as circuit, the amplifying circuit and so on, are provided for the panel design and data visualization analysis of the detecting and maintenance equipment. Through the simulation of the circuit, the input and output signal of the simulation circuit at all levels and even all the components are displayed on the computer, and the signal characteristics of the circuit working normally are obtained, which lays the foundation for the actual troubleshooting of the gun control system.

4.2   *Test the IVI program development of function module*

Test the functions of control and communication of the function module on the PCI board, is the key to realize the interaction of the hardware in the whole system. The application of IVI driver library developed by NI Company can easily help to develop the test function module. When the new instrument driver is installed in the system[6,3], the drive information, location information and instrument hardware address information belonging to IVI will be updated. The instrument can be updated only by modifying the specific driver associated with the defined logic name. After the system is configured, the IVI driver is available to writing the test program completely independent from hardware of the equipment.

4.3 *Design of fault diagnosis expert system*

The ultimate objective of this testing platform is to realize the analysis of the working state of the tested components, fault diagnosis and maintenance, and improve the capacity of the army to repair equipment, out of which a set of fault diagnosis expert system oriented to the object is established by the detection and control platform.

This system uses the VC++ language as an application development language, and for the knowledge base and database using ACCESS as the database of basic facts. The VC++ programming is applied to access to the knowledge base and the database of basic facts, as well as the location, judgment, reasoning of the faults, in order to realize the timely and accurate diagnosis of fault and the repair maintenance program and make the plan of the maintenance.

The object-oriented dialog box procedure Written by VC++ language is embedded in the main program of the detecting and maintenance equipment. Upon completion of the current interface test, click on the dialog box of the fault diagnosis, call the knowledge base according to the order of the test data and make a match between the IF field in knowledge base and the user's choice. If it is a successful match, call the solution of troubleshooting in the fault database to direct the users in fault processing, and ask the user whether the fault is clear. If no rules match, it means that there exist the new causes of malfunction and the user is required to update the original fault tree and database. When running the program next time, put the cause of the fault as the known one, namely the "black" incident turned to be "transparent" event, to fulfill the improvement and updating of the database. Through the test, the fault diagnosis expert system can make an accurate judgment of the faults and provide the methods of examination and treatment for their maintenance.

5 CONCLUSION

The virtual instrument technology, as an innovation combined computer technology with instrument technology, has very broad application prospects. The detection and maintenance equipment for x-type fire control system based on virtual instrument technology, shortens the development time and reduces development costs by applying IVI technology with the features of high performance, flexibility, simulation, interchangeability, which is suitable to the development trend of intelligence and networking of modern test system. The virtual instrument technology and fault diagnosis expert system are combined with the actual equipment detection and maintenance, to meet the requirement of different operational environment and support mode, and to simplify the procedures of the detection and maintenance. The diagnosis expert system can successfully analyze and judge the working status of each main module, and give guidance to fault maintenance, improving the ability of maintenance and support to x-type fire control system.

REFERENCES

[1] Wang Yifeng, Wen Xidong. 2005. Design of the Data Collection Module Based on CAN Bus. *Control & Automation.* 11(2):58–60.
[2] Xu Aijun. 1995. *Principles and Design of the Intelligent Measurement and Control Instrument.* Beijing: Beihang University press.
[3] Yang Xiaochuan, Xie Qinghua, He Jun. 2009. Fuzzy fault diagnosing method based on fault tree. *Journal of Tongji University.* 29(9):1058–1060.
[4] Zhang Debao. 2004. *X-type Electrical Equipment.* Beijing: General Equipment Support Department.
[5] Zhang Zhiyong, Wang Peng. 2003. The Development of Embedded PC104 Tester. *Industrial Control Computer.*
[6] Zhang Debao. 2009. Design of Online Detector of the Communication System of X-type Self-Propelled Artillery. *Control & Automation.* 6–16.

Advances in Engineering Materials and Applied Mechanics – Zhang, Gao & Xu (Eds)
© 2016 Taylor & Francis Group, London, ISBN 978-1-138-02834-0

The gas leak locating detection based on the improved ultrasonic transducer array group

Y. Pei

School of Automation, Beijing Institute of Technology, Beijing, China

ABSTRACT: Compressed air leak is a serious problem in the process of being used. This paper introduces a new type of ultrasonic air leak locating detection method, in the analysis of the leakage based on the principle of ultrasonic, we used the improved array-type ultrasonic transducer group and TDOA location algorithm to accurate positioning the location of the leak hole. In order to improve the recognition accuracy of the weak leak ultrasonic signal in various pressure channels and sealed containers, using the theory and method of data fusion, strengthen the ability to determine the presence of tiny leaks, therefore make the premise and foundation for the leak location.

1 INTRODUCTION

In the process of gas storage and transportation, once the leak hole exists, it will cause gas leaks[1]. The impact of the leak has not only produced losses of materials and energy, but also polluted the air[7, 3]. Locating the leak hole timely and accurate, having the great significance to improve the production efficiency and energy conservation. Therefore, gas-leak detection is an extremely important fault-detection technology, and to reduce the losses and the hazards of leakage accidents as much as possible, it is vital to detect a leak[2] immediately after the leakage accident has occurred, to specify the location of the leak, and to estimate the amount of leakage. Therefore, the research and development of leak location technology and related instrumentation have an important application value[4].

Ultrasonic leak location and detection is widely used in the field of non-destructive testing, as a kind of non-destructive testing methods, this method has the advantage of high efficiency, low cost and available online testing[5]. In the traditional ultrasonic leak location, the location of the leak detect is based on a single ultrasonic transducer, so the location accuracy depends on the directivity of the transducer[6]. But the problem of using a single ultrasonic transducer is not only the low detection efficiency, but also in many of the test results rely on the experience of testing personnel to judge. Therefore, this paper analyzes the principle of gas leak creating ultrasonic, on the basis of the array of ultrasonic transducer group signals through data fusion to receive the judgment result, and then advance the new ultrasonic leak detection and location method based on the TDOA location algorithm. The method does not need to set the background noise in the detection, and it can detect the slight leak hole within 1000 mm (0.05 mm diameter) for the accurate detection and location.

2 THE TESTING PRINCIPLE

2.1 *Propose the concept of ultrasonic transducer group*

When the pressure inside of the detected pipes or vessels is stronger than the external pressure and the leak hole is small enough, due to the large pressure difference between the inside and outside of the vessels, when the Reynolds number is high, the outflowing gas will generate turbulence, and the turbulence will produce a certain frequency of sound waves in the vicinity of the leak hole. So, we could say that the slight leak generates the ultrasonic. Considering the relationship between the ultrasonic signal leakages and noise of the frequency spectrum, a big difference value with respect to background noise around 40 kHz as can be seen in Figure 1. Therefore, we usually use an ultrasonic sensor with a rated frequency of 40 kHz to detect compressed air leak. In this experiment, FUS-40CR made by Fuji Ceramics Corporation is adopted as the ultrasonic sensor.

This paper presents the concept of the array ultrasonic transducer group; we could take the ultrasonic transducer group as a whole, and get more channels signals of ultrasonic leak detection. Using ultrasonic transducer group to detect leaks, since the detect objective is the same leak hole, each pair of the leak signals showing a significant correlation. According to the correlation theory, other signals are not correlated. So using the cross-correlation result of two transducers in an ultrasonic transducer group can serve as a leak-evaluation method that could improve the location accuracy.

2.2 *The leak location method based on the Time Difference Of Arrival (TDOA)*

Time difference of arrival leak location using the ultrasonic transducers array to receive the ultrasonic signals generated by the leak, and estimating the leak position by calculating the leak signal time delay between the different transducers. Therefore, the design of the ultrasonic transducer array has a very important implication on the time difference of arrival leak location technique. According to the results of analysis the principle of time delay estimation, the ultrasonic transducer group requires at least four transducers to determine where the positions of the leak signal of the pressure vessels in the actual detection. The planar array is not only easy to arrange, but also has the smaller array redundancy, so become the array form of widely studied.

When using the TDOA algorithm, since the two ultrasonic transducers could determine one pair hyperboloid, so it requires at least four transducers to locate the leak hole. The ultrasonic transducer group consisting of four ultrasonic transducers, and put the four transducers compact placed in the same plane of the four vertices of a square, thus forming a new ultrasonic transducer group. The arrangement is shown in the Figure 2.

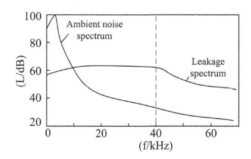

Figure 1.   The ultrasonic signal leakages and noise of the frequency spectrum distribution.

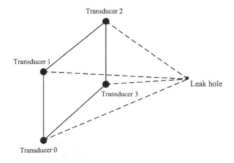

Figure 2.   The arrangement of the ultrasonic transducer group.

In Figure 3 the coordinates of the leak hole is (x, y, z), transducer 1, 2, 3, 4 located at the four vertices of a square in the same plane as the sensor for receiving signal, D is the distance between the two transducers of the diagonal cross-type array, so the coordinates of the respective transducers are transducer 1 $(D/2, 0, 0)$, transducer 2 $(0, D/2, 0)$, transducer 1 $(-D/2, 0, 0)$, transducer 1 $(0, -D/2, 0)$. The distance of the leak hole to the respective transducers is R_1, R_2, R_3, R_4. Since the ultrasonic signal generated by the slight hole can be seen as a point source, and ultrasonic signals transmit in the form of a spherical wave, t_1 is the acoustic travel time from the leak hole to the transducer 1. The time delay of transducer 1 to transducers 2, 3, 4 is τ_{12}, τ_{13}, τ_{14}, respectively. C is the velocity of sound. Equations by the geometric relationship are listed as follows:

$$\begin{cases}(x - D/2)^2 + y^2 + z^2 = R_1^2 \\ x^2 + (y - D/2)^2 + z^2 = R_2^2 \\ (x + D/2)^2 + y^2 + z^2 = R_3^2 \\ x^2 + (y + D/2)^2 + z^2 = R_4^2\end{cases} \text{ where } \begin{cases}R^2 = x^2 + y^2 + z^2 \\ R_2 = R_1 + C\tau_{21} \\ R_3 = R_1 + C\tau_{31} \\ R_4 = R_1 + C\tau_{41}\end{cases}$$

Solving the above equations can obtain:

$$\begin{cases}R_1 = \dfrac{C(\tau_{21}^2 + \tau_{41}^2 - \tau_{31}^2)}{2(\tau_{31} - \tau_{21} - \tau_{41})} \\[2mm] x = \dfrac{2(C\tau_{41} - C\tau_{21})R_1 + (C\tau_{41})^2 - (C\tau_{21})^2}{2D} \\[2mm] y = \dfrac{2(C\tau_{41} - C\tau_{21})R_1 + (C\tau_{41})^2 - (C\tau_{21})^2}{2D}\end{cases}$$

The spherical coordinates are
$$\begin{cases}R = \sqrt{R_1^2 + Dx - (D^2/4)} \\[2mm] \theta = \arccos\left(\dfrac{z}{\sqrt{R_1^2 + Dx - (D^2/4)}}\right) \\[2mm] \varphi = \arctan\dfrac{y}{x}\end{cases} \qquad (1)$$

Therefore, if get the values of τ_{12}, τ_{13}, τ_{14}, we can get the location of the leak hole, while the time delay will determine the accuracy of the location of the leak hole.

In Figure 4 shows the location process, through this process the leak location could be determined. Select as the reference point the center point of transducer 1 in the transducer array. First, determine the location of the leakage hole, and then test by comparing the real

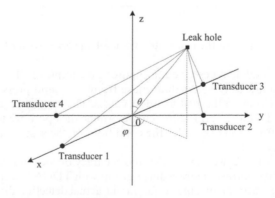

Figure 3. The modeling diagram of TDOA location algorithm.

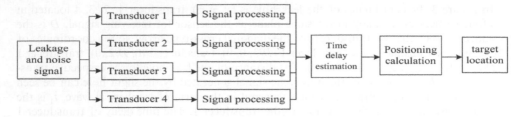

Figure 4. The location process of the system.

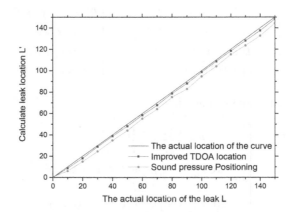

Figure 5. Comparison of the calculated and the true value curve.

Table 1. Distance 80 mm true value and measured value of leak location.

Group	Parameter distance (mm)	Azimuth	Pitch angle
Source target	80.00	1.319	0.764
Estimation target	77.91	1.286	0.753

Table 2. Distance 100 mm true value and measured value of leak location.

Group	Parameter distance (mm)	Azimuth	Pitch angle
Source target	100.00	1.396	0.785
Estimation target	101.53	1.427	0.762

value of the leakage point with the calculated value of the error to verify the effectiveness of the methods.

Using the experimental data described in Figure 5, the improved TDOA calculation algorithm gave results significantly better than those from the sound pressure calculation, with calculated positions closer to the real location values. The results were combined with the data from Table 1 and Table 2 and Figure 5 to determine the appropriate transducer spacing distance and satisfy Equation (1); under the conditions of the system, a maximum error of less than 2 mm was detected.

From Table 1 and Table 2, we can see that the accuracy of locating the leak hole that uses the improved array-type ultrasonic transducer group with TDOA location algorithm is relatively high. When the detection distance is far, in our actual detection distance, the maximum error is less than 2 mm, meet our expected requirements.

3 CONCLUSIONS

The results show the good performance of TDOA location algorithm in array ultrasonic gas leak detection and location, it could be extended to other related industries in leak detection and location. The main advantage of this method is different from the other detection methods: using the concept of the ultrasonic transducer group with four ultrasonic transducers then fuses the signal data, this method makes the results more accurate than a single ultrasonic transducer signal. The TDOA location algorithm has relatively high location accuracy and small amount of computation, which meets the requirements of leak detection and location for pressure container.

REFERENCES

[1] Alaska. 1999. Technical review of leak detection technologies. *Industry Preparedness and Pipeline Program*.
[2] Fuchs, H.V., Riehle, R. 1999. Ten years of experience with leak detection by acoustic signal analysis. *Applied Acoustic*, 33(1): 1–19.
[3] Pal-Stefan Murvay, Ioan Silea. 2012. A survey on gas leak detection and localization techniques. Journal of Loss Prevention in the Process Industries, 25, 966–973.
[4] F. Mohammad, A. Abdulhameed and N. Kahm (2007). Application of ultrasonic technology for well leak detection. *International Petroleum Technology Conference*, 2, pp. 1348–1352.
[5] Majid Ahadi, Mehrad Sharif Bakhtiar. 2010. Leak detection in waterfilled plastic pipes through the application of tuned wavelet transforms to Acoustic Emission signals. *Applied Acoustics*, vol. 71, pp. 634–639.
[6] Fernando Seco, Antonio Ramón Jiménez and María Dolores del Castillo. 2006. Air coupled ultrasonic detection of surface defects in food cans. *Meas. Sci. Technol*. 17 1409 doi: 10. 1088/0957-0233/17/6/019.

Advances in Engineering Materials and Applied Mechanics – Zhang, Gao & Xu (Eds)
© 2016 Taylor & Francis Group, London, ISBN 978-1-138-02834-0

Design of population statistics system based on real-time color video images

Q.Y. Zhang, X.P. Meng, H.Z. Xu & K.X. Chen
School of Automation, Wuhan University of Technology, Wuhan, Hubei Province, China

ABSTRACT: The system that can perform the human detection and population statistics real-timely according to the video scene of the fixed color camera in intelligent video monitoring system is realized. The system's interface is designed by the MFC, and we extract moving foreground to detect a human movement through the background difference method. The HOG is used to employ the human head and shoulders' feature extraction. The linear support vector machine as classifier realizes the head and shoulder model classifying. Dynamic human body counting is finally realized through targets mark and the real-time tracking. The experimental results show that the system can accurately realize the quantity statistics of floating population.

1 INTRODUCTION

With the continuous development of computer vision and pattern recognition technology, and the continuous improvement of hardware performance, and video monitoring system has been in constant development to intelligence. Detecting, tracking and recognizing the character, and then understanding and describing its behavior from the video image is becoming a hot spot of research in recent years. Population statistics is a vital step to human abnormal behavior symptom detection as well as public population control. Therefore, the population statistics in visual scene is of great significance.

In order to improve the practicability of population counting in video monitoring. A software, whose parameters can be changed real-timely, has been developed. The software can be debugged by the user in the process of monitoring according to the real-time effect, effectively avoiding the adaptability issues caused by the system algorithm itself, which makes its practicability greatly improved.

2 SYSTEM DESIGN

The population statistics system based on real-time color video images has been successfully developed by OPENCV, its interface is shown as Figure 1. It consists of four parts: control real-time video of original image area, set parameter area, display the effect of real-time parameter adjusting area, and display image about final processing result area.

2.1 Image control area

The area includes real-time video frame display area, video control area and video information display area.

The real-time image of monitored area is displayed in the real-time video frame display area and its information will show in the video information display area. We realize the function of monitoring the most original video and the image will be compared with the effect of all kinds of image as a finally reference.

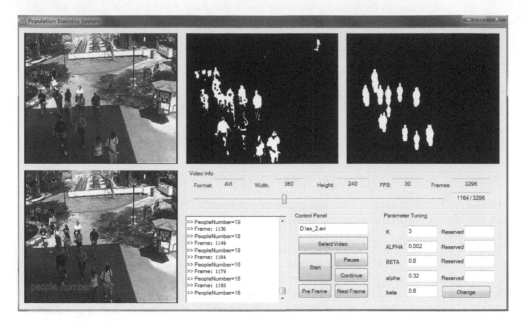

Figure 1.   System interface.

Video control area can realize the functions including choosing the position in which the video frame is stored in PC, opening, pausing, continuing and running the image in which the video frame exist and choosing the number of image frames (the number of frame can be read in the video information display area).

2.2   *Parameters tuning*

The correct execution of the system scheme depends on the following aspects mainly.

2.2.1   *Moving object detection based on the Gaussian background model*

The system uses the Gaussian background model to separate foreground and background. Because of several factors including the light and fluctuation, the same pixel's RGB channels will change in the video frames, so we should use the mixture Gaussian model that has the variable Gaussian components to describe different pixels at the same time and the same pixel at different time. It is necessary to change the Gaussian component according to the value of each pixel over time in order to make the mixture Gaussian model in the background modeling has the fastest convergence rate, minimal computational complexity, the most accurate and fast background updating. Based on the above analysis, the user can change the Gaussian component K real-timely in the parameter tuning area. If the K is taken too large, the calculation complexity is increased. If the K is too small, when it has a new moving object, the Gaussian background model will not fully updated and it will affect decay rate of the model. By testing, it is ideal for K to get 3~5.

In this algorithm, adjustable parameters also include the weight updating rate ALPHA and mean and variance updating rate BETA. When the two parameters are set too large, the learning rate is so fast that background model updating rate is too high, and it will lead to slow moving targets are detected as a background image, conversely, it is easy to appear smear or ghost phenomenon. Therefore, in order to avoid the occurrence of such phenomena, the users can adjust the value of this two parameters in 0~1 by themselves.

The original image is subtracted from the background image through the Gaussian background modeling method can obtain the foreground image. But the resulting target image often exists a serious shadow problem, so we need to deal with the problem. In the process of solving the shadow problem, the system provides two parameters alpha and beta to adjust the depth of shadow. The adjustment range is 0 < beta < alpha < 1.

Monitoring persons can adjust the above parameters to obtain the optimal effect according to observing the pictures in the parameter adjustment display area.

2.2.2   *Using HOG for feature extraction of human head and shoulder*

This system uses HOG feature as the feature descriptor to detect human head and shoulders. First, we need to construct the sample set for training classifier model, including positive and negative samples. Using human head and shoulder model instead of the traditional head model as the positive samples, not head and shoulder but suspected human head and shoulder as the negative samples. Because the human head and shoulders model is more characteristic than the traditional head model, so we can reduce the number of required samples greatly. Then we extract HOG feature vector, and through the geometric attack, we calculate gradient, calculate every cell's gradient orientation histogram, normalized for cell in each of overlapping block, assembling all of blocks' histogram vectors to get a large HOG feature vector, also called HOG feature descriptor.

2.2.3   *The linear SVM as classifier realize the head and shoulder model classifying*

We input a large number of positive and negative training samples' HOG feature descriptor and labels to SVM model for training, so we can get the optimal classifier according to this training samples. The video sequence to be detected will input to the trained classifier, and it will be scanned by the testing window at different positions. We can judge the window whether it is the region of human head and shoulders on the basis of the classifier's decision result. Finally, we can track and mark the position of human head and shoulder in the video sequence.

In this part, the sample training is realized finally by using SVM algorithm in OPENCV. The default parameters in the software package can solve the practical training problem, so users need not to adjust any of parameters.

2.3   *Parameter adjustment display area*

This area consists of foreground-stripping-image and sample-module-image. We can observe the effect to the algorithm according to adjusting above parameters. The sample-module-image will display the training samples' head and shoulder image. By observing this area, we can adjust the parameters to obtain the best effect.

2.4   *The final result display area*

This area consists of the marking of human head and shoulder and the displaying of counting value, system will mark the head and shoulder detected and display the amounts. These amounts mean the count of final number of people.

3   SYSTEM TEST RESULTS

The system was tested in the videos on Common scene and Overlooking scene, respectively. The video with overlooking scene contains 3296 frames and its experimental results are shown in Table 1.

The video on common scene consists of 908 frames. Its experimental results are shown in Table 2.

This system is suitable for common scene to quantity statistics of floating population, but it is not for overlooking scene. The error in this system is mainly caused by missing moving target in bidirectional projection detection or seriously image blocking. False detection will not occur in this system in the situation of fixed camera and relatively fixed background. The experimental results show that this system is accuracy, real-timing and robustly.

Table 1. Overlooking scene's experimental results.

Frame count	Actual number	Detect number	Accuracy
94	16	31	7%
314	9	17	12%
908	4	3	75%

Table 2. Common scene's experimental results.

Frame count	Actual number	Detect number	Accuracy
476	6	5	80%
1341	9	10	90%
2394	7	7	100%

ACKNOWLEDGEMENT

The paper is supported by WHUT National Undergraduate Training Programs for Innovation and Entrepreneurship (20141049711004).

Author to whom correspondence should be addressed. E-mail: qyzhang@whut.edu.cn.

REFERENCES

[1] Wang Yu-Ting. 2014. *The Vehicle Identification Technology Based On BP And HOG Apply To The Wireless Devices*. Jilin: Jilin University.
[2] Fu Zhi-yong. 2012. *Implementation and Optimization of Pedestrian Detection Algorithm based on HQG+SVM is DM6437*. Guangdong: South China University of Technology.
[3] Huang Dong-li, Dai Jian-wen, Feng Chao, et al. 2012. Three Linear Interpolation Algorithm in HOG Feature Extraction. *Computer Knowledge and Technology*, 8(31):7548–7551.
[4] Lu Hu-chuan, Zhang Ming-xiu, Zhang Ji-xia, et al. 2008. Effectual Real-time Method for Crowd Counting. *Computer Engineering*, 34(5):222–225.
[5] Hu Yao-wu. 2012. *Research on Video Analysis for People Flow Automatic Counting System*. Zhejiang: Zhejiang Sci-Tech University.
[6] Han Ting-mao. 2012. Counting people in crowds based on independent motion. *Electronic Design Engineering*, 20(20):55–57.
[7] Zhang Ying. 2013. *Research on Key Techniques Real-time People Counting Based in Video*. Hangzhou Dianzi University.

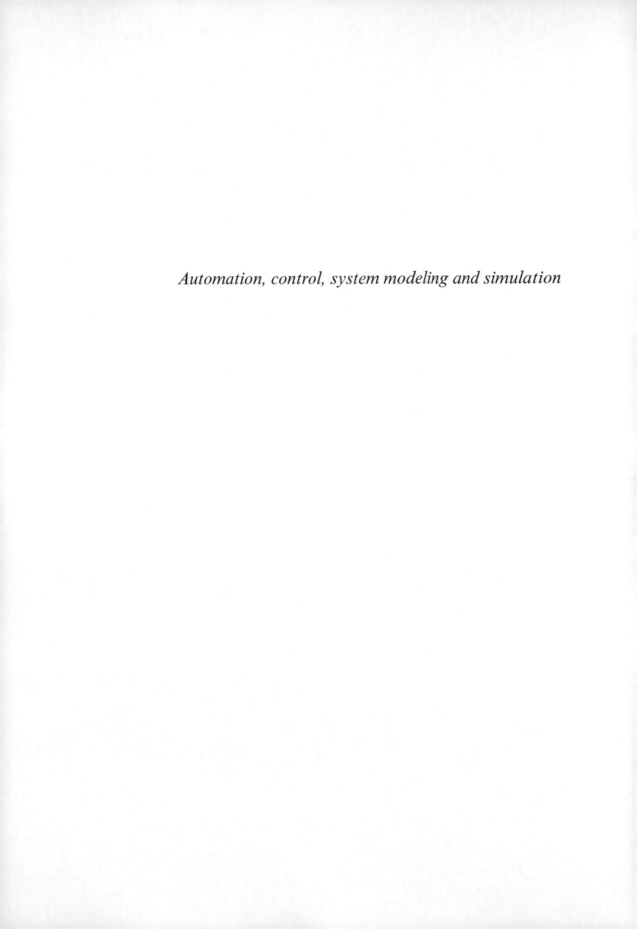

Automation, control, system modeling and simulation

Advances in Engineering Materials and Applied Mechanics – Zhang, Gao & Xu (Eds)
© 2016 Taylor & Francis Group, London, ISBN 978-1-138-02834-0

Path planning with PSO for autonomous vehicle

L. Cai & J.P. Jia
Mechanical and Electrical Engineering College, Wuhan Donghu University, Wuhan, China

ABSTRACT: Vehicle routing problem is a combinatorial problem, which has a practical engineering value in logistics distribution. In this paper, we intend to propose a path planning for vehicle model based on a particle swarm optimization algorithm, which aiming to achieve the optimization of path planning. The particle swarm optimization algorithm is a new swarm intelligent optimization method, which is applied to the vehicle routing problem, constructed by the particle representation methods for vehicle routing problem. The results show that the particle swarm optimization algorithm can obtain the solution for the vehicle routing problem quickly and effectively. It is a good method for solving the vehicle routing problem.

1 INTRODUCTION

Along with improvements of the market economy and logistics technology, the logistics industry has been developed rapidly. There are many optimization problems in the logistics business[1][2]. The reasonable selection of distribution route has a great influence on the quality of services such as the delivery speed, distribution costs and economic benefits. Many path planning methods for vehicle routing have been developed, such as global artificial potential field method, genetic algorithm method, and neural networks approach[3]-[5]. However, each method has its own limitations in some aspects. In recent years, intelligent soft computing techniques such as particle swarm optimization, artificial neural network, and interacting multiple model system are proven to be efficient and suitable when they are applied to a variety of systems[6][7]. A Particle Swarm Optimization (PSO) algorithm is a new evolutionary algorithm developed in recent years. In order to find the optimal solution, it uses the iterative method from the view of the random solutions and evaluates the quality of the solution through the fitness. And it is more simple than genetic algorithm because it has no genetic algorithm "crossover" and "mutation". It can find the global optimum by following the searched optimal values. Thus, it can be seen that the PSO algorithm is acceptable to solve routing problem, which has flexibility and collaboration features.

In view of improving the limitations, this paper proposes a path planning for autonomous vehicle based on the particle swarm optimization algorithm. Obstacle detecting and avoidance method for autonomous vehicle are implemented with the sensors. Then, the PSO algorithm for path planning is introduced and the simulation results in the MATLAB software show that the method of introducing PSO into autonomous vehicle is convenient. By this means, the path planning is well optimized in real-time way for vehicle routing problem. The chapters are as follows: Section 2 explains the system design of autonomous vehicle based on a single chip microcomputer. Section 3 introduces the principle and schematic diagram of the PSO algorithm, and then gives the simulation results in the MATLAB software.

2 STRUCTURE PRINCIPLE OF AUTONOMOUS VEHICLE

In fact, the autonomous vehicle is mainly made up of the mechanical part and the control part. The main mechanical part is designed for wheeled vehicle. The system of the vehicle has four wheels, in which two front wheels can control the direction, and two rear wheels are responsible for support. Thus it is relatively easy to control. The vehicle attached to a battery box, is equipped with four pieces of battery, so the operation of wheeled vehicle is more convenient. The single chip microcomputer is the core as the control part. The entire control circuit is composed of four parts: a motor drive module, obstacle avoidance module, tracking module, and extinguishing module.

So, the autonomous vehicle can control the steering through the servo motor by detecting obstacles on the road. The core chip can adjust the duty cycle of the signal through the signal line. Thus, it realizes the adjustment direction because the motor angle is different. In the tracing detection, infrared sensors are used for road detection. After the comparison, analysis and processing of the detection signals, the single chip microcomputer can issue corresponding instruction execution to drive motor, and control the entire operation of wheeled vehicle.

According to the characteristics of infrared detection, infrared rays have different reflection on different color surfaces. The wheeled vehicle continuously emits the infrared light to the ground. If infrared light meets the white line, it will produce a diffuse reflection, and the reflected light is received by the receiving tube arranged on the robot. While infrared light meets the black line, the light will be absorbed, and the receiving tube will not receive the infrared light. The single chip microcomputer can determine the position of black according to the reflected infrared light with three infrared detectors, which are respectively arranged on the robot at the lower part of the left, middle, and right.

3 THE PRINCIPLE AND SIMULATION OF PSO ALGORITHM

3.1 The principle of PSO algorithm

The PSO algorithm has a significant difference from other intelligent computing methods that few parameters need to be adjusted. However, the key parameters have a significant impact on the accuracy and efficiency of the algorithm. In general, the feasible solutions of each optimization problem are to search for a particle in space. Each particle has two parameters, one is the position in the search space, and the other is flying speed of a particle, which can be represented by the following equations:

$$v_{id} = v_{id} + c_1 * rand( )*(p_{id} - x_{id}) + c_2 * rand( )*(p_{gd} - x_{id}) \tag{1}$$

$$x_{id} = x_{id} + v_{id} \tag{2}$$

where v_{id} is the velocity component of the NO. i particle; c_1, c_2 are the acceleration constants; $rand( )$ is the uniform distribution random function between the interval; p_{id} is the current searched optimal solution component of the NO. i particle; x_{id} is the position component of the NO. i particle; and p_{gd} is the current optimal solution component of the whole community. Equation (1) determines the trajectories of particles in the population. The first part in Equation (1) is the previous speed, which determines the ability of particle in the new search area. The second part is "cognition", which can strengthen an enhanced learning process. The third part is "the agency", which represents the information sharing and cooperation. Schematic diagram of the PSO algorithm is shown in Figure 1. Specific steps of the PSO algorithm are as follows:

First, the parameters about the particle population are initialized, including the position and flight speed of the particles, and the variables are initialized such as current searched optimal solution of each particle, and current optimal solution of the whole community.

Second, it evaluates the fitness of each particle.

Third, it compares the fitness value of each particles: if the particle is better than current initialized optimal solution of each particle, set the particle to the current value. Meanwhile, if the particle is better than current initialized optimal solution of the whole community, set the particle to the current value.

Fourth, the particle position and velocity are changed according to above the Equations (1) and (2).

Finally, it will automatically end and return to step two if it reaches the end conditions, such as good location or the maximum number of iterations.

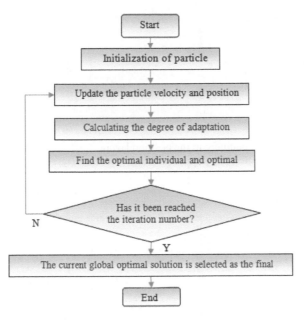

Figure 1.　Schematic diagram of the PSO algorithm.

In general, velocity must be limited to the maximum speed. The more maximum speed can improve the global search ability of the algorithm, while the smaller maximum speed can improve the local exploitation ability. But if the maximum speed is too high, the particles may fly over the good solution; if the maximum speed is too small, the particle will not be enough to explore in local interval. The maximum speed is usually set to 10–20% of variable range according to experience, which has a better effect.

Acceleration constants c_1, c_2 represent the statistical acceleration weight of current solution position of each particle and searched optimal solution of the group. So, c_1, c_2 is usually set to 2. Low values will allow particles wander outside the target area before being pulled back. While the high value will cause the particles over the target area suddenly. Population size is generally set to 20–50, which is related to the specific problems.

3.2　Simulation of PSO

Given n sites and the distances between any two sites, it seeks a shortest path through each site. The graph theory can be described as follows: given a weighted graph G = (V, E), where V is the vertex set, E is for the arc set connected with each vertex. The distances between the vertices are known. Let d_{ij} be the location of NO. i, j, which is the distance between the arc length. The introduction of variable x_{ij}, then the objective function is as follow:

$$Z_{\min} = \sum_{i=1, j=1}^{n} x_{ij} d_{ij} \qquad (3)$$

In Figures 2(a) and 3(a), the position layouts are different although the total number of places is the same, where n is set 50. And the maximum number of iterations is set to 200 and 150 respectively in the iterative process, as shown in Figures 2(b) and 3(b). There are many possible paths and it need to seek the optimal solution by comparing all possible cases. Then the final results about path planning will run out by searching the minimum value according to the Equation (3) under the different parameter settings, as shown in Figures 2(c) and 3(c). No matter from which side, it can draw the conclusion that the system with PSO has better tracking performance than the system without it and the method of introducing PSO into autonomous vehicle is convenient.

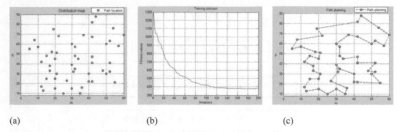

(a)   (b)   (c)

Figure 2.   Simulation of path planning in the position layout.
a—A position layout; b—The iterative process; c—Path planning diagram.

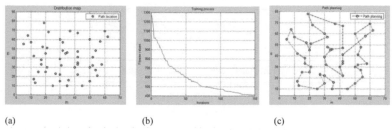

(a)   (b)   (c)

Figure 3.   Simulation of path planning in a new position layout.
a—A new position layout; b—The iterative process; c—Path planning diagram.

4   CONCLUSION

This paper introduces a simple design of an autonomous vehicle design based on the PSO algorithm, using a single chip microcomputer as the core, with photoelectric sensors to detect the path goals. Then the principle of the PSO algorithm is introduced and the simulation is developed by the MATLAB software. The simulation results show that the PSO algorithm into autonomous vehicle is feasible and has better performance of the path planning compared with the traditional method.

ACKNOWLEDGEMENT

This paper was supported by the Scientific Research Project of Hubei Province (No. B2013198).

REFERENCES

[1] Yin, X., Yang, C., Zhou, W., Xiong, D. 2014. Energy-efficient tracking control for wheeled mobile robots based on bio-inspired neurodynamic. *Comput. Inf. Syst.* 10(6): 2533–2541.
[2] P. Vadakkepat, O. Miin, X. Peng, and T. Lee. 2004. Fuzzy behavior based control of mobile robots. *IEEE Trans. Fuzzy Syst.* 12(4): 559–565.
[3] Rahim, Nawi. 2008. Path Planning Automated Guided Robot. *Engineering and Computer Science*, (4): 22–24.
[4] Chao, M. 2002. A Tabu Search Method for the Truek and Trailer Routing Problem. *Computers & Operations Researeh*, vol. 29: 33–51.
[5] Wang Quan, Wang Wei, etc. 2012. Wheeled robot path planning method based on hybrid strategy. *Computer Engineering and Applications*, vol. 9: 2–5.
[6] Francesco M. Raimondi, Maurizio Melluso. 2005. *Robotics and Autonomous Systems.* vol. 52: 115–131.
[7] Adem Tuncer, Mehmet Yildirim, Kadir Erkan. 2012. A Motion Planning System for Mobile Robots. *Advances in Electrical and Computer Engineering.* 12(1): 57–62.
[8] Lei Juan. 2012. *Principle and application of single chip micro.*

Advances in Engineering Materials and Applied Mechanics – Zhang, Gao & Xu (Eds)
© 2016 Taylor & Francis Group, London, ISBN 978-1-138-02834-0

Research on high-efficiency rotating screen printing control system of ceramic tile

L.G. Cao, H.P. Pan, L. Zhang & H. Feng
School of Mechanical and Electronic Engineering, Jingdezhen Ceramic Institute, Jingdezhen, Jiangxi, China

ABSTRACT: To improve the screen printing efficiency of ceramic tile, a kind of control system based on PLC and embedded system that has four operating position, is presented in this paper. The logic and sequence control system is designed with PLC, which controls the outputs of three AC motor, one server motor and one step motor and checks the signals of sensors in the system; the embedded system is constructed with the kernel of arm9, which implements functions of communication with PLC and other network devices and provides convenient user interface to operate the system. The embedded system communicates with PLC by RS232C serial port. Practice indicates that the system has merits of high efficiency, good operability and satisfies the requirement of production.

1 INTRODUCTION

The printing techniques of ceramic tile mainly include screen printing, roller printing and jet ink printing. In contrast to the other printing techniques, the screen printing technique has the merits of good quality, more multi-color, and slight pollution. Since 1980, the techniques about screen printing in the screen printing process and control, have been researched. Yongqin Liu presented a kind of screen printing technology of archaize tile; Zhiming Zhang designed a kind of screen printing control system that has eight color operating positions, with PLC that can control nine frequency converters; Huaizhong Chen built a monitor system of screen printing with PC-LINK communication network; Xiaowen Zha designed a new control method about screen printing and modified an image process algorithm of jet ink printing to improve efficiency; Fangbo Fu built an automatic control system of screen printing with PLC and touch screen.

Control systems of screen printing of ceramic tile in those researches are mainly in a batch mode, that is the system waits until the ceramic tile is transmitted to the printing position by belts or driving devices. The printing efficiency is low. Considering the problem, a kind of screen printing control scheme with four operating position is presented, in which kernel is PLC and embedded system. The system implements a logic and sequence control of screen printing process. Additional functions such as remote communication and convenient operation are designed to the system. The efficiency and operability of screen printing of ceramic tile is improved.

2 SCREEN PRINTING SCHEME OF CERAMIC TILE

The system mainly includes three parts: feeding control part, rotating work position control part, and output control part, as shown in Figure 1. The feeding control part that includes one AC motor for feeding ceramic tile, one step motor for adjusting vertical position of the cylinder, and conveyer belt, control position and speed of ceramic tile for feeding and insures that the supply of ceramic tile is not interrupt. The rotating work position control part that includes one AC motor of printing-squeegee, one AC server motor, rotating platform and one cylinder, are responsible for printing continuously, controlling the position of the rotating

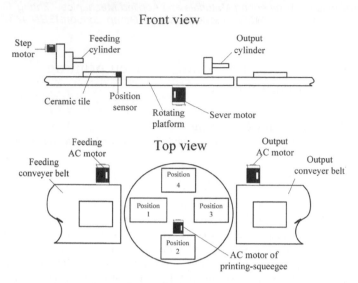

Figure 1. The screen printing scheme of ceramic tile.

platform and transmitting ceramic tiles to the output conveyer belt. The output control part that mainly includes one AC motor and conveyer belt, is used for transmitting the ceramic tile which has been printed, to the next operating position.

The working principle is that: the position sensor is trigged when the ceramic tile is transmitted to the end of the conveyer belt by the feeding AC motor. Then, the feeding AC motor stops and the step motor make a given turns, the plate goes down and the feeding cylinder act and push the ceramic tile on the feeding conveyer belt into the first operating position labeled as position 1 in Figure 1 on the rotating platform. There is a positioning fixture for ceramic tile at every operating position on the rotating platform. At the same time, the second AC motor for printing-squeegee starts to print on the operating position 2 and the cylinder on the operating position labeled as position 3 in Figure 1 push the ceramic tile on the operating position 3 to the output control part. The above three actions start at the same time, the period of screen printing is determined by the action among the three actions that need the longest time. After the actions complete, the AC server motor under the rotating platform turns 90 degree, and then the next work period starts.

The control of rotating platform is crucial for the system. There are four operating positions on which there is a positioning fixture, on the rotating platform. The work mode of the system is a three-stage pipeline that is feeding and fixture of ceramic tile, screen printing and output of the ceramic tile. The system can get high efficiency in this mode.

3 CONTROL SYSTEM OF SCREEN PRINTING OF CERAMIC TILE

To improve the stability and operability of screen printing of ceramic tile, two layer control structure is determined, as shown in Figure 2. The first layer is PLC control system that controls all motors, switches, and sensors in the system. The kind of motors of the system is divided into three class: three-phase asynchronous motor that includes feeding AC motor, output AC motor and AC motor of printing-squeegee, step motor which control the height of the feeding cylinder and the AC server motor that includes the reducer and encoder.

The second layer is an embedded system that connects PLC with RS232C serial port, and provides functions such as setting PLC running parameter, receiving data from PLC, providing UI, communicating with remote device, etc. The second layer provides additional functions to control the system and can installed not together with PLC. The second layer improves the operability of the system.

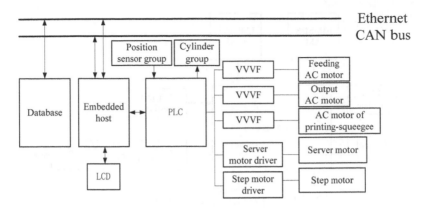

Figure 2. Control structure of the screen printing of ceramic tile.

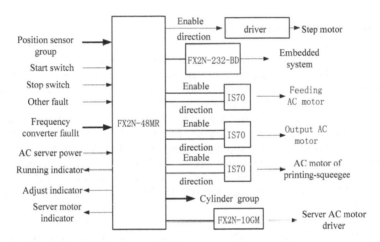

Figure 3. PLC connection diagram.

4 THE STRUCTURE OF THE PLC SYSTEM

The connection diagram is shown in Figure 3. The model of PLC is FX2 N-48MR that is manufactured by Mitsubishi Electric. The inputs of the PLC system include master switches, which can start, stop the system and other functions, fault signal from the three VVVF, position sensor signals and other fault signals. The position sensor signals include one limit switch of feeding, two limit switches on the cylinders, switch position signals about four operating positions on the rotating platform, and switch signal of printing-squeegee position. The output signals of PLC include two active signals of cylinders, activity and direction signals of AC motors and step motors. As shown in Figure 3, FX2 N-10GM, which outputs position set signals of the AC server motor and occupies eight I/O points, is an extension unit of FX2 N PLC and is a location unit of one axis. To allow some I/O point margins, the model of PLC FX2 N-48MR is selected, which includes 24 input points and 24 output points.

FX2 N-232-BD, which is a RS232C communication extension board of FX2 N PLC, connected with embedded system. IS70 is a common AC frequency converter that can be accessed by RS485 bus or extern analog voltage. Considering that the motor speed of the three AC motor in the system need not be adjusted continuously, the extern analog voltage access mode is adopted.

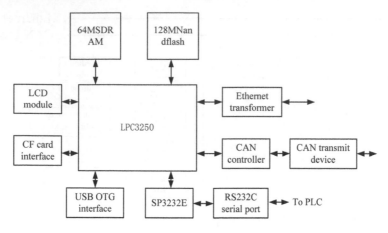

Figure 4. The structure of the embedded system.

5 EMBEDDED SYSTEM STRUCTURE

To improve the operability and functions of the system, the embedded system is designed with arm9 kernel. The embedded system can access database, provide UI, transfer data through CAN bus, Ethernet and RS232C bus, etc. The structure of the embedded system is shown in Figure 4. The model of the arm9 is LPC3250, which integrates ARM926EJS processor that can work at the frequency up to 266MHZ, a Vector Floating Point (VFP) coprocessor, 32 kB instruction cache and 32 kB data cache, two NAND flash controllers, external memory controller, eight channel general purpose DMA controller, 10/100 Ethernet MAC, USB interface supporting either device, LCD controller and Secure Digital (SD) memory card interface, etc. For Wince operating system to be ported, 128M NAND flash and 64M SDRAM memory structure are designed through external memory controller of LPC3250. The interface of CF card is used to store data, and in the process of the system porting, it is used to store wince system temporarily that has been complied. LCD module, USB port, RS232 port, CAN interface, and Ethernet interface are designed in the embedded system.

6 CONCLUSIONS

A kind of screen printing control system of ceramic tile, which include three stages pipeline, is presented in this paper. The PLC system is designed to control logic and sequence of screen printing of ceramic tile; it can improve reliability and stability of the system. The embedded system, which includes LCD module, Ethernet interface, USB interface, CF card interface, extern memory module, and CAN bus interface, is designed with arm9 kernel to improve the capability of remote communication and convenient operation. The embedded system communicates with PLC system by RS232C cable. The system improves the efficiency and operability of screen printing of ceramic tiles.

ACKNOWLEDGEMENTS

This work was financially supported by Key Technology R&D Program of JIANGXI China (20111BBE50020).

REFERENCES

Biaoguo Yao., Pengfei Li., Qin Chen. 2007. Multiple flat screen printing machine based on CAN bus line monitoring system. *Journal of Xi an Polytechnic University* 21(6): 858–863.

Daqin Zhou., Taosha Jiao. 004. A new type of double servo driving flat screen Printing Machine Electric control. *Textile Machinery* 4 (2): 28–30.

Fangbo Fu., Shunqi Mei., Lei Zhao., Zhiming Zhang. 2008. A study of automatic silk screen printing machine's control system based on function module serial communication. *Modern Machinery* 4 (4): 65–67.

Huaizhong Chen. 2014. Design of networked and electric control system for flat screen printing unit. *Journal of Textile Research* 35 (2): 78–83.

Junfeng Jing., Xuejuan Kang., Pengfei LI. 2010. digital PM AC servo in flat screen printing unit design. *Textile Machinery* 24 (1): 28–31.

Xiaobo Zha., Qiang Han. 2008. Control method for improving efficiency of digital inkjet printing. *Manufacturing Automation* 35 (6): 66–68.

Xiaohua Wang., Pengfei Li., Yuanxiang Ding. 2007. Design and Realization of Site Bus Control System of Flat-web Printing Machine. *Electric Engineering* 15 (9): 50–52.

Yongqin Liu. 2013. Antique ceramic screen printing. *Screen Printing*. 3 (3): 24–27.

Zhiming Zhang., Qiao Xu., Shunqi Mei. 2013. Research on automatic Screen Printing control system based on protocol macro. *Journal of Wuhan Textile University* 26 (6): 71–73.

Zhisong Zhu., Xiaozhao Yan., Jianwei Gong., Yiming Zhu. 2008. The design of the elevating mechanism of revolving printing machine. *Machinery Design & manufacture* 1 (10): 36–37.

The design of material sorting system based on S7-200

H.X. Wu
Wuhan Donghu University, Wuhan City, Hubei Province, China

ABSTRACT: The goods can be separated according to different materials by the automatic sorting system, it has the advantages of high efficiency, low cost etc., and PLC has the advantages of complete functions, versatility, high reliability, strong anti-interference ability, etc. The application of Siemens PLC in material sorting control system is introduced in this paper. Composition, control principle, control flow and control program were emphatically described, it meets the actual production requirements. The field experience shows the design of the system is reasonable, stable and reliable, and high automation.

1 INTRODUCTION

The automatic sorting system can pick up the goods in enormous and continuous quantities, it is widely used because of high efficiency, low cost, etc.[1]. And PLC has been widely used because of small volume, complete functions, versatility, high reliability, strong anti-interference ability, etc.[2, 3].

This paper puts forward a kind of material sorting system based on S7-200, it makes the different materials sorting available by the way of motor operation control, signal processing of sensors and magnetic valve control, etc. The system can realize the sorting of different materials according to the material texture and color. It has the advantages of high efficiency, low cost, high precision, and so on.

2 STRUCTURE AND WORKING PRINCIPLE OF MATERIAL SORTING SYSTEM

2.1 The structure

The workpiece can be separated according to the different materials, and the detection of different materials is mainly achieved through the sensor, different types of sensors can detect different kinds of material[5].

In this paper, the material sorting system composed of AC motor, cylinder, solenoid valve, rotary encoder and the color recognition sensor, capacitive sensor, inductive sensor and photoelectric sensor. The structure of the material sorting system is shown in Figure 1.

In Figure 1, YV1–YV4 are solenoid valve of the pushing cylinder 1–4, YV5 is solenoid valve of feed cylinder, SFW1–SFW4 are action limit switch of the pushing cylinder 1–4, SFW5 is action limit switch of the feed cylinder, SBW1–SBW4 are return limit switch of the pushing cylinder 1–4, SBW5 is return limit switch of the feed cylinder. SN is a feed sensor, it adopts photoelectric sensor and it converts the change of measured into optical signal change first, then convert optical signal into electrical signal with the help of photoelectric element, here it is used to detect the arrival of the goods.

SA is an inductive sensor, it is composed of high frequency oscillator and amplification circuit. The eddy currents generated in the object when the metal object approaches the vibration sensing head which can produce electromagnetic field, it reactions to the sensor

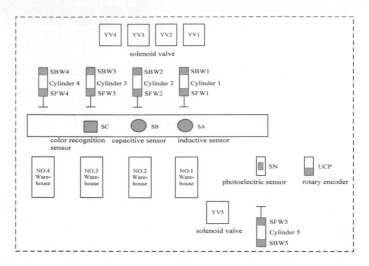

Figure 1.   The structure diagram of the material sorting system.

and makes the sensor ability of oscillation damping, the internal parameters of the circuit changing and so on[6]. It used to detect iron material in the system.

SB is a capacitive sensor. The measuring head is a plate of the capacitor, another plate is the measured object itself, the dielectric constant of the sensor will change when the object is close to the sensor[7]. It changes the state of the circuit that is connected with the measuring head, so as to control the on and off. It used to detect aluminum material in the system.

SC is a color recognition sensor. It is composed of photodiode and color filter. When the light intensity which incident to photodiode to keep a certain, the photoelectric current output is changes dependent on the wavelength of incident light[8]. It can reflect the difference of color due to the different wavelengths of light.

2.2   *The working principle*

The feeding sensor generates an output signal when the goods arrive, the goods are pushed to the conveyor belt by feed cylinder, and then the conveyor belt sends the goods to sensor detection area. It is judged as an iron material when the inductance sensor outputs signal, and it will be put into the no. 1 warehouse; It is judged as an aluminum material when the capacitive sensor outputs signal, and it will be put into the no. 2 warehouse; And the color recognition sensor is set to identify yellow color in the system, so it is judged as yellow plastic material when the color recognition sensor outputs signal, and it will be put into the no. 3 warehouse; It will be put into the no. 4 warehouse when there is no sensor output.

3   THE CONTROL PROCESS AND REQUIREMENTS OF THE SYSTEM

3.1   *The control process and requirements*

The control process and requirements of the material sorting system are as follows:

1. When the system is powered, the action mechanism will return to the initial position, all of the cylinder will return to the limit state and the conveyor belt will start running.
2. When goods are loaded to the discharge tower, feed sensor outputs signal and the conveyor belt will stop running. The goods are pushed to the conveyor belt by the feed cylinder, then the conveyor belt sends the goods to sensor detection area.
3. It will be put into the no. 1 warehouse when the inductance sensor outputs signal.
4. It will be put into the no. 2 warehouse when the capacitive sensor outputs signal.

5. It will be put into the no. 3 warehouse when the color recognition sensor outputs signal.
6. It will be put into the no. 4 warehouse when there is no sensor output.
7. The system reset after a period of operation when there are no goods on the conveyor belt.
8. The conveyor belt will stop running during the period of push cylinder working, it will be running until the cylinder back to return state.

3.2 *The flow chart*

The flow chart of control system is shown in Figure 2.

4 THE DESIGN OF MATERIAL SORTING SYSTEM BASED ON S7–200

4.1 *The I/O address allocation*

According to the analysis of the control requirement, the input signal of the system includes 4 sensor signals, 1 photoelectric encoder signal and 10 limit switch signal of the cylinder, output control object contains 5 solenoid valves of the four pushing cylinder and one feed cylinder, 1 conveyor belt motor.

This control system uses the Siemens S7–200 CPU226 to achieve control. And its I/O address allocation is shown in Table 1.

4.2 *Control program design*

The control program of the system is shown in Figure 3.

Network 1 is high counter initialized. The mode of high-speed counter was set by transmits 16 # F8 to SMB37 and the high speed counter HSC0 starts by high speed counter defined instruction HDEF.

In Network 2, the current value of high-speed counter was transmitted to VW20.

Network 8 is used to control the feed cylinder. When the feed cylinder in the location of return, it will start when the feed sensor outputs signal, and it will stop when it action to the limit state.

Networks 9 to 14 are used to control three push cylinder respectively. Networks 9 and 12 are used to control the No. 1 push cylinder. When the iron material passes, the inductance sensor outputs signal and No. 1 cylinder will moved from the return location to limit state.

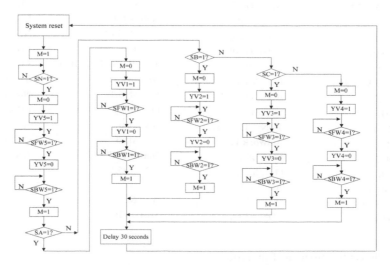

Figure 2.   The flow chart of control system.

275

Table 1. I/O address allocation.

Input device	PLC address	Output device	PLC address
SFW1	I0.0	YV1	Q0.0
SFW2	I0.1	YV2	Q0.1
SFW3	I0.2	YV3	Q0.2
SFW4	I0.3	YV4	Q0.3
SFW5	I0.4	YV5	Q0.4
SA	I0.5	M (The conveyor belt motor)	Q0.5
SB	I0.6		
SC	I0.7		
SBW1	I1.0		
SBW2	I1.1		
SBW3	I1.2		
SBW4	I1.3		
SBW5	I1.4		
SN	I1.5		
UCP	I2.0		

Network 1
LD   SM0.1
O   M5.2
MOVB 16#F8, SMB37
HDEF 0, 0
MOVD +0, SMD38
HSC 0
Network 2
LD   SM0.0
DTI HC0, VW20
Network 3
LD   SM0.0
MOVW VW20, VW24
DIV   77, VD22
Network 4
LDW= VW22, 0
AW>   VW24, 0
=   M5.1
Network 5
LD   M5.1
AW>   VW20, 400
=   M5.2
Network 6
LD   I1.4
AN   C0
=   Q0.5
Network 7
LD   M5.1
LD   SM0.1
LD   I1.4
EU

OLD
LD   I1.5
EU
OLD
CTU   C0, 7
Network 8
LD   M5.1
A   I1.5
O   Q0.4
AN   I0.4
=   Q0.4
Network 9
LD   I0.5
O   M1.0
AN   Q0.0
=   M1.0
TON   T33, +20
Network 10
LD   I0.6
O   M1.1
AN   Q0.1
=   M1.1
TON   T34, +50
Network 11
LD   I0.7
O   M1.2
AN   Q0.2
=   M1.2
TON   T35, +50

Network 12
LD   T33
O   Q0.0
AN   I0.0
=   Q0.0
Network 13
LD   T34
O   Q0.1
AN   I0.1
=   Q0.1
Network 14
LD   T35
O   Q0.2
AN   I0.2
=   Q0.2
Network 15
LD   SM0.1
O   C0
MOVB 16#FF, VB50
Network 16
LD   M5.1
O   M5.3
AN   C0
=   M5.3
Network 17
LD   M5.1
EU
A   M5.3
SLB   VB50, 1

Network 18
LD   M5.1
A   I1.4
S   V50.0, 1
Network 19
LD   I1.4
EU
R   V50.0, 1
Network 20
LD   I0.5
EU
S   V50.2, 1
Network 21
LD   I0.6
EU
S   V50.3, 1
Network 22
LD   I0.7
EU
S   V50.4, 1
Network 23
LDN   V50.5
A   M5.1
O   Q0.3
AN   I0.3
=   Q0.3

Figure 3.   The control program of the system.

Due to the motion of the cylinder speed is very fast, a short period of time delay is set in order to avoid the gap between sensor testing point and warehouse entry position.

5 CONCLUSION

The control system with PLC as the core, combined with pneumatic device, sensing technology, position control and other technology, which makes the whole control system has the advantages of high degree of automation, stable running, high precision, easy to control, and so on. Finally, this system also can realize the sorting of different materials by changing different sensor and it has broad prospects.

REFERENCES

[1] Qiang Wang, Ming-zhu Zhang. 2011. Application of PLC in Material Automatic Sorting System. *Journal of Capital Normal University (Natural Science Edition) Vol. 32 NO. 6*: 4–9.
[2] Chang-chu Liao. 2008. *PLC programming and applications (3rd edition)*. Beijing: Machinery Industry Press.
[3] Yong-hua Wang. 2008. *Modern electrical control and PLC application technology (2nd edition)*. Beijing: Beijing University of Aeronautics and Astronautics Press.
[4] Wei-hua Bian. 2008. *Research on automatic sorting material and import database system based on PLC*. Shanghai: Shanghai University.
[5] Hao Yu. 2011. A Study on PLC in Material of Automatic Sorting System. *Coal Technology Vol. 30*: 31–34.
[6] Bao-chun cui, Peng Shan. 2009. Design of control system for material sorting based on PLC. *Computer and Network Vol. 33*: 222–226.
[7] Ya Zhou, Lang Lang, Chang-jie Su. The design of material sorting system based on PLC and MCGS. *Journal of Nanyang Institute of Technology Vol. 5 NO. 6*: 6–10.
[8] Zhi-gang Jia. 2012. Auto sorting system for materials based on Mitsubishi PLC. *Journal of Liaoning Teachers College Vol. 14 NO. 2*: 83–86.

Advances in Engineering Materials and Applied Mechanics – Zhang, Gao & Xu (Eds)
© 2016 Taylor & Francis Group, London, ISBN 978-1-138-02834-0

Research on high precision linear motor control system based on adaptive fuzzy-PID control

J.F. Xiao & F.H. Zhu
College of Electrical Engineering, University of South China, Hengyang, Hunan, China

Q.M. Xiao
Foxconn Wireless Business Group, Hengyang, Hunan, China

ABSTRACT: This paper describes a new speed regulating method based on adaptive fuzzy-PID control, which used in the speed loop of linear motor control system. The design of fuzzy-PID speed controller is completed. The table of best fuzzy-PID parameters under different load conditions is given, which used the Permanent Magnet Linear Synchronous Motor (PMLSM) LSMF100302 as concrete research object. The simulation research on the new fuzzy-PID speed regulating system of linear motor is carried out. The simulation results show that the system based on adaptive fuzzy-PID control has better dynamic and static performance than the system based on conventional PID control. The permanent magnet linear synchronous motor LSMF100302 servo system experiment platform is set up. The British Renishaw XL-80 laser interferometer is used in the experiment to test the positioning accuracy of linear motor. The four curves of error between the actual displacement and the command displacement are given. The results of experiment show that the system is of high precision.

1 INTRODUCTION

AC servo motor speed control system is a complex system with multi-variable, strong coupling, nonlinear and time-varying. Without intermediate transmission structure, the permanent magnet linear synchronous motor is more easily influenced by external disturbance and parameter changes. At the same time, the "edge effect" will lead to the nonlinear characteristics. The traditional PID control can not meet the dynamic response, high performance and accuracy requirements[1–2,4,6,8].

Various parameter self-adjusting PID control methods of design and simulation for linear Motor have been proposed[3,5,8], which optimize the PID controller parameters to improve the precision of system. Those methods can meet the practical requirements of production, but are complex to control and the performance of linear motor control system is not ideal when the load changes[3,5–7].

Considering the high precision and steady performance of control system, a new linear motor control system is designed in this paper. The linear motor control system is a speed, current and position closed loop control system. The P adjustment is used in the position loop, the PI adjustment is used in the current loop and the adaptive fuzzy-PID control is used in speed loop. The designed new speed control method is applied to the PMLSM LSMF100302. A system based on two dimensional linear permanent magnet synchronous motor platform and XL-80 laser interferometer is designed to test the positioning accuracy of linear motor. The good feasibility of the system is proved by the experiments.

2 THE CONTROL PRINCIPLE OF LINEAR MOTOR CONTROL SYSTEM

In this design, linear motor control system is a speed, current and position closed loop control system. The P adjustment, PI adjustment and adaptive fuzzy-PID control are used in

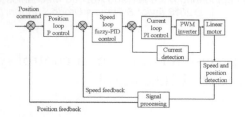

Figure 1. The system control principle diagram.

different loop of the linear motor control system. The system control principle diagram is shown in Figure 1.

3 ADAPTIVE FUZZY-PID CONTROL OF SPEED LOOP

3.1 *Fuzzy controller*

The inputs of fuzzy controller are E and EC, the outputs are Δk_p, Δk_i and Δk_d. The fuzzy language subset of E, EC and Δk_p, Δk_i and Δk_d are {NB (negative big), NM (negative middle), NS (negative small), Z0 (zero), PS (positive small), PM (positive middle), PB (positive big)}. Based on the variable language quantization level and the practical experience of expert, the fuzzy control inference rules are built as Tables 1–3.

3.2 *Adaptive fuzzy-PID control*

The objective of control method is to adjust the parameters of PID controller to improve the performance in a real time application. The K_p, K_i, K_d is described by (1), (2) and (3) .

$$K_p = K_{p0} + G_p \times \Delta k_p \tag{1}$$

$$K_i = K_{i0} + G_i \times \Delta k_i \tag{2}$$

$$K_d = K_{d0} + G_d \times \Delta k_d \tag{3}$$

where G_p, G_i, G_d are the scale factor, and K_{p0}, K_{i0}, K_{d0} are the initial parameters of K_p, K_i, K_d.

3.3 *Design example of the fuzzy-PID controller parameters in speed loop of linear motor*

3.3.1 *The selection of linear motor*
In this paper the permanent magnet linear synchronous motor LSMF100302 produced by Shenzhen Dazu Motor Technology Co. is the research object. And motor parameters: 340 N continuous thrust, 680 N peak thrust, continuous current 3.2 A (air cooling), 6.4 A (air cooling) peak current, 107 N/A thrust constant, 36 V/m/s back EMF constant, 7 ohm resistor, 30 mH inductor, 21 mm distance, position feedback encoder uses RGH22D Renishaw grating, reading head resolution 0.5 um and maximum no-load speed is 1 m/s.

3.3.2 *Fuzzy-PID controller parameters in speed loop*
In this design, their values are obtained as:

$$\begin{cases} G_p = 200 \\ G_i = 100 \\ G_d = 0.01 \\ K_{p0} = 520 \\ K_{i0} = 380 \\ K_{d0} = 0.018 \end{cases} \tag{4}$$

Table 1. Fuzzy control rules of Δk_p.

E	EC						
	NB	NM	NS	ZO	PS	PM	PB
NB	PB	PB	PM	PM	PS	ZO	ZO
NM	PB	PB	PM	PM	PS	ZO	NS
NS	PM	PM	PM	PS	ZO	NS	NM
ZO	PM	PS	NS	ZO	ZO	NS	NM
PS	PS	PS	ZO	NS	NS	NM	NM
PM	PS	PS	NS	NM	NM	NB	NB
PB	NS	ZO	NM	NM	NB	NB	NB

Table 2. Fuzzy control rules of Δk_i.

E	EC						
	NB	NM	NS	ZO	PS	PM	PB
NB	NB	NB	NM	NM	NS	NS	ZO
NM	NB	NB	NM	NS	NS	NS	ZO
NS	NB	NM	NS	NS	ZO	PS	PM
ZO	NM	NM	NS	ZO	ZO	PS	PM
PS	NM	NS	ZO	PS	PM	PM	PB
PM	ZO	ZO	ZO	PS	PM	PM	PB
PB	ZO	ZO	PS	PM	PM	PB	PB

Table 3. Fuzzy control rules of Δk_d.

E	EC						
	NB	NM	NS	ZO	PS	PM	PB
NB	PS	NS	NB	NB	NB	NM	PS
NM	PS	NS	NB	NM	NS	NS	ZO
NS	ZO	NS	NM	NM	NS	ZO	ZO
ZO	ZO	ZO	NS	NS	NS	NS	ZO
PS	NS	NS	ZO	ZO	ZO	ZO	ZO
PM	PB	PB	PM	PM	PS	PS	PB
PB	PB	PB	PM	PM	PS	PS	PB

The lookup table of fuzzy-PID controller is different when the load and speed of motor vary. So the PID controller parameters (K_p, K_i, K_d) are different when load changes, and the parameters also change when the speed changes. The parameters of adaptive fuzzy-PID controller are different under the same speed and different load conditions. The best parameters of fuzzy-PID controller when speed is 100 revolutions per minute are given as an example in Table 4.

The typical advantage of the control algorithm is that parameters of PID controller can be adjusted under different load conditions or different given speed.

3.4 *Simulation comparison of PID and fuzzy-PID control*

Using M-file of matlab, control system of linear synchronous motor is simulated. The capabilities of conventional PID controller and the fuzzy-PID controller in speed loop are compared.

Table 4. The best fuzzy-PID parameters of speed loop when speed is 100 r/min.

Different load	Parameters		
	K_p	K_i	K_d
0	450	350	0.015
10 KG	520	380	0.018
20 KG	600	450	0.020
30 KG	750	580	0.021
40 KG	920	720	0.023

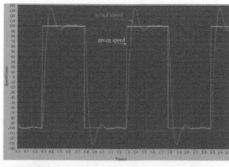

(a)PID control in speed loop

(b) Adaptive fuzzy-PID control in speed loop

Figure 2. The response of speed when load is 40 KG.

In the simulation test, stair case speed transients from −100 to 100 revolutions per minute, and load is 40 KG. The simulation results of the control system based on conventional PID and fuzzy-PID control in speed loop are shown in Figure 2.

Simulation results indicate that system based on fuzzy-PID controller in speed loop has better performance than the system based on PID controller in speed loop, such as the small overshoot, the short time of reaching steady, and the improved dynamic performance, robustness.

4 LINEAR MOTOR CONTROL SYSTEM DESIGN AND EXPERIMENT RESULTS

4.1 *The composition of linear motor control system*

This system based on fuzzy-PID control produces PWM signal from TMS320F2812. The control object is permanent magnet linear synchronous motor LSMF100302. The composition of linear motor control system is shown as Figure 3.

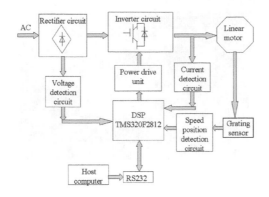

Figure 3. The composition of linear motor control system.

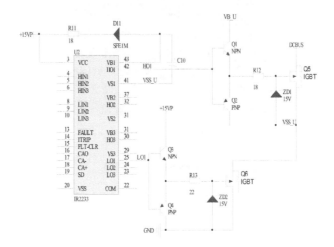

Figure 4. The inverter bridge circuit.

4.2 The system design

4.2.1 Hardware design

DSP TMS320F2812 is selected as the main control chip, FP30R06KE3 is selected as rectifier bridges, and 600 V IGBT modules in FS150R12KT3 series are selected to construct the three-phase inverters. The inverter bridge circuit is given as Figure 4.

The incremental grating RGH22D produced by the British Renishaw is used as position sensor. The output signal of incremental grating is 5 V voltage, and the maximum input voltage of DSP I/O is only 3.3 V. Therefore, AM26LS32AID is used as the driver chip.

The drive power board and the control board is designed. Figure 5 is the drive power board and Figure 6 is the control board.

4.2.2 Software design

Software is consisted by main program, interrupt service program and other related subprogram. The important subprogram is fuzzy-PID control program in speed loop. The flow chart is shown in Figure 7.

4.3 Test and analysis of motor position experiment

4.3.1 Instruments

The servo system is composed of XY two dimensional linear permanent magnet synchronous motor, servo driver (Hans Driver) and PC machine. The British Renishaw XL-80 laser

Figure 5.   The drive power board.

Figure 6.   The control board.

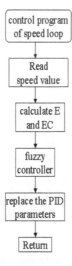

Figure 7.   Flow chart of control program in speed loop.

interferometer is used in the experiment to test the positioning accuracy of linear motor. It is an instrument has the advantages of high close, high precision up to ±0.5 PPM (0–40°C), large measuring range (linear measure up to 80 m), fast measuring speed (the speed is up to 4 m/s), high resolution (up to 1 nm).

The linear motor positioning test system is given in Figure 8.

Figure 8. Linear motor positioning test system.

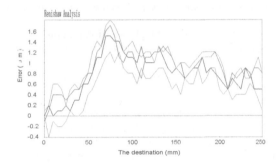

Figure 9. The error curves.

4.3.2 *Test method*

After the laser interferometer is installed and debugged, setting the driver makes the motor run in stepping mode, 5 mm step length, 250 mm the whole trip, 4 times back and forth. Laser interferometer and the motor run synchronously. This system captures the actual displacement data and calculates the error between the actual displacement and the command displacement.

4.3.3 *The test results*

The test data is analyzed using a special-purpose software, the positioning accuracy is ±1.1 um/250 mm and the repeat positioning accuracy is ±0.4 um. The test date is merged into the error curves through the special-purpose software too. The error curves are shown as Figure 9. The times of reciprocating motion of motor rotor are four, so there are four error curves.

High precision is shown in the experiment, the results also demonstrate that the direct drive system without reverse gap.

5 CONCLUSION

In this paper, a new speed control method based on adaptive fuzzy-PID control is applied to the speed loop for linear motor control system. The simulation results show that the system based on this speed regulating has better dynamic and static performance than the system based on conventional PID control. A permanent magnet linear synchronous motor LSMF100302 speed control system based on this new method is introduced. The results of experiment show that the system is of high precision.

ACKNOWLEDGEMENT

This paper is supported by Hunan National Science and Technology fund Program (2013GK3160). We would like to express our appreciation to all teachers and students of motor control and drives laboratory in the University of South China for their help and encouragements.

REFERENCES

[1] M. Abroshan et al. 2008. An optimal control for saturated interior permanent magnet linear synchronous motors incorporating field weakening. *13th Power Electronics and Motion Control Conference, EPE-PEMC*:1117–1122.

[2] A. Amthor et al. 2010. High precision position control using an adaptive friction compensation approach. *IEEE Transactions on Automatic Control* 55(1):274–278.

[3] Sun Hua & Yue-hong Dai. 2010. Fuzzy PID control and simulation experiment on permanent magnet linear synchronous motors. *2010 International Conference on electrical and Control Engineering* 39(5):1047–1049.

[4] Z. Z. Liu et al. 2005. Robust high speed and high precision motion control system of linear-motor direct drive for high-speed X–Y table positioning mechanism. *IEEE Trans. Ind. Electron* 52(5): 1357–1363.

[5] Kun Tao et al. 2014. Study of control system of permanent magnet linear synchronous motor. *Coal Mine Machinery* 35(5):14–16.

[6] K. Tsuruta et al. 2011. High-speed and high-precision position control using a sliding mode compensator. *Electrical Engineering in Japan* 174(2):65–71.

[7] Dao-jin Xu et al. 2014. Influence of varying load on linear motor control system of CNC Machine Tool [J]. *Mechanical Engineering & Automation1* 89(4):169–171.

[8] Li-juan Yu et al. 2014. A Summary on modern control strategy of permanent magnet synchronous linear motor. *Electric Drive Automation* 36(6):1–5.

Dispose of communication signals

Advances in Engineering Materials and Applied Mechanics – Zhang, Gao & Xu (Eds)
© *2016 Taylor & Francis Group, London, ISBN 978-1-138-02834-0*

Adaptive recognition algorithm for digital modulation communication signal

C.Y. Wang, W.L. Yu & M.Y. Hou
The Campaign and Command Department, Air Force Aviation University, Changchun, Jilin, China

ABSTRACT: Recognition of digital modulation schemes plays an important role in communication signal analysis. This paper presents an automatic computer simulation and adaptive recognition algorithm of six kinds of digital modulation types such as 2ASK, 4ASK, 2PSK, 4PSK, 2FSK, and 4FSK, which is applicable to the signal that changes within the scope of the Signal to Noise Rate (SNR) 5~30 dB. It is opinion that the adaptive algorithm operates rapidly and reliably in automatic recognition of communication signal and it's performance has been evaluated by simulating different types of bind-limited digital signals corrupted by white Gaussian noise. It is found that the overall success rate is not lower than 92% when SNR is over 7 dB.

1 INTRODUCTION

Automatic recognition of modulation schemes of the communication signals and computer simulation is a new research direction in signal analysis field, and it has a prodigious application foreground, especially in the realm of military communication. Along with the more attention paying to the study of electronic countermeasure, it cries for the study of the technology of automatic identification of modulation signal, it is widely used in the fields such as signal confirmation, interference identification, radio interception, signal monitor, analysis and so on.

By far it is generally considered that there are two basic measures of automatic identification of the communication signals modulation schemes. The first one is maximum likelihood hypothesis inspection measure basing on the decision-making academic, and the other one is statistical modulation recognition method basing on the character distilling. The basic frame of the avenue of decision-making academic accords with the hypothesis inspection measure, it is seasoned with the condition which the categories to be identified are limited, and that the basic frame of the avenue of statistical modulation recognition is to distill the character selected from the signal firstly, and then identify the mode.

E.E. Azzouz and A.K. Nandi proposed a modulation computer simulation and recognizer for digital modulations based on the decision-theoretic approach which uses five character parameters to recognize six modulation modes, and the accurate rate of the arithmetic for the six signals can reach 90% when the SNR is 10 dB. This article presents an adaptive recognition algorithm basing on this algorithm, it can reach better identification rate in lower SNR with only four parameters, and can be easily calculated.

2 DIGITAL MODULATION RECOGNITION ALGORITHM

Statistical modulation recognition method based on the character distilling involves two steps such as parameter distilling and identification as shown in Figure 1.

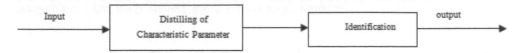

Figure 1. Structure of Statistic mode identification measure basing on the characteristic parameter distilling.

2.1 *Characteristic parameter distilling of digital modulation signal*

In the part of parameter distilling, the instantaneous amplitude $\alpha(t)$, instantaneous phase $\varphi(t)$ and instantaneous frequency $f(t)$ of the signal received by the receiver should be estimated firstly, then four parameters as followed should be distilling.

1. γ_{max} is the maximum of spectral density of the normalized-centered instantaneous amplitude and it is defined by

$$\gamma_{max} = \max\left|\mathbf{FFT}\left(a_{cn}(i)\right)\right|^2 / N_s \tag{1}$$

where N_s is the sample number of every length, $\alpha_{cn}(i)$ is the instantaneous amplitude of the signal normalization at the time instants $t = i/f_s$ and it is defined by

$$a_{cn}(i) = a_n(i) - 1$$

$$a_n(i) = \frac{a(i)}{m_a} \tag{2}$$

$$m_a = \frac{1}{N_s}\sum_{i=1}^{N_s} a(i)$$

2. σ_{aa} is the standard deviation of the normalized-centered instantaneous amplitude of a signal and it is defined by

$$\sigma_{aa} = \sqrt{\frac{1}{N_s}\left(\sum_{i=1}^{N_s} a_{cn}^2(i)\right) - \left(\frac{1}{N_s}\sum_{i=1}^{N_s}\left|a_{cn}(i)\right|\right)^2} \tag{3}$$

3. σ_{ap} is the standard deviation of the absolute value of zero-centered in the non-weak segment of instantaneous phase and it is defined by

$$\sigma_{ap} = \sqrt{\frac{1}{C}\left(\sum_{a_n(i)>t_a} \phi_{NL}^2(i)\right) - \left(\frac{1}{C}\sum_{a_n(i)>t_a}\left|\phi_{NL}(i)\right|\right)^2} \tag{4}$$

where $\phi_{NL}(i)$ is the non-linear part of the instantaneous phase after centralization. When the carrier wave is entirely in-phase,

$$\phi_{NL}(i) = \varphi(i) - \varphi_0 = \varphi(i) - \frac{1}{N}\sum_{i=1}^{N}\varphi(i) \tag{5}$$

where $\varphi(i)$ is instantaneous phase, and t_a is the amplitude threshold that using for distinguish the weak signal segment, when the amplitude of the signal is under the threshold, it's phase is sensitive to the noise, and the as a result that there will be a evident phase distortion. C is the number of the signal which is $a_n(i) > t_a$ in the data combination $\{\phi_{NL}(i)\}$. It is that the number of the non-weak signal.

4. σ_{af} is the standard deviation of the absolute value of the zero-centered normalization non-weak signal segment instantaneous frequency and it is defined by

$$\sigma_{af} = \sqrt{\frac{1}{C}\left(\sum_{a_n(i)>t_a} f_N^2(i)\right) - \left(\frac{1}{C}\sum_{a_n(i)>t_a}|f_N(i)|\right)^2} \tag{6}$$

$$f_N(i) = \frac{f_m(i)}{r_s} = \frac{f(i) - m_f}{r_s} = \left(f(i) - \frac{1}{N}\sum_{i=1}^{N}f(i)\right)\Big/ r_s \tag{7}$$

where r_s is the code rate of the digital signal, $f(i)$ is instantaneous frequency of the signal, $f_N(i)$ is instantaneous frequency after zero-centered normalization.

2.2 The classification of the signal

In this algorithm, it is basing on the fact that four parameters of γ_{max}, σ_{aa}, σ_{ap}, σ_{af} are chosen as followed.

1. γ_{max} is used to compartmentalize 2FSK and 4FSK to the first subset, and 2ASK and 4ASK to the second subset, 2PSK and 4PSK to the third subset. So the parameter γ_{max} can be used to divide the signal with amplitude information (2ASK, 4ASK; 2PSK and 4PSK) and without amplitude information (2FSK an 4FSK).
2. σ_{aa} is used to distinguish 2ASK that has no absolute amplitude information from 4ASK that has.
3. σ_{ap} is used to divide 4PSK signal that has absolute phase information from 2PSK signal that has no absolute phase information.
4. σ_{af} is used to divide 4FSK signal with absolute frequency information from 2FSK signal without absolute phase information.

3 COMPUTER SIMULATION

3.1 Flowchart for the adaptive recognition algorithm

The flowchart showed by Figure 2.

3.2 The setting of the adaptive threshold

The setting of the threshold plays an important role in recognizing. The best threshold is different with different SNR. Because of the unknown SNR of the pre-identified signal, the threshold can't be changed according to the SNR, so this paper puts forward the constitution of the best threshold which is applicable to the signal that changes within the scope of the signal to noise from 5 dB to 30 dB. Set the original thresholds of $t(\gamma_{max1})$, $t((\gamma_{max2})$, $t(\sigma_{aa})$, $t(\sigma_{ap})$, $t(\sigma_{af})$ to be 10, 0.8, 0.305, 0.75, 0.32 at first. Then the best settings of adaptive threshold are showed as followed.

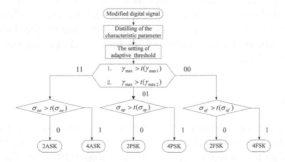

Figure 2.   The flowchart for the adaptive recognition algorithm.

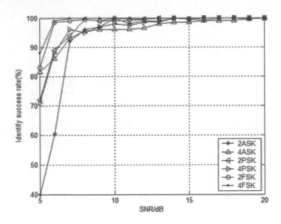

Figure 3. The relation between identify success rate and SNR.

$$t(\gamma_{\max 2}) = \begin{cases} 0.8, & \sigma_{af} \in [0:0.2) \\ 1.2, & \sigma_{af} \in [0.2:0.25) \\ 1.6, & \sigma_{af} \in [0.25:0.29) \\ 2.6, & \sigma_{af} \in [0.29:0.31) \\ 2.8, & \sigma_{af} \geq 0.31 \end{cases} \quad t(\sigma_{af}) = \begin{cases} 0.15, & \gamma_{\max} \in [0:0.2) \\ 0.18, & \gamma_{\max} \in [0.2:0.4) \\ 1.22, & \gamma_{\max} \in [0.4:0.8) \\ 0.30, & \gamma_{\max} \in [0.8:1.6) \\ 0.32, & \gamma_{\max} \geq 1.6 \end{cases}$$

$$t(\sigma_{ap}) = \begin{cases} 0.75, & \sigma_{af} \in [0:0.28) \\ 0.80, & \sigma_{af} \in [0:28:0.29) \\ 0.85, & \sigma_{af} \geq 0.29 \end{cases}$$

The simulation results shows that when the SNR is changing between 5 dB and 30 dB, the success rate for recognition of the modulation is enhanced with the setting of the best threshold.

3.3 *Results of the simulation*

The system simulation experiment is finished by using MATLAB (V6.5). The carrier frequency is 100 kHz and the code frequency is 12.5 kHz. Standard rectangle pulse figuration function is chosen to be the modulation signal, and the noise is the additive gauss noise. And make the frequency of the carrier frequency and code speed correspond with the real signal. The sample frequency is 400 kHz, and there are Ns = 1024 sample data in every frame. Identify per 10 frames, and test the signal in every modulation between SNR from 5 dB to 20 dB, and the interval is 1 dB. The simulation result of modulation recognition is showed by Figure 3.

The results show that: when the SNR is above 7 dB, the total recognition success rate is not lower than 92%. But when the SNR is below 6 dB, he overall success rate is very low, especially for some modulation kind (2ASK). When the SNR is below 5 dB, the system can't work. So it is concluded that time domain parameter is sensitive to additive noise, especially for the instantaneous amplitude and it cause the deformation of the distilling of the character parameter. It also shows the localization of the method that using time domain parameter for modulation recognition.

4 CONCLUSION

Pass through the result of computer simulation, we can discover this modified digital modulation automatic recognition algorithm needs less parameters, and less calculation. It is very useful. The recognition success rate is not lower than 92% when the SNR is above 7 dB.

The algorithm can be used to identify 2ASK, 4ASK, 2PSK, 4PSK, 2FSK and 4FSK. The simulation experiment also gives the best threshold value for different SNR.

REFERENCES

[1] Azzouz E.E., Nandi A.K. 1995. Automatic identification of digital modulation types. *Signal Processing*. 47(1):55–69.
[2] A.K. Nandi, E.E. Azzouz. 1998. Algorithms for automatic modulation recognition of communication signals. *IEEE Transaction on Communication*. 46(4):431–436.
[3] Cheng Lei, Ge Lindong, Peng Hua, etc. 2005. Modern and Developing Status of Communication Signals Modulation Classification. *Journal of Computer information*. 21(10–1);154–156.
[4] Chen Peiqing. 2007. *Digital signal processing course. Tsinghua University Press*.
[5] Li Yang, Li Guotong, Yang Genqing. 2005. Automatic digital modulation recognition algorithm of communication signal. *Journal of Electronics & Information Technology*. 27(2):197–201.
[6] Li Li. 2007. *Mechanical signal processing and application*. Huazhong university of science and technology press.
[7] Lin Hongbin, Xie Ping, Wang Na. 2009. *The principle and application of signal processing*. China Machine Press.
[8] Xu Chengbo, Tao Hongyan. 2008. *Digital signal processing and MATLAB implementation*. Tsinghua University Press.
[9] Xue Nianxi. 2003. *MATLAB in the application of digital signal processing*. Tsinghua University Press.
[10] Zhou Haomin, Wang Rui. 2005. *Test signal processing technology*. Beijing university of aeronautics and astronautics press.

Electronic design engineering

Advances in Engineering Materials and Applied Mechanics – Zhang, Gao & Xu (Eds)
© 2016 Taylor & Francis Group, London, ISBN 978-1-138-02834-0

Design of digital multi-channel temperature acquisition system

Lili Wan
Wuhan Donghu University, Wuhan City, HuBei Province, China

ABSTRACT: The digital multi-channel temperature acquisition system is consisted of main controller, temperature gathering electric circuit, temperature display circuit, alarm circuit and keyboard control circuit. MCU AT89C51 is used as the controller and the data processor. The intelligent temperature sensor DS18B20 is used as the temperature detector. The LED digital display tube is used as the temperature display device. The hardware circuit is simple, and has the advantages of low cost, large measuring range, high measuring precision, displaying intuitive and easy to use.

1 INTRODUCTION

Temperature is one of the most basic environmental parameters, and the environment temperature is closely related to people's lives. So, the study of the temperature measuring method and device is of great significance. The key for the temperature measuring device is the temperature sensor. The development of temperature sensor has experienced three stages of development: (1) The traditional separation of temperature sensor; (2) The analog integrated temperature sensor; (3) The intelligent integrated temperature sensor. At present, the new temperature sensor develops rapidly from analog to digital, from the integrated to intelligent and networked directions. A digital multi-channel temperature gathering system is designed in this paper.

The main technical index of the system:

1. Two channel temperature acquisition circuit at least;
2. The acquisition of temperature range is from −50°C to +110°C;
3. The temperature precision is within 0.1°C;
4. The display module, using LED digital tube to display.

2 GENERAL DESIGN

The system consists of 5 modules: the main controller, the temperature acquisition circuit, the temperature display circuit, the alarm circuit and the keyboard input circuit. The overall block diagram of the digital multi-channel temperature acquisition system circuit is shown in the Figure 1.

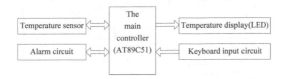

Figure 1. The structure diagram of multi-channel temperature acquisition system.

3 HARDWARE DESIGN

The temperature collecting and processing circuit is composed of temperature sensor, amplifier circuit, A/D conversion circuit etc, which can collect data and realize simple processing. It can meet the technical requirements of the design. In this system, the intelligent temperature sensor DS18B20 is used as the core device for the temperature acquisition circuit. It composed of DS18B20 and auxiliary circuit.

MCU is the core of the control circuit. AT89C51 is chosen for the system.

The input control circuit consists of keyboard and the interface. The independent type keyboard and interface circuit is used.

The LED digital tube is applied in Display circuit. Four pieces of 74 LS164 are used to convert the serial data of the controller output into parallel data to output, which used to drive 4 LED digital display tube display data.

The alarm circuit uses the piezoelectric buzzer as the sounding body. The buzzer sound is controlled by triode. The alarm circuit is controlled by P3.7 of the AT89C51, which is connected with the base tube of the triode C945 through a current limiting resistor. The collector of the triode C945 is connected with one end of the piezoelectric buzzer (BUZZER), and another end is connected with the power supply.

The power supply circuit consists of transformer, single-phase bridge rectifier circuit, filter circuit and three terminal voltage regulator circuit.

The circuit principle diagram of the digital multi-channel temperature gathering system is shown in Figure 2.

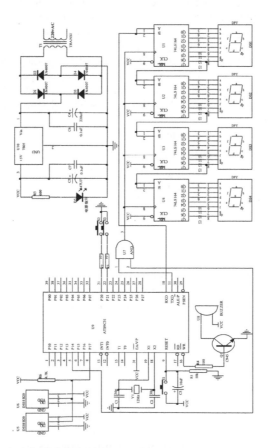

Figure 2.   The circuit principle diagram.

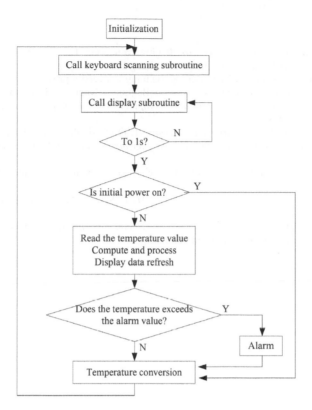

Figure 3.   The main program flow diagram.

4   SOFTWARE DESIGN

C51 is selected because the programming shows high efficiency, accuracy and stability. The main function of the main program is to call the subroutine, to read and process the DS18B20 value (temperature measured once every 1 second). And then realize the real-time temperature display. The main program flow diagram is shown in Figure 3.

5   PERFORMANCE ANALYSIS

Using the temperature measurement system designed and thermometer to measure the environmental temperature at the same time. By comparing and analyzing the results, it shows high precision of the digital temperature acquisition system. The error index can be limited within 0.1°C. In addition, the system with a −55°C to 125°C measuring range is suitable for general applications.

This system is mainly applicable to the collection and display of temperature, but it cannot adjust temperature. It can be extended by hardware and software. After the hardware extension, it can realize controlled function extension by software programming. For example, when the temperature reaches a certain value, judging by the software programming, the main controller outputs a command from a I/O port to drive the external control circuit to work. Thereby regulate temperature to meet people's requirements. There will be a good market in the social that popularizing prospect in the health consciousness of people is increasing day by day today.

REFERENCES

Bomin Zou. *The principle of automatic control*. Beijing: Machinery Industry Press.

Guangfei Li. 2004. *MCU Design Guide*. Beijing: Beihang University press.

Hangci Zhou. *SCM process design basis*. Beijing: Beihang University press.

Huaguang Kang. *Electronic Technology Foundation*. Beijing: Advanced Education press.

Jian Hu. *The practise course of single chip microcomputer principle and interface technology*. Beijing: Machinery Industry Press.

Mingying Chen. *8051 MCU design training materials*. Beijing: Tsinghua University press.

Quanli Li. *Single chip microcomputer principle and interface technology*. Beijing: Advanced Education press.

Yigang Zhang. 2004. Principle and application of single-chip microcomputer. Beijing: Advanced Education press.

Yongquan Yu. 2002. *Application technology of ATMEL Series MCU*. Beijing: Beihang University press.

Yufeng Li. *MCS-51 series single chip microcomputer principle and interface technology*. Beijing: People's Posts and Telecommunications Publishing House.

Machinery

A new method for topology optimization of tip/tilt mirror

C.L. Wang
The School of Computer Science, Xi'an Shiyou University, Xi'an, China

H.W. Wang
Engineering University of APF, Xi'an, China

ABSTRACT: In order to meet the strict requirement of surface-shape and realize the ultra-lightweight design of a tip/tilt mirror, the topology optimization method is introduced to the design of a φ110 mm round mirror for support part of the mirror. First, the theory and the approach of topology optimization is introduced, then absorbing the meanings of SIMP (Solid Isotropic Material with Penalization) and RAMP (Rational Approximation of Material Properties), a new approach ASMP (Adjusted Solid Material with Penalization) is proposed, and talking about its advantage. Finally, topology optimization for the mirror is processed, and it proved the new ASMP approach for topology optimization is correct and reliable. Compared with other traditional design method, it can improve surface profile remarkably. Furthermore, it lays a solid foundation for next parameters optimization.

1 INTRODUCTION

The elimination of atmospheric agitation in space stabilization image system is an important and difficult problem in the field of high-precision astronomical observation. This paper uses the rapid vibration of tip/tilt mirror to realize rapid correction of atmospheric agitation. The swing of tip/tilt mirror is implemented by correlation tracking system based on pzt driver, the processor in correlation tracking system can calculate the offset of the image. Then it drives the tiny wobble of mirror in two directions according to the offset, thus correct the offset in real-time in the field of view and achieve precise modification atmospheric agitation problem.

The fundamental frequency and surface-shape error of mirror are the key factors of this technology, so we must focus on the analysis and research of tip/tilt mirror. The relationship between system resonance frequency and moment of inertia of the mirror after the mirror is installed on the platform of correlation tracking system has been fully demonstrated in the relevant paper [1]. Therefore, in order to improve the resonant frequency of the whole system, the torque and quality of the mirror must be reduced, so the maximum lightweight design is very necessary in the process that ensures the mirror is sine swing.

The most effective method to complete lightweight design is structural optimization for the mirror is first, that is topology optimization for conceptual level is first, and then the parameter optimization after topology optimization results is improved. The difference algorithm for topology optimization design in this paper, absorbing the thought of the SIMP (Solid Isotropic Material with Penalization) interpolation algorithm and the RAMP (Rational Approximation of the Material Properties) interpolation algorithm of the variable density method, and proposes a new ASMP (Adjusted Solid Material with Penalization) interpolation algorithm which punishment center can be adjusted, and the topology optimization of the mirror is made by using the new algorithm. And the results are compared with the other optimization design, the results show that the new difference algorithm proposed in this paper can largely improve the surface-shape accuracy and is more flexible and convenient to use.

2 METHOD OF TOPOLOGY OPTIMIZATION

Topology optimization can be divided into two basic categories: evolutionary method and degradation method [2]. The basic idea of evolutionary method is that biological evolution is introduced into topology optimization, through evolution to obtain the optimal topology. The evolutionary method has higher global convergence, but the convergence speed is slow. And the basic idea of degradation method is that add all possible materials before optimization, then unnecessary materials are eliminated step by step through optimization method, finally get the optimal topology structure. Currently, the common degradation method used in topology optimization is mainly basic structure method, homogenization method, variable thickness method and variable density method, etc. The variable density method is a popular and effective method, compared with the size variable, it can reflect the essential characteristics of topology optimization [3,4,5]. It finds the structure with the minimum structural flexibility. Formula 1 is the mathematical model of topology optimization that uses structure supple degree as the objective function, volume as the constraint and pseudo density as the design variable.

$$
\left.
\begin{aligned}
\min : c(X) &= U^T K U = \sum_{e=1}^{N} x_e^p u_e^T k_0 u_e \\
\text{subject to}: \frac{V(x)}{V_0} &= f \\
: KU &= F \\
: 0 < x_{\min} &\le x < 1
\end{aligned}
\right\}
\tag{1}
$$

where c is the structure supple degree, that is the total strain energy, U is the displacement vector of structure, K is the stiffness matrix of structure, x_e is the pseudo density of unit, p is the punishment factor, k_0 is the stiffness matrix of unit, u_e is the displacement matrix of unit, V is the material volume (generally for density and unit volume weighted sum of real) should be preserved, and V_0 is the volume before the structure is topological optimized.

3 DENSITY–STIFFNESS INTERPOLATION MODEL

The variable density method uses pseudo density as design variables to replace the structure of the microstructure in the homogenization method. In order to eliminate the material with middle density, the design variables are punished between 0 and 1 by punishing function p, and it makes continuous topological optimization can approach topology optimization problem of discrete variable well. So the density–stiffness interpolation model is involved. The common interpolation model used by variable density method mainly includes: the SIMP (Solid Isotropic Microstructure with Penalization) interpolation model and the RAMP (Rational Approximation of the Material Properties) interpolation model.

SIMP stiffness–density interpolation model is shown in Figure 1(a), the mathematical model is formula 2. E is the elasticity modulus after interpolation, E_0 and $E_{\min}$ represent the elasticity modulus when materials are real and virtual. Generally, $E_{\min} = E_0/1000$, and P is the penalty factor.

$$
E = E_{\min} + x_e^p (E_0 - E_{\min})
\tag{2}
$$

The RAMP stiffness–density interpolation model as shown below, the mathematical model as shown in formula 3.

$$
E = E_{\min} + \left(\frac{x_e}{1 + p(1 - x_e)} \right)(E_0 - E_{\min})
\tag{3}
$$

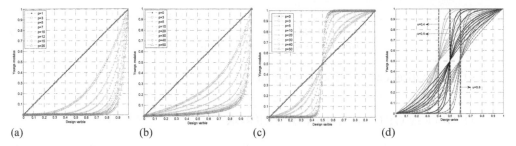

(a)   (b)   (c)   (d)

Figure 1.   (a) SIMP interpolation scheme; (b) RAMP interpolation scheme; (c) ASMP interpolation scheme; (d) ASMP interpolation scheme with different center.

These two interpolation models, both make the elasticity modulus of the most middle density units trend to zero after interpolation. As shown in Figure 1(a) and 1(b), as long as cell density is less than 1, the elastic modulus after interpolation will quickly tend to zero. Only density is equal to or close to 1 unit is considered as the real. This is unfavorable for the topology optimization process. The new interpolation model constructed in the paper is Adjusted Solid Material with Penalization (ASMP). The mathematic expression of the model is:

$$
E = \begin{cases}
E_{\min} + \dfrac{x_e}{2u}(x_e + 1 - u)^p \Delta E & 0 \le x_e < u \\[3mm]
1/2 & x_e = u \\[3mm]
E_{\min} + \left[-\dfrac{x_e - 1}{2(u-1)}(x_e + 1 - u)^{-p} + 1 \right]\Delta E & u < x_e \le 1
\end{cases}
\tag{4}
$$

Among the expression, $\Delta E = E_0 - E_{\min}$.

The punishment effect of this new type of interpolation model when $u = 0.5$ as shown in Figure 1(c). Which makes curve must pass through three point $(0, 0)$, $(u, 0.5)$, $(1, 1)$, by adjusting the penalty coefficient p to achieve the goal that let the Young's modulus whose pseudo density is higher than u quickly tends to 1, and quickly tends to zero if it is lower than u. Figure 1(d) shows the interpolation curve when u is 0.4, 0.5 and 0.6.

4   OPTIMIZATION ALGORITHM

At present there are mainly two types of commonly used optimization topology algorithms: the Optimization Criterion method (OC) and Mathematical Programming (MP). OC is mainly based on a heuristic update plan to update the design variables. It is efficient for the problems with many design variables and less constraints. To multi-constraint problem, response Lagrange multiplier of constraint problems should be introduced at a time. This paper uses the optimization criterion method. The variable update formula of optimization criterion method is as follows:

$$
x_e^{new} = x_e \left(\frac{p(x_e)^{p-1} q_c}{\lambda_0 V_e} \right)^{\varsigma} = x_e B_e^{\varsigma}
\tag{5}
$$

$$
B_e = \frac{p(x_e)^{p-1} q_c}{\lambda_0 V_e}
\tag{6}
$$

where λ_0 is Lagrange multiplier, and can be calculated by dichotomy, Newton iteration method and other optimization algorithms. To stabilize the iterative process and limit the variation of optimized variables in optimization process, m is introduced.

$$x_e^{new} = \begin{cases} \max(x_{\min}, x_e - m) & x_e B_e^\varsigma \le \max(x_{\min}, x_e - m) \\ x_e B_e^\varsigma & \max(x_{\min}, x_e - m) < x_e B_e^\varsigma < \min(1, x_e + m) \\ \min(1, x_e + m) & \min(1, x_e + m) \le x_e B_e^\varsigma \end{cases} \qquad (7)$$

ς is the numerical damping coefficient, and value is 0.5.

5 TOPOLOGY OPTIMIZATION OF MIRROR

The mirror whose research object is space stabilization image system, its caliber diameter is 110 mm, material is SiC. According to the requirements of overall size and the size of the assembly interface, three supporting way on the back is adopted, topology optimization model as shown in Figure 2(a). Red part is the non-design area, whereas green part is the design area. To make the topological shape more exquisite, unit will be divided into smaller units, a total of 92544 units and 103311 nodes. The non-design area is to ensure that the integrity of mirror surface, and the design area is to depict the shape of the back supporting. Under the comprehensive condition of gravity and the back supporting topology optimization analysis is taken.

The optimized lightweight form after using the new difference algorithm to optimize mirror as shown in Figure 2(b). According to the need to make a little change, because the mirror body is not very big, and SiC materials have limitations on whether it can chamfer and stiffener in the sintering process. So the topology model is transformed to stiffener form that is easy to sintering shaping and secondary processing. The model is built as in Figure 2(c).

For comparison of PV value and RMS value of the mirror after the topology optimization, this paper not only establishes the finite element model of topological optimization results (Fig. 2(d)), but also establishes the two kinds of model used by conventional empirical design that are finite element model of the fan lightening hole (Fig. 2(e)) and finite element model of triangle lightening hole (Fig. 2(f)) with same quality lightening hole. Under the condition of three-point support in the back and axial gravity, we can get deformation nephogram by calculating three mirror respectively as shown in Figure 3(a), (b) and (c).

Exporting the deformation results of finite element node on the mirror in the form of data, and then use the Matlab software to calculate the PV value and RMS value of the node that are shown in Table 1.

From Table 1 it can be seen that the mirror with triangle lightening hole and fan lightening hole based on experience and the mirror based on topology optimization both can meet the design requirements (PV < 10, lambda/RMS < lambda/40). But the mirror surface-shape based on topology optimization is better than that based on experience design. The PV value of topology mode is 31.45% lower than fan model and 17.92% lower than triangular model.

(a)   (b)   (c)   (d)   (e)

(f)

Figure 2. (a) Tilt/tip mirror for optimization; (b) optimization result (c); 3D model; (d) topology model; (e) fan model; (f) triangle model.

306

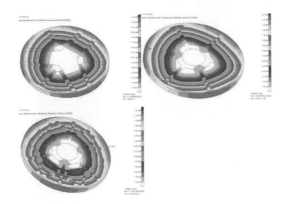

Figure 3. (a) Displacement contour of the topology optimized model; (b) displacement contour of the fan model; (c) displacement contour of the triangle model.

Table 1. The comparison of three kinds of lightweight swing mirror.

	Topology/mm	Fan/mm	Triangle/mm
PV	1.0722×10^{-5}	1.5640×10^{-5}	1.3063×10^{-5}
RMS	5.9579×10^{-6}	8.4706×10^{-6}	7.2366×10^{-6}

The RMS value of topology model below fan model and the triangle model is 29.66% and 17.71%, respectively.

6 CONCLUSION

Based on a new interpolation algorithm, this paper introduces a topology optimization analysis algorithm for the analysis of tip/tilt mirror for space image stabilized system, resulting in a great mirror model meeting with surface precision. The surface precision is up to 5.95 nm. Comparing with the lightweight method designed with traditional experience, it is characterized with reasonable mechanical transmission path, clear destination, and others. Even though the optimization design process for topology optimization has not redundantly referred to mirror and more specific optical parameter, it subsequently provides a solid foundation for parameter optimization calculation of these parameters involved with. It will have a very broad prospect in the large aperture optical design fields.

REFERENCES

[1] Wang Hong-wei, Ruan Ping, Xu Guang-zhou, et al. 2009. Design and Research of Tip/Tilt Mirror in a Space Telescope. *Acta Photonica Sinica*, 38(9):2368–2371.
[2] Eschenauer H.A. 2001. Topology optimization of continuum structures: A review. *Applied Mechanics Review*, 54(4):331–389.
[3] Liu Yin-dong, Bian Gang. 2006. Structure topology optimization and second development based on Ansys. *Journal of Ship Mechanics*, 10(2):120–125.
[4] Li Anhu, Liu Li-ren Sun Jan-feng. 2009. Large-aperture High-accuracy Optical Scanner. *Journal of Mechanical Engineering*, 45(1):200–204.
[5] Song Li-qiang, Yang Shi-mo, Chen Shi-yuan. 2009. Optimization design and analysis of the structure of beryllium mirrors of astronomical instruments in space. *Infrared and Laser Engineering*, 38(5):882–888.
[6] Zhuang Xue-jun, Li Zhi-lai, Zhang Zhong-yu. 2007. Space telescope a spherical mirror structure design based on SiC material. *Infrared and Laser Engineering*. 36(5):577–582.

Advances in Engineering Materials and Applied Mechanics – Zhang, Gao & Xu (Eds)
© 2016 Taylor & Francis Group, London, ISBN 978-1-138-02834-0

Realization method of NURBS path generation for 5-axis CNC controller

L.J. Chen, H.Y. Li & H.J. Zhang
Zhengzhou Institute of Aeronautical Industry Management, Zhengzhou, China

ABSTRACT: A real-time path generation method with Non Uniform Rational B Spline (NURBS) technology in the Computer Aided Design (CAD) field is presented and realized in a five-axis Computer Numerically Controlled (CNC) system. In this method, the tool path was represented into two NURBS curves, tool-tip and tool-orientation spline curve, based on the same knot vector. The Cutter Location (CL) data was firstly calculated and then transformed to five motion commands of five axes of machine tool through the real-time post-procession algorithm. The acceleration/deceleration controlling method was also presented to avoid the impact of machine tool. The proposed five-axis spline interpolation method is realized in our developing five-axis CNC machine tool and the result of machining shows that the method is feasible.

1 INTRODUCTION

Complex parts such as aeronautical turbine blades, impellors, dies, molds are machined on five-axis Computer Numerical Controlled (CNC) machines. The core of the machines is the contour controlling system which decides machining efficiency and accuracy of a part. However, the current five-axis CNC system mostly provide only line interpolators, that is, tool motion along straight lines is supported. To machine a surface, the tool paths, which are also known as the Cutter Location (CL) paths, are typically approximated with piecewise linear segments by CAD/CAM systems. The approximately machining method has its own drawbacks in advancing efficiency and accuracy of a machined part. To overcome the disadvantages, it must be taken into account that the five-axis control system has intelligent ability to generate the tool paths of the machined surfaces.

Non-Uniform Rational B-Spline (NURBS), as the only standard of data-exchange of product sanctioned by ISO, has been mostly used in almost all CAD/CAM systems to describe work-piece's surface. Based on the reality, the CNC system should be provided with the ability to directly generate five-axis NURBS curves or surfaces in real-time.

Some NURBS interpolation methods had been proposed by several investigators[1–6]. Cheng et al.[2] had proposed a real-time NURBS curve motion command generator for CNC machines. Zhiming et al.[5] developed a NURBS curve interpolator for CNC machining based on the geometric properties of the tool path. However, most of them had concentrated their attentions on 3-axis ball-end machining, but for five-axis NURBS surface interpolator, little has been done.

When developing and realizing a five-axis CNC system with function of NURBS interpolation, related study must be carried out. In this paper, a real-time five-axis NURBS curve generation method is given.

2 NURBS CURVE REPRESENTATION

A *p*-degree NURBS curve with parameter *u* can be defined as follows:

$$C(u) = [x(u)\ y(u)\ z(u)]^{\mathrm{T}} = \frac{\sum\limits_{i=0}^{n} N_{i,p}(u)W_i P_i}{\sum\limits_{i=0}^{n} N_{i,p}(u)W_i} \tag{1}$$

where P_i is the ith 3-D control point; W_i is the corresponding weight factor of P_i; $(n+1)$ is the number of control points; $N_{i,p}(u)$, B-spline basis function with degree of p, can be calculated by the following formula:

$$\begin{cases} N_{i,0} = \begin{cases} 1 & (u_i \le u < u_{i+1}) \\ 0 & (\text{otherwise}) \end{cases} \\ N_{i,p}(u) = \dfrac{u - u_i}{u_{i+p} - u_i} N_{i,p-1}(u) + \dfrac{u_{i+p+1} - u}{u_{i+p+1} - u_{i+1}} N_{i+1,p-1}(u) \\ \dfrac{0}{0} = 0 \quad (\text{prescribed}) \end{cases} \tag{2}$$

where $[u_i,\ \ldots,\ u_{i+p+1}]$ is the knot vector. More information about NURBS can be found in Ref.[7].

3   NURBS PATH REAL-TIME GENERATION METHOD

The task of real-time NURBS spline interpolation is to calculate next interpolation period's CL data that including tool-tip location coordination and tool-axis orientation and then transfer the CL data into machine tool's motion commands such as X, Y, Z, A, C through post-procession.

3.1   *Real-time calculation of CL data*

As shown in Figure 1, $C_0(u)$ and $C_1(u)$ is the cutter center point spline and the orientation spline, respectively. The two splines are constructed in the same knot vector. Assumed that parameter u is function of time t, that is, $u = u(t)$. By using Taylor's expansion of the parameter u with respect to time t to obtain the first order approximation interpolation algorithm, the first order approximation up to the first derivatives is

$$u_{k+1} = u_k + T_s \frac{\mathrm{d}u}{\mathrm{d}t}\Big|_{t=t_k} \tag{3}$$

where T_s is interpolation period, u_k and u_{k+1} are corresponding parameters of current and next time t_k and t_{k+1}.

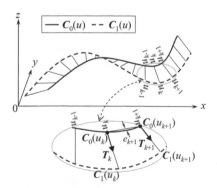

Figure 1.   Principle of five-axis NURBS interpolation.

The feedrate of tool-tip point along $C_0(u)$ is defined by

$$V(u) = \frac{dC_0(u)}{dt} = \left(\frac{dC_0(u)}{du}\right)\left(\frac{du}{dt}\right) \tag{4}$$

Since the curve speed

$$V(u_k) = \left\|\frac{dC_0(u)}{dt}\right\|_{u=u_k} \tag{5}$$

The first derivative of u with t is obtained as

$$\frac{du}{dt}\bigg|_{t=t_k} = \frac{V(u_k)}{\left\|\dfrac{dC_0(u)}{du}\right\|_{u=u_k}} \tag{6}$$

Therefore, the first-order interpolation algorithm is obtained by substituting Eq. (4) into (5), (5) can be processed as follows:

$$u_{k+1} = u_k + \frac{T_s V(u_k)}{\left\|\dfrac{dC_0(u)}{du}\right\|_{u=u_k}} \tag{7}$$

The first and second derivative of $C_0(u)$ with u is obtained as

$$\frac{dC_0(u)}{du} = \frac{\sum\limits_{i=0}^{n} N'_{i,p}(u)W_i P_i}{\sum\limits_{i=0}^{n} N_{i,p}(u)W_i} - \frac{\sum\limits_{i=0}^{n} N'_{i,p}(u)W_i \sum\limits_{i=0}^{n} N_{i,p}(u)W_i P_i}{\left(\sum\limits_{i=0}^{n} N_{i,p}(u)W_i\right)^2} \tag{8}$$

where the general algorithm for 1st order derivative of $N_{i,p}(u)$ is

$$N'_{i,p}(u) = p\left[\frac{N_{i,p-1}(u)}{u_{i+p-1} - u_i} - \frac{N_{i+1,p-1}(u)}{u_{i+p} - u_{i+1}}\right] \tag{9}$$

$C_0(u_{k+1})$ and $C_1(u_{k+1})$ are obtained by substituting the calculated u_{k+1} into $C_0(u)$ and $C_1(u)$. Suppose the tool-axis unit vector is T_{k+1}, so

$$T_{k+1} = \frac{C_1(u_{k+1}) - C_0(u_{k+1})}{\|C_1(u_{k+1}) - C_0(u_{k+1})\|} \tag{10}$$

3.2 Real-time post-procession of CL data

By now, we have the following CL data:

$$C_0(u_{k+1}) = (o_x, o_y, o_z) \qquad T_{k+1} = (t_x, t_y, t_z)$$

For five-axis machine of table tilting/rotating type (Fig. 2), the following post-procession method is presented to transfer the CL data into the machine's motion commands in real-time.

311

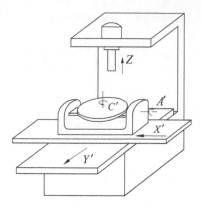

Figure 2.　A type of five-axis machine with tilting/rotating table.

a. if $t_x = t_y = 0$, $t_z = 1$, then $A' = 0$, $C' = 0$
b. else if $t_x = 0$, $t_y \neq 0$, then $A' = \arcsin(-t_y)$, $C' = 0$
c. else if $t_y = 0$, $t_x \neq 0$, then $A' = \arcsin(t_x)$, $C' = \pi/2$
$\qquad A' = \arccos(t_z)$
d. else, $C' = \begin{cases} \arccos(-t_y/\sin A') & t_x > 0 \\ 2\pi - \arccos(-t_y/\sin A') & t_x < 0 \end{cases}$

For the calculations of the three translation coordination, the Cramer rule is used here.

4　ACCELERATION/DECELERATION CONTROLLING METHOD

Assumed that the acceleration and deceleration is a and the destination federate is v as shown in Figure 3.

For the acceleration process, the following method can be used

$$V(u_k) = at_k$$

For the acceleration process, the key problem to be solved is evaluation of deceleration point. In Figure 3, S_D, distance of deceleration, can be calculated as follows

$$S_D = \frac{v^2}{2a} \tag{11}$$

The distance of deceleration, which is also the length between deceleration point and destination point along the spline curve, can be calculated as follows

$$\int_{u_D}^1 \left\| \frac{d\mathbf{C}_0(u)}{du} \right\| du = S_D \tag{12}$$

where u_D is corresponding parameter of the deceleration point. The following Newton-Rapson method is used to calculate u_D.

If

$$f(u) = \int_u^1 \left\| \frac{d\mathbf{C}_0(u)}{du} \right\| du - S_D \tag{13}$$

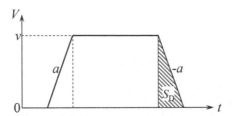

Figure 3.   Model of linear acceleration/deceleration.

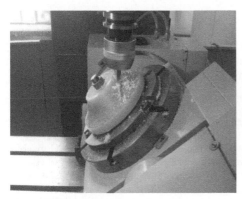

a) machining process   b) machining result

Figure 4.   Five-axis NURBS path generation.

Then we have the following result:

$$f(u_D) = 0$$

Assumed that the initial value of u_D is 1, that is, $f(u_0) = -S_D$, then we have the following equation

$$u_{i+1} = u_i - f(u_i)/f'(u_i) \quad i \geq 0 \tag{14}$$

where i is an integer.

Eq. (14) can be equivalent to the following formula

$$u_{i+1} = u_i + \frac{f(u_i)}{\left\| \dfrac{dC_0(u_i)}{du} \right\|} \quad i \geq 0 \tag{15}$$

$f(u_i)$ in Eqs. (14) and (15) can be obtained using Simpson integral method.

5   REALIZATION IN FIVE-AXIS CNC SYSTEM

The proposed five-axis NURBS interpolation method has been realized in the developing five-axis CNC system. The type of the five-axis machine is table tilting/rotating. Figure 4 shows that the CNC system is controlling cutting tool along a NURBS curve path to machine an impeller.

6 CONCLUSIONS

On the basis of analyzing the defects of the existing five-axis linear interpolation method used in the machining of sculptured surface, five-axis NURBS interpolation method with ACC/DEC controlling is presented and realized in a CNC system. Difference from discretization machining method of the five-axis linear interpolation, the proposed method directly interpolates a curve on a free-form surface. Consequently, the number of the NC code for the same machining path can be greatly decreased and machining efficiency and accuracy is also advanced.

ACKNOWLEDGMENTS

This research is supported by the National Natural Science Foundation of China (No. 51275485) and the program for Science & Technology Innovation Talents in Universities of Henan Province (No. 13HASTIT036). The authors thank for their support for this research.

REFERENCES

[1] Tikhon, M. 2004. NURBS interpolator for constant material removal rate in open NC machine tools. *International Journal of Machine Tools & Manufacture* 44:237–245.
[2] Cheng, M. Tsai, M. & Kuo, J. 2002. Real-time NURBS command generators for CNC servo controllers. *International Journal of Machine Tools & Manufacture* 42:801–813.
[3] Qiyi, G. Zhang, R. & Greenway, B. 1998. Development and implementation of a NURBS curve motion interpolator. *Robotics and Computer-Integrated Manufacturing* 14:27–36.
[4] Tsai, M. Cheng, C. & Cheng, M. 2003. A real-time NURBS surface interpolator for precision three-axis CNC machining. *International Journal of Machine Tools & Manufacture* 43:1217–1227.
[5] Zhiming, X. Jincheng, C. & Zhengjin, F. 2002. Performance evaluation of a real-time interpolation algorithm for NURBS curves. *International Journal of Advanced Manufacturing Technology* 20:270–276.
[6] Bahr, B. Xiao, X. & Krishnan, K. 2001. A real-time scheme of cubic parametric curve interpolations for CNC systems. *Computers in Industry* 45:309–317.
[7] Piegl, L. 1991. On NURBS: a survey. *IEEE Computer Graphics & Application* 11(1):55–71.

Advances in Engineering Materials and Applied Mechanics – Zhang, Gao & Xu (Eds)
© *2016 Taylor & Francis Group, London, ISBN 978-1-138-02834-0*

The strength analysis of non-power bogie frame for the passenger high-speed train

H. Wu & C.S. Yue
State Key Laboratory of Traction Power, Southwest Jiaotong University, Chengdu, China

ABSTRACT: This paper uses the large-scale general finite element analysis software Workbench to analyze the frame. First, the finite element analysis model of the welding frame is established; the load cases on the welding frame according to the leaflet UIC 515-4 "Passenger bogie structure strength and the test method" are counted; and then the static strength on the frame is carried out and the strength calculation is performed. The calculation results indicated that this bogie frame satisfies intensity request under the superior and operation load.

1 INTRODUCTION

The history of high-speed railway operation is more than 50 years. Even though it went through the tortuous process of development, it boomed rapidly because of its irreplaceable advantages. In order to adapt to high-speed railway rapid construction, we have to put the design and manufacturing at the forefront, while bogie design is the key part of high-speed railway design. This paper testifies that the bogie frame satisfies strength requirements under both supernormal and normal operating conditions through applying loads and static calculations.

2 BASIC STEPS OF FINITE ELEMENT TO DEALING WITH PROBLEMS

1. Establish models: According to the frame structure size, physical models are established by the software Pro/e, and then import it into the software Workbench.
2. Calculation standard research: Through the research of computational load and varieties of working conditions in the UIC 515-4 "Passenger cars bogie structure strength experimental method", determine the boundary conditions, loads and a variety of the working conditions required by infinite element models.
3. The evaluation of working conditions results: Calculate all working conditions, compare the results with evaluation specification to assess the structure intensity of the frame.

3 MODELING AND MESH DIVIDING

This paper uses the Workbench for structural strength analysis. Define the trains' forward motion direction as the X-axis positive direction, direction vertical the track plane upwards as Z-axis positive direction, the Y-axis is vertical XY plane, set the Y-axis positive direction according to the right-hand screw rule. Define the neutral plane of frame upper cover plate as XOY plane, the origin is set at geometric center of the frame.

After establishing geometric models, finite element meshing is performed for modeling finite element. Three steps are included in the mesh dividing process.

1. Define element properties: Element properties include element types, defining real constants and material features. Workbench do not limit the units of analysis system.

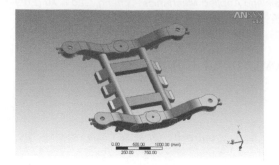

Figure 1.   Workbench model.

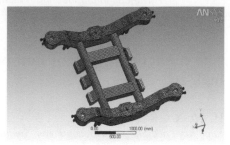

Figure 2.   The frame infinite element analysis model meshing.

Defining material properties, geometric size, lode values and other input values merely according to a unified system of units. This paper uses s, kg and mm as basic units, the unit of input force is N, the unit of calculation results displacement is mm, surface force is N/mm, and stress unit is Mpa. Element types are all physical models. Adopting Q345 steel and the yield limit is 345 Mpa.

2. Define the grid generation. We need to define mesh size and boundary conditions. Considering the bogie actual size, the mesh is defined as 35. The Workbench model is shown in Figure 1.

3. Generate grids. When the mesh is completed, the whole frame is divided as 40524 elements and 175137 modes, satisfying calculation requirement. The frame infinite element analysis model meshing is shown in Figure 2.

4   LOADING CONDITIONS AND CALCULATION

According to first two strength experiments in UIC 515-4 "Passenger cars bogie structure strength experimental method", supernormal load static strength experiment, it aims to simulate during the frame loading force to avoid the risk of elastic deformation. The aim of normal load static strength experiment is simulating that bogie will not crack for fatigue under the actions of main load (vertical, horizontal and track torsion).

The frame is applied with different combinations loads in supernormal conditions and all normal conditions. We use a vertical force percentage βF_z to describe the vertical force dynamic changes caused by car body vertical movements (rise and fall). We use a vertical force percentage αF_z (for normal running conditions: α values 0.1, β values 0.2) to describe vertical force dynamic changes caused by car body side-roll movements. Thirteen loading conditions including supernormal and normal loading conditions' calculation results are shown in Table 1.

According to the operating conditions, we need to apply vertical, lateral and skew-symmetric loads. Vertical loads need to be applied at air spring seat and they are plane loads. Lateral load is applied at the node where is in the middle of lateral stop and air spring seat steel tube. We apply the plane loads at lateral stop and nodal force at the node. Distorted loads are applied by elastic element node displacement.

4.1   *Boundary conditions*

According to car body actual operation conditions, we need to apply bogie with longitudinal, lateral and vertical elastic restraint.

Vertical restraint is applied at prime spring seat, and it is uniformly distributed at plane grid nodes. We make axle box spring vertical stiffness 1 MN/m.

Lateral restraint is applied at rotary arm seat from both sides of the frame. We make the lateral stiffness at rotary arm elastic node 5.49 MN/m, respectively at rotary arm positioning seat node.

Table 1. Fourteen combination loads operating conditions calculation formula.

Loading conditions	Vertical force		Lateral loads	Longitudinal loads	Distorted loads
	Left side beam	Right side beam			
Supernormal	F_{z1max}	F_{z2max}	F_{ymax}	0	10‰ track distortion
01	F_Z	F_Z	0	0	0
02	$(1+\alpha-\beta)\,F_Z$	$(1-\alpha-\beta)\,F_Z$	0	0	0
03	$(1+\alpha-\beta)\,F_Z$	$(1-\alpha-\beta)\,F_Z$	F_Y	0	0
04	$(1+\alpha+\beta)\,F_Z$	$(1-\alpha+\beta)\,F_Z$	0	0	0
05	$(1+\alpha+\beta)\,F_Z$	$(1-\alpha+\beta)\,F_Z$	F_Y	0	0
06	$(1-\alpha-\beta)\,F_Z$	$(1+\alpha-\beta)\,F_Z$	0	0	0
07	$(1-\alpha-\beta)\,F_Z$	$(1+\alpha-\beta)\,F_Z$	$-F_Y$	0	0
08	$(1-\alpha+\beta)\,F_Z$	$(1+\alpha+\beta)\,F_Z$	0	0	0
09	$(1-\alpha+\beta)\,F_Z$	$(1+\alpha+\beta)\,F_Z$	$-F_Y$	0	0
10	$(1+\alpha-\beta)\,F_Z$	$(1-\alpha-\beta)\,F_Z$	F_Y	0	5‰ track distortion
11	$(1+\alpha+\beta)\,F_Z$	$(1-\alpha+\beta)\,F_Z$	F_Y	0	5‰ track distortion
12	$(1-\alpha-\beta)\,F_Z$	$(1+\alpha-\beta)\,F_Z$	$-F_Y$	0	5‰ track distortion
13	$(1-\alpha+\beta)\,F_Z$	$(1+\alpha+\beta)\,F_Z$	$-F_Y$	0	5‰ track distortion

Table 2. Biggest prime stress of bogie.

Operating conditions	Biggest prime stress/MPa	Operating conditions	Biggest prime stress/MPa
Supernormal	115.73	7	100.80
1	107.29	8	139.01
2	96.10	9	138.26
3	100.83	10	108.83
4	139.02	11	138.21
5	138.26	12	107.18
6	96.10	13	138.18

Longitudinal restraint is applied from a wheel set at rotary arming positioning seat. We make rotary arm elastic node longitudinal stiffness 10MN/m respectively, at rotary arm positioning seat nodes.

4.2 Calculation results

We choose Q345 steel as the bogie frame material. The admissible stress in non-welding area under supernormal operating conditions is 345 MPa, whereas the admissible stress in welding area is 313 MPa. Admissible stress in non-welding area under normal operating conditions is 230 MPa, whereas the admissible stress in welding area is 209 MPa. The biggest prime stress of bogie is shown in Table 2.

As space is limited, we just show the frame vertical view plane stress distribution under operating condition 6 as shown in Figure 3. Biggest prime stress distribution is shown in Figure 4.

4.3 Analysis of the results

Under supernormal loads, none of biggest press reaches material minimum admissible stress and the admissible material stress in welding area is 313 MPa. Bogie frame satisfies requirements under supernormal loads.

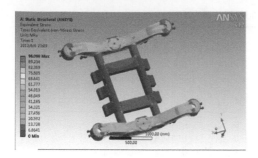

Figure 3. The frame vertical view plane stress.

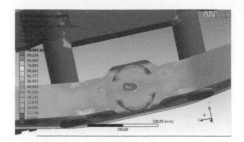

Figure 4. Biggest prime stress distribution under operating condition 6.

Since all of the biggest principle stress of all elements in different planes under 13 normal operating conditions do not surpass material minimum requirements with the maximum admissible stress 209 MPa, so the bogie satisfies the strength requirements.

Calculation results show that bogie frame comply with material admissible stress requirements regulated by "Provisional regulations" under both supernormal and normal operating loads conditions. So, the bogie proves to comply with design requirements.

5 CONCLUSIONS

1. Based on inadequate information about bogie, this paper modified and carefully reviewed the design process. Static strength analysis was performed on the bogie frame with the software Pro/e and Workbench in this paper. Strength analysis results show that maximum stress of bogie frame is all within the allowable range of base material and welding areas. So it can be safely operated.
2. As for box structure stress concentration especially in side beam and transverse beam, we do not increase thickness but set the partition, which can rationally relieve stress flow within the partition, reducing stress.
3. During the design process, sharp corners should be avoided as possible, we should use a larger arc radius to transit. As the model is modified, even though the joint parts are connected with circular transition, strength calculation shows stress concentration exists at sharp corners all the time. After we make improvements like decreasing connection stiffness and setting partition; however, there is no benefit so obviously we greatly draw our attention to improve the design process in the future.

ACKNOWLEDGEMENT

This work has been supported by the State Key Program of National Natural Science of China (61134002), the National Key Basic Research Program of China (973 Program) (2011CB711100) and Innovation Group of Ministry of Education funded project (IRT1178).

Author's brief introduction: Wu Hao (1988–), Harbin, HeiLongJiang. majorly research vehicle system dynamic.

REFERENCES

[1] Brenna F.P., Dover W.D., Kare R.F., Hellier A.K. (1999). Parametric equations for T-butt weld toe stress intensity factors. *Int J Fatigue*; 21:1051–1062.
[2] Fu B., Haswell J.V., Bettess P. (1993). Weld magnification factors for semi-elliptical surface cracks in fillet welded T-butt joints models. *Inc Fract*; 63:155–171.

[3] Gurney T.R., Maddox S.J. (1973). A re-analysis of fatigue data for welded joints in steel. *Welding Research Int*; 3(4):1–54.

[4] Hobbacher A. (1996). Recommendations for fatigue strength of welded components. *Cambridge: Abington Publishers.*

[5] Isao Okamoto (1998). Railway Technology Today 5-How Bogies Work Japan Railway. *Transport Review 18,* (12):52–61.

[6] J.F. arm (2003). Evaluation of wheel damperson an intercity train. *Journal of Sound and Vibration,* (5):734–747.

[7] J. Perez, J.M. Busturiaa, R.M. Goodallb (2002). Control strategies for active steering of bogie-based railway vehicles. *Control Engineering Practice,* (10):1005–1012.

[8] J.K. Mok, J. Yoo (2001). Numerical study on high speed train and tunnel hood interaction. *Journal of Wind Engineering and Industrial Aerodynamics,* (8):17–29.

[9] Marquis G. (1995). High cycle spectrum fatigue of welded components. *Espoo:Dissertation for the Degree of Doctor of VTT Manufacturing Technology.*

[10] Neuber H. (1958). Theory of notch stress. *Publ Springer Verlag, Berlin.*

[11] Radaj D. (1990). Design and analysis of fatigue resistant welded structures. *Cambridge:Abington Publishers.*

[12] R.K. Luo, B.L. Gabbitas and B.V. Brickle (1996). Dynamic Stress Analysis of an Open-shaped Railway Bogie Frame. *Engineering Failure Analysis,* (1):53–64.

[13] R.K. Luo, B.L. Gabbitas, B.V. Brickle, W.X. Wu (1998). Fatigue damage evaluation for a railway vehicle bogie using appropriate sampling frequencies. *Vehicle System Dynamics,* (28):405–415.

[14] Sung II Seo, Choon Park, Ki Hwan Kim (2005). Fatigue strength evaluation of the aluminum car body of urban transit unit by large scale dynamics load test. *JSME international journal,* Vol. 48(l).

[15] Wolfgang Fricke (2003). Fatigue analysis of welded joints: state of development. *Marine Structures.* 16:185–200.

Advances in Engineering Materials and Applied Mechanics – Zhang, Gao & Xu (Eds)
© *2016 Taylor & Francis Group, London, ISBN 978-1-138-02834-0*

Design of control system for smart window

Juan Lei
Wuhan Donghu University, Wuhan City, Hubei Province, China

ABSTRACT: Along with the development of the times, people demand for smart home is becoming more and more urgent. In the design of smart home, most people will first pay attention to the control of household appliances and the realization of the intelligent home appliances, but there are some other aspects that if transformed also will greatly enhance the intelligent degree of residence, such as the smart window. So, this paper tried to design a control system for smart window. The system has two main functions: automatic environment regulating function and automatic security protection function. The system uses a MCU, several kinds of sensors and signal processing circuit. According to the current actual situation, it can open windows or close windows by itself.

1   INTRODUCTION

With the development of the times, people demand for smart home is becoming more and more urgent. The system of smart home based on the house, uses electronic technology, network communication technology, computer technology and automatic control technology to centrally manage appliances, household environment, security facilities and other facilities related to the household life[1,2]. In order to achieve a comfortable, convenient, safe and energy-saving residential space with the ability of self-regulation.

In the design of smart home system, most people will first pay attention to the control of household appliances and the realization of the intelligent home appliances, but there are some other aspects that if transformed also will greatly enhance the intelligent degree of residence, such as the smart window designed in this paper.

2   DESIGN SCHEME

The smart window control system designed in this paper has two main functions: automatic environment regulating function and automatic security protection function. According to the current actual situation, the system can open the window or close the window by itself. For example, when it rains, the system can automatically close the window, so that the house will not be wetted even if there is no one at home; when it is too hot or there are harmful gases in room, the system can open the window to get fresh air; when there is someone want to approach the window, the system can close and lock the window.

So, the design of smart window control system will include two main functions, the environment regulating function and the security protection function, as shown in Figure 1.

In the design of the whole system, we use different types of sensor and processing circuit for each detection circuit. And also use MCU to analyse, process the output signals of these detection circuit, to detect the current state of the window, and then output a control signal to the window's control circuit. The system's hardware structure diagram is shown in Figure 2.

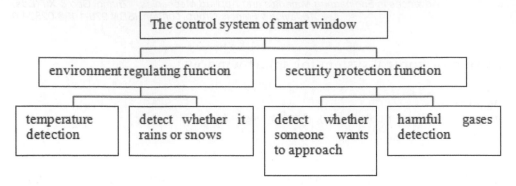

Figure 1. The system structure diagram.

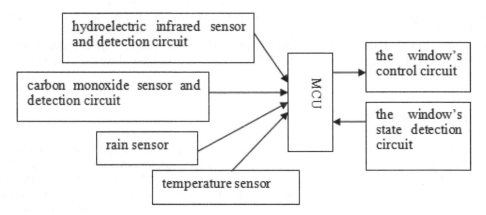

Figure 2. The system hardware structure diagram.

2.1 *Security protection function*

The security protection function has two main parts: detection of whether someone wants to approach, and detection of whether there are harmful gases (such as CO), then according to the result of detection, output a signal for closing window or opening window.

In this paper, we use the pyroelectric infrared sensor to detect the human body that approach a window. The temperature of human body is generally constant, at 36 to 37 degrees, so human body sends a specific infrared, and the infrared's wavelength is around 10UM. The pyroelectric infrared sensor detects the infrared in non-contact form, and then converts it into voltage signal. In addition, this sensor has two detection units that have the opposite polarity, characteristics consistent, and equal size. Environmental radiation will have almost the same effect on the two units, the voltage signal converted from them cancel each other out. But once someone come into the area detected, the two detection units will receive the infrared radiation emitted from the intruder, and the heat they received is different, the voltage signal converted from them cannot be offset, the sensor will output a voltage signal[3,4]. Therefore, the sensor can identify the moving biological and non biological by itself. If the pyroelectric infrared sensor works with Fresnel lens, it can separate light, sunlight and the other kinds of infrared radiation, and its effective detection range can be up to 12 m, horizontal angle will be 120 degrees[5].

As shown in Figure 3, the system has two pyroelectric infrared sensors. When the two sensors find that someone wants to approach the window at the same time, the MCU output a signal of closing and locking window, and then the window will close and lock automatically.

322

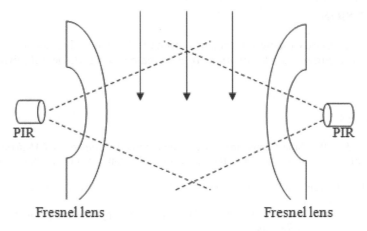

Figure 3.   Detection of whether someone wants to approach the window.

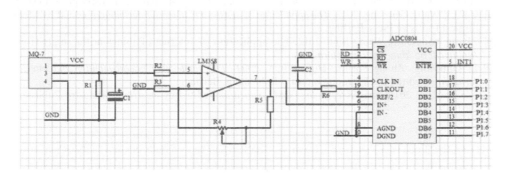

Figure 4.   The circuit that used to detect indoor CO concentration.

In addition, because the main harmful gas in room is carbon monoxide, carbon monoxide is colourless and odourless, it is not easy to find, but it can make people dead. It is very dangerous. The system use the MQ-7 carbon monoxide sensor to detect the indoor CO concentration[6,7], and then combine with the signal processing circuit and A/D conversion circuit, output to MCU. The circuit is shown in Figure 4. If the indoor CO concentration is higher than the allowable value, MCU outputs a signal of opening window.

2.2   *Environment regulating function*

In this paper, we use a rain sensor to detect whether it rains[8], use two temperature sensors to measure the indoor and the outdoor temperature. The rain sensor we chose is FM-YX, this sensor has a good stability. The temperature sensor we chose is DS18B20, because its output signal is digital. The two kinds of sensors output signals to MCU directly. According to the signals, the MCU outputs different signals to the window's control circuit. Then when it rains, the window can be closed, and when the outdoor temperature is too high or too low, the window can be closed automatically.

2.3   *The window's control circuit and window's state detection circuit*

The two kinds of circuit are used to open and close the window. MCU can detect some special judge circuits on and off to judge the window's state. Then MCU controls several motors to open or close the window.

3 CONCLUSION

In this way, the smart window system can automatically open windows or close windows according to the current actual situation. It can create a comfortable, safe living space for human.

REFERENCES

[1] Lv Li, Luo Jie. 2007. Smart home and its development trend. *Computer and Modernization.*
[2] Qi ZhenXing. 2009. Introduction to the smart home development of our country. *Guangxi light industry.*
[3] Meng LiFan, Lan JinHui. 2011. *The principle and application of sensor*, edited by electronics industry Publishing, BeiJing.
[4] Dong Ji-hong, Bai Ming, Lang Pei. 2007. Alarm system with pyroelectric infrared sensors. *Journal of Tianjin University of Technology.*
[5] LinXueMei. 2005. Pyroelectric infrared sensor and its application. *Industrial Science and technology.*
[6] Xu KeJun. 2011. *Sensor and detection technology*, edited by electronics industry Publishing, BeiJing.
[7] Jin FaQin. 2012. *Sensor technology and Application*, edited by Mechanical industry press, BeiJing.
[8] Cai Li, Wang Guorong. 2013. *The application of sensor and detection technology,* edited by Metallurgical Industry Press, BeiJing.
[9] Zhang YongGang, Wang Bing. 2012. The application of Internet of things in the smart Home. *Intelligent building and city information.*
[10] Li QunFang, Zhang ShiJun, Huang Jian. 2009. *The single chip microcomputer and interface technology*, edited by electronics industry Publishing, BeiJing.
[11] Nie HuiHai. 2012. *Sensor technology and Application*, edited by electronics industry Publishing, BeiJing.

Advances in Engineering Materials and Applied Mechanics – Zhang, Gao & Xu (Eds)
© *2016 Taylor & Francis Group, London, ISBN 978-1-138-02834-0*

Study of squeeze film damping characteristics in damper of linear rolling guide

L.L. Li, J.J. Yang & W.W. Liu
School of Mechanical Science and Engineering, Huazhong University of Science and Technology,
Wuhan City, Hubei Province, China

ABSTRACT: Squeeze film damper can attenuate vibration in the linear rolling guide of CNC machine tool. The damping effect occurs when the damper filled with very thin film approaches the rail. In this paper, squeeze film damping characteristics between the damper and rail are investigated. The mathematical models of squeeze film pressure and damping coefficient are established based on the two-dimensional Reynolds equation which is solved by using the finite difference method. Numerical simulations are performed to analyze influence of the damper parameters on the film pressure and damping coefficient. The results show that the film pressure and damping coefficient increase with the decreasing values of film thickness and increasing values of film length-width ratio of the damper, which can be as a reference for designers of squeeze film damper.

1 INTRODUCTION

When two plates filled with thin film approach each other at a normal velocity, it yields the resistance for presence of the film pressure, and then squeeze film damping occurs. Squeeze film damping can dissipate vibration energy and attenuate the vibration amplitude (Bicak & Rao 2010). Griffin et al. (1966) discussed experimentally and theoretically squeeze film damping effect for long parallel plates and circular parallel plates of the gas squeeze film damper. Pandey et al. (2007) presented analytical model of squeeze film damping by using Green's function in MEMS structure. Bicak & Rao (2012) presented the finite element solution of squeeze film damper considering coupled fluid-structure effect. Wang et al. (1993) modelled the rolling guide system with damping film and presented the identification method of squeeze film damping coefficient. Jiang et al. (1995a) derived the formula of the damping coefficient by analyzing damping effect of parallel-gap film during angular vibration for rectangular plates, which was validated experimentally through simulative test rig of the rolling guide. Tian et al. (1985) investigated relation between the damping coefficient and feed velocity of CNC machine tool guide by experiments. Jiang et al. (1995b) performed analysis of the film damping effect when inserting film into the rolling guide. It is found that a significant effect of attenuating vibration was produced for films of different forms.

In this paper, numerical analysis of squeeze film damping characteristics in damper of the linear rolling guide is presented based on two-dimensional Reynolds equation which is solved by using the finite difference method. The effects of the film thickness and length-width ratio of damper on squeeze film pressure and damping coefficient are investigated.

2 MATHEMATICAL MODEL

Figure 1 shows a schematic of squeeze film damper with the length l and the film width b. The damper is approaching the rail normally with a velocity $\partial h/\partial t = 2\pi f A \cos(2\pi f t)$, where A is the amplitude and f is the excitation frequency. The film damping between the damper and

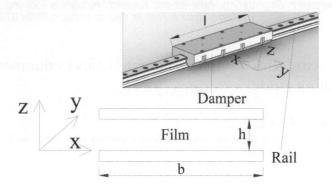

Figure 1. Schematic of squeeze film damper of linear rolling guide.

rail occurs due to squeeze film effect. The Reynolds equation considering squeeze film effect alone for the incompressible fluid with the constant viscosity and density can be expressed as (Wen & Huang 2002)

$$\frac{\partial}{\partial x}\left(h^3 \frac{\partial p}{\partial x}\right) + \frac{\partial}{\partial y}\left(h^3 \frac{\partial p}{\partial y}\right) = 12\eta \frac{\partial h}{\partial t} \tag{1}$$

where p is the film pressure distribution, η is the viscosity, and h is the film thickness of the damper. Using the dimensionless variables and parameters:

$$X = \frac{x}{b}, \ Y = \frac{y}{l}, \ H = \frac{h}{h_m}, \ T = \frac{t}{t_c}, \ \beta = \frac{l}{b}, \ P = \frac{p h_m^2 t_c}{12\eta b^2} \tag{2}$$

Since the film thickness h is a function of the time t alone, the dimensionless form of Equation (1) becomes

$$\frac{\partial^2 P}{\partial X^2} + \frac{1}{\beta^2} \frac{\partial^2 P}{\partial Y^2} = \frac{1}{H^3} \frac{\partial H}{\partial T} \tag{3}$$

with the boundary conditions

$$P = 0 \ \text{at} \ X = 0, \ X = 1 \tag{4}$$

$$P = 0 \ \text{at} \ Y = \pm 1/2 \tag{5}$$

The dimensionless damping force F is obtained by integrating the dimensionless pressure distribution P

$$F = \int_0^1 dX \int_{-1/2}^{1/2} P(X,Y) dY \tag{6}$$

The dimensionless damping coefficient C of the damper can be written as

$$C = \frac{F}{\partial H/\partial T} \tag{7}$$

3 SOLUTION USING THE FINITE DIFFERENCE METHOD

Equation (3) is solved numerically by using the finite difference method (Wen & Huang 2002). The first and second derivatives of Equation (3) are approximated by using the central difference method. For a given mesh point (i,j) in the discretized domain, the dimensionless film pressure can be written as

$$P_{i,j} = \frac{\Delta Y^2 \left(P_{i+1,j} + P_{i-1,j}\right) + \frac{\Delta X^2}{\beta^2}\left(P_{i,j+1} + P_{i,j-1}\right) - \Delta X^2 \Delta Y^2 \frac{1}{H^3}\frac{\partial H}{\partial T}}{2\left(\frac{\Delta X^2}{\beta^2} + \Delta Y^2\right)} \tag{8}$$

where $\Delta X = 1/m$ and $\Delta Y = 1/n$ are the space length in X and Y directions. The convergence criterion in the kth step iteration is defined by

$$Err(k) = \frac{\sum\limits_{i=2}^{m}\sum\limits_{j=2}^{n}\left|P_{i,j}^k - P_{i,j}^{(k-1)}\right|}{\sum\limits_{i=2}^{m}\sum\limits_{j=2}^{n}\left|P_{i,j}^k\right|} \leq \delta \tag{9}$$

where δ is the relative error. The dimensionless damping force F and damping coefficient C can be obtained in finite difference form as follows

$$F = \sum\limits_{i=1}^{m}\sum\limits_{j=1}^{n} P_{i,j}\Delta X \Delta Y \tag{10}$$

$$C = \frac{\sum\limits_{i=1}^{m}\sum\limits_{j=1}^{n} P_{i,j}\Delta X \Delta Y}{\partial H/\partial T} \tag{11}$$

4   NUMERICAL SIMULATIONS

In this section, numerical simulations of squeeze film damping characteristics are presented. The parameters used in simulations are as follows: the excitation frequency $f = 12$ HZ, the amplitude $A = 1$ μm, the relative error $\delta = 10^{-6}$, $m = n = 60$ and $h_m = 40$ μm.

Figure 2 presents the variation of the dimensionless film pressure P of damper with the axial coordinate X for the different dimensionless film thickness H and the different film length-width ratio β respectively at the dimensionless time $T = 4.522$ and axial coordinate

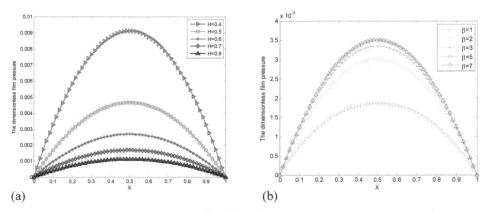

Figure 2.   Variation of the dimensionless film pressure P with the axial coordinate X at the dimensionless time $T = 4.522$ and axial coordinate $Y = 1/6$ for (a) the different dimensionless film thickness H and (b) the different film length-width ratio β.

$Y = 1/6$. It is observed that the values of the dimensionless film pressure P decrease significantly as the values of the dimensionless film thickness H increase. And the values of P increase as the values of the film length-width ratio β increase. Therefore, the film thickness and film length-width ratio are the important parameters affecting the film pressure distribution.

It is found that the damping coefficient C is negative by numerically solving Equation (3). For convenience C is assumed to denote the magnitude alone. The variation of the dimensionless damping coefficient C of damper with the dimensionless film thickness H at the film length-width ratio $\beta = 6$ is shown in Figure 3. It is observed that the values of the dimensionless damping coefficient C decrease significantly as the values of the dimensionless film thickness H increase. Thus, the film thickness has a significant influence on the damping coefficient, which should be considered for designers of the squeeze film damper.

The variation of the dimensionless damping coefficient C with the film length-width ratio β at the dimensionless film thickness $H = 0.55$ is shown in Figure 4. It is observed from Figure 4 that the values of the dimensionless damping coefficient C increase as the values of the film length-width ratio β increase. The variation of damping coefficient C is slow when $\beta > 6$, which denotes that the effect of length-width ratio is not so significant in this condition.

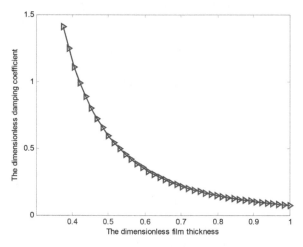

Figure 3. Variation of the dimensionless damping coefficient C with the dimensionless film thickness H at the film length-width ratio $\beta = 6$.

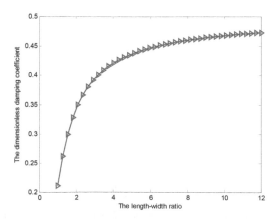

Figure 4. Variation of the dimensionless damping coefficient C with the film length-width ratio β at the dimensionless film thickness $H = 0.55$.

5 CONCLUSIONS

This paper presents numerical study of squeeze film damping characteristics in damper of linear rolling guide of CNC machine tool. Numerical simulations of squeeze film pressure and damping coefficient between the damper and rail are performed based on two-dimensional Reynolds equation which is solved by using the finite difference method. It is found that the values of film pressure and damping coefficient increase with the decreasing values of film thickness H and increasing values of film length-width ratio β. Therefore, H and β are the important parameters influencing squeeze film damping characteristics of the damper. This paper can be as a reference for designers of damper of linear rolling guide.

ACKNOWLEDGMENT

This study was supported by National Science and Technology Major Project of the Ministry of Science and Technology of China, under Grant no. 2013ZX04008-031.

REFERENCES

Bicak, M.M.A. & Rao, M.D. 2010. Analytical modeling of squeeze film damping for rectangular elastic plates using Green's functions. *Journal of Sound and Vibration* 329(22): 4617–4633.

Bicak, M.M.A. & Rao, M.D. 2012. Coupled squeeze film analysis by Reissner-Mindlin plate elements. *Journal of Vibration and Control* 18(5): 632–640.

Griffin, W.S., Richardson, H.H. & Yamanami, S. 1966. A study of fluid squeeze-film damping. *ASME Journal of Fluids Engineering* 88(2): 451–456.

Jiang, Z., Wang, Q.Y. & Qian, W.M. 1995a. Damping effect and dynamic coefficient of oil films on rolling guideways system. *Journal of Northeastern University (Natural Science)* 16(3): 248–252.

Jiang, Z., Wang, Q.Y. & Qian, W.M. 1995b. Vibration attenuation due to oil-film damping in rolling guideway systems. *Journal of Northeastern University (Natural Science)* 16(4): 417–419.

Pandey, A.K., Pratap, R. & Chau, F.S. 2007. Analytical solution of the modified Reynolds equation for squeeze film damping in perforated MEMS structures. *Sensors and Actuators A: Physical* 135(2): 839–848.

Tian, H.Q., Liao, X.G., Liu, E.C., Zhu, Q.Q., Li, G.B. & Wang, X.M. 1985. The frictional damping behavior of guideway on a CNC machine tool. *Journal of Huazhong University of Science and Technology* 13(5): 139–146.

Wang, Q.Y., Jiang, Z., Qian, W.M., Qi, H.S. & Zheng, H.W. 1993. Research on machine tools rolling slideways assembly with damping oil-films. *CIRP Annals-Manufacturing Technology* 42(1): 441–444.

Wen, S.Z. & Huang, P. 2002. *Principles of Tribology*. Beijing: Tsinghua University Press.

Advances in Engineering Materials and Applied Mechanics – Zhang, Gao & Xu (Eds)
© *2016 Taylor & Francis Group, London, ISBN 978-1-138-02834-0*

Pre-tension analysis of the tractor anchor wire in anchorage construction

Z.L. Liu, P.P. Ma & H. Zhou
School of Mechanical Engineering, Zhengzhou University, Zhengzhou, China

ABSTRACT: At present, a tractor is widely used in the overhead line construction. During the working process, a tractor mainly relies on the anchorage of its anchor wire and the inter-action between their legs and foundation to maintain its balance, thus, the pre-tension of the tractor anchor wire is crucial to smooth traction work in the anchorage construction. For the existing arrangement of tractor anchor wire, we take a certain type of tractor for the study in this paper, and by using of finite element method, the demand pre-tension of each tractor anchor wire in initial anchorage scheme under full load working condition is analyzed. On this basis, a comparative analysis of the tractor anchor wire pre-tension is carried out by changing the anchorage scheme, and we get the importance of each tractor anchor wire in maintaining tractor balance when it is in the process of traction. This paper provides useful guidance for the design and anchorage construction of the tractor.

1 INTRODUCTION

Wiring job is the key step in transmission line construction and it requires to maintain certain tension in the cable, as a consequence, cable tractor and tensioner are all critical equipment in wiring construction. Tensioner puts the cable at one side of the construction and at another side of the construction tractor pulls and recycles the cable during the wiring construction[1]. During wiring construction, the tractor is supported on the ground by its front and rear legs, at the same time, in order to fix tractor on the ground, anchor wires are installed on the tractor. In the process of anchorage construction, pre-tension should be kept in tractor anchor wire in case the location of tractor changes after loading. The security of the anchor wire is significance to the normal wiring construction[2].

In this paper, taking a self-driven traction as the research object, we analyzed the tension in each anchor wire when the tractor is under full load working condition in ANSYS. In addition, a pre-tension comparative analysis for removing anchor wire 2 or removing anchor wire 3 is carried out. On this basis, we get the pre-tension demand in each anchor wire during the anchorage construction when it is under every anchorage scheme. This paper provides useful guidance for the design and anchorage construction of the tractor.

2 ANCHORING SITUATION AND EXISTING PROBLEMS OF THE TRACTOR

The structure of the self-driven tractor that this paper refers to is shown in Figures 1 and 2. This type of tractor integrates two functions of traction and driving. As a result of that, the tractor is flexible in transitions and positioning, and it can effectively reduce the manpower and construction costing[1]. As a consequence, the self-driven tractor has been widely applied in the current wiring construction.

As shown in Figures 1 and 2, five anchor wires are arranged on both sides of the tractor. During designing, only textual requirements are raised for anchorage construction. As a consequence, it exists security risks. Therefore, it is necessary to determine the pre-tension

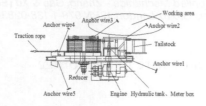

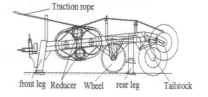

Figure 1. Structure schematic (a) of the tractor.   Figure 2. Structure schematic (b) of the tractor.

requirements in anchor wire. In addition, the anchor wire 2 and anchor wire 3 are just located in the working area. It brings inconvenience to workers; in addition, it cases a great number of threats to the safety of the worker. Therefore, removing anchor wire 2 or anchor wire 3 is considered during the design process, and then quantitative pre-tension analysis of the reserve anchor wire is indispensable.

3 PRE-TENSION ANALYSIS AND CALCULATION OF THE ANCHOR WIRE IN ORIGINAL ANCHORAGE SCHEME

3.1 *Structure simplification and finite element model establishment*

In this paper, combining with traditional mechanical analysis, we analyzed the tractor which is installed anchor wires by using the classical mechanics analysis module in ANSYS, and calculated the tension in each anchor wire during working process[3,4].

As shown in Figures 1 and 2, the engine, reduce and open gears drive the traction wheel to rotate during working process. Then, relying on the friction between traction wheel and traction rope, combining with the coil that mounted on the tailstock, we can recycle the pulling rope. The front and rear legs adopt hydraulic legs. We should adjust the height of the legs before the anchorage construction and fix the legs height when it reaches the desired position.

Because the tractor legs have little effect on the analysis of tractor anchor wire when it is working, we make the tractor as a whole. Simplify the parts that have no effect on the overall stress distribution, for example, ignoring threaded holes and lifting lug. Hydraulic tank, reducer and other components are treated as the weight load. In this paper, the solid45 is selected to mesh the model and the link10 is selected to mesh the anchor wire. At the position of anchor wire connects with anchor ear, the paper couples the node[5]. Model after meshing is shown in Figure 3.

The main material used for all parts of tractor frame beam and legs is Q235A. The performance parameters of frame material are[6,7]: density 7.85×10^3 Kg/m^3, yield limitation 235 MPa, elastic modulus 2.0×10^{11} Pa, Poisson's ratio 0.3; the performance parameters of anchor wire material are: elastic modulus 1.06×10^{11} Pa, Poisson's ratio 0.3, real constant of section 1.256×10^{-3} m^3.

3.2 *Load model and constraint, load applied*

Analysis of the tractor anchor wire is under the case of full load working condition. Because the weight of the tractor is large, we consider the tractor weight during finite element analysis. In addition, the other load and its parameters that tractor suffered is shown in Table 1. The loading position of each part is shown in Figures 1 and 2. The load is applied at each respective site in the form of uniform load.

In the direction of perpendicular to the main girder (X direction) and perpendicular to the ground (Y direction), tractor has no displacement during wiring construction, as a consequence, freedom of X, Y direction is limited. Tractor has displacement in the direction of main girder (Z direction), freedom of Z direction is unrestraint. When imposing constraint to the anchor wire, we should limit all the freedom of the tractor anchor wire at ground end, and another end that connects with the anchor ear should be coupled with anchor ear.

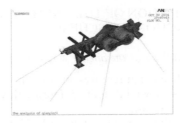

Figure 3. Model of the tractor after meshing.

Table 1. Various parts parameters of the tractor.

Load name	Load value/kN	Loading area/m^2	Uniform load/Pa
Hydraulic tank, meter box	4.5	0.385	11688.31169
Reducer	22	0.4243412	51845.07184
Engine weight	23	0.25	92000
Tailstock	70	0.20979056	333666.1097
Rated traction	250	0.00751753	33255603.9

```
PRINT REACTION SOLUTIONS PER NODE

 ***** POST1 TOTAL REACTION SOLUTION LISTING *****

 LOAD STEP=     1  SUBSTEP=     1
  TIME=    1.0000      LOAD CASE=   0

 THE FOLLOWING X,Y,Z SOLUTIONS ARE IN THE GLOBAL COORDINATE SYSTEM

   NODE       FX          FY          FZ
 173992    24057.     -57802.    -0.14200E+06
 173994   -15677.     -17674.     -44201.
 173996   -55844.     -20885.     -52214.
 173998   -19817.     -5329.4     -2740.2
 174000    20609.     -5775.7     -2978.3

TOTAL VALUES
VALUE    -46873.    -0.10747E+06-0.24414E+06
```

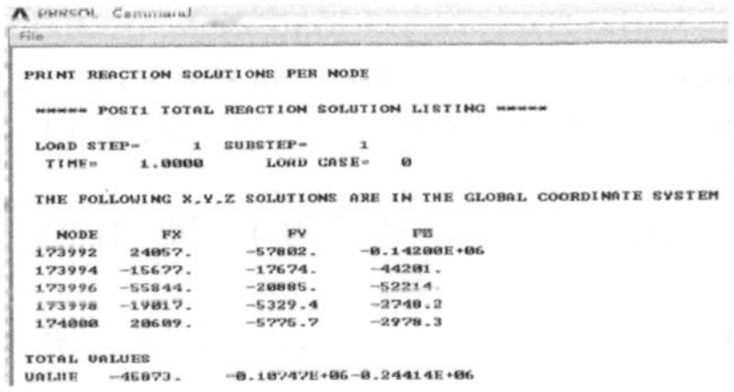

Figure 4. Results of reaction force in original anchorage scheme.

3.3 *Pre-tension results and analysis of the anchor wire in original anchorage scheme*

By calculation, we can get the component load in X, Y, Z direction of the joint reaction force that anchor wire connected with the ground which is shown in Figure 4 when tractor has completed anchor wire.

According to the component load in X, Y, Z direction, the paper calculated the total tension in each anchor wire and all the tension direction are all along the anchor wire[8,9].

Total tension in anchor wire 1:

$$F_1 = \sqrt{F_x^2 + F_y^2 + F_z^2} = \sqrt{24057^2 + 57802^2 + 142000^2} = 155189.595\ \text{N}$$

In the same way, total tension in anchor wire 2: $F_2 = 50118.53$ N.

Total tension in anchor wire 3: $F_3 = 79252.99$ N; Total tension in anchor wire 4: $F_4 = 19939.95$ N.

Total tension in anchor wire 5: $F_5 = 21609.25$ N.

Sequence of total tension in each anchor wire in original anchorage scheme: $F_1 > F_3 > F_2 > F_5 > F_4$.

The result shows that: the total tension in anchor wire 1 is the most severe and the value is 155189.595 N. The total tension in anchor wire 3 takes the second place. The total tension in anchor wire 4 is minimum. Pre-tension in anchor wire 1 should be ensured mainly.

4 PRE-TENSION CALCULATION OF THE ANCHOR WIRE IN IMPROVEMENT ANCHORAGE SCHEME

4.1 *Pre-tension results and analysis of the anchor wire in anchorage scheme 2*

Anchorage scheme 2 is the scheme that removes anchor wire 2 but keeps other anchor wire, and the finite element model is shown in Figure 5. During this condition, the load and constraint that tractor suffered is constant.

Under anchorage scheme 2, the component load in X, Y, Z direction of the joint reaction force that anchor wire connected with the ground is shown in Figure 6. During this anchorage scheme, the total tension in each anchor wire is as follow and all the tension direction are all along the anchor wire.

Total tension in anchor wire 1: F_1 = 183418.70 N; Total tension in anchor wire 3: F_3 = 105196.70 N.

Total tension in anchor wire 4: F_4 = 25094.57 N; Total tension in anchor wire 5: F_5 = 25727.83 N.

Sequence of total tension in each anchor wire in anchorage scheme 2: $F_1 > F_3 > F_5 > F_4$.

4.2 *Pre-tension results and analysis of the anchor wire in anchorage scheme 3*

Anchorage scheme 3 is the scheme that removes anchor wire 3 but keeps other anchor wire, and the finite element model is shown in Figure 7. During this condition, the load and constraint that the tractor suffered is constant.

Under anchorage scheme 3, the component load in X, Y, Z direction of the joint reaction force that anchor wire connected with the ground is shown in Figure 8. During this anchorage scheme, the total tension in each anchor wire is as follow and all the tension direction are all along the anchor wire.

Total tension in anchor wire 1: F_1 = 180270.40 N; Total tension in anchor wire 2: F_2 = 81840.31 N.

Total tension in anchor wire 4: F_4 = 29722.81 N; Total tension in anchor wire 5: F_5 = 20431.20 N.

Figure 5.   Finite element model in anchorage scheme 2.

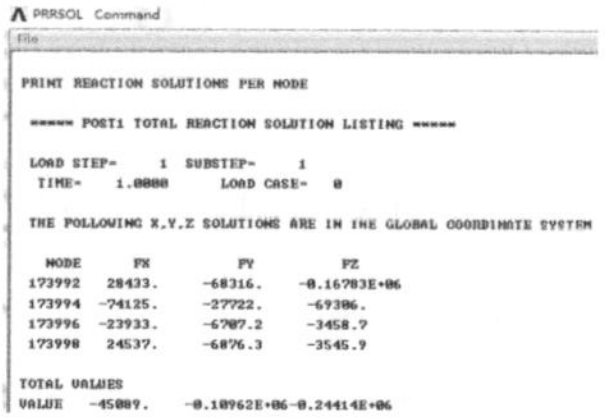

Figure 6.   Results of reaction force in anchorage scheme 2.

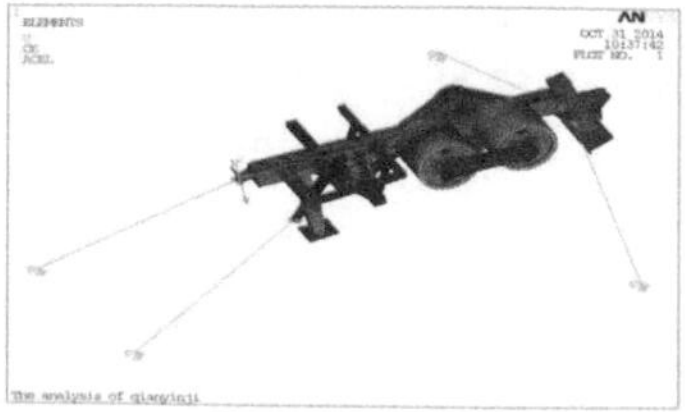

Figure 7.   Finite element model in anchorage scheme 3.

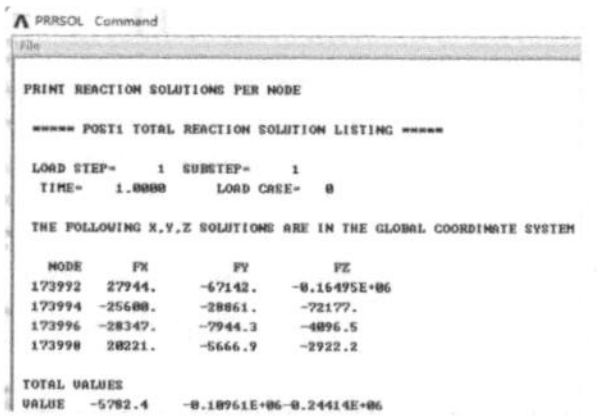

Figure 8.   Results of reaction force in anchorage scheme 3.

Table 2. Pre-tension comparison of anchor wire in each anchorage scheme.

Anchor wire	Original anchorage scheme	Anchorage scheme 2	Anchorage scheme 3
Anchor wire 1	155189.595 N	183418.702 N	180270.402 N
Anchor wire 2	50118.530 N		81840.312 N
Anchor wire 3	79252.996 N	105196.704 N	
Anchor wire 4	19939.945 N	25094.574 N	29722.813 N
Anchor wire 5	21609.254 N	25727.831 N	20431.196 N

Sequence of total tension in each anchor wire in anchorage scheme 3: $F_1 > F_2 > F_4 > F_5$.

5 PRE-TENSION COMPARATIVE ANALYSIS OF EACH ANCHOR WIRE IN DIFFERENT ANCHORAGE SCHEME

The pre-tension comparative result in each anchor wire under three different anchorage schemes is shown in Table 2.

The Table 2 shows that: the total tension in anchor wire 1 is the most severe when it is under three different anchorage schemes. Among them, the total tension in anchor wire 1 is 155189.595 N when it is under anchorage scheme 1. The total tension in anchor wire 1 is 183418.702 N when it is under anchorage scheme 2. The total tension in anchor wire 1 is 180270.402 N when it is under anchorage scheme 3. Compared with the original anchorage scheme, when it is under anchorage scheme 2 or anchorage scheme 3, the total tension in anchor wire 1 increased 18.19% and 16.17% respectively. This is due to the removing of the anchor wire 2 or anchor wire 3 and consequently increases the load in anchor wire 1.

Anchoring of each anchor wire is carried through lever hoist. In order to satisfy the anchorage demand of each tractor anchor wire, the paper selects different rated lifting weight lever hoist and the selection is made according to the pre-tension value that listed in the Table 2[10].

6 CONCLUSIONS

We analyzed the tractor anchor that is under three different anchorage schemes wire by using the finite element method. In the three different anchorage schemes, we obtained the tension in each anchor wire when it is under full load working condition, and on this basis, the pre-tension comparative analysis of each anchor wire is carried out. The result shows that: in the original anchorage scheme, the total tension in anchor wire 1 is the most severe and the value is 155189.595 N. The total tension in anchor wire 3 takes the second place. The total tension in anchor wire 1 maintains the most severe when it is under anchorage scheme 2, and the value is 183418.702 N. Compared with the original anchorage scheme, the total tension in anchor wire 1 increased to 18.19%. The total tension in anchor wire 3 takes the second place. The total tension in anchor wire 1 is still keeping the most severe when it is under anchorage scheme 3 and the value is 180270.402 N. Compared with the original anchorage scheme, the total tension in anchor wire 1 increased to 16.17%. The total tension in anchor wire 2 takes the second place. On this basis, we can get that: when it is under original anchorage scheme or anchorage scheme 2, the anchor wire 1 plays the most important role in maintaining the tractor stability, and the function of anchor wire 3 takes the second place. Therefore, the pre-tension in anchor wire 1 and anchor wire 3 should be ensured. In addition, when it is under anchorage scheme 3, the anchor wire 1 plays the most important role, and the function of anchor wire 2 takes the second place. Therefore, the pre-tension in anchor wire 1 and anchor wire 2 should be ensured.

REFERENCES

[1] Jin-ning Nie. 2012. *Finite Element Analysis and Structure Optimization of Key Components on the Towing Machine for Cable*. Jinan: University of Jinan.

[2] Hai-ping Jiang. 2004.*Tension stringing equipment and application*. Beijing: China electric power press.

[3] Yan-shen Xu, Xue-ling Zhang. 2003. *Mechanical structural static and dynamic performance optimization design that based on the finite element analysis*. Precision Manufacturing and Automation. 56–58.

[4] Xu-cheng Wang, Min Shao. 2006. *Basic principles of the finite element method and numerical methods*. Beijing: Tsinghua university press.

[5] Shu-guang Gong, Gui-lan Xie. 2009. *ANSYS Parameter Programming and Command Reference*. Beijing: Machinery Industry Press.

[6] Hua-rong Rao, Xin-peng You. 2010. *Force Mechanism of Installing Drag System with Flexible-stiff Combination in Ultra-long cable in Stayed-cable Bridge*. China Harbour Engineering, 167.3.

[7] Properties data sheet of mechanical engineering materials Editorial board.1995. *Properties data sheet of mechanical engineering materials*. Beijing: Mechanical industry press.

[8] Theoretical Mechanics Department of Harbin Institute of Technology. 2009. *Theoretical Mechanics*. Beijing: Higher education press.

[9] Liang-gui Pu, Ming-gang Ji. 2006. *Design of Machinery*. Beijing: Higher education press.

[10] Hao Xu. 2004. *Mechanical design handbook*. Beijing: Machinery industry press.

Advances in Engineering Materials and Applied Mechanics – Zhang, Gao & Xu (Eds)
© *2016 Taylor & Francis Group, London, ISBN 978-1-138-02834-0*

Serial port expansion technologies based on the VK series chip

Y.X. Lu & Z.Y. Cai
School of Electrical Engineering, Shenyang University of Technology, Shenyang, China

ABSTRACT: In recent years, the number of serial module has been rapidly increased in the application of electronic engineering. This paper introduces a kind of serial port expansion technologies based on the VK series chip, the author used VK3366, which is one of the VK series serial port expansion chips of Viken Technology, to extend four serial ports for a kind of electric fire monitoring device. In this paper, the design of electric fire monitoring device system and the basic characteristics of VK3366 have been introduced, and the hardware interface circuit of VK3366 and the key program flow chart of initialization program and receiving data program have been designed. Serial port expansion technologies have realized multi-serial port multiplexing and data transmission, and solved the problem of microcontroller lack of serial port.

1   THE REQUIREMENT BACKGROUND OF SERIAL PORT EXPANSION

Serial port has many advantages, such as convenient, easy to realize, anti-interference, and long distance transmission. In recent years, with the rapid development of MCU technology, using MCU as the core of the control system is applied more and more widely. The interfaces of these devices are mostly serial ports. At the present stage, such as the 51 series microcontroller, PIC series, AVR series and MSP430 series, most of them only produce one or two serial ports. The engineer often used the timer and the I/O to imitate the timing of asynchronous serial communication. But this method has many disadvantages, such as, sampling results are not accurate, and anti-interference ability is poor. The better approach is to use serial port expansion chip to expand serial port.

2   THE DESIGN OF ELECTRIC FIRE MONITORING DEVICE SYSTEM

When the protected circuit parameter exceeds a set value, electrical fire monitoring system can send out alarm signal and control signal, and alarm the location of the system. Electrical fire monitoring includes electric fire monitoring device and electrical fire monitoring detector. The logic diagram of electric fire monitoring device is shown in Figure 1, the chip of the monitoring equipment is PIC16F4620. Electrical fire monitoring detector is connected to the monitor through the RS485 bus. According to the design requirement, we used the serial port expansion chip VK3366 to expend serial ports for the electric fire monitoring device. SD card modules, touch screen module, and serial port Ethernet module are connected to the monitoring equipment through the serial port.

2.1   *The basic characteristics of VK3366*

VK3366 is one of the VK series serial port expansion chips of Viken Technology. VK3366 supports a variety of host interface, the user can choose UART, IIC, SPI or 8 bit parallel port. The baud rate, parity bits, UART format of each sub channel can be set up respectively. Each sub channel provides the maximum communication rate at 1 Mpbs. Each sub channel

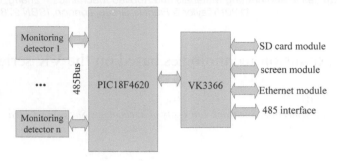

Figure 1. Electric fire monitoring device logic diagram.

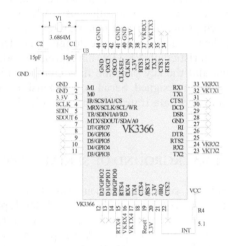

Figure 2. Circuit diagram of VK3366.

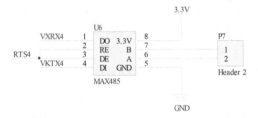

Figure 3. RS-485 interface circuit.

has independent 16 bit FIFO (First in First Out) buffer, which has 4 level programmable interrupt conditions trigger point [1]. The VK3366 serial port expansion chip occupied only 3 ports of microcontroller. In this design, we used the SPI interface of VK3366 to extend serial ports. VK3366 work in slave mode of SPI, and it support the SPI model 0. It has the maximum speed at 5 Mbit/s when the crystal oscillator is 20 MHz.

2.2 Circuit diagram of VK3366

As shown in Figure 2, the power supply voltage of VK3366 is 2.5 ~ 5.5 V, we used the 3.3 V power supply in this design. E0 pin and E1 pin are connected to the low level, and we set VK3366 to SPI mode. Sub serial port baud rate is determined by the external crystal oscillator and software, and we chose 3.6864 MHz. The SPI interface of VK3366 comprises of the following three signals: SDIN, SDOUT and SCLK, the three pins are respectively connected to the SPI pin of MCU. IRQ is the global interrupt pins, the notification of data transceiver to the microcontroller through the IRQ pin outputs a low level signal, so the pin should be connected to pull-up resistors, and the typical value is 5.1 k. The sub channels of VK3366 are connected respectively to touch screen module, SD card module, and Ethernet module. Because each sub channel of VK3366 can be set to RS-485 for automatic sending and receiving mode, the fourth serial port is connected to MAX485 interface chip, and RTS4 is the control pin. As shown in Figure 3.

3 MICROCONTROLLER SOFTWARE DESIGN

MCU software design includes the main function, MCU initialization, VK3366 initialization, data transmission program, and VK3366 interrupt receive program. We used C language as the program language that has a good readability. It is easy to modify and expand, that language can greatly shorten the software development cycle. VK3366 software design idea and frame are as follows.

3.1 *The design of VK3366 initialization program*

The initialization of VK3366 through the SPI interface includes the interface baud rate, data format and working mode. First, the SPI driving function of MCU should be compiled, which used to read and write VK3366 register. In the write operation, the MCU sends a control byte, and then send a byte of data. In the read operation, the MCU sends a command byte, and then the data bytes read from the register. In RS485 mode, VK3366 only supports 9 data bits, if you want to transfer 8 bit data bits which have one stop bit, it needs to be set to RS232 mode. Through the RTS signal, we can send or receive data. The initialization procedure of VK3366 as shown in Figure 4.

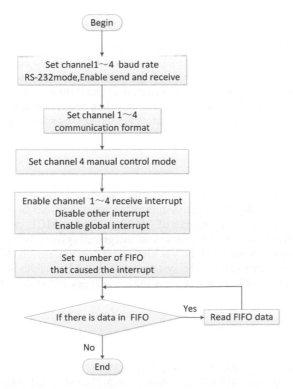

Figure 4. The initialization program flow chart of VK3366.

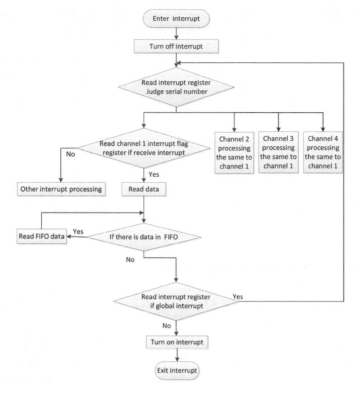

Figure 5. The receive data program flow chart of VK3366.

3.2 *The receiving data program of VK3366*

It is easy to realize that microcontroller send data to VK3366 by SPI, when the sub channel data of VK3366 have been received, it will be stored in FIFO. If the number of receiving bite is up to receiving the trigger point, there will generate an interrupt, which notify the MCU to read data. Therefore, the IRQ pin is connected to the RB0 pin of PIC[2]. After interrupt generation, the MCU read the interrupt status, and then determine which channels generate the interrupt. The receiving data of procedure VK3366 are shown in Figure 5.

4 CONCLUSIONS

As the market for serial communication interface module is increased, industrial control, data acquisition, and intelligent home system often need multi-serial port communication. Because of the restrictions on the number of MCU serial port, MCU cannot connect to multiple serial port devices to constitute a variety of integrated system. In the design of electrical fire monitoring equipment, we used the VK3366 serial port expansion chip to connect to the microcontroller. After verification, the system is stable and reliable. Serial port expansion technologies based on the VK series chip has the advantages of high reliability, low cost, and easy implementation.

REFERENCES

[1] Chengdu Viken Technology VK3366 Data Sheet [Z]. 2011.
[2] Microchip Technology Inc. PIC18F2525/2620/4525/4620 Data Sheet [Z]. 2006.
[3] Magda, Yury. 2009. Serial port tests digital circuits. *Measurement World*, 293: 48.

[4] Jing Zhang, Xi-tao Zheng, Ye-hua Yu, Yong-wei Zhang, Kun Yang, Jun Shi. 2013. Network structure and reliability analysis of a new integrated circuit card payment system for hospital. *Journal of Shanghai Jiaotong University (Science),* 185: 630–633.

[5] Liakot Ali, Roslina Sidek, Ishak Aris, Alauddin Mohd. Ali, Bambang Sunaryo Suparjo. 2004. Design of a micro-UART for SoC application. *Computers and Electrical Engineering,* 304: 257–268.

[6] You Zhi, Chen Shu-zhen, Huang Ji-wu. 2005. Serial peripheral interface communication between MCU EM78P447A and RF reader IC MF RC530. *Wuhan University Journal of Natural Sciences,* 103: 550–554.

[7] Ke Peng, Ji-da Chen, Lan Chen. 2002. Hierarchical hybrid control network design based on LON and master-slave RS-422/485 protocol. *Journal of Central South University of Technology,* 93: 202–207.

[8] Jia Liu, Guangmin Sun, Dequn Zhao, Xu Yao, Yihang Zhang. 2012. MCU-Controlling Based Bluetooth Data Transferring System. *Procedia Engineering,* 29: 2109–2115.

[9] Gao Guanwang, Wang Yanpeng, Sha Zhanyou. 2012. The Design of Embedded MCU Network Measure and Control System. *Energy Procedia,* 17: 983–989.

[10] Shaobao Li, Juan Wang, Xiaoyuan Luo, Xinping Guan. 2010. A new framework of consensus protocol design for complex multi-agent systems. *Systems & Control Letters,* 601: 19–26.

[11] A. Poursaee, W.J. Weiss. 2010. An Automated Electrical Monitoring System (AEMS) to assess property development in concrete. *Automation in Construction,* 194: 485–490.

Advances in Engineering Materials and Applied Mechanics – Zhang, Gao & Xu (Eds)
© 2016 Taylor & Francis Group, London, ISBN 978-1-138-02834-0

Cutting tool wear monitoring by Empirical Mode Decomposition method based on wavelet packet analysis

M.H. Shao, W. Li, S.C. Li & S.Y. Wang
School of Mechanical and Electrical Engineering, Jiangsu Normal University, Xuzhou, China

L. Wei
Tangshan Key Laboratory of Mechantronics, Tangshan College, Tangshan, Hebei, China

ABSTRACT: While we analyze the tool wears signal in turning process, the mode mixing problem caused by Empirical Mode Decomposition (EMD) is a great challenge. In this paper the wavelet analysis combined with the EMD was chosen to extract the features of the cutting tool wears signal. Firstly, the EMD method was used to decompose the wear signal into several Intrinsic Mode Functions (IMF), and wavelet analysis was employed to decompose the IMF and to make it more relevant to the tool wear, the IMF was rebuilt. After calculating the correlation coefficient of the rebuilt IMFs and cutting tool wear, sensitive singles based on the rebuilt IMFs are selected as the inputs of Support Vector Machine (SVM). Finally, we can identify the tool wear condition.

1 INTRODUCTION

Nonlinear and non-stationary signals were sampled in the process of tool cutting. Fourier Transform was mainly used to process the stationary signal [1–4]. EMD was applied to process it because the method has many advantages [5–7]. The wavelet bases of wavelet and wavelet packet were chose according to the signal features [8–11]. However, EMD overcomes this problem, and signal is adaptively decomposed into each IMF component with the physical sense. EMD is effective in the diagnosis of rotating machinery and tool fault, but the main drawbacks are the modal aliasing. Signal with noise was decomposed by EMD, and then similar scale signal may be distributed in different IMF or an IMF contains several large-scale difference signals, which can not make various IMF signals well decomposed into IMF with physical signification in the case of noise. In the diagnosis of tool wear, it can not extract the sensitive IMF which reflects the state of wear and its characteristics.

The EMD based on wavelet packet analysis was used in this paper. The Aliasing IMF was decomposed through wavelet Packet after conducting EMD decomposition. Wavelet packet coefficients which have high correlation with tool wear were reconstructed and they were used as the new IMF variable. Then, the eigenvalues were extracted from the new IMF and used as the inputs of SVM, and then the state of tool wear was identified. The results show that it is effective using the method to improve the recognition accuracy of tool wear state.

2 THE EMD METHOD BASED ON THE WAVELET PACKET

The application of the EMD method based on wavelet packet in the tool wear monitoring is shown in Figure 1.

1. Several IMF components is acquired by using EMD to decompose the vibration signal acquisition;
2. Calculating correlation coefficients of IMF components and tool wear status;

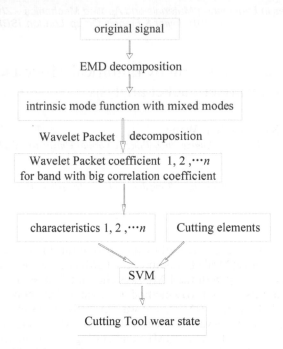

original signal

↓

EMD decomposition

↓

intrinsic mode function with mixed modes

Wavelet Packet ↓ decomposition

Wavelet Packet coefficient 1, 2 ,···n
for band with big correlation coefficient

↓

characteristics 1, 2 ,···n Cutting elements

SVM

↓

Cutting Tool wear state

Figure 1.   Schematic diagram of tool wear monitoring.

3. IMFs of large correlation coefficients are decomposed by 4-layer wavelet packet;
4. IMF is reconstructed through extracting wavelet packet coefficients which have high cor-relation coefficients with tool wear state;
5. The features of the reconstructed IMF, which can reflect the states of tool wear, are extracted;
6. The features extracted in the previous step and three elements of cutting are regarded as the input of SVM;
7. The tool wear states are identified according to SVM after training.

3   RESULTS AND DISCUSSION

3.1   *Vibration signal characteristics*

The cutting tool vibration signal power spectrums in different wearing degree were collected at the cutting speed of 1900 mm/s, feed rate of 0.8 mm/min and cutting depth of 0.4 mm conditions and the results are derived as shown in Figure 2.

As can be seen in Figure 2, features are in the frequency range of 0–500 Hz, 1000–1600 Hz and 5500–6500 Hz. Therefore, sensitive features which are associated with the cutting tool wear in the vibration signal can be obtained through using EMD method to decompose the signal adaptively.

3.2   *EMD decomposition of vibration signal*

The signals are decomposed into the finite IMF and a residual term by the EMD method, and each IMF characterizes the local information of signals, but the residual term reflects the overall trend of signal.

Energy change of different bands in different wear conditions is characterized by root mean square value of IMF, and it can effectively reflect the wear state of tool. In this paper, the root mean square value of each IMF was chosen as the feature vector, and the correlation coefficients of it and tool wear status are shown in Figure 3.

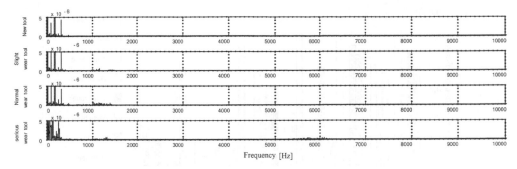

Figure 2.   Vibration power spectrum of z direction.

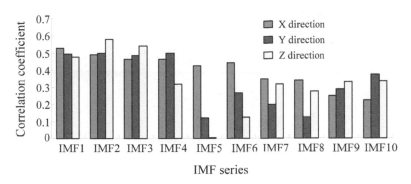

Figure 3.   Correlation coefficients of the root mean square values with the tool wear status.

3.3   *Vibration signal based on EMD decomposition of wavelet packet*

As can be seen in Figure 3, the root mean square values of IMF1–IMF4 had a better correlation with tool wear status. Meanwhile, the main frequency components of original signal were decomposed into IMF2~IMF5 components by using the Fourier transformation to process the original signal and IMF components. Therefore, IMF1–IMF5 were used as the IMF components of tool wear.

The IMF1–IMF5 of vibration in X direction were decomposed by the DB8 wavelet packet in this research, and then reconstructed, and the correlation coefficients of the root mean square value of new IMF and tool wear status were obtained, as shown in Figure 4.

As can be seen in Figure 4, the features obtained based on the EMD decomposition of wavelet packet can reflect the wear state of the tool better.

3.4   *SVM prediction to tool wears state*

The root mean square values of IMF1–IMF5 were acquired according to EMD decomposition, and the root mean square values of the new IMF based on EMD decomposition of wavelet packet were found out, and then they were chosen as the feature vectors to establish SVM model.

According to Figure 5, we know that the predicted accuracy of SVM training set based on EMD method is 92.9688% (119/128), while predicted accuracy of test set is 81.25% (52/64).

As can be seen Figure 6, the predicted accuracy of SVM training set of EMD method based on wavelet packets is 99.2188% (127/128), while predicted accuracy of test set is 92.1875% (59/64).

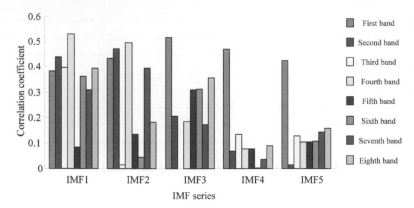

Figure 4.　Correlation analysis.

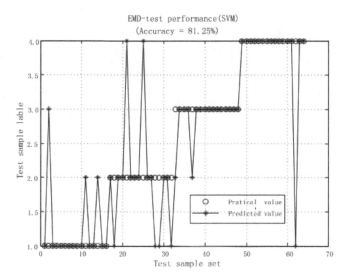

Figure 5.　Test performance of SUV based on EMD.

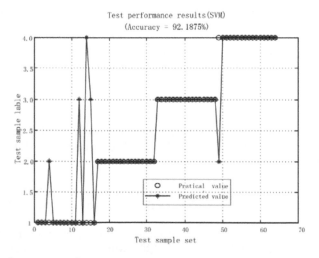

Figure 6.　Test performance results.

4 CONCLUSIONS

The experimental results demonstrated that EMD decomposition based on the wavelet packet is effective for avoiding mode mixing in the EMD decomposition. The sensitive features associated with the cutting tool wear were extracted for the diagnosis of cutting tool wear, and it establishes a good foundation to improve the accuracy of pattern recognition. Now, wavelet, wavelet packet and EMD method are applied to signal analysis of cutting tool wear respectively. The method proposed in this paper combines the advantages of the wavelet packet and the EMD method, and it provides a novel approach for characterization of the cutting tool wear.

ACKNOWLEDGEMENTS

This research is supported by Scientific Researching Fund Projects of JiangSu Normal University (Grant No. 14XLB09), and Natural Science Foundation of Jiangsu Higher Education Institutions (Grant No. 13KJB510009).

REFERENCES

[1] D.W. Cho, W.C. Choi, H.Y. Lee. 2000. Detecting tool wear in face milling with different work piece materials. *Key Engineering Materials*, 183(1): 559–564.
[2] J.C. Wang, S.Y. Li et al. 1995. Tool Wear Monitoring on the Basis of Multisensor Fusion by Neural Network. *Mechanical Science and Technology for Aerospace Engineering*, 6:125–130.
[3] X. Li et al. 1998. Tool wear monitoring with wavelet packet transform-fuzzy clustering method. *Wear*, 219: 145–154.
[4] R.J. Kuo, P.H. Cohen. 1998. Intelligent tool wear estimation system through artificial neural networks and fuzzy modeling. *Artificial Intelligence in Engineering*, 12:229–242.
[5] T. Wang & T. Xu. 2011. Tool Wear Fault Diagnosis Based on Empirical Mode Decomposition and Support Vector Machines. *Tool Engineering*, 45(263):63–67.
[6] S. Guan & L.S. Wang et al. 2011. Identification Method of Tool Wear Based on Empirical Mode Decomposition and Least Squares Support Vector Machine. *Journal of Beijing University of Aeronautics and Astronautics*, 37(2):144–148.
[7] P. Nie & H.Y. Xu et al. 2012. Application of EEMD method in state recognition of tool wear. *Transducer and Microsystem Technologies*, 31(5):147–149.
[8] B. Li & Y.F. Ding et al. 2010. Application of Wavelet Package Analysis in Tool Wear Monitoring System based Electric Current signals. *Mind and Computation*, 4(4):258–264.
[9] N. Fan et al. 2009. Characteristics of Cutting-Force Signals Based on Wavelet Analysis. *Machine Tool & Hydraulics*, 37(1):162–163.
[10] X.L. Zhao et al. 2001. Wavelet Measurement of Cutting-Tool Wear. *Tool Engineering*, 35(2):39–41.
[11] X.X. Wang et al. 2011. Study on Cutting Sate Monitoring Technology Based on the Wavelet Analysis of Vibration Signals. *Journal of Nanchang Hangkong University*, 25(3):42 –47.

Advances in Engineering Materials and Applied Mechanics – Zhang, Gao & Xu (Eds)
© 2016 Taylor & Francis Group, London, ISBN 978-1-138-02834-0

Reliability analysis of component in series system using masked data

X.L. Shi

School of Electronic Engineering, Xi'an University of Posts and Telecommunications, Xi'an, China

ABSTRACT: In this paper, reliability estimate of Rayleigh component in the series system are investigated using masked system life data under the Type-II censoring. The Maximum Likelihood Estimations (MLE) of the unknown parameters for Rayleigh components and reliabilities index are derived. In view of the limitation of MLE under complete masked case, we employ the Bayes approach. Bayes estimations of the Rayleigh parameters and reliabilities index are obtained under the Square Error loss function. Also, a numerical example is given by means of the Monte-Carlo simulation for comparing the effects of the MLE and the Bayes estimation, which shows that the Bayes estimation is better than MLE.

1 INTRODUCTION

Lifetime data from multi-component systems are often used to estimate the reliability of each component. Such estimates can be extremely useful since they reflect component reliability after assembly into an operational system. In engineering practice, most of high technology products are composed of many components, such as computer, circuit card and so on. Due to some certain environmental conditions, such as the constraints of cost and time, the exact component that causes the system to fail is often unknown. Instead, it is assumed that the component causes the system failure belongs to some subset of the components which is considered potentially responsible for the failure. In this case, the cause of failure is masked.

Recently, estimating component reliability using the masked data has been considered by several authors. Literature[1] firstly proposed the concept of masked data and obtained likelihood function of the life test sample. Literature[2-4] derived the maximum likelihood and Bayes estimates of components parameters and reliabilities in the series system when the lifetime of components follows Weibull, Pareto and Geometric distribution, respectively. Sarhan[5] extended the reliability analysis of masked data in the parallel case, while the reliability estimating of verse Weibull component was discussed in a parallel system by Zhang[6]. Lin & Guess[7] presented a proportional probability model for a dependent masked probability in the series system. Based on this model, reliability estimations of Geometric components from series system are studied by Sarhan[8]. Fan and Wang[9] studied Statistical inference in accelerated life tests for Weibull series systems with masked data. Literature[10] considered the parametric estimation in a wide class of parametric distribution families based on right-censored competing risks data and with masked failure cause. Literature[11,12] studied Bayesian analysis of masked data in step-stress accelerated life testing and Bayesian reliability analysis of Burr XII component in series system using masked life data.

Rayleigh distribution has wide applications in reliability analysis. In this paper, we mainly consider the estimates of parameters and reliabilities index for Rayleigh components in series system. Both MLE and Bayes approaches are utilized, and their results are compared in the simulation. The paper is organized as follows. In Section 2, MLEs of the unknown parameters for Rayleigh components and reliabilities index are derived. In Section 3, the Bayes estimations of the Rayleigh parameters and reliabilities index are obtained under the square error loss function. In Section 4, a numerical example is given for comparing the effects of the maximum likelihood estimation and the Bayes estimation. The conclusions are provided in section 5.

2 LIKELIHOOD FUNCTION AND MAXIMUM LIKELIHOOD ESTIMATION

Suppose that N identical systems are put on the life test, the test is stopped when the n-system failure (n is a pre-given constant $n < N$). Each system consists of k independent but non-identical components connected in series, and each component life distribution independent of each other. The random variables $T_{ij}(i = 1, 2, ..., N, j = 1, 2, ..., k)$ denote the lifetime of the j-th component in the i-th system, so the lifetime of the i-th system is $T_i = \min(T_{i1}, T_{i2}, ..., T_{ik}), (i = 1, 2, ..., N)$. T_{ij} follows the Rayleigh distribution, and its probability density as follows.

$$f_j(t) = 2t/\lambda_j \exp(-t^2/\lambda_j), \quad t > 0. \tag{1}$$

Reliability function and failure rate function of the j-th component in the i-th system are $R_j(t) = \exp(-t^2/\lambda_j)$ and $H_j(t) = 2t/\lambda_j, t > 0, j = 1, 2, ..., k$, respectively. After the life test is stopped, we can obtain the observed data $(t_1, S_1), (t_2, S_2), ..., (t_n, S_n)$, where $S_i(i = 1, ..., n)$ express the set of possible failure cause in i-th system. For non-failure system, we only observed censored lifetime $t_i = t_n$, $i = n+1, n+2, ..., N$. When masking is independent of failure cause, the likelihood function can be derived.

$$L(\lambda_1, \lambda_2) = \prod_{i=1}^{n}\left(\sum_{j \in S_i} f_j(t_i)\prod_{l \in M_j} R_l(t_i)\right)\left(\prod_{k=n+1}^{N}\prod_{l=1}^{k} R_l(t_k)\right), \tag{2}$$

where $(M_j = 1, 2, ..., j-1, j+1, ..., K)$. In this paper, we consider the case $k = 2$. Let n_1 and n_2 be the numbers of system failures for which the component 1 and 2 cause the failure, respectively. While n_{12} denotes the number of failed systems that the cause is not directly observed. That is, $n_j(j = 1, 2)$ is the number of observations when $S_i = \{j\}$ and n_{12} is the number of observations when $S_i = \{1, 2\}$. Note that $n = n_1 + n_2 + n_{12}$, the likelihood function is then expressed

$$L(\lambda_1, \lambda_2) = 2^n \left(\prod_{i=1}^{n} t_i\right)(1/\lambda_1)^{n_1}(1/\lambda_2)^{n_2}(1/\lambda_1 + 1/\lambda_2)^{n_{12}} \exp\left\{-(1/\lambda_1 + 1/\lambda_2)\sum_{i=1}^{N} t_i^2\right\}. \tag{3}$$

The maximum likelihood estimation of λ_1, λ_2 can be obtained by maximizing the likelihood function given by formula (3). One can derive the log-likelihood function.

$$\ln L(\lambda_1, \lambda_2) = n \ln 2 + \sum_{i=1}^{n}\ln t_i - (n_{12} + n_1)\ln \lambda_1 - (n_{12} + n_2)\ln \lambda_2$$
$$- (1/\lambda_1 + 1/\lambda_2)\sum_{i=1}^{N} t_i^2 + n_{12}\ln(\lambda_1 + \lambda_2).$$

Thus, by setting the derivative zero, the likelihood equations can be obtained as

$$\partial \ln L(\lambda)/\partial \lambda_1 = -(n_{12} + n_1)/\lambda_1 + \sum_{i=1}^{N} t_i^2/\lambda_1^2 + n_{12}/(\lambda_1 + \lambda_2) = 0$$
$$\partial \ln L(\lambda)/\partial \lambda_2 = -(n_{12} + n_2)/\lambda_2 + \sum_{i=1}^{N} t_i^2/\lambda_2^2 + n_{12}/(\lambda_1 + \lambda_2) = 0$$

Solving the equations above with respect to λ_1, λ_2, we can directly get

$$\hat{\lambda}_j = \sum_{i=1}^{N} t_i^2 \Big/ \left[n_j + n_{12} \cdot n_j/(n_1 + n_2)\right], (j = 1, 2). \tag{4}$$

According to the invariance property of MLE, by replacing the parameters λ_1, λ_2 with their MLE $\hat{\lambda}_1, \hat{\lambda}_2$ in $R_j(t)$ and $H_j(t)$, we can get the MLE of the reliability function and failure rate function. $\hat{R}_j(t) = \exp\{-t^2/\hat{\lambda}_j\}, \hat{H}_j(t) = 2t/\hat{\lambda}_j, j = 1, 2$.

3 BAYES ESTIMATION

From the above discussion we can conclude that when $n_1 = n_2 = 0$, the MLE of parameters can not be obtained from (4). Thus Bayes approach is considered in this section. In order to derive the Bayesian estimation, the prior distributions of the parameters λ_1, λ_2 should be considered initially. In this paper, the prior probability density function (pdf) is taken as following form.

$$\pi_j(\lambda_j) = \frac{\beta_j^{\alpha_j}}{\Gamma(\alpha_j)}\left(\frac{1}{\lambda_j}\right)^{\alpha_j+1} \exp(-\beta_j/\lambda_j), (\lambda_j > 0, \alpha_j > 0, \beta_j > 0, j = 1, 2), \tag{5}$$

where $\alpha_j, \beta_j(j = 1, 2)$ called hyper-parameters are given by historical data or experience of experts. Assume that the parameters λ_1 and λ_2 is independent of each other. Then the joint prior pdf of λ_1, λ_2 are $\pi(\lambda_1, \lambda_2) = \pi_1(\lambda_1)\pi_2(\lambda_2)$. By using binomial expansion, we have

$$(1/\lambda_1 + 1/\lambda_2)^{n_{12}} = \sum_{k=0}^{n_{12}}\binom{n_{12}}{k}\left(\frac{1}{\lambda_1}\right)^k\left(\frac{1}{\lambda_2}\right)^{n_{12}-k}.$$

Hence the likelihood function becomes

$$L(data \mid \lambda_1, \lambda_2) = 2^n\prod_{i=1}^n t_i \exp\left\{-(1/\lambda_1 + 1/\lambda_2)\sum_{i=1}^N t_i^2\right\}\sum_{k=0}^{n_{12}}\binom{n_{12}}{k}\left(\frac{1}{\lambda_1}\right)^{n_1+k}\left(\frac{1}{\lambda_2}\right)^{n_2+n_{12}-k} \tag{6}$$

Once the likelihood function and the joint prior pdf are constructed, we can derive the joint posterior pdf of λ_1, λ_2 as follows.

$$\pi(\lambda_1, \lambda_2 \mid data) = \frac{\pi(\lambda_1, \lambda_2)L(data \mid \lambda_1, \lambda_2)}{\int_0^{+\infty}\int_0^{+\infty}\pi(\lambda_1, \lambda_2)L(data \mid \lambda_1, \lambda_2)d\lambda_1 d\lambda_2}$$

$$= e^{-(\beta_1+T)/\lambda_1}e^{-(\beta_2+T)/\lambda_2}\sum_{k=0}^{n_{12}}\binom{n_{12}}{k}\left(\frac{1}{\lambda_1}\right)^{A+1}\left(\frac{1}{\lambda_2}\right)^{R+1} \bigg/ \sum_{k=0}^{n_{12}}\binom{n_{12}}{k}(\beta_1+T)^{-A}\Gamma(A)(\beta_2+T)^{-B}\Gamma(B)$$

$$= e^{-(\beta_1+T)/\lambda_1}e^{-(\beta_2+T)/\lambda_2}\sum_{k=0}^{n_{12}}\binom{n_{12}}{k}\left(\frac{1}{\lambda_1}\right)^{A+1}\left(\frac{1}{\lambda_2}\right)^{B+1} \bigg/ I_0,$$

where

$$I_0 = \sum_{k=0}^{n_{12}}C_{n_{12}}^k(T+\beta_1)^{-A}\Gamma(A)(T+\beta_2)^{-B}\Gamma(B), T = \sum_{i=1}^N t_i^2, A = n_1 + k + \alpha_1, B = n_2 + n_{12} - k + \alpha_2.$$

Thus the marginal posterior pdf of λ_1, λ_2 can be formulated as following equations

$$\pi_1(\lambda_1 \mid data) = \int_0^{+\infty}\pi(\lambda_1, \lambda_2 \mid data)d\lambda_2 = e^{-(\beta_1+T)/\lambda_1}\sum_{k=0}^{n_{12}}\binom{n_{12}}{k}\left(\frac{1}{\lambda_1}\right)^{A+1}(T+\beta_2)^{-B}\Gamma(B)/I_0,$$

$$\pi_2(\lambda_2 \mid data) = \int_0^{+\infty}\pi(\lambda_1, \lambda_2 \mid data)d\lambda_1 = e^{-(\beta_2+T)/\lambda_2}\sum_{k=0}^{n_{12}}\binom{n_{12}}{k}\left(\frac{1}{\lambda_2}\right)^{B+1}(T+\beta_1)^{-A}\Gamma(A)/I_0.$$

Under the square error loss function $L(\theta,d) = (d-\theta)^2$, the Bayes estimators of λ_1 and λ_2 are their posterior mean, that is, $\hat{\lambda}_j = E(\lambda_j \mid data)$ $j = 1, 2$. Hence the Bayes estimators of λ_1 and λ_2 are

$$\hat{\lambda}_1 = E(\lambda_1 \mid data) = \int_0^{+\infty}\lambda_1\pi_1(\lambda_1 \mid data)d\lambda_1$$

$$= J_1^{(1)}\bigg/\sum_{k=0}^{n_{12}}\binom{n_{12}}{k}(T+\beta_1)^{-A}\Gamma(A)(T+\beta_2)^{-B}\Gamma(B) = J_1^{(1)}/I_0,$$

$$\hat{\lambda}_2 = E(\lambda_2|data) = \int_0^{+\infty} \lambda_2 \pi_2(\lambda_2|data)d\lambda_2$$

$$= J_2^{(1)}\Big/\sum_{k=0}^{m_{12}}\binom{n_{12}}{k}(T+\beta_1)^{-A}\Gamma(A)(T+\beta_2)^{-B}\Gamma(B) = J_2^{(1)}/I_0,$$

where

$$J_1^{(1)} = \sum_{k=0}^{m_{12}}\binom{n_{12}}{k}(T+\beta_2)^{-B}\Gamma(B)(T+\beta_1)^{-(A-1)}\Gamma(A-1),$$

$$J_2^{(1)} = \sum_{k=0}^{m_{12}}\binom{n_{12}}{k}(T+\beta_1)^{-A}\Gamma(A)(T+\beta_2)^{-(B-1)}\Gamma(B-1)$$

Similarly, we can obtain the Bayes estimators for the reliability function R_1, R_2 below.

$$\hat{R}_1(t) = \int_0^{+\infty} e^{-t^2/\lambda_1}\pi_1(\lambda_1|data)d\lambda_1 = \sum_{k=0}^{m_{12}}\binom{n_{12}}{k}(T+\beta_2)^{-B}\Gamma(B)\int_0^{+\infty}\left(\frac{1}{\lambda_1}\right)^{A+1}e^{-(T+\beta_1+t^2)/\lambda_1}d\lambda_1\Big/I_0.$$

$$\hat{R}_2(t) = \int_0^{+\infty} e^{-t^2/\lambda_2}\pi_2(\lambda_2|data)d\lambda_2$$

$$= \frac{1}{I_0}\sum_{k=0}^{m_{12}}\binom{n_{12}}{k}(T+\beta_1)^{-A}\Gamma(A)\int_0^{+\infty}\left(\frac{1}{\lambda_2}\right)^{B+1}e^{-(T+\beta_2+t^2)/\lambda_2}d\lambda_2 = \sum_{k=0}^{m_{12}}\binom{n_{12}}{k}I_2\Big/I_0.$$

In the reliability analysis, Component failure rate function is calculated using the formula $H(t) = -R'(t)/R(t)$. Hence the Bayes estimators of component failure rate function H_1 and H_2 are respectively.

$$\hat{H}_1(t) = \frac{2t}{I_0}\sum_{k=0}^{m_{12}}\binom{n_{12}}{k}(T+\beta_2)^{-B}\Gamma(B)(T+\beta_1)^{-(A+1)}\Gamma(A+1) = 2t\sum_{k=0}^{m_{12}}\binom{n_{12}}{k}M_1\Big/I_0$$

$$\hat{H}_2(t) = \frac{2t}{I_0}\sum_{k=0}^{m_{12}}\binom{n_{12}}{k}(T+\beta_1)^{-A}\Gamma(A)(T+\beta_2)^{-(B+1)}\Gamma(B+1) = 2t\sum_{k=0}^{m_{12}}\binom{n_{12}}{k}M_2\Big/I_0.$$

4 NUMERICAL EXAMPLES AND ANALYSIS

In this section, we assume that there are 100 same systems are put on the life test at the same time. Each system is consisted of two independent components connected in series. The lifetimes of the components follows Rayleigh distribution with parameters λ_1, λ_2. The test is stopped when the n system fail ($n < N$). The hyper-parameters in prior distribution are taken as $\alpha_1 = 3, \alpha_2 = 4, \beta_1 = \beta_2 = 2$, and we Let $\lambda_1 = 1, \lambda_2 = 0.9, t_0 = 1.5$. According to the given masking level l and censoring number n, we obtain the observed data $(t_1, s_1), (t_2, s_2), \cdots (t_n, s_n)$ $(t_{n+1}, *), \ldots, (t_N, *)$ by using Monte-Carlo method, where $t_{n+1} = t_{n+2} = \cdots = t_N = t_n$. Then we can get the value of n_1, n_2, n_{12}. After take these data to the theoretical results in section 2 and section 3. The MLE and Bayes estimations of $\lambda_j, R_j, H_j(j=1,2)$ are calculated. Repeat the steps above 1000 times, and then the Mean Squared Errors (MSE) of these estimations are computed, and the results are presented in the Table 1 and the Table 2. (In the Table 1, N means for the Mean Square Error (MSE) can not be given).

From the results shown in Table 1, one can conclude that: (1) When the masking level l and censoring number n are given, the MSE of Bayes estimation for the parameters is always less than the MSE of the MLE. Therefore, the effect of the Bayes estimation is better than the effect of MLE. (2) For a given Censoring number n, the MSE of the MLE and Bayes estimation for the parameters decrease with increasing masking level l. However, for the cases in which masking level is 100%, the MSE of the MLE for the parameters can not be obtained.

Table 1. The MSE of the parameter estimations.

n	λ_i	$l=0$ MLE	Bayes	$l=20\%$ MLE	Bayes	$l=40\%$ MLE	Bayes	$l=60\%$ MLE	Bayes	$l=100\%$ MLE	Bayes
20	λ_1	0.2189	0.0568	0.3060	0.0585	0.3159	0.0651	0.3832	0.0682	N	0.1645
	λ_2	0.2083	0.0556	0.2583	0.0580	0.2701	0.0610	0.3002	0.0661	N	0.1365
40	λ_1	0.1080	0.0459	0.1219	0.0478	0.1278	0.0552	0.1390	0.0653	N	0.1128
	λ_2	0.1038	0.0388	0.1048	0.0438	0.1145	0.0473	0.1284	0.0522	N	0.0678
60	λ_1	0.0608	0.0380	0.0823	0.0396	0.0850	0.0479	0.0930	0.0597	N	0.0651
	λ_2	0.0576	0.0331	0.0775	0.0346	0.0798	0.0371	0.0864	0.0465	N	0.0609
80	λ_1	0.0496	0.0320	0.0589	0.0342	0.0621	0.0376	0.0743	0.0427	N	0.0516
	λ_2	0.0429	0.0274	0.0523	0.0301	0.0528	0.0311	0.0767	0.0324	N	0.0358

Table 2. The MSE of the Bayes estimations for the parameter and reliabilities index ($n = 60$, $t_0 = 1.5$).

l	$\hat{\lambda}_1$	$\hat{\lambda}_2$	$\hat{R}_1$	$\hat{R}_2$	$\hat{H}_1$	$\hat{H}_2$
20%	0.0387	0.0335	0.0301	0.0347	0.0325	0.0378
40%	0.0465	0.0365	0.0328	0.0428	0.0361	0.0416
60%	0.0586	0.0458	0.0446	0.0486	0.0464	0.0508
100%	0.0654	0.0615	0.0656	0.0678	0.0659	0.0674

But based on the discussion of section 3, the MSE of the Bayes estimation can still be calculated, and the results are good. The results presented in Table 2 shows, for a given Censoring number n, the MSE of the Bayes estimations for the parameter and components reliabilities index decrease with increasing masking level l. However, the MSE of the Bayes estimations are smaller.

5 CONCLUSIONS

In this paper, we discuss the estimations of Rayleigh components reliabilities index in a series system using Type-II censored and masked system life data. MLE and Bayes approach are exploited for estimating. The MLE and Bayes estimations of the Rayleigh parameters and reliabilities index are obtained. Also, a numerical example is given by means of the Monte-Carlo simulation for comparing the effects of the maximum likelihood estimation and the Bayes estimation, which shows that the MSE of the Bayes estimations are smaller than MLE.

ACKNOWLEDGMENTS

This work is supported by the National Natural Science Foundation of China (71401134) and Scientific Research Program Funded by Shaanxi Provincial Education Department (14JK1673).

REFERENCES

[1] Usher J.S., Hodgson T.J. 1998. Maximum likelihood analysis of component reliability using masked system life data. *IEEE Trans. on Reliability,* 37(5), 550–555.

[2] Sarhan A.M. 2003. Estimation of system components reliabilities using masked data. *Applied Mathematics and Computation,* 136:79–92.

[3] Sarhan A.M., Awad I. 2003. EI-Gohary. Estimations of parameters in Pareto reliability model in the presence of masked data. *Reliability Engineering and System Safety,* 82:75–83.

[4] Sarhan A.M. 2008. Debasis Kundu. Bayes estimators for reliability measures in geometric distribution model using masked system life test data. *Computational Statistics & Data Analysis,* 52:1821–1836.

[5] Sarhan A.M. 2003. El-Bassiouny Ahmed H. Estimation of components reliability in a parallel system using masked system life data. Applied Mathematics and Computation, 138(1):61–75.

[6] Fan Zhang, Yimin Shi. 2009. Parameter estimation of aerospace power supply system using masked lifetime data. *Aerospace Control,* 27(4):96–100.

[7] Lin D.K.J, Guess F.M. 1994. System life data analysis with dependent knowledge on the exact cause of system failure. *Microelectronics Reliability,* 34:535–544.

[8] Sarhan A.M., Guess F.M., Usher J.S. 2007. Estimators for reliability measures in geometric distribution model using dependent masked system life test data. *IEEE Trans.on Reliability,* 56(2): 312–320.

[9] Tsai-Hung Fan and Wan-Lun Wang. 2011. Accelerated life tests for weibull series systems with masked data. *IEEE Trans. on Reliability,* 60(3):557–569.

[10] Qiqing Yu, G.Y.C. Wong, Hao Qin. 2012. Jiaping Wang, Random partition masking model for censored and masked competing risks data. *Ann Inst Stat Math,* 64:69–85.

[11] Ancha Xu, Yincai Tang. 2014. Bayesian analysis of masked data in step-stress accelerated life testing. *Communications in Statistics-Simulation and Computation,* 43(8):2016–2030.

[12] Yimin Shi, Xiaolin Shi. 2015 Bayesian reliability analysis for Burr XII component in series system using dependent masked life data. *ICIC Express Letters,* 9 (1):283–288.

Advances in Engineering Materials and Applied Mechanics – Zhang, Gao & Xu (Eds)
© 2016 Taylor & Francis Group, London, ISBN 978-1-138-02834-0

Failure calculating and risk assessment method of crane parts

Z.Y. Xie, F.Z. Qu & D.G. Pan
School of Mechanical Engineering, Dalian University of Technology, Liaoning, China
School of Transportation and Mechanical Engineering, Shenyang Jianzhu University, Liaoning, China

Q.L. Yang
China Zhonghua Geotechnical Engineering (DALIAN) Co. Ltd., Beijing, China
School of Mechanical Engineering, Dalian University of Technology, Liaoning, China

X. Zhao
China Liaoning Academy of Safety, Liaoning, China

ABSTRACT: As a special equipment, the crane is widely used in building, metallurgy, petrochemical fields and other fields. The failure degree and risk of crane parts has a significant influence on crane operation performance. According to the typical failure modes and working characteristics of crane parts, a method that using the reliability theory and coefficient correction method to compute the failure probability and then combining the risk matrix method to assess risk of crane parts is proposed. In this paper, a mathematical model of failure probability is built, and the influence weights of failure parts are determined by the Analytic Hierarchy Process. The failure correction coefficient of crane parts is calculated. And then, the process of the failure probability computation and risk assessment for the crane parts is given. Took a typical crane part as an example, the results indicated that the method can solve the problem of computing failure probability and assessing the risk of common crane parts, and it provides the basis and reference for the risk assessment of crane.

1 INTRODUCTION

With the rapid development of industry, the crane is widely used in building, petrochemical, metallurgy, mining, chemical, wind power and other industry fields. As a main lifting equipment, crane has significant advantages, such as, big work strength, high use frequency, wide work area, and complex work environment. However, the accidents caused by the failure of parts are getting increased. The current method for preventing the accidents is inspecting the structure and the parts of the crane regularly, but the problem of inspection is insufficient or excessive, and even waste of resource exists. For that reason, it is essential to analyze the failure of the parts, make a risk assessment for the crane in service, evaluate the failure state of the parts, analyze the risk situation of the parts reasonable, and make a reasonable detection project or cycle to avoid the failure accidents.

The failure and risk assessment of the crane structure has aroused attention from the scholars. (David J. Edwards. 2010) made a risk assessment and analyzed the potential risk factors[1], when the excavator used as a crane in construction. (Duan & Wang. 2012) proposed a method to analyze bridge crane reliability of structure based on time-variant failure[2]. Based on the principles of quantified risk assessment, (I.A. Papazoglou & M. Konstandinidou) presented a methodology for managing occupational risk owing to contact with moving parts while working with machines[3]. (Guo & Xu, 2011) proposed a method for structure safety measurement using the Bayesian network[4]. (Liu & Yang, 2008) introduced the Analytic Hierarchy Process (AHP) into the unit risk comprehensive assessment of special equipment M, and built unit risk comprehensive assessment Hierarchy structure model of the special equipment[5].

(Lu & Zhu, 2010) proposed an approach to the reliability-based risk assessment of cracked structures, take the relationship between design parameters and reliability into consideration, so as to optimize the design parameters in accordance to the required target reliability index[6]. (Sun & Zhao, 2011) proposed to combine the Monte Carlo method and the fault tree analysis to study the reliability of the crane girder[7]. These studies always focus on the risk detection, management and control, while rarely mentions about the failure and the risk assessment of crane parts.

This paper proposes a method of using the reliability theory and coefficient correction method to compute the failure probability and gives the process of the failure probability computation and risk assessment for the common crane parts. By taking a general crane as an example, the failure probability of the general crane parts are calculated, the failure consequences are determined, the risk assessment matrix is established, and the risk assessment of the crane parts is implemented.

2 FAILURE PROBABILITY MODEL

The failure mechanism of different types of crane parts is different, and it mainly includes the parts of luffing mechanism, lifting mechanism, rotating mechanism, such as roller, wire rope, hook, pulley, brakes, coupling and so on. The failure of crane parts is affected by many factors, so we took the working conditions, the degree of corrosion, the working level of the crane, the impact of human factors into account, and modified the failure probability of the crane parts.

According to the property of the conditional probability formula and the time inclusion relation, the instantaneous failure rate is as follows:

$$p = \frac{P(t \le T \le t + \Delta t)}{P(T > t)} = \frac{F(t + \Delta t) - F(t)}{R(t)} \tag{1}$$

According to the principle of instantaneous failure rate, the failure probability model of a certain failure mechanism is established as follows:

$$\lambda(t) = \lim_{\Delta t \to 0} \frac{1}{\Delta t} p(t < T < t + \Delta t | T > t) = \lim_{\Delta t \to 0} \frac{F(t + \Delta t) - F(t)}{\Delta t} \times \frac{1}{R(t)} = \frac{f(t)}{R(t)} = \frac{dF(t)}{R(t)} \tag{2}$$

Since the failure life distribution of crane parts is different, according to the modes of failure mechanism, the failure life distribution in a certain failure mechanism is inspected through corresponding inspection methods. And according to the different failure life distribution, corresponding probability density function is chosen. This paper used the chi-square test method to check the distribution type of the failure life data. As most of the strength and life of the crane structure and parts are fit in Normal distribution and Weibull distribution, this paper uses the common lognormal distribution as the failure life distribution of the parts to make analytic explanation. The probability density function of the lognormal distribution is as follows:

$$f(t) = \frac{1}{\sigma x \sqrt{2\pi}} \exp\left[-\frac{1}{2}\left(\frac{\ln t - \mu}{\sigma} \right)^2 \right], \quad t > 0 \tag{3}$$

The distribution function:

$$F(t) = \int_0^t \frac{1}{t\sigma\sqrt{2\pi}} \exp\left[-\frac{(\ln t - \mu)^2}{2\sigma^2} \right] dt = \Phi\left(\frac{\ln t - \mu}{\sigma} \right) \tag{4}$$

Based on the method of maximum likelihood to determine the parameters of the lognormal distribution function, then uses the Chi square test method to check whether the life data fit the lognormal distribution, if it do not conform to the assumption, we reassume the distribution function and test again. The failure probability function of the parts fits lognormal distribution:

$$\lambda(t) = \frac{f(t)}{R(t)} = \frac{\phi\left[\dfrac{(\ln t - \mu)}{\sigma}\right]}{t\sigma\left(1 - \Phi\left[\dfrac{\ln t - \mu}{\sigma}\right]\right)} \tag{5}$$

where:
$f(t)$—the probability density function of the parts
$R(t)$—the reliability function of the parts
t—work time of the parts.
The failure probability of the parts under a certain failure mode can be obtained by placing the parameters of into the expression 5.

3 CORRECTION COMPUTATION OF THE FAILURE PROBABILITY OF THE PARTS

Except its corresponding failure mode, the failure of crane parts is also affected by environmental factors, material factors, human factors, and their work level. There are lots of failure correction factors. According to the degree of various factors impact on parts failure, a hierarchical model of parts correction factor is established. Based on the materials, parts, work level and the working environment, human factors, and static equipment requirements, parts correction factors are divided into three layers. The first layer is the correction factor of parts; the damage factor, environment factor, mechanical factor and work factor are divided into the second layer; and the third layer is the sub-factors of the corresponding secondary factors. The failure correction factor of the parts as shown in (Fig. 1).

Using Three Standard Degree Method to compare the importance of each factor, each factor weight is determined, and compare matrix C_{ij} is established, C_{ij} represents the importance of i elements to j elements. The importance sorting index of the compare matrix is

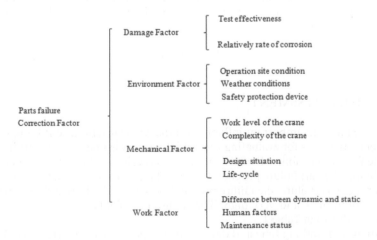

Figure 1.   Failure correction factor of part.

calculated as $R_{ij} = \sum C_{ij}$. And then the importance of sorting index is determined, and the judgment matrix is established.

$$
j_{ij} = \begin{cases} \dfrac{R_i - R_j}{R_{max} - R_{min}}(I_m - 1) + 1 & (R_i \geq R_j) \\[4mm] \left[\dfrac{R_j - R_i}{R_{max} - R_{min}}(I_m - 1) + 1\right]^{-1} & (R_i < R_j) \end{cases}
\tag{6}
$$

Take logarithm of the judgment matrix $A = Log\ j_{ij}$, and calculate the average of the A_{ij}, then structure the new optimal transfer matrix based on the new elements of average, and finally take $10O_{ij}$ as new elements to define the matrix Q,

$$
q_{ij} = 10^{\frac{1}{n}\sum\limits_{k=1}^{n}(a_{ik} - a_{jk})}
\tag{7}
$$

The importance of each factor to the upper layer factor ordered by eigenvector matrix Q, the weight of each factor value is:

$$
\omega^* = \left(1\Big/\sum\limits_{i=1}^{n} q_{i1}, 1\Big/\sum\limits_{i=1}^{n} q_{i2}, \cdots, 1\Big/\sum\limits_{i=1}^{n} q_{im}\right)
\tag{8}
$$

The evaluation matrix of each factor is calculated by using hierarchy grey comprehensive evaluation method, and then experts make a score on the correction factors. Grading standard is divided into five classes, and they are very low, low, medium, high, and very high. The corresponding assignments are given to each gray class (9—very low; 7—low; 5–3—high; 1—high), and the lower of the rank, the greater of the effect on parts. Based on the grey class, whiten weight function is structured, and the whitening value of weight function is calculated. The results of the grey class whiting values divided by the sum of five grey class whiting of values each factor compose a new element evaluation matrix. And then comprehensive evaluation matrix M is obtained through the multiplication of evaluation matrix and weight matrix. The correction factor is determined as follows:

$$
F = \omega^* \cdot M \cdot (9, 7, 5, 3, 1)^T
\tag{9}
$$

The specific failure probability of the parts consists of failure probability and the corresponding correction factor in one certain failure mode. The failure probability of the crane parts is shown in formula 10.

$$
P = \lambda(t) \times F
\tag{10}
$$

4 PARTS RISK ASSESSMENT

The risk of the parts has important influence on the whole machine, so it is important to get the risk level of the parts for evaluating the risk of the whole machine. According to the formula 10, the failure probability P of the crane parts is calculated and 5×5 order risk matrix of parts is built. The part failure probability interval is [0, 1], combining with the characteristics of the crane parts failure, the failure probability is divided into five classes, which is very low, low, medium, high, and very high. The failure probability and the level of the failure probability are shown in Table 1.

According to the working time and damage of the parts during service period, the failure consequence is divided into five parts: very low, low, medium, high, and very high. The specific failure consequences class can be divided referring to the parts failure standard

Table 1. Level of the failure probability.

Level	Very low	Low	Medium	High	Very high
Interval	<0.0001	0.0001–0.001	0.001–0.01	0.01–0.1	0.1–1

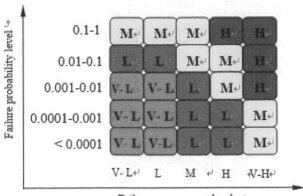

Figure 2. Risk assessment matrix of part.

and regulation. Taking the wear of lifting cable wire as an example, under the condition of normal maintenance and use, the wear loss less than 10% of its' diameter, the level of failure consequences is defined as very low. When the wear loss reaches 10%–15%, the level is defined as low. When the wear loss reaches 15%–25%, the level is defined as medium, and when the wear loss reaches 25%–35%, the level is defined as high. When the wear loss reaches 35%, the level is defined as very high.

The parts risk assessment matrix is established based on the parts failure probability level and failure consequence level, as shown in Figure 2. The vertical axis represents the level of failure probability, and the horizontal axis represents the level of failure consequence.

According to the calculation of the parts failure probability, the level of failure probability is determined. The failure consequence level is determined based on the service situation and failure degree of the parts, and according to the Figure 2 the parts risk level can be determined.

5 EXAMPLE ANALYSIS

According to the calculating method and risk assessment method of the crane parts failure mentioned above, taking the wear of lifting cable wire on a tower crane in service as an example, failure calculation and risk assessment is analyzed. The failure mode of lifting cable wire includes fatigue bending, wear, wire breaking, strand breakage, and some of the wire rope have more than one failure mode. A statistic is carried on the failure life of the crane lifting cable wire, and data are shown in Table 2.

By chi-square test, the data of failure life fit the lognormal distribution. The general failure probability of the wire rope is calculated by formula 5, and the result is shown in Figure 3.

The weight correction coefficient of each factor is calculated by formula 8,

$W1 = (0.75,0.25)$;
$W2 = (0.63,0.26,0.11)$;
$W3 = (0.56,0.26,0.12,0.06)$;
$W4 = (0.64,0.26,0.10)$;
$W = (0.25,0.10,0.50,0.15)$.

Table 2.   Failure life of the wire rope (unit: h).

9600	10900	10000	11100	10206	10503
11249	11345	11441	11358	11236	11432
11414	10713	10712	9800	10715	10716
11725	11728	11761	11762	11763	11764
11775	11778	11782	11788	11791	11793
11812	11816	12420	12115	12231	12435
11180	11255	11345	12380	12428	11428
11319	12560	12680	12825	14717	10721
12498	12398	12480	11768	11772	12580
12900	12820	11795	11797	12228	12589
12850	12338	12341	12620	–	–

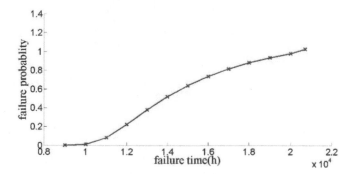

Figure 3.   Failure probability of the wire rope.

Weight vectors are:

r1 = {(0.33,041,0.26,0,0), (0.41,040,0.19,0,0)};
r2 = {(0.30,0.39,0.31,0,0), (0.29,0.37,0.33,0,01,0), (0.31,0.40,0.30,0,0)};
r3 = {(0.33,0.41,0.26,0,0), (0.48,0.38,0.14,0,0), (0.26,0.33,0.36,0.05,0), (0.27,0.34,0.3,5,0.04,0)};
r4 = {(0.45,0.40,0.19,0,0), (0.25,0.32,0.37,0,07,0), (0.30,0.38,0.33,0,0)}.

Evaluation matrix:

$$R = \{(0.39,0.40,0.21,0,0),(0.30,0.39,0.31,0.002,0);$$
$$(0.4,0.38,0.31,0.01,0),(0.33,0.38,0.28,0.01,0)\}$$

Comprehensive evaluation matrix is achieved through multiplying evaluation matrix by weight matrix r. The comprehensive evaluation matrix: M = (0.38,0.39,0.23,0.007,0). Expert estimation of gray classifications: $(9,7,5,1)^T$. Correction coefficient of wire rope:

$$F = M (9,7,5,1)^T = 7.27$$

According to the formula (5) and Figure 3, the general failure probability of wire rope is calculated. By using the correction coefficient modification, the failure probability of wire rope is obtained. For example, a wire rope has been used 11000 h, the general failure probability is 0.072667, and its failure probability is:

$$P = \lambda(t) \times F = 0.072667 \times 7.27 = 0.528$$

So, the failure probability of the wire rope which has been used 11000 h is 0.528, failure probability level is H, the wear loss reaches 30% of the diameter, and failure consequence level is H. According to the Figure 2, the risk level of the wire rope can be obtained through the risk assessment matrix.

6 CONCLUSION

In this paper, based on the reliability theory, the general failure probability model of the crane parts is established, the failure impact factor hierarchy model of crane parts is set up, to calculate the failure probability of the crane parts is calculated by correction coefficient method, the levels of failure probability and failure consequence are determined, and the risk assessment analysis of crane parts is realized. The main conclusions are as follows:

1. This paper proposed a method to calculate the failure probability of crane parts based on reliability theory and coefficient correction method, established the failure probability mathematical model, and gave methods to analyze failure probability model under different failure modes;
2. Gave the method to calculate failure impact factors of the crane parts, obtained the parts failure correction coefficient by calculation, and then, calculated the failure probability with the correction coefficient;
3. The failure probability and failure consequence were divided into different levels. Using 5×5 order risk matrix to realize the risk assessment of crane parts, which provides a reference for assessing the risk of the whole machine and testing the crane parts. Taking the hoisting wire rope of a certain crane in service as an example, the failure probability, failure consequence and risk grade of the crane wire rope by calculation are obtained, and the calculation of failure probability and risk assessment of crane parts is realized.

ACKNOWLEDGEMENTS

The project is supported by the National Nature Science Foundation of China (No. 51275070).

REFERENCES

[1] David J. Edwards. & Gary D. Holt. 2010. Case study analysis of risk from using excavators as cranes. *Automation in Construction, 19(2)*:127–133.
[2] Duan Zhi-bin. Wang Jian-hua. Xu Ge-ning, 2012. Analyzing Reliability of Bridge Crane Steel Structure Based on Time-varying Failure Theory. *China Safety Science Journal, 22(8)*:64–69.
[3] I.A. Papazoglou, M. Konstandinidou, O.N. Aneziris. 2013. etc. Quantification of occupational risk owing to contact with moving parts of machines. *Safety Science, 51(1)*:382–396.
[4] Guo Wenxiao. Xu Gening. 2011. Structure Safety Assessment of General Overhead Traveling Crane Based on Bayesian Network. *Hoisting and Conveying Machinery (10)*:79–83.
[5] Liu Jin-lan. Yang Zhen-lin. 2008. Study on Special Equipment's Risk Evaluation Based on AHP. *Pressure Vessel Technology 25(9)*:28–33.
[6] Lu Hao. Zhu Li-sha. 2010. An Approach to Reliability-Based Risk Assessment of Crackable Structures. *Journal of Northeastern University (Natural Science) 31(11)*:1599–1602.
[7] SUNZhi-li. Zhang Lei. Zhao Xin. 2011. Application of Monte Carlo Simulation and Fault Tree Analysis to the Reliability Calculation of the Crane Girder. *Journal of Northeastern University (Natural Science), 32(6)*:843–845.
[8] The State Administration of quality inspection and quarantine. *Report on the 2012 national special equipment safety status*, 2013.6.

Advances in Engineering Materials and Applied Mechanics – Zhang, Gao & Xu (Eds)
© *2016 Taylor & Francis Group, London, ISBN 978-1-138-02834-0*

Research on rubber damper dynamic model

J. Yang, M.R. Chi, M.H. Zhu, J. Zeng & P.B. Wu
State Key Laboratory of Traction Power, Southwest Jiaotong University, Chengdu, China

ABSTRACT: The paper presents a new method of rubber damper modeling. In this paper, the friction force element was used to describe the damping dependence of displacement amplitude, the fractional derivative model based on the standard linear solid model was used to describe the damping dependence of frequency. By the comparison with the experimental results, it shows that the newly proposed model can describe the dynamic characteristics of rubber damper better than the equivalent model, the standard linear solid model and the fractional derivative model based on the standard linear solid model.

1   INTRODUCTION

As the polymer material, rubber has complex dynamic characteristics of nonlinear stiffness and damping. The dynamic performance is related with the hardness, the exciting frequency, the exciting amplitude and the ambient temperature. Several models are more widely used to describe the dynamic performance of rubber damper, like Maxwell model, Kelvin model, standard linear solid model and generalized Maxwell model, generalized Kelvin model, etc. Maxwell model consists of a linear spring and dashpot in series, and it can simulate dynamic stiffness in the high frequency region very well, however it can't simulate the damping because the dissipative energy presents a downward trend in the high frequency region[1]. Kelvin model, also can be called the equivalent model, consists of a linear spring and dashpot in parallel, this model can't reflect the trends of dynamic characteristics during the whole frequency region accurately. Standard linear solid model consists of a Kelvin model and a linear spring in series, and it can describe the relationship between the stiffness and the damping varies with frequency simultaneously.

Compared with the traditional integer differential constitutive model, fractional derivative model has been taken more seriously for it can characterize the viscoelasticity of rubber damper in the wide frequency region with less parameters[2]. Fractional derivative constitutive model was first suggested by Gemant[3] in 1936, soon afterwards, Rabotnov[4] put forward the fractional derivative constitutive equation based on the abundant experiments, then Bagley[5] found that the stress relaxation and the creep of rubber material could be described by the fractional derivative constitutive relation. Pritz[6] put forward the so-called five-parameters fractional derivative model based on the Zener model. Besides, the carbon black filler which is added to the rubber damper during preparation process results in a friction property. To characterize this friction property, M Berg[1] raised a piecewise function which was proved to explain the strain amplitude dependence very well.

This paper combined the standard linear solid model with the fractional calculus theory, and put forward the standard-linear-solid-fractional-derivative constitutive model, which was used to describe the viscoelasticity of rubber damper. Meanwhile, this paper introduced the Berg friction force unit to describe the elastoplasticity. The new non-linear model was compared with the equivalent model, the traditional standard linear solid model and the fractional derivative model by the vertical dynamic experiment of axle-box rubber mat.

2 DYNAMIC EXPERIMENT OF RUBBER DAMPER

In order to verify the new proposed model, sufficient test on vertical dynamic property has been carried out. The test adopted the MTS vertical vibration test-bed. The bottom of rubber damper is fixed, while the actuator applying a vertical displacement exciting on the top of rubber damper with the maximum amplitude of ±52.5 mm. The maximum vertical load which the MTS vertical vibration test-bed can apply is 500 kN, and the maximum frequency it can apply is 100 Hz.

CSR Zhuzhou Times New Material Technology Co., Ltd provided the prototype of rubber mat with 4 kinds of Shore hardness (50HS, 55HS, 61HS, 65HS). The thickness of the rubber mat is 30 mm, and the inner diameter is 110 mm, while the outer diameter is 270 mm. The test was carried out under the ambient temperature of 20 Celsius with the preload of 85 kN. Sinusoidal exciting displacement $x = x_0 \sin \omega t$ is adopted in this test, in which the displacement amplitudes are 0.1, 0.3, 0.5, 0.7, 0.9 mm, and the exciting frequency ranges from 0.5 Hz to 20 Hz.

3 DYNAMIC MODEL OF RUBBER DAMPER

The dynamic models of rubber damper are shown in Figure 1(a–d). We can find that, Figure 1(a) indicates the equivalent model which consists of a linear spring and dashpot in parallel, Figure 1(b) indicates the Standard Linear Solid model (SLS model) which consists of a Kelvin model and a linear spring in series, Figure 1(c) indicates the fractional derivative model (SLSFD model) which substitutes the dashpot in the SLS model for the fractional element. Figure 1(d) indicates the new proposed non-linear dynamic model (TPL model), which consists of the SLSFD model and the friction element in parallel.

3.1 The equivalent model

The constitutive equation of the equivalent model is:

$$\sigma = E_{eq}\varepsilon + \eta_{eq}\dot{\varepsilon}. \tag{1}$$

where E_{eq} is the elastic modulus of linear spring, η_{eq} is the viscosity coefficient of the dashpot. In order to make the comparative analysis more convenient, this paper set

$$q_0 = E_{eq}, q_1 = \eta_{eq}$$

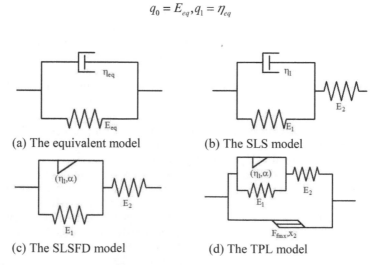

(a) The equivalent model

(b) The SLS model

(c) The SLSFD model

(d) The TPL model

Figure 1. Dynamic models of rubber damper.

The equivalent dynamic stiffness and equivalent damping are:

$$k_{d\_eq} = \frac{A}{h}\sqrt{q_0^2 + q_1^2\omega^2} \tag{2}$$

$$c_{d\_eq} = q_1 A / h. \tag{3}$$

Eq. 2 and Eq. 3 show that the dynamic stiffness of the equivalent model will maintain a rapid growth rate in the high frequency region, and the damping has nothing to do with the frequency.

3.2 The SLS model

By connecting a linear spring in series, we can obtain the SLS model which can reflect the characteristics of transient, relaxation and creep. The constitutive equation of the SLS model is:

$$\sigma + p_1\dot{\sigma} = q_0\varepsilon + q_1\dot{\varepsilon} \tag{4}$$

where σ, ε are the stress and the strain of the model; p_1 equals $\eta_1/(E_1+E_2)$ which describes the characteristic of relaxation; q_0 equals $E_1E_2/(E_1+E_2)$ which describes the characteristic of elasticity and creep transient; q_1 equals $\eta_1 E_2/(E_1+E_2)$ which describes the characteristic of viscosity; E_1, E_2 represent the elastic modulus of the spring in series and the spring in parallel respectively.

The dynamic stiffness and equivalent damping of the SLS model are:

$$k_{d\_S} = (A/h)\sqrt{E_S'^2(\omega) + E_S''^2(\omega)} \tag{5}$$

$$c_{d\_S} = E_S''(\omega)A/(\omega \cdot h) \tag{6}$$

3.3 The SLSFD model

By substituting the fractional element for the dashpot, we can obtain the fractional derivative model of the conventional SLS model. The fractional element model, which can be called the Scott-Blair[7–8] model, is one kind of physical model with the structure of self-similarity, and this model can reflect the viscoelasticity characteristic deeply. The stress is proportional to the fractional derivative of the strain in the simplified fractional element. The constitutive equation of the SLSFD model is:

$$\sigma + p_1\frac{d^\alpha\sigma}{dt^\alpha} = q_0\varepsilon + q_1\frac{d^\alpha\varepsilon}{dt^\alpha} \tag{7}$$

where α means the αth order derivative, and $0 < \alpha < 1$. The common used definition of fractional derivative is given by Riemann-Liouville. According to the fractional derivative Fourier transform, it's easy to get the storage modulus E_{SF}' and the loss modulus E_{SF}'':

$$E_{SF}'(\omega) = \frac{q_0 + \cos(\alpha\pi/2)(p_1q_0+q_1)\cdot\omega^\alpha + p_1q_1\omega^{2\alpha}}{1 + 2p_1\omega^\alpha\cdot\cos(\alpha\pi/2) + p_1^2\omega^{2\alpha}} \tag{8}$$

$$E_{SF}''(\omega) = \frac{(q_1 - p_1q_0)\omega^\alpha\sin(\alpha\pi/2)}{1 + 2p_1\omega^\alpha\cos(\alpha\pi/2) + p_1^2\omega^{2\alpha}}. \tag{9}$$

The dynamic stiffness and damping of the SLSFD model can be calculated referring to Eq. 5 and Eq. 6.

3.4 *The TPL model*

It's known that the rubber damper will have a larger stiffness under a small displacement amplitude exciting, and the stiffness will turn smaller with the amplitude getting larger. This is well-known as the Payne effect. To describe the amplitude dependence, M Berg put forward the piecewise function of friction force. In this model, friction force is independent on the strain rate. The definition is:

If $x = x_s$

$$F_f = F_{fs} \tag{10a}$$

If $x > x_s$

$$F_f = F_{fs} + \frac{(x - x_s)(F_{f\max} - F_{fs})}{x_2(1 - \mu) + (x - x_s)} \tag{10b}$$

If $x < x_s$

$$F_f = F_{fs} + \frac{(x - x_s)(F_{f\max} + F_{fs})}{x_2(1 + \mu) - (x - x_s)} \tag{10c}$$

where $\mu = F_{fs}/F_{f\max}$;
the initial reference state $(x_{s0}, F_{fs0}) = (0,0)$;
$F_{f\max}$ is the maximum friction force, x_2 is the displacement needed to reach the half of $F_{f\max}$.
Assuming a sinusoidal displacement exciting is imposed on the element, the additional dynamic stiffness and the energy loss per cycle of the friction force element can be calculated:

$$k_{d\_f} = \frac{F_{f\max}}{2x_2 x_0}\left(\sqrt{x_2^2 + x_0^2 + 6x_2 x_0} - x_2 - x_0\right) \tag{11}$$

$$W_{f0} = 2F_{f\max}\left(2x_0 - x_2(1 + \mu_0)^2 \cdot \ln\frac{x_2(1 + \mu_0) + 2x_0}{x_2(1 + \mu_0)}\right) \tag{12}$$

where $\mu_0 = F_{f0}/F_{f\max}$.
The TPL non-linear model consists of the SLSFD model and the friction force element in parallel. The total dynamic stiffness and damping of the TPL non-linear model is:

$$k_{d\_TPL} = k_{d\_SF} + k_{d\_f} \tag{13}$$

$$c_{d\_TPL} = c_{d\_SF} + W_{f0}/(\omega\pi x_0^2) \tag{14}$$

4   EXPERIMENTAL RESULTS AND SIMULATION OF MODELS

Considering the friction force element, static vertical stiffness test is needed. The parameters $F_{f\max}$ and x_2 can be solved by Eq. 12, while the viscoelastic force is ignored. However, the effect of the friction force should be eliminated when we start to solve the parameters of the fractional derivative element. Suppose that the friction element stiffness is $\hat{K}_f$ and the stiffness of the dynamic test is $\hat{K}_m$, we can get the stiffness of fractional derivative element $\hat{K}_d = \hat{K}_m - \hat{K}_f$. The damping of the dynamic test is $\hat{C}_{dm}$. The objective function defined by the least-square method is:

$$e^2 = \sum_n W_1\left[\hat{K}_d(\omega_n) - \hat{K}_{d\_SF}(\omega_n)\right]^2 + \sum_n W_2\left[\hat{C}_{dm}(\omega_n) - \hat{C}_{d\_TPL}(\omega_n)\right]^2. \tag{15}$$

where n is the number of frequency, W_1, W_2 are weight coefficient.

By optimization program, the parameters of models mentioned above can be calculated, which are listed in Table 1. The stiffness and damping can be calculated respectively.

4.1 Dynamic stiffness

Figure 2 shows the experimental results of rubber mat stiffness under the exciting amplitudes of 0.1 mm, 0.5 mm and 0.9 mm, from which we can find that the rubber damper has a large dynamic stiffness at the small displacement amplitude of 0.1 mm, and then the dynamic stiffness decreases rapidly. This phenomenon, which is called the Panye effect, has been considered in this new proposed non-linear model since we introduced the Berg friction force element. Figure 3–5 shows the dynamic stiffness of different models under the exciting amplitudes of 0.1 mm, 0.5 mm and 0.9 mm respectively, including the experimental results. The simulation of the TPL non-linear model fits the experimental results very well, except the region of 5–7 Hz, where the TPL non-linear model overestimates the dynamic stiffness.

Besides, the equivalent model, the SLS model and the SLSFD model can't reflect the amplitude dependence of the dynamic stiffness. In terms of the frequency dependence, the SLSFD model behaves better than the SLS model under the exciting amplitudes of 0.5 mm and 0.9 mm, which means that the fractional derivative model can describe the viscoelasticity more accurately.

4.2 Dynamic damping

Figure 6 shows the experimental results of rubber mat damping under the exciting amplitudes of 0.1 mm, 0.5 mm and 0.9 mm, from which we can find that only slight changes of the damping take place when the exciting displacement amplitude turns larger. But with the exciting frequency increasing, the damping decreases rapidly until the frequency reaches to middle and high frequency region. Figure 7–9 shows the dynamic damping of different models under the exciting amplitudes of 0.1 mm, 0.5 mm and 0.9 mm respectively, including the experimental results. The simulation of the TPL non-linear model underestimates the dynamic damping, especially in the low frequency region. However, the simulation of the TPL non-linear model

Table 1. Parameters of each model.

Name	The equivalent model	The SLS model	The SLSFD model	The TPL non-linear model
F_{fmax} (kN)	–	–	–	1.718
x_2 (mm)	–	–	–	0.105
q_0 (Mpa)	36.0	34.4	0.074	0.039
p_1 (s^α)	–	0.037	0.303	0.115
q_1 ($Mpa \cdot s^\alpha$)	0.125	1.400	43.4	33.6
α	–	–	0.030	0.042

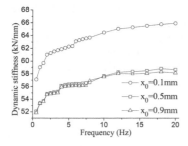

Figure 2. Experimental results of dynamic stiffness.

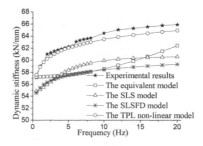

Figure 3. Stiffness of different models under the exciting amplitude of 0.1 mm.

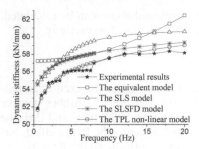

Figure 4. Stiffness of different models under the exciting amplitude of 0.5 mm.

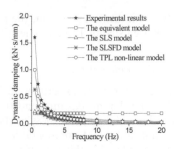

Figure 7. Damping of different models under the exciting amplitude of 0.1 mm.

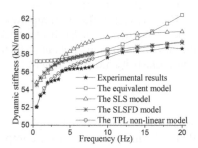

Figure 5. Stiffness of different models under the exciting amplitude of 0.9 mm.

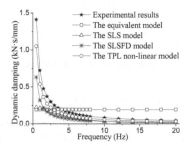

Figure 8. Damping of different models under the exciting amplitude of 0.5 mm.

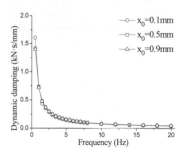

Figure 6. Experimental results of dynamic damping.

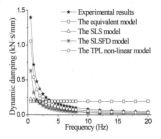

Figure 9. Damping of different models under the exciting amplitude of 0.9 mm.

fits the experimental results very well in the high frequency region. For we focus more on the high frequency excitation, the new proposed non-linear model is advisable.

Besides, the equivalent model, the SLS model and the SLSFD model can't reflect the amplitude dependence of the dynamic damping. The equivalent model stays unchanged under different conditions. In terms of the frequency dependence, the SLSFD model behaves better than the SLS model under the exciting amplitudes of 0.5 mm and 0.9 mm, especially in the low frequency region, which means that the fractional derivative model can describe the viscoelasticity more accurately.

5 CONCLUSIONS

This paper has built a non-linear model which consists of the fractional derivative model of the standard linear solid model and the Berg friction force element in parallel. Abundant experiments on the vertical vibration test-bed have been carried out to make comparative analysis of the equivalent model, the SLS model, the SLSFD model and the TPL non-linear model. The experimental results show that:

1. The TPL non-linear model can describe the dynamic stiffness characteristic of rubber damper, which is dependent both on the exciting frequency and the exciting displacement amplitude obviously; the equivalent model, the SLS model and the SLSFD model can't describe the amplitude-dependent characteristic of dynamic stiffness at all, and they can't reflect the frequency-dependent characteristic of dynamic stiffness accurately.
2. The TPL non-linear model can describe the dynamic damping characteristic of rubber damper, which is dependent on the exciting frequency obviously and on the exciting displacement amplitude slightly; the equivalent model, the SLS model and the SLSFD model can't describe the amplitude-dependent characteristic of dynamic damping at all, and they can't reflect the frequency-dependent characteristic of dynamic damping accurately.

ACKNOWLEDGEMENTS

This work has been supported by the National Key Basic Research Program of China (2011CB711100); Chinese National Science Foundation Key Program (61134002) and Innovation Group of Ministry of Education funded project (IRT1178).

REFERENCES

[1] Sjöberg M. 2002. Non-linear Behavior of a Rubber Isolator System Using Fractional Derivatives. *Vehicle System Dynamics*, 37(3):217–236.
[2] Zhou Yun. 2006. Structural Design of Viscoelastic Damper. Wuhan, Wuhan University of Technology Press.
[3] Gemant A. 1936. A method of analyzing experimental results obtained from elasto-viscous bodies. *Physics*, 7(8):311–317.
[4] Hao Songlin, Lao Liang. 1990. Theory of Viscoelasticity—An Introduction. Beijing, Sicence Press.
[5] Bagley R.L, Torvik P.J. 1986. On the fractional calculus model of viscoelastic behavior. *Journal of Rheology*, 30(1):133–155.
[6] Pritz T. 2003. Five-parameter fractional derivative model for polymeric damping materials. *Journal of Sound and Vibration*, 265(5):935–952.
[7] Chen Ming, Jia Laibing, Yin Xiezhen. 2011. The Fractional Zener Model to Describe The Relaxation Characteristics of Fin Material. *Chinese Journal of Theoretical And Applied Mechanics*, 43(1):217–220.
[8] Zhu Ke-qin, Yang Di, Hu Kai-xin. 2007. Fractional Element of Viscoelastic Fluids and Start-up Flow in a Pipe. *Chinese Quarterly of Mechanic*, 28(4):521–527.

Advances in Engineering Materials and Applied Mechanics – Zhang, Gao & Xu (Eds)
© 2016 Taylor & Francis Group, London, ISBN 978-1-138-02834-0

Bed–chair integration—new developing trend of helpage assistive robot

L.Z. You, S.D. Zhang & L.D. Zhu
College of Mechanical Engineering and Automation, Northeastern University, Shenyang, China

ABSTRACT: Bed–chair integration robot uses the multi-mode human–computer interaction technology, such as the modular mechanism configuration design, the separation of bed-chair, etc, to construct a set of deformable multifunctional bed–chair integration systems. As a branch of the service robot, the helpage assistive robot is mainly used to assist the aged and the disabled people's daily life. This paper analyzes the needs for the life of the aged and the lower limb disabled, puts forward the optimization strategy, and then summarizes and compares with various existing helpage assistive robots and the bed–chair integration products based on the robot developing plan outline. Finally, we analyze the developing trend of the bed–chair integration industry, which would provide related theories and methods for the design and manufacture of the home service robot.

1 INTRODUCTION

According to the data of "The Second China National Handicapped Person Sample Survey" in 2006, the total number of people with disabilities is 82.96 million that accounts for 6.34% of the total population, including 24.12 million physical disabilities that accounts for 29.07%[1]. "The 2013 Development Statistics Bulletin of the Social Services" showed that, by the end of 2013, the number of the elderly people in China who aged over 60 years had reached 202.43 million, accounting for 14.9% of the population[2]. As the most populous country, with population aging developing and increasing lower limbs dyskinesia patients caused by disease and disasters, the helpage assistive problem is becoming a major social problem, and how to make the aged and the disabled live better is drawing more attention.

This paper is based on the demand of the age and the lower limb disabled, putting forward the optimization strategy, analyzing and comparing the existing helpage assistive robots and the products of the bed–chair integration, and then having a further cognition about the accuracy of optimization.

2 DEMAND FOR THE AGE AND THE LOWER LIMB DISABLED PEOPLE

The number of aging, disease, disability, empty nest, and childless elderly population are rising year by year. In addition, the faster rhythm of life makes many people do not have more time to take care of the age and the disabled people at home. What worried us is that the safety and the living conditions of the elderly with weak ability of self-care and the lower limb disabled people. Through the investigation, we analyze that the main demand for the age and lower limb disabled people is mainly divided into service, health care and safety monitoring. On this basis, their needs of auxiliary move, rehabilitation training and emergency processing are worthy for our urgent attention. Therefore, the design of helpage assistive product should start from the demand for the age and the disabled.

3 PERFORMANCE COMPARISON AMONG THE EXISTING HELPAGE ASSISTIVE ROBOTS

"National Middle/Long Term Plan for Science and Technology Development (2006–2020)" explicitly pointed out that the service robot would be regarded as a priority development of strategic high technology in the future, and then put forward that with an emphasis on service robot application demand, to research common foundation technology such as design, manufacturing process, intelligent control, application system integration, etc[3]. With the further study in fields of function structure and automation, the domestic and foreign experts have put forward various helpage assistive robot design concept. The main function of helpage assistive robots and products is to assist the age and the disabled people's daily life, and the wheelchair, the nursing bed and the rehabilitative robot are typical products. The following is a description of the several existing helpage assistive robots.

3.1 The traditional helpage assistive robots

The power wheelchair is the most common helpage assistive robots at present. According to the controlling modes of power wheelchair, it can be divided into two kinds: general power wheelchair controlling with joystick, which is used for the lower disabled and the aged people, while intelligent wheelchair controlling with human–computer interaction such as voice, head, biological signal, which is used for the severe disabled people[4]. The core technologies of the power wheelchair are electric drive assembly, the embedded controller, sensor and its information processing technology and power supply technology[5].

In addition to the electric wheelchair, there are some intelligent wheelchairs that have special functions such as stair climbing, obstacle avoidance, etc. The obstacle crossing mechanism of current stair-climbing wheelchair has experienced courses of wheel, crawler, multiple crawler, wheel leg compound and wheel crawler compound[6]. The designing purpose of obstacle negotiation mechanism: on the one hand is to adapt to the better terrain, increase the mobile mechanism and the contact area, and improve the traction force and the stability of the body. On the other hand is to ensure continuity, improve the ability of obstacle negotiation and energy utilization efficiency[7].

In recent decades, the stair-climbing wheelchair was studied by many universities, companies and research institutions. Then, they put forward a variety of solutions and designed different products. Overall, depending on the climbing structure, there are three kinds of principle structures: planet gear that rotates around both axis and a common axis, crawler, and leg-foot, as shown in Figure 2.

In order to increase the intelligence, some designers install manipulator to the wheelchair to grasp objects, open the door, feed, etc. Operating robots are used as daily walking tools as well as perform specific actions by manipulator, which can make the age and disabled people to restore to the largest degree of self-care ability.

As shown in Figure 4, the helpage assistive robot system integrates multiple devices such as operation arm, wheelchair or mobile robot, design and install sensor system, control system, navigation system and human–computer interface system. Each system collaborates with each other to help the aged and disabled people to complete a variety of complex movements, greatly improving their self-care ability, and this is the new pattern and concept for the aged and the disabled in the future social life.

With the renewal of the intelligent wheelchair, the development of multifunctional nursing bed is also very quickly. After the intelligent wheelchair proposed, some nursing bed also increased sorts of new functions to meet what helpage assistive products need. Now there are different functional nursing bed on the market, such as SS18 Computer Far-infrared Physical Therapy Bed that has the function of massage, Chinese Medicine Massage Robot that has achieved operation of 10 kinds of typical traditional Chinese medicine massage[9], T-1 multifunctional side-tumbling nursing bed, MYD, which owns split type and whose bed body can be adjusted to meet the user's requirements of various states, etc. The occurrence of these nursing beds has provided new ideas for the helpage assistive robots industry.

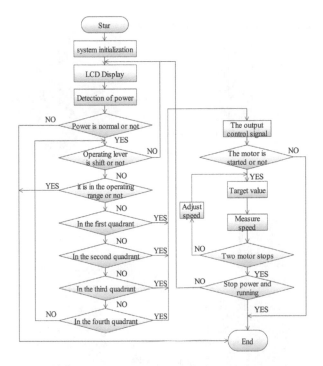

Figure 1.　The flow chart of common power wheelchair system's main program.

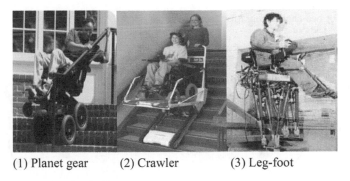

(1) Planet gear　　(2) Crawler　　(3) Leg-foot

Figure 2.　Three types of stair climbing.

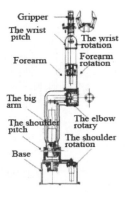

Figure 3.　7-DOF light operation arm mechanical structure[8].

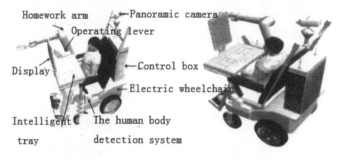

Figure 4.   Whole structure of manipulator wheelchair.

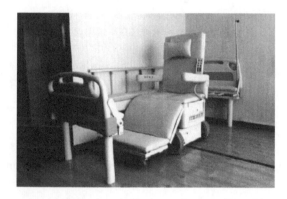

Figure 5.   *E-bed A* bed-chair-integration robot.

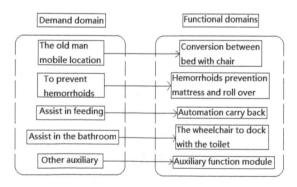

Figure 6.   Separation process of bed and chair.

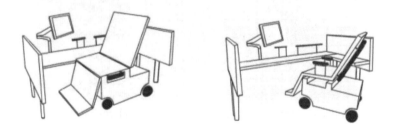

Figure 7.   The transformation of the demand to functional domains.

3.2 The bed–chair integration products

To the age with weak ability of self-care and the lower limb disabled people, bed and wheelchair will occupy a large space of the bedroom. Besides, the aged and the lower limb disabled are hard to perform the action of getting up and leaving from the bed. What worries us most is that these actions are extremely dangerous. If there is no one else to take care of the aged and disabled people in home, it will cause catastrophic consequences to be eager to do all these actions. Then, the concept of the bed–chair integration appeared to construct a set of deformable multifunctional bed–chair integration system by using the modular organization configuration design and the separation of bed–chair multi-mode human–computer interaction technology. Currently, Robotic Bed developed by Panasonic and the series of e-bed Bed–Chair-Integration Robot designed by Beihang University are relatively proficient in the bed–chair integration field.

E-bed multifunctional bed–chair integration robot consists of bed body, seat, information and control systems[10]. E-bed series products use the original separation methods of open type, rotating seat type, rotating chassis type and the docking methods of laser radar type and line-tracking type, to achieve automatic conversion between nursing bed and wheelchair. In the state of nursing bed, it can realize the basic function of multifunctional nursing bed such as turning over, lifting back and curving legs through the verbal control. Setting electric elevating system at the bottom of the seat to solve the problem of people staying in bed for the toilet. In the state of the wheelchair, users can move the wheelchair in the bathroom, corridor and any indoor space by controlling joystick, to relieve themselves, and they can also make it back to bed automatically. Moreover, the series of products own network information interaction system, to realize medicine-take reminding, remote video communication, equipped with monitoring detection and alarm system of physiological parameters such as blood pressure, pulse and body temperature, etc.

E-bed bed–chair integration system is equipped with multifunctional nursing bed, intelligent power wheelchair, urine processing equipment and information system, which can greatly reduce the burden of nursing staff, expand user's scope of activities and improve their ability to live independently as well.

4 THE DEVELOPING TREND OF BED–CHAIR INTEGRATION

Because bed–chair integration is a new concept that was proposed in recent years, the related research is still in its infancy, and the combination of bed and chair does not reach the designated position. Bed–chair integration products still have promoted space, such as installing a manipulator, adding function of stair climbing to wheelchair, etc. With the development of intelligent wheelchair, multifunctional nursing bed and rehabilitation robot, the bed–chair integration product will be more perfect.

At present, most studies in the field of intelligent wheelchair are still in the laboratory stage, and intelligent service robot products are mostly designed for some specific conditions, which cannot produce into flow line. And it is hardly to achieve full independence when the aged and the disabled use the helpage assistive robots. In the future, the research on the bed–chair integration helpage assistive robot should mainly focus on the following aspects:

1. Low price. Mostly aged and disabled people who use wheelchair spend lots of money on their treatment and care. The existing intelligent helpage assistive robot products rarely put into flow line, so they are relatively so expensive that most people cannot afford them. Also the lack of independence on the use of the products, which makes users still need a lot of human resources, which does more harm than good.
2. Modularization. Due to the different using environment, it often requires different modules of products to service for some special sites. It must realize modular of sensor interface and human–computer interface. Many modules have been designed for the toilet in e-bed series products, which can easily apply to different people. So, intelligent service robots will own different modules to meet different needs in the future.

3. Lightness. At present, the stair-climbing wheelchairs are restricted on the market because of the weight problem. Most products with the electric drive always take much space due to the corresponding accessories, which has great influence on popularity. As a result, lightness is an inevitable outcome of the development of the robot in the future.
4. Multifunction. The future intelligent service robots will not simply limited to some of the basic requirements of the life. In the past period of time, the wheelchair has become from simple walking tool to the equipment with a rehabilitation treatment function, and now it has been more intelligent. The functions of helpage assistive robot have constantly improved, and in recent years, the raise of bed–chair integration is to combine the bed and the wheelchair. In terms of functional design, they should meet both users' physical and psychological demand, and let them not to play a role of rehabilitation tools.
5. Security. Wheelchair design for the aged and the disabled people should meet to the requirement of safety and comfort. For example, intelligent obstacle avoidance systems use sensors to give users a lot of convenient. Grasping the details can make the users feel reassuring.

5 CONCLUSION

With the development of technology and the improvement of living standards, people put forward with higher requirements for the social issue on assisting the aged and the disabled people. It attracts more attention on how to exploit high-tech means to improve the living space of the aged and the disabled people. Bed–chair integration can save space and expand assistive function to the maximum extent, through the combination of intelligent wheelchair and multifunctional nursing bed. If the Bed–chair integration products are able to get to the flow line production in the practical application, it will greatly reduce the burden of medical staff, and improve the independent living ability of the aged and the disabled people. Therefore, it has momentous realistic meaning on the bed–chair integration system's further research and improvement, and the home-service robot industry.

REFERENCES

[1] X. Chen. 2008.The Second National Handicapped Person Sampling Survey Data. *Chinese Journal of Reproductive Health*, 19 (2): 68.
[2] Ministry of Civil Affairs of the People's Republic of China. 2014. Social Service Development Statistical Bulletin in 2013.
[3] T.M. Wang, Y. Tao, Y. Chen. 2012. Service Robot Technology Research Status and Development Trend. *Chinese Science: Information Science* 42: 1049–1066.
[4] T. Lu, K. Yuan, H.B. Zhu. 2008. The Research Status and Development Trend of Intelligent Wheelchair. *Robot Technique and Application*, (3): 1–5.
[5] Z.D. Deng, Z.B. Cheng. 2009. The Helpage Assistive Robot Industry and Technology Development Present Situation Research. *Robot Technique and Application*, (2): 20–24.
[6] Y.W. Ma, C. Zhao. 2008. Research on Stair Climbing Wheelchair Based on Planet Gear. Tianjin University.
[7] C.N. Sui, Z.W. Wu. 2010. A Kind of Multiple Mobile Robots Based on Differential Mechanism of Leg: CN, 201010563622.7.
[8] Y.F. Yao, Z. Wang, L. Zhou. 2008. Homework Helpage/Assistive Robot System Research. *Journal of Mechanical Design* 25 (2): 18–20.
[9] H.B. Gao, S.Y. Lu, T. Wang, etc. 2011. Research and Development of Traditional Chinese Medicine Massage Robot. *Robot* 33 (5): 553–562.
[10] M.H. Hu, J.H. Liu, etc. 2013. Bed-Chair-Integration of Multifunction Nursing Bed—Dream of the Old Man's Self Care. *Robot Technique and Application*: 42–46.

Advances in Engineering Materials and Applied Mechanics – Zhang, Gao & Xu (Eds)
© *2016 Taylor & Francis Group, London, ISBN 978-1-138-02834-0*

Research on decelerator based on Planetary Roller Screw

D.Y. Yu, D.X. Sun & X.W. Tong
Harbin Institute of Technology, Harbin, China

ABSTRACT: In this article, a new decelerator design scheme based on the planetary roller screw is proposed, which has the characteristics of small friction, high efficiency, long life time, small volume, strong bearing capacity, etc. After introducing the working principle of above-mentioned new decelerator design, equivalent mechanism method is used to analyze the movement of main components in decelerator. The formulas are deduced for the design of structure parameters and transmission ratio. Then virtual prototype was produced in ADAMS environment for simulation. Finally, the simulation results verify the rationality of design scheme, research method, and analysis on movement characteristics.

1 INTRODUCTION

Robots have been widely used in automobiles, machinery processing, electrical and electronic engineering and other fields. The main transmission device in the joint of robot has become an important problem to be solved. At present, precision gear transmission are mainly used in robot with the application of harmonic drive, cycloidal pinwheel transmission, RV transmission, etc.

In the aspect of materials and manufacturing processes, the development of precision transmission in Japan, Germany, American and other countries has reached international leading level. In foreign countries, the RV reducer has achieved serialization, while our country is still in the development process. Thus, we should strengthen the research of new type of precision decelerator, in addition to systematically study the RV transmission gear optimization design theory and manufacture technology, thus to reduce our dependence on foreign technology.

Planetary Roller Screw (PRS) is similar to ball screw transmission, which is a mechanical transmission that can realize the mutual conversion between rotary and linear motion, with roller act as intermediate, using rolling friction instead of sliding friction.

In this paper, the working principle will be analyzed according to the structure characteristics of decelerator. The kinematic relationship between these components will be parsed level by level through equivalent mechanism method. The theory formula will be derived to calculate the structural parameters. And the virtual prototype simulation will be accomplished in the ADAMS software to verify the correctness of analysis method and structural design scheme.

2 WORKING PRINCIPLE ANALYSIS

The decelerator is composed of two-stage PRS connected in radial, which converts the motion of motor into the spiral movement of intermediate nut. Figure 1 shows the kinematic sketch. The output shaft of the motor is connected with input screw through coupling, and then rotary motion of motor translates into axial movement of intermediate nut after first-stage PRS. The axial velocity is low owes to the small lead of first-stage PRS. Part of the axial velocity translates into the rotate speed of the intermediate nut when the second-stage

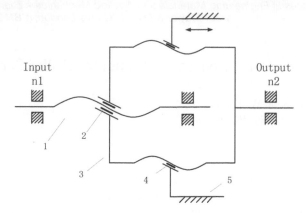

Figure 1.   Kinematic sketch of decelerator 1—input screw, 2—roller of first-stage PRS, 3—intermediate nut, 4—roller of second-stage PRS, 5—shell.

PRS works. Accordingly, the rotate speed is low because of the big lead of second-stage PRS. Thus, the reducer implements the speed reduction effect.

2.1   Motion analysis

According to the structure characteristics, it parses the kinematic relationship between these components step by step. What's more, the discussion of rotation and axial direction is separated.

The circumferential movement of decelerator can be viewed as 2K-H planetary gear train. Then turn it into ordinary gear train by equivalent mechanism method.

In first-stage PRS, the calculation formula can be expressed as follows:

$$i_{23}^H = \frac{\omega_R - \omega_H}{\omega_N - \omega_H} = \frac{d_N}{d_R} \tag{1}$$

$$i_{13}^H = \frac{\omega_S - \omega_H}{\omega_N - \omega_H} = -\frac{d_N}{d_S} \tag{2}$$

where ω_S = screw rotation speed; ω_R = roller rotation speed; ω_H = holder rotation speed; ω_N = nut rotation speed; i_{23}^H = transmission ratio from roller to nut; and i_{13}^H = transmission ratio from screw to nut.

During working process, axial movement exists between the main components. Assume the nut is fixed in the circumferential direction, i.e. $\omega_N = 0$, considering $\omega = \theta/t$, the axial displacement of roller relative to nut is x_{23}:

$$x_{23} = \frac{\theta_H}{2\pi} \cdot q_N + \frac{\theta_R - \theta_H}{2\pi} \cdot q_R = \frac{q_N \cdot d_R - q_R \cdot d_N}{d_S + d_N} \cdot \frac{d_S}{d_R} \cdot \frac{\theta_1}{2\pi} \tag{3}$$

The axial displacement of roller relative to screw is x_{21}:

$$x_{21} = \frac{\theta_S - \theta_H}{2\pi} \cdot q_S + \frac{\theta_R - \theta_H}{2\pi} \cdot q_R = \frac{-q_S \cdot d_R - q_R \cdot d_S}{d_S + d_N} \cdot \frac{d_N}{d_R} \cdot \frac{\theta_1}{2\pi} \tag{4}$$

The axial displacement of nut relative to screw is x_{31}:

$$x_{31} = x_{21} - x_{23} = \frac{-q_S \cdot d_N - q_N \cdot d_S}{d_S + d_N} \cdot \frac{\theta_1}{2\pi} \tag{5}$$

where q_S = lead of screw; q_R = lead of roller; and q_N = lead of nut.

All the above motion relationships apply to the second-stage PRS.

2.2 *Transmission ratio*

According to formula (5), the leads of PRS are available as q_1 and q_2:

$$q_1 = \frac{-q_{1s} \cdot d_{1N} - q_{1N} \cdot d_{1S}}{d_{1S} + d_{1N}} \tag{6}$$

$$q_2 = \frac{-q_{2s} \cdot d_{2N} - q_{2N} \cdot d_{2S}}{d_{2S} + d_{2N}} \tag{7}$$

where subscript 1 = first-stage PRS, and subscript 2 = second-stage PRS.

Considering that the characteristics of PRS transmission can be expressed as formula (8), the transmission ratio of decelerator is available as formula (9).

$$\begin{cases} \dfrac{2\pi}{q_1} = \dfrac{2\pi n_1 + 2\pi n_2}{60v} \\ \dfrac{2\pi}{q_2} = \dfrac{2\pi n_2}{60v} \end{cases} \tag{8}$$

$$i = \frac{n_1}{n_2} = \frac{-q_{2s} \cdot d_{2N} - q_{2N} \cdot d_{2S}}{d_{2S} + d_{2N}} \left/ \frac{-q_{1s} \cdot d_{1N} - q_{1N} \cdot d_{1S}}{d_{1S} + d_{1N}} - 1 \right. \tag{9}$$

where v = moving speed of nut; n_1 = input angular velocity; and n_2 = output angular velocity.

3 STRUCTURE DESIGN

The decelerator structure design is based on PRS, the PRS parameter selection mainly includes pitch diameter of screw, length, number and pitch of screw, roller and nut, etc.

It can be drawn from the above formula that the relative motion between the screw, roller and nut is determined by the relationship of diameter ratio and thread leads. There is no axial displacement between roller and screw in the decelerator discussed here, i.e. $x_{21} = 0$, $x_{31} = x_{23}$, thus $q_S/q_R = -d_S/d_R$, where '$-$' means thread rotate in opposite directions.

According to the roller screw thread relationship, set screw and roller thread number of s/S respectively, thus $q_S/q_R = -s/S$. Since the motion relationship of roller screw is known, the axial displacement of roller $\Delta x = p \cdot (s + d \cdot S/D)$. Figure 2 shows the axial displacement of

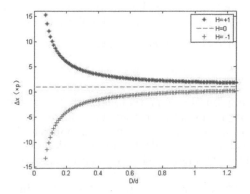

Figure 2. Lead of roller-screw.

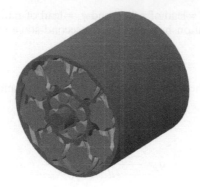

Figure 3.   3D model of decelerator.

roller relative to screw in the situation that roller in right hand, smooth and left hand when screw in right hand. It is obvious that when the diameter ratio is bigger, the relative screw roller axial displacement becomes smaller.

Based on the above principles for determining the reducer structural parameters, 3D modeling is accomplished by using the Cero 2.0 software; the virtual assembly is shown in Figure 3.

4   SIMULATION ANALYSIS

ADAMS software has a powerful postprocessor, and the simulation results can be used to grasp the motion relationship between the main components systematically, which is in accord with modern design idea.

4.1   *Create a virtual prototype*

The 3D model is established in Creo 2.0 software, saved as *.x_t file, then imported in Adams.

Redefine the following motion pair constraint in ADAMS.

1. Create a fixed pair. Fix the second-stage nut to ground.
2. Create a rotation pair between the input screw and ground.
3. Create cylindrical pair during intermediates, screw, roller and nut, respectively.
4. Create screw pair during intermediates, screw, roller and nut, respectively.
5. Create marker points as the meshing point between screw, roller and nut.
6. Create gear pair at the marker point created in step (5).
7. Add the driver. Click the rotation drive button in the toolbar, and then add a speed of 30°/sec rotary actuation at the rotation pair created in the above step (2).

So far, the decelerator virtual prototype is created.

4.2   *Simulating calculation*

1. Screw driver motor rotation function is set to $\theta(t) = 47.5 \times 2\pi \times \sin(2\pi \times t/100)$, take angular velocity of Motion_1 and nut as measurement of target. Simulation result is shown in Figure 4. The rotation of input and output is in opposite direction, which is consistent with the theoretical analysis. While during the working procedure, the ratio of input angular velocity and output angular velocity is constant, and it is equal to the theoretical calculation result.
2. Screw driver motor rotation function is set to $\theta(t) = 2\pi \times \sin(2\pi \times t/100)$, take angular velocity and axial velocity of intermediate nut as a target. Figure 5 shows the result.

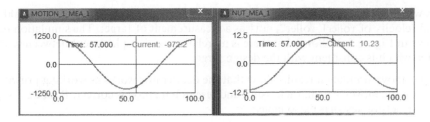

Figure 4. Result of simulation No. 1.

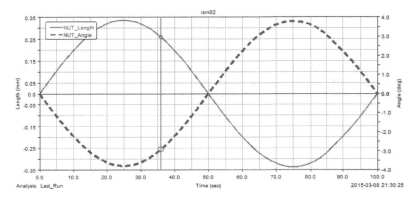

Figure 5. Result of simulation No. 2.

Table 1. Comparison of simulation result and theoretical value.

	Simulation (mm/rad)	Theory (mm/rad)	Relative error
First-stage lead	0.3368	1/3	1.03%
Second-stage lead	31.9953	32	0.01%

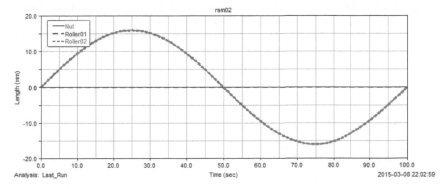

Figure 6. Result of simulation No. 3.

The simulation result at t = 36.0s is compared with the theoretical value as shown in Table 1.

There is deviation between simulation value and theoretical value, and there are varies reasons, such as the errors exist in geometrical model and assembly. But the relative error is lesser, thus the result is still validating the lead theory calculation formula.

3. Screw driver motor rotation function is set to $\theta(t) = 47.5 \times 2\pi \times \sin(2\pi \times t/100)$, take axial velocity of roller1, roller2, nut1 as a measurement of target. The simulation result is shown in Figure 6. It shows that there is no relative axial displacement between screw and roller, which is consistent with the design target.

All the above simulation results show that the decelerator works as working principle, the analysis on the movement characteristics and research method is correct. Besides, it suggests that the structure design is reasonable.

5 CONCLUSIONS

In this paper, the motion relationship between the components is analyzed and derived by equivalent mechanism method. And it discusses the influence of diameter ratio on decelerator movement space, then the design basis and principle of parameter selection is determined. The simulation result shows that the analysis on the movement characteristics and research method is correct, and the structure design is reasonable.

ACKNOWLEDGEMENT

The research was supported by National High-tech Research and Development Projects (863): 2013AA040902-2.

REFERENCES

He, Yayin. 2009. Kinematic simulation of gear reducer based on ADAMS. *Mechanical Management Development* 24(3): 165–166.

Jing, Qianzhong & Yang, Jiajun & Sun, Jianli. 1998. Kinematic characteristic and parameter selection of planetary roller screw. *Manufacturing Technology & Machine Tool* (5): 13–15.

Liu, Geng & Ma, Shangjun & Tong, Ruiting. 2012. Development and key technology of planetary roller screw. *Mechanical Transmission* 36(5): 103–108.

Ni, Jie. 2002. Research on roller-screw meshing transmission. Tianjin University.

Wu, Suzhen, & Chen, Dan. 2014. Advances of robot joint drive precision decelerator. *Journal of Henan Institute of Science and Technology* 42(6): 58–63.

Velinsky, S.A. & Chu, B & Lasky, T.A. 2009. A kinematics and efficiency analysis of the planetary roller screw mechanism. *Journal of Mechanical Design* Vol. 131: 1–8.

Yousef Hojjat & M. Mahdi Agheli. 2009. A comprehensive study on capabilities and limitations of roller-screw with emphasis on slip tendency. *Mechanism and Machine Theory* (44): 1887–1899.

Advances in Engineering Materials and Applied Mechanics – Zhang, Gao & Xu (Eds)
© 2016 Taylor & Francis Group, London, ISBN 978-1-138-02834-0

Combustion simulation of opposite axial piston engine with unsymmetrical cam

L. Zhang, C.Y. Pan, R. Zhen & J.Y. Tan
College of Mechatronics Engineering and Automation, National University of Defense Technology, Changsha, China

S.X. Li
Engineer Military Representative Office of the General Armaments Department, Xi'an, China

ABSTRACT: Opposite axial piston engine could improve the power density of the engine effectively by increasing the number of cylinders and the working frequency. As the main power transmission component of engine, the unsymmetrical cam is invented to increase combustion efficiency. Based on the structure introduction of the axial piston engine, the working principle of engine is presented. Then the mathematical model of unsymmetrical profile of cam is built and analyzed. Under the assumption of no heat loss during the whole working process, the thermodynamic model of the engine is established. The Vibe function is applied to simulate the heat release during the combustion process. The results show that the unsymmetrical cam profile can improve the combustion efficiency by increasing the maximal combustion pressure. The maximal pressure in cylinder can be increased by 28% when the unsymmetrical coefficient is set to 2.

1   INTRODUCTION

Axial piston engine, which has a significant high power density, is originally designed to power the aircraft. In the early twentieth-century, several physical prototypes of axial piston engine have been manufactured and tested. The Almen aero engine (1921), the Dyna-Cam engine (1941), etc are the most famous axial piston engines invented about one hundred years ago. The Duke engine company and the Dyna-cam engine company have proposed some innovative axial piston engines recently. Some tests on the axial piston engine have approved that the advantage of arranging the cylinders around the output shaft. But the poor emission performance and the high abrasion of cam restrain the usage of the axial piston engine. However, with the innovation of new material with high abrasive resistance and breakthrough in the new emission control strategy, the potential of axial piston engine in power density catches attention for the scientists once again. Especially in some special applications area, for example the portable electrical power generating system in small scale, axial piston engine shows some promising advantages.

The structure and working principle of innovative axial piston engine is presented, furthermore a combustion model based on the Vibe function is established to analyze the combustion process of axial piston. In this paper, the energy balance of engine is studied in detail. A mathematic model is built to analyze the profile of cam. Then the relationship between the unsymmetrical coefficient and the combustion pressure is studied. The results show that the unsymmetrical cam can increase the maximal combustion pressure. The combustion efficient can be increased at the same time.

2 COMBUSTION MODEL OF AXIAL PISTON ENGINE

2.1 *Structure and working principle introduction of axial piston engine*

Axial piston engine applies a cam-roller mechanism as the main power transmission components, which is used to transfer the reciprocating motion of piston to the revolution of cam. Eight cylinders that are divided into two groups are laid on both sides of cam. Four cylinders including two power cylinders and two charge cylinders are on each side. Power cylinder and its corresponding charge cylinder are nested next to each other. The fresh air flows into charge cylinder through carburetor during which the gasoline is included. The fresh mixture is compressed under the force of charge piston, and then the compressed mixture is used to blow away the waste gas in power cylinder. This explains the innovation of opposite axial piston engine. The out view and mechanical schematic of axial piston engine is shown Figure 1.

The axial piston engine applies a cam-roller mechanism as the primary power transmission component, which allows the axial arrangement of eight cylinders. The innovative cam-roller mechanism is the origin of the high power density of axial piston engine. The reciprocating motion of piston leads to the volume variation of cylinder, which could be used to convert the internal energy of combusted gas to the mechanical energy of cam. Axial piston engine is designed to work as a two-stroke engine, which has a higher power density. The cams with different profiles have different combustion performances, which will be studied below.

2.2 *Energy balance analysis*

In order to simulate the combustion process of the axial piston engine, a combustion model is built to calculate the total heat release during the working cycle. Till now several combustion models have been proposed to predict the combustion heat release, such as the single Vibe function, the double Vibe function, the two zone model, the Woschni/Anisits model, etc. Except for being influenced by the empirical coefficients heavily, the single Vibe function is the simplest among all those models. Therefore, the single vibe model that does not consume too much computer resource is applied to validate the advantage of axial piston engine with an unsymmetrical cam in combustion. In Vibe model, the burned mass fraction can be regarded as a function of the angle of output shaft 'α', the start angle 'α_o' of combustion, the duration '$\Delta\alpha_c$', the shape parameter 'm' and the Vibe parameter 'a', as shown in Equation (1).

$$\frac{dx}{d\alpha} = \frac{\alpha}{\Delta\alpha_c} \cdot (m+1) \cdot \left(\frac{\alpha-\alpha_o}{\Delta\alpha_c}\right)^m \cdot e^{-ay(m+1)} \tag{1}$$

where x is the mass fraction burned, a is set to 6.9, m is set to 2.1.

The thermodynamic state of the cylinder can be calculated based on the first law of thermodynamics. If the gas leakage and the heat loss can be ignored, the internal energy in

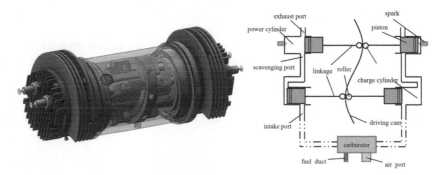

Figure 1.   Out view and schematic of axial piston engine.

cylinder only varies with the work done by the piston, the heat released by combustion process, the mass flowing into cylinder and the mass flowing out of cylinder.

$$\frac{d(m_c \cdot u)}{d\alpha} = -p_c \cdot \frac{dV}{d\alpha} + \frac{dQ_F}{d\alpha} + \sum \frac{dm_i}{d\alpha} \cdot h_i - \sum \frac{dm_e}{d\alpha} \cdot h \tag{2}$$

where $d(m_c \cdot u)$ is the internal energy variation in cylinder, p_c is the pressure, and m_i is the mass.

The mass in cylinder is the sum of the mass flowing in cylinder and the mass flowing out cylinder, which can be expressed using the following equation:

$$\frac{dm_c}{d\alpha} = \sum \frac{dm_i}{d\alpha} - \sum \frac{dm_e}{d\alpha} \tag{3}$$

Since the gas exchange between cylinder and ambient lasts for a very short period, mass flow through the intake and exhaust ports can be regarded as isentropic orifice process. Therefore the mass flow can be determined using the following Equation (4), the unknown coefficient ψ can be calibrated by experiments.

$$\frac{dm}{dt} = \psi \cdot A_{eff} \cdot p_{o1} \cdot \sqrt{\frac{2}{R_o \cdot T_{o1}}} \tag{4}$$

where A_{eff} is the effective flow area, p_{o1} is the upstream stagnation pressure, T_{o1} is the upstream stagnation temperature, and R_o is the gas constant.

3 PROFILE DESIGN AND CHARACTERISTICS ANALYSIS OF CAM

3.1 *Mathematical model of cam profile*

A piecewise function is applied here to model the profile of the unsymmetrical cam, as shown in Equation (5). The shape of the cam curve depends on unsymmetrical parameter n. With any given parameter n, the profile is expressed as the function of angle displacement of output shaft, as shown below:

$$y = \begin{cases} f(\alpha), [\pi, 2\pi] \\ 1 - f(\alpha + \pi), [0, \pi] \end{cases} \tag{5}$$

where $f(\alpha)$ is the function of cam profile in section $[\pi, 2\pi]$, it descents slowly after the TDC (short for top dead center), and ascents quickly before TDC. $f(\alpha)$ can be expressed like below:

$$f(\alpha) = \frac{(n^2 - 1)\sin(\alpha - \pi/2)}{2n[n + \sin(\alpha - \pi/2)]} + \frac{n+1}{2n} \tag{6}$$

3.2 *Evaluation of model*

The 3D model of the cam and the relative position with respect to output shaft angle are displayed in Figure 2. The cam is made up of cam body and the support; the cam surface is well grinded to ensure the accuracy. As shown in Figure 1, in the expanding stroke piston stays around the TDC for a longer interval, and then the piston goes down to BDC (short for bottom dead center) quickly. The new profile of cam can keep the piston staying around the TDC longer, which would increase the maximal pressure and temperature in cylinder. According to the thermodynamic principle, higher combustion efficient could be achieved.

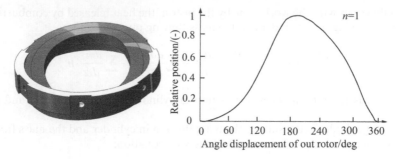

Figure 2. 3D model and relationship between relative position and output shaft angle of cam.

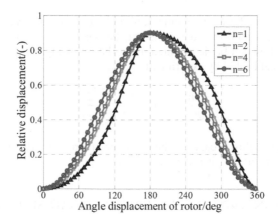

Figure 3. The relationship between relative position and output shaft angle for different profiles.

Table 1. Parameters defined in simulation model.

Parameter	Bore	Stroke	Compress ratio	Intaking pressure	Exhausting pressure	Air/fuel ratio
Value	14 mm	13.5 mm	10	1.1 Bar	1.0 Bar	12.5

In order to analyze the influence of unsymmetrical coefficient n in Equation (7) to the profile of the cam, profiles with different coefficients are deprived respectively, which are displayed in Figure 3. The asymmetry of the profile increases with the decrease of the parameter n, which results a longer interval around the TDC.

In order to evaluate the thermodynamic model of two-stroke axial piston engine that takes combustion, gas scavenges process and gas flow loss into consideration at the same time, some parameters like cylinder diameter, stroke length, and compress ratio and ports area are need to define at first. The needed parameters are listed in Table 1.

4   SIMULATION RESULTS

The pressure curves in cylinder with different cam profiles are shown in Figure 4. The combustion start angle is 5 degree before TDC, the combustion last angle is 40 degree. The lower heating value of combustion mixture is set to 43500 kJ/kg. The simulation of combustion

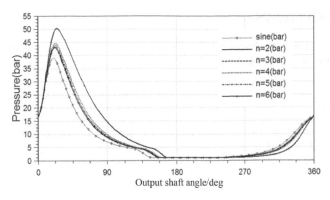

Figure 4.   Pressure in cylinder with respect to output shaft angle.

process in cylinder is carried out after the relative parameters are properly set. The simulation results are shown in Figure 4.

It can be concluded that the maximal pressure in cylinder increases to 39 Bar when the cam has a sine profile. However, when the unsymmetrical parameter n is set to 2 the maximal pressure in cylinder could reach 50 Bar, which is 1.28 times of the pressure in cylinder with a sine cam profile. The maximal pressure in cylinder achieves an obvious increase when cam has an unsymmetrical profile. The simulation results also show that the mean pressure in cylinder with an unsymmetrical profile is larger than pressure with the sine profile, which means more work done during the expansion stroke. More work done by piston means higher energy converting efficiency. The asymmetry of cam is determined by the parameter n. When n gets bigger the profile of cam gets closer to sine curve.

Compared with the sine profile, the maximal pressure angle when n is 6 is delayed. And with the decrease of n, the delay phenomenon becomes more obvious. The delay of maximal pressure angle is influenced by the heat release of combustion gas in cylinder and the work done by the piston at the same time. As to the cam with sine profile, the pressure in cylinder decreases when the rate of work done by piston becomes larger than the heat release rate. At the same time as to the unsymmetrical cam, the rate of work done by piston is much smaller for longer interval of staying around the TDC. As a result, the pressure and temperature can increase higher.

During the compression stroke pressure in cylinder in the situation of unsymmetrical cam is much smaller compared with the sine cam. Similar to the TDC situation, the piston also stays around the BDC for a longer period, which enlarges the scavenge interval. More fresh mixture can be boosted into cylinder that will increase the combustion quality and decrease the production of pollution. The pressure in cylinder when cam is unsymmetrical is smaller compared with the sine cam, as a result the energy the piston consumes during the compression stroke to compress the mixture is much smaller.

5   CONCLUSIONS

1. Cam with an asymmetrical profile can increase the maximal pressure and temperature in cylinder that will improve the combustion efficient of engine. During the compression stroke, the mean pressure in cylinder is also decreased; as a result the energy used to compress gas in cylinder gets smaller. The total energy efficient of engine is increased correspondingly.
2. The longer period of piston staying around the BDC means the higher scavenge efficiency. As a result, the pollution exhausted by the two-stroke engine can be effectively decreased.

ACKNOWLEDGEMENTS

The authors would like to thank the funding of National Natural Science Foundation of China with Grant No: 51475464 and 51175500.

REFERENCES

Cantore, G. et al. 2014. A new design concept for 2-Stroke aircraft Diesel engines. *68th Conference of the Italian Thermal Machines Engineering Association* 45: 739–748.

Feng, L.Q. et al. 2008. Research Status of Free Piston Generator Engine, *Small internal combustion engine and motorcycle* 37(4): 91–96.

Hermann H. & Peter P. 2007. Charging the Internal Combustion Engine Powertrain Lavanttal. Austria: Springer Wien New York.

Perini F.P.F. et al. 2010. A quasi-dimensional combustion model for performance and emissions of SI engines running on hydrogen–methane blends. *International Journal of Hydrogen Energy* 35: 687–701.

Saffa, B.R. & Gan, G.H. 2000. Numerical determination of energy losses at duct junctions. *Applied Energy* 67: 331–340.

Verhelst, R.S.S. et al. 2007. A quasi-dimensional model for the power cycle of a hydrogen-fuelled ICE, *International Journal of Hydrogen Energy* 32: 3545–3554.

Verhelst, S. et al. 2009. Multi-zone thermodynamic modelling of spark-ignition engine combustion—An overview, *Energy Conversion and Management* 50:1326–1335.

Wang Y. et al. 2011. Numerical analysis of two-stroke free piston engine operating on HCCI combustion. *Applied Energy* 88: 3712–3725.

Yong L. et al. 2014. An evaluation of a 2/4-stroke switchable secondary expansion internal combustion engine. *Applied Thermal Engineering* 73: 323–332.

Zhang, L. et al. 2014. Analysis of Unrestrained Curve of Rectangular Piston Ring based on Energy Principle, *International Conference on Control, Automation, Robotics and Vision*, Singapore.

Zhang, L. et al. 2015. Combustion simulation and key parameter optimization for opposite axial piston engine in small scale. *Journal of Central South University*. (under publish).

Simulation analysis of basic motion performance of a tracked efficient steering platform

Y.N. Zhang, J. Zhang, T. Huang & H.B. Yang
Academy of Armored Forces Engineering, Beijing, China

ABSTRACT: The problem of tracked platform's difficulty in steering is caused by the high lateral resistance on rigid pavements. To solve this, a new type of tracked mobile platform-tracked efficient steering platform has been designed. In order to study the basic motion performance of a tracked efficient steering platform, a brushless DC motor control system model was built by Matlab/Simulink and multi-body dynamics model of platform was built by virtual prototyping technology software ADAMS/View were used. The motor control system model and multi-body dynamics model composed co-simulation model. The basic motion performance such as longitudinal steer performance, central steering performance and climbing performance were simulated and analyzed. The simulation result verified the correctness of the model and indicated the basic motion performance of efficient platform is same as traditional tracked platform.

1 INTRODUCTION

The steering resistance of traditional tracked mobile platform is high, which mainly includes the ground deformation resistance and lateral resistance. The lateral resistance only includes sliding frictional resistance when driving on the cement road, tar road, and other rigid pavements[1,2]. For the problem of tracked platform's difficulty in steering caused by the high lateral resistance on rigid pavements, a new type of tracked mobile platform, tracked efficient steering platform, is designed. The tracked efficient steering platform can improve the steering performance of tracked platform fundamentally. To verify the correctness of the model and lay a foundation for the development of the platform, a simulation analysis will be carried out on the basic movement performance of the platform.

2 THE DESIGN OF EFFICIENT STEERING CATERPILLAR

The principle structure of efficient steering caterpillar is shown in Figure 1, which is mainly composed of drive wheel, track shoe, roller, road wheel, carrier wheel, and inducer. This caterpillar is hanged several rows of track shoes, mounting bracket is outside the track shoes and roller is fixed on the mounting bracket. Mesh axis is inside the track shoes spaced and meshing with drive wheel and inducer. The roller installed on the grounding of track shoe is cylindrical and can roll freely. The angle between axis of the roller and axis of pin hole of the track shoe is known as offset angle γ, wherein $-\pi/2 < \gamma < \pi/2$. Roller in two different offset angles is arranged on longitudinal center line of the track shoes symmetrically. Efficient steering caterpillar has two degrees of freedom in direction, one is longitudinal motion i.e. the winding direction of the caterpillar, and the other is perpendicular to the axis of roller. The overall structure of efficient steering caterpillar is same as traditional caterpillar, just need to install roller on the track shoes, so the refit is very convenient and simple.

The layout structure of tracked efficient steering platform is similar to traditional tracked platform. Compared with the traditional mechanical transmission structure, the electric drive

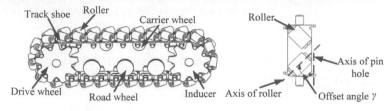

Figure 1. The principle structure of efficient steering caterpillar.

structure has the advantages of high maneuverability, flexible overall arrangement characteristic, and so on. So, the platform is driven by dual motors independently.

3 MODELING OF MOTOR CONTROL SYSTEM AND PLATFORM

3.1 *Modeling of control system for brushless DC motor*

Based on maintaining the superior speed control performance of traditional DC motor, brushless DC motor overcomes a series of problems caused by the original mechanical reversing and brush. So, this paper selects brushless DC motor to build motor control system model[3]. In the brushless DC motor, the induced electromotive force is trapezoidal wave, current is square wave, and the inductance is nonlinear. Therefore, a phase variable method is used to analyze the brushless DC control system. And used the piecewise straight line to represent the induced electromotive force based on the position of rotor[4,5]. The brushless DC motor control system model established in this paper adopts double loop PI control in which the outer loop is speed loop, whereas the inner is current loop. The brushless DC motor control system model built by Matlab/Simulink is shown in Figure 2.

3.2 *Multi-body dynamics modeling of platform*

In order to emulate the basic performance of platform conveniently, this paper builds the virtual prototype model using the multi-body dynamics software ADAMS. Considering the particularity of the caterpillar structure, the ADAMS/ATV module is not suitable to build model. Therefore, all parts of the model are built by the Solidworks. Then, the assembly model in Solidworks is imported into ADAMS/View in Parasolid. Define name, material properties, constraints, applied load, relevant motion functions for all parts[6,7,8]. Thus the platform can be driven to steer. This paper cannot use road surface spectrum to build a road model, so the rigid pavement of which adhesion coefficient is 0.8 is used as the simulation road. The virtual prototype model of platform built in this paper is shown in Figure 3.

The structure parameters of prototype in Figure 3 are shown in Table 1.

4 CO-SIMULATION ANALYSIS

In order to simulate the practical movement of the platform, this paper uses the ADAMS/Control software interface technology to combine multi-body dynamics simulation model with motor control system model. Set the environment parameters of co-simulation, drive wheel angular velocity of the platform prototype is used as feedback to input brushless DC motor control system model and the output torque of motor control system is used as drive wheel torque of platform prototype, which forms a closed-loop control system[9].

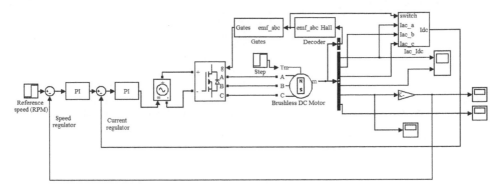

Figure 2.   The brushless DC motor control system model.

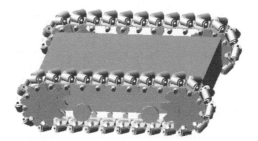

Figure 3.   Virtual prototype model of tracked efficient steering platform.

Table 1.   Structure parameters of prototype.

Parameters	Value
Size (length × breadth × height) (mm)	$1050 \times 850 \times 200$
Mass of platform m (kg)	150
Offset angle of roller γ (°)	±45°
Track-center distance (mm)	900
Radius of drive wheel (mm)	80

4.1   *Simulation on longitudinal steer performance*

Set 10 rad/s speed for the dual motors, and do longitudinal steer simulation on horizontal rigid pavement whose adhesion coefficient is 0.8. The track of the platform centroid and the angular velocity of drive wheel are shown in Figure 4.

4.2   *Simulation on central steering performance*

Set ±10 rad/s speed for the dual motors to do simulation on central steering. The track of drive wheel centroid and angular velocity of dual drive wheel are shown in Figure 5. The result shows track of drive wheel is approximate to a circle. The amplitude of drive wheel angular velocity is same and direction of angular velocity is opposite.

4.3   *Simulation on climbing*

Set 20 rad/s speed for the dual motors, and do climbing simulation on slope rigid whose slope angle is 32° and adhesion coefficient is 0.8. The track of the platform centroid and speed of

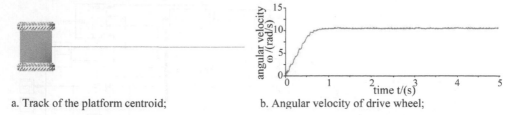

a. Track of the platform centroid; b. Angular velocity of drive wheel;

Figure 4.   Track of the platform centroid and the angular velocity of drive wheel.

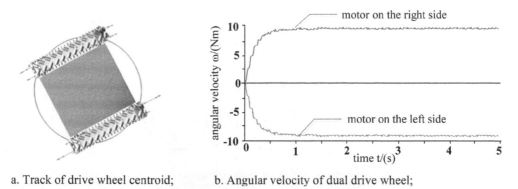

a. Track of drive wheel centroid; b. Angular velocity of dual drive wheel;

Figure 5.   Track of drive wheel centroid and angular velocity of dual drive wheel.

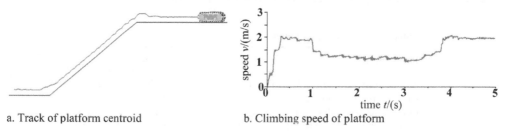

a. Track of platform centroid b. Climbing speed of platform

Figure 6.   Track of the platform centroid and speed of climbing.

climbing are shown in Figure 6. From the figure, we can conclude that tracked efficient steering platform has the same climbing performance as traditional tracked platform.

5 CONCLUSIONS

The simulation on basic motion performance using the virtual prototyping technology can arrive at the following conclusions:

1. The simulation verified the correctness of the model, indicating that the use of virtual prototyping technology for platform basic motion performance simulation and assessment is feasible and effective.
2. The basic performance of tracked efficient steering platform is similar to traditional tracked platform.

REFERENCES

[1] Wang, M.D. & Zhao, Y.Q. et al. 1983. *Driving principle of tank*. Beijing: National Defense Industry Press.

[2] Cao F.Y. & Zhou Z.L. et al. 2007. Research on simulation of turning performance of tracked vehicle. [J]. *Transactions of the Chinese Society for Agricultural Machinery* (01):184–188.

[3] Xia C.L. 2009. *Brushless DC motor control system*. Beijing: Science Press.

[4] Yin Y.H. & Zheng B. et al. 2008. A method for modeling and simulation of brushless DC motor control system based on Matlab. *Journal of System Simulation*. 20(2):293–298.

[5] Zhang L. 2012. *Brushless DC motor control system design and research*. Heng Yang: Nan Hua University.

[6] Zheng J.R. 2008. ADAMS—*Virtual prototyping technology introduction and rise*. Beijing: Machinery Industry Press.

[7] Wang X.S. & Shen Y.F. 2012. Dynamics simulation analysis of crawer walking mechanism of caterpillar bulldozers based on ADAMS. *Journal of Shandong University of Technology (Natural Science Edition)* 26(3):34–37.

[8] Huang T. & Zhang Y.N. et al. 2014. Design & kinematics Analysis a tracked omnidirectional mobile platform. *Chinese Journal of Mechanical Engineering* (21):206–212.

[9] Wang S.S. 2012. *Research on simulation and motion control of omnidirectional platform*. Beijing: Academy of Armored Forces Engineering.

Design and implementation of the tensioning device in the underwater diamond wire saw machine

H. Zhang

School of Naval Architecture and Ocean Engineering, Zhejiang Ocean University, Zhoushan, China

ABSTRACT: In order to improve the cutting accuracy of the underwater diamond wire saw, a new device, namely the tensioning device of the underwater diamond wire saw machine with adjustments, was proposed. Through the analysis of the underwater diamond wire saw machine working principle and structure feature of the tensioning device, shortcomings of existing tension device are presented. According to the design principle of transforming rotary motion into linear motion by the screw, the adjustment institutions of the axle position of the tensioning wheel and the tension testing structures are designed. The main parts are calculated. The result shows that the device is feasible.

1 INTRODUCTION

On the research of diamond wire saw cutting technology at home and abroad, the diamond wire saw is used in nuclear power plant dismantling and maintenance of subsea components (Qi, Q. 2011). Today, it is toward efficient heavy-duty cutting and underwater cutting development (Buttner, N. et al. 1999). Xu, X.T. (1998) introduced a single rope of diamond wire saw machine. It was used for plastic, mineral exploitation, trimming, removal, and demolition of houses. Liao, Y.S. (1997) presented a multi rope diamond wire saw machine. It was mainly used for cutting marble and granite plates. By replacing the different types of diamond string bead wire or changing the linear speed of the wire, it was easily to cut and greatly improved production efficiency. Liao, Y.S. & Tao, C.G. (2011) introduced a CNC diamond wire saw machine. It used the flexible features of cutting tool to cut stone. It was mainly used for complex designs. Wang, H. (2008), Yong, H.K. (2013) & Zhou, B. (2011) described a kind of underwater diamond wire saw machine for cutting. It had high precision and high efficiency. In addition, the cutting equipment was light weight, easy to transport, simple and compact structure, and low construction equipment. It can cut and used for the pile foundation of offshore platform jacket or underwater large diameter pipe and pipe cutting jobs and the nuclear power industry. Diamond rope saw machine cutting is mainly relies on flexible diamond beaded rope and rigid diamond particles to cut pipelines with the different material or shape. The tensioning force of the diamond beaded rope must be required in order to keep constant of the tensioning force and guarantee normal of cutting job. The reliable security of the tensioning institution is a premise for cutting work, so the tension institution is very important for whole system. Therefore, the diamond wire saw machine needs a proper tensioning device to maintain optimal cutting efficiency and long service life. However, existing tensioning institutions of the underwater diamond wire saw cannot take-up tension detection and does not guarantee the moving precision of tensioning wheel. The stress effect is not good, more tension cannot be taken.

To overcome the above disadvantages, a tensioning device of underwater diamond rope saw machine is proposed based on the existing approach. It consists of a tensioning wheel, a tensioning force detection institution, an adjustment handle, a screw and a guide. It can guarantee beaded rope initial the tensioning force and processing process in the stable of the tensioning force, and convenient disassembly. Through tension sensor for real-time monitoring and control beaded rope radial into speed, it ensures cutting job security and efficient. The method is verified by an example in this paper. Corresponding of conceptual design is proposed. The tensioning device of underwater diamond rope saw machine is designed and calculated.

The paper is organized as follows. In the next section, we propose the problems of the tensioning device of underwater diamond rope saw machine. In section 3, conceptual design of the tensioning device of underwater diamond rope saw machine is presented. Finally, we conclude our paper in section 4.

2 STRUCTURAL ANALYSIS OF THE TENSIONING DEVICE

The underwater diamond wire saw is mainly composed of the wire, the motion systems, the feed system, the control system, the tensioning device and the guide organization parts. The diamond wire used as cutting is typically a closed-ring form. It is mounted on the driving wheel. The idlers and the tensioning wheel are driven by the driving wheels at an ideal rate.

2.1 Characteristics of current tensioning devices

Wire of the underwater diamond wire saw machine needs tension and the tension is constant. Because tensioning of the wire plays an important role in the process of cutting, it guarantees the service life of the wire and cutting efficiency. When the tension force is too small, the wire is small friction and low cutting efficient. When tension is too large, the wire is low life expectancy. When it encounters a strong impact, the wear can be accelerated, and severe rope phenomenon occurs (Zhang, L. et al. 2005). In order to facilitate assembly and disassembly of diamond wire, the initial installation distances between the driving wheels and the wheel of the diamond wire saw machine should be within the length of the error of the wire. After the completion of processing, the wire do not need to keep too much tension, namely the tension is adjustable.

2.2 Analysis of the tensioning device of underwater diamond wire saw machine

Now commonly used for tensioning device of the underwater diamond wire saw machine is manual tensioning and hydraulic tensioning device. Manual tensioning device uses a common screw, and its structure is shown in Figure 1. Manual tensioning device consists of the nut, the screw, and the tensioning wheel component parts. The tensioning wheel is fixed on the nut connected by a key component. As the screw rotates, the nut moves for linear and drives the tensioning wheel moving around. The effect of tension and relaxation is achieved.

The tensioning device shown in Figure 2 is to rely on hydraulic cylinders to achieve tension and relaxation. The automatically filling oil and holding pressure of the tensioning system are measured by the hydraulic system with pilot operated check valve and electric-contact pressure gauge. This approach is not only steady pressure but also has long holding time. The

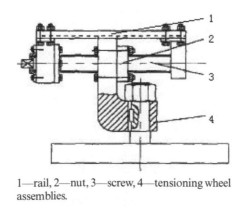

1—rail, 2—nut, 3—screw, 4—tensioning wheel assemblies.

Figure 1.   Manual tensioning device.

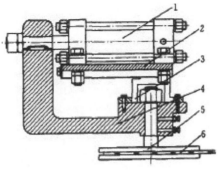

1—tensioning cylinders, 2—bow plate, 3—rail,
4—tensioner, 5—tensioning wheel, 6—wire.

Figure 2.   Hydraulic tensioning device.

rail set on the tensioning wheel is to avoid tension cylinder piston rod to subject to bending moment. When the tensioning cylinder piston rod extends to the left, the tensioning wheel moves left, the wire is tensioning and cutting. When the tensioning cylinder moves right, the wire relaxes and removes the wire.

Because the diamond rope saw machine works in underwater 60M, the hydraulic drive has advantage for underwater machines. The hydraulic tensioning device has following advantages: hydraulic has high keep capacity. The device still can maintain a stable of pressure even stop the oil. In the underwater work special seal is not needed, and it can make mechanical structure of design simplified.

It can be found by comparing that the principle of the manual tensioning device relied on nut screw and the tensioning device relied on hydraulic is the same. All is to achieve the effect of tension and relaxation through adjusting position of the tensioning wheel. Only one is to change through the nut screw, and the other is changed by a hydraulic cylinder. The manual tensioning device has a simple structure, but does not automatically adjust the tension. The hydraulic tensioning device can automatically adjust the tension though the structure is relatively complex. Combining the advantages of each, the tensioning device is designed.

3   DESIGN OF THE TENSIONING DEVICE

3.1   *Determination of the tensioning device*

Through the analysis of existing tensioning devices of underwater diamond wire saw machine, it is found that the structure of the manual tensioning device is simple, but the tension does not automatically adjust, which affected the stability of tensioning. The key to cutting is to guarantee work stability. The hydraulic tensioning device is realized by hydraulic cylinders. The structure is relatively complex, but stable tension. Even the oil supply is stopped, and the device can still maintain a steady pressure. Two tensioning devices are relatively more worthwhile to apply in practice for current technologies.

The design is determined according to relative movement between objects. The spiral drive is used. It can change rotary movement to line movement. The regulation of tensioning force needs the mobile of the tensioning wheel. Through rotating of the screw, the screw nut is set on the screw. If the screw nut is not moved, the screw will move. Thus the tensioning axle seat is moved. The tensioning wheel is also moved. In order to ensure the reliable and the stability of the institution, a pedestal is put on between screws and the tensioning wheel. The base is connected to the screw through the screw nut. The rail is laid on the base to accurately ensure the moving of the tensioning wheel. The base also can be arranged connecting the other mechanism of diamond wire saw machine through the above installation holes to ensure the stability of the institution as a whole. As shown in Figure 3.

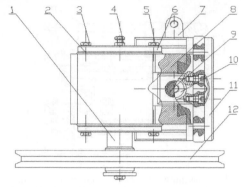

1—axis, 2—adjust washer, 3—screws, 4—nut,
5—elastic washers, 6—end caps, 7—grip,
8—tensioning wheel axle retaining, 9—screw,
10—ring, 11—end plate, 12—tensioning wheel.

Figure 3. The tensioning device connected with the guide.

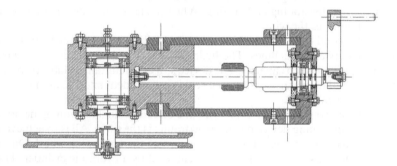

Figure 4. Tensioning device.

3.2 *Structural characteristics of the tensioning device*

The tensioning device of the underwater diamond rope saw machine is shown in Figure 4. It consisted of the tensioning wheel, the tensioning axle seat, tensioning force detection institution, the adjustment handle, the screw, the side board, the end board and the base. The screw is located in the side board. Its end wears out end board and is fixed by the adjustment handle. The nut sets in screw and is fixed with the base. The base is tied with the tensioning wheel axle seat guide. The end plate is connected with the side plate by the screw. The base and screw is connected with the screw. Holes in the side panels and the base are mounting holes of the tensioning device and the main points of diamond wire saw machine.

The device used to adjust the radial position of the tensioning wheel can be also clearly shown in Figure 4. The screw is used to adjust the location of the outer ring through the bearing outer ring. The nut is locked after adjusting. Generally, it is essential operations when need to adapt. As the radial location needed to adjust is comparatively tiny, so this mechanism is feasible. The tensioning wheel and the handle are fixed by screws through the retainer ring locking. The circular ring screw located on the left end of the screw and the bulge sites located on the right side of the screw are used to limit the nut. Thus the moving distance of the tensioning wheel is limited and the nut is prevented rotary out of the screw. The side board and the end board are connected by the inner six angle cylindrical head screws. The tensioning force sensor is set on the screw near handles end. It both ensures the initial tensioning force of the beaded rope and the stable of tensioning force processing in the process. The beaded rope is conveniently loaded and split. The tensioning force sensor can be the real-time control radial speed of the beaded rope and ensure the efficient and accurate cutting job.

3.3 Design and calculation of the screw

Considering self-locking, the screw selects the trapezoidal threads. The trapezoidal thread profile as isosceles trapezoid, tooth-fillet $\alpha = 30°$. Inner and outer thread with conical trapezoidal screw threads snapping is not easy to loose.

The self-locking condition of the thread is $\psi \leq \varphi_v$, in the formula, ψ for pitch diameter angle; φ_v is the equivalent friction angle. In order to ensure the lock, that is $\psi \leq \varphi_v$. Because the tensioning wheel is rotating in a high-speed when cutting, it needs to be fixed. Namely thread is self-locking, so screw need to check self-locking conditions. The equivalent of the friction formulas of the helix angle:

$$\varphi_v = \arctan\left(\frac{f}{\cos \beta}\right) \tag{1}$$

where β = flank angle; f = friction coefficient; symmetric tooth flank angle $\beta = \dfrac{\alpha}{2} = 15°$ and friction coefficient = 0.1.

$$\varphi_v = \arctan\left(\frac{f}{\cos \beta}\right) = \arctan\left(\frac{0.1}{\cos 15°}\right) = 5.91°$$

Rise angle formula is:

$$\psi = \arctan\left(\frac{p}{\pi d_2}\right) \tag{2}$$

where p = pitch (mm); d_2 = diameter (mm).

Designed thread nominal diameter is 34 mm, look-up table pitch $p = 6$ mm, and $d_2 = 31$ mm.

$$\psi = \arctan\left(\frac{p}{\pi d_2}\right) = \arctan\left(\frac{6}{31\pi}\right) = 3.53°$$

$\psi \leq \varphi_v$, so screw the self-locking condition is met. The screw plays a role in tension transfer, the check is very necessary. Therefore, the screw strength should be checked according to the fourth strength theorem, the check formula is:

$$\sigma_e = \sqrt{\sigma^2 + 3\tau^2} = \sqrt{\left(\frac{4F}{\pi d_1^2}\right)^2 + 3\left(\frac{T}{\pi d_1^3/16}\right)^2} \leq [\sigma]\text{MPa} \tag{3}$$

where d_1 = pitch diameter (mm); $[\sigma]$ = screw the allowable stress (MPa). The torque T is:

$$T = \frac{F \cdot \tan(\psi + \varphi_v)d_2}{2} \tag{4}$$

where F = axial force (n); ψ = lead angle; φ_v = equivalent friction angle; d_2 = pitch diameter (mm);

$$T = \frac{F \cdot \tan(\psi + \varphi_v)d_2}{2} = \frac{1000 \times \tan(3.53° + 5.91°) \times 31}{2} = 2577.13 \text{ N} \cdot \text{mm}$$

$$\sigma_e = \sqrt{\sigma^2 + 3\tau^2} = \sqrt{\left(\frac{4F}{\pi d_1^2}\right)^2 + 3\left(\frac{T}{\pi d_1^3/16}\right)^2} = \sqrt{\left\{\frac{4 \times 1000}{(34-6)^2}\right\}^2 + 3\left\{\frac{2577.13}{(34-6)^3 \pi/16}\right\}^2} = 2.5 \text{ MPa}$$

$$[\sigma] = \frac{\sigma_s}{S} = \frac{980}{4} = 245 \text{ MPa}$$

$\sigma_e \leq [\sigma]$, so to meet the requirements.

Figure 5.   Model of the tensioning device of the underwater diamond wire saw.

4   SIMULATION EXPERIMENT

According to determined design parameters above chapters, parts of the institution are drawn by using the Pro-E software. The parts of the institution are assembled. The three dimensional model of the tensioning device in the underwater diamond wire saw machine is established in Figure 5. The movement path of screw drives institution for the tensioning device is simulated with the help of the designed specific size and force size on the tensioning device. The strength and lock function are tested. The experimental results show that the device is simple and feasible.

5   CONCLUSIONS

The underwater diamond wire saw is a device for cutting metals, non-metallic and composite materials in the submarine working environment. Using the screw theory of transforming rotary motion into linear motion, the device has a simple structure of the tensioning device of the underwater diamond wire saw machine. The device solves the current problems of underwater diamond wire saw, and with good results. Next, the mechanical characteristics of the rotating handle and the tensioning wheel need to analyze through simulation software to determine the optimal design parameters.

REFERENCES

Buttner, N. Schwan, G. Wunderlich, P. 1999. Travelling diamond wire saw for the processing of natural stone. *Industrial Diamond Review.* 59:221–224.
Liao, Y.S. 1997. Poly rope diamond wire-saw. *Stone.* 17(6):7–9.
Liao, Y.S. Tao, C.G. 2011. Overview of diamond wire-saw developments. *Stone.* 21(6):22–29.
Qi, Q. 2011. The diamond wire saw cutting steel. *Superhard Materials and Engineering.* 21(4):36–41.
Wang, H. 2008. Experimental research of underwater diamond wire saw. *Harbin Engineering University.*
Xu, X.T. 1998. Development and application of diamond wire saw the new. *Stone.* 18(10):13–15.
Yong, H.K. 2013. Use diamond wire sawing salvage. *Industrial Diamond.* 23(2):36–38.
Zhang, L. Meng, Q.X. Wang, L.Q. 2005. Analysis on technical scheme of underwater diamond wire saw machine and design. *Mechanical Engineering.* 22(8):8–10.
Zhou, B. Guo, H. 2011. Domestic development present situation and development prospect of wire saw. *Stone.* 21(6):31–33.

Advances in Engineering Materials and Applied Mechanics – Zhang, Gao & Xu (Eds)
© 2016 Taylor & Francis Group, London, ISBN 978-1-138-02834-0

Design of underwater cover opening mechanism for autonomous underwater vehicle

H. Zhang

School of Naval Architecture and Ocean Engineering, Zhejiang Ocean University, Zhoushan, China

ABSTRACT: In order to ensure the accuracy of the mechanism, an underwater cover opening mechanism based on an escaping device is presented in this paper. First, the characteristics, structure and working principle of the underwater covers are analyzed. The underwater cover opening mechanisms with different releasing devices are summarized. Then, according to the characteristic structure and the working process of the releasing device, the advantages and disadvantages of it are analyzed. The conceptual design is determined. Finally, the structural design for key components is completed. The results show that the design is feasible.

1   INTRODUCTION

The underwater cover opening mechanism refers to the special work environment. After running out of energy when underwater, the original settings of the mechanical structure are used for completing the open action (Du, L.N. 2012 & Chen, W. 2007). Lv, S.L. (2010) addressed the needs of an armored tank open and close lid and designed a hydraulic opening cover body to realize the weight of larger slot cover open-close control. Wang, H.P et al. (2002) established the dynamics model of the lid body on the storage missiles canister using the finite element method, and automatically calculated the pressure distribution on the cover, and analyzed characteristics of bearing. Jin, Y.B. (2010) analyzed the gas pressures, structural strength and structure of launcher box in the opening moment instantaneous release of gas pressure in the box. Miao, P.Y. & Yuan, Z.F. (2004) discussed anisotropic material canister cover to achieve the concentric launch canister missiles self launch front cover automatically opens to the side, and simulated complex three dimensional flow field inside the tube. Li, J.W. (2006) set up the corresponding motion model after payload separation for underwater characteristics were studied, and simulated and analyzed the separatist movement. Xiao, B. (2012) used the research methods of ANSYS ADAMS, virtual prototype technology and finite element analysis, combined with design guidelines for pressure vessels, designed the escape tube for fast buoyancy ascent escape device. Yu, W.B. et al. (2002) applied mechanics and theory of mechanics to ship missile launcher cover opening mechanism, simulated mathematical model and find out ways to resolve the effects. Liu, F. Yu, & W.B. (2004) used VB language to develop software for ship simulator mathematical model for simulation of an open institution, and made a comparative analysis of the simulation results with the test results (Liu, F. Yu et al. 2004). These researchers provided some methods used in the different area, but they do not resolve the accuracy of the underwater cover opening mechanism.

In this paper, based on the existing approaches, a new method, namely a new mechanism for underwater cover opening, was proposed. It analyses the structures of the cover opening, according to the principle of underwater cover opening, a novel of cover opening mechanism for underwater cover opening is designed based on underwater principle, and the method is verified by an example of in this paper. Corresponding of conceptual design is proposed.

The paper is organized as follows. In the next section, we propose the problems of underwater cover opening for underwater work. In section 3, conceptual design of cover opening mechanism and the structural design for key components are presented. Finally, we conclude our paper in section 4.

2 ANALYSIS OF THE UNDERWATER COVER OPENING MECHANISM

2.1 *Working principles of the underwater cover opening mechanism*

The underwater cover opening mechanism is used for releasing the airbag in the craft, so to facilitate the recovery of the aircraft. When power has been consumed, the control system will give instructions. The electric control valves are opened. The high pressure gas is filled in balloon through a high pressure pipeline. The balloon begins to swell and to put pressure on the board. When the inflation pressure is more than the withstand pressure of the push plate, the slider begins to move and the pin is cut off. The connecting device is released. Finally, under the combined effect of the expansive force of the airbag and the preload of the tighten screws, two semicircular cover is opened. In accordance with the set requirements, air bags continue to provide buoyancy. The aircraft floats. As shown in Figure 1.

2.2 *Types of the cover opening mechanism*

2.2.1 *Explosive bolt-type cover opening mechanism*
The interior of the explosive bolt has a cavity. It can accommodate a certain amount of blasting explosives and detonators. Incendiary device ignited charge after the action signal is received. The entire bolt is cut into 2 3 parts. The explosive bolt must be an adequate mechanical strength, while also providing a reliable explosive separation. The explosive separation must be no, or only a few fragments. In order to ensure this, stress tanks are suppressed on the explosive bolt. When explosion, the bolt ruptured from the stress of tank. Meanwhile, reaction time of the explosive bolt has strict requirements. Due to the special work environment, the explosive bolt must also work in a large temperature range. Generally, reliable work from dozens of degrees below zero to zero near the temperature range is required. As shown in Figure 2.

The cover opening mechanism of the explosive bolt has a simple structure, high reliability characteristics. But explosive bolts cover opening mechanism existed certain danger in the practical application.

2.2.2 *Electromagnetic relay cover opening mechanism*
Such cover opening mechanism is relatively simple. Open cover bodies are divided into left and right doors. The electromagnetic relay in aircraft is fixed with the left cover. The armature is fixed with the right cover. When the aircraft has energy, electromagnetic relays power will flow through a certain amount of current in the coil. Electromagnetic effects would generate.

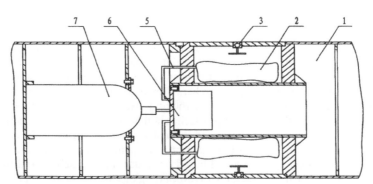

1-underwater aircraft 2-balloon 3-cover opening mechanism 5-high pressure pipes 6-control system 7-high pressure gas cylinders.

Figure 1.   Underwater cover opening mechanism.

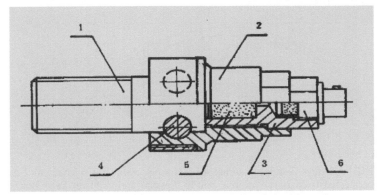

1-screw 2-bolt 3-seat 4-shear pin 5-detonator 6-electric ignition tubes.

Figure 2. The explosive bolt.

The armature will attract return towards the core by overcoming the spring tension under the action of electromagnetic forces. Thus moving contact and fixed opened contact of the armature are attracted. When aircraft power and coil power are exhausted, magnetic suction will disappear. The armature returns to the original position reaction. The dynamic contact point and the original static closed contact are attracted. High pressure gas cylinders inflated the balloon. Left and right doors are separated under thrust of airbags. The airbags continue to swell. When the buoyant force acted on airbags is greater than the value of aircraft, the aircraft floats for easy recycling. Electromagnetic relay-type cover opening mechanism with limited contact life and produce electromagnetic interference faults, effect underwater normal operation of other equipment and so on.

To overcome above disadvantages, a new underwater opened cover institution is proposed. It is composed with the airbag, the pushed board, the free device, the broken pin, the mobile pin, the spring, and the sliding rod. It can guarantee the cover process under the action of the pre-tight force of the spring, the pre-tight force of the bolt and the expansion force of the airbag. It provides the needed buoyancy of the sailing device floating. It ensures security reliable, and using convenient, and positioning accurate of the opened cover process.

3 DESIGN AND CALCULATION OF THE UNDERWATER COVER OPENING MECHANISM

3.1 *Design of the underwater cover opening mechanism*

By analyzing the original open cover bodies, it is found that an explosive bolt cover opening mechanism and electromagnetic relay type cover opening mechanism have obvious short-comings. So, a new type of underwater cover institutions is designed to address security, stability and control inaccurate problems.

This design uses force to push the lid and the pusher. The compression springs are pushed on both sides of mobile pins and screw. The releasing head screws arise out of a force on the right cover from the left cover by overcoming the preload, as shown in Figure 3.

3.2 *Characteristics of the underwater cover opening mechanism*

On the releasing device of the cover opening mechanism, the left cover and the base are fixed connection. The right cover is fixed by screws and the releasing head. The left cover and the right cover are connected with the ball. On the left half of the releasing head has holes that match ball. It belongs to the transition fits that can ensure not loosen up on the x-y plane. And it can guarantee that when the steel ball in the direction of the z-axis the spaces exits

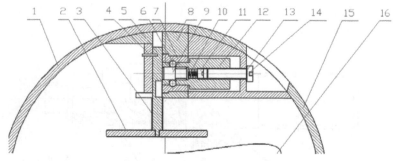

1-left lid 2-trays 3-support spring 4-broken pin 5-slip rod 6-smallball 7-base 8-connected fixed part 9-mobile pin 10-pressure spring 11-no head bolt 12-releasing head 13-washers 14-screw15-right cover 16-airbag.

Figure 3.   The underwater cover opening mechanism front view.

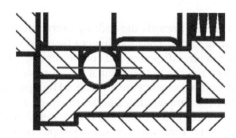

Figure 4.   The ball position.

lines fit on. The ball can move along the z-axis. The releasing head and the base are sliding coupling. The ball makes the head and the base could not be moved. This allows the left and right door close coincidence and does not open automatically.

3.3   *Parameters of key parts*

3.3.1   *Design and calculation of small steel balls*
According to aircraft parts proportions, the diameter of the ball is selected as 40 mm. It is made of steel. The number is 4. Ball size as:

$$V = (4/3)\pi R^3 \tag{1}$$

where r = radius of the sphere, steel ball size is:

$$V = (4/3)\pi R^3 = (4/3)\ \pi 24^3 = 33510.32 \text{ mm}^3$$

According to the machine manual, the density of the steel is 7.85, the quality of the steel ball $M = \rho V = 57.904 \times 7.85 = 263.06$ g.

In accordance with the underwater cover opening mechanism, on the inner surface of the chassis opens 4 hemispherical concave. The radius is 10 mm. When the upper three-fourths part of the steel ball is in the holes, the one-fourth part of the ball is in the groove of the base, as shown in Figure 4. When there is a gap at the top of the ball, if the releasing head exerts a force on the ball, the ball will move up under the action of external forces, lower one-fourth parts of the steel ball will be separated from the chassis.

3.3.2   *Design and calculation of the releasing head*
The left half of the releasing head has 2 round holes, which the radius is 20 mm and the depth is 30 mm, namely for three-fourths of the steel ball diameter.

The internal thread in the right half of the releasing head is fit with screw thread, and the length is set to 180 mm. The spring is placed in the middle of the releasing head. The length is the length of spring compression, and is set as 60 mm. The left half length of the releasing head is equal to the length of mobile pin. So, the total length of the releasing head is 180 + 60 + 160 = 400 mm. The inner diameter of the releasing head is equal to the left half of the move pin, and is 80 mm. The inner diameter of internal threads of the releasing head is equal to the outer diameter of the pressure spring and for 58.5 mm. The outer diameter of internal threads of the releasing head is 68.5 mm.

3.3.3 *Design and calculation of the spring*
According to Hooke's law:

$$\Delta f = k \times \Delta x \tag{2}$$

where Δf = pressure or tension of the spring, k = the spring constant, Δx = spring-shaped variable, then spring constant formula:

$$K = (G \times d4)/(8 \times Dm3 \times Nc) \tag{3}$$

where G = the modulus of rigidity of the wire, d = wire diameter, Dm = diameter, Nc = effective number of laps, the data show that $DM = Do - d$, Do = outside diameter, $Nc = N - 2$ and N = total number of laps. Spring common materials for making available according to the data are the piano wire $G = 8000$. Set the spring wire diameter = 6.50 mm, out diameter = 58.50 mm, total lap count = 9 laps, the steel wire material is the piano wire, then:

$$K = (G \times d4)/(8 \times Dm3 \times Nc) = 1.81 \text{ kgf/mm}$$

The spring length is set to 120 mm. The depth of the hole on the sliders is 40 mm. The deformation of the spring is at least 40 mm. The spring deformation is set to 60 mm. When the cover opening mechanism does not work, the pressure of the spring is:

$$\Delta f = k \times \Delta x - 1.81 \times 60 = 108.6 \text{ kgf} = 1064.28 \text{ N}$$

When the cover opening mechanism works, the push force that the spring moves the pin into the hole of the slider is:

$$\Delta f = k \times \Delta x = 1.81 \times 20 = 36.2 \text{ kgf} = 354.76 \text{ N}$$

So, the spring force used for closing the lid will be 354.76 N.

3.3.4 *Design and verification of the bolt and the screw*
The length of the bolt is set to 35 mm. The inner diameter of the thread and the releasing head is same as 58.5 mm. The outer diameter of the thread and the releasing head is same as 68.5 mm. The releasing head is fixed to the right cover by the screw. Because of the need to ensure the tight junction of the right cover and the releasing head, a preload is required to the screw and set to 0.5 N. The inner diameter of the screw and the releasing head is same as 58.5 mm. The outer diameter of the screw and the releasing head is same as 68.5 mm. The strength check of screw is:

$$\sigma_y = 1.3 F_0 / (\pi d_1^2 / 4) \tag{4}$$

where Type: σ_y = screw the yield strength, F_0 = bolt preload, d_1 = screw diameter, the strength of screws according to the set conditions is:

$$\sigma_y = 1.3 F_0 / (\pi d_1^2 / 4) = 1.3 \times 0.5 / (\pi \times 58.5^2 / 4) \approx 242.00 \text{ MPa}$$

The screw material is 45th steel, the yield strength of the 45th steel $\sigma_s = 355$ MPa according to look up table, and then the screw is qualified.

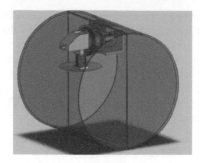

Figure 5.   The model of the underwater cover opening mechanism.

4   SIMULATION EXPERIMENT

Based on the parameters identified in the previous section, a three dimensional model of the underwater cover opening mechanisms is set up by the Solid Works software as shown in Figure 5. The structure of the various parts and parts work together are simulated. An underwater opening force provided by the air bag and screw in the body to complete the entire opening process is simulated. The separation process of the release device in the cover opening mechanism is analyzed. The sizes of the mobile pin and the releasing head are verified. The experimental results show that the mechanism is simple and safe.

5   CONCLUSIONS

The underwater cover opening mechanism is mainly used for underwater recovery and utilization. The relief device requirement is used for underwater cover opening, so as to achieve the purpose of underwater vehicle recycling. Aiming at the shortcomings of existed open institutions, a mechanical structure of underwater cover institutions is designed in order to ensure accurate work of underwater cover bodies. Subsequent work to be done is the calculation of screw pre-tightening force and the pin. With the acceleration of marine development and application of autonomous underwater vehicle, the underwater open bodies as an important structure of the autonomous underwater vehicle will be more and more attention.

REFERENCES

[1] Du, L.N. 2012. Study on the potential development of ocean engineering equipment manufacturing industry in China. *Dalian Maritime University Master's thesis.*
[2] Chen, W. Dou, Y.S. 2007. Ocean engineering industry development present situation analysis of the world. *China Water Transport (Academic Version)*, 7(08): 199–200.
[3] Lv, S.L. 2010. Armored tank lid mechanism design. *Hydraulic and Pneumatic.* 12: 7–8.
[4] Wang, H.P. Luo, Y.Z. Zhang, B.S. 2002. Missile storage canister fault replay and institutions of the open cover before improving. *Missile and Space Launch Vehicle Technology.* 255 (01): 36–40.
[5] Jin, Y.B. 2010. Missile launcher instant problem analysis and solving measures of the impact opening. *Tactical Missile Technology.* 30(05): 75–78.
[6] Miao, P.Y. Yuan, Z.F. 2004. Concentric launch canister gas open end technology. *Journal of Beijing Institute of Technology (Natural Science Edition).* 24(04): 283–285.
[7] Li, J.W. 2006. Load separation for underwater maneuverability. *Northwestern Polytechnic University Master's thesis.*
[8] Xiao, B. 2012. Breathing fast buoyancy ascent escape tube design of the device. *Hubei University of Technology Master's thesis.*
[9] Yu, W.B. Xu, G.H. Liu, Y.C. Chen, B.F. 2002. Ship Simulator mathematical model of missile launcher cover opening mechanism. *Ship Science and Technology.* 24(06): 34–36,39.
[10] Liu, F. Yu, W.B. 2004. VB-based ship Simulator mathematical model for simulation of an open institution. *Ship Science and Technology.* 26(Z1), 59–61.

Advances in Engineering Materials and Applied Mechanics – Zhang, Gao & Xu (Eds)
© 2016 Taylor & Francis Group, London, ISBN 978-1-138-02834-0

Analysis on cooling air of engine compartment inlet section

H.K. Cai, Y.L. Zhang, Y. Xiang, W.Z. Yang & D.X. Deng
ShenZhen Research Institute of Xiamen University, Shenzhen City, Guangdong Province, China
Department of Mechanical and Electrical Engineering of Xiamen University, Xiamen City,
Fujian Province, China

ABSTRACT: Most of the heat generated in engine compartment is taken away by cooling air from inlet grids suctioned by cooling fan. The analysis of cooling air in the inlet section is important to the thermal balance of engine. Therefore, this paper is addressed to air inlet grids and cooling fan by model establishment and numerical simulation. The attention is paid to the effect of characteristic parameters on the performance of air inlet grids and cooling fan, including their geometry parameters and operating parameters. The research will be helpful in guiding future design and optimization of vehicle cooling system and engine compartment.

1 INTRODUCTION

A good cooling cycle should ensure the vehicle engine to work at an appropriate temperature. With the development of vehicle, its engine power and heat flux are increasing, consequently the cooling system is becoming larger and larger, whereas the engine compartment has a limited space for all components[1-2]. One way to resolve this contradiction is to raise cooling efficiency by improving the cooling system. The vehicle cooling system almost includes engine, thermostat, water pump, expansion tank, fan, radiator, air inlet grids, pipes and other units[3]. The cooling air is suctioned into the compartment from inlet grids by a fan rotation, and flows through the radiator to refrigerate down cooling fluid from the engine. It can be followed that air inlet grids and fan greatly influence the thermal balance of engine compartment.

To analyze the thermal balance of engine compartment is usually by experiments or simulations. As many CFD softwares are arising and booming, researchers have used them to solve many questions of cooling fan and air inlet grids[4-6]. For instant, Khaeld has studied the effects of ground vehicle inclination on compartment cooling[7]. Pradeep[8] and Kim[9] have combined a simulation with experiments and proposed a non-linear control method for the components of cooling system to improve engine cooling capacity and fuel economy. However, there are still few papers for a throughout study about air inlet grids. Plus, the new noise limitation requirement in GB/T 25982-2010 will be carried out and it will ask for a further noise reduction of cooling fan. A fan with large mass flow volume and low noise is needed a more study based on present work. Therefore, this paper is addressed to thermal analysis of cooling air in engine compartment inlet section by focusing on air inlet grids and cooling fan, which will be helpful for the design and optimization of vehicle cooling system.

2 COOLING AIR INLET GRIDS

2.1 *Numerical model*

In practice, the amount of cooling air inlet grids could be up to thousands. This is difficult to simulate if all grids are meshed. As a result, it is simplified to be a flat plate keeping real characteristics. Usually, there are three types of air inlet grid: circle aperture, long aperture and square aperture, which are shown in Figure 1. The plates own the same aperture opening ratio. In order to analyze heat dissipation of different apertures, the initial temperature is supposed to be 60 °C and heat convection coefficient is supposed to be 310 W/(m² · K).

2.2 *Simulation results*

For different air inlet grids, to evaluate the effect of heat dissipation is to compare the temperature it can reach. Among the three types of aperture, the lowest temperature of circle, long and square aperture is 52.4, 48.4 and 53.3 °C respectively. The temperature distribution of long aperture is shown in Figure 2, which will be discussed next.

2.3 *Discussion*

The discussion is centered on the parameters of long aperture to see the effect on lowest temperature. Its geometry dimension varies from 66 × 2 × 2.5, 66 × 2 × 3.5 (as shown in Fig. 2), 66 × 2 × 5.5 and 68 × 2 × 2.5 mm (length × width × distance of long aperture). The simulation results are shown in Table 1. With different distance of long aperture, the closer the long aperture is, the better its heat dissipation is. However, when its length gets larger, its

Figure 1.   Air inlet grids (from left to right: circle, long and square aperture).

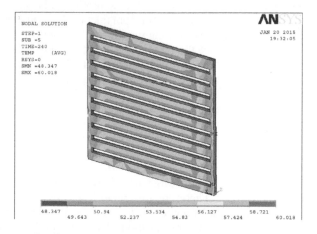

Figure 2.   Temperature distribution of long aperture.

Table 1.   Simulation results of the four long apertures.

	No. 1	No. 2	No. 3	No. 4
Dimension/mm (length × width × distance)	66 × 2 × 2.5	66 × 2 × 3.5	66 × 2 × 5.5	68 × 2 × 2.5
Temperature/°C	43.8	48.4	53.3	46.6

effect of heat dissipation does not get better according to the simulation. This may be due to the turbulent flow. Since the length increases, the aperture is close to the edge of plate gradually, which will increase the turbulence of air flow and this could induce a heat stack near the aperture edge. This phenomenon will be further studied in the Fluent software.

3 COOLING FAN

3.1 *Numerical model*

The air suctioned through inlet grids is blown into engine compartment by cooling fan, so the fan is another important component in cooling system and its flow volume is a key parameter for the fan. Meanwhile, as environment pollution is more and more drawn the attention of researchers and the new noise policy is carried out, the noise produced by fan needs a further study. Therefore, flow volume and noise are both the aims for fan simulations.

The fan model is simulated in Fluent and its geometry dimension of the whole model is shown in Table 2. The flowing area is divided into four parts: inlet section, rotating fluid section, outlet section and pipe section as shown in Figure 3. Here, RNG k-ε two equations model is used to solve fluid control equations and second order upwind scheme is adopted for the turbulent flow of the fan. The boundary condition of inlet section is pressure-inlet and the outlet section is pressure-outlet condition, where a monitor point of mass flow volume is also set. After 790 iterations, the simulation converges and mass flow rate keeps at 0.484 kg/s as shown in Figure 4.

The sound field simulation is similar to that of flow described above. Here, the Large Eddy Model is adopted. The time step is set to be 0.00005 s and step number to be 4000. The number of maximum iterations for each step is 25. After 332941 iterations, the simulation converges as shown in Figure 5. The spectrogram of 1/3 octave is shown in Figure 6, where attention is paid to the frequency interval between 2000 and 6000 Hz.

3.2 *Simulation results*

Figure 7 shows the effect of revolving speed on flow volume and overall sound pressure of A type. Obviously, with the increase of revolving speed, both flow volume and sound pressure

Table 2. Geometry dimension of fan simulation model.

Fan model	The hub radius is 50 mm, hub thickness is 50 mm, hub ratio is 0.236, outer felloe radius is 190 mm and blade number is 4.		
Fluid area	Radius (mm)	Length (mm)	Mesh spacing (mm)
Inlet section	220	3000	20
Rotating section	200	60	10
Outlet section	220	4500	20
Pipe section	205	70	16

Figure 3.   Fluid area by meshed.

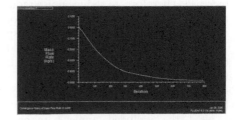

Figure 4.   Mass flow rate convergence curve.

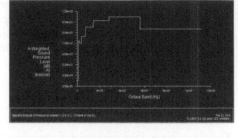

Figure 5.   Mass flow rate convergence curve.

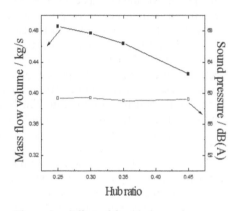

Figure 6.   Spectrogram of 1/3 octave.

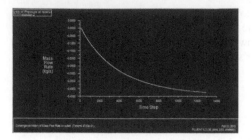

Figure 7.   Effect of fan revolving speed.

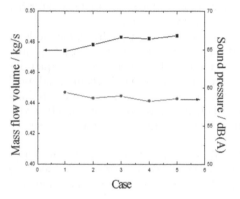

Figure 8.   Effect of fan blade number.

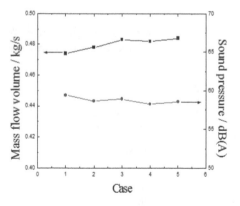

Figure 9.   Effect of blade included angle.

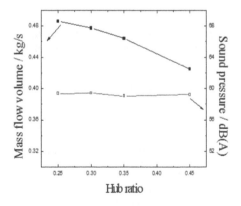

Figure 10.   Effect of fan hub ratio.

increase. Therefore, to reduce noise radiation, the first measure is to reduce revolving speed and slow down cooling fan in the precondition of fulfilling mass flow volume and cooling capacity of engine. Here, the revolving speed of 2000 rpm is chosen in next research.

Then the number of blade is changed from 2 to 5. As shown in Figure 8, the flow volume of fan increases with blade number, but the amount of increasing mass is slowing down. The way to enlarge cooling wind volume by increasing blade number is restricted. On the other hand, for the sound pressure, the effect of blade number is little in the simulation for the wide blade. Here, the blade number of 4 is still adopted in further simulation.

Next the included angle between two blades is changed from 70°-110°-70°-110° (Case 1), 75°-115°-75°-115° (Case 2), 80°-100°-80°-100° (Case 3), 85°-95°-85°-95° (Case 4)

410

to 90°-90°-90°-90° (Case 5). As shown in Figure 9, the flow volume of fan does not vary greatly with the change of included angle. The sound pressure is the same, except for Case 1. For Case 1, sound pressure increases to be 59.39 dB(A). Therefore, for the wide blade in the simulation, the included angle is suggested to be not smaller than 75°. Here, the Case 5 is still used for final study.

Finally, the hub ratio is changed from 0.25, 0.3, 0.35 to 0.45. As shown in Figure 10, with hub ratio increase, fan flow volume is decreasing, while sound pressure does not change greatly, which is keeping at the value of about 59.2 dB(A). To reduce the noise radiation by changing hub ratio is not to be suggested.

4 CONCLUSION

This paper is addressed to the analysis of cooling air in engine compartment inlet section. For air inlet grids, three different types of grid are studied and it is found out that long aperture with $66 \times 2 \times 2.5$ mm (length × width × distance) is the best for heat dissipation in the simulations. For cooling fan, revolving speed, blade number, the blade included angle and hub ratio play important role in flow volume and sound pressure of cooling fan, among which the revolving speed is the most. The research will be helpful in guiding future design and optimization of thermal balance in engine compartment and vehicle cooling system.

ACKNOWLEDGEMENT

The work was supported by Knowledge Innovation of Shenzhen City of China (JCYJ20140417162429675), Natural Science Foundation of Fujian Province of China (Grant No. 2014J01210) and Collaborative Innovation Center of High-End Equipment Manufacturing in FuJian.

REFERENCES

[1] Robert, W.P. & Wsewolod, J.H. et al. 2005. Thermal management for the 21st century—improved thermal control and fuel economy in an army medium tactical vehicle. SAE01: 2068.
[2] Kyoung, S.P. & Jong, P.W. et al. 2002. Thermal flow analysis of vehicle engine cooling system. *KSME International Journal*. 975–985.
[3] Defraeye, T. & Blocken, B. et al. 2010. Aerodynamic study of different cyclist positions: CFD analysis and full-scale wind-tunnel tests. *Journal of Biomechanics.* 43(7):1262–1268.
[4] Bayraktar, I. 2012. Computational simulation methods for vehicle thermal management, *Applied Thermal Engineering.* 36: 325–329.
[5] Mahmoud, K.G. & Loibner, E. 2003. An overview of the vehicle thermal management simulation tools, *Veh. Therm. Manage. Syst.* VTMS, Brighton, United Kingdom, 585–597.
[6] Khaled, M. & Alshaer, A. et al. 2012. Effects of ground vehicle inclination on under-hood compartment cooling, *International Journal of Automotive Technology*. 13(6), 895–904.
[7] Pradeep, S. & John, R.W. et al. 2005. An Advanced Engine Thermal Management System: Nonlinear Control and Test. *Transactions on Mechatronics.* 10(2): 210–220.
[8] Kim, K.B. & Chor, K.W. et al. 2010. Active Coolant Control Strategies in Automotive Engines. *International Journal of Automotive Technology.* 11(6): 767–772.

Advances in Engineering Materials and Applied Mechanics – Zhang, Gao & Xu (Eds)
© 2016 Taylor & Francis Group, London, ISBN 978-1-138-02834-0

The study of characteristics in flat-tube radiator based on multiple rows

H.K. Cai, Y.L. Zhang, T.F. He, C. Li & Y.X. Zhang
School of Physics and Mechanical and Electrical Engineering, Xiamen University, Xiamen, China
Shen Zhen Research Institute, Xiamen University, Shenzhen, China

ABSTRACT: According to numerical simulation to get thermal performance for radiator with different numbers of tube rows 2, 3, 4, 5, and 6. Through the numerical simulation of uni-directional fluid-structure coupling by ANSYS Workbench platform, the heat transfer coefficient, air-side outlet temperature and air pressure drop are obtained. When the numbers of tube rows are two, three, and four, they can match the demand of cooling performance for the diesel engines well, when the numbers of tube rows increases to five or six rows, they can not meet the cooling performance of diesel engines. Finally, the rules of heat transfer and flow resistance characteristics of radiator with different numbers of tube rows are compared, to provide basis design for a radiator product of engineering machinery.

1 INTRODUCTION

Due to the high speed and big power of engineering machinery diesel engine, thermal load is extremely great. Meanwhile, in order to achieve the engineering machinery NVH requirements, other energy in the close engine compartment is also transformed into heat. High efficient and compact louvered fin radiators are widely used to meet the demand of cooling performance for diesel engine, due to good heat transfer and flow resistance characteristics[1–8].

In recent years, Lyman, A.C[1], Witry A[2], Oliet, C[3] investigated air-side heat transfer and flow behavior for louvered fin radiators structure by numerical simulation, Yuan, Z.Q[4], Li, K.N[5] also studied the louvered fin radiators with the heat transfer and flow characteristics by experimental. However, due to the increase in number of tube rows (radiator core thickness), certain aluminum material will lead to the cooling capacity decrease[6]. So, most researchers are focused on the thermal performance of radiator with single tube row, only few researchers studied the radiator with multi-tube rows. Liu, J.X[7] carried on the research for three, four, five tube rows radiator by numerical simulation method, Tong, Z.M[8] studied the single and two tube rows radiator thermal performance, and discovered radiators with double tube rows cost more than radiators with single tube row, but the heat transfer performance has not improved significantly. However, multi-row tube radiators can play an important role in large scale engineering machinery, therefore, a radiator of mechanical engineering with different numbers of tube rows is necessary to study and analysis.

In the FLUENT or CFX soft, it is difficult to calculate the complex structure of radiator with multi-row tube. They need a high computer configuration, it is easy to make a mistake sometimes. However, when we use the ANSYS Workbench (AWB) platform including the CFX and Steady-state Thermal coupling simulation analysis, it can provide convenient and time-saving analysis method for the radiators' study[9]. During the observation of surface temperature distribution and air pressure drop based on a radiator with four tube rows, the characteristics of heat transfer and flow resistance of tube arrangement by comparing the two tube rows to the six tube rows, and study the rules of thermal performance for radiators with different numbers of tube rows.

2 THE GEOMETRIC MODEL OF THE RADIATOR

2.1 *Louvered fin radiator geometric parameters*

According to the parameters provided by radiator manufacturers and actual mapping, got the main structure of an engineering machinery radiator size, see Table 1 and Figure 1.

2.2 *The establishment of louvered fin radiator model*

The Figure 2 shows 3-D structure of the model with four tube rows, and each row of louvers were arranged just the opposite to next one. Air can flow through the fin and the louver, and the cooling water only flows in the tubes. To analyze fluid resistance and heat transfer performance, we imported the structure into the AWB platform. But the whole radiators model is too large, the radiators has some similar characteristics, so the imports only include half of tubes and two fins. To minimize the effect of the air-side flowing, we designed a 10 mm structure to the import and a 20 mm structure to the export, we built the radiators model from two tube rows to six tube rows.

Table 1. Backhoe loader radiator structure parameters.

The form structure of radiator	The core structure size width high thickness/mm	The cooling water tube number	The cooling water tube size/mm	Belt number	Belt pitch/mm	Tube pitch/mm	Row pitch/mm
Flat-tube	$460 \times 550 \times 120$	152	3×23	38	2	12	7

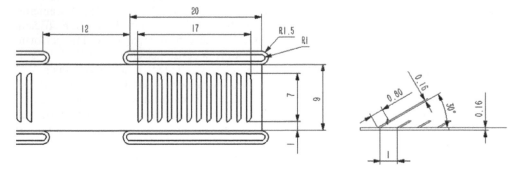

Figure 1. Radiator basic size.

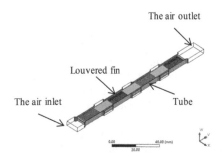

Figure 2. Simulation structure.

414

3 STUDY ON THE METHOD OF RADIATOR

3.1 The setting of boundary conditions and grid mesh

Setting boundary conditions: in CFX,

1. the air inlet was set as velocity boundary condition, the air temperature was set to 27°C;
2. the air outlet was set as pressure boundary condition;
3. the fluid-structure coupling surface was set as adiabatic wall;
4. the other surface was set as periodic boundary condition;

   In Steady-state Thermal,

5. the tube walls were set to 90°C as a fixed temperature;
6. the fin surface was imported convective heat transfer coefficient from CFX.

The radiators adopt the hexahedral mesh in ANSYS Meshing. Table 2 shows the mesh independent of air-side domain for four tube rows radiator. From the Table 2, we can conclude that it could achieve good mesh quality when the size of face mesh and body mesh were set to 0.4 mm, the calculation results of air pressure drop and heat transfer coefficient error are less than 5%. So the air-side domain grid number is 161238, the node number is 146086. The face mesh size and body mesh of structure domain also were set to 0.4 mm, but the quality of grid is poorer than the air domain mesh quality, the grid number is 52531, the node number is 67716.

3.2 Matching the thermal performance for radiator with different numbers of tube rows

Different engineering machinery power of diesel engines, small size loaders: power is less than 74 kW; medium size loaders: power at 74~147 kW; large size loaders: power at 147~515 kW; very large size loaders: power is more than 515 kW. According to the formula[10], it can calculate the heat dissipation for the power of the engines:

$$Q = \frac{A g_e P_e h_u}{3600} \tag{1}$$

Type: fuel quantity of heat, Q kJ/s; the heat to the cooling system of the fraction A, 0.25; the fuel consumption rate g_e, $kg/(kW \cdot h)$, 0.27; engine effective power P_e, kW; low calorific value of fuel h_u, kJ/kg, 41870 kJ/kg.

At the same time, according to the empirical formula:

$$Q_w = \frac{KF\Delta t_m}{\varphi} \tag{2}$$

Type: heat exchange coefficient K, $kCal/m^2/h$, take 102; radiation area required F, m^2; mean temperature difference, the cooling water and air Δt_m, $°C$, take 26; radiator reserve coefficient φ, scale and clay influence, generally = 1.1~1.5, get 1.5. Calculate the heat dissipation of various radiators to match the heat dissipation for different power of engines.

Table 2. Different sizes of mesh.

Face mesh size/mm	Body mesh size/mm	Grid quality (skewness)	Elements	Nodes	Iteration steps	Time/s	Pressure drop/Pa	Heat transfer coefficient/
0.5	0.5	General	107210	90360	45	780	774.25	135.84
0.4	0.6	General	116790	91673	47	900	797.87	134.8
0.4	0.4	Good	161175	146327	45	1200	793.87	137.87

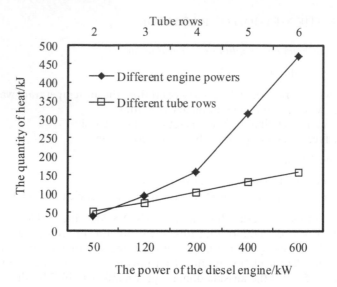

Figure 3.   Different number of tube rows and different powers diesel engines matching.

It can be seen from Figure 3: the numbers of water tube rows were six, they can meet the powers of diesel engines below 220 kW; the numbers of water tube rows were five, they can meet the powers of diesel engines below 170 kW; the numbers of water tube rows were four, they can meet the powers of diesel engines below 120 kW; the numbers of water tube rows were three, they can meet the powers of diesel engines below 80 kW; the numbers of water tube rows were two, they can meet the powers of diesel engines below 50 kW. So, this can predict different powers of diesel engines need numbers of tube rows from the design of the radiators with different numbers of tube rows, and make some theoretical analysis and instruction according to the quantity of heat. Of course, in the actual situation, engineering machinery working environment is complex, the radiators can match powers of diesel engines will below the theoretical value.

4   SIMULATION AND ANALYSIS OF RADIATOR

4.1   *The results of numerical simulation*

When the inlet velocity was 10 *m/s*, the air was flow from left to right, and we can get different numbers of tube rows of air pressure drop and temperature distribution map by the fluid-structure coupling analysis, such as in Figure 4 to Figure 8 (In the figure, the left is the temperature scaling, and on the right is the air pressure drop scaling, the above is the air pressure drop of the radiator, the below is the temperature distribution in the surface of the radiator).

It can be seen from the picture, the heat of tube wall surface is transferred to the fin surface, due to the direct contact with the louver area and tube wall surface. Most of the heats are transmitted to the louver area, then, the air take away of the heat on the surface of the fin. A small part of the heats are transferred to the fin area among the water tubes. So it can be seen from the figure, the louver surface temperature is higher than the fin surface temperature among the water tubes. With numbers of tube rows increases, the surface temperature of radiators and air pressure drop are becoming high, meanwhile, the convection heat transfer coefficient between air and fin surface are being reduced. Especially, it can be seen from the Figure 8 the temperature distribution on the surface of the radiator with six tube rows, the fin surface temperature distribution of sixth tube row is almost reached 90°C.

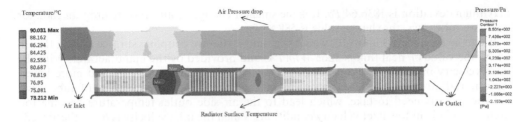

Figure 4.   Four rows tube radiator air pressure drop and surface temperature.

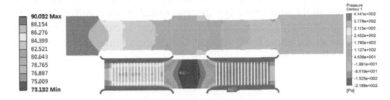

Figure 5.   Two rows tube radiator air pressure drop and surface temperature.

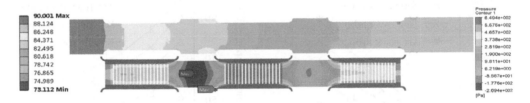

Figure 6.   Three rows tube radiator air pressure drop and surface temperature.

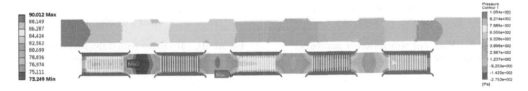

Figure 7.   Five rows tube radiator air pressure drop and surface temperature.

Figure 8.   Six rows tube radiator air pressure drop and surface temperature.

4.2   *Thermal performance analysis*

To get further analysis, we changed the inlet velocities from 6 *m/s* to 14 *m/s* to simulation engineering machinery work under different conditions. It can be seen from Figure 9, the air flow resistance is becoming bigger with the increase of numbers of tube rows. Then, the air pressure drop between import and export become more and more bigger. The air pressure drop of radiator with six tube rows increases rapidly as the inlet velocity increases, the maximum deviation is 1836.64 *Pa*. It is the smallest change of air pressure drop for radiator with two tube rows, the change value is 588.83 Pa.

It can be seen from Figure 10, we set the fluid-structure coupling surface as heat flux in CFX, the value of heat flux is 3265 $W/m^2/°C$ and provided by the manufacturer. With the inlet velocity increases, the air-side outlet temperature is become more and more lower. Among radiators with different numbers of tube rows, the more numbers of tube rows, the more heats need to take, which lead to the air-side outlet temperature become too high, especially in low inlet velocity conditions. When the inlet velocity is 6 *m/s* the air-side outlet temperature of radiator with five tube rows and six tube rows is nearly 90°C, which is not allowed in actual working condition of engineering machinery.

It can be seen from Figure 11, with the air flow speed increases, the average heat transfer coefficient is become more and more big among radiator with different numbers of tube rows. But the change of average heat transfer coefficient is extremely small as the numbers of tube rows increase at the same inlet velocity, the change range of values is basically maintained within 2%.

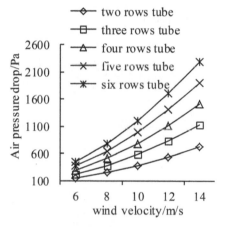

Figure 9. Radiator with different numbers of tube rows air pressure drop changes.

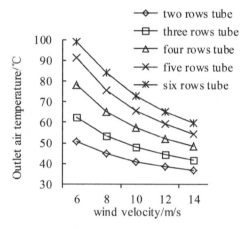

Figure 10. Radiator with different numbers of tube rows air-side outlet temperature changes.

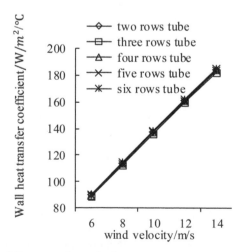

Figure 11. Radiator with different numbers of tube rows average wall heat transfer coefficient changes.

5 CONCLUSIONS

Under the different inlet velocity conditions, with the increases of numbers of tube rows, the air pressure drop become more and more bigger, the air-side outlet temperature become more and more higher, and the average wall heat transfer coefficient have no big change among radiators with the different numbers of tube rows. In summary, it can get thermal performance of radiators with different numbers of tube rows to match the need of heat dissipation of various engine powers: when the numbers of tube rows are two, three, four, they can match the demand of powers for the diesel engines under 50 kW, 80 kW, 120 kW respectively; when the numbers of tube rows increase to five or six rows, especially six tube rows, the air outlet temperature is too high. Although the air pressure drop can take heat away, but it will affect the cooling of engine compartments.

ACKNOWLEDGEMENTS

The work was supported by the Natural Science Foundation of Fujian Province of China (Grant No. 2014J01210), Knowledge Innovation of Shenzhen City of China (Grant No. JCYJ20140417162429675), Collaborative Innovation Center of High-End Equipment Manufacturing in FuJian (Grant No. 2013BAF07B04), and National Key Technology R&D Program of China (Grant No. 2013BAF07B04).

REFERENCES

[1] Lyman, A.C. & Stephan, R.A. et al. 2002. Scaling of heat transfer coefficients along louvered fins. *Experimental Thermal and Fluid Science* 26: 547–563.
[2] Wilry A. & Al-hajeri M.H. et al. 2005. Thermal performance of automotive aluminum plate radiator. *Applied Thermal Engineering* 25: 1207–1218.
[3] Oliet, C. & Olive A. et al. 2007. Parametric studies on automotive radiators. *Applied Thermal Engineering* 27: 2033–2043.
[4] Li, K.N. & Zhou, W. et al. 2014. Experiment and simulation of the performance of automotive radiator. *Mechanical Science and Technology for Aerospace Engineering* 33: 1079–1082.
[5] Yuan, Z.Q. & Gu, Z.Q. et al. 2008. Study on Numerical Simulation of the Airflow Pressure Drop for Automotive Fin-tube Radiator with Various Structural Parameters, *Science & Technology Review* 26(21): 52–56.
[6] Chen, G.J. 2011. *Hydraulic excavator*. Huazhong University of Science & Technology press.
[7] Liu, J.Y. 2013. Research on heat transfer performance of heat-dissipation module for construction machinery. *Jilin University*. 43–46.
[8] Tong, Z.M. & Zhang, Y.F. et al. 2014. Research on the structure optimization of automobile radiators. *Energy Research and Information* 30(2): 108–112.
[9] Song, X.G. & Cai, L. et al. 2012. ANSYS fluid and structure coupling analysis and engineering examples. *China Water & Power Press*.
[10] Zhang, J. 2004. Design and calculation of heat quantity of the cooling system of engine. *Equipment Manufacturing Technology*. 21–23.

Research on the acquisition of steering performance parameters of armored vehicle based on experiments

H.Y. Wang, B. Chen, S.L. Li, Q. Rui & Q.L. Wang
Department of Mechanical Engineering, Academy of Armored Force Engineering, Beijing, China

J. Guo
Science and Technology on Vehicle Transmission Laboratory, China North Vehicle Research Institute, Beijing, China

ABSTRACT: With the combination of National Instruments measurement system, memory revolving speed and torque tester and GPS-based steering track test instrument, uninterrupted experiments of multiple steering radiuses for steering kinematics and dynamic parameters have been implemented, and the test efficiency and precision of the steering performance parameters have been improved significantly. Then, the usage of test equipment and the process of test data have been depicted in detail, and the critical question of truncation and synchronization of experimental data measured from the multi-set of test instrument have been resolved. Finally, the test results are analyzed and compared with the calculation results of the theoretical model. The research provides an important technical method for steering performance test and theoretical model validation for tracked vehicles.

1 INTRODUCTION

Research on the Acquisition of the Steering Performance of Armored Vehicle is consisted of a theoretical research and experiment research. To make sure the veracity of the result, the experiment research is necessary. In this paper, the research will test actual steering radius, actual steering angular velocity, offset of steering center point, the slip and spin, traction and barking power, etc.

Wong, Said did research according to date of Ehlert experiments, and they validated the model of steering on a firm ground and soft ground. Fang Zhiqiang and Cheng Junwei measured the steering performance parameters on a series of road condition. At the present condition, getting all the acquisition in one research, several sensors and test devices are necessary. Our study is proposed according to the GPS-based steering track test instrument. With the combination of National Instruments measurement system, memory revolving speed and torque tester, uninterrupted experiments of multiple steering radiuses for steering kinematics and dynamic parameters have been implemented. Finally, the test results are analyzed and compared with the calculation results of the theoretical model.

2 RESEARCH METHOD ABOUT THE ACQUISITION OF STEERING PERFORMANCE PARAMETERS OF ARMORED VEHICLE

Several parameters of armored vehicle are shown in the Figure 1. O_L, O_S and O_C are theoretical steering center, test steering center, and geometry center. O_2 and O_1 are high-speed steering

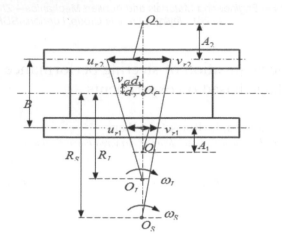

Figure 1. Parameters while steering.

center and low-speed steering center. u_{r2} and u_{r1} are the implicative velocity. B is for the center distance of the two tracks. d_1 and d_2 are the location parameters of the GPS, and V_G is the velocity of GPS, respectively.

2.1 Steering radius and the modified coefficient of steering radius

2.1.1 Actual steering radius and actual relatively steering radius
According to longitude and latitude location of the armored vehicle while steering, a GPS can measure the detail location, velocity and steering angle, finally it can figure out the steering trajectory by difference calculations, or using the velocity and steering angle to draw the steering trajectory.

2.1.2 The modified coefficient of steering radius
The modified coefficient of steering radius is the ratio of actual steering radius and theoretical steering radius as Equation 1 below

$$K_\rho = \frac{2}{B} \frac{u_{r2} - u_{r1}}{u_{r2} + u_{r1}} \sqrt{R_G^2 - d^2} \qquad (1)$$

2.2 Steering angular velocity and the modified coefficient of steering angular velocity

While measuring the vehicle steering angle, steering angular velocity can be calculated through difference calculations.

2.2.1 The modified coefficient of steering angular velocity
The modified coefficient of steering angular velocity is the ratio of actual steering angular velocity and theoretical steering angular velocity.

3 STEERING PERFORMANCE TEST SYSTEM

The measurement system is consist of National Instruments measurement system, memory revolving speed and torque tester and GPS-based steering track test instrument, the following Figure 2 shows the location of the sensors.

Memory revolving speed and torque tester is used for collecting the memory revolving speed and torque, the Figure 3 shows the location of it.

(a) Sensor on the driver   (b) Electronic guider   (c) National Instruments
                                                               measurement system

Figure 2.   Sensor and NI system.

Figure 3.   Memory revolving tester.

4 MEASUREMENT OF THE DATE AND THE RESULT ANALYSIS

4.1 *Synchronization and truncation of test date*

Four steps are useful for dealing with the date as follows:

1. Comparing the rotate speed signals, dates from memory revolving speed and torque tester and National Instruments measurement system are not synchro. Figure 4 shows that before synchronization, one of the date left behind, while after synchronization, they fit well.
2. Truncation of test date
   Figure 5 shows movement trajectory of the vehicle, we choose the velocity and angular velocity of armored vehicle which is symmetrical as the date for measurement.
3. Calculating the theoretical steering velocity, theoretical steering radius, theoretical steering angular velocity, actual steering radius and actual steering velocity. Figuring out the actual steering angular velocity using steering angle and sampling frequency.
4. Using the date we calculated, figure out the modified coefficient of steering radius, the modify coefficient of steering angular velocity, the slip, the spin, the traction, and the barking power.

4.2 *Result analysis of test date*

An example is that an armored vehicle is steering on sand testing road in the speed of 3 m/s. Figure 6 shows that while in different steering radiuses, the date curves are imitated though the modified coefficient of steering radius and the modified coefficient of steering angular velocity. Because of the spin and the slip, the actual steering radius is 1.4 to 1.6 times of the theoretical steering radius. The actual steering angular velocity is from 60% to 75% of the theoretical steering angular velocity.

Figure 7 shows that the imitated curves of steering radius changes according to the spin on the high-speed track and the slip on the low-speed track. The spin and slip are getting

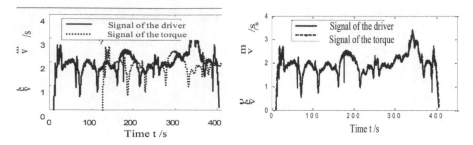

Figure 4.   Measurement of the test date.

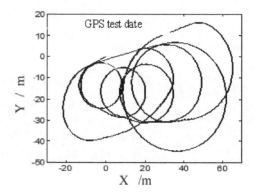

Figure 5.   Movement trajectory of the test.

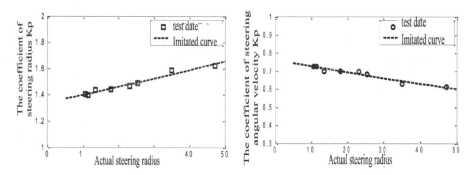

Figure 6.   Modified coefficient of steering radius and angular velocity.

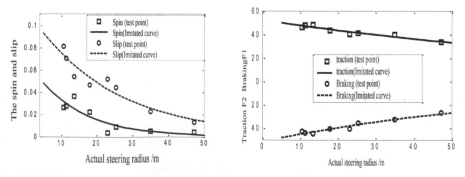

Figure 7.   The spin, slip, traction, barking power and actual radius.

424

weak while the steering radius getting bigger and the spin is smaller than the slip. That means, while the steering radius is small the spin and slip are much severe, the relationship between the traction, barking power and actual steering radius. The traction grows with the steering radius, while the barking power reduces with the steering radius. Furthermore, absolute value of the traction is bigger than the barking power. That means, while armored vehicle steering, the traction should be bigger than the barking power, and the smaller the steering radius is the bigger the traction and barking power are needed and steering became difficult.

5   CONCLUSION

1. Using the steering performance test system movement trajectory based on GPS, uninterrupted experiments of multiple steering radius for steering kinematics and dynamic parameters have been implemented, and the test efficiency and precision of the steering performance parameters have been improved significantly.
2. With the combination of National Instruments measurement system, memory revolving speed and torque tester and GPS-based steering track test instrument, finished the research on the acquisition of steering performance parameters of armored vehicle.
3. By analyzing steering kinematics and dynamic parameters, this paper provides a technique method and test date for the research of armored vehicle steering performance and theoretical steering model.

REFERENCES

Cheng, Jun wei. 2007. Study on steering of military tracked vehicle equipped with power-shift steering transmission. Beijing: *Academy of Armored Force Engineering*.

Cheng Jun wei, Gao,Lian hua, Wang Hong yan. 2006 Analysis of tracked vehicles steering considered tracks' skid and slip. *Journal of Mechanical Engineering*, 42 (B05):192–195.

Chi Yuan. 2013. Differential steering technology and theory of tracked vehicle. Beijing: *Chemical Industry Press*.

Ehlert W., Hug B., Schmid I.C. 1992. Field measurements and analytical models as a basis of test stand simulation of the turning resistance of tracked vehicles. *Journal of Terramechanics*, 29(1):57–69.

Fang, Zhi qiang, Gao Lian hua, Wang Hong yan. 2005. Indexed analysis and experimental research on the steering performance of tracked vehicles. *Journal of Academy Force Engineering*, 19(4): 47–50.

Said, A.M., Kaspar A., Lakmal D.S. 2008, Maneuverability performance of tracked vehicles on soft terrains [C]//2008 IEEE/RSJ Internal Conference on Intelligent Robots and Syntems Acropolis Convention Center Nice, France, September, 22–26:107–112.

Wei, Chen guan. 1980. Research on steering issues of tracked vehicle. *Tractor & Farm Transporter*, (1):17–35.

Wong, J.Y. 2001.Theory of Ground Vehicles. 3rd edition. New York: *John Wiley & Sons*: 388–430.

Wong, J.Y., Chiang C.F. 2001 A general theory for skid steering of tracked vehicles on firm ground. *Proceedings of the Institution of Mechanical Engineers, Part D: Journal of Automobile Engineering*, 215:343–355.

Zhu, Yan, Li Heyan, Ma Biao. 2011 Theoretical analysis and experimental research on renewable power hydraulic steering of tracked vehicle. *Machine Tool &Hydraulics*, 39(13):45–47.

Advances in Engineering Materials and Applied Mechanics – Zhang, Gao & Xu (Eds)
© 2016 Taylor & Francis Group, London, ISBN 978-1-138-02834-0

Research and development of on-line visual inspection and traceability system for robot glue spreading quality of car engine hood

R.F. Li, Y.Y. Wang & K. Wang
State Key Laboratory of Robotics and System, Harbin Institute of Technology, Harbin, China

G.L. Sun
FAW Car Co. Ltd., Changchun, China

ABSTRACT: An entire on-line visual inspection and traceability system was developed in order to increase the inspection speed of robot glue spreading quality of car engine hood and achieve the traceability of quality problem. Based on the robot glue spreading production line of welding workshop of FAW Car, this system built up a hardware platform and designed the on-line inspection and traceability software. A specific camera calibration algorithm was proposed aimed at the characteristic that the position and orientation of robot changed in real time during spreading the glue. The experimental results show that the detection precision can reach to the technical index, 0.3 mm, with the glue spreading speed, 10 m/min, and the detection time can satisfy the requirements of on-line real-time inspection.

1 INTRODUCTION

At the moment, automobile industry has become one of the most important application fields of industrial robots. Meanwhile, glue spreading process, as one major part of automobile production, is accomplishing the switch from handwork to robots.

The traditional inspection methods of glue spreading quality include manual inspection and off-line visual inspection. At present, the manual inspection has gradually quitted the historical stage, whereas the off-line visual inspection has dominated the inspection methods. Shanghai GM developed a set of off-line visual inspection system and applied it to the body workshop.

With increasingly intense competition among automobile industry, the traditional inspection technology has already been unable to satisfy the practical production requirement. Therefore, this paper develops a set of system that can achieve on-line real-time inspection of glue spreading quality to increase production efficiency, and adds quality information inquiry function to the inspection system so as to realize the control of glue spreading quality.

2 ON-LINE INSPECTION AND TRACEABILITY SYSTEM

On-line visual inspection and traceability system for robot glue spreading quality of car engine hood is mainly made up of visual hardware platform, and on-line inspection and traceability software. When system running normally, real-time cameras capture glue images and send them to system software for image processing. With technical requirements and relevant data of glue-robot system, the quality information including diameter of dot damping glue, width of strip damping glue and folding glue is analyzed. Then judge whether alarm signal is given out or not, and save necessary information to local database for quality tracing. Figure 1 shows the system framework of on-line inspection and traceability system for robot glue spreading quality.

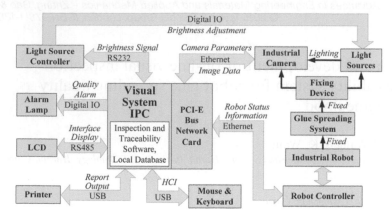

Figure 1.   On-line inspection and traceability system.

2.1   *Hardware platform of visual inspection system*

As shown in Figure 2, visual hardware platform mainly consists of two parts. One is the equipment and wiring circuit that are installed inside control cabinet. The other includes industrial cameras, optical lens, light sources and their fixing device which are fixed on the glue guns located at the end of robot. The equipment inside control cabinet contains IPC, LCD, printer, light source controller, alarm lamp, and the equipment of power circuit and trigger circuit.

Unlike the off-line visual inspection, in order to inspect in real time, cameras and light sources are fixed on the glue guns, and change their position and orientation along glue spreading track all the time. In addition, because the surface of the car engine hood inspected is complex and has warping areas, double cameras are used and detachable, adjustable flexible fixing device is designed to solve the problem that the glue is blocked by the warping areas or glue guns at some turning points during inspecting.

The fixing device of light sources, cameras and lens uses the supporting structure of dial gauge for reference. And based on it, the fixing device is improved a lot aimed at specific requirement. Ultimately, the demand of practical stability of light sources, cameras and lens in the process of glue spreading is satisfied while keeping the flexibility and adjustability of original structure.

2.2   *On-line inspection and traceability software*

On-line inspection and traceability software mainly consists of man-machine interactive interface, database operation module, communication module between IPC and robot controller, camera calibration program, and image processing and analysis program. Multithreading structure is adopted to achieve multitask parallel processing for higher reliability and efficiency.

Main interface of system software is designed with **MS Visual Studio 2010** for man–machine interaction. The interface is made up of two windows, visual inspection display and configuration window as well as glue spreading database query window.

MS SQL Server 2008 is used to establish and manage the local database. Meanwhile, **ADO** technology is adopted to access the SQL database established and conduct record operations. Some information at specific moments such as original images, processed images and analysis result needs to be saved in database for use of quality tracing and querying. Brightness of every light source and camera calibration information of every mark point also need to be stored for use when system running.

During the system running, current glue-robot position, mark point number and glue spreading status need to be learned in real time to call calibration data accurately for acquiring, processing and analyzing images with high speed. Because the information above can be obtained

Figure 2.   Hardware platform of visual inspection system.

only through the communication between IPC and robot controller, the robot communication module is developed based on DLL provided by the MOTOCOMES software.

Camera calibration program based on OpenCV first preprocesses the images of strip and dot calibration objects captured in the calibration process, then analyzes the ratios between number of pixels and physical distance of different areas in every mark point image, and saves relevant information into glue spreading database at last. The calibration algorithm in details is explained in Section 3.

Combining with corresponding data saved when calibrating, image processing and analysis program processes the images captured during system running and analyzes size information of folding and damping glue. Then judge whether glue spreading quality is up to standard and system main thread sends alerts or not. The algorithm in details is explained in Section 3.

3   IMAGE PROCESSING AND ANALYSIS ALGORITHM

The core goal of this system is to accomplish on-line real-time inspection of glue size information. As the theoretical basis to achieve the above goal, image processing and analysis algorithm is mainly made up of three parts including camera calibration algorithm, image preprocessing algorithm and glue size analysis algorithm.

3.1   *Camera calibration algorithm*

When analyzing physical size of the inspected objects in the images, the ratio between number of pixels and physical distance (namely projection ratio) of every image needs to be learned through camera calibration in advance.

Because the surface of car engine hood is complex and the glue is three-dimensional, specific calibration objects, whose sizes are known and which are similar to the glue, are made to improve calibration accuracy. And fix calibration objects in standard glue spreading position and acquire their images at every mark point of glue spreading track.

When processing images, as the background of calibration objects images is complex, it is adopted to clip ROI manually. And there are several kinds of interferences on the glue spreading scene and they might result in the loss of calibration accuracy, so filtering processing is necessary. The image test result shows that the Gaussian filter belonging to smoothing filter has the best filtering effect. As calibration objects are dark black, image binarization can be used to segment calibration objects from background. OTSU belonging to threshold method has the best effect for image segmentation. Binary ROI of calibration objects might contain defects like holes or fracture and need close operation belonging to morphological operation for modification.

As shown in Figure 3, for binary ROI of strip calibration objects, click centers of centroid extracting regions and set region sizes to extract the centers of calibration objects. Then conduct cubic Hermite interpolation and fitting of the obtained centroids. Search the edge

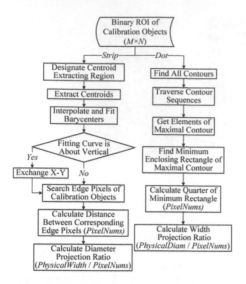

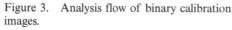

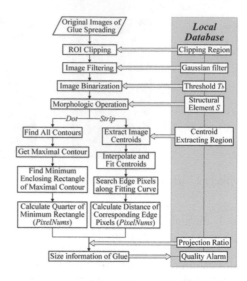

Figure 3.   Analysis flow of binary calibration images.

Figure 4.   Analysis flow of glue spreading images.

pixels of calibration objects by Sobel operator along normal direction of the fitting curves, and calculate number of pixels between corresponding edge points every other fixed distance. Finally, the ratios between pixel distance and physical width (namely width projection ratios) of different areas in every mark point image are figured out according to physical width of calibration objects.

When analyzing binary ROI of dot calibration objects, find maximal contour of the region. If ROI is chosen properly, the maximal contour is just peripheral contour of this calibration object. And then seek out minimum enclosing rectangle of the maximal contour. A quarter of perimeter of the rectangle is about the number of diameter pixels of this object. According to physical diameter of dot calibration objects, the ratios between pixel distance and physical diameter (namely diameter projection ratios) in every mark point image are worked out finally.

3.2 *Glue spreading image analysis algorithm*

Combining the ratios between pixel distance and physical distance in every mark point image, an image analysis algorithm can obtain size information of strip glue or dot glue just through figuring out the number of corresponding pixels in width or diameter direction. Meanwhile, as glue and calibration objects have similar size and shape and lie in the basically same position, the information obtained by calibration algorithm, such as ROI location, gray threshold, and centroid extracting region, can be stored in the database for use when processing and analyzing glue images in the normal running process. Therefore, the image analysis algorithm of glue spreading is greatly simplified, and its flow chart is roughly as shown in Figure 4.

4   RESULT AND ANALYSIS OF GLUE SPREADING TEST

In order to verify the effect and accuracy of this system during visual inspection, 10 of a set of glue spreading images (totally 195 images) acquired when system running normally are randomly chosen to conduct a test. First, according to shooting time and mark point serial number of the images, find size detection result of every image through database query window. Then use vernier caliper to accomplish the precision measurement of glue size on corresponding glue spreading positions of the images (3 measuring points per strip glue

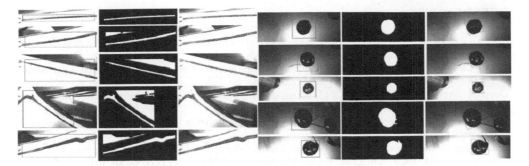

Figure 5. Glue spreading images for test.

Table 1. Inspection results and errors of test images (unit mm).

Image No.	1			2			3			4		
Measurement	1.85	2.12	1.86	1.75	2.33	2.40	1.72	1.98	2.20	2.41	1.58	1.76
Inspection	1.73	1.89	1.77	1.90	2.44	2.65	1.65	1.82	1.99	2.67	1.77	1.88
Error (Absolute)	0.12	0.23	0.09	0.15	0.11	0.25	0.07	0.16	0.21	0.26	0.19	0.12

Image No.	5			6	7	8	9	10
Measurement	2.23	3.21	4.27	16.82	17.02	12.05	19.73	17.42
Inspection	2.18	3.12	4.17	17.06	16.82	13.24	19.62	17.63
Error (Absolute)	0.05	0.09	0.10	0.24	0.20	0.19	0.11	0.21

image). Finally, make a comparison between measuring result and visual inspection result, analyze the detection error of visual system and judge whether its detection precision reaches technical requirement or not. Figure 5 shows the images for test, including original images, binary images and analysis images.

The results of precision measurement and visual inspection as well as their errors on corresponding glue spreading positions of the test images are shown in Table 1. After calculating the data of Table 1, it is known that the average absolute error between precision measurement and visual inspection is 0.1575 mm, the limit error is ±0.26 mm and the standard deviation of detection errors is 0.0647 mm. The analysis result shows that detection precision can satisfy the technical index, ±0.3 mm.

5 CONCLUSION

Aimed at the mission of on-line inspection and tracing of robot glue spreading quality, this paper has designed hardware platform mainly consisting of IPC, industrial cameras and machine vision light sources, and developed on-line inspection and traceability software containing user interactive interface, database operation module, camera calibration program and image processing and analysis program. The test result shows that this system can ensure real-time on-line visual inspection of glue spreading quality, improve detection efficiency, achieve tracing and inquiry of quality problem and has a good application prospect.

ACKNOWLEDGEMENT

The research was supported by National High-tech Research and Development Projects (863): 2013 AA040902-2.

REFERENCES

Cao, Tielin., Jiang, Tao & Guo, Zhiheng. 2006. Successful application of robot in windscreen gelatinizing system. *Robot Technique and Application* (2): 12–15.

Chen, Zailiang., Zou, Beiji & Huang, Minzhi. 2012. Influence of intensity feature on ROI extraction. *Journal of Central South University* 43(1): 208–214.

Deep, K., Arya, M & Thakur M. 2013. Stereo camera calibration using particle swarm optimization. *Applied Artificial Intelligence* (27): 618–634.

Huang, Jiwei. 2011. Application of robot bodywork gelatinizing system. *Automobile Technology & Material* (2): 51–56.

Wu, Jigang & Bin, Hongzan. 2010. Research and development of dimensional inspection system for thin sheet part based on machine vision. *Machine Tool & Hydraulics* (17): 86–88, 101.

Yin, Guofu. 2007. Image segmentation technology based on thresholding. *Modern Electronics Technique* (23): 107–108.

Zhu, Zhengde., Yang, Hong & Fang Lin. 2010. Machine vision: the 3rd eye of quality monitoring. *Modern Components* (10): 56–63.

Advances in Engineering Materials and Applied Mechanics – Zhang, Gao & Xu (Eds)
© 2016 Taylor & Francis Group, London, ISBN 978-1-138-02834-0

Research on the control method of hydro-viscous speed control system based on system identification

L.G. Ma & C.L. Xiang
School of Mechanical Engineering, Beijing Institute of Technology, Beijing, China

T.G. Zou & J.W. Zhang
China North Vehicle Research Institute, Beijing, China

ABSTRACT: In this paper, the transfer function of the system is identified with the least squares method. Based on the system identification theory and open loop characteristics of the hydro-viscous speed control system. A self-tuning PID controller parameter adjustment rule of the system is established based on the double optimization design method, using variable γ as a measure index of the response speed and the control smoothness. According to the requirements of the actual system, the optimal PID controller can be obtained by selecting the appropriate value of γ following the controller design method in this paper. The superiority of the control method is verified by the simulation analysis and bench test. The test results show that robustness of the system response and stability of the control is better when the speed input is higher.

1 INTRODUCTION

Hydro-viscous clutch is the inevitable choice for cooling fan system of tracked vehicle that has the advantages of stepless speed regulation, small size, high power transmission and high reliability. At present, fan speed control system of tracked vehicle is lack of a set of intelligent and practical control system to achieve its stepless speed regulation and high reliability. The ultimate goal is to obtain a stable output speed fan by adjusting the input current of the proportional valve to control the hydro-viscous clutch in the hydro-viscous speed control system. A mathematical model of hydro-viscous clutch is the most complex, which is generally difficult to establish its precise control model in the system. The dynamic characteristics are also relatively poor, so it is difficult to control the entire system.

To solve these problems, this paper adopts the system identification method to obtain the transfer function of the actual system and designs PID parameters adjustment rules using the bi-level optimization method. Furthermore, the response characteristics of the system are simulated using the AMESim. Then the control method is verified by the test bench.

2 SYSTEM PRINCIPLES

The hydro-viscous speed control system working principle diagram is shown in Figure 1. The system mainly consists of an oil supply mechanism, a regulating oil pressure mechanism, and a speed control mechanism. The supply oil mechanism includes tank 1, pumps 2, oil filter 3, constant pressure valve 4, regulating oil pressure mechanism includes electro-hydraulic proportional valve 5, ECU 7, speed control mechanism includes hydro-viscous clutch 6. The basic principle of hydro-viscous speed control system is a stable supply oil pressure is supplied by pump and constant pressure valve of supply oil mechanism, regulating oil pressure

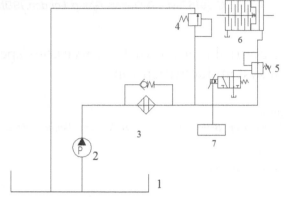

1. tank; 2. pumps; 3. oil filter; 4. constant pressure valve; 5. electro-hydraulic proportional valve; 6. hydro-viscous clutch; 7. ECU.

Figure 1.   Working principle diagram in hydro-viscous speed control system.

mechanism through the ECU control input current of electro-hydraulic proportional valve, thereby regulating the output oil pressure of electro-hydraulic proportional valve, speed control mechanism change output speed by changing the control oil pressure of hydro-viscous clutch.

3   SYSTEM IDENTIFICATION

In this paper, the system identification data mainly come from the range of the output speed controllable. The specific value has little effect to determine the system transfer function, so assuming the input speed to a certain value. The system open-loop response is obtained by setting the input speed of hydro-viscous clutch to 800 rpm and setting the amplitude of the input ramp voltage to 10 V.

In this paper, the mathematical model of hydro-viscous speed control system is identified by the least squares method. Assuming that it is the First Order Plus Time Delay (FOPTD). The input step amplitude is h, so the step response is:

$$y(t) = hk\left(1 - e^{-\frac{t-L}{T}}\right) + w(t), t \geq L \tag{1}$$

where $w(t)$ is white noise.

Integrate the formula (1) from $t = 0$ to $t = \tau$, the next equation can be obtained:

$$\int_0^\tau y(t)dt = hk\left(t + Te^{-(t-L)/T}\right)\Big|_L^T + \int_0^\tau w(t)dt \tag{2}$$

Assume:

$$A(\tau) = \int_0^\tau y(t)dt, \delta(\tau) = [Tw(t)]\Big|_L^T + \int_0^\tau w(t)dt$$

Then, the formula (2) can be written as:

$$\left[h\tau - h - y(\tau)\right]\begin{bmatrix} K \\ LK \\ T \end{bmatrix} = A(\tau) - \delta(\tau) \ t \geq L \tag{3}$$

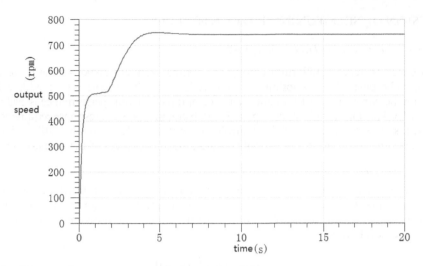

Figure 2. The open-loop step response characteristics of the system.

Then, formula (3) is written as a linear equations group:

$$\psi\theta = \Gamma + \Delta, \ t \geq L \tag{4}$$

Among formula (4):

$$\theta = \begin{bmatrix} K \\ LK \\ T \end{bmatrix}, \psi = \begin{bmatrix} hmT_s & -h & -y[mT_s] \\ h(m+1)T_s & -h & -y[(m+1)T_s] \\ \vdots & \vdots & \vdots \\ h(m+n)T_s & -h & -y[(m+n)T_s] \end{bmatrix}$$

$$\Gamma = \begin{bmatrix} A[mT_s] \\ A[(m+1)T_s] \\ \vdots \\ A[(m+n)T_s] \end{bmatrix}, \delta = \begin{bmatrix} -\delta[mT_s] \\ -\delta[(m+1)T_s] \\ \vdots \\ -\delta[(m+n)T_s] \end{bmatrix}$$

where T_s is the discrete sampling time; mT_s must be greater than or equal L.
Finally, the estimated value of θ is determined by a least square method:

$$\bar{\theta} = (\psi^T\psi)^{-1}\psi^T\Gamma \tag{5}$$

And then find the value of the parameter k, T, L.
The estimated value of θ is obtained by the least squares method combined with the open-loop step response curve.

$$\bar{\theta} = \begin{bmatrix} 2450 \\ 3535.4 \\ 0.267 \end{bmatrix}$$

Therefore, the identification of transfer function of the system is:

$$G(s) = \frac{2450}{0.267s+1}e^{-1.443s} \tag{6}$$

435

4   DESIGN OF SELF-TUNING PID CONTROLLER

4.1   *Design method of bi-level optimization*

In this paper, to reach equilibrium among the optimal transient response, the protection of actuator and robust, a bi-level optimization controller design framework is proposed, which can determine the parameters of **PID** controller. The optimal transient performance is evaluated in the lower level problem, while at the top of the problem, the transient response requirements are decreased, so that the changer of the controller output is smaller, thus when still maintain the transient performance within a pre-specified range, the impact effect of the actuator is reduced.

4.1.1   *Problem form*

In order to balance the trade-offs between transient performance and actuator preservation and to guarantee the robustness of the system, the controller design scheme is formulated into a two-step optimization problem, or a bi-level optimization problem, as follows:

The upper level problem:

$$\begin{cases} \rho^* = \arg\min_\rho \int_0^\infty \left[tu'(t) \right]^2 dt \\ \quad M_S < M_S^* \\ \quad \int_0^\infty \left[te(t) \right]^2 dt \le \lambda ISTE^* \end{cases}$$

The lower level problem:

$$\begin{cases} ISTE^* = \min_\rho \int_0^\infty \left[te(t) \right]^2 dt \\ \quad M_S < M_S^* \end{cases} \tag{7}$$

where $\rho = [K_p, T_i, T_d]$ is the vector of controller parameters.

In the lower level problem, the transient response is optimized to obtain the minimal value of the *ISTE* index. In the upper level problem, the total variations in the control variable are minimized to obtain the balanced controller. A constraint on the *ISTE* index is introduced in the upper level to ensure that the resultant balanced transient performance only degenerates within a specified extent, which is controlled by the weighting factor $\lambda > 1$, which also ensures that the transient response speed is still within a reasonable range. In fact, owing to the existence of stabilizing **PID** controllers, the lower level problem will be feasible and have finite optimal solution if M_s^* is properly selected. In this paper, M_s^* is chosen as 2, to ensure the feasibility of the problem, which also allows a 'faster' controller to be obtained in the lower level.

4.1.2   *Discussions on the weighting factor*

To achieve uniform trade-off between transient performance and actuator preservation for different processes, the following approach is employed. First, the properties of the *ISTE* optimal response and balanced response are exploited. These properties allow piecewise-affine approximations of the responses to be derived and the uniformness measures of the transient responses to be defined. Then, for any fixed value of the measure and for any process, the *ISTE* index of the balanced response with uniform extent of relaxation is predicted based on the approximation method and the corresponding optimal transient performance. Finally, the tuning rules of the weighting factor λ in (7) are proposed to restrict the actual *ISTE* index within this predicted bound.

Minimizing the *ISTE* index with respect to a step set point change usually corresponds to an approximate-monotone response with small overshoot and short settling time. Also, because the variations in the controller output are minimized and the *ISTE* index is restricted, the balanced controller obtained by solving (7) also corresponds to an approximate monotone step response with acceptable settling time. These properties provide the possibility to approximate the corresponding step responses using piecewise affine curves, which is illustrated in Figure 3.

In Figure 3, curve OABC denotes the approximation of *ISTE* optimal response and curve OADC denotes the predicted balanced response to be obtained by solving (7). The slopes of AB and AD are denoted as k_1 and k_2 respectively.

Owing to the approximate-monotonousness properties, the transient performance can be characterized by the rising speed of the response, which is defined as the average slope of the rising part, and the uniformness of the relaxation in transient performance can be evaluated by defining a dimensionless quantity $\gamma = \arctan k_2/\arctan k_1$ which can be regarded as the relative rising speed of the balanced response against the rising speed of the *ISTE* optimal response. Since $k_2 < k_1$, $\gamma \in (0,1)$, for set point tuning, by specifying the value of g, curve OADC can be predicted, and the maximal *ISTE* value allowed can thus be calculated, which leads to the tuning formula of λ in

$$\lambda = ISTE_{OADC}/ISTE_{OABC} = \frac{10L^3 + \dfrac{10L^2}{k_2} + \dfrac{5L}{k_2^2} + \dfrac{1}{k_2^3}}{10L^3 + \dfrac{10L^2}{k_1} + \dfrac{5L}{k_1^2} + \dfrac{1}{k_1^3}} \qquad (8)$$

where L denotes the time delay of the system, $ISTE_{OABC}$ denotes the *ISTE* value of the approximated *ISTE* optimal response and $ISTE_{OADC}$ denotes the predicted maximal *ISTE* value allowed in the upper level problem. Note that k1 can be estimated from the step response by solving the differential equations of the system using numerical methods.

4.2 *PID parameter adjustment rules*

For this model by system identification, PID tuning rules with respect to set point changes and load disturbances are separately studied. The transfer function of the FOPTD model is:

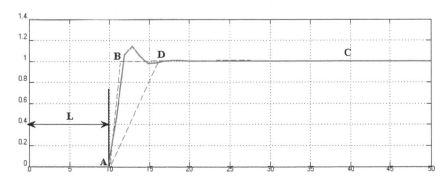

Figure 3.   Piecewise affine approximation of set point.

Table 1.   The PID parameter adjustment rules.

γ	Parameter adjustment rules		
0.4	$k_p = \begin{cases} 0.59\tau^{-0.92}/k; \tau \le 1 \\ 0.55\tau^{-0.53}/k; \tau > 1 \end{cases}$	$T_i = \begin{cases} T(1.1 + 0.15\tau + 0.55\tau^2); \tau \le 1 \\ T(1.09 + 0.23\tau); \tau > 1 \end{cases}$	$T_d = \begin{cases} 0.01T(0.4 + 0.35\tau - 0.13\tau^2); \tau \le 2 \\ 0^s; \tau > 2 \end{cases}$
0.6	$k_p = \begin{cases} 0.797\tau^{-0.92}/k; \tau \le 1 \\ 0.773\tau^{-0.47}/k; \tau > 1 \end{cases}$	$T_i = \begin{cases} T(0.979 + 0.32\tau + 0.124\tau^2); \tau \le 1 \\ T(1.015 + 0.396\tau); \tau > 1 \end{cases}$	$T_d = \begin{cases} T(0.02 + 0.17\tau - 0.05\tau^2); \tau \le 2 \\ T/(1 + e^{1.557\tau - 1.414}); \tau > 2 \end{cases}$
0.8	$k_p = \begin{cases} 0.797\tau^{-0.86}/k; \tau \le 1 \\ 0.773\tau^{-0.47}/k; \tau > 1 \end{cases}$	$T_i = \begin{cases} T(0.979 + 0.302\tau + 0.124\tau^2); \tau \le 1 \\ T(1.015 + 0.396\tau); \tau > 1 \end{cases}$	$T_d = \begin{cases} T(0.02 + 0.17\tau - 0.05\tau^2); \tau \le 2 \\ T/(1 + e^{1.557\tau - 1.414}); \tau > 2 \end{cases}$

$$G(s) = \frac{k}{Ts+1} e^{-Ls} \tag{9}$$

where K is the process gain, T is the time constant and L is the time delay. To derive simple tuning formulae for this model, the normalized time delay $\tau = L/T$ is employed, which has been found to have direct relations with controller parameters in the literature. To find the specific relations, problem (7) is solved repeatedly by varying the value of τ from 0.1 to 10 and the value of λ among 0.4, 0.6 and 0.8, for set point changes and load disturbances, respectively.

Using the least-square fitting technique, the obtained data are fitted to obtain the tuning formulae, which are summarized in Table 1. Note that each interpolation function has been piecewise devised by hand based on the above relations and the simulation results; the parameters of the functions are then optimized to minimize the total estimation error, so that the performance and stability properties obtained by solving (7) could be inherited by the tuning rules.

5 SIMULATION ANALYSIS

According to the model, $\tau = L/T = 1.443/0.267 = 5.4$. When $\tau > 2$, T_d is very small, so setting it to 0 in the paper. The control method of the system becomes the proportional-integral control. The parameters of the PI controller in different situations can be get based on different γ values as shown in Table 2 according to PID parameter adjustment rules in the Table 1.

According to the controller parameters in Table 2, self-tuning PID controller has added in the simulation system. The response characteristics of the output speed and control voltage can be obtained by the simulation of the system as shown in Figure 4.

As shown in Figure 4-a, when γ is larger, the response speed of the system is faster, while the system may be overshooting in a certain degree. As shown in Figure 4-b, with the γ decreasing, the intensity of changing in system control voltage is gradually reduced. In the actual system, it can make the amplitude and frequency of the electromagnetic valve spool

Table 2. The parameters of the controller.

γ	K_p	T_i	T_d
0.4	$9.3 \times e^{-5}$	0.6229	0
0.6	$1.4 \times e^{-4}$	0.8424	0
0.8	$1.7 \times e^{-4}$	0.7789	0

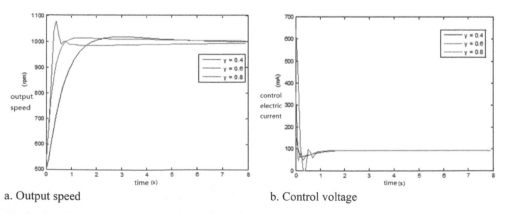

a. Output speed   b. Control voltage

Figure 4. The closed-loop step response simulation curve.

movement become smaller in order to increase the life of the elements and working stability of the system.

According to the simulation results in Figure 4, the superiority of self-tuning PID controller can be seen obviously. The measurement factor γ of the transient response speed and the stability of controller can be determined easily according to the response performance needed after identifying the mathematical model of the system. Then corresponding to different γ, the controller parameters of the system are calculated according to the formula in Table 1.

6 EXPERIMENTAL VERIFICATION

In this paper, the system step response is tested with a different γ value when the input speed is 1600 rpm and 2532 rpm respectively. The response characteristics of the system are verified by the γ value selected according to the test.

6.1 When the input speed is 1600 rpm

In this paper, the closed-loop control test is carried out when $\gamma = 0.8$, 0.6, 0.4 respectively and setting the target output speed to step signal at t = 0 which amplitude is 1200 rpm. The mathematical model is obtained after filtering and mathematical fitting through the software, as shown in Figure 5.

As shown in Figure 5, with the value of γ decreasing, the response speed of the system becomes slower, while the stability of the system is improved.

Because of the small control parameters of the system, the weak control performance and the low switching frequency, the smoothness of control is increased and a more balanced performance of the control system is obtained. The response of the system is verified by the given sine signal after getting the step response characteristic when $\gamma = 0.6$. The sine response of the system is shown in Figure 6.

As shown in Figure 6, with a given precision of the controller, the response speed and stability of system have been greatly improved. The response error mainly comes from the hysteresis phenomenon of hydro-viscous speed control system itself. The stability and desired response of the control system can be improved by improving the control algorithm only.

6.2 When the input speed is 2532 rpm

When the input speed is high, the controllable area of the system output speed is large, so the adjustment range of controller parameter is relatively large. The test is carried out still

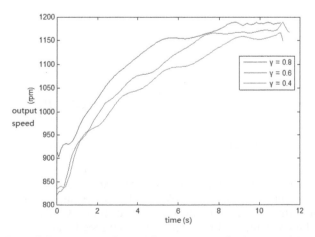

Figure 5.   Comparison of three groups of controller parameters in the step response.

439

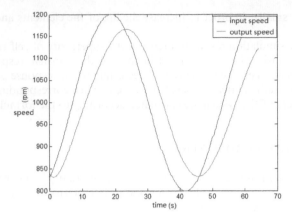

Figure 6. Sine signal response characteristics.

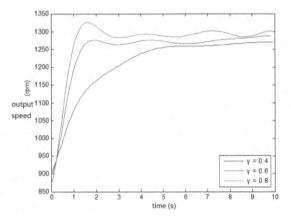

Figure 7. Comparison of three groups of controller parameters in the step response.

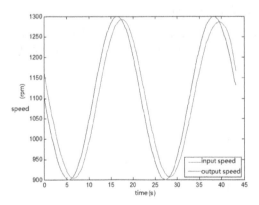

Figure 8. Sine signal response characteristics.

at $\gamma = 0.8, 0.6, 0.4$. The closed-loop control test is carried out by setting the step signal as the target output speed which amplitude is 1200 rpm. The test results are obtained after filtering and mathematical fitting through the software, as shown in Figure 7.

By comparing Figure 5 and Figure 7, when the input speed is at a high speed, the response speed of the system is greatly improved, even in the $\gamma = 0.8$ conditions which is the highest demand for transient performance, the control performance of the system is ideal, and the overshooting is small. It can meet the system requirements.

In this case, the system response is verified with sinusoidal signal when the controller parameters $\gamma = 0.8$, as shown in Figure 8.

As shown in Figure 8, when the input speed is at a high speed, the system response with sinusoidal signal is relatively better, the hysteresis phenomenon has reduced and the control stability is also improved. The effect of controller is obvious.

7 CONCLUSION

1. Transfer function of the system is obtained by the open-loop characteristic test and using the least square method. Then, a system simulation model is established on the basis of the transfer function.
2. Based on the model identification, the parameter adjustment rules of a self-tuning PID controller for hydro-viscous speed control system were developed by using a controller design method with bi-level optimization and proposing the variables γ can be used as a measure index of response speed and control smoothness which are verified through experiment.
3. The tuning rules of controller in the paper can provide a variety of freed choice. The controller can be designed according to the requirements of actual system for the transient performance and switching frequency. Selecting the appropriate γ values according to the actual system requirements, the best PID controller can be obtained following the controller design method in this paper, so as to achieve the best control effect.

REFERENCES

[1] Martin Lund. 2009. *Wet clutch performance and durability*. Lulea University of Technology. Sweden.
[2] Robert D. & Chalgren A. 2002. *Controlled EGR Cooling System for Heavy Duty Diesel Applications Using the Vehicle Cooling System Simulation*. SAE paper. (2):58–76.
[3] Ngy Srun Ap. 2005. *Ultimate Cooling TM System for New Generation of Vehicle*. SAE Paper. (1): 1–5.
[4] Seok-BoemLee. 2006. Closed-Loop Estimation of permanent Magnet Synchronous Mortor parameters by PI Controller Gain Tuning. *Transactions on energy conversion*. (13):863–870.
[5] Rikard Maki. 2005. *Wet clutch tribology-friction characteristics in limited slip differentials*. Lulea University of Technology. Sweden.
[6] P. Marklund & R. Larsson. 2007. Wet clutch under limited slip conditions-simplified testing and simulation. Proc. IMechE Vol. 221 Part J: *J. Engineering Tribology*.
[7] P. Marklund. 2008. *Wet clutch tribological performance optimization methods*. Lulea University of Technology. Sweden.
[8] Liu, R. Alleyne. 2000. A. Nonlinear force tracking of an electro-hydraulic actuator. *SAME Journal of Dynamic Systems, Measurement, and Control*. (122):232–237.
[9] SeokBoemLee. 2006. Closed-Loop Estimation of permanent Magnet Synchronous Mortor parameters by PI Controller Gain Tuning. *Transactions on energy conversion*. (13):863–870.
[10] Bin Yao & M. Tomizuka. 2001. Adaptive robust control of MIMO nonlinear systems in a semi-strict feedback form. *Automatica*. 37(9):1305–1321.

Advances in Engineering Materials and Applied Mechanics – Zhang, Gao & Xu (Eds)
© 2016 Taylor & Francis Group, London, ISBN 978-1-138-02834-0

Calculation method and testing verification of ground traction of tracked vehicle

S.L. Li, H.Y. Wang, B. Chen & Q. Rui
Department of Mechanical Engineering, Academy of Armored Force Engineering, Beijing, China

J. Guo
Science and Technology on Vehicle Transmission Laboratory, China North Vehicle Research Institute, Beijing, China

ABSTRACT: In this paper, the driving characteristics of tracked vehicles on soft ground, especially the properties of relationship between traction force and slip ratio when driving straightly on soft ground were analyzed. Aiming at the result of the ground pressure distribution of a tracked vehicle, a simplified pressure model was built and a calculation method about the ground traction exerted by the tracked vehicle is presented, the ground traction of each road wheel and the whole vehicle can be achieved with the parameters of soil carried out by test. The method is proved credible by the consistency between the result of on-vehicle experiment and the calculation result, the coherence suggests the feasibility of models, which makes a foundation for the calculation of driving load in the tracked vehicle.

1   INTRODUCTION

Ground traction of tracked vehicles is a key factor, which has an important influence on maneuverability of tracked vehicles. Research[1] shows that a load distribution under the track is the foundation to study the ground traction. Domestic and overseas researchers assume that the load distribution is a simple function[2–5]. These assumptions make the calculation easier, but the results differ greatly from the actual results. Researcher Rowland[1] did a large number of tests for tracked vehicles in different structure. From the testing results, we know the study of the ground traction is based on the test of load distribution. In this paper, a simplified track-ground pressure distribution is built based on actual testing. The ground traction can be achieved with the parameters of soil and vehicle. Tracked vehicle with same parameters is experimented to verify the model.

2   SIMPLIFIED MODEL OF LOAD DISTRIBUTION

2.1   *Test of load distribution on the track-ground interface of tracked vehicles*

The testing process and sensor are shown in Figure 1 and the parameters of the testing vehicles are shown in Table 1.

A typical ground pressure distribution is gained in the test as the blue line shown in Figure 2, we can see that max pressures are under the road wheels and the pressure distribution presents like a discrete and approximate triangle. A simplified track-ground load distribution is built based on the testing result as the black line in Figure 2.

(a) Pressure sensor (b) Laying ways

Figure 1. Pressure sensor and laying ways.

Table 1. Vehicle parameters.

Parameter	Value
Vehicle mass W/kg	20 380
Track-ground contact length L/cm	451
Track width b/cm	29
Track pitch l/cm	14.8
Numbers of road wheels on one side n	6
Road wheel diameter D/cm	60

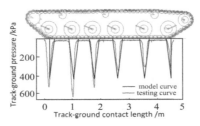

Figure 2. Contrast picture of model and testing result.

2.2 Model of track-ground pressure

Figure 2 is the contrast figure of the testing result and pressure centralizes in two track shoe width under road wheels. From the figure, we can see the pressure centralizes in the two track shoe width under road wheels is in accordance with the testing result.

Ground pressure in vertical direction balances gravity with tracked vehicle in low speed, so

$$\frac{1}{2} \times 2ab \times P_m \times 10 = Wg \tag{1}$$

where W = vehicle mass (kg); $2a$ = the base length of the triangle (cm); b = track width (cm); P_m = max pressure under road wheels (kPa); and g = acceleration of gravity (m/s²).

From Equation 1, we get $P_m = 474.9$ kPa.

From the Figure 2 and above analysis, we get $a = 1 = 14.8$ cm.

3   CALCULATION MODEL OF TRACTION FORCE

3.1   *Expression of traction force*

Soil shear stress and ground traction are a pair of acting force and reacting force. The testing process of them are same essentially, this indicates the curve of soil shear can change into the curve of traction force-slip ratio. According to soil shear stress-shear displacement equation put forward by Janosi and Hanamoto[6], soil shear stress can be expressed as follows:

$$F = F_m\left\{1-\exp\left(-\frac{j}{k}\right)\right\} = A(c + P_t\tan\varphi)\left\{1-\exp\left(-\frac{j}{k}\right)\right\}$$ (2)

where F = shear stress (N); F_m = maximum shear stress (N); j = shear deformation (cm); K = deformation constant of soil shear curve (cm); A = shear area (cm^2); c = cohesion (kg/cm^2); φ = angle of internal friction (degrees); P_t = normal pressure (kPa); τ = shear stress (kPa); and τ_m = maximum shear stress (kPa).

According to the equation of ground traction put forward by Komandi in his research, the traction curve's equation $T_i = f_i(i_0)$ under the ith road wheel is represented as follows:

$$T_i = T_{im}\left\{1-\exp\left(-\frac{i_0}{K_i}\right)\right\}$$ (3)

where T_i = ground traction due to soil shear under the ith road wheel (N); T_{im} = maximum value of T_i (N); i_0 = slip ratio of vehicle; K_i = deformation constant of ground traction curve.

3.2   *Ground traction under the 1th road wheel*

According to the ground traction curve shown in Figure 3, ground traction T_1 is given as follows:

$$T_1 = 2bc\left\{a - \frac{K}{i_0} + \frac{K}{i_0}\exp\left(-\frac{i_0 a}{K}\right)\right\} + 2bE\left\{\frac{a^2}{2} - \frac{Kl}{i_0} + \frac{K^2}{i_0^2}\exp\left(-\frac{i_0 a}{K}\right)\right\}$$ (4)

From Equations (3) and, (4), we get

$$\frac{T_{1m}}{K_1} = \frac{ba^2(3c + Ea)}{3K}$$ (5)

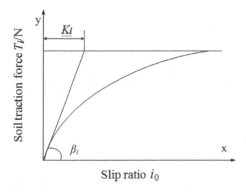

Figure 3.   Ground traction curve.

The maximum ground traction T_{1m} may be produced at $i_0 = 1$. Assuming that the exponential term of Equation (6) is negligible when $i_0 = 1$, T_{1m} is obtained approximately as follows:

$$T_{1m} = 2bc(a - K) + bE(a^2 - 2Ka + 2K^2) \tag{6}$$

From Equations (7) and (8),

$$K_1 = \frac{3K\{2c(a - K) + E(a^2 - 2Ka + 2K^2)\}}{a^2(3c + Ea)} \tag{7}$$

Equation (6) provides the relationship between deformation constants of the shear curve and ground traction curve. The ground traction curve can be obtained by substituting Equations (5) and (6) in Equation (3). Where K_1 is deformation constant of ground traction curve under the 1th road wheel.

3.3 Ground traction under the ith road wheel

According to the ground pressure shown in Figure 2 and as is calculated under the 1th road wheel, ground traction under ith road wheel is expressed as

$$T_i = 2bc\left\{2a + \frac{K}{i_0}\exp(-X) - \frac{K}{i_0}\exp(-Y)\right\} + 2bE\left\{a^2 + \frac{2K^2}{i_0^2}\exp(-Z) - \frac{K^2}{i_0^2}\exp(-X) - \frac{K^2}{i_0^2}\exp(-Y)\right\} \tag{8}$$

$$K_i = \frac{K}{x_i} \tag{9}$$

where $i == 2 \ldots 5$.

3.4 Ground traction under the nth road wheel

According to the ground pressure shown in Figure 2 and as is calculated under the 1th road wheel, ground traction under the nth road wheel can be expressed as

$$T_n = 2bc\left\{a + \frac{K}{i_0}\exp(-Z') - \frac{K}{i_0}\exp(-Y')\right\} + 2bE\left\{\frac{a^2}{2} + \frac{Ka}{i_0}\exp(-Z') + \frac{K^2}{i_0^2}\exp(-Z') - \frac{K^2}{i_0^2}\exp(-Y')\right\} \tag{10}$$

$$K_n = \frac{3K(2c + Ea)}{3c(2x_n - a) + El(3x_n - a)} \tag{11}$$

From the above-mentioned equations, and parameters of soil $c = 5.638$ kPa, $\varphi = 30.84°$, $K = 5.23$ cm the total ground traction is given and calculated as a function of slip ratio i_0 as follows:

$$T = \sum_{i=1}^{n} T_i = T_{1m}\left\{1 - \exp\left(-\frac{i_0}{K_1}\right)\right\} + \sum_{i=2}^{n-1} T_{im}\left\{1 - \exp\left(-\frac{i_0}{K_i}\right)\right\} + T_{nm}\left\{1 - \exp\left(-\frac{i_0}{K_n}\right)\right\} \tag{12}$$

4 TESTING VERIFICATION

Testing process of ground traction-slip ratio is shown in Figure 4. As shown in the figure, the testing vehicle is in the front and the brake vehicle is in the rear. Tensile sensor is linked between the two vehicles; changing the brake load in order to obtain different slip ratio of the

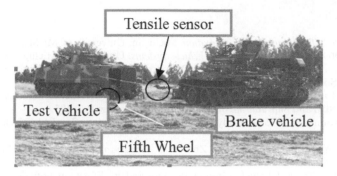

Figure 4.   Schematic diagram of testing process.

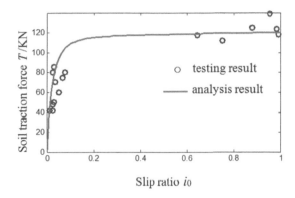

Figure 5.   Relationship between total ground traction and slip ratio i_0.

test vehicle. In the testing, the Fifth Wheel is used to measure the velocity of the test vehicle, optical electronic sensor is used to measure the speed of sprocket, and the tensile sensor is used to measure the traction force of the two vehicles. The slip ratio of the test vehicle is computed from the vehicle speed and the circumferential speed of sprocket. Motion resistance is 9.81 kN and the equation used in the testing is expressed as follows:

$$T = T_l + R \tag{13}$$

where T is ground traction of test vehicle; R is the motion resistance of the test vehicle; T_l is the traction load between the two vehicles.

Figure 5 shows the theoretical analysis and experimental results. From the results, we can see the result of on-vehicle experiment and the calculation result are in great consistency.

5   CONCLUSION

1. A simplified pressure model is built and a calculation method about the ground traction force exerted by a tracked vehicle is presented, the ground traction of each road wheel and the whole vehicle can be achieved with the parameters of soil and vehicle.
2. The model is proved credible by the consistency between the result of on-vehicle experiment and the calculation result, the coherence suggests the feasibility of models, which makes a foundation for the calculation of driving load in the tracked vehicle and can be used to calculate the driving performance of tracked vehicles.

REFERENCES

Bekker, M.G. 1962. *Theory of Land Locomotion*. Michigan: University of Michigan Press.

Bekker, M.G. 1969. *Introduction to Terrain-Vehicle Systems*. University of Michigan Press.

Janosi, Z, Hanamoto, B. 1961. The Analytical Drawbar-Pull as Function of Slip for Tracked Vehicle in Deformable Soil. *Mechanics of Soil-Vehicle Systems*: 707–736.

Jun-wei Cheng, Lian-hua Gao, Hong-yan Wang. 2006. Analysis of Tracked Vehicles Steering Considered Tracks' Skid and Slip. *Journal of Mechanical Engineering*: 192~195.

Kogure, K, Sugiyama, N. 1975. A Study of Soil Thrust Exerted by a Tracked Vehicle. *Terramechanics* 12: 225–238.

Ming-de Wang, Yu-qin Zhao, Jia-guang Zhu. 1983. *Theory of Tank locomotion*. Beijing: National Defense Industry Press.

Rowland, D. 1972. Tracked Vehicle Ground Pressure and its Effect on Soft Ground Performance. *Mechanics of Soil-Vehicle Systems*: 353–384.

Wong, J.Y. 2001. *Theory of Ground Vehicles*. New York: John Wiley & Sons Inc.

Condition assessment of hydraulic pump based on fuzzy comprehensive evaluation

C. Wu, Y.F. Peng, X.Y. Mao, H.Y. Ding, J.H. Hu & Y. Liu
Wuhan Second Ship Design and Research Institute, Wuhan, China

ABSTRACT: As it is difficult to assess the hydraulic pump condition accurately due to its affect by a number of factors, based on the analysis of the main fault forms of hydraulic pump, which shows its fuzziness. A fuzzy comprehensive evaluation method for the condition assessment of the hydraulic pump is proposed with the fuzzy theory. With the evaluation set V, the factor set U and the weight vector A, the mathematical model for the fuzzy comprehensive evaluation is built, which can assess the technical condition of hydraulic pump with the test data. The application examples show that the fuzzy comprehensive evaluation can deal with several fuzzy information, which can assess the technical condition of the hydraulic pump accurately.

1 INTRODUCTION

It is frequent to assemble and disassemble equipments at different project sites for construction enterprise. If the important equipments cannot work normally during installation, the project construction period will extend, which seriously impacts the economic benefits. As the output conversion device, the pump is the core component of the entire hydraulic system, thus, the hydraulic pump failure will make the entire hydraulic system to work abnormal[1]. The newly installed pump often fails to work in the enterprise, making it so important to assess accurately the technical condition of the hydraulic pump before installing for the construction enterprise.

The formation and development of the faults of the hydraulic pump is a gradual process. There is a long intermediate state in the equipment from good condition to fault condition, and this process of change is fuzzy[1]. There are many forms of pump faults, such as small output flow, low output pressure, large vibration and high oil temperature, while the size of output flow, the level of output pressure, the size of vibration and the level of oil temperature have no strict correspondence with the technical condition of the pump. When the test data of a hydraulic pump is just in the fuzzy interval between normal and fault values, it is difficult to make an accurate assessment of the technical condition of the pump directly. A mechanism to express and deal with a fuzzy logic is provided by the fuzzy mathematics theory, which can deal with the uncertain information and incomplete information in data, making it to become the important method and a tool to solve the fuzzy information[2].

2 MATHEMATICAL MODEL OF FUZZY COMPREHENSIVE EVALUATION

Based on the fuzzy set theory, a fuzzy comprehensive evaluation can quantify the various fuzzy information. Using a linear transformation principle and the principle of maximum membership to consider the various factors associated with the evaluation objects, the fuzzy comprehensive evaluation can unify the indicators to quantify and assign weight coefficients depending on the degree of influence on indicators to evaluation objects, which can make a reasonable and comprehensive evaluation[3]. When applying this method to deal with practical engineering problems, a reasonable mathematical model has to be established to use all kinds

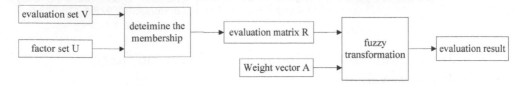

Figure 1.   Evaluation process of fuzzy comprehensive evaluation.

of information, because the incorrect mathematical model can cause errors in evaluation[4]. The mathematical model of fuzzy comprehensive evaluation is composed of the evaluation set V, the factor set U, the weight vector A and the evaluation matrix R, and its evaluation process is shown in Figure 1.

1. Assuming that there are n evaluation rating and m considerations, the evaluation set $V = \{v_1, v_2 \ldots v_n\}$, and the factor set $U = \{u_1, u_2 \ldots u_m\}$.
2. The fuzzy relationship between evaluation set V and factor set U is shown in the evaluation matrix R as follows:

$$R = \begin{bmatrix} r_{11} & r_{12} & \cdots & r_{1n} \\ r_{21} & r_{22} & \cdots & r_{2n} \\ \vdots & \vdots & & \vdots \\ r_{m1} & r_{m2} & \cdots & r_{mn} \end{bmatrix} \qquad (1)$$

In formula (1), r_{ij} represents the membership of the i-th factor to the j-th rating, requiring $0 \leq r_{ij} \leq 1$, $1 \leq i \leq m$ and $1 \leq j \leq n$.
3. According to the influence of each factor to the evaluation results, the weight vector A is determined as follows:

$$A = [a_1, a_2, \ldots, a_m] \qquad (2)$$

In formula (2), a_i is the weight coefficient, which represents the importance of the i-th factor, requiring $\sum_{i=1}^{m} a_i = 1$, and $0 \leq a_i \leq 1$.
4. The synthesis algorithms based on weighted average is used to carry out the fuzzy transformation. The evaluation result is calculated as follows:

$$B = A \circ R = [a_1 \quad a_2 \quad \cdots \quad a_m] \circ \begin{bmatrix} r_{11} & r_{12} & \cdots & r_{1n} \\ r_{21} & r_{22} & \cdots & r_{2n} \\ \vdots & \vdots & & \vdots \\ r_{m1} & r_{m2} & \cdots & r_{mn} \end{bmatrix} = [b_1 \quad b_2 \quad \cdots \quad b_n] \qquad (3)$$

In formula (3), $b_j = \min\left\{1, \sum_{i=1}^{m} a_i r_{ij}\right\}$.
5. According to the principle of maximum membership of fuzzy mathematics, identify the largest b_j in the evaluation results matrix B, which can determine the evaluation rating.

3   FUZZY COMPREHENSIVE EVALUATION OF HYDRAULIC PUMP CONDITION

3.1   *Establish the evaluation set V*

Too many evaluation rating of hydraulic pump may cause that there are only a little differences among the b_j in the evaluation result matrix B, resulting in inaccurate evaluation of their technical condition, which illustrates a reasonable number of evaluation rating is very important. Combined with the actual situation of this enterprise, the number of technical condition rating of the hydraulic pump is determined as 3, as shown in Table 1.

So, the evaluation set $V = \{v_1, v_2, v_3\}$.

3.2 *Establish the factor set U*

By accessing the relevant information, and combining with actual experience of the pump failure in the enterprise, the main fault forms of a hydraulic pump are low output pressure, small output flow, large vibration and high oil temperature. Therefore, the factor set U is composed of output pressure u_1, output flow u_2, vibration u_3 and oil temperature u_4, which makes $U = \{u_1, u_2, u_3, u_4\}$.

3.3 *Establish the membership function*

Membership function is the basic concept of fuzzy mathematics, the definition of which is that: assuming that A is a fuzzy subset of the domain U, and for any $x \in U$, specify a real value $f_A(x) \in [0,1]$, which is the membership of $\times$ to A, and the mapping $f_A(x)$ is called the membership function[5]. The fuzzy subset is entirely described by its membership function, which can quantify the fuzzy set and analyze and deal with fuzzy information using precise mathematical methods. Several ways are used to determine the membership function, which should be determined separately based on the characteristics of each factor. Take the most common type of a pump in this enterprise for example, the membership function of each factor is determined with its factory technical parameters as the fundamental basis, as shown in Table 2.

Set $f_{ij}(x)$ denotes the membership function of the i-th factor to the j-th rating, according to the data in Table 2, each membership function is as follows:

1. Membership function of the output pressure u_1:

$$f_{11}(x) = \begin{cases} 1 & x \geq 18.5 \\ 1 + \dfrac{1}{(18.5-18)}(x-18.5) & 18 < x < 18.5 \\ 0 & x \leq 18 \end{cases}$$

$$f_{12}(x) = \begin{cases} 1 - \dfrac{1}{(18.5-18)}(x-18) & 18 < x < 18.5 \\ 1 & 17 \leq x \leq 18 \\ 1 + \dfrac{1}{(17-16.5)}(x-17) & 16.5 < x < 17 \\ 0 & x \geq 18.5, x \leq 16.5 \end{cases}$$

$$f_{13}(x) = \begin{cases} 0 & x \geq 17 \\ 1 - \dfrac{1}{(17-16.5)}(x-16.5) & 16.5 < x < 17 \\ 1 & x \leq 16.5 \end{cases}$$

Table 1. Technical condition rating of hydraulic pump.

Rating	Condition
Good (v_1)	Can work normally, and need common maintenance
General (v_2)	Can work normally, but need increased maintenance efforts
Faulty (v_3)	Work abnormally, need repair

451

Table 2. The relationship of the main parameters and technical condition.

Condition	Parameter			
	Output pressure (u_1)/MPa	Output flow (u_2)/L/min	Vibration (u_3)/mm/s	Oil temperature (u_4)/°C
Good (v_1)	≥ 18.5	≥ 145	≤ 2	≤ 50
General (v_2)	17~18	130~140	3~5	55~65
Faulty (v_3)	≤ 16.5	≤ 125	≥ 6	≥ 70

2. Membership function of the output flow u_2:

$$f_{21}(x) = \begin{cases} 1 & x \geq 145 \\ 1 + \dfrac{1}{(145-140)}(x-145) & 140 < x < 145 \\ 0 & x \leq 140 \end{cases}$$

$$f_{22}(x) = \begin{cases} 1 - \dfrac{1}{(145-140)}(x-140) & 140 < x < 145 \\ 1 & 130 \leq x \leq 140 \\ 1 + \dfrac{1}{(130-125)}(x-130) & 125 < x < 130 \\ 0 & x \geq 145, x \leq 125 \end{cases}$$

$$f_{23}(x) = \begin{cases} 0 & x \geq 130 \\ 1 - \dfrac{1}{(130-125)}(x-125) & 125 < x < 130 \\ 1 & x \leq 125 \end{cases}$$

3. Membership function of the vibration u_3:

$$f_{31}(x) = \begin{cases} 1 & x \leq 2 \\ 1 - \dfrac{1}{(3-2)}(x-2) & 2 < x < 3 \\ 0 & x \geq 3 \end{cases}$$

$$f_{32}(x) = \begin{cases} 1 + \dfrac{1}{(3-2)}(x-3) & 2 < x < 3 \\ 1 & 3 \leq x \leq 5 \\ 1 - \dfrac{1}{(6-5)}(x-5) & 5 < x < 6 \\ 0 & x \geq 6, x \leq 2 \end{cases}$$

$$f_{33}(x) = \begin{cases} 0 & x \leq 5 \\ 1 + \dfrac{1}{(6-5)}(x-6) & 5 < x < 6 \\ 1 & x \geq 6 \end{cases}$$

4. Membership function of the oil temperature u_4:

$$f_{41}(x) = \begin{cases} 1 & x \le 50 \\ 1 - \dfrac{1}{(55-50)}(x-50) & 50 < x < 55 \\ 0 & x \ge 55 \end{cases}$$

$$f_{42}(x) = \begin{cases} 1 + \dfrac{1}{(55-50)}(x-55) & 50 < x < 55 \\ 1 & 55 \le x \le 65 \\ 1 - \dfrac{1}{(70-65)}(x-65) & 65 < x < 70 \\ 0 & x \ge 70, x \le 50 \end{cases}$$

$$f_{43}(x) = \begin{cases} 0 & x \le 65 \\ 1 + \dfrac{1}{(70-65)}(x-70) & 65 < x < 70 \\ 1 & x \ge 70 \end{cases}$$

According to the above equations, the membership function curves are shown in Figures 2 to 5.

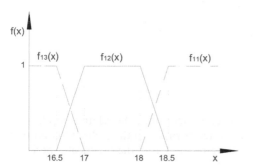

Figure 2. Membership function of u_1.

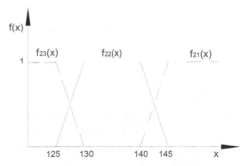

Figure 3. Membership function of u_2.

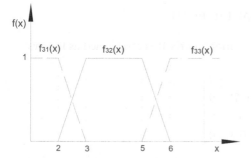

Figure 4. Membership function of u_3.

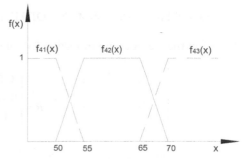

Figure 5. Membership function of u_4.

Table 3. The average value of test data.

Parameter	Out pressure (MPa)	Output flow (L/min)	Vibration (mm/s)	Oil temperature (°C)
Average value	18.3	143.6	1.5	51.5

3.4 Obtain the membership

In the special test facilities, one hydraulic pump is selected to carry out the technical condition assessment, the data of the output pressure, output flow, vibration and oil temperature are collected with a sampling frequency of 1 Hz. The average value of the above data is used to assess the condition of the hydraulic pump. After the mathematical calculation, the test data are shown in Table 3.

Take the data in Table 3 into the membership functions in section 2.3, and the membership are obtained as follows:

The membership of the output pressure to each evaluation rating:

$$r_{11} = 0.6, r_{12} = 0.4, r_{13} = 0;$$

The membership of the output flow to each evaluation rating:

$$r_{21} = 0.72, r_{22} = 0.28, r_{23} = 0;$$

The membership of the vibration to each evaluation rating:

$$r_{31} = 1, r_{32} = 0, r_{33} = 0;$$

The membership of the oil temperature to each evaluation rating:

$$r_{41} = 0.7, r_{42} = 0.3, r_{43} = 0.$$

3.5 Determine the weight vector

Experts with rich experience in engineering are invited to determine the relative importance of the output pressure, output flow, vibration and oil temperature in the technical condition assessment of hydraulic pump with the expert scoring method, and the practical experience in the enterprise is combined to determine the weight vector.

$$A = [0.305\ 0.336\ 0.202\ 0.157]$$

4 CONDITION ASSESSMENT OF HYDRAULIC PUMP

Based on the membership in section 2.4, the evaluation matrix R is established as follows:

$$R = \begin{bmatrix} 0.6 & 0.4 & 0 \\ 0.72 & 0.28 & 0 \\ 1 & 0 & 0 \\ 0.7 & 0.3 & 0 \end{bmatrix}$$

The fuzzy transformation is carried out with Equation (3):

$$B = A \circ R = [0.305 \quad 0.336 \quad 0.202 \quad 0.157] \circ \begin{bmatrix} 0.6 & 0.4 & 0 \\ 0.72 & 0.28 & 0 \\ 1 & 0 & 0 \\ 0.7 & 0.3 & 0 \end{bmatrix} = [0.737 \quad 0.263 \quad 0]$$

According to the principle of maximum membership of fuzzy theory, the technical condition of this hydraulic pump belongs to rating v_1, which means the condition is good, so the hydraulic pump can be transported to the construction site for installation directly, and just need common maintenance. Through keeping track with the construction site, the hydraulic pump keeps working in good condition all the time, which shows that the conclusion made by the fuzzy comprehensive evaluation and the actual technical condition of the pump is consistent. With this method, the technical condition of the other 10 pumps with the same model in the enterprise is assessed, and the accurate evaluation results are presented.

5 CONCLUSIONS

1. In this paper, the fuzzy comprehensive evaluation method is used to quantify the fuzzy information in hydraulic pump fault, which can give comprehensive analysis and accurate assessment of the technical condition of hydraulic pump, indicating a large application value in this field.
2. Through establishing the membership functions of the main parameters, the final evaluation matrix is determined with the membership calculated by the test data. The membership is determined by the means of functions, which can guarantee the reliability of the value in evaluation matrix in order to accurately reflect the real technical condition of the hydraulic pump.
3. The factor set contains the four most important parameters of the hydraulic pump, which can make a correct assessment. In order to meet the higher demand assessment, the pressure pulsation, the amount of leakage and other technical parameters can be added in the factor set U to make a more comprehensive and accurate evaluation.

REFERENCES

[1] Hu K.L. 2003. The study of the fuzzy synthetical judgment in the axial piston pump. *Lanzhou University of Technology.*
[2] Cao F. 2010. Power transmission projects decision making based on two-level fuzzy comprehensive evaluation. *Water resources and power* (9):145–147.
[3] Wang Z.G. 2004. Fuzzy comprehensive evaluation method of wind power generation unit. *Acta energiae solaris sinica* (2):177–181.
[4] Ji L.X. 2012. The research on application of fuzzy mathematics in purchasing large medical equipment. *China Medical Equipment* (1):22–25.
[5] Zhang S.Z. 2011. Fuzzy comprehensive evaluation for songhua river basin appraisal of the quality. *Environmental science and management* (3):163–165.
[6] Zhong, W.F. 2004. Fuzzy Control of Electro-hydraulic Position Servo System. *Machine Tools & Hydraulics* (07):86–87.
[7] Fu Y.L. 2007. Application of Sliding Mode Fuzzy Control in Robot. *China Mechanical Engineering* (10):1168–1170.
[8] Pan Y.P. 2007. A Novel Method of Model Reference Adaptive Fuzzy Control in Hydraulic Servo System. *Machine Tools & Hydraulics* (04):120–122.
[9] Dong Y.H. 2004. The Fuzzy Control Simulation of Hydraulic Force Servo Manipulator Based on Matlab. Machine Tools & Hydraulics (08):85–87.
[10] Zhang F.J. 2010. Application of Genetic Algorithm in Hierarchical Fuzzy Control of Electro-hydraulic Servo System. *Process Automation Instrumentation* (02):39–42.

Advances in Engineering Materials and Applied Mechanics – Zhang, Gao & Xu (Eds)
© 2016 Taylor & Francis Group, London, ISBN 978-1-138-02834-0

Application of fault tree analysis diagnosis on the gearbox of certain type of self-propelled artillery

J.M. Yang, Z. Zhang, Re.B. Zhou & C.F. Liu
Wuhan Mechanical College, Wuhan, Hubei, China

ABSTRACT: In this paper, an analysis has been made on the causes of difficulty and failure in gear engaging of certain type of self-propelled artillery before constructing the fault tree. Based on the results of the qualitative analysis of the fault through calculation of minimum cut sets an effective diagnosis, and implementation plan is introduced by following the principles, that is "Inspection before the test, complexity before simplicity, judgment before disassembly and outside before inside". This method has the advantages of simplicity, visualization, and availability.

1 INTRODUCTION

The gear box is an important component of a certain type of self-propelled gun, which is composed of a box body, a driving shaft assembly, the intermediate shaft assembly, spindle assembly, reverse gear shaft assembly and gearshift etc. (1), which integrates machine, electricity and fluid technology with high technology and complex structure. Coordination among each part and component is required, for the bad performance of any assembly will lead to the fault. Therefore, it is more economically beneficial and secure to make timely and accurate judgment of fault and its position, while finding the cause and suitable solution without disintegrating the gearbox case, which can reduce the blindness of repair and enjoy a promising application.

2 FAULT TREE; FAULT DIAGNOSIS[2–3]

Fault tree, also called negative analysis tree, is a kind of logical causality diagram. Logic analysis should be made on various factors (including hardware, software, environment, etc.) that may cause failure symptoms before forming a dendritic graphic refined from the whole to the part layer by layer, which is called as fault tree. And the fault tree analysis is used a method to identify the basic fault, determine the fault reason, fault impact and probability of occurrence and to analyze the reliability of the system in form of the fault tree.

Fault tree analysis is a systematic method of deduction with studying lest favorable fault condition (top event) in the system as the goal of fault analysis. All the possible reasons will be tracked back layer by layer downward until those original factors (basic events) that no longer need to be delved into. All of these events are represented by appropriate symbols, and through logic doorknob top events, intermediate events and basic events are connected into an inverted tree graphic.

3 FAULT ANALYSIS OF TRACKED VEHICLE GEARBOX[1]

The gearbox is a very complex system, in which many faults often appear such as difficulty or failure in gear engagement, automatic gear disengagement, double gear engagement, oil spilling, and abnormal sound within gear box. In practice, the gearbox failure is often caused by one or more than two fault, so fault diagnosis should start by analyzing main phenomenon of the faults based on the main technical performance standards for gear box. And there are too many malfunctions to make a whole analysis on the system. Therefore, a systematic analysis is undertaken only on the common and typical fault phenomena.

4   ESTABLISHING A FAULT TREE

In this paper, a certain type of self-propelled gun gearbox is taken for an example with difficulty or failure in gear engagement as the top event, assuming there are only two states (working and fault) for each part of the system for analysis, and the basic fault events of each component is independent to each other. Through manual manner, the fault tree establishment starts from top events till the end event. Figure 1 is the fault tree for the gear-engaging malfunctions of a type of self propelled artillery.

Intermediate event including M1: fault in speed change device of gearbox, M2: Breakdown in functional unit of gearbox, M3: Incompleteness in gear releasing of the main clutch.

The basic event X_1: some pins of the synchronizer break off, X_2: a slide shaft on the shifting fork break off, X_3: burr in inner and outer ring gear of driven gear and sliding connection gear sleeve, X_4: bending or break off of locator spring or spring tube wearing down, X_5: burrs or serious wear and tear in the shaped hole of Synchronizer, X_6: the wire rope of the locking device is too long, the X_7: burrs or bending in drive rod, X_8: vertical shaft bracket becomes flexible or prop breaks, X_9: fixation of transmission becomes flexible, X_{10}: the vertical axis is short of oil or dirty, X_{11}: rod is jammed, X_{12}: bearing of transmission drive shaft becomes severely worn, X_{13}: the fixed nut of main clutch becomes flexible, X_{14}: warp and sintering in friction plate, X_{15}: increasing false stroke of the main clutch control device, X_{16}: too large free stroke and too small pressure plate stroke due to improper installation and adjustment.

5   QUALITATIVE ANALYSIS OF FAULT TREE

In fault tree analysis, if the collection of a few bottom events of the fault tree happens at the same time, and leads to a top event, this collection is then called cut sets, a kind of system failure mode. The minimum set of failure modes that will lead to system faults is called a minimum cut set[4]. In other words, the minimum cut set is the set containing the minimum

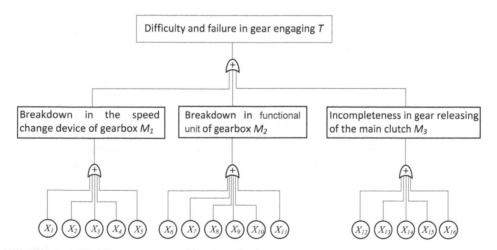

Figure 1.   A certain type of self-propelled gun gearbox fault tree blocking hanging difficulties or do not hang on the block.

number of essential bottom events, which depict for us the basic fault that must be repaired in the system with malfunctions. Through analysis on the minimum cut sets, the weak links of the system can be found for improving the reliability of the system. Therefore, the qualitative analysis by means of fault tree analyzing, first of all, requires all the minimum cut sets of fault trees, which could be worked out through ascending method and descending method[1]. In this paper, by using the ascending method in this case we find all the minimum cut sets are all the bottom events, including all the minimal cut sets

$$\{X_1\},\{X_2\},\{X_3\},\{X_4\},\{X_5\},\{X_6\},\{X_7\},\{X_8\},\{X_9\},\{X_{10}\},\{X_{11}\},\{X_{12}\},\{X_{13}\},\{X_{14}\},\{X_{15}\},\{X_{16}\}.$$

The minimum cut sets are independent to each other. No matter where there is one cause of the malfunction, the top event will occur.

6 DESIGN ON FAULT DIAGNOSTIC SCHEME

Through establishing the fault tree, all possible reasons of the faults that cause difficulty and failure in gear engagement may be found, on the basis of which we can work out practical fault diagnostic scheme. And implementing plan of fault diagnosis is made in line with the structure and principle of the gearbox and the diagnosis principle namely "Inspection before the test, complexity before simplicity, judgment before disassembling and outside before inside".

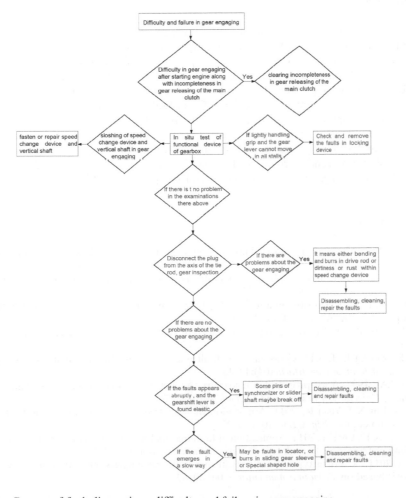

Figure 2. Process of fault diagnosis on difficulty and failure in gear engaging.

If the fault is only difficulties in gear engagement at engine starting or working with incomplete releasing of the main clutch at the same time, we should first eliminate the possibility of incomplete releasing of the main clutch. Make an in situ check on functional device of gearbox. Lightly hold the grip handle, check and repair the faults in locking device if the gear lever cannot move in all gears; watch the transmission and the vertical axis at gear engagement, and fasten or repair the one with movement. If there are no problems in the check there above, then remove the tie rod from the fork shaft arm and make a check on gear engagement. If the fault continued, it shows that the possibility of burr or bending on the transmission rod or the transmission is too dirty or rusty, which calls for disassembly, cleaning and repair; if not the vertical axis must be short of oil or dirty, calling for oiling or cleaning. If the gear works after disconnecting the tie rod, the fault should be inside the gear box. If the faults appear abruptly and the gearshift lever is found elastic, some pins of synchronizer or slider shaft may break off, which requires suspending and repair. If the fault emerges in a slow way, the faults may lie in locator, or there are burrs in sliding gear sleeve or special shaped hole, which requires disassembling, check and repair.

Based on the above analysis we establish a "process tree" of fault diagnosis on difficulty or failure in gear engagement, shown in Figure 2. According to which, we can locate the fault position and complete troubleshooting in a rapid and accurate manner.

7 CONCLUSION

Through establishing the fault tree, a list can be made on all reasons causing difficulty or failure in gear engagement of the gearbox of a certain type of self-propelled artillery, combined with the actual experiences in diagnosis and the corresponding auxiliary examination, accurate judgments can be made on the location of the faults, which helps us master the rules of faults and system features, providing comprehensive and reliable information for timely troubleshooting and effective practice in preventing accidents. Other faults in the gearbox of the self-propelled artillery can also apply same measure by establishing fault tree, listing all the reasons of the fault and grasping the rules and characteristics of the faults in gearbox. Moreover, through the fault tree analysis theory further organic combination can be achieved between conventional fault diagnosis method and computer technology, forming expert system, providing more effective fault diagnosis on gearbox or a certain type of self-propelled artillery.

REFERENCES

[1] Bai Y. 2010. Fault diagnosis on the abnormal sound in automatic gearbox of automotive based on fault tree analysis. *Journal of Changzhou Institute of Technology* (12):30–34.
[2] Jiang Y.N, Lou Y.H. 2001. Application of fault tree analysis in fault diagnosis of gasoline engine. *small internal combustion engine* (01):45–48.
[3] Prashad H. 1984. Diagnostic monitoring of rolling-element bearings by high-frequency resonance technique. *Asle Transactions* (04):439–448.
[4] Tang L.W. 1998. Self detection and fault diagnosis of gear box gun. *of Ordnance Engineering College* (1):3–12.
[5] Qu Y.K, Zhong L.H. 2011. Guns based on fault tree analysis method for analysis of servo system fault risk. *Sichuan armamentaria* (04):40–43.
[6] Zhang Z.M. 1997. A self-propelled gun gearbox and method for fault diagnosis of engine research. *Huazhong University of Science and Technology* (01):2–7.
[7] Chen B, Liu X.J, Yuan R. 2011. Application of fault tree analysis method in fault diagnosis of marine main engine. *Ship & Boat* (03):35–37.
[8] He Y, Cao X.D, Liu H.F. 2011. Application of fault tree analysis method in intelligent electromagnetic descaling instrument in fault diagnosis. *Instrument technology* (07):61–70.
[9] Hu L.J. 2011. A certain type of antiaircraft gun fault tree analysis method of servo system fault diagnosis based on. *Sichuan armamentaria* (04):40–43.

Advances in Engineering Materials and Applied Mechanics – Zhang, Gao & Xu (Eds)
© 2016 Taylor & Francis Group, London, ISBN 978-1-138-02834-0

Design of automobile active suspension controller based on genetic algorithm

D. Zhao & Z.Q. Tang
School of Automobile, Chang'an University, Xi'an, China

ABSTRACT: In the view of linear quadric optimal control method exists insufficiency in designing the active suspension control, this paper based on the theory of linear quadratic optimal control method, putting forward a design method that puts a genetic algorithm and the linear quadratic together to design and optimize the active suspension controller, namely the automobile body vertical acceleration, automobile suspension dynamic schedule and wheel dynamic load as the control target. And to optimize the optimal control rules of vehicle active suspension, so as to improve the control effect of active suspension controller. In order to show the feasibility of the optimization method, the design is based on the automobile single wheel model of active suspension control strategy, the simulation results show that an active suspension based on the genetic algorithm for the improvement of car's comfort and stability is significant.

1 INTRODUCTION

Vehicle active suspension system is closely related to vehicle's reliability, comfort, stability and safety factor[1]. After Federpid-Labrosse puts forward how to design the vehicle active suspension problems in 1995, study on the active suspension system and how to achieve better control performance have attracted many scholars to study it in the community[2]. In recent years, with the rapid development of the control theory, industrial automation, power electronics and computer technology, the performance of automotive active suspension has been significantly improved, but how to design stability, reliability better active suspension system controller is still one of the core technology. At present, how to apply the theory to practice, scholars have done a lot of research, for example, the linear quadratic optimal control, genetic algorithm control, adaptive control, synovial variable structure control, neural network control and the traditional PID control, etc[3].

Now the stochastic linear quadratic optimal control theory has been widely used in the process of vehicle active suspension; however, according to the stochastic linear quadratic optimal theory, local optima rather than the global optimal solution can be obtained[4]. Because the control methods use a state feedback control, this method requires that the system has a high sensitivity and it should use high-frequency servo components, but due to random interference it also may produce oscillations and misuse. Therefore the design of the control system is costly, less reliable and stable, which is according to this theory. In order to improve the deficiencies, this paper will study compound control mode, the compound control mode is combined the linear quadratic regulator (Linear Quadratic Regulator, LQR) control with other optimization algorithms (genetic algorithms, fuzzy control, neural network control, the traditional PID control, adaptive control, synovial variable structure control, etc.)[5]. The genetic algorithm is a parallel random search optimization algorithm, which is formed according to the simulation of natural biological and genetic mechanisms and biological evolution. This paper combines GA with LQR control, and gets a new optimal control algorithm, which can be referred to as GALQR (Genetic Algorithm Linear Quadratic Regulator) optimal control methods. In this paper, GALQR control method is applied to the design of active suspension optimal control, GALQR control method weight matrix optimization can be optimized for weight matrix, this feature makes up for lack of active suspension system which is designed by LQR control theory, allowing the system to achieve better control effect[6]. Finally, the simulation analysis carried out in the MATLAB, using simulation results and genetic algorithms optimize the LQR design parameters, and find the global optimal solution.

2 ESTABLISHMENT OF VEHICLE ACTIVE SUSPENSION MATHEMATICAL MODEL

Single wheel vehicle model is treated as the research object, and single wheel vehicle model is shown in Figure 1. m_1 is the quality of single wheel vehicle body; m_2 is the wheel quality; x_1 is the vehicle body dynamic displacement; x_2 is the wheel dynamic displacement; x_3 is the road surface vertical displacement; k_1 is the elasticity coefficient of suspension; k_2 is the elasticity coefficient of tire; U is the input of control force. Parameters of single wheel vehicle model are shown in Table 1.

According to Newton's motion laws, differential equation of system movement is established as follows:

$$m_1\ddot{x}_1 = -k_1(x_1 - x_2) + U \tag{1}$$

$$m_2\ddot{x}_2 = k_1(x_1 - x_2) - k_2(x_2 - x_3) - U \tag{2}$$

The state variables is for

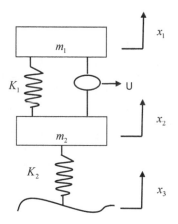

Figure 1.   Single wheel vehicle model.

Table 1.   Parameters of single wheel vehicle model.

Item	Value
Vehicle body quality m_1 (kg)	320
Wheel quality m_2 (kg)	40
Elasticity coefficient of suspension k_1 (N/m)	2000
Elasticity coefficient of tire k_2 (N/m)	20000
Road roughness coefficient G_0 (m³/cycle)	5×10^{-6}
Vehicle rated velocity v (m/s)	20
Lower cutoff frequency f_0 (Hz)	0.1
Elastic modulus of suspension K_{pas} (N/m)	22000
Damping coefficient C_{pas} (Ns/m)	1000

$$x = (\dot{x}_1, \dot{x}_2, x_1, x_2, x_3)^T \tag{3}$$

The input of road is filtered white noise

$$\dot{x}_3(t) = -2\pi f_0 x_3(t) + 2\pi\sqrt{G_0}vw(t) \tag{4}$$

In Equation (4), G_0 is the road roughness coefficient; v is the vehicle speed; w is the Gaussian white noise; f_0 is the lower cutoff frequency.

System State Equation is as Equation (5).

$$\dot{x} = Ax + Bu + Gw \tag{5}$$

In Equation (5), $w = (w(t))$ is white noise matrix; $u = (U(t))$ is enter the matrix.

$$A = \begin{bmatrix} 0 & 0 & -\dfrac{K_1}{m_1} & \dfrac{K_1}{m_1} & 0 \\ 0 & 0 & \dfrac{K_1}{m_2} & -\dfrac{K_1+K_2}{m_2} & \dfrac{K_1}{m_2} \\ 1 & 0 & 0 & 0 & 0 \\ 0 & 1 & 0 & 0 & 0 \\ 0 & 0 & 0 & 0 & -2\pi f_0 \end{bmatrix} \quad B = \begin{bmatrix} \dfrac{1}{m_1} \\ -\dfrac{1}{m_2} \\ 0 \\ 0 \\ 0 \end{bmatrix} \quad G = \begin{bmatrix} 0 \\ 0 \\ 0 \\ 0 \\ 2\pi\sqrt{G_0}v \end{bmatrix}$$

3 LQR CONTROLLER

Evaluation of vehicle active suspension performance indicators are body vertical acceleration which represents riding and driving comfort, suspension travel which affects vehicle stability and dynamic displacement of tire which affects driving reliability[7]. Therefore, design performance indicators of LQR controller are:

$$J = \int_0^\infty [q_1(x_2 - x_3)^2 + (q_2(x_1 - x_2)^2) + q_3\ddot{x}_1]dt \tag{6}$$

Combining Equation (5) into Equation (6), there is:

$$J = \int_0^\infty (x^T Q x + u^T R u + 2x^T N u)dt \tag{7}$$

In the Eq. (7)

$$Q = \begin{bmatrix} 0 & 0 & 0 & 0 & 0 \\ 0 & 0 & 0 & 0 & 0 \\ 0 & 0 & q_2 + \dfrac{k_1^2}{m_1^2} & -q_2 - \dfrac{k_1^2}{m_1^2} & 0 \\ 0 & 0 & -q_2 - \dfrac{k_1^2}{m_1^2} & q_1 + q_2 + \dfrac{k_1^2}{m_1^2} & -q_1 \\ 0 & 0 & 0 & -q_1 & q_1 \end{bmatrix}, \quad N = \dfrac{1}{m_1^2}\begin{bmatrix} 0 \\ 0 \\ -k_1 \\ k_1 \\ 0 \end{bmatrix} \quad R = \dfrac{1}{m_1^2}$$

Optimal control force of actuator is

$$U = -Kx(t) \tag{8}$$

In the Equation (8), $x(t)$ is the feedback state variable of any time[5]; k is the feedback matrix of optimal control process; S is the solution of algebraic riccati equation; E is the poles and zeros of closed-loop system. The function of linear quadratic optimal controller design can be obtained by using MATLAB.

$$[K, S, E] = L \, Q \, R \, (A, B, Q, R, N) \qquad (9)$$

According to Equation (6) to the Equation (9), it is found that the design of optimal control entirely depends on the choice of the weighting function coefficients ($q_1 \, q_2 \, q_3$). In previous designs, the weighting coefficient is obtained based on the designer's experience and constant repeated experiments. Although in a sense the optimal control rule has been got, but the optimal results, which are obtained in this way, are great uncertainty. Then on the basis of genetic algorithms, the optimal weighting coefficient of LOR controller is globally found.

4 OPTIMAL DESIGN OF LINEAR QUADRATIC OPTIMAL CONTROLLER BASED ON GA

Performance indicators include three components, and they are vehicle body vertical acceleration, suspension travel and tire dynamic displacement, and the unit orders of magnitude of three components are inconsistent. Therefore, they are divided by the corresponding performance indicator of passive suspension, the optimization problem is expressed as follows:

$$L \min = \frac{BA(X)}{BA_{pas}} + \frac{sws(X)}{SWS_{pas}} + \frac{DTD(X)}{DTD_{pas}} \qquad (10)$$

$$X = (q_1, q_2, q_3), 0.1 < X_i < 10^6, i = 1, 2, 3 \qquad (11)$$

$$s.t. \begin{cases} BA < BA_{pas} \\ SWS < SWS_{pas} \\ DTD < DTD_{pas} \end{cases} \qquad (12)$$

In the Equations (10)–(12), BA, SWS, DTD respectively represent the body vertical acceleration, suspension travel and rms value of tire dynamic displacement. BA_{pas}, BA_{pas}, DTD_{pas} represent response performance of automotive passive suspension; K_{pas} represents the elastic modulus of suspension; C_{pas} represents the damping coefficient. Other conditions are the same as the active suspension, optimization variables X represents weighting factors q_1, q_2, q_3.

Constraint processing methods are as follows: for each set of weighting coefficients X, according to equation (10) L is calculated, judging whether it satisfy the constraints (12), if it meets the conditions, then the fitness function value is L; Otherwise, in order to punish the weighting coefficients that does not satisfy the constraints, considering the optimization problem is the minimum optimization problem, to make the fitness value of weighting coefficient L plus 10, away from the minimum, thus guiding the population to evolve which meets the constraints.

Based on GA, to find the optimal solution of weighting coefficients for LOR controller, specific steps are as follows:

1. Set initial population size of genetic algorithm is 100, evolutionary generation is 20, and the proportion of cross progeny is 0.4.
2. The population of each individual is in turn assigned to the weighting coefficients of LQR controller q_1, q_2, q_3, according to Equation (9) K is obtained, according to Equation (8) U is obtained and acting on single wheel model, then obtain performance indicators J of automobile suspension.
3. According to Equation (10) each individual population value of the fitness function is obtained, and determine whether it meets the conditions to terminate GA. If it meets, then quit and obtain the best individual; otherwise go to step (4).
4. Genetic algorithm selection, retention elite, crossover and mutation to produce new species, and go to step (2).

5 IMPLEMENTATION AND SIMULATION ANALYSIS OF CONTROL MODEL

In the course of the simulation, the white noise signal of the road input is generated by function wgn (M, N, P), M represents the sample point of signal, N represents the number of signal, P represents the noise intensity of signal, and taking $M = 10000$, $N = 1$, $P = 20$. The settings of improved genetic algorithm parameter are shown in Table 2.

Variation of optimizing the fitness function and the best individual function value of GA are shown in Figure 2. Body vertical acceleration, suspension travel and tire dynamic displacement of MATLAB simulation results are respectively shown in Figures 3–5.

Comparison of the corresponding indicator obtained from MATLAB with indicator passive suspension is shown in Table 3. According to Table 3, it can be seen that the tire dynamic displacement is 6.2526 before optimization, it is 5.4842 after optimization, and the performance improves to 12.29%. The body acceleration of LQR controller controls is 1.7816 before optimization, it is 1.6251 after optimization, and the performance improves to 5.2481%.

Table 2. Parameter settings of genetic algorithm.

Parameter	Explanation	Parameter	Explanation
Coding scheme	Real number coding	Cross-function	Decentralized cross
Initial population	Randomly generated within the upper and lower range	Maximum evolution algebra	20
Population size	100	Variation function	Constrained adaptive mutation
The number of elite	10	Deviation of the fitness function values	1e-100
Cross rate	0.4	Stopped algebra	20

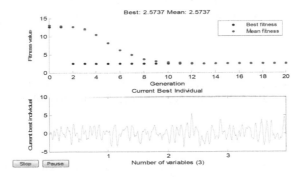

Figure 2. Optimal fitness value.

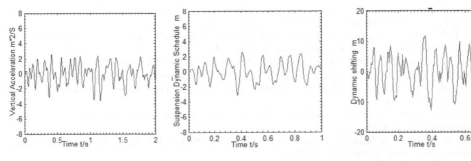

Figure 3. Body vertical acceleration.

Figure 4. Suspension travel.

Figure 5. Tire dynamic displacement.

Table 3. Comparison of genetic algorithm optimization controller of active suspension and passive suspension.

Performance indicators	Results of this paper	Passive suspension
Tire dynamic displacement	5.4842	6.2526
Body acceleration	1.6251	1.7816
Suspension travel	11.9626	17.1284

The maximum suspension dynamic travel is 17.1284 before optimization, it is 11.9626 after optimization, and the performance improves to 30.16%.

6 CONCLUSION

By establishing a mathematical model of vehicle active suspension, and to analyze and design the active suspension process of LQR controller, the optimal selection weighting coefficients importance is illustrated. Due to inadequacies of the LQR controller, this paper uses the genetic algorithms to optimize controller and to find the global optimization of weighting factor, and the simulation analysis was performed. The results show that the LQR controller based on genetic algorithm makes performance indicators of active suspension have been significantly improved, and optimization effect is very effective. But when the system model is built this paper only considers the linear relationship, ignoring nonlinear factors of the suspension system such as friction and other, so this paper only ensures the expected performance of ideal mathematical model. The actual suspension system containing many uncertain factors, nonlinear time-varying, and high-end power system, it is difficult to reach the predetermined performance requirements with steady feedback system. Therefore, the uncertain factors should be taken full account in the actual design of the controller.

ACKNOWLEDGEMENTS

Project supported by the special fund project of central university basic scientific research business expenses (No. 2013G3322009, No. 2014G1321040, No. 2013G1321048).

REFERENCES

[1] Sang, N. 2012. Design of Active Suspension Controller Based on Stochastic Linear Optimal Control Theory. *Journal of Zhengzhou University (Engineering Science)* (3):112–116.
[2] Sang, N. & Bai, Y. 2013. Optimal Design of Active Suspension Controller Based on Genetic Algorithm. *Mechanical Science and Technology for Aerospace Engineering* (9):1400–1004.
[3] Meng, J. 2013. Simulation research of the PID controller of vehicle suspension based on the genetic algorithm. *Modern Manufacturing Engineering* (6):141–143.
[4] Liu, H.S. 2008. Application of LQG control simulation in vehicle active suspensions based on simulink. *Machinery Design & Manufacture* (8):315–319.
[5] Hu, F. & Zhao, Z.G. 2011. Optimization of Weighting Factors for LQR Controller of Active Suspension Based on Multi-objective Genetic Algorithm. *Machinery & Electronics* (2):21–25.
[6] Zhang, R. 2013. The Optimal Control of Active Suspension Based on Simulink. *Automobile applied technology* (2):89–94.
[7] Zhao, Z.N. 2010. *Research on land Vehicle's ride Comfort Based on Suspension Control.* Harbin: Harbin Institute of Technology.
[8] Zhou, K. & Han, Z.N. 2010. Design of LQG Controller of Automobile Active Suspension Based on Simulink. *Automobile technology* (2):97–102.
[9] Liu, X.P. 2013. The Optimal Design of Automobile Suspension with LQR Controller Based on Genetic Algorithm. *Modular machine tool & automatic manufacturing technique* (6):51–55.

Machinery design

Advances in Engineering Materials and Applied Mechanics – Zhang, Gao & Xu (Eds)
© 2016 Taylor & Francis Group, London, ISBN 978-1-138-02834-0

Design of a novel dual-band metamaterial absorber

C.S. Tian, K. Zhou & Y.L. Guan

College of Information and Communication Engineering, Harbin Engineering University,
Harbin City, Heilongjiang Province, China

ABSTRACT: A dual-band metamaterial absorber was designed based on the electromagnetic resonance. The absorber consists of electrical metal ring resonators, lossy dielectric and metal backboard. Using the electromagnetic simulation software simulated, and different polarized vertical incident electromagnetic wave and oblique incident electromagnetic wave were applied on the surface of the absorber respectively. The simulation results show that the absorptivity for different polarized vertical incident electromagnetic waves is 99.7% at 6.3 GHz and 99.2% at 13.01 GHz respectively. When the incident angle increases to 40 deg, the absorptivity of the absorber remains above 91.5% at two frequencies. The absorber exhibits a wide incident angle characteristic. By studying distribution of electric on the surface of structure, we analyzed its working mechanism. Using absorbing characteristic, we can attach absorbers to the antenna. By reducing antenna's RCS, we can achieve its stealth purpose.

1 INTRODUCTION

Radar absorbing material is a kind of functional material that can effectively absorb the incident electromagnetic wave, and significantly reduce the target echo intensity. It can also substantially reduce radar cross section of targets to improve its stealth capability. Traditional radar absorbing materials are restricted in the application for the defections in thick, heavy and poor stability. Therefore, seeking and designing of the radar absorbing material with thin thickness, light weight, high absorptivity and good stability has been the focus in study of material science. Metamaterial is an artificial composite structure or composite material with extraordinary physical properties which natural materials don't have. In 2002, Engheta first proposed the thought of designing ultra-thin absorbing material based on characteristic which the sub-wavelength structure of metamaterial that can break through the limitation of the material intrinsic nature laws and implemented by frequency selective surface (Engheta 2002). In 2008, Landy first put forward Perfect Metamaterial Absorber (PMA) which consists of electric resonator, lossy substrate and metal microstrip basing on the electromagnetic resonance properties of metamaterial (Landy 2008). Compared with traditional absorbing material, PMA has the advantages of simple structure, ultra-thin, small volume and high absorptivity, etc. So as it was put forward, it has aroused extensive attention of scientific researchers, and related research also began on it.

With in-depth research, the new absorbing structure units have been put forward continually, the electromagnetic properties of structure have been significantly improved. And various insensitive-polarization (Landy 2009, Sepiadeh 2012), wide angle (Olli 2009, Jiang 2010) and multi-band (Somak 2013, Theofano 2012) metamaterial absorbers are constantly being proposed. In this paper, a dual-band metamaterial absorber with high absorptivity, insensitive-polarization and wide angle is proposed. Compared with previous dual-band structures, the proposed absorber achieves compact design.

2 MECHANISM ANALYSIS

Reflectance, transmittance and absorbance are three frequency dependent parameters to measure the performance of metamaterial absorber. Absorbance is equal to

$$A = 1 - R - T \tag{1}$$

where, R is the reflectance and is equal to $|S_{11}|^2$ and T is the transmittance and is equal to $|S_{21}|^2$ (Somak 2013).

To achieve absorbing must have two conditions: (1) achieving impedance matching with free space to ensure that the incident electromagnetic wave can all enter into the internal structure; (2) having a loss characteristic to ensure that the internal electromagnetic wave is absorbed absolutely. When the electromagnetic wave incidents from free space to absorber interface, a part of the electromagnetic wave is reflected, another part of the electromagnetic wave turns into the interior of the structure. The absorber structure reflection coefficient can be expressed as

$$\Gamma = \frac{Z_0 - Z_i}{Z_0 - Z_i} \tag{2}$$

where, Z_0 is the impedance of free space and $Z_0 = \sqrt{\mu_0/\varepsilon_0}$, Z_i is the impedance of absorber structure and $Z_i = \sqrt{\mu_i/\varepsilon_i}$, ε_0 and μ_0 are the permittivity and permeability of free space respectively, ε_i and μ_i are the permittivity and permeability of the structure, respectively (Yoo 2014). Simplifying the above equation, we can get

$$\Gamma = \frac{1 - \sqrt{\mu_r/\varepsilon_r}}{1 + \sqrt{\mu_r/\varepsilon_r}} \tag{3}$$

where, $\varepsilon_r = \varepsilon_i/\varepsilon_0$ and $\mu_r = \mu_i/\mu_0$, ε_r and μ_r are the effective permittivity and permeability of the structure, respectively. Then if $\varepsilon_r = \mu_r$, $\Gamma = 0$, that is, absorber structure achieves the impedance matching with free space. The internal electromagnetic waves under the influence of dielectric loss and ohmic loss will transform into heat and achieve perfect absorption of incident electromagnetic waves.

3 DESIGN AND SIMULATION

The proposed structure has been shown in Figure 1 that (a) and (b) respectively, which are front view and side view of the cell structure. The absorber consists of three parts: circular copper ring and square copper ring at the top of lossy FR4 substrate and a square metal film at the bottom of substrate. The FR4 substrate has dielectric constant of 4.4 and tangent loss of 0.02. The upper circular ring and square ring constitute electrical resonator which can produce electrical resonance to electric field component of the incident electromagnetic waves. The upper rings and bottom metal film constitute magnetical resonator which can produce magnetical resonance to magnetic field component of the incident electromagnetic waves. By proper design of the structure dimensions, electric resonance and magnetic resonance appear simultaneously at a given frequency. Thus the absorber can separately absorb the electric field component and magnetic field component of the incident electromagnetic waves, and then achieve perfect absorption.

In this work, we use electromagnetic simulation software Ansoft HFSS 13.0 for simulation modeling. By setting Master/Slave boundary conditions and Floquet port, we simulate an infinite periodic unit. Specific parameters of structural unit are obtained through simulation and optimization and are shown as follows: periodical cell length L = 10.2 mm, square ring length w = 2.95 mm, width d = 1 mm, circular ring inner radius r1 = 4 mm, outer

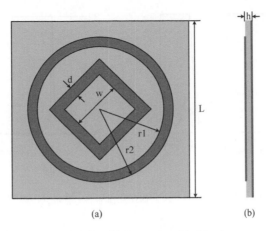

Figure 1. A unit cell of dual-band absorber (a) front view (b) side view.

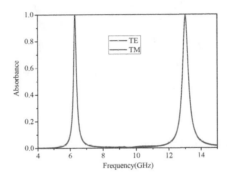

Figure 2. Absorbance for TE polarization vertical wave and TM polarization vertical wave.

radius r2 = 4.5 mm, the overall thickness of the structure h = 0.61 mm, the inclination of square ring is 45°. In this work, the absorber is centrosymmetric. Using this structure, we can eliminate the incident waves' anisotropy. So the absorber has polarization insensitivity to incident waves. At the same time, the bottom of structure is all-metal. This can ensure that no electromagnetic transmission and the transmittance of the structure T is equal to zero. Therefore, absorbance $A = 1 - R - T = 1 - R$. With TE polarized wave and TM polarized wave vertical beaming on absorber surface, we can get the absorbance as shown in Figure 2. As can be seen form the figure, TE polarized wave curve and TM polarized wave curve are basically the same. The absorber exhibits excellent absorption characteristic at the two frequencies of 6.3 GHz and 13.01 GHz. The absorbance of 99.7% is achieved at 6.3 GHz for TE wave and TM wave and the absorbance of 99.2% is achieved at 13.01 GHz for TE wave and TM wave.

Figure 3 shows the absorbance for different polarization oblique electromagnetic waves. As can be seen from the figure, when the incidence angle $\theta = 20°$, TE polarized wave curve and TM polarized wave curve are basically the same and absorbance remains above 98% at 6.3 GHz and 13.01 GHz. With the increase of the incidence angle, when the incidence angle increases to 40°, absorbance decreases. However, absorber absorbance remains above 91.5% in two frequencies. Especially for TM polarized wave, absorbance for oblique wave remains unchanged, above 99%. This indicates that the absorber shows perfect polarization insensitivity and wide incidence angle characteristic. The main reason for decreasing absorbance is that the equivalent impedance of absorber becomes smaller with increasing incidence angle, and absorber unable to achieve impedance matching with free space, which leads to lower absorbance.

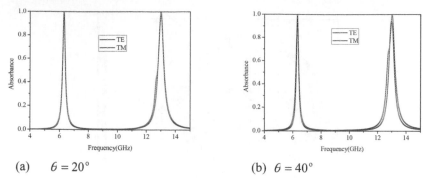

(a) $\theta = 20°$ (b) $\theta = 40°$

Figure 3. Absorbance for TE polarization oblique wave and TM polarization oblique wave.

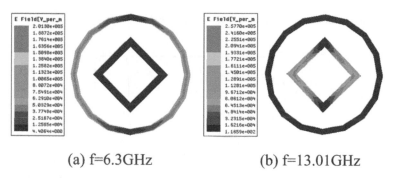

(a) f=6.3GHz   (b) f=13.01GHz

Figure 4. Absorber electric field distribution.

In order to understand the absorber working mechanism to achieve dual-absorbing more clearly, Figure 4 shows the electric field distribution (Liu 2014) for the electromagnetic waves of different frequencies irradiated to the surface of the structure. From the figure we can see that the electric field around circular ring is stronger than the electric field around square ring at the frequency of 6.3 GHz. This shows that the absorption at 6.3 GHz of the absorber is mainly produced by circular. At the same time, the electric field around square ring is stronger than the electric field around circular ring at the frequency of 13.01 GHz. This indicates that the absorption at 13.01 GHz of the absorber is mainly produced by circular.

4 CONCLUSION

Based on the electromagnetic resonance characteristics of metamaterial, we design a novel dual-closed-ring miniaturization metamaterial absorber. The absorber shows polarization insensitive, wide incidence angle and high absorbance. By analyzing distribution of electric on the surface of structure, we understand the absorber dual-working mechanism. This design has potential applications in electromagnetic shielding, stealth, detection and other fields.

ACKNOWLEDGMENTS

The authors would like to thank the anonymous reviewers of this manuscript for their constructive comments and suggestions. This paper is funded by the International Exchange Program of Harbin Engineering University for Innovation-oriented Talents Cultivation.

REFERENCES

Engheta N. 2002. Thin Absorbing Screens Using Metamaterial Surfaces. *IEEE Antennas and Propagation Society International Symposium*, 2: 392–395.

Jiang Zhihao, Wu Qi, Wang Xiande, et al. 2010. Flexible wide-angle polarization-insensitive mid-infrared metamaterial absorbers. *IEEE Antennas and Propagation Society International Symposium*, 1–4.

Landy N.I., Sajuyigbe S., Mock J.J., et al. 2008. A Perfect Metamaterial Absorber. *Physical Review Letters*, 100, 207402.

Landy N.I., C.M. Bingham, T. Tyler, et al. 2009. Design, Theory and Measurement of A Polarization-Insensitive Absorber for Terahertz Imaging. *Physical Review B* 79, 125104.

Liu Yahong, Zhao Xiaopeng. 2014. Perfect absorber metamatetial for designing low-RCS patch antenna. *IEEE Antennas and Wireless Propagation Letters*, 13: 1473–1476.

Olli Luukkonen, Filippo Costa, Constantin R. Simovski, et al. 2009. A Thin Electromagnetic Absorber for Wide Incidence Angles and Both Polarizations. *IEEE Transactions on Antennas and Propagation*, 57(10): 3119–3125.

Sepiadeh Fallahzadeh, Keyvan Forooraghi, Zahra Atlasbaf. 2012. A Polarization-Insensitive Metamaterial Absorber with A Broad Angular Band. *20th Iranian Conference on Electrical Engineer*, 1540–1543.

Somak Bhattacharyya, Saptarshi Ghosh, Kumar Vaibhav Srivastava. 2013. A dual band metamaterial absorber using electric field driven LC (ELC) and cave ELC structures. *IEEE Antennas and Propagation Society International Symposium*, 1452–1453.

Somak Bhattacharyya, Kumar Vaibhav Srivastava. 2013. An ultra thin electric field driven LC resonator structure as metamaterial absorber for dual band applications. *2013 International Symposium on Electromagnetic Theory*, 722–725.

Theofano M. Kollatou, Alexandros I. Dimitriadis, Nikolaos V. Kantartzis, et al. 2012. A class of multi-band, polarization-insensitive, microwave metamaterial absorbers in EMC analysis. *2012 International Symposium on Electromagnetic Compatibility*: 1–4.

Yoo Minyeong, Lim Sungjoon. 2014. Polarization-independent and ultrawideband metamaterial absorber using a hexagonal artificial impedance surface and a resistor-capacitor layer. *IEEE transactions on antennas and propagation*, 62(5): 2652–2658.

Advances in Engineering Materials and Applied Mechanics – Zhang, Gao & Xu (Eds)
© *2016 Taylor & Francis Group, London, ISBN 978-1-138-02834-0*

Coiled tubing small turbo drill cascade design optimization and numerical simulation of the flow field

Q. Zhang, Z. Chen, X.H. Dong, F. Zhang & K.J. Luo
College of Mechanical Engineering, Yangtze University, Jingzhou, Hubei, China

ABSTRACT: In this paper, a coiled tubing small turbo drill design is proposed for coiled tubing in ultra-deep slim-hole drilling process, the presence of a conventional screw drill has poor temperature performance and on sensitive issues such as oil-based drilling fluids. By using the Quintic polynomial method to construct multiple-sectional linear turbine blades and by using the Bezier curve to link the leading and trailing edges of each section in order to transit the liner of each section we devised a small size, high efficiency three-dimensional Cascade turbo drill. The flow field simulation helps to explore the feasibility of using the three-dimensional modeling method Cascades.

1 INTRODUCTION

With the explorations and developments of the gas and oil move into a higher level, coiled tubing as a high efficiency and low cost technology has raised wide attention. Due to the stress of the ultra-deep slim-hole drill string is complicate, it needs high quality mini size drill and conventional screw drill can no longer be suitable for ultra-deep slim-hole drilling, it is also difficult to select the drill bit, however, there is still no report on the coiled tube mini size turbo drill in our country. According to the working technical characteristics of the coiled tubing, the turbo drill should meet these requirements: (1) To prevent the hurt from the coiled tubing itself, it needs thermo stability and fewer transverse vibrations. (2) The drill should adapt the low drilling pressure, low anti-torque and the high speed. Turbo drill is an important downhole power drill. Plane cascade is the most widely used turbo drill cascade. It means when you cut it in liquid flow line, it can spread like a plane. The flow of the mud fluid in drill is complicate, the sections of the cuts of the cascade liquid flow lines are different, so we provided a dimensional cascade modeling to increase the efficiency of single-stage turbine.

2 OPTIMIZATION OF CASCADE

We use Φ127 mm small size turbo drill as the prototype of the design. According to Neyfor turbo drill manual, we can make sure the performance parameters of the Φ127 mm single-stage turbo drill: working speed 1300 r/min, working torque 6.1 Nm, output volume 12 L/S. Turbine blade height 10 mm, turbine blade leading edge taper angle 20°, the rear edge of the cone angle 11°, the leading edge radius 0.8 mm, the trailing edge radius 0.4 mm, and the relative pitch 9.07 mm.

$$\left[\begin{cases} \cot\alpha_{1k} = \dfrac{1}{\overline{C}_z}\left(m_a + \dfrac{1}{2}\overline{C}_u\right) \\[2mm] \cot\alpha_{2k} = \dfrac{1}{\overline{C}_z}\left(m_a - \dfrac{1}{2}\overline{C}_u\right) \end{cases} \begin{cases} \cot\beta_{1k} = \dfrac{1}{\overline{C}_z}\left[1 - \left(m_a + \dfrac{1}{2}\overline{C}_u\right)\right] \\[2mm] \cot\beta_{2k} = \dfrac{1}{\overline{C}_z}\left[1 - \left(m_a - \dfrac{1}{2}\overline{C}_u\right)\right] \end{cases} \right.$$

In the formula, circulation coefficient M_a, axial velocity coefficient of the turbo drill $\bar{C}_z$, and the turbine circulation coefficient $\bar{C}_u$.

We can get the structure angle and the setting angle of the blade. Stator outlet structure angle α_{1k}, rotator outlet structure angle β_{2k}, stator entrance structure angle α_{2k}, rotator entrance structure angle β_{1k}, and setting angle β_m. We used the quintic polynomial method to build the linear state of the pressure surface and the suction surface. With the help of Matlab software to solve the system of linear equations, we can get the equation of the state of the pressure surface and the suction surface of the blade. Plane cascade blade of the turbo drill is shown in Figure 1.

The blades of the turbo drill mostly are plane cascade, so the outlines of each section are the same in radial direction. Plane cascade causes the low water efficiency of the single-stage turbo drill, and its long length does not fit coiled tubing. In fact, the flows of drilling fluid are very different in each section of the turbine runner, so under the same rotation and output torque, the outlines of each section should be different in theory. We use the section diameter of the blade as optimum design parameter, creating 3-dimensional cascade turbo drill with the Bezier curve transits the leading and trailing edge of the arc and the center of the circle to optimize the small-size turbo drill cascade. We divide the fluid surrounds the cascade into several flow layers in same thickness, use the diameter of the section we have got from the plane cascade design as median, pick five sections symmetrically from both sides. Table 1 shows the relevant parameters of

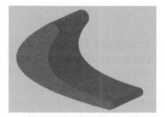

Figure 1. Plane cascade blade of turbo drill.

Table 1. The semi-diameters of the blade in different sections, dimensionless coefficient.

Section	R	M_a	$\bar{C}_z$	$\bar{C}_u$
The first section	35.00	0.5	0.901	3.316
The second section	38.00	0.5	0.764	2.813
The third section	41.00	0.5	0.656	2.416
The fourth section	44.00	0.5	0.570	2.098
The fifth section	47.00	0.5	0.499	1.839

Table 2. Structure angles and setting angles of stators and rotators in different sections.

Section	α_{1k}	α_{2k}	β_{2k}	β_{1k}	β_m
The first section	22.66°	142.11°	22.66°	142.11°	60.97°
The second section	21.84°	139.88°	21.84°	139.88°	56.80°
The third section	21.01°	137.18°	21.01°	137.18°	52.69°
The fourth section	20.20°	133.92°	20.20°	133.92°	48.74°
The fifth section	19.37°	130.05°	19.37°	130.05°	44.94°

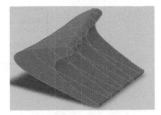

Figure 2. 3-dimensional turbo drill blade.

the sections. We also can get the structure angle and the setting angle of the blade and it is shown in Table 2.

According to the parameters from the 5 sections, we can use the quantic polynomial method to work out the pressure surface and suction surface equation from each section and build up the alphabet of lines of the section. We also can ensure the coordinates of the center of the circle of the leading and trailing edge. Then with the coordinates of the center of the circle of both edges and the semi-diameter from different sections, we can build a smooth Bézier curve to transit the 5 sections in to a 3-dimensional turbo drill blade. It is shown in Figure 2.

3 ANALYZE TURBO DRILL CASCADE IN CFD

3.1 *Modeling and simulation calculation*

We can use fluid simulation software to simulate the hydraulic performance of the cascade. First, build up the geometric models of the plane cascade and 3-dimensional cascade turbo drill. Then we can use FLURNT to do the fluid simulation, we pick turbulence model to simulate the flow of turbo drill, pick regularization model in $k - \varepsilon$ model as control equation. In the hydraulic analysis, we choose pure water as the medium. Stator's entrance condition is set as velocity entrance. Stator's outlet is set as pressure outlet, rotator's entrance is set as pressure entrance, these two surfaces contact with each other and they need coupling. The outlet condition of the rotator is set as pressure outlet, and we use normal atmosphere as initial pressure. Stator is still and has no speed, we give rotator a speed circles around the central axis. The other parts we make it non-slither. We choose flow 12 l/s, the speed of rotator gradually increase from 0 r/min.

3.2 *Calculation results and analysis*

According to the results of the calculation, we can draw torque-speed regulation and voltage drop-speed regulation curves. On the whole situation from the simulation of the blade flow field, it shows the design is logical. The voltage drop of the 3-dimensional cascade blade decreased obviously, the function increased compared with plane cascade blade in same displacement and meets the performance parameter of the turbo drill. Also, we can get the efficiency of turbo drill-speed regulation curves of 3-dimensional cascade and plane cascade. They are shown in Figure 3.

When the designed parameter meet, the peak efficiency of plan cascade turbo drill is about 59.27%, the displacement is about 12 l/s, the speed is about 1800 r/min and the voltage drop is about 0.119 MPa. The peak efficiency of 3-dimensional cascade turbo drill is about 62.88%, the displacement is about 12 l/s, the speed is about 2000 r/min, and the voltage drop is about 0.113 MPa. The 3-dimensional method in designing blade has increased the efficiency of single-stage turbo drill. It will do contribute to small-size, high-efficiency turbo drill in coiled tubing.

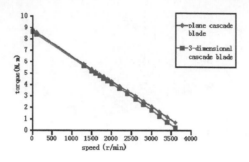

Torque-speed regulation

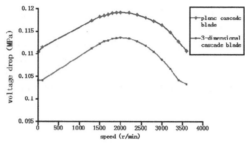

Voltage drop-speed regulation

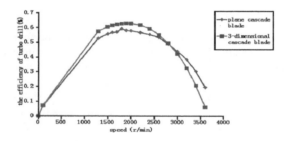

Efficiency-speed regulation

Figure 3. The results of the calculation curves.

4 CONCLUSION

Due to the structure characteristics of regular small-size cascade turbo drill and the requests in coiled tubing, we carry out the quantic polynomial method to build up the linear of several sections of the blades. By using the Bézier curve to optimize the 3-dimensional cascade.

We use the CFD software to perform the flow field simulation analysis and testified the practicability of using 3-dimensional modeling method to design turbo drill. The 3-dimensional cascade increases the hydraulic efficiency compared with the plane cascade. It has guiding significance towards solving the problems in low efficiency single-stage turbo drill. The 3-dimensional cascade increases the speed of the turbo drill compared with the plane cascade, which can fit the requirement in coiled tubing.

REFERENCES

Feng, D. 2007. Domestic turbo drill structure and performance analysis. *China Petroleum Machinery.* 35(1):59–61.

Feng, J. 2000. Turbodrill turbine blade design a new method. *Petroleum Machinery.* 28(11):9–12.

Feng, D. 2004. Turbodrill turbine section of the modular design method. *Jianghan Petroleum Institute.* 26(1):19–122.

He, H.Q. 2009. Coiled tubing drilling technology and equipment. *China Petroleum Machinery.*

Liu, X.G. 2006. CFD in turbo drill performance prediction analysis application. *Manufacturing and information technology.* (1):21–28.

Xie, L.H. 2012. *ANSYS CFX fluid analysis and simulation.* Beijing: Electronic Industry Press.

Lin, Y.H., 2004. Other new research tool turbine blade profile and computer-aided design. Chongqing University 27(2):24–27.

Wang, L.B. 1999. *Petroleum Engineering Fluid Machinery.* Beijing: Petroleum Industry Press.

Xu, F.D. 2004. *Reducer turbo drill work with synchronous mechanics and performance simulation.* Beijing: China University of Geosciences Press.

Zhao, H.B. 2012. Hydraulic calculation turbodrill turbine cascade analysis. *China University of Geosciences.* 39(11):29–32.

Advances in Engineering Materials and Applied Mechanics – Zhang, Gao & Xu (Eds)
© 2016 Taylor & Francis Group, London, ISBN 978-1-138-02834-0

Optimal design on dual-front-axle steering mechanism of a heavy vehicle based on simulated integration technology of iSIGHT and ADAMS software

H.N. Zhou, Y. Feng & Q. Hu
Hubei University of Automotive Technology, Hubei, China

ABSTRACT: In this paper, a simulation model of dual-front-axle steering mechanism is established by using the application ADAMS/View, which aimed on the tire abnormal wearing problem for a heavy vehicle with the dual-front-axle steering system. The kinematics simulation results show that the actual steering angle of steering wheel is different from the theoretical results. The integration of kinematics simulation with ADAMS/View is achieved and the optimal analysis process of the steering mechanism is established based on Multi-disciplinary Design Optimization (MDO) software iSIGHT. With Orthogonal Arrays DOE analysis, the steering mechanism is optimized by applying multi-objective optimization genetic algorithm NSGA-II. The optimization results show that the steering angle absolute error is decreased greatly after the optimization, i.e. the tire abnormal wearing problem could be solved effectively in engineering.

1 INTRODUCTION

In recent years, with the wide use of dual-front-axle steering vehicles in transportation, some typical problems usually occur, mainly in the abnormal wearing problem of steering wheels. To reduce tire's wearing, the wheels of each two steering axle should first satisfy the Ackermann steering theory. Extensive researches on designing and optimizing of dual-front-axle steering mechanism have been done, and many methods and theories have been proposed both at home and abroad[1–8][11–12]. And with the application of various special software of engineering calculation, it brings convenience for these studies. In aspect of kinetic and dynamic simulation and optimization of mechanism or system, multi-body dynamics simulation software ADAMS and Insight module is widely applied[2–5][11–12]. Meanwhile, the iSIGHT software that is based on Multi-disciplinary Design Optimization (MDO) platform has covered in the fields of aviation, shipping, electronics, etc. By using the iSIGHT, the integration of simulation and analysis process of complicated system is achieved, as well as the design analysis and optimization of multi-disciplinary and multi-objective engineering problem[6]. Therefore, it shorts the product development cycle and reduces the cost of research and development greatly.

This paper aimed on the tire abnormal wearing problem of a heavy vehicle with dual-front-axle steering system, the simulation model of the dual-front-axle steering mechanism is made by using the application ADAMS/View, and the simulation and optimal analysis process is established based on the simulated integration technology of iSIGHT and ADAMS. With Orthogonal Arrays DOE analysis, the steering mechanism is optimized by applying the multi-objective genetic algorithm NSGA-II. This paper could offer a new effective way of the design and development on automotive steering system, and the chassis as well.

2 KINETIC SIMULATION OF DUAL-FRONT-AXLE STEERING MECHANISM

Figure 1 shows the 3D chassis model of a prototype vehicle, which is obtained by the 3D scanning and reverse modeling technology. So, the coordinate values of the key points in dual-front-axle steering mechanism are obtained, and the simplified simulation model is created by the ADAMS/View software package, as shown in Figure 2. In the simulation model, all the coordinate values of key points have been parameterized in preparation for the optimization.

To avoid the tire wearing, each steering wheel should be in the state of pure rolling when turning the wheels, that is to say, the steering wheels' motion of the first and second axle should meet the coordinative relations. So, in theory, the basic design of the steering kinematics must satisfy the Ackermann conditions. According to the Ackermann steering theory of multi-axle steering mechanism, the theoretical steering angle relations of each steering wheel which is mentioned in many references can be obtained in the study[1-3].

In the ADAMS/View, the absolute error of the actual and theoretical steering angle is obtained when the other three steering wheels turning with the left wheel of the first axle, as shown in Figure 3. From the Figure 3, it can be observed that the steering angle absolute

Figure 1.   3D chassis model of prototype vehicle.

Figure 2.   ADAMS/view model.

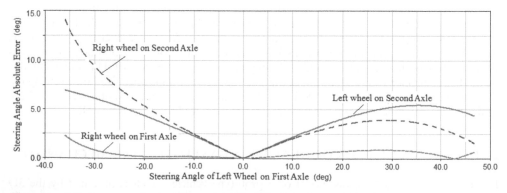

Figure 3.   Steering angle absolute error of steering wheels before optimization.

error of the actual and theoretical exists in the steering mechanism, and the steering angle error of left and right wheels on second axle is relatively larger. The large error may lead to the tire wearing abnormally, so it is necessary to carry out the optimal design on the steering mechanism.

3 ESTABLISHING SIMULATION PROCESS IN ISIGHT

The iSIGHT has widely CAD/CAE software integrated interface, and can establish complicated simulation process quickly. It also provides comprehensive methods and algorithms for optimal design and analysis, such as Design of Experiment (DOE), Gradient Optimization, Direct Search, Global Optimization, Multi-Objective Optimization, Approximation Models, Monte Carlo Simulation, etc. According to the design variable and optimization objective, it can carry out circulation analysis and optimal calculation automatically in iSIGHT[6]. Integrating ADAMS/View application program with iSIGHT is in two ways: Simcode program and iSIGHT Adams. In this paper, the first method is selected, because it is common to use and it is an interface that can integrate input, execution, and output of the application program completely. The Simcode component contains three modules essentially, the Data Exchanger of input file, the OS Command to carry out the application program and the Data Exchanger of output file. So, three files of ADAMS/View, runadams.bat batch file, *.cmd command file and *.bin model file, should be prepared before integration. The integration of ADAMS/View kinetic simulation with iSIGHT is achieved after defining the input, command and output in Simcode component and through testing the program. The integrated simulation process frame in iSIGHT is shown in Figure 4.

4 DESIGN OF EXPERIMENT (DOE)

4.1 *Design variable and optimization objective*

The parameters, such as the wheelbase, position of kingpin center, wheel alignment parameters and hinged joint position of drop arm and frame etc., can not be adjusted in large scale in this study, or it will affect the vehicle performance and arrangement. Considering the feasibility of the actual steering mechanism in adjusting and improving, in this paper, the steering mechanism is optimized in the main way of changing the length and angle of the rods. Table 1 shows the primary selected design variables.

The actual and theoretical steering angle of steering wheels should correspond closely as far as possible when the parameters of the steering mechanism are optimized. According to the actual driving condition, the actual and theoretical steering angle error in initial position of the steering mechanism should be small than in the limit position to reduce the tire wearing in common position or common wheel steering angle (high vehicle velocity). In some reference, the weighting function is cited to describe the requirement of different steering angle error with different steering angle application, or the objective function is the mean value of steering angle absolute error of the steering wheels, and or the maximum of steering angle

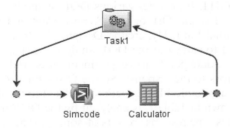

Figure 4. Integrated simulation process in iSIGHT.

Table 1. Initial values and deviation of design variable.

Name and definition	Initial value	Upper deviation	Lower deviation
DV_1 Upper drop arm length of first axle [mm]	210	+15	−15
DV_2 Drop arm initial angle of first axle [°]	90	+5	−5
DV_3 Upper drop arm length of second axle [mm]	238	+15	−15
DV_4 Drop arm initial angle of second axle [°]	90	+5	−5
DV_5 Bottom angle of steering trapezoid of first axle [°]	77	+1	−1
DV_6 Bottom angle of steering trapezoid of second axle [°]	73	+1	−1
DV_7 Lower drop arm length of second axle [mm]	325.2	+15	−15
DV_8 Steering trapezoid length of first and second axles [mm]	183.3	+15	−15
DV_9 Caster angle of first axle [°]	2	+1	−1
DV_10 Kingpin inclination of first axle [°]	7	+1	−1
DV_11 Kingpin inclination of second axle [°]	2	+1	−1
DV_12 Caster angle of second axle [°]	7	+1	−1
DV_13 Lower drop arm length of first axle [mm]	315.8	+15	−15
DV_14 Kingpin length of first and second axle [mm]	200.2	+15	−15

absolute error of the steering wheels is the objective[2–5][11–12]. Because the method of weighting function may be affected greatly by the designer's subjective factors, and the weighting value varies with the types of vehicle, this method is not used in this paper. The third method with the maximum of steering angle absolute error as the objective function is not used either, because it does not consider the common driving condition, the actual meaning of the optimal result needs further research. After comprehensive consideration, this paper uses the mean value of steering angle absolute error of the steering wheels as the objective function when the other three steering wheels turning with the left wheel of the first axle, so there are three optimal objectives.

4.2 DOE analysis of deign variable

Adding the DOE component to the task component in Figure 4, the DOE analysis process can be established in iSIGHT. It provides various DOE methods, such as Parameter Study, Full/Fractional Factorial Design, Orthogonal Arrays, Central Composite Design, Latin Hypercube Design, Optimal Latin Hypercube Design, etc. In this paper, the Orthogonal Arrays method is selected to carry out the DOE analysis, because it has the advantages of the uniform distribution of data points, less experiment times, and it can also carry out inter-active analysis of all factors to the response. So the Orthogonal Arrays method is a highly efficient, high-speed and economical DOE method. Related to the three objective responses, the design parameters shown in Table 1, is analyzed by the DOE method, in order to select the proper design variables. The response results of some parameters to the three objectives are presented in Figure 5.

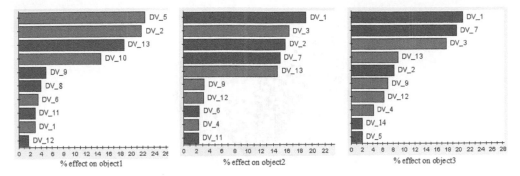

Figure 5.   Pareto figure of DOE analysis.

In DOE analysis, the Pareto figure reflects the contribution rate of all the design factors (Design variables) to the responses, and the blue bars indicate positive effect, while the red bar indicate negative effect. According to the response results in Figure 5, the factors to different objectives are not exactly the same, even may conflict each other. The factors, DV_1, DV_2, DV_3, DV_5, DV_7, DV_10, and DV_13, are selected as the optimal variable because of the greater response results to the three objectives.

4.3   Multi-objective optimization

Adding the optimization component to the task component in Figure 4, the optimal calculation process can be established in iSIGHT. Selecting the above seven design variables as the optimal variables and the mean value of steering angle absolute error of the steering wheels as the objective function when the other three steering wheels turning with the left wheel of the first axle, it forms a multi-objective optimization problem.

Most actual engineering optimization problems are multi-objective optimization problems, and the sub-objectives may conflict with each other, that is to say the improvement of one sub-objective may lead to the reduction of other sub-objective, so all the objectives achieve to optimum result is impossible at the same time. The final purpose to solve the multi-objective problems is that each objective achieve to the optimum as far as possible by the methods of coordinating, judging and compromising with each objective. Because the non-scalar method do not need to transverse the multi objectives to single objective, but it can avoid the disadvantages of scalar method, and the frontier of optimal solution set could approximate to the Pareto frontier as far as possible[6][9].

Multi-Objective Genetic Algorithm (MOGA) is the typical method to solving the non-scalar problem. In iSIGHT, it provides three types of MOGA, such as NCGA, NSGA-II and AMGA. In this paper, the NSGA-II (Non-Dominated Sorting Genetic Algorithm) which is the improvement of NSGA, is applied to optimize the dual-front-axle steering mechanism, because it adds the conception of crowding distance, and the individuals close to Pareto frontier is selected in non-dominated sorting, so the algorithm of NSGA-II has great advanced power, and excellent exploring performance. In iSIGHT optimization component, the population size is set to 32, the number of generations is set to 30 and the crossover probability is set to 0.9, after 960 times iteration. The Pareto optimal solution set of multi objectives is obtained as shown in Figure 6. In the Figure 6, the blue individuals that compose the Pareto optimal solution set form the Pareto frontier, and the red individual is the final Pareto optimal solution selected by the program automatically. Table 2 shows the change of the seven optimal variables before and after the optimization.

Figure 7 is the variation of steering angle absolute error of other three steering wheels when turning with the left wheel of the first axle. Comparing with the Figure 3 (before optimization), the steering angle absolute error is reduced greatly, especially the two wheels on second axle, even at large steering angle the maximum error is also reduced greatly. So, when the left wheel

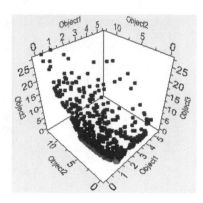

Figure 6.   3D scatter diagram of optimal objects and Pareto solution set.

Table 2.   The change of optimal variables before and after the optimization.

	DV_1 [mm]	DV_2 [°]	DV_3 [mm]	DV_5 [°]	DV_7 [mm]	DV_10 [°]	DV_13 [mm]
Before optimization	210	90.00	238	77.00	325.2	7.00	315.8
After optimization	196.3	88.3	250.2	77.9	311.2	7.9	326.6
Change rate	−6.52%	−1.99%	5.13%	1.17%	−4.31%	12.86%	3.42%

Figure 7.   Steering angle error of wheels after optimization.

turning from middle position to the two sides of the limit position, the actual steering angle relation of the three wheels can approximate the Ackermann theoretical steering angle relation as far as possible and the problem of tire abnormal wearing is solved effectively at last.

5   CONCLUSIONS

Based on iSIGHT and ADAMS/View software, the integration of kinematics simulation of dual-front-axle steering mechanism is achieved in this paper. By using the Orthogonal Arrays DOE method, the steering mechanism is optimized by applying multi-objective optimization genetic algorithm NSGA-II. The optimal result shows that the tire abnormal wearing problem is solved effectively. As there are many reasons to cause the tire wearing, considering multi-conditions and multi-factors, the multi-objective optimization of steering system needs further researches. The application of integration technology of iSIGHT and ADAMS software could offer a new effective way of the design and development of automotive steering system, and the chassis as well.

ACKNOWLEDGEMENTS

This study is supported by Education Science Research Project of Hubei Provincial Department (D20122302) and the Open Fund Project of Hubei Province Key Laboratory of Modern Automotive Technology in Wuhan University of Technology (2014-08), in China.

REFERENCES

[1] Miller, G. & Reed, R. & Wheeler, F. 1991. Optimum Ackerman for Improved Steering Axle Tire Wear on Trucks. *SAE paper* 912693, 1991:572–578.
[2] Hu, M.F. 2013. Optimization of automobile dual-front-axle steering mechanism. *Master Degree Paper of Hunan University*, 2013, 06.
[3] He, Y.S. & Xiong T.C.1993. Design and Analysis of Multi-axle Steering System of Heavy Duty Vehicle. *SAE paper* 931919, 1993:160–166.
[4] Liu, Z.S. & Zhao, L.L. 2013. Motion compatibleness analysis and optimization of double front axle steering mechanism and suspension system. *China Mechanical Engineering*, 24(16):2164–2167.
[5] Li, Q.H. & Zhang, D.S. & Lv, S.Q. 2006. Design and optimization of the double-rocker of heavy vehicles with the double-front-axle based on the ADAMS models. *Journal of Hefei University of Technology*, 29(1):80–83.
[6] Lai, Y.Y. 2012. Isight Parameter optimization Theory and Example Detail Introduction. *Beijing: Beijing University of Aeronautics and Astronautics Press*.
[7] Mu, C.Y & Yu, J.N. & Yang, Y.Y. 2010, Wu Kai. Design for Dual-front axle Steering Angle of the Heavy Truck. *International Conference on Educational and Network Technology*, 2010, 185–187.
[8] Ning, J.X. & Zhou, H.X. 2007. Mathematical modeling and optimization on rocker mechanism of automobile. *Journal of Machine Design*, 24(2):65–67.
[9] Srinivas, N. & Deb, K. 1995. Multi-objective Function Optimization Using Non-dominated Sorting Genetic Algorithms. *Evolutionary Computation*, 2(3):221–248.
[10] Wei, F.T. & Song, L.L. & Yan, L.P. 2011. Application of iSIGHT in Solving Multi-objective Optimal Design Problems *International Conference on Consumer Electronics, Communications and Networks*, 2011, 110–113.
[11] Yuan, X.L. & Wang, J.Y. 2013. Kinematics simulation and optimization design of multi-axis steering system. *Mechanical Science and Technology for Aerospace Engineering*, 33(12):1795–1797.
[12] Zhu, L. & Feng, Y. & Yan, Y.B. 2013. Parameters optimization of dual-front-axle steering based on response surface methodology. *Hubei University of Automotive Technology*, 27(1):1–4.

Material science and advanced materials

Advances in Engineering Materials and Applied Mechanics – Zhang, Gao & Xu (Eds)

Research and analysis of giga-fatigue life of FV520B

Y.L. Zhang, J.L. Wang, Q.C. Sun & H. Zhang
Dalian University of Technology, Dalian, P.R. China

ABSTRACT: It is an important research topic in remanufacturing field to investigate the fatigue property and make up the lack of the experimental data of FV520B. In this paper, fatigue phenomenon and data were obtained from the results of the Ultrasonic fatigue experiment. The fatigue phenomenon showed that the inclusions could lead to fatigue failure. And with the fatigue data (life and inclusion size), a life prediction model could be obtained by using a fitting algorithm, then the suitability of the model was verified by the comparison between experiment data and predicted life, and errors were in acceptable range. This model could be used to predict the fatigue life of FV520B and this not only made a contribution to the theoretical study of FV520B.

1   INTRODUCTION

FV520B is a very important engineering material and has been widely used in the manufacture of mechanical equipments such as centrifugal compressor impellers or some other expensive, mechanical units. FV520B has good corrosion resistance, high hardness, and good welding property. However, during the usage there are many factors will affect the fatigue property of FV520B, one of these factors that researchers most commonly choose to study on is known as fatigue damage which is also called fatigue crack. An external cyclic loading will cause a high stress concentration in the position of inclusion and create cracks because of the significant property differences between inclusion and matrix material[1, 2]. The high stress concentration will lead to the crack nucleation, and this influence is more significant for high strength material as FV520B.

However, the investigation of the fatigue property of FV520B is still shallow and lacks enough fatigue life data. So in this paper, the fatigue experiment for FV520B was carried out to obtain the fatigue data and phenomenon to establish the specific model of FV520B. FV520B fatigue life prediction is difficult in theory, but useful in practical engineering. It has a high value in the study of remanufacturing and also the remanufacturing engineering.

2   FV520B FATIGUE EXPERIMENT

In this paper, the fatigue experiment for FV520B was carried out according to ASTM (American Society for Testing Material) standards, using an Ultrasonic Fatigue Test System operating at 20 KHz with zero mean stress that allows us to obtain fatigue properties completely reliable. The experiment observations, especially on the fracture characteristics, were investigated and measured by a Scanning Electron Microscope (SEM) and Optical Microscope (OM). The experimental results included inclusion size and fatigue life data.

2.1   *Fatigue experimental phenomenon*

From the experiment, several fatigue experimental phenomenon of FV520B had been observed, such as "Fish-eye" and Granular Bright Facet (GBF), as shown in Fig.1a and Fig.1b[3].

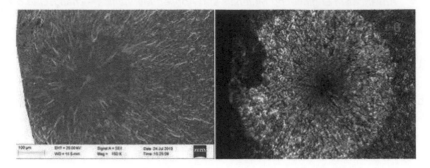

Figure 1.   a. A "Fish-eye" example (SEM). b. GBF caused by inclusion (OM).

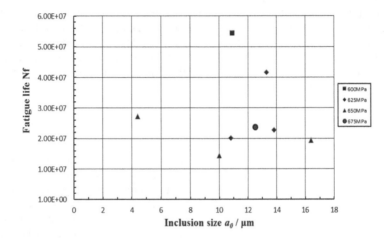

Figure 2.   Scatter plot of experimental data.

"Fish-eye" boundary can be clearly observed as reflected in Fig.1a, which proved that giga-cycle fatigue really happened in the experiment. It is the most significant characteristic of the giga-cycle fatigue fracture. GBF was clearly captured through the OM as shown in Fig.1b, it is the most significant characteristic of subsurface crack propagation and it has been widely accepted that the occurred giga-cycle fatigue failure is due to subsurface crack propagation.

2.2   *Fatigue experimental data*

According to the SEM images the inclusion size that was considered to be the crack size. The fatigue data obtained from the experiment of FV520B is shown in Fig. 2. The experimental data contains three parts: load amplitude, inclusion and life. But there is no clear relationship between inclusion and life.

3   FATIGUE LIFE PREDICTION MODEL

3.1   *Analysis of life prediction model*

In the past a few years, researchers have established many life prediction models. Professor Q.Y. Wang summarized a comprehensive model for metal fatigue life prediction, which contains many factors that will affect the fatigue life such as crack size, load and strength[4, 5].

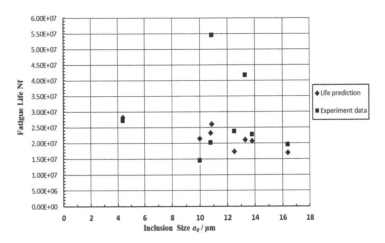

Figure 3. Experiment data and life prediction.

For high strength material, the long crack propagation life accounts for a very small proportion of the whole fatigue life, can be neglected. So the basic model can be simplified as shown in Eq. (1):

$$N_f = \frac{9\Delta K_{th}^2 G}{2E(\Delta\sigma - \Delta\sigma_D^R)^2 a_o} + \frac{a_o^{(1-n/2)}}{C\Delta\sigma^n \beta_1^n \pi^{n/2}(\frac{n}{2}-1)} \qquad (1)$$

where N_f is fatigue life; a_0 is inclusion size (equal to crack size); $\Delta\sigma$ is stress amplitude, $\Delta\sigma_D^R$ is fatigue strength limit when stress ratio is R; both E (elastic modulus) and G (shear modulus) are constant; β_1 is the geometry constant.

But this model cannot be used to predict fatigue life of all kinds of metals because there are three unknown parameters related to the FV520B property, including ΔK_{th}, C, and n, and these three parameters are constant to a particular metal. So the specific life prediction model for FV520B can be established with determination of the unknown parameters related to FV520B. According to the least square method and the iterative computation, the unknown parameters can be calculated based on the fatigue data, as shown in Eq. (2):

$$\Delta K_{th} = 6.3, \quad C = 3.14 \times 10^{-14}, \quad n = 2.95 \qquad (2)$$

The life prediction model can be obtained by substituting the three parameters into the basic model as shown in Eq. (3):

$$N_f = \frac{9 \times 6.3^2 G}{2E(\Delta\sigma - \Delta\sigma_D^R)^2 a_o} + \frac{a_o^{(1-2.95/2)}}{3.14 \times 10^{-14} \times \Delta\sigma^{2.95} \beta_1^{2.95} \pi^{2.95/2}(\frac{2.95}{2}-1)} \qquad (3)$$

3.2 The results and errors

As mentioned earlier, fatigue life prediction model for FV520B can be obtained by the substitution of three parameters, ΔK_{th}, C and n, that is model (3). These three parameters are deduced with the fatigue experimental data and the basic model is deduced by the theoretical formulas, so this model has both practicalness and theory. According to the inclusion size measured in the experiment, theoretical life can be calculated; Fig. 3 shows the relationship between the theoretical results and the experimental data.

Obviously, errors can be found between theoretical life and experimental data, such as 100.95%, 98.32%, 10.13%, 1.32%, 3.58%, 14.7%, 32.69%, 37.67, six sets of errors are in an acceptable range, that exclude two large errors.

Although the error exists, but they are still in acceptable range. Therefore, the life prediction can still make a meaningful prognostic for preventive maintenance decision can be considered in use of prediction of the fatigue life of FV520B in applications.

4 CONCLUSIONS

In this paper, fatigue experiment was carried out, FV520B specimen is tested in the Ultrasonic Fatigue Test System and the fitting model, based on the experiment data, is used to estimate the parameters of FV520B fatigue life model. By analyzing the errors between analytical results and experimental data, a modified fatigue life prediction model for FV520B can be obtained, which can be used to predict the fatigue life of FV520B and benefit the study of FV520B's fatigue characteristics and the prognostic of performance fatigue failure. Therefore, the research on remanufactured-part life and fatigue life prediction still faces many challenges in theory and applied modeling, further investigation and effort are urgently needed in remanufacturing engineering theory development.

ACKNOWLEDGEMENT

The authors gratefully acknowledge the financial support from the National Key Basic Research and Development Program (No. 2011CB013401).

REFERENCES

[1] Bao Rui and Zhang Xiang. 2010. Fatigue crack growth behaviour and life prediction for 2324-T39 and 7050-T7451 aluminium alloys under truncated load spectra, *Int J Fatigue,* 32,1180–1189.
[2] Mahadevan S, Yongming Liu and Stratman B. 2006. Fatigue crack initiation life prediction of railroad wheels, *Int J Fatigue*, 2847–756.
[3] Zhang Min, Wang Weiqiang, Wang Pengfeia, Liu Yana and Li Jianfenga. 2014. Fatigue behavior and mechanism of FV520B-I in ultrahigh cycle regime, *20th European Conference on Fracture (ECF20)*, 2035–2041.
[4] Wang Q.Y., Bathias C., Kawagoishi N and Chen Q. 2002. Effect of Inclusion on Subsurface Crack Initiation and Giga-cycle fatigue Strength, *Int J Fatigue*, 24, 1269–1274.
[5] Tanaka K and Mura, T. 1981. A Dislocation Model for Fatigue Crack Initiation, *J APPL MECH-T ASME*, 48, 97–103.

Advances in Engineering Materials and Applied Mechanics – Zhang, Gao & Xu (Eds)
© 2016 Taylor & Francis Group, London, ISBN 978-1-138-02834-0

Synthesis of $CaCO_3$ particles on stearic acid template

M. Cai, H.Y. Zeng, M.R. Jiao, Q.T. Cheng & S.P. Zhu
School of Materials Science and Engineering, Hubei University of Technology, Wuhan, China

J.X. Jiang
School of Materials Science and Engineering, Hubei University of Technology, Wuhan, China
Hubei Provincial Key Laboratory of Green Materials for Light Industry, Hubei University of Technology, Wuhan, China
Collaborative Innovation Center for Green Light-weight Materials and Processing, Hubei University of Technology, Wuhan, China

ABSTRACT: $CaCO_3$ powder was prepared by dispersing CO_2 gas into the $Ca(OH)_2$ slurry through a micropore plate with different addition of stearic acid as template. The polymorph and morphology of $CaCO_3$ particles were investigated by using XRD patterns and TEM photographs, respectively. The XRD patterns indicate that the addition of stearic acid cannot change the crystalline phase of $CaCO_3$ particles. The remarkable difference in morphology for different addition of stearic acid was observed in the TEM photographs. The formation mechanism of $CaCO_3$ powder on stearic acid template was proposed.

1 INTRODUCTION

Based on microdispersion, one of micromixing technology that has a tremendous mass transfer due to the path of molecular diffusion greatly shortens and homogenous mixing will take place in extremely short time (millisecond to microsecond grade) only depending on molecular diffusion. Wu et al.[1] have developed a new reactor to intensify gas–liquid mass transfer, in which a micro-pore glass plate with average pore size about 10 μm was adopted as gas disperser and this method was defined as micropore dispersion.

There are some literatures on the application of stearate in the preparation of $CaCO_3$ and in the surface modification of $CaCO_3$ particles. Ukrainczyk et al.[2] and García-Carmona et al.[3] have investigated the different impact of sodium stearate on the morphology of $CaCO_3$ particles at different concentration of dissolved Ca^{2+}. Trana et al.[4] have successfully modified nano-$CaCO_3$ particles with sodium stearate by adding at the end of carbonation stage and well dispersed nano-$CaCO_3$ particles and rod-like or spindle-like aggregates of primary particles are obtained.

However, the effect of the different amount of stearic acid on the morphology of $CaCO_3$ particles has rarely involved. Therefore, the effect of different amount of stearic acid on the polymorphs and morphologies of $CaCO_3$ particles is focused on and the formation mechanism of different morphology of $CaCO_3$ is proposed in the present work.

2 EXPERIMENT

$Ca(OH)_2$ and absolute ethyl alcohol are of analytical grade, stearic acid is of chemical grade, and the above regents are made by Sinopharm Chemical Reagent Company. Water is distilled water and the purity of CO_2 exceeds 99.5%. The schematic of reaction equipment refers to our previous research[5].

To prepare 100 mL of $Ca(OH)_2$ slurry, 5 g of pre-ground $Ca(OH)_2$ was added in the distilled water. Stearic acid was dissolved in absolute ethyl alcohol to form a solution with the concentration of 0.1 mol/L. A certain amount of this solution was added in the $Ca(OH)_2$ slurry to make the weight ratio of stearic acid to $CaCO_3$ was selected as 0, 1.0, 1.5, 2.0, 2.5, 3.0, 3.5 and 4.0% and the corresponding sample number was marked as CS0, CS1.0, CS1.5, CS2.0, CS2.5, CS3.0, CS3.5 and CS4.0, respectively. CO_2 gas was introduced into the mixture and reaction was not stopped until PH value dropped to 7. The final suspension was centrifuged at 8000 rpm for 10 min and the supernatant solution was discharged, and then the solid was redispersed in distilled absolute ethyl alcohol. This process was repeated and sedimented particles were dried in the oven at 110°C for 4 h.

The polymorphs of $CaCO_3$ were detected by the XRD (D/MAX–RB, RIGAKU). The morphology was characterized by the TEM (Tecnai, G20).

3 RESULTS AND DISCUSSION

To study the effect of different addition of stearic acid on the polymorphs of $CaCO_3$, XRD experiments were carried out and the results are shown in Figure 1. From the XRD patterns, it can be found that calcite is the only crystalline phase in all samples, which indicates that the different amount of stearic acid cannot alter the crystal type of $CaCO_3$.

Rhombic calcite, needle-like aragonite and spherical vaterite are the three typical polymorphs of $CaCO_3$. Different crystalline phases can be obtained in different preparation method. Generally, calcite is the only crystalline phase during the carbonation of $Ca(OH)_2$ aqueous slurry[1–3,6–9]. The result in the present work showed that stearic acid cannot change the crystalline phase of $CaCO_3$ particles as in accordance with the case of other surfactants.

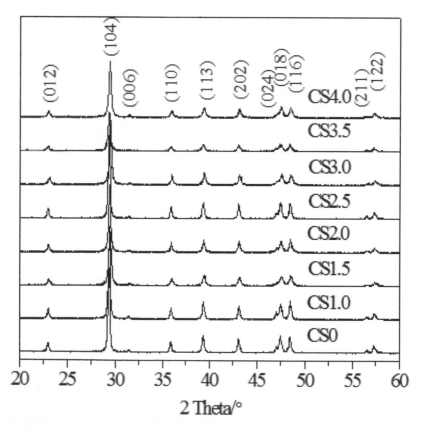

Figure 1.   XRD patterns of $CaCO_3$ with different addition of stearic acid.

TEM images of CaCO$_3$ particles at the absence and in the presence of stearic acid with the amount of 1.5, 2.5 and 3.5 wt% are shown in Figure 2. From Figure 2(a) and (b), although some particles with size less than 100 nm exist in the sample without addition of stearic acid, lots of large particles and irregular aggregation can be found. In the sample with addition of 1.5 wt% stearic acid, small CaCO$_3$ particles gather into regular rod-like aggregation, as shown in Figure 2(c). From Figure 2(d) and (e), most of the small CaCO$_3$ particles gather into imperfect spindle-like aggregation when 2.5 wt% of stearic acid is added. When the addition of stearic acid reaches to 3.5 wt%, the profile of spindle-like aggregation becomes more integrated, as shown in Figure 2(f). In addition, the single CaCO$_3$ particle in the sample with addition of stearic acid is smaller than that of in the sample at absence of stearic acid.

It is well understood that the decrease in the particle size of CaCO$_3$ in the presence of surfactant is due to the adsorption of surfactant on the surface of CaCO$_3$ particles caused by the amphiphilic structure of surfactant. Spherical micelles in tail-to-tail arrangement form in the hydrophilic polar solvents when the amount of stearic acid, an amphiphilic molecule, reaches to a certain value[10], and the micelles deform into ellipsoid when CaCO$_3$ particles exist in the system. Ca(OH)$_2$ powder added in distilled water cannot completely dissolve and the undissolved particles suspend in slurry. When stearic acid is added in the slurry, the amphiphilic molecules first adsorb to form a monolayer with the hydrophilic headgroup bonded to the surface of Ca(OH)$_2$ particles lining up with one another having the tails perpendicular to the surface and the adsorption is chemisorption. The amphiphilic molecules of insoluble calcium sterate, formed by stearic acid and Ca^{2+} ions, then adsorb on the top of this first monolayer in tail-to-tail arrangement to form a hemimicelle[11] and the adsorption belongs to physisorption. The appearance of the whole micelle consisted of the two adsorptions is also ellipsoidal, as illustrated in Figure 3(a). When CO$_2$ gas is introduced, extremely small CaCO$_3$ particles form and Ca(OH)$_2$ particles are gradually consumed, during which CaCO$_3$ particles aggregate under the driving force of high surface energy. The aggregation process results in the attraction between micelles and line contact is the contact type due to the maximum reduction in surface energy, as shown in Figure 3(c). When the concentration of stearic acid is enhanced,

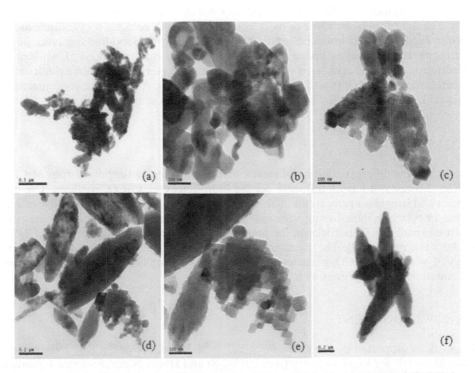

Figure 2. TEM photograph for the samples (a)–(b) CS0, (c) CS1.5, (d)–(e) CS2.5 and (f) CS3.5.

497

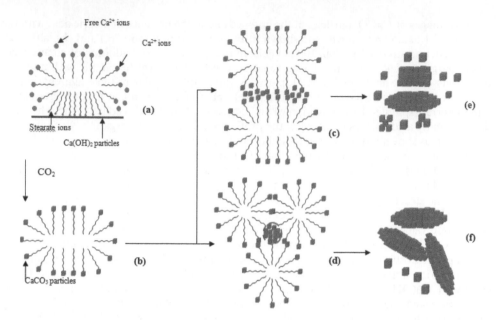

Figure 3. (a) Adsorption of stearic acid on surface of $Ca(OH)_2$ particles, the formation of micelles consists of stearic acid and calcium stearate. (b) Formation of extremely small $CaCO_3$ particles and consumption of $Ca(OH)_2$ particles after the introduction of CO_2. (c) Aggregation of micelles forced by surface energy of $CaCO_3$ particles at low addition of stearic acid. (d) Aggregation of micelle forced by surface energy of $CaCO_3$ particles at increased addition of stearic acid (sectional view). (e)–(f) Crystal growth or/and aggregation of $CaCO_3$ grains at different addition of stearic acid.

the distance between micelles shortens and the contact opportunity of more micelles (three or four micelles) increases. The contact type does not change, however, the result is some different. The area formed by three or more ellipsoidal micelles becomes a cylindrical area, as shown in Figure 3(d). At low concentration of stearic acid, small rod-like aggregations form centering on the intersecting line of two ellipsoidal micelles (in sample CS1.5), while at high concentration of stearic acid, lager spindle-like aggregations form around the cylindrical area between three or more ellipsoidal micelles (in sample CS3.5). At moderate concentration of stearic acid, rod-like and spindle-like aggregations coexist (in sample CS2.5).

4   CONCLUSIONS

In summary, the different addition of stearic acid cannot change the polymorphs of $CaCO_3$ particles, however, it has an influence on the morphology of $CaCO_3$ particles. The larger particles and irregular aggregations at absence of stearic acid become regular rod-like aggregations, imperfect spindle-like aggregations and relative integrated spindle-like aggregations consist of small $CaCO_3$ particles in the presence of stearic acid with the amount of 1.5, 2.5 and 3.5 wt%, respectively. The interattraction of ellipsoidal micelles forced by high surface energy of very small $CaCO_3$ particles adsorbed on the hydrophilic end of micelles results in the formation of aggregations with different shapes.

ACKNOWLEDGEMENT

This work is supported by the Open Foundation of Hubei Provincial Key Laboratory of Green Materials for Light Industry (No. 2013-2-8) and Hubei Natural Science Foundation (No. 2009CDA030).

REFERENCES

[1] Wu, G.H. Wang, Y.J. Zhu, S.L. et al. 2007. Preparation of ultrafine calcium carbonate particles with micropore dispersion method. *Powder Technology*: 172 82–88.

[2] Ukrainczyk, M. Kontrec, J. Kralj, D. 2009. Precipitation of different calcite crystal morphologies in the presence of sodium stearate. *Journal of Colloid Interface Science*: 329 89–96.

[3] García-Carmona, J. Gómez-Morales, J. Rodríguez-Clemente, R. 2003. Rhombohedral—scalenohedral calcite transition produced by adjusting the solution electrical conductivity in the system $Ca(OH)_2$–CO_2–H_2O. *Journal of Colloid Interface Science*: 261 434–440.

[4] Trana, H.V. Tranb, L.D. Vua, H.D. et al. 2010. Facile surface modification of nanoprecipitated calcium carbonate by adsorption of sodium stearate in aqueous solution. *Colloids Surface A*: 366 95–103.

[5] Jiang, J.X. Liu, J. Liu, C. Zhang, G.W. et al. 2011. Roles of oleic acid during micropore dispersing preparation of nano-calcium carbonate particles. *Applied Surface Science*: 257 7047–7053.

[6] Han, Y.S. Hadiko, G. Fuji, M. et al. 2006. Factors affecting the phase and morphology of $CaCO_3$ prepared by a bubbling method. *Journal of the European Ceramics Society*: 26 843–847.

[7] Wang, C.Y. Piao, C. Zhai, X.L. et al. 2010. Synthesis and characterization of hydrophobic calcium carbonate particles via a dodecanoic acid inducing process. *Powder Technology*: 198 131–134.

[8] Liu, Q. Wang, Q. Xiang, L. 2008. Influence of poly acrylic acid on the dispersion of calcite nano-particles. *Applied Surface Science*: 254 7104–7108.

[9] Zhao, Z.X. Zhang, L. Dai, H.X. et al. 2011. Surfactant-assisted solvo- or hydrothermal fabrication and characterization of high-surface-area porous calcium carbonate with multiple morphologies. *Microporous and Mesoporous Materials*: 138 191–199.

[10] De Mul, M.N.G. Davis, H.T. Evans, D.F. et al. 2000. Solution Phase Behavior and Solid Phase Structure of Long-Chain Sodium Soap Mixtures. *Langmuir*: 16 8276–8284.

[11] Chen, S.H. Frank, C.W. 1989. Infrared and fluorescence spectroscopic studies of self-assembled n-alkanoic acid monolayers. *Langmuir*: 5 978–987.

Advances in Engineering Materials and Applied Mechanics – Zhang, Gao & Xu (Eds)
© *2016 Taylor & Francis Group, London, ISBN 978-1-138-02834-0*

Study on the manufacturing technology of optical materials

Y. Chen
*Key Laboratory of Optical System Advanced Manufacturing Technology, Changchun Institute of Optics,
Fine Mechanics and Physics, Chinese Academy of Sciences, Changchun, China*

ABSTRACT: In this paper, the performance comparison between glass ceramics (Zerodur) and Silicon Carbide (SiC), which are two kinds of common materials used for space optical reflector, is carried out, and several lightweight structure forms are analyzed. The oval plane reflector is applied in ultra-low temperature environment of space, by taking this kind of reflector as an example, its lightweight structure is optimized by CAD, then through finite element analysis, deformation of the planar lightweight mirror is 0.014λ (rms) in gravity condition, and deformation is 0.002λ (rms) in ultra-low temperature of -150 degrees environment. The actual lightweight processing is controlled by the CNC system in a graphical way, and a chemical method is used to eliminate the stress and micro crack generated during processing, its final surface shape precision reaches 0.022λ (rms). Finally, this paper introduces the manufacture method of novel Silicon Carbide (SiC), and analyzes the current situation and development trends of the spatial lightweight reflector manufacturing technology.

1   INTRODUCTION

Recently, national optics technology has been developed rapidly, and the optical system is the important payload of the space remote sensor, whose quality determines the transmission cost. The camera should have a minimum size and quality under the premise of ensuring better imaging quality of the optical system. This requires that the camera should have sufficiently high dynamic and static stiffness and strength, meanwhile its optical system has the maximum degree of lightweight. Lightweight process of the camera primary reflector is the most basic and important key techniques.

2   REFLECTOR MATERIAL

2.1   *Requirements for materials*

Lightweight mirror should be able to meet the quality requirements of optical surface shape through optical processing system; and when the mirror is working in the space environment, the surface accuracy of which should be stable. Therefore, the selection of light reflector material should primarily consider several aspects such as rigidity, thermal stability, chemical stability, operational safety, and processing quality. The material of mirror blank must be able to adapt to the requirements of the optical surface shape processing, the microstructure of it should be homogeneous, at the same time, it should be able to withstand the process conditions of optical coating and easy to combine with the reflective film. The material thermal stability including its low coefficient of expansion and good thermal conductivity, a combination of these two properties help the body of the mirror to eliminate its temperature gradient, so that the reflector quickly achieves thermal equilibrium. The high specific stiffness of materials is conducive to achieve higher rates of lightweight of the mirrors, and also has a stronger ability to resist vibration and shock. The existing materials are difficult to simultaneously achieve these performance requirements; selection of materials

Table 1. The performance parameters of two commonly used materials.

Material	Performance					
	Density ρ (g/cm³)	Elastic modulus E (GPa)	Coefficient of thermal expansion α (10⁻⁶/K)	Thermal conductivity λ (W/m·K)	Specific heat Cp (J/g·K)	Ratio of stiffness E/ρ
Zerodur	2.53	91	0.05	1.64	0.821	35.97
SiC	3.13	400	2.5	141	0.69	127.8

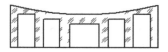

Figure 1. The honeycomb structure with back openings.

(a) Triangle holes; (b) Quadrilateral holes; (c) Hexagonal holes; (d) Round holes;(e) Fan-shaped holds

Figure 2. Commonly used lightweight holes.

must be a comprehensive comparison, selected as far as possible to meet system performance requirements.

2.2 *Light reflector materials*

The common materials used for mirrors include glass ceramics (Zerodur) and Silicon Carbide (SiC). The following is a list about performance parameters of these two kinds of materials.

The Table 1 shows, glass ceramics with low thermal expansion coefficient and small thermal deformation, can obtain high quality optical surface, so it is known as zero-expansion glass ceramic, its technology is more mature and the material is relatively cheap, because all of above reasons, applications in the modern space optical systems are more widely; with higher stiffness, silicon carbide can be made for reflector with a higher rate of lightweight, but its processing efficiency is lower than the glass ceramic. However, with the manufacture technology of silicon carbide is mature, light reflector made of silicon carbide has been more and more used in the engineering, so silicon carbide is becoming the most potential material used to make space optical system reflector.

3 LIGHTWEIGHT STRUCTURE

At present, the structure of the space light reflector body is in the form of the honeycomb with back opening. A schematic diagram is shown in Figure 1.

The shapes of the honeycomb holes on the back of the lightweight reflectors are various, such as triangle holes, quadrilateral holes, hexagonal holes, fan-shaped holes, round holes, hole and hole mix. Their forms are shown in Figure 2.

The density and layout of the lightweight holes should be considered according to the requirements of the optical system on the primary mirror surface. The circular holes are easy to process among the holes, but the lightweight rate is relatively low; Fan-shaped hole is generally used for lightweight of round mirrors with a center hole, which can be divided into continuous and discontinuous ribs lightweight holes according to the different types of the ribs arranged. The process of triangular holes, hexagonal holes and rectangular holes is similar; however, the longest side length is required to be a triangle to enclose a hollow area, quadrilateral is second, hexagonal is minimum, for the same diameter of the circular lens body, the mirror that adopts the way of triangle lightweight has the largest mass and lowest level of lightweight, so its area density is maximum, geometry remains unchanged in the other case, the mirror with the same mass density to the triangular lightweight mirror has thicker lens and higher surface precision.

4 WEIGHTLIGHT MANUFACTURING AND RESULTS

4.1 Lightweight structure analysis

In this paper, based on the oval planar reflector as an example to explain the lightweight manufacturing process of the reflector made of glass-ceramic, the reflector dimensions are: long axis 730 mm × short axis 525 mm, thickness 90 mm, surface shape accuracy better than 0.025 λ (rms, λ = 633 nm), lightweight rate greater than 30%, the lowest application temperature −150°C.

Through the CAD optimized design, the lightweight rate of lightweight reflector is 33%, the thickness is 20 mm, the rib thickness of triangular hole is 14 mm, and distance from the center of mass to the reflector surface is 39.95mm. It can be observed in Figure 3.

Because of the large size of reflector, we are more concerned about the effect its weight and low temperatures on the surface accuracy, so finite element analysis should be made, the finite element model of the overall structure of the reflector is shown in Figure 4.

Obtained by calculating, the results of the light reflector deformation under gravity are shown in Table 2, the mirror deformation at −150°C temperature environment is shown in Table 3.

4.2 Light manufacturing

Polished lightweight mirror is shown in Figure 5. Due to the larger caliber, the long axis of the oval reflector can be reached 730 mm, while the diameter of ZYGO 24″ flat interferom-

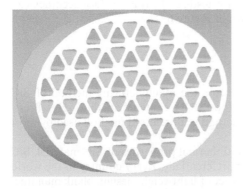

Figure 3. CAD three-dimensional map of light reflector.

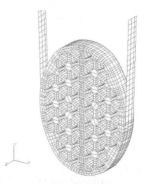

Figure 4. The finite element model of overall structure.

Table 2. Results of reflector deformation.

Evaluation method	PV (nm)	RMS (nm)
Results	32.96	8.68

Table 3. Mirror deformation at −150°C temperature.

Temperature (°C)	PV (nm)	RMS (nm)
−150	8.00	1.14

Figure 5.   Polished lightweight mirror.

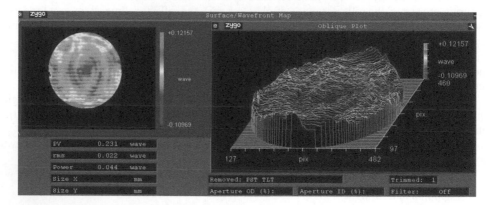

Figure 6.   The surface test results of reflector.

eter existed in laboratory is only Φ610 mm, it can not test the full aperture of the reflector, therefore, by using the Ritchey–Common method for testing, the final surface accuracy can be reach 0.022 λ (rms, λ = 633 nm), which is shown in Figure 6.

5   SILICON CARBIDE MATERIALS

SiC is a novel material used for space optical reflector, the manufacturing method of SiC reflector blank mainly are: Reaction Bonded method (RB), Sintered method, Hot Press/Hot Isostatic Press (HP/HIP), etc., in which RB method is a preferred method used by most countries to manufacture SiC reflector blank.

Because of the excellent physical, mechanical and thermal properties of SiC material, SiC is more and more chosen to be used for reflector preferentially. Every country in the world have accumulated a wealth of experience in the respect of lightweight design; blank manufacturing; mirror optical processing and testing, technology is more mature. Domestic research in this field starts later, a lot of research about integration of SiC reflector design, manufacturing, evaluation and testing has to be done.

6   CONCLUSIONS

With the improvement of the space optical remote sensor resolution in the future, the reflector diameter of the space optical remote sensor increases, the demand for lightweight of the structure and adapting to the changes of the space environment gets corresponding increases, which further leads to the choice of materials, structural design, processing methods, and many other innovations, in the future the following three aspects of research should be strengthened.

1. From the aspect of material selection, currently the most commonly used material is glass-ceramic (Zerodur) and Silicon Carbide (SiC), both have their own characteristics, whether there are novel reflector materials with more excellent properties are what we need to explore actively;
2. Doing more in-depth study for the existing lightweight design, make sure the space optical reflector guarantees the stability of the surface accuracy, on this premise, make the lightweight rate further improve to reduce costs and improve production efficiency, to meet the need for space optical engineering;
3. The space for the manufacturing methods of lightweight reflector also can be improved; the manufacture of advanced equipment and the improvement of processing technology are all worth us to devote more effort.

ACKNOWLEDGEMENTS

The authors would like to thank the large lightweighted mirror lab in Changchun Institute of Optics, Fine Mechanics and Physics, Chinese Academy of Sciences, for the technique supports.

REFERENCES

Chen, H.N., Wang, J.L. Liu, J. 2013. Uniform Removal of the Surface Material of the Optical Components in Ultra-smooth Technology. *Acta Photonica Sinica*, 42(4):417–422.

Li SH Y., Dai Y.F., Xie X.H., et al. 2011. New Technology for Manufacturing and Measurement of large and Middle-scale aspheric surfaces, *National Defense Industry Press*,:37–40.

Lee, Y.J., Joo, H.J. 2004, Ablation characteristics of carbon fiber reinforced carbon (CFRC) composites in the presence of silicon carbide (SIC) coating. *Surface and Coatings Technology*, 180–181.

Liu, H.W., Zhang, Q., et al. 2003. Design of strip primary mirror supporting structure based on finite element analysis. *Optics and Precision Engineering*. 11(6):556–558.

Preston, F.W. 1927. Glass technology. *Journal of the society of Glass Technology*. 11(10):277–281.

Sen Han, Erik Novak and Mike Schurig. 2001. Application of Ritchey-Common Test in Large Flat Measurements. *Proc. of Spie*. 4399:131–136.

Wu, Q.B., Chen, SH J., Dong, S.H. 2003. Optimization of parameters structural design of lightweight reflector. *Optics and Precision Engineering*. 11(5):467–470.

Y.M., Zhang J.H., et al. 2004. Large-scale fabrication of lightweight Si/SiC ceramic composite optical mirror. *Materials Letters*. 58:1204–1208.

Yan, Y., Jing, G., Yang, H.B. 2008. Lightweight structural design of space mirror. *Infrared and Laser Engineering*, 2(1):98–100.

Yang, L. 2001. *Advanced optical manufacture technology*. Bei Jing: Science Press. 195–198.

Zhao, W.X., Zhang, G., Zhao, R.C., Bao, J.X. 2011. Fabrication of silicon carbide lightweight mirror blank. *Optics and Precision Engineering*, 11(11):2609–2617.

Nucleation, crystallization and growth of $CaCO_3$ grains on lauric acid template during micropore dispersion carbonation

Q.T. Cheng, X.P. Gan, Z.L. Yan, M. Cai & H.Y. Zeng
School of Materials Science and Engineering, Hubei University of Technology, Wuhan, China

J.X. Jiang
School of Materials Science and Engineering, Hubei University of Technology, Wuhan, China
Hubei Provincial Key Laboratory of Green Materials for Light Industry, Hubei University of Technology, Wuhan, China
Collaborative Innovation Center for Green Light-weight Materials and Processing, Hubei University of Technology, Wuhan, China

ABSTRACT: Nano $CaCO_3$ powder was prepared by dispersing CO_2 into $Ca(OH)_2$ slurry *via* a micropore plate with different addition of lauric acid as template, and the nucleation, crystallization and growth of $CaCO_3$ grains on lauric acid template were investigated in the present work. The polymorph and morphology of $CaCO_3$ particles was characterized by XRD and TEM, respectively. XRD patterns indicated that calcite is the only crystalline phase for different addition of lauric acid. From TEM images, decreased particle size and improved dispersity were found when lauric acid was added and the best dispersity was at 2.5 wt% of lauric acid. The mechanism of nucleation, crystallization and growth of $CaCO_3$ grains on lauric acid template were also proposed.

1 INTRODUCTION

In traditional carbonation of $CaCO_3$, CO_2 gas was blown into $Ca(OH)_2$ slurry *via* a tube with diameter of millimeters. Because of the inefficiency of mass transfer between liquid and gas, the carbonation rate is low, the carbonization time is long and the size of particles is big and not even. Wu et al[1] developed a new reactor to intensify gas–liquid mass transfer, in which a micro–pore glass plate with average pore size about 10 μm was adopted as gas disperser and this method was defined as micropore dispersion.

Although other carboxylic acids have been employed as surfactant in the preparation of $CaCO_3$[2-6] and in the surface modification of $CaCO_3$ particles to improve their dispersibility[7], the application of lauric acid in this aspect has rarely been reported. C.Y. Wang et al[8] have synthesized $CaCO_3$ powder consists of most rod–like particles with diameter of 200–400 nm and length of 2–4 μm and few ellipse–like particles with diameter of about 100 nm when lauric acid is used as organic substrate. Nevertheless, the action mechanism of lauric acid on the morphology of $CaCO_3$ particles has not been involved. Therefore, nano $CaCO_3$ powder is prepared by dispersing CO_2 into $Ca(OH)_2$ slurry *via* a micropore plate with assistance of lauric acid as surfactant, the effect of different addition of lauric acid on the polymorphs and morphologies of $CaCO_3$ particles is investigated and the mechanism of nucleation, crystallization and growth of $CaCO_3$ grains on lauric acid template are focused on in the present work.

2 EXPERIMENT

$Ca(OH)_2$ and absolute ethyl alcohol are of analytical grade, lauric acid is of chemical grade, and the above regents are purchased from Sinopharm Chemical Reagent Company. Water is distilled water and the purity of CO_2 exceeds 99.5%. The schematic of reaction equipment refers to our previous research[2].

5 g of pre–ground $Ca(OH)_2$ was added in distilled water to prepare 100 mL $Ca(OH)_2$ slurry. Lauric acid was dissolved in absolute ethyl alcohol to form a solution with the concentration of 0.1 mol/L and a certain amount of this solution was added in $Ca(OH)_2$ slurry. The weight ratio of lauric acid to $CaCO_3$ was selected as 0, 1.0, 1.5, 2.0, 2.5, 3.0, 3.5 and 4.0% and the corresponding sample No was marked as CL0, CL1.0, CL1.5, CL2.0, CL2.5, CL3.0, CL3.5 and CL4.0, respectively. CO_2 gas was introduced into the mixture and reaction was not stopped until PH value dropped to 7. The final suspension was centrifuged at 8000 rpm for 10 min and the supernatant solution was discharged, and then the solid was redispersed in distilled absolute ethyl alcohol. This process was repeated and sedimented particles were dried in the oven at 110 °C for 4 h.

The polymorphs of $CaCO_3$ were detected by XRD (D/MAX–RB, RIGAKU). The morphology was characterized by TEM (Tecnai, G20).

3   RESULTS AND DISCUSSION

XRD patterns of $CaCO_3$ powder prepared with different addition of lauric acid are shown in Figure 1, from which it can be found that calcite is the only crystalline phase in all samples, indicating that the addition of lauric acid cannot alter the polymorphs of $CaCO_3$.

Calcite, aragonite and vaterite are three polymorphs of $CaCO_3$. Different crystalline phases can be obtained in different preparation method. Generally, calcite is the only crystalline phase during the carbonation of $Ca(OH)_2$ slurry[1–4,9]. The result in this work that lauric acid cannot change the polymorphs of $CaCO_3$ particles is in accordance with previous conclusions.

TEM images of $CaCO_3$ particles at absence and in the presence of lauric acid with the amount of 1.5, 2.5 and 3.5 wt% are illustrated in Figure 2. From Figure 2a and 2b, although some particles with size less than 100 nm exist in the sample without addition of lauric acid, lots of large particles and serious aggregation can be found. In the sample with addition of 1.5 wt% of lauric acid, as shown in Figure 2c and 2d, the average particle size is about 30–40 nm, the dispersity is markedly improved and the shape of aggregation is irregular. Although there is little difference in morphology of single particle, the dispersity is further improved when the amount of lauric acid is increased to 2.5 wt%, as shown in Figure 2e and f, and the dispersity become slightly poor when the amount of lauric acid reached to 3.5 wt%, as shown in Figure 2g and h.

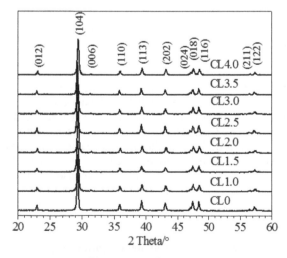

Figure 1.   XRD patterns of $CaCO_3$ with different addition of lauric acid.

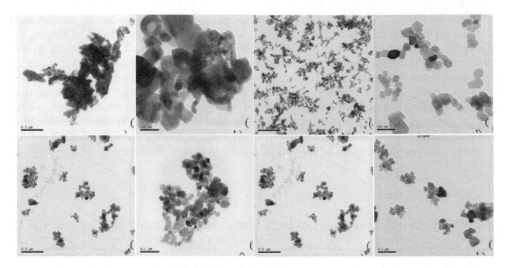

Figure 2.　TEM photograph of sample of (a)–(b) CL0, (c)–(d) CL1.5, (e)–(f) CL2.5 and (g)–(h) CL3.5.

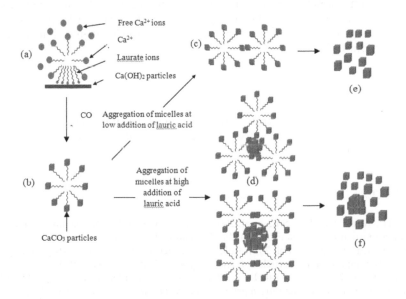

Figure 3.　(a) Adsorption of lauric acid on surface of $Ca(OH)_2$ particles, the formation of micelles consist of lauric acid and calcium salt of fatty acid. (b) Formation of extremely small $CaCO_3$ particles and consumption of $Ca(OH)_2$ particles after the introduction of CO_2. (c) Aggregation of micelles forced by surface energy of $CaCO_3$ particles at low addition of lauric acid. (d) Aggregation of micelle forced by surface energy of $CaCO_3$ particles at increased addition of lauric acid (sectional view). (e) and (f) Crystal growth or/and aggregation of $CaCO_3$ grains at different addition of lauric acid.

$Ca(OH)_2$ powder cannot completely dissolve when it is added in distilled water and the undissolved particles suspend in slurry. When lauric acid is added in slurry, the amphiphilic molecules first adsorb to form a monolayer with the hydrophilic headgroup bonded to the surface of $Ca(OH)_2$ particles lining up with one another having the tails perpendicular to the surface and the adsorption is chemisorption. The amphiphilic molecules of insoluble calcium laurate, formed by lauric acid and Ca^{2+} ions, then adsorb on the top of this first monolayer in tail–to–tail arrangement to form a hemimicelle[10] and the adsorption belongs to physisorption. The appearance of the whole micelle consisted of the two adsorptions is

509

spherical, as illustrated in Figure 3a. When CO_2 gas is introduced, extremely small $CaCO_3$ particles form and $Ca(OH)_2$ particles are gradually consumed, during which $CaCO_3$ particles aggregate under the driving force of high surface energy. Point contact is the contact type for spherical micelle because of the spherical symmetry, as shown in Figure 3b and c. When the concentration of lauric acid is increased, the distance between micelles shortens and the contact opportunity of more micelles grows. Point contact is still the contact type, however, three or more spherical micelles gather together to form a spherical space, in which extremely small $CaCO_3$ particles aggregate and grow up, as illustrated in Figure 3d. When the carbonation finished, different morphologies of $CaCO_3$ particles and aggregations, with spherical micelles adsorbed on their surface, can be formed, as shown in Figure 3e and f.

4 CONCLUSIONS

In summary, the different addition of lauric acid cannot change the polymorphs of $CaCO_3$, however, it has influence on the morphology of $CaCO_3$ particles. Comparing with the case at absence of lauric acid, the average particle size markedly decreases and the aggregation is remarkably improved when lauric acid is added. Although there is no evident difference in particle size with different amount of lauric acid, the sample with 2.5 wt% lauric acid has the best dispersity. The formation of spherical micelles and the interattraction between micelles under the driving force of high surface energy of extremely small $CaCO_3$ particles adsorbed on the polar group of lauric acid results in the different particles aggregation in the presence of different amount of lauric acid.

ACKNOWLEDGEMENT

This work is supported by the Open Foundation of Hubei Provincial Key Laboratory of Green Materials for Light Industry (No. 2013-2-8) and Hubei Natural Science Foundation (No. 2009CDA030).

REFERENCES

[1] Wu, G.H., Wang, Y.J., Zhu, S.L., et al. 2007. Preparation of ultrafine calcium carbonate partcles with micropore dispersion method. *Powder Technology*: 172 82–88.
[2] Jiang, J.X., Liu, J., Liu, C., et al. 2011. Roles of oleic acid during micropore dispersing preparation of nano-calcium carbonate particles. *Applied Surface Science*: 257 7047–7053.
[3] Ukrainczyk, M., Kontrec, J., Kralj, D., 2009. Precipitation of different calcite crystal mophoologies in the presence of sodium stearate. *Journal of Colloid Interface Science*: 329 89–96.
[4] García-Carmona, J., Gómez-Morales, J., Rodríguez-Clemente, R., 2003. Rhombohedral–scalenohedral calcite transition produced by adjusting the solution electrical conductivity in the system $Ca(OH)_2$–CO_2–H_2O. *Journal of Colloid Interface Science*: 261 434–440.
[5] Chen, Y.X., Ji, X.B., Wang, X.B., 2010. Facile synthesis and characterization of hydrophobic vaterite $CaCO_3$ with novel spike-like morphology via a solution route. *Materials Letters*: 64 2184–2187.
[6] Wang, C.Y., Piao, C., Zhai, X.L., et al. 2010. Synthesis and character of super-hydrophobic CaCO3 powder in situ. *Powder Technology*: 200 84–86.
[7] Trana, H.V., Tranb, L.D., Vua, H.D., et al. 2010. Facile surface modification of nanoprecipitated calcium carbonate by adsorption of sodium stearate in aqueous solution. *Colloids Surface A*: 366 95–103.
[8] Wang, C.Y., Piao, C., Zhai, X.L., et al. 2010. Synthesis and characterization of hydrophobic calcium carbonate particles via a dodecanoic acid inducing process. *Powder Technology*: 198 131–134.
[9] Wang, C.Y., Sheng, Y., Bala, H., et al. 2007. A novel aqueous-phase route to synthesize hydrophobic CaCO3 particles in situ, *Materials Science and Engineering C*: 27 42–45.
[10] Chen, S.H., Frank, C.W., 1989. Infrared and fluorescence spectroscopic studies of self-assembled n-alkanoic acid monolayers. *Langmuir*: 5 978–987.

Advances in Engineering Materials and Applied Mechanics – Zhang, Gao & Xu (Eds)
© 2016 Taylor & Francis Group, London, ISBN 978-1-138-02834-0

Preparation of onion-based composite and study on its antimicrobial activity

X.N. Cheng, R.Y. Song, Y.F. Qian & J. Wei
School of Textile and Material Engineering, Dalian Polytechnic University, Dalian, China

ABSTRACT: The preparing method of onion-based composite was studied in this project. These materials took onion fiber as the base and took tussah silk as the reinforcement. The composite material showed the largest tensile strength when the weight contain of tussah silk is 31.75% and the length of tussah silk is 2 mm. Onion skin contains natural flavonoids that has good antibacterial properties. The ethanol method was adopted in this study for extracting flavonoids from abandoned onion skins, and then made it as the antibacterial agent for the composite materials. The results showed that the best extraction technology was ethanol concentration 60%, extraction time 2 h, extraction temperature 60°C, and ratio of solid to liquid 1:20. After antibacterial finishing, the produced composite showed some antimicrobial activity against both *Staphylococcus aureus* and *Escherichia coli*, and the antibacterial effect on *Staphylococcus aureus* was better than *Escherichia coli*.

1 INTRODUCTION

The majority of packaging material is usually made of high molecular compound. These packaging materials contain chemical substances which is harmful to human body. This composite not only causes environmental pollution, but also causes great harm to people's health[1].

The method for preparing onion-based composite was studied in this project. These materials took onion as the basis and took tussah silk as the reinforcement. The composite material has good mechanical properties, which can be used as a kind of degradable green packaging materials.

Bacteria are every where in our daily life, which contains a lot of harmful pathogenic bacteria. With the improvement of people's life level, the antimicrobial resistance of daily necessities is increasingly attracting the attention of people. The traditional antibacterial finishing agent has harmful on the environment and human body[2]. More and more people like the antimicrobial agent that extracted from plants, because it is safe, non-toxic side effects, environmental protection.

Onion skin contains a lot of natural flavonoids, which is a kind of antibacterial substance. Ethanol extraction method was adopted in this study for extracting the flavonoids from abandoned onion skin, and then made it as the antibacterial agent for composite materials.

2 EXPERIMENTAL

2.1 Materials and chemical

Fresh onions were purchased from the vegetable market. Tussah silk was originated from abandoner silk guilt. Rutin, ethanol, sodium hydroxide, sodium nitrite, aluminum nitrate, peptone, beef extract, agar and sodium chloride were purchased from Kemiou Chemical Inc. (Tianjin, China).

2.2   The molding method of composite materials

According to the process of the experiment[3], method for preparing onion-based composite was studied in this project. These materials take onion fiber as basics and take tussah silk as reinforcement. This paper researches the influence of added quantity and fiber length on the mechanical properties of composite materials.

2.3   Extraction of flavonoids

According to the previous study[4–5], the influence factors of ethanol extraction of flavonoids from onion skins are the concentration of ethanol, extraction of solid–liquid ratio, extraction of temperature, and extraction of time. The orthogonal experiment was done on these four factors in this paper, and we studied the influence of various factors on the extraction rate of flavonoids.

2.4   The antimicrobial activity test

The antimicrobial performance was tested by agar plate diffusion method (GB/T 20944.1-2007) using *Staphylococcus aureus* and *Escherichia coli*. The antibacterial effect was evaluated by the size of inhibition zone[6–8].

3   RESULTS AND DISCUSSION

3.1   Mechanical properties of composite materials

Take 15 cm long, 1.5 cm wide strip from the onion-based composite material as the experimental sample. Then test the tensile strength of experimental sample. The effect of silk content and silk length on the mechanical properties of the composites was showed in Figure 1 and Figure 2.

It can be seen from Figure 1, with the increasing of tussah silk content in composite materials, the thickness of the composite materials is more and more thin. With the increase of tussah silk content, the tensile strength of composites increased gradually. When the tussah silk content in the composite material is about 31.75%, the tensile strength of composite

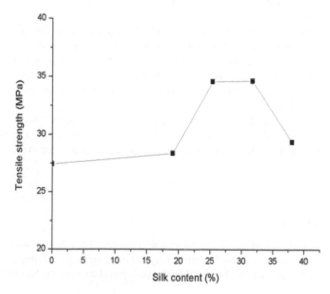

Figure 1.   The effect of silk content on the tensile properties of composites.

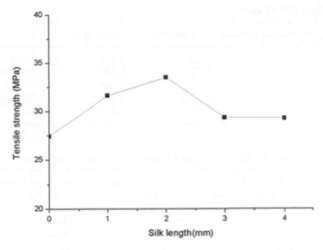

Figure 2. The effect of silk length on the tensile properties of composites.

Table 1. The results of orthogonal experiment and the analysis of range.

Experiment number	Factor				The flavonoid content (mg)	Extraction rate (%)
	A	B (h)	C (°C)	D		
1	30%	1	50	1:10	96	3.20
2	30%	2	60	1:15	120	4.00
3	30%	3	70	1:20	156	5.20
4	60%	1	60	1:20	255	8.50
5	60%	2	70	1:10	208	6.93
6	60%	3	50	1:15	192	6.40
7	90%	1	70	1:15	140	4.67
8	90%	2	50	1:20	222	7.40
9	90%	3	60	1:10	142	4.73
K1	12.4	16.4	17.0	14.9		
K2	21.8	18.3	17.2	15.7		
K3	16.8	16.3	16.8	21.1		
k1	4.13	5.46	5.67	4.95		
k2	7.28	5.43	5.73	5.02		
k3	5.60	6.10	5.60	7.03		
R	3.15	0.67	0.13	2.08		

material is the maximum. When the tussah silk content in the composite material exceeds 31.75%, the strength of the composite materials becomes gradually smaller.

It can be seen from Figure 2, with the increase of the length of silk, the tensile strength of the composite material increases gradually. When the length of silk is 2 mm, the tensile strength of the composite material is maximum. When the length of tussah silk in the composite material exceeds 2 mm, the strength of the composite materials becomes gradually smaller.

3.2 *Flavonoids extraction rate*

This topic uses rutin as a reference substance to test the flavonids content of extraction liquid. The results and analysis were showed in Table 1 and Table 2.

According to the experimental results, the primary and secondary order of the influence factors: ethanol concentration > solid–liquid ratio > extraction time > extraction temperature.

Table 2. The analysis of variance.

Factor	Sum of squared deviations	Freedom	Mean square	F	Significance
A	5.301	2	2.651	530.2	**
B	0.190	2	0.145	29.00	*
C	0.010	2	0.005		
D	2.773	2	1.387	227.4	**
E	0.010	2	0.005		

Table 3. Composite antibacterial activity against *Staphylococcus aureus*.

Sample label	Concentration of flavonoids (mg/ml)	Size of inhibition zone (cm)	Bacterial reproduction under the specimen
1	0.0	0	Much
2	0.1	1.56	None
3	0.2	1.63	None
4	0.3	1.68	None
5	0.4	1.71	None

Table 4. Composite antibacterial activity against *Escherichia coli*.

Sample label	Concentration of flavonoids (mg/ml)	Size of inhibition zone (cm)	Bacterial reproduction under the specimen
1	0.0	0	Much
2	0.1	1.09	None
3	0.2	1.21	None
4	0.3	1.28	None
5	0.4	1.34	None

The results showed that the best extraction technology were ethanol concentration 60%, extraction time 2 h, extraction temperature 60°C, and ratio of solid to liquid 1:20.

3.3 *Antibacterial properties*

The antimicrobial performance was tested by agar plate diffusion method using *Staphylococcus aureus* and *Escherichia coli*. The antibacterial effect was evaluated by the size of inhibition zone.

It can be seen from Table 3 and Table 4, the composite materials after finished by flavonoids have some antibacterial property, and the antibacterial effect on *Staphylococcus aureus* was better than *Escherichia coli*.

4 CONCLUSIONS

1. This project produces the degradable composite made of onion fiber and tussah silk. The composite material has the largest tensile strength when the containing of tussah silk is 31.75% and the length of tussah silk is 2 mm.
2. Ethanol method was adopted in this study for extracting of flavonoids from abandoned onion skin. The best extraction technology was ethanol concentration 60%, extraction time 2 h, extraction temperature 60°C, and ratio of solid to liquid 1:20.

3. The composite materials after finished by flavonoids have some antibacterial property, and the antibacterial effect on *Staphylococcus aureus* was better than *Escherichia coli*.

Corresponding Author: Ruo-yuan SONG, sry@hotmail.co.jp, 18640872045.

ACKNOWLEDGMENT

This work was financially by the science and Technology Department of Liaoning Province (NO. 20131005) and the Education Department Liaoning Province educating (NO. L2014221), which is gratefully acknowledged.

REFERENCES

[1] Yan Wang, Hui Yu, Yuhong Zhu, Jingzhuo Guo. 2012. Outline of Bio-based Composite Materials and the Standards. [J]. *China standardization* (3): 57–60.

[2] Chang-gen Feng, Wu-xian Wu, Xia Liu, Sheng-cai Li. 2003. Research Advancement of Onion's Chemical Ingredient and Pharmaceutical Action. [J]. *Shanghai Journal of Traditional Chinese Medicine* 37(7): 63–65.

[3] Ruo-yuan Song, Lai-jiu Zheng, Yu-ping Zhao, Yong-fang Qian. 2014. Preparation and Mechanical Properties of Short Antheraea pernyi Silk Fiber Reinforced Onion Composite. [J]. *Advanced Materials Research* (842): 110–113.

[4] Tian-pi Xu, Ji-xian Gong, Zheng Li, Jian-fei Zhang, Li Wang, Fan-jie Meng. 2013. Flavonoid Extraction from Apocynum Venetum Linn and its Functional Finishing on Cotton Fabrics. [J]. *Knitting Industries* (1): 29–32.

[5] Yan Xie, Lin-ping Cui, Yun-jie Cao, He-ping Zeng. 2011. Progress in Research on Extraction and Pharmacological Activities of Flavoniod in Onion. [J]. *Hubei agricultural sciences* 50(2): 233–236.

[6] Xiang-bo Ma, Bei Wang, Rui-xue Rong, Zhi-yong Dang. 2013. Optimization of Uitrasonic Extraction Technology of Flavonoids from Herba Verbenae and Study on the Antimicrobial Activity. [J]. *Hubei Agricultural Sciences* 52(3): 645–648.

[7] Dan Zhu, Guang-cai Niu, Xi-yun Niu, Xian-jun Meng. 2006. Study on Antimicrobial Effect of Flavonodis from Fortulace Olernacea. *Journal of Anhui Agricultural Sciences* 34(1): 7–8.

[8] Anjali Karolia & Umang Khaitan. 2012. Antibacterial Properties of Natural Dyes on Cotton Fabrics. [J]. *RJTA* 16(2): 55–61.

Advances in Engineering Materials and Applied Mechanics – Zhang, Gao & Xu (Eds)
© *2016 Taylor & Francis Group, London, ISBN 978-1-138-02834-0*

Acoustic absorption of MRF under different external field

Y.Z. Huang
College of Mechanical Engineering, Guangxi University, Nanning, P.R. China

S.D. Luo, H.B. Mao & T.L. Yang
State Key Laboratory of Digital Manufacturing Equipment and Technology,
Huazhong University of Science and Technology, Wuhan, P.R. China

ABSTRACT: This paper presents an equivalent of the shear moduli of MRF, which is an important parameter in the Biot theory based on the relaxation spectrum. The Biot's model is used to calculate the complex wave numbers of the two compressional waves propagating in MRF at normal incidence. The surface impedances and the acoustic absorption of MRF under different external field are simulated. The results agree with the test by Guicking and the absorption of MRF can be changed and improved with the application of external field.

1   INTRODUCTION

Magnetorheological Fluid (MRF) is a kind of smart material, which dramatically changes the rheological properties in the presence of an external magnet field (Rabinow 1948). The suspension particles of MRF interact with each other and form a fiber structure and its other properties except rheological property, such as acoustic property (Nahmad-Molinari et al. 1999, Nahmad-Molinari et al. 2000), change with the external field. The sound wave propagate in MRF can be modeled by the Biot theory (Biot 1956, Biot 1956, Biot 1961) that the MRF is treated as an equivalent porous material while an external field employed. For the controllable acoustic property of MRF, the composite structure involves MRF has a great potential as a smart acoustic active absorber, and this paper gives a basic study on the acoustic property of MRF. The key to implement the Biot theory is the parameters' equivalent and the equivalent of the shear modulus of MRF is investigated by using the relaxation spectrum theory (Mahjoob et al. 2012).

2   BIOT'S MODEL

The Biot's model considers that two compressional waves and one shear wave can propagate in a porous medium simultaneously. In the case of normal incidence, which is the case discussed throughout the paper, the shear wave is not excited. Note that a harmonic time dependence of the $e^{j\omega t}$ type will be assumed and will therefore not appear in the equations given. The Biot theory gives an approach to calculate the complex wave numbers of the compressional waves:

$$\delta_1^2 = \omega^2 \left[P\rho_{22} + R\rho_{11} - 2Q\rho_{12} - \sqrt{\Delta} \right] / 2(PR - Q^2) \tag{1}$$

$$\delta_2^2 = \omega^2 \left[P\rho_{22} + R\rho_{11} - 2Q\rho_{12} + \sqrt{\Delta} \right] / 2(PR - Q^2) \tag{2}$$

$$\Delta = \left[P\rho_{22} + R\rho_{11} - 2Q\rho_{12} \right]^2 - 4(PR - Q^2)(\rho_{11}\rho_{22} - \rho_{12}^2) \tag{3}$$

where ρ_{11}, ρ_{12} and ρ_{22} are the modified Biot's densities, ω is the circular frequency, P, Q and R are the elastic coefficients of the porous material, which can be evaluated from the "gedanken experiments" suggested by (Biot 1957). In the case, the compressibility modulus of the elastic solid from which the frame is made K_s, is much greater than the compressibility modulus of the porous frame K_b, the simplified expressions of the coefficients are as follows:

$$P = 4N/3 + K_b + K_f (1-\varphi)^2/\varphi$$
$$Q = K_f (1-\varphi), \ R = \varphi K_f \tag{4}$$

where the dynamic elastic moduli are:

$$K_b = 2N(\nu+1)/3(1-2\nu) \tag{5}$$

where K_f is the bulk modulus of the fluid, N is the shear modulus, and ν is the Poisson ratio.

The expression of the surface impedance of a single layer of porous material placed against a rigid wall as shown in Figure 1 was worked out by (Allard 2009):

$$Z = -j\left(Z_1^s Z_2^f \mu_2 - Z_2^s Z_1^f \mu_1\right)/D \tag{6}$$

where

$$D = (1-\varphi+\varphi\mu_2)\left[Z_1^s - (1-\varphi)Z_1^f \mu_1\right]\tan(\delta_2 l)$$
$$+ (1-\varphi+\varphi\mu_1)\left[Z_2^f \mu_2 (1-\varphi) - Z_2^s\right]\tan(\delta_1 l) \tag{7}$$

in which Z_i^s (analogously, Z_i^f) denotes the characteristic impedance of the ith compressional wave in the solid (fluid) part of the porous medium. The parameters that describe the acoustical properties of the acoustical materials and structures are the reflection coefficient and absorption coefficient and they can be directly obtained from the surface impedance. While a plane wave impinges upon a layer of acoustical medium, the reflection coefficient R_s and absorption coefficient α_s of the layer can be expressed as follows:

$$R_s = (Z_s - Z_c)/(Z_s + Z_c), \ \alpha_s = 1 - |R_s|^2 \tag{8}$$

where Z_c is the characteristic impedance of the medium neighboring to the layer, and Z_s is the surface impedance at normal direction of the layer.

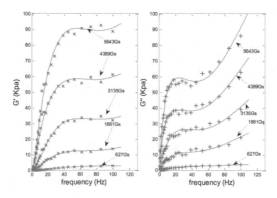

Figure 1.   Test and estimated storage/loss modulus of MRF under different external magnetic field.

3 SHEAR MODULUS EQUIVALENT

It can be found that there are eleven parameters in the Biot model, which are the density ρ_0, the dynamic viscosity η and the bulk modulus K_f of the fluid phase; the density ρ_1 and the bulk modulus K_s of the solid phase; the shear modulus N, the Poisson ratio v, the static permeability q_0, the porosity φ, the tortuosity α_∞ and the viscous characteristic length Λ of the porous frame. It is not easy to obtain the frame parameters during the application of the Biot theory. Therefore, the equivalent of the parameter is very important and the equivalent of the frame shear modulus by the relaxation spectrum theory is discussed in this part.

The rheological dynamic moduli (i.e. loss and storage moduli) can be measured using a rheometer, which is the shear modulus from the definition of the dynamic moduli.

$$G(i\omega) = (\cos\theta + i\sin\theta)\sigma_0/\gamma_0 = G'(\omega) + iG''(\omega) \tag{9}$$

where $G'(\omega)$ is the storage modulus, $G''(\omega)$ is the loss modulus, and $\tan\theta$ is the loss factor.

The dynamic moduli are measured in the oscillation mode of rheometer while the applicable angular frequency range is 0.01–100 rad/s. Therefore, the dynamic moduli of MRF can only be measured directly up to 100 rad/s. Nevertheless, the frequency range involved in acoustic application is higher, such as 20Hz–20KHz for the audible sound. Herein, the relaxation spectrum is used to estimate the dynamic moduli of MRF indirectly beyond the measured range.

The parameters associated with the relaxation spectrum of a viscoelastic fluid are connected to experimental results obtained in rheology test by the (Fredholm) first-order integral equation (Mohammad et al. 2012). The relationship between the relaxation spectrum parameters and the storage/loss moduli (test results) are as follows:

$$G'(\omega) = \sum_{i=1}^{n} H_i(\lambda_i)\omega^2\lambda_i^2/(1+\omega^2\lambda_i^2) \tag{10}$$

$$G''(\omega) = \sum_{i=1}^{n} H_i(\lambda_i)\omega\lambda_i/(1+\omega^2\lambda_i^2) \tag{11}$$

where λ_i is the ith discrete relaxation time and $H_i(\lambda_i)$ is the corresponding discrete relaxation spectrum, and n is the number of the relaxation time point. The discrete relaxation spectrum can be calculated by using test data by the minimizing the function given below:

$$\sum_{j=1}^{m}\left(\left(1-\sum_{i=1}^{n}\frac{1}{G'(\omega_j)}H_i(\lambda_i)\frac{\omega_j^2\lambda_i^2}{1+\omega_j^2\lambda_i^2}\right)^2 + \left(1-\sum_{i=1}^{n}\frac{1}{G''(\omega_j)}H_i(\lambda_i)\frac{\omega_j\lambda_i}{1+\omega_j^2\lambda_i^2}\right)^2\right) = \min \tag{12}$$

where m is the testing number of the moduli. The relaxation time should be appointed before solving Eq. (12). Generally, the reciprocal of the testing upper and lower limit frequency are chosen as the original relaxation time. Then the relaxation time value linear distributed from the minimum time of test to the max time at the logarithmic coordinates,

$$\lambda_i = \lambda_{\min}\left(\lambda_{\max}/\lambda_{\min}\right)^{(i-1)/(n-1)} \tag{13}$$

where $\lambda_{\min}$ and $\lambda_{\max}$ are the lower and upper limit relaxation time.

Figure 1 show the experiment results (Mohammad 2012) of storage and loss moduli of MRF and their estimated value under different external magnetic field. While the magnetic field is wake, the viscosity of MRF dominates the whole testing frequency band compared with the elasticity. The domination of viscosity and elasticity of MRF is governed by

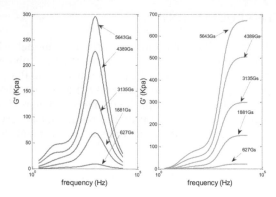

Figure 2.   Estimated storage/loss modulus of MRF in global frequency range.

Table 1.   Parameters of MRF (SI).

ρ_0	ρ_1	R	c_0	c_1	η
997.04	6889	4×10^{-6}	1496.7	5900	0.042
l	φ	α_∞	q_0	Λ	
0.025	0.746	1.38	1.91×10^{-7}	110×10^{-6}	

frequency while the magnetic field is moderate. While the magnetic field is high, the viscosity dominates at low frequency, otherwise the elasticity dominates. Altogether, the storage and loss moduli of MRF are growing with the external magnetic field.

The estimated storage and loss moduli of MRF in global frequency range are plotted in Figure 2. The shear modulus of MRF is growing with the external field while the storage modulus gradually increases to a steady state, and the loss modulus increases to a max and then decreases rapidly.

Other parameters of the frame, i.e. the static permeability q_0, the porosity φ, the tortuosity α_∞ and the viscous characteristic length Λ value refer to similar acoustic media. For example,

$$\varphi = 1 - V_c \tag{14}$$

where V_c is the volume ratio of the suspension particles, α_∞ is in the range of 1 to 2. The parameters of MRF are listed in Table 1.

4   RESULTS AND DISCUSSION

The helixes of the normalized impedance of MRF under different external magnetic field are plotted in Figure 3. Those helixes rotate clockwise and gradually move to the location (1, 0), which indicates that the absorption coefficient of MRF approach to one with the increase of frequency. Each circle of the helixes represents a peak of absorption curve. While there is no external field, the center location of the first circle of the helix is about (13, 0) while the left limit is (0, 0). The peaks and troughs of the absorption are not obvious then.

Zooming Figure 3(a) and the comparison of the helixes becomes clearer, as shown in Figure 3(b). The centers of those helixes approach the location (1, 0) while the radius decrease which indicates that the troughs become flatter with higher external field. Those helixes cutting their horizontal axis at about the same location (0.4, 0) which means that the peak values

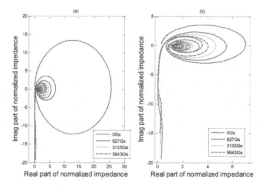

Figure 3. Helix of normalized impedance of MRF under different magnetic field. Dots: 0 Gs; Solid: 627 Gs; Dashed: 3135 Gs; Dash-Dot: 5643 Gs.

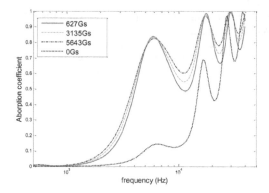

Figure 4. Absorption coefficients of MRF under different magnetic field. Dots: 0 Gs; Solid: 627 Gs; Dashed: 3135 Gs; Dash-Dot: 5643 Gs.

of different external field are the same. These simulated characteristics agree with the testing results by (Guicking et al. 2002).

The acoustic absorption coefficients of MRF under different magnetic field are given in Figure 4. The absorption of MRF is improved by the application of external field that changes the internal micro-structure, and the resonance of the layer is more obvious. With the increase of the field, the peak value almost not change while the trough becomes flatter and the absorption becomes better then.

5 CONCLUSION

The acoustic absorption of MRF is studied based on the Biot theory in this paper while the shear modulus of MRF under an external magnetic field is equivalent by the relaxation spectrum theory. The simulation of this work shows that the absorption coefficients of MRF improved by the applying of an external field. The impedance of MRF layer changes with the external field so that the MRF may be used as an active sound absorber material.

ACKNOWLEDGMENT

This work was supported by the National Natural Science Foundation of China (No. 51175195).

REFERENCES

Allard J.F. & Atalla N. 2009. Propagation of Sound in Porous Media: Modelling Sound Absorbing Materials, *John Wiley & Sons Inc, Chichester*.

Biot, M.A. 1956. Theory of Propagation of Elastic Waves in a Fluid-Saturated Porous Solid. I. Low-Frequency Range. *The Journal of the Acoustical Society of America*, 28(2): 168–178.

Biot, M.A. 1956. Theory of Propagation of Elastic Waves in a Fluid-Saturated Porous Solid. II. Higher-Frequency Range. *The Journal of the Acoustical Society of America*, 28(2): 179–191.

Biot, M.A. 1961. Generalized Theory of Acoustic Propagation in Porous Dissipative Media. *The Journal of the Acoustical Society of America*, 34(2): 1254–1264.

Biot, M.A., Willis, D.G. 1957. The Elastic Coefficients of the Theory of Consolidation. *J. Appl. Mechanics*, 24: 594–601.

Guicking, D. & Wicker, K. & Eberius, C. 2002. Electrorheological Fluids as an Electrically Controllable Acoustic Medium: I. Experimental Arrangement and Application to an Absorber of Underwater Sound. *Acta Acustica United with Acustica*, 88: 886–895.

Mahjoob M.J. & Mohammadi, N. & Malakooti, S. 2012. Analytical and experimental evaluation of magnetic field effect on sound transmission loss of MR-based smart multi-layered panels. *Applied Acoustics*, 73(6): 614–623.

Nahmad-Molinari, Y. & Arancibia-Bulnes C.A. & Ruiz-Suárez J.C. 1999. Sound in a Magnetorheological Slurry. *Physical Review Letters*, 82(4): 727–730.

Nahmad-Molinari, Y. & Ruiz-Suárez, J., 2000. Reply. *Physical Review Letters*, 84(2): 397.

Rabinow, J. 1948. The Magnetic Fluid Cluth. *AIEE Transaction*, 671308-6.

Quantitative analysis of aflatoxins content in sweet corn at early stages using an improved electronic nose

K.J. Huang
State Key Laboratory of Materials Processing and Die & Mould Technology,
Huazhong University of Science and Technology, Wuhan, P.R. China
Key Laboratory of Advanced Technology for Materials Synthesis and Processing,
Wuhan University of Technology, Wuhan, P.R. China

J. Song & C. Zhu
State Key Laboratory of Materials Processing and Die & Mould Technology,
Huazhong University of Science and Technology, Wuhan, P.R. China

ABSTRACT: Aflatoxins content in sweet corn at early stages was quantitatively detected by an improved electronic nose. After storing 96 h at 30°C and 75% RH, the aflatoxins concentration measured by the competitive direct enzyme-linked immunosorbent assay (CD-ELISA) was 4.32 ppb that was over the legal limit (4 ppb). The loading analysis and Cluster Analysis (CA) were applied to optimize sensor arrays, and the Multilayer Perceptron Neural Network (MLPNN) was also applied to aflatoxins content. After optimizing the sensor arrays, the correlation coefficient of the prediction and real aflatoxins content increased from 0.99636 to 0.99871, and the average relative error decreased from 1.58% to 0.92%. The excellent performance of the improved electronic nose makes it a good choice to play as a rapid, easy operation, low cost and effective method to quantify aflatoxins content in sweet corn at early stages.

1 INTRODUCTION

Sweet corn has become one of the most popular vegetables in Europe, America and China, because of the high level of sugar and the property of anti-cancer. However, sweet corn can be easily contaminated by toxigenic mould, especially *Aspergillus* in relatively high humidity and high temperature or other stressful conditions. Aflatoxins (e.g., B_1, B_2, G_1 and G_2) are the second metabolites of *Aspergillus*, most notably *A. flavus* and *A. parasiticus*. The aflatoxins cause serious harm to human and animal health because of its high toxicity and carcinogenicity. Since 1993, aflatoxin B_1 has been classified as Group 1 carcinogen, which primarily affects the liver by the International Agency for Research on Cancer (IARC). Totally, 76 countries regulated limit levels for the sum of the aflatoxins B_1, B_2, G_1 and G_2 in 2003 and the most frequently occurring limit for total aflatoxins is at 4 ppb which applied by 29 countries. Besides, sweet corn contaminated by aflatoxins caused tremendous waste and economic loss, for example, aflatoxin-contaminated maize products result in the loss of between US$270 million to US$500 million annually (Abbas et al., 2011).

Many methods had been developed for quantitative analysis of aflatoxins and those methods were ranged from directly detecting the toxins themselves, based upon physical characteristics of the toxins, to methods for indirectly detecting the toxins. Microfluidic chip-based nano-liquid chromatography (chip nanoLC) coupled to triple quadrupole mass spectrometry (QpQ-MS) was applied for quantification of aflatoxins in peanut with limits of detection of 0.004–0.008 ppb and good intra-day/inter-day precision (2.3%–9.5%/2.3%–6.6%) and accuracy (96.1%–105.7%/95.5%–104.9%) (Liu et al., 2013). Spectrofluorescence coupled with parallel factor analysis was used to detect aflatoxin B_2 and G_2 in peanuts with a detection limit of 0.05 and 0.04 ppb, respectively (Luna et al., 2013). And most of the novel publications had arisen around the use of antibody-based technologies, such as indirect competitive Enzyme-Linked Immunosorbent Assay (icELISA) with limit of detection of 0.52 + 0.36 ppb (mean + 3SD) for aflatoxin B1 in eight agricultural products (Jiang et al., 2013), lateral flow immunoassay with detection limit of 1 ppb and good intra-day/inter-day accuracy (70%–118%/90%–130%) (Anfossi et al., 2011). The above-mentioned methods usually have high accuracy, sensitivity, recovery and precision. However, most of those methods require sample preparation steps that are complicated, time-consuming and need large volumes of organic solvent to enhance sensitivity and selectivity. Therefore, those methods are usually high-cost, time-consuming, limited to particular laboratories and required the services of highly trained operators.

Electronic nose that provides qualitative, semi-quantitative or quantitative results has been paid attention to monitor mycotoxins in food, such as wheat and corn. The underlined hypothesis for the use of electronic nose for mycotoxins is that several Volatile Organic Compounds (VOCs) potentially detected by the instrument will be produced or their concentration will be changed due to the growth and metabolism of the mycotoxin-producing fungi. An electronic nose was used to qualitatively discriminate maize samples contaminated and non-contaminated fumonisin B_1. The best classification scores of k-Nearest Neighbours (kNN) and Linear Discriminant Analysis (LDA) were 98.8% and 94%, respectively. Afterwards, the electronic nose was applied to semi-quantitatively predict low fumonisin content (below 1.6 mg/kg) and high fumonisin content (above 1000 mg/kg) in maize cultures. For the detection of aflatoxins, commercially available electronic nose (PEN2) was able to qualitatively classify maize samples contaminated and non-contaminated with aflatoxins by the use of LDA and Principle Component Analysis (PCA). However, none of those works had conducted to quantitatively detect aflatoxins content in sweet corn at early stages that was very important for early prevention of aflatoxin contamination.

Therefore, the aim of this study is to develop an improved electronic nose that is rapid, easy operation and cost effective for quantitatively detecting the aflatoxins content in sweet corn at early stages. The improved electronic nose is intended as a preliminary screening method before the samples were detected by the conventional quantitative methods to save much time and cost. Therefore, the improved electronic nose will be noticed by the corn farmers, corn processors, corn quality testing institutions and some other people, like instrument manufacturer.

2 EXPERIMENTAL METHODS

2.1 *Sweet corn samples*

Once the harvest of sweet corn, they were bought for our study without further processing. The fresh sweet corn evenly divided into 11 samples (80 g) were analyzed by the electronic nose, and several samples (2.5 g) were analyzed by using the Competitive Direct Enzyme-Linked Immunosorbent Assay (CD-ELISA). Then, all of the samples were put into petri dishes (Φ = 150 mm) and kept at 30 °C and 75% RH in high-low temperature damp-heat testing chamber (Beijing hong da tian ju test equipment Co., Ltd) to promote the formation of aflatoxins. CD-ELISA and electronic nose were applied to detect aflatoxins content and fingerprint of VOCs of sweet corn samples every 24 hours, respectively.

2.2 *CD-ELISA analysis*

The total aflatoxins (B1, B2, G1, and G2) contents of fresh and stored sweet corn were measured by the PriboFast® aflatoxin total ELISA kit following the manufacturer's instructions.

2.3 *Electronic nose*

A total of 12 nano-SnO_2-based gas sensors were fabricated to compose an electronic nose. Nano-SnO_2 powders were prepared by hydrothermal method. Then, mixed with several organic solutions, the screen-printing technique was used for printing the paste onto four different locations of an alumina substrate which had already printed Au interdigital electrodes and RuO_2 heater. Afterwards, the alumina substrates were financed at 550°C for 2 hours. Subsequently, 0.1 μL 0.02 mol/L $Y(NO_3)_3$, $LaCl_3$, $Ce(NO_3)_3$, $PdCl_2$, $CdCl_2$, $TiCl_4$, $Gd(NO_3)_3$, $SrCl_2$, blank, $AlCl_3$, WCl_6, $FeCl_3$ solutions were injected onto SnO_2 gas sensor films to form 12 nano-SnO_2-based gas sensors from S1 to S12. Then, the gas sensor films were financed at 600 °C for 2 hours. After some other processes, such as soldering and ageing, 12 nano-SnO_2-based flat-type coplanar gas sensors were fabricated (Fig. 1).

The schematic diagram of the electronic nose measurement during experiments is shown in Figure 1. Before the test, petri dish with sweet corn was closed tightly and held at 30 °C for 0.5 h to reach equilibrium of the headspace. Then, the petri dish and sampling window were opened and the petri dish was directly put into the measurement chamber through the sampling window. The primary material and size of the measurement chamber were polymethyl methacrylate (PMMA) and 230 mm × 190 mm × 80 mm, respectively. Afterwards, the sampling window was quickly closed. Through the signal conditioning circuit, the resistance changes that the sensors experiences were recorded by a computer which properly stored the acquired data for later use. The measurement process lasted 450 s, time enough for those sensors to reach a stable value. The collected data interval was 0.1 s. When the measurement was completed, the sampling window was opened about 300 s to eliminate the remaining volatiles and the tested sample were put back to the high-low temperature damp-heat testing chamber, waiting for the next test. When those sensors returned to their baseline, the next set of measurements would begin. Each sample had two replicates.

2.4 Data analysis

The responses (S) of the two replicates of the sample were taken average and the average value was chosen as representative feature. The response (S) was defined as Eq. (1).

$$S = \frac{R_a}{R_g} \tag{1}$$

In Eq. (1), R_a and R_g represent the resistance of the sensors in the air and in the test gases, respectively. Then, the extracted feature was processed by the loading analysis and Cluster Analysis (CA) to optimize the sensor array. Subsequently, the total aflatoxins content in sweet corn at early stages were quantitatively predicted by conducting the Multilayer Perceptron Neural Network (MLPNN) on the original sensor arrays and the optimized sensor arrays, respectively. All calculations were carried out with the IBM® SPSS® Statistic Version 19.0 (SPSS, Inc., Chicago, IL).

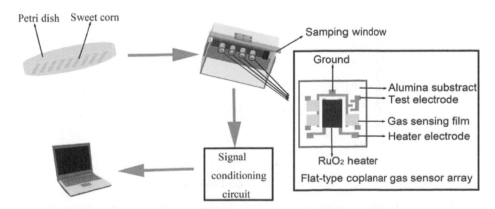

Figure 1.   Schematic diagram of electronic nose measurement during experiments.

3 RESULTS AND DISCUSSION

3.1 *CD-ELISA analysis and photographs of sweet corn samples*

The aflatoxins concentrations measured by CD-ELISA in sweet corn stored from 0 h to 96 h were 0.74 to 4.32 ppb, shown in Figure 2. From Figure 2, the aflatoxins content increased with the extension of the storage times and increased more sharply after 48 h. From 0 h to 96 h, the increasing trend of aflatoxins concentrations agreed with the growth trend of *A. flavus* (Li et al., 2007). After storing 96 h, the aflatoxins content is 4.32 ppb that is higher than the legal limit (4 ppb). Because the main purpose of this study is to quantitatively detect aflatoxins content in sweet corn at early stages, so the measurements did not continue after storing 96 h.

At the same time, the photographs of sweet corn samples stored at different times were also taken, shown in Figure 3. With the extension of storage time, the color of the sweet corn kernels changed from bright to dark, the plumpness of the sweet corn kernels reduced and the mildew degree increased, especially the storage time exceed 48 h. The phenomenon was similar to the reference (Zhou, 2004). The appearance changes of the sweet corn kernels were consistent with the changes of aflatoxins content measured by CD-ELISA.

3.2 *Electronic nose analysis of sweet corn samples*

3.2.1 *Sensor arrays response to sweet corn samples*
Figure 4 shows the typical response curve of the sensor arrays to the sweet corn stored 48 h, other response curves of the sensor arrays to sweet corn stored at different times were similar. From Figure 4, it is obvious that each sensor has a relatively high response and reaches a stable value after collection time 250 s. Each sensor signal was considered to be used in the analysis.

3.2.2 *Loading analysis and Cluster Analysis (CA)*
Loading analysis and Cluster Analysis (CA) were applied to optimize the sensor arrays to provide maximum diversity in response to different levels of aflatoxins contamination of sweet corn according to larger diversity of sensors leads to better classification perform-ance. The loading analysis will help to identify the importance of sensors responsible for discrimination in the current pattern file. Sensors with loading parameters near to zero for a particular principal component have a low contribution to the total response of the arrays, whereas high values indicates a discriminating sensor. If the sensors have almost identical

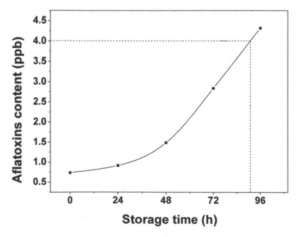

Figure 2. The aflatoxins content in sweet corn samples at different storage times detected by CD-ELISA.

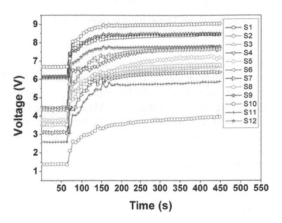

Figure 3. The images of sweet corn samples sensor arrays to the sweet corn stored 48 h

Figure 4. The typical response curve of stored at the different times. a 0 h; b 24 h; c 48 h; d 72 h; e 96 h.

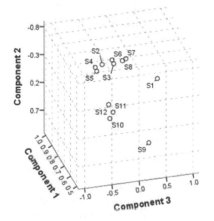

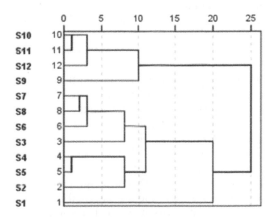

Figure 5. Loading analysis of the gas sensor arrays.

Figure 6. Cluster analysis of the gas sensor arrays.

loading parameters, their effect for analysis could be represented by just one sensor. The loading analysis was performed and a loading plot of the loading parameters for sweet corn was shown in Figure 5. Figure 5 shows that the sensors S3 and S6 have similar loading factor; sensors S4 and S5 have similar loading factor; sensors S7 and S8 have similar loading factor; sensors S10 and S11 have similar loading factor, and therefore could be represented by just one sensor.

The cluster analysis is a statistical method grouping objectives according to their similarity. Each group or cluster is homogeneous in connection with certain characteristic, so observations in each group or cluster are similar to each other and observations of different groups are different. In general, similarity is determined by comparing the distances in some space: the smaller the distances between observations, the more similar they are. Cluster analysis can be classified into hierarchical and non-hierarchical cluster analysis. In this study, hierarchical cluster analysis was used to decide the number of sensors constituting an array. In detail, a squared Euclidean distance was used in the CA with a between-groups method of linkage and Z scores standardized method of transforming values. The CA dendrogram is shown in Figure 6. The dendrogram indicates that sensors S4 and S5 are classified in a group; sensors S10 and S11 are classified in a group; sensors S7 and S8 are classified in a group. Combined with the results of loading analysis and cluster analysis, sensors S5, S6, S8 and S11 should be eliminated from the sensor arrays, that is, the optimized sensor arrays contain S1, S2, S3, S4, S7, S9, S10, and S12.

3.2.3 *Comparison of the results of Multilayer Perceptron Neural Network (MLPNN)*

The Multilayer Perceptron Neural Network (MLPNN) that has feedforward architecture produces a predictive model for one or more dependent (target) variables according to the values of the predictor variables. The MLPNN was first applied for the signals of the original gas sensor arrays. Before running of the multilayer perceptron procedure, the initialization value for the random number generator was set at 20000039 to reproduce the same randomized results in the future. The proportion of training samples and holdout samples was 7:3. The "best" structure of the network was selected by the procedure automatically. The type of training was batch and the optimization algorithm was scaled conjugate gradient. Therefore, 39 samples were randomly assigned to the training set and 16 samples were randomly assigned to the holdout set to build and estimate the prediction model. The network had 12 neurons in input layer, 4 neurons in hidden layer, and 6 neurons in output layer. "12" represents the response signals of twelve gas sensors and "6" represents the five different levels of aflatoxins content in sweet corn and the aflatoxins content in sweet corn. The activation functions of hidden layer and output layer were hyperbolic tangent and identity. The training process was stopped due to the maximum number of epochs (100) exceeded. Both of the training and holdout data had correct identification rates of 100%. The liner fitting of the prediction and real aflatoxins content was shown in Figure 7(a). The liner fitting equation is Eq. (2).

$$M_{predicition} = 0.03683 + 0.98611 M_{real} \qquad (2)$$

where $M_{prediction}$ and M_{real} represent the prediction and real aflatoxins content, respectively. The correlation coefficient (R) is 0.99636. The relative error for storing 0 h, 24 h, 48 h, 72 h and 96 h is 3.59%, 2.61%, 1.08%, 0.09% and 0.54%, respectively. The average relative error is 1.58%.

Then, the MLPNN was applied for the signals of the optimized gas sensor arrays. The random number generator was set at 20000028. The proportion of training samples and holdout samples was 7:3. The "best" structure of the network was selected by the procedure automatically. The type of training was batch and the optimization algorithm was scaled conjugate gradient. Therefore, 40 samples were randomly assigned to the training set and 15 samples were randomly assigned to the holdout set to build and estimate the prediction model. The network had 8 neurons in input layer, 4 neurons in hidden layer and 6 neurons in output layer. "8" represents the response signals of eight gas sensors and "6" represents the five different levels of aflatoxins content in sweet corn and the aflatoxins content in sweet corn. The activation functions of hidden layer and output layer were hyperbolic tangent

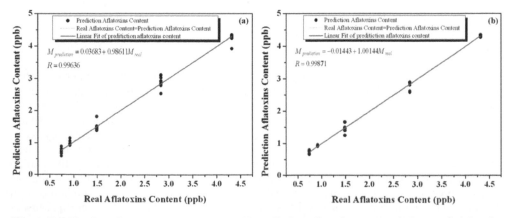

Figure 7.   The liner fitting curve of the real and prediction aflatoxins content before optimizing the sensor arrays (a) and after optimizing the sensor arrays (b).

and identity. The training process was stopped due to the maximum number of epochs (100) exceeded. Both of the training and holdout data had correct identification rates of 100%. The liner fitting of the prediction and real aflatoxins content was shown in Figure 7(b). The liner fitting equation is Eq. (3).

$$M_{predicition} = -0.01443 + 1.00144\ M_{real} \tag{3}$$

The correlation coefficient (R) is 0.99871. The relative error for storing 0 h, 24 h, 48 h, 72 h, and 96 h is 1.80%, 1.42%, 0.82%, 0.36%, and 0.19%, respectively. The average relative error is 0.92% which is lower than the average relative error (1.58%) of the original sensor arrays. Therefore, the optimized gas sensor arrays have stronger prediction ability than the original gas sensor arrays. The low relative errors show that the electronic nose has good ability to quantitative prediction of aflatoxins content in sweet corn at early stages.

From the results of MLPNN based on the optimized sensor arrays, sweet corn samples contaminated by different concentrations of aflatoxins at early stages can be well identified and the aflatoxins concentrations can be quantitatively predicted accurately. The reason is that several volatile compounds, such as, 3-methyl-1-butanol, 2-methylpropan-1-ol, octanal, octane and nonanal, are produced by aflatoxins producing fungi cultured on corn and the quantities of these volatiles changed with storage time (De Lucca et al., 2010). More details should be further investigated. Besides, the pattern recognition algorithms also contribute to better results. Compared with CD-ELISA and some other quantitative methods, the improved electronic nose composed of 12 nano-SnO_2-based gas sensors is a rapid, easy operation, without pretreatment and cost-effective tool to quantitatively detect aflatoxins content in sweet corn at early stages. Therefore, the improved electronic nose could act well as a preliminary screening tool to screen sweet corn samples contaminated by different levels of aflatoxins at early stages.

4 CONCLUSIONS

The CD-ELISA results shown that only the sweet corn samples stored 96 h had a higher aflatoxins content (4.32 ppb) than the legal limit (4 ppb), indicating that sweet corn samples contaminated by different levels of aflatoxins at early stages were obtained. The appearance changes of the sweet corn kernels were consistent with the changes of aflatoxins content measured by CD-ELISA. The sensors S5, S6, S8, and S11 were eliminated to optimize the sensor arrays by using the loading analysis and CA. The optimized gas sensor arrays had better performance, for the correlation coefficient of the prediction and real aflatoxins content increased from 0.99636 to 0.99871 and the average relative error decreased from 1.58% to 0.92% by using the MLPNN. The excellent performance of the electronic nose indicates it could be used for quantitatively detecting aflatoxins content in sweet corn at early stages.

ACKNOWLEDGEMENTS

This work was supported by the Natural Science Foundation of Hubei Province, China (Grant No. 2014CFB1009), and the Opening Project (2015-KF-6) of State Key Laboratory of Advanced Technology for Materials Synthesis and Processing (Wuhan University of Technology).

REFERENCES

Abbas, H.K., Zablotowicz, R.M., Horn, B.W., Phillips, N.A., Johnson, B.J., Jin, X. & Abel, C.A. 2011. Comparison of major biocontrol strains of non-aflatoxigenic *Aspergillus flavus* for the reduction of aflatoxins and cyclopiazonic acid in maize. *Food Addit. Contam. A* 28(2): 198–208.

Anfossi, L., D'Arco, G., Calderara, M., Baggiani, C., Giovannoli, C. & Giraudi, G. 2011. Development of a quantitative lateral flow immunoassay for the detection of aflatoxins in maize. *Food Addit. Contam. A* 28(2): 226–234.

De Lucca, A.J., Boué, S.M., Carter-Wientjes, C.H., Bland, J.M., Bhatnagar, D. & Cleveland, T.E. 2010. Volatile profiles of toxigenic and non-toxigenic *Aspergillus flavus* using SPME for solid phase extraction. *Ann. Agric. Environ. Med.* 17(2): 301–308.

Jiang, W.X., Wang, Z.H., Nölke, G., Zhang, J., Niu, L.L. & Shen, J.Z. 2013. Simultaneous determination of aflatoxin B1 and aflatoxin M1 in food matrices by enzyme-linked immunosorbent assay. *Food Anal. Method* 6(3): 767–774.

Li, R.F., Han, B.Z. & Chen, J.Y. 2007. Establishment of a predictive model for the growth of *Aspergillus flavus* and content variation of aflatoxin. *Food Res. Dev.* 128(12): 129–132.

Liu, H., Lin, S., Chan, S., Lin, T. & Fuh, M. 2013. Microfluidic chip-based nano-liquid chromatography tandem mass spectrometry for quantification of aflatoxins in peanut products. *Talanta* 113: 76–81.

Luna, A.S., Luiz, R.A., Lima, I.C.A., Março, P.H., Valderrama, P., Boqué, R. & Ferré, J. 2013. Simultaneous determination of aflatoxins B2 and G2 in peanuts using spectrofluorescence coupled with parallel factor analysis. *Anal. Chim. Acta* 778: 9–14.

Zhou, J.X. 2004. The biochemistry of moldy grain. *Grain Storage* 33(1): 9–12.

Advances in Engineering Materials and Applied Mechanics – Zhang, Gao & Xu (Eds)
© 2016 Taylor & Francis Group, London, ISBN 978-1-138-02834-0

Mold filling and defects of Ti-6Al-4V alloy during centrifugal casting

L.M. Jia, R.L. Liu, J.X. Li, S.H. Wei & Y.Z. Li
Hebei Key Laboratory of Material Near-Net Forming Technology, Hebei University of Science and Technology, Shijiazhuang, P.R. China
School of Materials Science and Engineering, Hebei University of Science and Technology, Shijiazhuang, P.R. China

ABSTRACT: The present investigation combined with the experimental and numerical method mainly focuses on the effects of centrifugal force on mold filling of Ti-6Al-4V alloy for wedge and stepped castings, and it also analyzes the influence of mold rotating rates on defects of the alloy. The results show that, the melt fills the furthest of horizontal runner immediately while it arrives at the bottom of sprue under a centrifugal force, and then completes the reverse filling. It also found that the mold-filling time is shorter in centrifugal force than in gravity, and the former has a better filling behavior. The numerical simulation indicates that shrinkages in stepped castings decrease exponentially with the increasing of mold-rotating rate from 0 rpm to 510 rpm, which has the similar changing trend with the previous experimental research.

1   INTRODUCTION

Low density and high strength make titanium and its alloys very attractive in the applications for the aerospace industry as structural materials[1], and Ti-6Al-4V is one of the most widely used titanium alloy due to its excellent comprehensive mechanical properties. Centrifugal casting process has been adopted for most titanium alloys to improve their mold-filling capability, since it is difficult to fill the mold completely for titanium castings by conventional method in gravity field[2]. However, the behaviors of the alloy mold-filling and solidification are more complex under centrifugal force than gravity, and it is difficult to investigate the whole casting process only with experimental methods. Therefore, for the centrifugal casting of titanium alloy, it is especially important with numerical simulation to research its mold filling and solidification behaviors. Process parameters used in the simulation can be obtained with experimental methods to get the practical data of Ti-6Al-4V centrifugal castings in order to improve the accuracy of calculation results.

The main defects of titanium centrifugal castings are shrinkage, segregation, pores and so on, and some investigations have been done on these defects[3–9], however, the effects of centrifugal force on titanium shrinkages need to be investigated deeply. Based on this issue, the present investigation mainly focuses on the effect of centrifugal force on mold filling of Ti-6Al-4V alloy, and it also analyzes the influence of mold-rotating rates on shrinkage defects by the means of combining with experimental and numerical simulation.

2   RESEARCH METHOD

The Ti-6Al-4V alloy was remelted in copper crucible in evacuated chamber, with the vacuum degree of 0.1 Pa. The melt would be poured into graphite mold with the casting system shown as Figure 1 under gravity, and cooling curves collected by computer with ECON

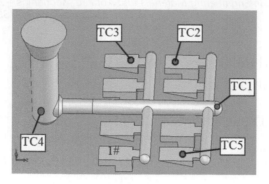

Figure 1.   Centrifugal casting system and locations for data acquisition.

Temperature Data Acquisition. Meanwhile, ProCast was adopted to calculate cooling rate at different locations, and compared with the measured curves. The Equivalent Heat Transfer Coefficient (EHTC) between casting and mold are adjusted for many times until the error between calculation and experiment has the minimum value to determine the optimal process parameters of EHTC and the melt pouring velocity, which are adopted by the ProCast to simulate the centrifugal casting process for Ti-6Al-4V alloy.

2.1   *Experiment of Ti-6Al-4V alloy cast in graphite mold*

As mentioned above, the casting system used in this paper is shown in Figure 1, the maximum radius, measured by the farthest end of the horizontal runner from the turning axis, is about 500 mm. And, four stepped and four wedge-shaped castings alternately distributed in a high-strength graphite mold and the axis of the jar-shaped sprue is taken as the rotating axis. The two secondary horizontal runners, with the length of about 440 mm, are 170 mm apart. The four stepped castings are 95 mm long, 56 mm wide, and 20 mm and 40 mm in thicknesses, respectively. The other four wedge castings have the same length and width as the stepped castings but with a maximum thickness of 20 mm.

While the alloy casts under gravity, five groups of cooling curves at locations of thermal couples numbered as TC1 to TC5 shown in Figure 1 are measured.

2.2   *Determination of casting process parameters for Ti-6Al-4V alloy*

The physical parameters of Ti-6Al-4V alloy can be obtained by the calculation of ProCast once the compositions are provided. The casting process parameters needed to be determined are pouring velocity of melt and EHTC between the casting and mold. The process of determination the two parameters are described as follows.

Determination of pouring velocity: mold-filling time at different locations should be determined, which can be represented by melt contact time difference between TC4 and other thermal couples, expressed as time intervals with TC4. First, setting an initial pouring velocity, and then calculating the filling time at different positions from TC1 to TC5. The filling time calculated should be compared with the one measured by the experiment to adjust pouring velocity, and calculate again until the appropriate pouring velocity is found. Figure 2 shows part of the comparison of mold-filling time by calculation and experiment at different pouring velocities. Obviously, the differences between two values are the smallest, while pouring velocity is 0.722 m/s. Therefore, the pouring velocity of Ti-6Al-4V casting in present investigation is 0.722 m/s.

Determination of EHTC between Ti-6Al-4V cast and graphite mold: at the beginning of calculation, an initial EHTC value should be given, and the comparison would be done between calculation and experiment cooling curves at different locations where thermal couples have been deposited. If large difference between the two values, manual adjustment was needed to change EHTC values until the two values have the best approximation for five

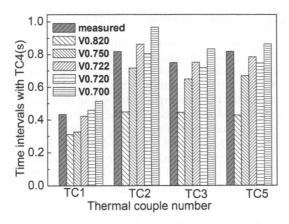

Figure 2. Time intervals of fluid contacting for the different thermal couples relative to TC4.

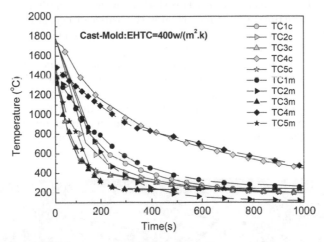

Figure 3. Comparisons between the calculated and measured cooling curves at different thermal couple locations while EHTC is 400 w/(m²·k).

groups of data. In present simulations, eight EHTC values have been tried: 240, 300, 400, 450, 500, 800, 1000 and 1180 w/(m²·k). Result indicates that, calculation and experiment values have the best approximation when EHTC is 400 w/(m²·k), as shown in Figure 3, in which, "TCim" represents measured cooling curves and "TCic" represents calculated cooling curves. Therefore, the EHTC value between Ti-6Al-4V casting and graphite mold in present investigation can be determined as 400 w/(m²·k).

Pouring temperature of Ti-6Al-4V alloy can be obtained according to the measured cooling curves from Figure 3, and the pouring temperature is 1800°C in present investigation.

3 NUMERICAL SIMULATION OF TI-6 AL-4V ALLOY CENTRIFUGAL CASTING

According to the above analysis, it is known that the casting process parameters of Ti-6Al-4V alloy are the follows: pouring temperature 1800°C, pouring velocity 0.722 m/s and EHTC 400 w/(m²·k). These parameters are imported into ProCast to simulate centrifugal casting process of Ti-6Al-4V. In present simulation research, six mold-rotating rates are considered, 0, 110, 210, 310, 410 and 510 rpm to investigate the alloy mold filling and solidification behavior.

3.1 Effect of centrifugal force on mold filling of Ti-6Al-4V alloy

Among the calculation results of six mold-rotating rates mentioned above, the results of 0 rpm and 110 rpm are adopted to analyze the melt mold-filling behavior and the effect of centrifugal force on filling process. Figure 4 shows mold-filling process of the melt cast under gravity and centrifugal force with the mold-rotating rate of 110 rpm. The mold-filling time given in the figures starts while the melt arrives at horizontal runners.

It can be seen from Figure 4, compared with the alloy cast under gravity, the melt has a quick mold-filling process under centrifugal force. For example, the mold-filling time of 110 rpm is about one half of 0 rpm. And, the melt flows quickly to the end of the runner as soon as it arrives at the horizontal runner, and then accomplishes the reverse mold-filling process layer by layer and has a smooth filling. Meanwhile, since the first stream of the melt, which contains pores and inclusions, flows into runner base, it will have sound castings.

3.2 Effect of centrifugal force on Ti-6Al-4V alloy defect of shrinkages

To compare with the experiment results that have been done in previous work[10], in this paper, the effect of centrifugal force on Ti-6Al-4V alloy defect mainly focuses on the shrinkages of the stepped castings. Generally, the gravity coefficient G is adopted to represent the centrifugal force[10], and G values will be obtained from 0 to 118.7 for present simulation research with the castings distributed as Figure 1.

A section in the middle of the casting in width should be done, and then the ratio of shrinkages on the section would be counted through the ProCast software to analyze the effect of centrifugal force on shrinkages, which can be expressed as shrinkage ratio (s), and s is the ratio of shrinkage area to the whole area on the section of the casting.

Figure 5 shows part of calculation results of shrinkages in stepped castings numbered as 1# in Figure 1 under different mold-rotating rates. It can be seen from Figure 5 that

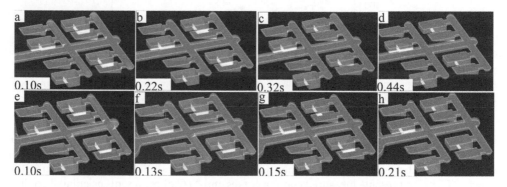

Figure 4. Mold-filling process of the melt cast under gravity and centrifugal force (a)–(d) 0 rpm; (e)–(h) 110 rpm.

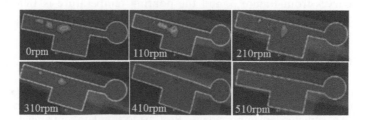

Figure 5. Shrinkages in stepped castings of 1# range from 0 rpm to 510 rpm.

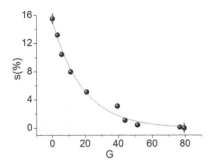

Figure 6.   Relationship between centrifugal force and shrinkage of Ti-6Al-4V alloy.

shrinkage decreases with the increasing of mold-rotating rate, and the shrinkages nearly disappear while the mold-rotating rate is 510 rpm. A statistical analysis result of shrinkage ratio at different gravity coefficient for stepped castings is shown in Figure 6, and both the calculation data and the fitted curve are shown in the figure. By exponential fitting, the relationship between shrinkage ratio and gravity coefficient was obtained, and can be expressed as:

$$s = 15.248 \exp(-G/18.589) - 0.035 \tag{1}$$

In previous work, the relationship between shrinkage ratio and gravity coefficient has been researched through experiment method[10]. It has already got the result that shrinkage ratio decreases exponentially with the increasing of gravity coefficient. However, the mold-rotating rates ranged from 0 rpm to 210 rpm for the limited experiment conditions. Therefore, the simulation results again prove the effect of centrifugal force on casting defects when the mold-rotating rates increases. Difference between the simulation and the experiment fitting formulas is the coefficient, which may be caused by the difference of boundary conditions, and it should be investigated further.

According to the simulation results, the shrinkages in wedge castings decrease more obviously than they are in stepped castings, and there is almost no pores in castings when mold-rotating rate increase to 310 rpm.

4   CONCLUSION

1. It is feasible to determine the interface equivalent heat transfer coefficient and pouring velocity through multiple step trials based on the measured cooling curves, and this can provide necessary parameters for numerical simulation of the casting process.
2. Under centrifugal force, the alloy melt fills the furthest of runner first, and then completes the reverse filling; mold-filling time is shorter in centrifugal force than in gravity, and the former has a better filling behavior.
3. The shrinkages in stepped castings decrease exponentially with the increasing of mold-rotating rate from 0 rpm to 510 rpm according to the numerical simulation results, and this exhibits the similar changing trend with the previous experimental research.

ACKNOWLEDGEMENT

The authors would like to thank the financial support of Natural Science Foundation of Hebei Province (E2013208181), and the funding agency of Project 973 (No. 51319).

REFERENCES

[1] Xu, Z.S. 1988. Status and development of titanium and its alloys. *Journal of materials* 19:17–22.

[2] Watanabea, K. et al. 2003. Casting behavior of titanium alloys in a centrifugal casting machine. *Biomaterials* 24(10):1737–1743.

[3] Atwood, R.C. & Leep, D. 2003. Modeling of microstructure and porosity formation during solidification. *Acta Materiallia* 51:2988–2998.

[4] Sui, Y.W. 2009. Infiltration flux and similarity criterion during centrifugal casting titanium alloy melts Feeding. *Rare Metal Materials and Engineering* 4(38):594–598.

[5] Delgado-Ruiz, R.A. et al. 2014. Porous titanium granules in critical sized effects of rabbit tibia with or without membranes. *International Journal of Oral Science* 6:105–110.

[6] Xu, D.S. et al. 2006. Point defects and mechanical behavior of titanium alloys and intermetallic compounds. *Journal of Physics* 29:220–227.

[7] Xu, R.H. 1995. Vertically centrifugal casting of aluminum heater plate. Foundry 6:28–29.

[8] Suzuki, K.I. et al. 1996. Mold filling and solidification during centrifugal precision casting of Ti-6Al-4V alloy. *JIM* 37(12):1793–1801.

[9] Dimitris, E. & Zinolis, S. 2004. Papadopoulos T. Porosity of cpTi casting with four different casting machines. *J Prosthetic Dent* 10(4):377–381.

[10] Jia, L.M. et al. 2009. Effects of centrifugal force on the pore and shrinkage of TC4 alloy. *Special Casinging & Nonferrous Alloys* 29(9):827–829.

Design of HBc-based nanoparticles for efficient delivery of MEF2D siRNA and induction of apoptosis of hepatoma cells

J. Kong & X.P. Liu

Faculty of Life Science and Chemical Engineering, Huaiyin Institute of Technology, Huai'an, Jiangsu Province, P.R. China

ABSTRACT: *Objective:* Hepatitis B virus nucleocapsid protein (HBc) can be able to assemble into nano-particulate structures when expressed in *E.coli*. Aim to use non-virus-like particles HBc as a vector to investigate the impact of knockdown of MEF2D gene expression in the apoptosis of hepatocellular cells.

Methods: Stable si-MEF2D/HBc nanoparticles were prepared and transfected into hepatocellular cells. The MEF2D RNAi effects on cell apoptosis were detected by flow cytometry analysis, DNA fragmentation assay and Caspase 3/7 enzyme activity assay.

Result: In contrast to control cells, the apoptosis of treated hepatocellular cells were increased significantly. Pretreatment of HepG$_2$ cells with si-MEF2D/HBc complex enhanced apoptosis rate in a dose-dependent manner.

Conclusion: As a new biological nano-materials, HBc particles have the capability of delivery of siRNA. It will give a great contribution on later applied to other aspects of research.

1 INTRODUCTION

Virus Like Particles (VLPs) is a class of nanoparticles and non-viral vectors, which are made up of all or part of the virus structural protein self-assembly. Virus-like particles can be generated by heterologous expression of viral structural proteins and their spontaneous self-assembly. The core protein of hepatitis B virus expressed in bacteria forms shells resembling those in hepatitis B virus infected liver cells. E.coli expressed core has been demonstrated to assemble icosahedral shells with 180 (T = 3) or 240 (T = 4) subunits. Due to its advantageous features, hepatitis B virus core has been extensively exploited as a carrier for foreign epitopes. Most importantly, core particles have been demonstrated to drastically improve the immunogenicity of foreign protein segments presented on their surface (Buchner et al., 2014). Moreover, on the basis of hepatitis B virus core, highly promising vaccine candidates have been generated for influenza and malaria. The immunostimulatory properties of virus-like particles have been exploited recently for the generation of highly immunogenic chimeric particles. The core antigen of hepatitis B virus in particular, has been used for the presentation of various foreign epitopes. As a pivotal member of the virus-like particles family, core-shell or core of the mammalian hepatitis B virus core protein (HBc) contains a sequence which is composed of 183 or 185 amino acid and its protein molecular weight is about 21 kDa (Delfino et al., 2014). Its depolymerization self-assembly function lays foundation for HBc particles as a small molecule carrier. HBc particles can efficiently express, folding and self-assembly to form particles effectively in all known homologous and heterologous expression of the system and it has good biological compatibility, no direct toxic to human cells. HBc is one of the first viral proteins expressed heterologously in bacteria. Recombinant hepatitis B virus nucleocapsid protein, even with deletions at the carboxyl-terminal end or with small internal insertions, maintains the ability to form particles. Two kinds of three-dimensional HBc particles have been reported. The diameter of HBc particles infectious virus particles is about 30–34 nm. The structure of the HBc particles has been extensively analysed (Somvanshi et al., 2009). Present study showed, the core antigen of hepatitis B virus in particular, has been used for the presentation of various foreign epitopes. This paper aims to use HBc particles package wrapping specific siRNA to protect degradation and transfer it to target position in hepatoma cells.

Myocyte Enhancer Factor 2 (MEF2) is a specific transcription factor which is found in skeletal muscle tube nucleus. Myocyte enhancer factor 2 is a family of transcription factors that is composed of different isoforms, such as MEF2A, MEF2B, MEF2C and MEF2D, which are highly expressed in developing neurons during early dendritic maturation and synapse formation (Andzelm et al., 2015). These transcription factors play critical roles in several key intracellular pathways, including neuronal survival and apoptosis. It has been postulated from different studies that MEF2 family members influence neuronal differentiation by supporting the survival of newly formed neurons and regulate excitatory synapses in hippocampal and cerebellar granule neurons (Omori et al., 2015). The MEF2 family is expressed in distinct but overlapping temporal and spatial expression patterns in the embryo and adult. Both MEF2C and MEF2D are implicated in myogenesis. MEF2 factors alone do not possess myogenic activity, but work in combination with the MRFs to drive the myogenic differentiation program. MEF2 proteins control differentiation, proliferation, survival and apoptosis in a wide range of cell types (Ogembo et al., 2015). The N-terminus of the MEF2 proteins contains a highly conserved MADS box and an immediately adjacent motif termed MEF2 domain. Together, these motifs mediate dimerization, DNA binding and co-factor interactions. They can regulate the expression of related genes in cell proliferation, differentiation, survival and apoptosis process. MEF2D expresses abnormal in some types of cancer, proving that it has a certain impact in the occurrence and development of malignant tumor (Ham et al., 2013).

RNA interference (RNAi) or the degradation of specific mRNA species in response to cytosolic presentation of sequence identical dsRNA molecules is a widespread phenomenon among eukaryotes. Following its discovery, RNAi has been widely adopted as powerful loss of function gene analysis tool. Components of the intracellular RNAi pathway machinery like the dsRNA processing enzyme Dicer, and the RNA Induced Silencing Complex (RISC) have been found in many eukaryote model organisms. RNA interference technology is vividly called gene knock-down or gene silencing. This technique has proven to be a powerful method to investigate gene function when genetics or biochemical methods are difficult (Li et al., 2015). It is a typical method of post-transcriptional gene regulation which is also called post-transcriptional gene silencing (Li et al., 2009). siRNA contains 19–21 nucleotides and can degrade a target mRNA in a sequence-specific manner. RNA interference technology has opened up new avenues for the study of gene function and gene therapy. A variety of therapeutic siRNAs had been used to treat numerous diseases including cancer and genetic disorders (Sioud, 2015b). siRNA is usually introduced into tumor cells. However, siRNA cannot achieve the desired effect because of low permeability and easy to decompose. Therefore, in this study, we used non-virus-like nanoparticles HBc as a vector of siRNA to inhibit the expression of MEF2D in human hepatocellular carcinoma cell lines, and investigate the impact of knock-down of MEF2D gene expression on the apoptosis of hepatoma cells.

2 MATERIALS AND METHODS

2.1 Materials

Human hepatocellular carcinoma cell line HepG$_2$ cell was purchased from Takara Biotechnology Company. Lipofectamine 2000 was purchased from Invitrogen Company. All other reagents were of analytical grade.

2.2 Design and synthesize MEF2D-siRNA

We designed and screened the effective siRNA of gene MEF2D (Gene bank NM_001271629.1) using online design software *http://sirna.wi.mit.edu/home.php*. To ensured that the candidate siRNA targeted sites would not inhibit other human genes, we used sequence homology alignments (BLAST analysis) (Sioud, 2015a). After designing the complete appropriate sequence, we entrusted Shanghai Sangon Biotechnology Company which is specialized in nucleic acid synthesis.

2.3 Warp MEF2D-siRNA with HBc virus-like nanoparticles

About 40 µl solution containing HBc particles at a concentration of 20 µg/µl was mixed with an organic solvent in 50 µl HBS buffer (pH 7.4), vortexes for 30 min at room temperature. Solution containing 2 µl siRNA was added drop by drop at a concentration of 500 ng/µl under vortex conditions (Di Nicola et al., 2014). Then vortex for 40 min and waited for 50 min to make full package of si-MEF2D/HBc particles in order to provide sufficient time for HBc self-assemble into nanoparticles.

2.4 Cell cultures

Human hepatocellular carcinoma cell line $HepG_2$ cell was purchased from Takara Biotechnology Company. Cells were incubated in RPMI 1640 medium supplemented with 10% heat-inactivated fetal bovine serum, 100 U/ml penicillin, and 100 mg/ml streptomycin at 37°C in 95% air/5% CO_2 atmosphere. For all experiments, cells were plated at a density of 1.0×10^5 cells/ml.

2.5 si-MEF2D/HBc nanoparticles were transfected into $HepG_2$ cells

Using Invitrogen's Lipofectamine 2000 method in accordance with its instructions. $HepG_2$ cells in the logarithmic growth phase were prepared into cell suspension with trypsin enzyme digesting technique (cell number is about 5×10^4/ml) (Safaralizadeh et al., 2009). Cultured cells could be used for transient transfection when their cell fusion degree reached about 90%–95% in 6-well plate. Then added 4 µl si-MEF2D/HBc complex and 8 µl Lipofectamine 2000 in each well. Cells were harvested after 72 h for the cell apoptosis assay.

2.6 Detection of MEF2D RNAi effects on cell apoptosis

2.6.1 Flow cytometry analysis
Treated cells were trypsinized to form single-cell suspensions. After being rinsed twice with PBS, cells were fixed with PBS containing 90% methanol for 1 h at 4°C. Subsequently, they were centrifuged (12000 rpm, 5 min) and washed twice with cold PBS. Flow cytometry analysis technique was applied after cells were stained with 50 µg·mL[-1] Propidium Iodide (PI) for 30 min. Apoptosis Index (AI) = the number of apoptosis cells/the number of total cells × 100% (Wagenblast et al., 2013).

2.6.2 DNA fragmentation assay
Analysis of DNA fragmentation by agarose gel electrophoresis was performed as described previously. Briefly, cells were washed with PBS and resuspended in ice-cold lysis buffer (10 mmol Tris-HCl pH 7.4, 1 mmol EDTA, 0.2% Triton X-100). After incubating for 15 min at 4°C, cell lysates were centrifuged at 12000 rpm for 15 min. The supernatant was treated with 0.2 mg·mL[-1] of proteinase K in a buffer containing NaCl 150 mmol, Tris-HCl 10 mmol (pH 8.0), EDTA 40 mmol and 1% SDS for 6 h at 37°C. The DNA preparations were extracted with phenol/chloroform twice, to remove proteins. DNA was precipitated with 140 mmol NaCl and two volumes of ethanol at −20°C overnight. DNA precipitates were recovered by centrifugation at 12000 rpm for 15 min at 4°C, washed twice in cool 75% ethanol and air dried. DNA pellets were completely resuspended in 20 µl of TE (10 mmol Tris-HCl (pH 8.0), 1 mmol EDTA) and electrophoresed in 1.2% agarose gels. Gels were placed on a UV light box to visualize the DNA ladder pattern.

2.6.3 *Caspase 3/7 enzyme activity assay*

Apoptosis was also evaluated by Enzyme-Linked Immunosorbent Assay (ELISA) using Cell Death Detection ELISA Plus Kit according to the manual instructions. The relative amounts of apoptotic mono- and oligo-nucleosomes generated from the apoptotic cells were quantified using monoclonal antibodies directed against DNA and histones by ELISA (Bi et al., 2013). Detection of apoptosis was verified by measuring caspase 3/7 enzyme activity as described in previous study.

2.7 *Statistical analysis*

Comparison between groups was performed by the *t*-test. Results are expressed as means ± Standard Deviation (SD) of at least 3 independent experiments. Statistical significances were calculated for comparisons of treated groups and controls using Dunnett's test and one-way ANOVA (Taneja et al., 2010). SPSS version 12.0 for Windows was used for the analysis (SPSS Inc., IL, USA), and P values of <0.05 were considered statistically significant.

3   RESULTS

3.1 *Design of specific MEF2D-siRNA*

The screened siRNA target sequences were AACCCACACTCTAGACTGAAACC.
  Two MEF2D interference sequences for the target sequence were designed as follows:

siRNA1   sense: 5′-GUCCUCGAGAUUCUUAUGAUU-3′
         antisense: 3′-UUCAGGAGCUCUAAGAAUACU-5′;
siRNA2   sense: 5′-GAACUGGGCAUUUGCCAAAUU-3′
         antisense: 3′-UUCUUGACCCGUAAACGGUUU-5′.

Selected a negative control sequence that was not homologous to the target sequence of 16 consecutive bases as follows:

sense: 5′-UUCUCCGAACGUGUCACGUTT-3′
antisense: 5′-ACGUGACACGUUCGGAGAATT-3′.

3.2 *Detection of MEF2D RNAi effects on cell apoptosis*

si-MEF2D/HBc complex induced apoptosis in HepG$_2$ cells, even at low concentration. Data showed an obvious difference in cell morphology between the test group and controls. The morphology of cells that were treated with si-MEF2D/HBc complex for 2 h appeared to have some of the characteristics of apoptosis, such as compaction and cytoplasm leakage into membrane bound-vesicles. But control cells had none of these characteristics.

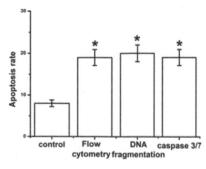

Figure 1.   Apoptosis analysis of hepG$_2$ cells treated by si-MEF2D/HBc complex using flow cytometry, DNA fragmentation assay and caspase 3/7 enzyme activity assay. *$P < 0.05$.

We confirmed this result by DNA fragmentation assay. Agarose gel electrophoresis of DNA showed the characteristic ladder pattern of apoptosis in HepG$_2$ cells incubated in the presence of si-MEF2D/HBc complex.

In order to confirm the apoptosis rate of cells, besides flow cytometry analysis and DNA fragmentation analyses, we examined the changes in caspase 3/7 enzyme activity. The results revealed that there was a dose-dependent increase in caspase 3/7 enzyme activity by si-MEF2D/HBc complex. Pretreatment of HepG$_2$ cells with si-MEF2D/HBc complex enhanced apoptosis by increasing caspase 3/7 enzyme activity.

4 DISCUSSION

The incidence and mortality of hepatocellular carcinoma are the second-highest among the cancer patients in China. Hepatocellular cancer is hard to cure. Compared with many other cancers, achieving a response to therapy in this cancer is much more intractable, and hepatocellular cancers are among the least responsive. In fact, the annual mortality rate of hepatocellular cancer almost approximates the annual accident rate (Rotem et al., 2015). On a case to case basis, hepatocellular cancer has the shortest median survival time for all cancer types. There is a big difference between the early and advanced gastric cancer patients on 5-year survival rate. Neoplastic disease was identified by its very nature as a "genetic disease"; therefore, biotherapy targeted at genetic level might be a better choice. siRNA has highly specific gene silencing function, which can make the target genes stable silencing without affected the expression of other normal gene, and has the characteristics of specificity, high efficiency and simplicity.

During recent years, different virus like particle carrier systems have been developed. For the generation of chimeric virus like particles as potential vaccines, the insertion capacity of the carrier is a critical issue. Carriers like bacteriophage coat proteins have been found to have a very limited insertion capacity. In contrast, hepatitis B virus core protein can folded and assembled themselves correctly in both eukaryotic cells and prokaryotic cells, and thus has significant advantages as a particle-like carrier. Th1 and Th2 cytokines can be detected when HBc was used as a carrier. Non-virus-like nanoparticles HBc can protect siRNA degradation without any toxicity to normal human cells. It is well known to us that MEF2D play an important regulatory role in the growth of liver cancer which are highly expressed in developing neurons during early dendritic maturation and synapse formation. These transcription factors play critical roles in several key intracellular pathways, including neuronal survival and apoptosis (Runfola et al., 2015). It has been postulated from different studies that MEF2 family members influence neuronal differentiation by supporting the survival of newly formed neurons and regulate excitatory synapses in hippocampal and cerebellar granule neurons (Yu et al., 2015). Apoptosis, as one of the biological behavior of cancer, which is an important step in malignant tumor growth process. RNAi is a process in which a sequence-specific pool of mRNA can be eliminated from the cytoplasm of cells by an endogenous surveillance mechanism triggered by the introduction of exogenously supplied dsRNA (Wang and Ye, 2015). In this paper, siRNA was transfected into liver cancer cells with HBc virus-like nanoparticles as the vector to induce RNA interference effect. In order to get more accurate results, we used three experimental methods to detect apoptosis rate of hepatoma cells. The results showed that the apoptosis rate of HepG$_2$ cells increased significantly by knockdown of MEF2D gene expression. It showed that HBc virus-like nanoparticles could be transferred siRNA into hepatoma cells effectively, and it would be one of promising biological nanomaterials.

CORRESPONDING AUTHOR

Corresponding author: Dr. Jing Kong. Email address: cpukj@163.com.

ACKNOWLEDGEMENT

This project is funded by Jiangsu Provincial Special Program of Medical Science (BL2013021) and Huai'an Science foundation (SN12037).

REFERENCES

Andzelm, M.M., Cherry, T.J., Harmin, D.A., Boeke, A.C., Lee, C., Hemberg, M., Pawlyk, B., Malik, A.N., Flavell, S.W., Sandberg, M.A., Raviola, E. & Greenberg, M.E. 2015. MEF2D Drives Photoreceptor Development through a Genome-wide Competition for Tissue-Specific Enhancers. *Neuron,* 86, 247–63.

Bi, L., Chen, J., Yuan, X., Jiang, Z. & Chen, W. 2013. Salvianolic acid A positively regulates PTEN protein level and inhibits growth of A549 lung cancer cells. *Biomed Rep,* 1, 213–17.

Buchner, A., Omar, F.E., Vermeulen, J. & Reynders, D.T. 2014. Investigating hepatitis B immunity in patients presenting to a paediatric haematology and oncology unit in South Africa. *S Afr Med J,* 104, 628–31.

Delfino, C.M., Gentile, E.A., Castillo, A.I., Cuestas, M.L., Pataccini, G., Canepa, C., Malan, R., Blejer, J., Berini, C., Eirin, M.E., Pedrozo, W., Oubina, J.R., Biglione, M.M. & Mathet, V.L. 2014. Hepatitis B virus and hepatitis D virus in blood donors from Argentina: circulation of HBsAg and reverse transcriptase mutants. *Arch Virol,* 159, 1109–17.

Di Nicola, E., Tavazza, M., Lucioli, A., Salandri, L. & Ilardi, V. 2014. Robust Rna silencing-mediated resistance to Plum pox virus under variable abiotic and biotic conditions. *Mol Plant Pathol,* 15, 841–7.

Ham, M.S., Lee, J.K. & Kim, K.C. 2013. S-adenosyl methionine specifically protects the anticancer effect of 5-FU via Dnmts expression in human A549 lung cancer cells. *Mol Clin Oncol,* 1, 373–8.

Li, F.F., Zheng, G.H., Xu, Y.H. & Luo, Q. 2009. [Effect of siRna targeting Vegf on cell apoptosis and the expression of survivin in K562 cells.]. *Zhonghua Xue Ye Xue Za Zhi,* 30, 825–8.

Li, Y., Tian, H., Ding, J., Lin, L., Chen, J., Gao, S. & Chen, X. 2015. Guanidinated Thiourea-Decorated Polyethylenimines for Enhanced Membrane Penetration and Efficient siRNA Delivery. *Adv Healthc Mater.*

Ogembo, J.G., Muraswki, M.R., Mcginnes, L.W., Parcharidou, A., Sutiwisesak, R., Tison, T., Avendano, J., Agnani, D., Finberg, R.W., Morrison, T.G. & Fingeroth, J.D. 2015. A chimeric EBV gp350/220-based VLP replicates the virion B-cell attachment mechanism and elicits long-lasting neutralizing antibodies in mice. *J Transl Med,* 13, 50.

Omori, Y., Kitamura, T., Yoshida, S., Kuwahara, R., Chaya, T., Irie, S. & Furukawa, T. 2015. Mef2d is essential for the maturation and integrity of retinal photoreceptor and bipolar cells. *Genes Cells.*

Rotem, A., Janzer, A., Izar, B., JI, Z., Doench, J.G., Garraway, L.A. & Struhl, K. 2015. Alternative to the soft-agar assay that permits high-throughput drug and genetic screens for cellular transformation. *Proc Natl Acad Sci U S A,* 34, 235–7.

Runfola, V., Sebastian, S., Dilworth, F.J. & Gabellini, D. 2015. Rbfox proteins regulate tissue-specific alternative splicing of Mef2D required for muscle differentiation. *J Cell Sci,* 128, 631–7.

Safaralizadeh, R., Soheili, Z.S., Deezagi, A., Pourpak, Z., Samiei, S. & Moin, M. 2009. FcepsilonRI-alpha siRNA inhibits the antigen-induced activation of mast cells. *Iran J Allergy Asthma Immunol,* 8, 177–83.

Sioud, M. 2015a. Overcoming the Challenges of siRNA Activation of Innate Immunity: Design Better Therapeutic siRNAs. *Methods Mol Biol,* 1218, 301–19.

Sioud, M. 2015b. Strategies for siRNA Navigation to Desired Cells. *Methods Mol Biol,* 1218, 201–16.

Somvanshi, P., Singh, V. & Seth, P.K. 2009. High throughput prediction and analysis of small interfering RNA from the 5'UTR and capsid genes of flavivirus through in silico strategies. *Interdiscip Sci,* 1, 298–302.

Taneja, P., Maglic, D., Kai, F., Sugiyama, T., Kendig, R.D., Frazier, D.P., Willingham, M.C. & Inoue, K. 2010. Critical roles of DMP1 in human epidermal growth factor receptor 2/neu-Arf-p53 signaling and breast cancer development. *Cancer Res,* 70, 9084–94.

Wagenblast, J., Hirth, D., Eckardt, A., Leinung, M., Diensthube R, M., Stover, T. & Hambek, M. 2013. Antitumoral effect of PLK-1-inhibitor BI2536 in combination with cisplatin and docetaxel in squamous cell carcinoma cell lines of the head and neck. *Mol Clin Oncol,* 1, 286–290.

Wang, H. & YE, Y.F. 2015. Effect of survivin siRNA on biological behaviour of breast cancer MCF7 cells. *Asian Pac J Trop Med,* 8, 225–8.

Yu, H., Sun, H., Bai, Y., Han, J., Liu, G., Liu, Y. & Zhang, N. 2015. MEF2D overexpression contributes to the progression of osteosarcoma. *Gene,* 563, 130–5.

Advances in Engineering Materials and Applied Mechanics – Zhang, Gao & Xu (Eds)
© *2016 Taylor & Francis Group, London, ISBN 978-1-138-02834-0*

The X-ray powder diffraction data and crystal structure of Al$_2$Cu$_3$Dy ternary compound

D.G. Li, L.Q. Liang, S.H. Liu, M. Qin, B. He, C.S. Qin, C.B. Li & Y.Z. Huang
Department of Physics and Communication Engineering, Baise University, Baise, Guangxi, P.R. China

L.M. Zeng
School of Materials Science and Engineering, Guangxi University, Nanning, Guangxi, P.R. China

ABSTRACT: The ternary compound Al$_2$Cu$_3$Dy has been synthesized by stoichiometric elemental constituents. The X-ray powder diffraction data for the Al$_2$Cu$_3$Dy compound are presented. The structure type and precise lattice constants have been determined by pattern indexing and Rietveld refinement. The Al$_2$Cu$_3$Dy compound tends to crystallize systemmatically into hexagonal structure type with the space group P6/mmm and the lattice parameters of a = b = 5.1524 Å, c = 4.1481 Å, V = 95.37 Å^3, Z = 1, Density = 7.091 g/cm^3, and the RIR value is 1.70.

1 INTRODUCTION

The phases of Al-Cu-Dy ternary system were studied widely, and a large number of compounds were reported, such as Al$_{0.55}$Cu$_{0.45}$Dy, AlCu$_4$Dy, Al$_{1.64}$Cu$_{0.36}$Dy (*Kuz'ma Yu B & Milyan V V,* 1989)[1], AlCuDy (*Tsvyashchenko A V & Fomicheva L N,* 1987)[2], Al$_{10}$Cu$_7$Dy$_2$ (*Prevarskii A P & Kuz'ma Yu B,* 1989)[3], Al$_3$CuDy (*Kuz'ma Yu B & Stel'makhovich B M,* 1988)[4], Al$_6$Cu$_6$Dy (*Felner I,* 1980)[5], Al$_7$Cu$_{16}$Dy$_6$ (*Stel'makhovich B M & Kuz'ma Yu B,* 1990)[6], Al$_8$Cu$_4$Dy (*Felner I & Nowik I,* 1979)[7], Al$_{8.4}$Cu$_{2.6}$Dy$_3$ (*Stel'makhovich B M, et al.* 2000)[8], etc. The crystal structure of the new compound Al$_2$Cu$_3$Dy has not been reported. In this work, the crystal structure of Al$_2$Cu$_3$Dy has been studied by the X-ray powder diffraction technique and Rietveld refinement.

2 EXPERIMENTAL

2.1 *Sample preparation*

The new compound of Al$_2$Cu$_3$Dy with the total mass of 2 gram was synthesized by arc melting in argon atmosphere with the composition of 13.26 wt% Al, 46.83 wt% Cu and 39.91 wt% Dy. The raw material of high purity metals with 99.99 wt% Al, 99.99 wt% Cu and 99.9 wt% Dy were from China New Metal Materials Technology Co., Ltd. The mixed metal was melted at least three times in order to ensure the metals fused together completely and the composition was well distributed, and the mass loss was less than 1 wt% after melting process. The alloy was sealed in an evacuated quartz glass tube and annealed at the temperature of 1103 K for 30 days, and then cooled down to room temperature at the rate of 0.2 K/min. The alloy sample was grounded into powder so that the particle size is less than 10 μm in an agate mortar.

2.2 *Data collection and analysis*

The X-ray powder diffraction data of Al$_2$Cu$_3$Dy compound powder were collected by the Rigaku Smart Lab diffractometer with the Cu Kα radiation at room temperature. The X-ray powder diffractometer was operated in the condition of voltage 40 kV and current 150 mA, the scan range of Bragg angle was from 10° to 100°. The stepping scanning mode with the step size of 0.02° and the speed of 2 s per step was adopted. The internal standard method had been used in order to correct the systematical errors for observed peak positions, and the X-ray powder diffraction data for the mixture of Al_2Cu_3Dy and the internal standard material silicon were collected. The X-ray powder diffraction data of the mixture of 50 wt% Al_2Cu_3Dy and 50 wt% corundum were collected in order to calculate the RIR value. And then, the pattern of Al_2Cu_3Dy was refined by DBWS9807 program (*Young R A*, 2000)[9].

3 RESULTS

The all diffraction peaks in the whole X-ray diffraction pattern of Al_2Cu_3Dy were successfully indexed by MDI Jade 6.5 program (*Materials Data Inc.*, 2002)[10] with the hexagonal structure type of the space group P6/mmm, and the Smith-Snyder figure-of-merit (*Smith G S & Snyder R L*, 1979)[11] F_{30} is 162.1 (0.0068, 33). The X-ray powder diffraction data of Al_2Cu_3Dy are listed in Table 1. Lattice constants had been refined by the DBWS9807 program, and the precise lattice parameters were determined with a = b = 5.1524 Å, c = 4.1481 Å, Vol = 95.37 Å^3,

Table 1. The X-ray powder diffraction data of Al_2Cu_3Dy ($CuK\alpha_1$, with $\lambda = 1.5406$Å).

No.	h	k	l	$2\theta_{cal}$	$2\theta_{obs}$	$\Delta 2\theta^a$	I/I_0	d_{cal}	d_{obs}	Δd^b
1	1	0	0	19.887	19.878	0.009	4.1	4.4608	4.4627	−0.0019
2	0	0	1	21.407	21.398	0.009	26.5	4.1474	4.1492	−0.0018
3	1	0	1	29.381	29.378	0.003	34.5	3.0374	3.0378	−0.0004
4	1	1	0	34.805	34.8	0.005	30.9	2.5755	2.5759	−0.0004
5	2	0	0	40.407	40.399	0.008	12	2.2304	2.2308	−0.0004
6	1	1	1	41.227	41.222	0.005	100	2.1879	2.1882	−0.0003
7	0	0	2	43.610	43.618	−0.008	22.4	2.0737	2.0734	0.0003
8	2	0	1	46.174	46.178	−0.004	2.8	1.9644	1.9642	0.0002
9	1	0	2	48.362	48.357	0.005	1.5	1.8805	1.8807	−0.0002
10	2	1	0	54.369	54.357	0.012	0.9	1.686	1.6864	−0.0004
11	1	1	2	56.965	56.961	0.004	22.1	1.6152	1.6153	−0.0001
12	2	1	1	59.098	59.099	−0.001	6.1	1.5619	1.5619	0
13	2	0	2	60.954	60.959	−0.005	7.9	1.5187	1.5186	0.0001
14	3	0	0	62.400	62.399	0.001	3.4	1.4869	1.4870	−0.0001
15	3	0	1	66.778	66.779	−0.001	16.8	1.3997	1.3997	0
16	0	0	3	67.721	67.731	−0.010	0.5	1.3825	1.3823	0.0002
17	1	0	3	71.369	71.380	−0.011	2.1	1.3205	1.3203	0.0002
18	2	1	2	72.145	72.144	0.001	0.5	1.3082	1.3082	0
19	2	2	0	73.478	73.48	−0.002	5.3	1.2877	1.2877	0
20	3	1	0	77.012	77.023	−0.011	0.5	1.2372	1.237	0.0002
21	2	2	1	77.561	77.567	−0.006	0.8	1.2298	1.2297	0.0001
22	1	1	3	78.451	78.441	0.010	6.8	1.2181	1.2182	−0.0001
23	3	0	2	79.202	79.200	0.002	2.8	1.2084	1.2084	0
24	3	1	1	81.039	81.039	0	1.9	1.1856	1.1856	0
25	4	0	0	87.372	87.361	0.011	0.6	1.1152	1.1153	−0.0001
26	2	2	2	89.516	89.519	−0.003	7.6	1.0940	1.0939	0.0001
27	2	1	3	92.196	92.200	−0.004	1.8	1.0690	1.0690	0
28	3	1	2	92.935	92.941	−0.006	0.2	1.0625	1.0624	0.0001
29	0	0	4	95.958	95.960	−0.002	0.9	1.0369	1.0368	0.0001
30	3	0	3	99.067	99.061	0.006	2.3	1.0125	1.0125	0
31	1	0	4	99.406	99.399	0.007	1	1.0099	1.0100	−0.0001

$^a\Delta 2\theta = 2\theta_{cal} - 2\theta_{obs}$; $^b\Delta d = d_{cal} - d_{obs}$.

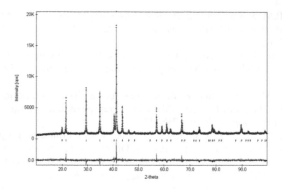

Figure 1.   Observed, calculated and residuals of X-ray powder diffraction pattern for Al_2Cu_3Dy after Rietveld refinement.

Table 2.   Atomic positions and occupancy of Al_2Cu_3Dy with Rietveld refinement.

Atom	Position	X	Y	Z	SOF
Dy	1a	0	0	0	1
Cu	2c	0.3333	0.6667	0	1
Cu	3g	0.5	0	0.5	0.3333
Al	3g	0.5	0	0.5	0.6667

Density $= 7.091$, $Z = 1$, with goodness-of-fit parameters: $R_p = 4.17\%$, $R_{wp} = 5.53\%$, $R_B = 8.66\%$, and $R_F = 6.72\%$. The observed, calculated and residuals of X-ray powder diffraction pattern for Al_2Cu_3Dy after Rietveld refinement have been shown in Figure 1, and the atomic positions and their occupancy have been presented in Table 2. It was found that the phases of Al_2Cu_3Dy and Al_2Cu_3 La have the same crystal structure type in the report (*Guo Yongquan, et al,* 1997)[12]. The RIR value is 1.70 that was calculated from the ratio of the strongest peak intensity between Al_2Cu_3Dy and corundum in the X-ray powder diffraction pattern for the mixture of Al_2Cu_3Dy and corundum.

ACKNOWLEDGEMENTS

This work was financially supported by the department of education of Guangxi Zhuang Autonomous Region (project numbers: YB2014386 and KY2015YB279).

REFERENCES

[1] Kuz'ma, Yu. B. & Milyan, V.V. 1989. Phase equilibria in the system Dy-Cu-Al at 500 degree C. *Russian Metallurgy* (1): 216–218.
[2] Tsvyashchenko, A.V. & Fomicheva, L.N. 1987. High-pressure synthesis and structural studies of rare earth (R) compounds RCuAl. *Inorganic Materials* 23(7): 1024–1027.
[3] Prevarskii, A.P. & Kuz'ma, Yu. B.1989. New Compounds with the Th_2Zn_{17}-type structure in REM-Al-Cu systems. *Russian Metallurgy*, 1989(1): 216–218.
[4] Kuz'ma, Yu. B. & Stel'makhovich, B.M. 1988. New compounds $RCuAl_3$ (R = Tb, Dy, Ho, Er, Tm, Yb) and their crystalline structure. *Dopovidi Akademii Nauk Ukrains'koi RSR, Seriya B: Geologichni, Khimichni ta Biologichni Nauki* 1988(11): 38–41.
[5] Felner, I. 1980. Crystal structures of ternary rare-earth-3d transition metal compounds of the RT_6 Al_6 type. *Journal of the Less-Common Metals* 72: 241–249.

545

[6] Stel'makhovich, B.M. & Kuz'ma, Yu. B. 1990. New compounds $Ln_6(Cu, Al)_{23}$ and their crystalline structure. *Dopovidi Akademii Nauk Ukrains'koi RSR, Seriya B: Geologichni, Khimichni ta Biologichni Nauki* 1990(6): 60–62.

[7] Felner, I. & Nowik, I. 1979. Magnetism and hyperfine interactions of 57Fe, 151Eu, 155Gd, 161Dy, 166Er and 170Yb in RMM compounds. *Journal of Physics and Chemistry of Solids* 40: 1035–1044.

[8] Stel'makhovich, B.M.; Gumeniuk, R.V.; Kuz'ma, Yu. B. 2000. Compounds $Dy_3 Ag_{2.3} Al_{8.7}$, $Ho_3 Ag_{2.1} Al_{8.9}$, $Dy_3Cu_{2.6} Al_{8.4}$ and $Ho_3Cu_{2.4} Al_{8.8}$ as new representatives of the $La_3 Al_{11}$-type structure. *Journal of Alloys and Compounds* 307: 218–222.

[9] Young, R.A.; Allen C.L.; Paiva-Santos C.O. 2000. User's Guide to Program DBWS9807a for Rietveld Analysis of X-ray and Neutron Powder Diffraction Patterns with a PC and Various Other Computers. *School of Physics, Georgia Institute of Technology, Atlanta, GA.*

[10] JADE Version 6.5. 2002. XRD pattern processing. *Materials Data Inc.*

[11] Smith G.S. and Snyder R.L. 1979. FN: A criterion for rating powder diffraction patterns and evaluating the reliability of powder-pattern Indexing. *J. Appl. Crystallogr* 12: 60–65.

[12] Guo Yongquan; Liang Jing Kui; Zhang Xinhui; Tang Wei Hua; Zhao Yanming and Rao Guanghui. 1997. Effects of Mn and Cu doping in $La (Ta, Al)_{13}(T = Fe, Co)$ on crystal structure and magnetic properties. *Journal of Alloys and Compounds* 257: 69–74.

Advances in Engineering Materials and Applied Mechanics – Zhang, Gao & Xu (Eds)
© 2016 Taylor & Francis Group, London, ISBN 978-1-138-02834-0

Molecular Dynamics simulation of Ni thin film growth on Cu (001) substrate

Y.J. Li, Y.J. Mo, J.N. Huang & S.J. Jiang
State Key Laboratory of Optoelectronic Materials and Technologies, Sun Yat-sen University, Guangzhou, P.R. China

ABSTRACT: Molecular dynamics simulations have been carried out to study the hete-roepitaxial growth of Ni thin film deposited on Cu (001) surface at the atomic scale. The results reveal that the growth mode of Ni thin film changes from island growth mode to layer-by-layer growth mode as the substrate temperature and incident energy increase. In addition, Ni atoms hardly penetrate into Cu substrate while Cu atoms easily diffuse into Ni deposition layers. The thickness of intermixing region depends on the incident energy and higher incident energy results in greater thicknesses. Specially, regardless of how much the incident energy is, the incremental value of the incident energy is about 1.6 eV due to local acceleration. Finally, the peaks of Radial Distribution Function (RDF) clearly indicate that the structure of Ni thin film tends to be amorphous as substrate temperature increases.

1 INTRODUCTION

With the development of modern technology, Understanding of thin film growth mechanism has been playing an important role in optimizing device characteristics for technological application[2,7]. The nanoscale magnetic thin films consisting of ferromagnetic/non-ferromagnetic multilayer for Giant Magnetoresistance (GMR) can be used in magnetic sensors and hard disk devices[1]. It was found that the giant magnetoresistance was related to microstructure, interface roughness and mixing of multilayer films[16, 15]. Therefore, it is of great interest in investigating the growth behaviors of thin film in details. In the past two decades, extensive researches about structure, magnetic and electrical properties of Ni/Cu thin films were studied by different experiment techniques had been reported[24,5,12].

However, a better understanding of the physical mechanism of thin films growth can be achieved via Molecular Dynamics (MD) simulation, which can provide information about not only the microprocess of thin film growth but also the evolution law of thin film at the atomic level. Soon-Gun Lee et al[17, 18]. have simulated thin films of transition metals (Fe, Ni and Co) deposited on Al substrates using MD simulation. They found that active surface alloying occurred easily in the initial stage for Fe-Al, Ni-Cu and Co-Cu systems, and that the intrinsic Fe-Al, Co-Al and Ni-Al compounds of B2 structure at the interfaces were formed in spite of low incident energy of 0.1 eV. Moreover, the growth mode changed from island growth to layer-by-layer growth for Ni-Al system when incident energy increased. M. F. Francis et al[13]. discovered the roughness of Cu films deposited on Ta substrate decreased as adatom energy increased. Due to crystalline symmetry between Cu film and Ta substrate, the crystalline grains had only two orientations. V Georgieva et al[22]. analyzed Radial Distribution Function (RDF) and the density of Mg–Al–O film. And they found that thin film transited from crystalline to amorphous when the Mg concentration decreases. In addition, molecular dynamics simulation was also used to investigate the effect of 3D ES barrier in texture selection[3], defects in thin films[26] and mechanism for fabricating Y-shaped nanorods[10].

The growth kinetics of thin film is related to the experimental growth conditions, such as substrate temperature, incident energy and deposition rate et al. In this paper, Ni atoms onto

Cu (001) substrate are taken as a prototype in order to study how the growth behaviors of heteroepitaxial film are affected by substrate temperatures and incident energies. The layer density, surface roughness, layer coverage, evolution of the kinetic energy and radial distribution function of Ni thin film will be analyzed by molecular dynamics method.

2   SIMULATION METHOD

The Embedded-Atom Method (EAM) potential can accurately describe the atomic many-body interactions in metallic systems and has been used to investigate the growth mechanism of thin film[14,6,4]. In this work, the EAM potential proposed by Bonny et al. for Cu/Ni system is adopted[8], which can be used to describe the alloy thermodynamic properties and the interactions between Cu and Ni atoms. The size of the Cu (001) substrate is set to 43.38Å × 43.38Å × 20.244 Å with the surface perpendicular to the z axis in Figure 1. The Cu substrate containing 3456 perfect Face Centered Cubic (FCC) atoms comprises 12 Cu atomic layers. The bottom three atomic layers called fixed area are frozen to their perfect lattice to prevent simulation cell moving. The middle three layers of Cu substrate called constant temperature area are introduced to maintain temperature of the substrate. And the velocities of these atomic layers are adjusted every ten time steps according to prescript substrate temperatures. In order to make the simulation more realistic and reduce other constraints affecting the interaction between the deposited atoms and the substrate atoms, the top six atomic layers of free area are in none of any constraint conditions except the Newton's laws of motion. Periodic boundary conditions are employed in both x-axis and y-axis, but not in z-axis direction. Ni atoms are placed at 34Å length above the substrate surface, randomly assigned in the x–y plane, and deposited perpendicular to the surface of Cu substrate. In addition, the time step is set to 1 femtosecond (fs) and the deposition rate is set to 2 atoms/ps. In order to eliminate the internal stress of the substrate and reach system equilibrium, substrate is relaxed 10ps before atom deposition. Then, 1500 Ni atoms will be deposited onto the Cu (001) substrate. The substrate temperature is set from 100 to 800 K and the incident energy of deposited Cu atoms varies in the range of 0.1–6 eV. All simulations are performed using LAMMPS, a classic MD package[19].

3   RESULTS AND DISCUSSION

Thin film growth can be divided into three growth modes: layer-by-layer growth, island growth and initially layer-by-layer growth to island growth. The layer density and surface

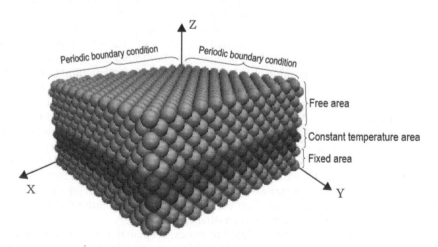

Figure 1.   Atomic configuration of Cu (001) substrate employed for MD simulation.

roughness are used to characterize the growth mode, which has an important impact on physical characteristics. The layer density of Ni thin film will be examined as a function of substrate temperature and incident energy. As shown in Figure 2(a), the layer density of the 2nd atomic layer is only accomplished 80%, whereas Ni atoms can be found at the 4th atomic layer at substrate temperature of 100 K. Consequently, the surface of Ni thin film is rough. However, the tendency changes as the substrate temperature increases to 800 K. The growth behaviors seem to change to layer-by-layer growth mode in which the 4th atomic layer begins filling after the layer density of the 2nd atomic layer reaches 100%. The similar phenomenon occurs as the incident energy increases from 0.1 eV to 5 eV (Fig. 2(b)). It is noted that the curve of layer density at the incident energy of 5 eV is steeper than 1 eV, and thus the surface of Ni thin film becomes smoother. In addition, the layer density of Ni thin film is also analyzed at the substrate temperature of 300 and 500 K and the incident energy of 1 and 3 eV. As a result, the effect on the layer density of Ni thin film is similar. The dependence of layer density on incident energy effect is similar to the Ni–Ni system[20]. The explanation about the above phenomenon is: high substrate temperature and incident energy are conducive to increase the thermal diffusion and atomic activity, so that the atoms can arrive at their thermodynamically optimum positions. Therefore, Ni thin film becomes denser and smoother, while growth behaviors change from island growth mode to layer-by-layer growth mode.

In an attempt to further study the growth behaviors of Ni thin film, the surface roughness is employed to indicate the dependence of the growth mode on the substrate temperature and incident energy. The surface roughness is defined as[25]:

$$\omega = \sqrt{\frac{\sum_{i=1}^{N}(h_i - \bar{h})^2}{N}}, \tag{1}$$

where N is the total number of exposed atoms on the surface, h_i is the height of the exposed atom i on surface, and $\bar{h}$ represents the average height of all the exposed atoms. The surface roughness is plotted in Figure 3(a) as a function of substrate temperature. It is found that the surface roughness of Ni thin film decreases and Ni thin film becomes smoother with increasing substrate temperature from 100 to 800 K. On the other hand, the insets display the surface morphologies of Ni thin films at 100 and 800 K, respectively, which imply that a transition from island growth mode to layer-by-layer growth mode is observed as the substrate temperature increases. Similarly, Figure 3(b) shows that when the incident energy increases from 0.1 to 5 eV, the surface roughness of Ni thin film descends asymptotically, thus the growth behaviors seemed to change to layer-by-layer growth. As a consequence, analyzing

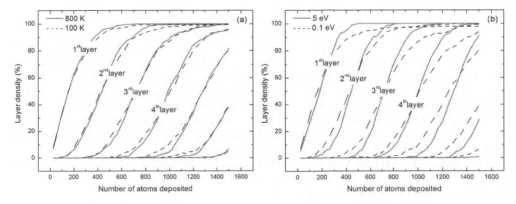

Figure 2. The layer density of Ni thin film on Cu (001) substrate as a function of number of atoms deposited corresponding to: (a) at substrate temperature of 100 and 800 K, (b) at incident energy of 0.1 and 5 eV.

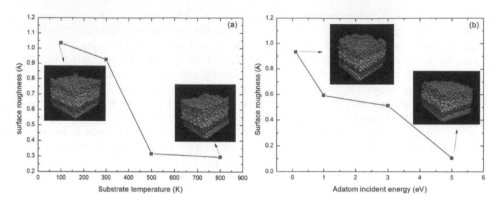

Figure 3. Calculation of the surface roughness of Ni thin film as a function of various deposition parameters: (a) substrate temperatures from 100 K to 800 K, (b) incident energies from 0.1 eV to 5 eV. The insets are simulated morphologies of Ni thin film growing on Cu (001) substrate at the substrate temperatures of 100 and 800 K and the incident energies of 0.1 and 5 eV, respectively.

the layer density and surface roughness reveal that high substrate temperature and incident energy could enhance the thermal diffusion and atomic activity, and promote the transition of the growth mode of thin film.

To quantitatively understand the mixing phenomenon in the interface region, it is necessary to analyze the layer coverage function which represents the relative proportion of the substrate atoms and deposited atoms in each atomic layer from the substrate layers to the deposited layers. The layer number 0 represents the top layer with Cu substrate, and layers with negative number on the left side of the dotted line are considered as Cu substrate layers, while layers with positive number represent the Ni deposited layers (Fig. 4). Figure 4(a) and (b) clearly illustrate that the intermixing phenomenon appears when the incident energy is 0.1 and 1 eV at the substrate temperature of 300 K, while atomic penetration only emerges between the top layer (layer number 0) of Cu substrate and the first deposited layer (layer number 1). However, intermixing degree aggravates as the incident energy increases to 3 and 5 eV, which is shown in Figure 4(c) and (d). Substrate Cu atoms are found to penetrate into three deposited layers (layer number 1, 2 and 3), but the deposited Ni atoms only diffuse onto the top layer of the substrate (layer number 0). The intermixing phenomenon for Ni/Cu (001) can be also observed in the experiments[11, 9]. Thus, we can conclude that the incident energy has significant effects on the intermixing between the substrate layers and deposited layers. Moreover, the diffusion depth of Cu atoms to the deposited layer is greater than Ni atoms to the substrate layer. The surface energy of Cu (1.52 J/m^2) is higher than that of Ni (1.94 J/m^2) [23], which lead to the asymmetry of the interfacial intermixing. The atoms of low surface energy tend to diffuse to the system surface so as to minimize the total energy of the system.

Evolutions of the kinetic energy of one deposited atom are plotted in Figure 5. Figure 5(a) reveals that the initial kinetic energy of the deposited atom is 0.1 eV at substrate temperature of 300 K, but the kinetic energy increases rapidly when the deposited atom is close to the substrate surface. Consequently, the maximum of the kinetic energy can reach 1.7 eV. The phenomenon of the local acceleration near the substrate is similar to the previous works[21]. The attracting force between the deposited atoms and substrate atoms lead to the local acceleration. Furthermore, it is interesting to find that the incremental value of the incident energy is about 1.6 eV regardless of the incident energy in Figure 5(a-d).

The Radial Distribution Function (RDF) describes the spatial organization of atoms around a central atom and provides a signature for identifying the lattice structure of a system. It is defined by the following equation[13]:

$$g(r) = \frac{n(r)}{\rho \pi r^2 \Delta r} \tag{2}$$

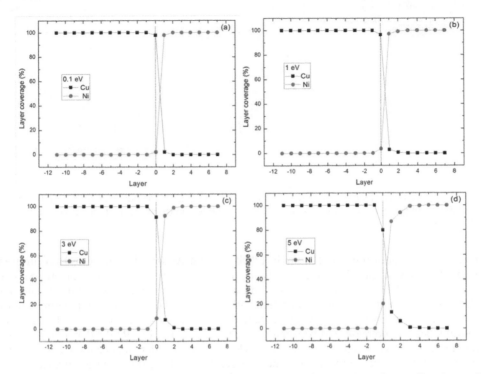

Figure 4. Layer coverage function of the deposited atoms and substrate atoms along z-direction at different incident energies: (a) 0.1 eV, (b) 1 eV, (c) 3 eV and (d) 5 eV. The layer number 0 represents the top layer of the substrate and the layers of negative number on the left side of dotted line are considered as Cu substrate layers, while layers of positive number represent the Ni deposited layers.

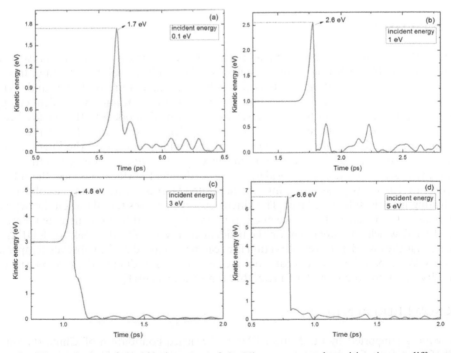

Figure 5. The evolution of the kinetic energy of the Ni atom versus deposition time at different incident energies: (a) 0.1 eV, (b) 1 eV, (c) 3 eV and (d) 5 eV.

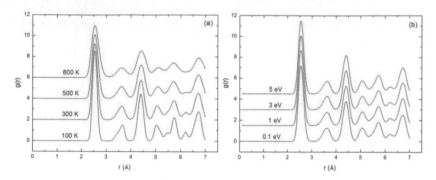

Figure 6. The radial distribution function g(r) of Ni thin film at various deposition conditions: (a) at substrate temperature of 100 K, 300 K, 500 K and 700 K and (b) at incident energy of 0.1 eV, 1 eV, 3 eV and 5 eV.

where $n(r)$ is the mean number of atoms in a shell of width, Δr is the distance away from an atom, and ρ is the mean atomic density. In a radial distribution function, sharp and narrow peaks are indicative of ordered structure, while low and wide peaks represent an amorphous state. The radial distribution functions of Ni thin film with respect to the substrate temperature and the incident energy are in Figure 6. When the temperature increases from 100 to 800 K, the peaks of RDFs become decrease and broaden gradually (Fig. 6(a)). The RDFs depending on substrate temperatures reveal that Ni thin film has long-range disorder and becomes amorphous at higher substrate temperature. Namely, the disorder degree of the thin films is enhanced resulting from an increasing substrate temperature, which gives rise to a higher chance for the regular Ni lattice to be destroyed and high perturbation to occur at the lattice position. In contrast, as the incident energy increases from 0.1 to 5 eV, the differences between these four cases is slight (Fig. 6(b)). The result reveals that the radial distribution function is independent of the incident energy. The similar phenomenon is found in Co/Cu and Fe/Cu systems[27].

4 CONCLUSION

In this work, molecular dynamics simulations have been performed to study the growth of Ni thin film on Cu (001) substrate at the atomic scale. The results reveal that the thin films grow approximately in a layer-island mode at low substrate temperature of 100 K and incident energy of 0.1 eV. However, for the substrate temperature to 800 K and the incident energy to 5 eV, growth behaviors seem to follow layer-by-layer growth mode and the surface become smoother since high substrate temperature and incident energy can enhance the thermal diffusion and atomic activity. The influences of the incident energy on the intermixing are explored quantitatively and the degree of intermixing aggravated is observed as the incident energy increases to 3 and 5 eV. Besides, owing to different surface energies, the diffusion thickness of Cu atoms penetrating into the deposited layer is greater than that of Ni atoms penetrating into the substrate layer. The kinetic energy increases rapidly when the deposited atom is close to the substrate surface and it is interesting to find that the incremental value is about 1.6 eV, which is independent on the incident energy. Finally, the peaks of RDF clearly indicate that the disorder degree of Ni thin film can be enhanced as the substrate temperature increases. We expect these results can offer a clearer understanding to the mechanism and a capability to control the growth of metal thin films experimentally.

ACKNOWLEDGMENTS

This work is supported by the National Natural Science Foundation of China (Grant No. 61275159) and the Fundamental Research Funds for the Central Universities (Grant No. 13lgjc02).

CORRESPONDING AUTHOR

Corresponding author: Shao-Ji Jiang, Email: stsjsj@mail.sysu.edu.cn, Tel: +86 020-84113306.

REFERENCES

[1] B.K. Kuanr. 2003. Interlayer exchange coupling of epitaxial Fe/Al/Fe trilayer films: Dynamic and static measurements. *J. Appl. Phys.* 93: 7232–7234.
[2] Cynthia L. Kelchner and Andrew E. DePristo. 1997. Molecular dynamics simulations of multilayer homoepitaxial thin film growth in the diffusion-limited regime. *Surf. Sci.* 393: 72–84.
[3] Christopher G. Johansen and Hanchen Huang. 2007. Effects of three-dimensional Ehrlich-Schwoebel barrier on texture selection during Cu nanorod growth. *Appl. Phys. Lett.* 91: 121914.
[4] C.M. Gilmore and J.A. Sprague. 1991. Molecular-dynamics simulation of the energetic deposition of Ag thin films. *Phys. Rev. B* 44: 8950–8957.
[5] Eva Pellicer, Aïda Varea and Salvador Pané et al. 2010. Nanocrystalline Electroplated Cu–Ni: Metallic Thin Films with Enhanced Mechanical Properties and Tunable Magnetic Behavior. *Adv. Funct. Mater.* 20: 983–991.
[6] E.B. Dolgusheva and V. Yu. Trubitsin. 2014. Study of size effects in structural transformations of bcc Zr films by molecular-dynamics simulation. *Comput. Mater.* Sci. 84: 23–30.
[7] Graeme Henkelman and Hannes Jónsson. 2003. Multiple Time Scale Simulations of Metal Crystal Growth Reveal the Importance of Multiatom Surface Processes. *Phys. Rev. Lett.* 90: 11610.
[8] G. Bonny, RC. Pasianot and N. Castin et al. 2009. Ternary Fe–Cu–Ni many-body potential to model reactor pressure vessel steels: First validation by simulated thermal annealing. *Philos. Mag.* 89: 3531–3546.
[9] H.L. Meyerheim and D. Sander et al. 2008. Buried Ni/Cu (001) interface at the Atomic Scale. *Phys. Rev. Lett.* 100: 146101.
[10] Jian Wang, Hanchen Huang and S.V. Kesapragada et al. 2005. Growth of Y-Shaped Nanorods through Physical Vapor Deposition. *Nano Lett.* 5: 2505–2508.
[11] J. Lindner, P. Poulopoulos and F. Wilhelm. 2000. Atomic exchange processes at the interface and their role on the magnetic moments of ultrathin Ni/Cu (001) films. *Phys. Rev. B* 62: 10431–10435.
[12] M. Hemmous, A. Layadi and A. Guittoum et al. 2014. Structure, surface morphology and electrical properties of evaporated Ni thin films: Effect of substrates, thickness and Cu underlayer. *Thin Solid Films* 562: 229–238.
[13] M.F. Francis, M.N. Neurock and X.W. Zhou. 2008. Atomic assembly of Cu/Ta multilayers: Surface roughness, grain structure, misfit dislocations, and amorphization. *J. Appl. Phys.* 104: 034310.
[14] Masao Doyama and Y. Kogure. 1999. Embedded atom potentials in fcc and bcc metals. Comput. *Mater. Sci.* 14: 80–83.
[15] N. Rajasekaran, J. Mani, B.G. Tóth, G. Molnár. 2015. Giant magnetoresistance and structure of electrodeposited Co/Cu multilayers: the influence of layer thicknesses and Cu deposition potential. *Journal of The Electrochemica Society.* 162:204–212.
[16] S. Rout, M. Senthil Kumar. 2013. Study of structure, microstructure and giant magnetoresistance in nanogranular FeCuAg thin films with wide concentration range. *Journal of Alloys and Compounds.* 563:197–202.
[17] Soon-Gun Lee and Yong-Chae Chung et al. 2006. Surface characteristics of epitaxially grown Ni layers on Al surfaces: Molecular dynamics simulation. *J. Appl. Phys.* 100: 074905.
[18] Soon-Gun Lee and Yong-Chae Chung et al. 2009. Molecular dynamics investigation of interfacial mixing behavior in transition metals (Fe, Co, Ni)-Al multilayer system. *J. Appl. Phys.* 105: 034902.
[19] S.J. Plimpton. 1995. Fast Parallel Algorithms for Short-Range Molecular Dynamics. *J.Comput. Phys.* 117: 1–19.
[20] Soon-Gun Lee and Yong-Chae Chung. 2005. Molecular dynamics investigation for thin film growth morphology of Ni/Ni(111). *IEEE Transactions on Magnetics* 41: 3431–3433.
[21] Sang-Pil Kim and Seung-Cheol Lee et al. 2008. Asymmetric surface intermixing during thin-film growth in the Co–Al system: Role of local acceleration of the deposited atoms. *Acta Materialia* 56: 1011–1017.
[22] V Georgieva, M Saraiva and N Jehanathan et al. 2009. Sputter-deposited Mg–Al–O thin films:linking molecular dynamics simulations to experiments. *J. Phys. D: Appl. Phys.* 42: 065107.
[23] V. K. Kumikov and Kh. B. Khokonov. 1983. On the measurement of surface free energy and surface tension of solid metals. *J. Appl. Phys.* 54: 1346–1350.

[24] X.L. Yan, J.Y. Wang. 2013. Size effects on surface segregation in Ni–Cu alloy thin films. *Thin Solid Films.* 529:483–487.

[25] Xing-bin Jing, Zu-li Liu and Kai-lun Yao. 2012. Molecular dynamics investigation of deposition and annealing behaviors of Cu atoms onto Cu (0 0 1) substrate. *Appl. Surf. Sci.* 258: 2771–2777.

[26] Z. S. Pereira and E. Z. da Silva. 2010. Study of defects in Pd thin films on Au (100) using molecular dynamics. *Phys. Rev. B* 81: 195417.

[27] Zheng-Han Hong, Shun-Fa Hwang and Te-Hua Fang. 2008. Effect of substrate temperature and deposition rate on alloyzation for Co or Fe onto Cu(001) substrate. *J. Appl. Phys.* 103: 124313.

Advances in Engineering Materials and Applied Mechanics – Zhang, Gao & Xu (Eds)
© 2016 Taylor & Francis Group, London, ISBN 978-1-138-02834-0

Application of dendrimer modified montmorillonite in EPDM composites

R.M. Li, C. Pan, J.M. Wu & J.C. Wang

College of Chemistry and Chemical Engineering, Shanghai University of Engineering Science, Shanghai, China

ABSTRACT: FR-DOMt was a type of effective filler for improving the mechanical and flame-retardant properties of EPDM composites. The experimental results revealed that incorporation of 1–15 phr of FR-DOMt into pure EPDM could change T10 and T90 of these composites. At loading of 3–5 phr of FR-DOMt, the tensile strength was improved from 10.5 to 12.0 MPa and elongation at break from 427 to 523%. Meanwhile, the abrasion loss of EPDM/FR-DOMt-1 was 0.32 cm^3, decreased 9% in comparison with that of pure EPDM, 0.35 cm^3. In addition, with the addition of 1 phr of FR-DOMt into EPDM system, the vertical burning time was increased from 250 s to 370 s.

1 INTRODUCTION

Due to the specialty and general-purpose applications, EPDM rubber is a type of synthetic rubber that has wide usage. It is valuable for its high resistance to heat, oxidation, and weather aging resulting from its stable polymer backbone. In addition, it has a good electrical resistivity and polar solvents resistance. However, inherently high flammability for this polymer is one of the typical setbacks that limit its usages[1].

The flame-retardant properties of organic polymers can be optimized by using the flame-retardant additives[2]. Two main types of additives were focused on, that is, halogenated additives or high loadings of Aluminium Hydroxide (ATH). But neither of which was satisfactory. The traditional halogens flame-retardant additives can obtain good flame-retardant property, but it will release poisonous hydrogen halogens and has harm to the human body and environment. In addition, ATH is green and good to the environment. However, high quantity of this additive is needed to satisfy the flame-retardant requirement of polymeric matrix and this may influence the mechanical properties of polymers. Montmorillonite (Mt) has now obtained a special attention in the field of flame-retardancy due to its good thermal stability, small particle size, and intercalation properties. Improved tensile properties, decreased gas permeability, and enhanced thermal stability can be endowed for the polymeric composites when Mt particles were intercalated[3,4].

Recently, a new type of macromolecules, dendrimers, has received more attention resulting from their special structures, such as plenty of functional end groups, three-dimensional networks, difficult to crystallize, and high compatibility with other polymers[5]. The dendrimers exhibited the features of both molecular and polymer chemistry. The step by step controlled synthesis of dendrimers belonged to molecular chemistry like properties. Meanwhile, dendrimers were synthesized from monomers and this behavior was ascribed to polymer chemistry[6,7]. In our previous studies, a novel Flame-Retardant Dendrimer modified Organic Mt (FR-DOMt) was prepared[8].

In this study, FR-DOMt reinforced EPDM composites were prepared by common mixing techniques. These EPDM composites exhibited better tensile properties, decreased hardness, and increased flame retardance.

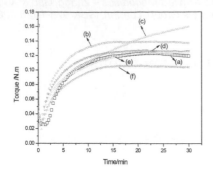
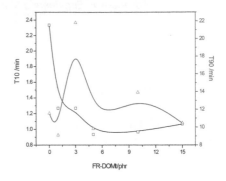

Figure 1. Cure curves of different EPDM composites. (a) EPDM, (b) EPDM/FR-DOMt-1, (c) EPDM/FR-DOMt-3, (d) EPDM/FR-DOMt-5, (e) EPDM/FR-DOMt-10, (f) EPDM/FR-DOMt-15.

Figure 2. T10 and T90 of different EPDM composites.

2   CURE PROPERTIES OF EPDM/FR-DOMT COMPOSITES

Figure 1 presented the cure curves of different EPDM composites. The curves were almost similar and the theoretical sulfurate time was about 30 min. From these curves, we can see that with the addition of 1–5 phr of FR-DOMt, the maximum torque of EPDM composites was increased. Moreover, the data of other cure characteristics, T10 and T90, for different EPDM composites were summarized in Figure 2. T10 represented the scorch time during which a rubber can be safely worked at a given temperature in rubber preparation before curing began. T90 referred to optimum cure time taken for the rubber to rise to its maximum point, and at 90% of this level. For T10, the longer was the scorch time, the safer the processing was. The addition of FR-DOMt (1~15 phr) can decrease scorch time and make processing unsafer. For T90, the longer was the cure time, the more energy was used. It can be concluded that almost no energy was saved in the vulcanization process deduced from the longer technical cure time.

3   TENSILE PROPERTIES OF EPDM/FR-DOMT COMPOSITES

The tensile properties of EPDM/FR-DOMt were given in Figure 3. It can be seen that different trends of tensile strength and elongation at break were shown for different EPDM composites. When FR-DOMt content was lower than 3 phr, their tensile strength was increased with the increasing amount of organoclays (Fig. 3(a)). At 3 phr, the EPDM/FR-DOMt composite showed the highest tensile strength, 12.0 MPa. This was 15% higher than that of pure EPDM, 10.5 MPa. However, some silicate layers aggregated when their contents were increased. This led to the decrease of the tensile strength. In addition, the elongation at break showed almost different trend when this additive was added. At loading of 5 phr of FR-DOMt, the elongation at break was 523%. This was 22% higher than that of pure EPDM, 427%. Moreover, the best elongation break, 556%, was obtained when the amount of FR-DOMt was increased to 15 phr. The improvement of tensile strength and elongation at break were attributed to two facts: (1) The silicate layers of FR-DOMt in the polymeric matrix showed uniformly dispersion and possessed a high stress bearing capability; (2) More effective constraint of the motion of rubber chains resulted from the stronger interactions between the additive and EPDM chains.

4   HARDNESS OF EPDM/FR-DOMT COMPOSITES

Figure 4 showed the hardness of different EPDM/FR-DOMt composites. With the increasing amount of FR-DOMt, the hardness was relatively decreased, especially for the composite

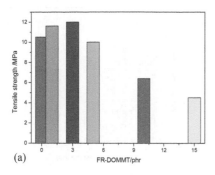
(a)

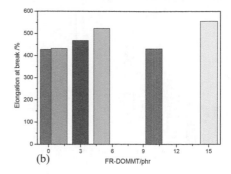
(b)

Figure 3.   Tensile properties of different EPDM composites (a) tensile strength, (b) elongation at break.

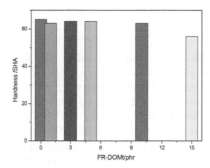

Figure 4.   Hardness of different EPDM composites.

with 15 phr of FR-DOMt. This was resulted from the less cross-linking networks in the dendrimers. This may be suitable for the application of these additives in the fields that has this requirement.

5   WEAR RESISTANCE OF EPDM/FR-DOMT COMPOSITES

Figure 5 presented the abrasion loss value of different EPDM composites. At loading of 1 phr of FR-DOMt, the EPDM composite showed the lowest abrasion loss value, 0.32 cm^3. This was 9% lower than that of pure EPDM, 0.35 cm^3. However, this composite showed the worst wear resistance after the addition of 10 phr of FR-DOMt. Compared with pure EPDM, the abrasion loss of EPDM/FR-DOMt-10 was 1.03 cm^3, nearly 194% increase. This was due to the worse compatibility between EPDM and the aggregated silicate layers.

6   FLAME RETARDANCE OF EPDM/FR-DOMT COMPOSITES

The vertical burning time of different EPDM composites were shown in Figure 6. The burning time of pure EPDM was 250 s. By adding 1~15 phr of FR-DOMt into pure EPDM, the burning time was changed. Due to the flame inhibition effect of FR-DOMt, the EPDM/FR-DOMt-1 exhibited a longer burning time, 370 s. However, with addition of 15 phr of FR-DOMt into EPDM, this composite presented a decreased burning time, 210 s. The more aggregated silicate layers in the polymeric matrix resulted more holes and a damaged structure. This structure was beneficial for the oxygen to flow into the rubber matrix. This phenomenon may increase the burning rate of this composite.

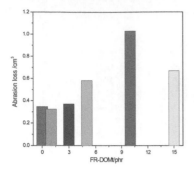

Figure 5. Abrasion loss of different EPDM composites.

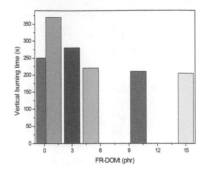

Figure 6. Vertical burning time of different EPDM composites.

7 CONCLUSIONS

FR-DOMt/EPDM composites were prepared using this flame-retardant organic silicate layers. The properties such as cure characteristics, tensile strength, elongation at break, hardness, abrasion loss, and vertical burning time were studied. The addition of FR-DOMt improved the mechanical properties of EPDM composites. The composites with 3 phr of FR-DOMt showed the highest tensile strength, 12.0 MPa. Meanwhile, the EPDM/FR-DOMt-15 showed the highest elongation at break, 556%. In addition, at loading of 1 phr of FR-DOMt, the abrasion loss was 0.32 cm^3 and was decreased 9% compared with that of pure EPDM, 0.35 cm^3. Moreover, the flame-retardant behavior of EPDM composites was improved after the addition of 1–3 phr of FR-DOMt. The vertical burning time was increased from 250 s to 370 s after addition of 1 phr of FR-DOMt into EPDM.

ACKNOWLEDGEMENT

This work was financially supported by "National Natural Science Funds (Project No. 51173102)".

REFERENCES

[1] Vikas, M., Jin, K.K., Kaushik, P., 2011. *Recent advances in elastomeric nanocomposites.* Springer-Verlag Berlin Heidelberg, New York. pp 160.

[2] Wang, J.C., Guo X., Zheng X.Y., Zhao Y., Li W.F. *Clays and Clay Minerals*, 2011, 59, 446–458.

[3] Acharya, H., Pramanik, M., Srivastava, SK, Bhowmick, A.K., 2004. Synthesis and evaluation of high-performance ethylene-propylene-diene terpolymer/organoclay nanoscale composites. *J. Appl. Polym. Sci.* 93, 2429–2436.

[4] Mousa, A., 2007. Evolution of the mechanical properties of EPDM vulcanizates by compounding with layered organo-montmorillonite. *Int. J. Polym. Mater.* 56, 355–363.

[5] Tan, H.M., Luo, Y.J., 2002. *Dendritic Polymers.* Chemical Industry Press, Beijing.

[6] Dasgupta, M., Peori, M.B., Kakkar, A.K., 2002. Designing dendritic polymers containing phosphorus donor ligands and their corresponding transition metal complexes, Coordin. *Chem. Rev.* 233–234, 223–235.

[7] Malik, A., Chaudhary, S., Garg, G., Tomar, A., 2012. Dendrimers: A Tool for Drug Delivery, *Advan. Biol. Res.* 6, 165–169.

[8] Wang, J.C., Sun, K., Hao, W.L., Du, Y.C., Pan, C., 2014. Structure and Properties Research on Montmorillonite Modified by Flame-retardant Dendrimer. *Appl. Clay Sci.* 90, 109–121.

Advances in Engineering Materials and Applied Mechanics – Zhang, Gao & Xu (Eds)
© 2016 Taylor & Francis Group, London, ISBN 978-1-138-02834-0

Mechanisms and microstructure evolution in pure iron by PM under tensile load

Lei Li
Department of Mechanics, Inner Mongolia University of Technology, Hohhot, China

Z.M. Shi
School of Materials Science and Engineering, Inner Mongolia University of Technology, Hohhot, China

ABSTRACT: The analysis and measurement of the damage variable of materials is important for research on damage mechanics. For this study, pure iron test samples were fabricated by powder metallurgy. The crack formation and propagation mechanism was investigated during a dynamic changing process under tensile load by environmental scanning electron microscopy. The results indicate that, at the initial stage of loading, micro-cracks easily appear at different positions around the prefabricated notch, connecting with surrounding micro-cracks and propagating along the grain boundaries, eventually forming a dominant crack. The dominant crack is formed at the notch and connects with neighboring micro-cracks while propagating along the grain boundaries in the middle period of loading. When the crack propagation is disrupted, new cracks form in front of the older cracks and connect with each other, again propagating along the grain boundaries. Finally the samples lose stability and fracture rapidly. The fracture surfaces do not show visible necking phenomena along the thickness direction. Most dimples are oval in shape, indicating that they are formed by normal stress. The damage evolution law of the microstructure of the pure iron samples is discussed and the evolution equation is established for future application to other materials fabricated by powder metallurgy.

1   INTRODUCTION

Many theories and experiments suggest a combined occurrence and formation of structural damage and the fracture of materials due to the presence of micro-defects. Metal materials generally undergo three stages of deformation: elastic deformation, plastic deformation and fracture[1–2]. Recently, several research groups investigated the structural damage and the fracture of materials at the micro-, meso- and macroscopic level, especially focusing on the mesoscopic scale. For instance, Rabotnov proposed the concept of the damage factor, and introduced a constitutive equation for damage coupling and a damage evolution equation[3]. Gurson later proposed an integral constitutive equation based on the results published by McClintock and Rice and Tracey[4], describing the effect of void damage on the plastic deformation behavior of a material[4], while Dhar *et al.* suggested direct and indirect methods for fracture damage assessment[5]. For most metal materials, fracture damage generally occurs in three stages: nucleation, extension, and coalescence of voids[6]. Void damage occurred on the mesoscopic scale and even physical changes on the microscopic scale were observed. In order to find a method to quantitatively describe the material's microstructure irregularities and randomness, a micro evolution[7–9] of the material damage was performed on the mesoscopic level. In this study, the area affected by the micro-cracks can be approximated by a square flake, with the square value of the length of the crack line representing the damage area. A selected crack length factor reflects the change in length of the micro-crack square, and is used to eventually derive the damage variable. The crack propagation and evolution of the damage variable were investigated under application of a tensile load. The analysis of mesoscopic images and the damage curves[10-13] showed that the mesoscopic damage of pure iron fabricated by powder metallurgy affected the formation and propagation of macroscopic cracks. Research on the damage propagation in pure iron fabricated by powder metallurgy is important for the future prediction of fracture in metal materials.

2   EXPERIMENTAL DETAILS

The samples of pure iron were fabricated by the powder metallurgy method, and were provided by the University of Wuppertal, Germany. They were subject to a rolling and annealing process prior to the experiments. The mechanical properties of the samples are listed in Table 1.

The test samples used for the *in situ* observations were fabricated by the wire cutting method. They were placed in an environmental scanning electron microscope (ESEM-Hitachi High technologies S-3400N). A tensile load was then applied to the samples and the resulting initiation and propagation of cracks were observed *in situ*. The maximum displacement of the ESEM was 3 mm, and the maximum load 2 kN. A unilateral precast notch with a depth of 1 mm was cut into the samples by the wire cutting method. The dimensions and the loading state of the samples used for the *in situ* ESEM observations are shown in Figure 1[14]. After the sample fractured, the fracture was investigated in detail in order to determine the crack propagation mechanism.

3   RESULTS AND DISCUSSION

The ESEM micrographs of the fracture surfaces of the iron samples are shown in Figure 2 revealing the crack formation and propagation in pure iron samples fabricated by powder metallurgy subjected to tensile stress. In the earlier stages of loading, micro-cracks easily appeared in the vicinity of the sample notch. When the applied stress had reached 272.8 MPa, micro-cracks vertical to the tension direction appeared in the notch region. These cracks propagated along the grain boundaries, which is typical for crack formation under tensile stress. A large number of new micro-cracks appeared in the sample along the grain boundaries and the interaction of these micro-cracks led to the formation of dominant cracks as the stress level increased. Previously appearing micro-cracks did not develop into

Table 1.   Mechanical properties of the pure iron samples fabricated by powder metallurgy.

Sample number	Mechanical properties of the pure iron samples		
	Density (g·cm⁻³)	σ_b (MPa)	ψ_k (%)
1	7.13	270	3.3
2	7.22	314	3.1

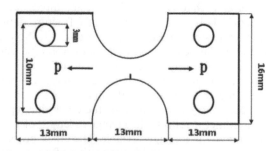

Figure 1.   Dimensions and loading state of the samples used for the in situ ESEM observations.

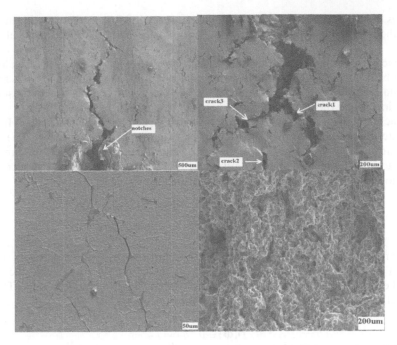

Figure 2. SEM micrographs revealing the formation and propagation of micro-cracks and the fracture surfaces.

the dominant cracks; instead the dominant cracks appeared at the sample notch. As the stress level continuously increased and the propagation of the dominant cracks was deterred, the dominant cracks connected to preceding micro-cracks and the ambient micro-cracks extended with the connection; the dominant cracks tended to propagate in a zigzag pattern and roughly vertical to the tension direction.

Some facts are suggesting that the sample fractured rapidly and the surface of the sample lost stability due to the rapid expansion of cracks since the stress decreased to 258.9 MPa. Since the grains did not deform easily, a passive area was produced in front of the cracks on a microscopic view, and then the cracks produced branches expanding along the boundaries between neighboring grains. New micro-cracks were formed in front of older cracks and expanded along the grain boundaries as the dominant cracks were expanding. Then, the new micro-cracks connected to adjacent cracks to form a new dominant crack. Therefore, the cracks in the test samples mainly appeared at the grain boundaries and expanded by connecting with neighboring cracks to form new cracks or new holes when the sample was subjected to tensile stress. The new cracks and new holes connected with each other to propagate towards the undamaged regions of the sample. The sample lost stability and fractured rapidly as the dominant crack reached a certain length. Figure 2 shows that there is no visible necking phenomenon.

Dimples appeared on the fracture surfaces as shown in Figure 2. Some dimples are very large, others very small, which can be explained by the formation of accumulative regions and scarce regions during the fracture experiment. Grain spacing in the scarce region is large, and the constraining force at the grain boundary is small. Therefore, the size of the dimples visible on the corresponding fracture surfaces is large. Grain spacing in the accumulative region is small, and the constraining force at the grain boundaries is large, and, therefore, the size of the dimples visible on the corresponding fracture surfaces is very small. Most dimples have an oval shape.

In order to quantitatively describe the deformation of the microstructure of the metal material and the metallic structure evolution, the crack length factor Φ was introduced, which is typically expressed by the following equation[15-17]:

$$\Phi = \Delta L / A_0 = (L_c - L_0)/ A_0 \qquad (1)$$

where L_c is the total length of the cracks in the plane section for any deformation state, which includes the length of the precracks, and A_0 denotes the length of the precracks at the initial state.

If the measured experimental data has enough statistical significance, the crack length factor Φ can be used as an indicator for the deformation inside the material, i.e., it is reasonable to quantitatively describe the shape evolution law and the variation characteristics of the plane section inside the material with the help of the crack length factor Φ.

The crack length factor Φ changes with the microstructure of the material. Φ It increases as the material deformation increases during the damage formation process. Therefore, the damage variable can be defined using the crack length factor Φ for a certain damage stage of the material. The relative crack length factor ξ gradually decreases as the microstructure deformation of the material increases. ξ It is defined as:

$$\xi = \frac{\Phi_f - \Phi}{\Phi_f} \tag{2}$$

where Φ and Φ_f correspond to the crack length factor of ε_p and ε_p, respectively. The above expression is a simple and convenient method for material testing and subsequent calculations, and can be used to sensitively characterize the degree of material damage. The relative crack length factor can then be used to define the damage variable D:

$$D(\Phi) = 1 - \frac{\Phi_f - \Phi}{\Phi_f} \tag{3}$$

where $D(\Phi)$ denotes the damage variable of the materials. It can express the degree of damage of the material during deformation, hardening, softening and fracture:

1. If $D(\Phi) = 0$, the material is undamaged;
2. If $0 < D(\Phi) < 1$, the material is damaged;
3. If $D(\Phi) = 1$, the material is fractured.

This study focused on the crack length as the measurement parameter and uses the calculation function of the analytics software to measure the total length of the cracks of the selected sample area.

As seen in Figure 3, the crack length factor Φ monotonically increases with the applied strain level. The crack length factor Φ expresses the variation per unit length of the crack. The larger the crack length factor Φ the larger the degree of plastic deformation. Therefore, the crack length factor Φ can be used as an indicator for the deformation inside the material, and it is feasible to use the crack length factor to establish the microstructure evolution law of the material Φ.

$$D = C_1 + C_2 e^{\varepsilon C_3}$$

Pure iron samples with two different densities were used for the fracture experiments. Figure 3 show that the damage variable D is monotonically increasing with the strain.

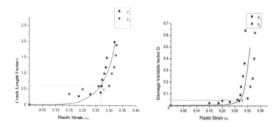

Figure 3. Crack length factor plotted versus the applied plastic strain (left) and damage variable plotted versus the applied plastic strain (right).

Point B in Figure 3 denotes an inflection point. At first, the crack length factor Φ increases very slowly and the measurement data points are basically aligned parallel to the strain axis. However, after point B is reached, the crack length factor sharply increases. The damage value also exhibits this apparent turning point and starts to increase significantly. Therefore, we can define the value ε at point B as the critical strain ε_c, and the damage variable D_c corresponding to ε_c as the critical damage factor. If this value is exceeded, the material rapidly becomes unstable and starts to fracture.

The curves in Figure 3 were obtained by fitting them to the experimental data, resulting in the following equation:

$$D = C_1 + C_2 e^{\varepsilon C_3} \tag{4}$$

where C_1, C_2 and C_3 are material constants.

From this damage evolution law for pure iron fabricated by powder metallurgy, we now know that the evolution of the plastic damage in the samples can be divided into two stages:

1. The initial stage of damage evolution
   If $0 < D \leq D_c$ and $0 < \varepsilon \leq \varepsilon_c$, damage evolution is very slow and uniform. At this stage, micro-cracks do not appear in the material or the micro-cracks are very small. The grain shape does not change significantly and the density of the sample does not have an impact at this stage, and the fitted curve is relatively smooth.
2. The rapid increase stage of damage evolution (the stage of hole formation and propagation)
   If $D_c < D \leq 1$ and $\varepsilon_c < \varepsilon \leq \varepsilon_f$, the degree of damage in the material increases rapidly. This stage is indicative of the prospective material damage, i.e., it is the fastest damage evolution stage of the material. As the deformation value approaches ε_c ($\varepsilon_c = 0.026$ to 0.03 for the tested pure iron samples), i.e., as the damage value D approaches D_c ($D_c = 0.046$ for the tested pure iron samples), the fitted curve rapidly increases. The dominant cracks start to appear in the sample and a large number of micro-cracks propagate around the dominant crack from a macroscopic point of view. However, grain deformation is not obvious on a microscopic level. The micro-cracks rapidly propagate along the grain boundaries and connect with each other, and micro-cracks preferably form in the dominant crack region. The amount of material damage increases sharply with the strain until the material fractures. The two types of samples with different densities show only a slightly different behavior as the strain reaches ε_c. The damage evolution curves for the sample 1 (lower density) and the sample 2 (higher density) start to rapidly increase at $\varepsilon_c = 0.026$ and $\varepsilon_c = 0.03$, respectively, i.e., the higher density sample requires a larger amount of strain in order to show the same degree of damage compared to the lower density sample.

4 CONCLUSIONS

1. The cracks in pure iron samples fabricated by powder metallurgy mainly appeared at the grain boundaries as the sample was subjected to tensile stress. The cracks propagate and connect with neighboring cracks to form new cracks or holes. Then, the newly formed cracks or holes connect with each other and propagate towards undamaged regions. Finally, the sample rapidly loses stability and fractures as the propagation of cracks exceeds a certain level. Most of the dimples are oval in shape, indicating that they are formed under normal stress.
2. The damage variable D is also monotonically increasing with the plastic strain. The equation describing the fitted curve was obtained and can be written as:

$$D = C_1 + C_2 e^{\varepsilon C_3} \tag{5}$$

where C_1, C_2 and C_3 are material constants.
3. From the obtained damage evolution law, it is clear that the plastic damage evolution can be divided into two stages. If $0 < \varepsilon \leq \varepsilon_c$, damage evolution is very slow and uniform. If, $\varepsilon_c < \varepsilon \leq \varepsilon_f$, the material damage is rapidly increasing. This stage is indicative of the

prospective material damage, i.e., it is the fastest damage evolution stage of the material. As the deformation value approaches ε_c ($\varepsilon_c = 0.026$ to 0.03 for the tested pure iron samples), i.e., as the damage value D approaches D_c ($D_c = 0.046$ for the tested pure iron samples), the fitted curve rapidly increases.

4. At the initial stage of damage formation, the sample density does not have an obvious impacted. It is clear that the damage evolution law for the iron samples is basically the same at the initial stage, and that the density has little impact on the damage evolution law. However, the higher density sample requires a larger amount of strain in order to show the same degree of damage compared to the lower density sample.

ACKNOWLEDGMENT

The authors gratefully acknowledge the support from the National Natural Science Foundation of China (grant no. 11002065). The supply of test material by the University of Wuppertal, Germany, is much appreciated.

REFERENCES

[1] Gao X., Wang H.G., Kang X.W. and Jiang L.Z. 2010. Analytic Solutions to Crack tip Plastic Zone under Various Loading Conditions. *Eur. J. Mech. A-Solid*, 29(4): 738–745.

[2] F. Delale and F. Erdogan. 1983. The Crack Problem for a Nonhomogeneous Plane. *J. Appl. Mech*, 50(3): 609–614.

[3] Rabotnov Y.N. 1963. On the Equations of State for Creep. *Progress in Applied Mechanics*, 178(1): 117–122.

[4] Gurson A.L. 1975. Continuum theory of ductile rupture by void nucleation and growth. Part I. Yield criteria and flow rules for porous ductile media. *Brown Univ., Providence, RI (USA): Div. of Engineering*.

[5] S. Dhar, P.M. Dixit and R. Sethuraman. 2000. A Continuous Damage Mechanics Model for Ductile Fracture. *International Journal of Pressure Vessels and Piping, Engineering Materials and Technology*, 77: 335–344.

[6] Z.M. Shi, H.L. Ma and J.B. Li. 2011. A Novel Damage Variable to Characterize Evolution of Microstructure With Plastic Deformation For Ductile Metal Materials Under Tensile Loading. *Engineering Fracture Mechanics*, 78(3): 503–513.

[7] Wang J., Ju D. and Yin F. 2011. Microstructure evaluation and crack initiation crack for AZ31 sheet under biaxial stress. *Procedia Engineering*, 10: 2429–2434.

[8] Wang Q., Sha A. and Wu G. 2012. Evolution of Microstructure During Tensile Deformation of TB5 Titanium Alloy. *Procedia Engineering*, 27: 840–846.

[9] Liu J.B., Meng L. and Zeng Y.W. 2006. Microstructure evolution and properties of Cu–Ag microcomposites with different Ag content. *Materials Science and Engineering* A, 435: 237–244.

[10] Yunxue J. and Jungang L. 2007. Crack forming and propagation mechanisms of TiC/Ti composites during dynamic tension. *Rare Metal Materials and Engineering*, 36(5): 764–768.

[11] Niu Rong Mei and Zhang Guo Jun.2006. SEM In-situ Observation on Microstructure of Pure Mo. *Rare metal materials and engineering*, 35: 559–563. (in Chinese)

[12] Lan L.Y., Qiu C.L. and Zhao D.W. 2011. Microstructural Evolution and Mechanical Properties of Nb-Ti Microalloyed Pipeline Steel. *Journal of Iron and Steel Research, International*, 18(2): 57–63.

[13] Yang G., Yang M., Liu Z., et al. 2011. Three-Dimensional Microstructures and Tensile Properties of Pure Iron During Equal Channel Angular Pressing. *Journal of Iron and Steel Research, International*, 18(12): 40–44.

[14] VI.V. Skorokhod and V.D. Krstic. 2000. Processing, Microstructure, and Mechanical Properties of B_4C-TiB_2 Particulate Sintered Composites. II. Fracture and Mechanical Properties. *Powder Metallurgy and Metal Ceramics*, 39(9–10): 504–513.

[15] J.A. Jimenez, G. Frommeyer. 2010. Analysis of the microstructure evolution during tensile testing at room temperature of high-manganese austenitic steel. *Materials Characterization*, 61: 221–226.

[16] D. Barbier, N. Gey, S. Allain, et al. 2009. Analysis of the tensile behavior of a TWIP steel based on the texture and microstructure evolutions. *Materials Science and Engineering A*, 500: 196–206.

[17] S. Bystrzanowski, A. Bartels, A. Stark, et al. 2010. Evolution of microstructure and texture in Ti-46 Al-9 Nb sheet material during tensile flow at elevated temperatures. *Intermetallics*, 18: 1046–1055.

Advances in Engineering Materials and Applied Mechanics – Zhang, Gao & Xu (Eds)
© 2016 Taylor & Francis Group, London, ISBN 978-1-138-02834-0

Simulation research on constant pressure system of synchronous generator based on two-dimensional cloud model PID

Z.Z. Li & C.J. Shi

School of Marine Engineering, Dalian Maritime University, Dalian, China

ABSTRACT: For the poor adaptive ability of controllable phase compound excitation system in the ship power system, a two-dimensional cloud model of PID excitation is designed and applied to the excitation system of synchronous generator. Combining the cloud theory and fuzzy theory, PID parameters are self-tuned. Through the establishment of models of synchronous generator controllable phase compound brushless excitation system and the two-dimensional cloud model PID excitation controllable system, the Matlab is used to simulate various situations of the synchronous generator, such as sudden load on and off. The simulation of experiments shows that the two-dimensional cloud model PID excitation controllable system model has better robustness and better adaptability than the controllable phase compound brushless excitation system.

1 INTRODUCTION

A ship power system needs to have static stability and dynamic stability, so the excitation system has to maintain constant generator terminal voltage and reactive power regulation. Controllable brushless excitation system can be seen as a conventional PID excitation controller that can response to the change of generator terminal voltage and achieve steady control easily. But compared with the large power grid on land, ship power plant capacity is much smaller, circuit of power plant is shorter and time constant is smaller. Due to the characteristics of nonlinear and time varying, parameters of synchronous generator change frequently. As the adaptive ability of controllable brushless excitation system is poor, synchronous generator terminal voltage cannot be controlled precisely[1–8]. Combining the cloud theory and fuzzy theory, a two-dimensional cloud model PID excitation system is designed. The simulations of sudden load on and off show that the latter has better robustness.

2 FIVE ORDER MODEL OF SYNCHRONOUS GENERATOR

Ignoring transient situation of stator windings, an ideal salient pole motor is applied.

$$T'_d \dot{E}'_q = -\left(X_d - X'_d\right)I_d - E'_q + E_f \tag{1}$$

$$T''_d \dot{E}''_q = -\left(X'_d - X''_d\right)I_d + E'_q - E''_q + T''_d \dot{E}'_q \quad T''_q \dot{E}''_d = \left(X_q - X''_q\right)I_q - E''_d \tag{2}$$

$$U_d = -R_a I_d + X''_q I_q + E''_d \quad U_q = -R_a I_q - X''_d I_d + E''_q \tag{3}$$

$$T_J \frac{dw}{dt} = T_m - \left[E''_q I_q + E''_d I_d - \left(X''_d - X''_q\right)I_d I_q\right] - D\left(w-1\right) \tag{4}$$

$$\frac{d\delta}{dt} = w - 1 \tag{5}$$

Static load model:

$$U_d = L\dot{I}_d - wLI_q + rI_d \qquad U_q = L\dot{I}_q - wLI_d + rI_q \qquad (6)$$

where E''_d is d-axis subtransient electromotive force; E''_q is q-axis subtransient electromotive force; E'_q is q-axis subtransient electromotive force; E_f is stator excitation electromotive force; I_d is d-axis electric current; I_q is q-axis electric current; X'_d is d-axis transient reactance; X_d is d-axis reactance; T'_d is transient time constant of d-axis short circuit; T''_q is subtransient time constant of q-axis; T''_d is subtransient time constant of d-axis; w is power angle; X_q is q-axis reactance; X''_d is d-axis subtransient reactance; X''_q is q-axis subtransient reactance; T_m is mechanical torque; T_J is inertial time constant of synchronous generator; R_a is the stator winding resistance; U_d is d-axis voltage; U_q is q-axis voltage; δ is electrical angle; r is the resistance of load; and L is the reaction of load.

3 CONTROLLABLE PHASE COMPOUNDING EXCITATION SYSTEM

Components of self-excitation current and phase compound current are overlapped to supply field windings of the brushless exciter through the rectifier link. As the load current changes, the exciter excitation current changes and dynamic voltage are adjusted. Then, the feedback signal of voltage difference is gained to accomplish static voltage adjustment.

$$U_t = \sqrt{\left(U_d - mI_qR\right)^2 + \left(U_q + mI_dR\right)^2} \qquad (7)$$

$$\Delta U = U_{ref} + U_0/k_a - U_{tf} + U_{stab} - U_{fr}; U_{tf} = \sqrt{\left(U_d\right)^2 + \left(U_q\right)^2} \cdot \frac{1}{1 + sT_d} \qquad (8)$$

$$U_h = \Delta U \frac{1 + sT_h}{1 + sT_c} \qquad (9)$$

$$U_a = U_h \frac{k_b}{1 + sT_a} \qquad (10)$$

$$E_f = U_a + U_t \qquad (11)$$

$$U_f = \frac{1}{T_b s + k_e} \cdot E_f \qquad (12)$$

$$U_{fr} = U_f \cdot \frac{sk_f}{1 + sT_{fr}} \qquad (13)$$

where U_t is output voltage of phase compound; m is component coefficient of voltage and current; R is phase-shifting reactance; U_{ref} is reference voltage of automatic pressure regulating device; U_{stab} is the grounding voltage; U_0 is the initial value of the excitation voltage; k_a is the

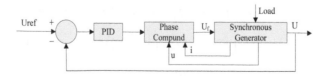

Figure 1. Controllable phase compound excitation brushless excitation system.

effective gain of AC exciter; T_d is the time constant of low pass filter; U_{fr} is the output voltage of feedback link; ΔU is the integrated voltage difference signal; U_h is output voltage of compensator; T_h is the lead compensation time constant of compensator; T_c is the lag compensation time constant of compensator; U_a is the output voltage of amplifier; k_b is the amplifier gain; T_a is the time constant of amplifier; E_f is the output voltage of regulating device; U_f is the AC exciter pressure; k_e is the AC exciter amplification coefficient; T_b is the time constant of AC exciter; k_f is the feedback link gain; and T_{fr} is time constant of feedback link.

4 TWO-DIMENSIONAL CLOUD MODEL PID EXCITATION SYSTEM

The fuzzy membership degree is often used to describe the fuzziness in the field of fuzzy theory. But if the membership is determined, a precise mathematical method is used to calculate without randomness. On the basis of fuzzy theory, the cloud theory is proposed with the characteristic of fusion of fuzziness and randomness. The fuzzy membership degree is probable distribution in the field of cloud theory. An arbitrary number e, belonging to the universe U and mapped to the interval [0,1], is a probability distribution instead of a fixed value. The three characteristic parameters of expectation Ex, entropy En and hyper entropy He are used to describe the normal cloud model.

Assumed that U is the quantitative universe of discourse, and C is the qualitative concept in U. If the quantitative value (x,y) is one of the random values of C belonging to U, and the certainty degree $u_{(x,y)}$ of (x,y) to C is a random number with stable tendency. Then if the quantitative value (x,y) satisfies with conditions: (x,y)-$N(E_x,E_y,En_x'^2,En_y'^2)$, (En_x',En_y')-$N(En_x,En_y,He_x^2,He_y^2)$, the certainty degree $u_{(x,y)}$ of (x,y) to C is calculated by the following equation:

$$u = e^{-\left[\frac{(x_i-Ex)^2}{2En_{xi}'^2} + \frac{(y_i-Ey)^2}{2En_{yi}'^2}\right]} \tag{14}$$

The double condition multi-rule normal cloud generator:
Input: $(E_x,E_y,En_x,En_y,He_x,He_y,n)$;
Output: drop (x_i,y_i,u_i), $i = 1,2,3...n$.
Algorithm steps: (1) One of the two-dimensional normal random data pairs (En_{xi}',En_{yi}'), which of expectation is (En_x,En_y) and variance is (He_x^2,He_y^2), is generated. (2) One of the two-dimensional normal random data pairs (x_i,y_i), which of expectation is (E_x,E_y) and variance is $(En_{xi}'^2,En_{yi}'^2)$, is generated. (3) The certainty degree $u_{(xi,yi)}$ of data pair (x_i,y_i) is calculated by the Equation (14). (4) The value (x_i,y_i,u_i) can be seen as a cloud droplet. (5) Repeat the step from (1) to (4) until the number of cloud drops reaches n.

The inference method of double condition multi-rule is adopted as follows.

The following block diagram is the control scheme of cloud model of PID excitation controller.

The input variables of two dimensional cloud model of PID excitation controller are difference voltage E between synchronous generator terminal voltage and reference voltage and change rate of difference voltage EC. Output variables are real-time parameters $K_{p,n+1}$, $K_{i,n+1}$, $K_{d,n+1}$ of PID. According to the established cloud inference rules, three parameters of PID are timely under control. The cloud model controller process is divided into three parts: positive cloud generator, rule constructor, and backward cloud generator.

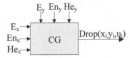

Figure 2.   Two-dimensional positive normal cloud generator.

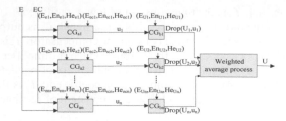

Figure 3.   Double condition multi-rule inference.

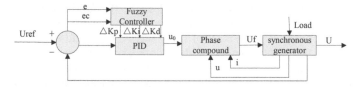

Figure 4.   Two-dimensional cloud model of PID excitation controller.

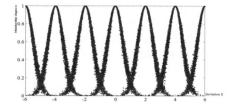

Figure 5.   Membership cloud of E and EC.   Figure 6.   Membership cloud of U_p, U_i and U_d.

Table 1.   Two-dimensional cloud model control rule of U_p, U_i and U_d.

E	EC						
	NB	NM	NS	Z	PS	PM	PB
NB	PB/NB/PS	PB/NB/PS	PM/NM/NB	PM/NM/NB	PS/NS/NB	PM/NM/NB	Z/Z/PS
NM	PB/NB/PS	PB/NB/PS	PM/NM/NB	PS/NS/NM	PS/NS/NM	Z/Z/NS	NS/Z/Z
NS	PM/NB/Z	PM/NM/NS	PM/NS/NM	PS/NS/NM	Z/Z/NS	NS/PS/NS	NS/PS/Z
Z	PM/NM/Z	PM/NM/NS	NS/PS/NS	Z/Z/NS	NS/PS/NS	NM/PM/NS	NM/PM/Z
PS	PS/NM/Z	PS/NS/Z	Z/Z/Z	NS/PS/Z	NS/PS/Z	NM/PM/Z	NM/PB/Z
PM	PS/Z/PB	Z/Z/PS	NS/PS/PS	NM/PS/PS	NS/PM/PS	NS/PB/PS	NB/PB/PB
PB	Z/Z/PB	Z/Z/PM	NM/PS/PM	NM/PM/PM	NM/PM/PS	NM/PB/PS	NB/PB/PB

The set of predefined linguistic values {PB, PM, PS, Z, NS, NM, NB} stand for positive maximum, positive middle, positive minimum, zero, negative minimum, negative middle and negative maximum, respectively. The cloud models of deviation E: E1(PB) = (6,0.5,0.04); E2(PM) = (4,0.5,0.04); E3(PS) = (2,0.5,0.04); E4(Z) = (0,0.5,0.04); E5(NS) = (−2,0.5,0.04); E6(NM) = (−4,0.5,0.04); E7(NB) = (−6,0.5,0.04).

The positive normal cloud models of change rate of deviation EC are the same with those of E. The cloud models of control variable Up: Up1(PB) = (3,0.5,0.01); Up2(PM) = (2,0.5,0.01); Up3(PS) = (1,0.5,0.01); Up4(Z) = (0,0.5,0.01); Up5(NS) = (−1,0.5,0.01); Up6(NM) = (−2,0.5,0.01); Up7(NB) = (−3,0.5,0.01); The parameters of control variables U_i and U_d cloud model are the same with those of U_p.

The membership cloud diagrams of all variables are shown in Figure 5 and Figure 6.

The cloud uncertain linguistic rule like "IF E_i and EC_j THEN Up_m" can be constructed, where the values i, j and m are from one to seven. The cloud linguistic rules of U_i and U_d are the same as that of U_p.

The real-time parameters of PID are controlled by output variables U_p, U_i, U_d of two-dimensional cloud model. Quantitative factors: Ke = 20, Kec = 10, scale factor: Kup = 5, Kui = 10, Kud = 1

$$K_{p,n+1} = K_{p,n} + \Delta K_p = K_{p,n} + K_{up}U_p; \ K_{i,n+1} = K_{i,n} + \Delta K_i = K_{i,n} + K_{ui}U_i; \ K_{d,n+1}$$
$$= K_{d,n} + \Delta K_d = K_{d,n} + K_{ud}U_d \tag{15}$$

where $K_{p,n}$, $K_{i,n}$ and $K_{d,n}$ are the parameters of cloud model PID controller. $K_{p,n+1}$, $K_{i,n+1}$ and $K_{d,n+1}$ are the real-time parameters of cloud model PID controller. Output variable u of two dimensional cloud model of PID controller can be calculated:

$$u_{(t)} = K_{p,n+1}e_{(t)} + K_{i,n+1}\int e_{(t)}dt + K_{d,n+1}\frac{de_{(t)}}{dt} \tag{16}$$

5 SIMULATIONS

The main parameters of synchronous generator are shown as follows: P = 1875 KW, U = 450 V, n = 720 r/min, f = 60 Hz, cosΨ = 0.8, R_a = 0.0098, X'_d = 0.204, T'_d = 1.141 s, $X_d = X_q$ = 2.06, $X''_d = X''_q$ = 0.157, T''_d = 0.113 s.

After ten seconds, the sudden 50% load take-on test is conducted. The overshoot of cloud model PID excitation system is 7.94% fewer than that of the controllable phase compound excitation. The adjustment time shortens 0.36 seconds and the steady state error is reduced by 0.51%. Twenty seconds later, the 50% load dump test is conducted. The overshoot of cloud model PID excitation system is 6.88% fewer than that of the controllable phase compound excitation. The adjustment time shortens 0.374 seconds and the steady state error is reduced by 0.049%. Due to the synchronous generator running for a long time, a series of factors result in change of synchronous generator parameters.

$$R_a = 0.008, \ X_d = X_q = 2.04, \ X'_d = 0.186, \ X''_d = X''_q = 0.146, \ T'_d = 1.23 \text{ s}, \ T''_d = 0.214 \text{ s}$$

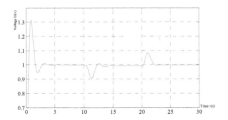

Figure 7.   Controllable phase compound.

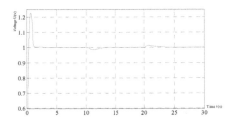

Figure 8.   Two-dimensional cloud model PID.

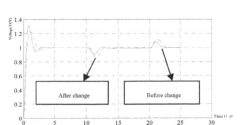

Figure 9.   Controllable phase compound.

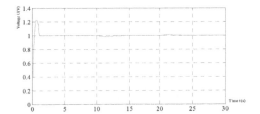

Figure 10.   Two-dimensional cloud model PID.

From Figures (9) and (10), the result of test shows that the moderating effect of cloud model PID excitation system is better.

6 CONCLUSION

On the basis of traditional fuzzy theory, combined with the cloud theory, a two-dimensional cloud model PID excitation controller is designed. In the nonlinear and time-varying ship power system, the cloud model PID excitation system is able to infer by cloud control rules and achieve PID parameters self-tuned. The mathematical models of controllable phase compound excitation system and two-dimensional cloud model PID excitation system are established. The result of simulation indicates that the two-dimensional cloud model PID excitation system in ship power system has better adaptability and robustness.

REFERENCES

[1] Z. Xu & P. Zhu. et al. 2013. Research on backward cloud model with uncertainty. *Journal of Networks*, 8(11):2556–2563.
[2] Awadallah & A. Mohamed. 2010. Fuzzy-based on-line detection and prediction of switch faults in the brushless excitation system of synchronous generators. *Electric Power Components and Systems*, 38(12):1370–1388.
[3] A.E. Leon & J.A. Solsona. et al. 2011. Optimization with constraints for excitation control in synchronous generators. *Energy*, 36(8):5366–5373.
[4] Farouk & Naeim. et al. 2012. Design and implementation of a fuzzy logic controller for synchronous generator. *Research Journal of Applied Sciences Engineering and Technology*, 4(20):4126–4131.
[5] W. Wang & Y.K. Zhang. et al. 2012. Study and simulation of fuzzy PID control system of BLDCM. *Journal of computer simulation*, 29(4):196–200.
[6] C.F. Zong. 2015. Simulation and research of fuzzy-PID excitation control based on double controller switching. *Electric Drive Automation*, 37(1):26–29.
[7] Y.C. Liu & Y.T. Ma. et al. 2012. Study on characters of cloud model based on high-order gaussian distribution with iterations. *Journal of Electronics*, 40(10):1913–1919.
[8] G.W. Yan & X.X. Gong. et al. 2014. Conceptual representation and measurement model of ball mill fill level based on cloud model. *Proceedings of the CSEE*, 34(14):2281–2287.
[9] S.B. Zhang & C.X. Xu. 2013. Study on the trust evaluation approach based on cloud model. *Chinese Journal of Computers*, 36(2):422–429.

Advances in Engineering Materials and Applied Mechanics – Zhang, Gao & Xu (Eds)
© 2016 Taylor & Francis Group, London, ISBN 978-1-138-02834-0

Numerical simulation and experimental verification of long-rod projectile penetration into A95 ceramic composite armor

R.Y. Li & Y.X. Sun

National Key Laboratory of Transient Physics, NUST, China

ABSTRACT: In this paper, the long-rod projectile penetrations into composite target vertically were simulated based on EPIC-2D. Combining the Johnson–Cook constitutive model, Lemaitre yield criterion and micro crack damage equations to describe the vandalism of ceramic material in the condition of high speed impact. The composite target was composed of cover plate, A95 ceramic and authenticate target. The relationship between penetration depth and impact velocity of projectile was achieved, simulation results were agreed with experimental data very well. The results showed that the A95 ceramic was better than steel A3 on thickness efficiency only when the impact velocity of projectile was above critical value. However, A95 ceramic could reduce armor mass greatly. The conclusions provide important reference value to the further study of ceramic target protection and long-rod penetration into target.

1 INTRODUCTION

Ceramic materials have the characteristics of low density (3.6 g/cm³), high compression strength (1.9~3.0 GPa), low tensile strength (0.2~0.3 GPa), and low fracture strain (0.001~0.003). Due to high compression and low density and the other excellent physical and mechanical performance, ceramic has been a frequently used lightweight armor materials. Donald A. Shockey[1] has studied the phenomenon of ceramic target's destruction under the penetration of long-rod projectile. Z. Rosenberg[7] has carried out 2D numerical simulation researches of tungsten alloy projectile penetration into ceramic. Li Yongchi[4] has established an elastic plastic hydrodynamic constitutive model that considers the evolution of ceramic damage and carried out 2D numerical simulation researches of ceramic cone. Du Zhonghua[2] has conducted experimental research on the ballistic performance of ceramic/GFRP/steel composite target and set up an engineering analysis model of long-rod projectile penetration into multi-layer composite armor. Sun Yuxin[8][9] has given a preliminary experimental analysis of the anti-penetration property of A95 ceramic and toughening ceramic.

EPIC-2D, the two-dimensional impact dynamics finite element code, was applied to carry out the numerical simulation of the long-rod projectile penetration into confined A95 ceramic. The contrast between simulated results and experimental data showed that the established model in this paper was correct and reliable. Change law between thickness efficiency, mass efficiency and velocity of projectile was achieved through the simulated and experimental results, which provides a significant basis to the correlation study on the anti-penetration property of ceramic.

2 NUMERICAL MODEL AND PARAMETERS

2.1 *Sketch of model and scheme*

Figure 1 is sketch of target model, in which 1(a) is the reference target and 1(b) is the ceramic composite target. δ_1 is the thickness of cover plate and δ_2 is the thickness of ceramic target. The

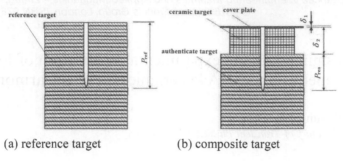

(a) reference target	(b) composite target

Figure 1. Sketch of target model[8].

ballistic experiment was conducted under the velocities of 1054 m/s, 1177 m/s and 1265 m/s, respectively. Reference penetration depth P_{ref} and penetration depth of authenticate target P_{res} will be achieved, and then obtained the efficiency of A95 ceramic with thickness δ_2.

2.2 Geometric dimension

The projectile used in the experiment has 7.6 mm diameter and 0.023 kg mass and with material of 35CrMnSi. The target is composed of cover plate (steel A3), A95 ceramic and authenticate target (steel A3). In which, the size of cover plate is 95 mm × 95 mm and its thickness is $\delta_1 = 2$ mm; the diameter of ceramic target is 80 mm and its thickness is $\delta_2 = 30$ mm; the size of authenticate target is 95 mm × 95 mm and its thickness is 95 mm. For purpose of constraint ceramic, sleeve tightly closed to ceramic target, which the material is steel 45 and thickness is 8 mm. Triangular elements were used to build two-dimensional axial symmetry model in EPIC-2D. The total elements were 28582, in which projectile occupied 582.

2.3 Material models and parameters

Johnson–Cook[3] constitutive model were used to describe metal materials:

$$\sigma_y = A\left(1 + B\varepsilon^n\right)\left(1 + C\ln\frac{\dot{\varepsilon}}{\dot{\varepsilon}_0}\right)\left[1 - \left(\frac{T - T_0}{T_m - T_0}\right)^m\right] \tag{1}$$

In equation (1), A, B, n, C and m are the material constants; σ_y, ε, $\dot{\varepsilon}$ and $\dot{\varepsilon}_0$ are the yield stress, effective strain, equivalent strain rate and reference strain rate respectively; T_m and T_0 are the melting temperature and reference temperature. Corresponding state equation is Gruneisen equation[5]. Lemaitre yield criterion[6] was applied on ceramic:

$$\sigma^* \equiv \bar{\sigma} R_v^{1/2} \tag{2}$$

In this equation, $\bar{\sigma} = \sqrt{3/2\, s_{ij}s_{ij}}$ and $R_v = 2/3\left(1 + v\right) + 3\left(1 - 2v\right)\left(P/\bar{\sigma}\right)^2$ are the Mises equivalent stress and triaxial stress factor; s_{ij}, P and v are the deviatoric stress, hydrostatic pressure and Poisson ratio, respectively. The constitutive model ceramic use is:

$$\sigma_y = Y_d(1 + B\varepsilon^n)\left(1 + C\ln\frac{\dot{\varepsilon}}{\dot{\varepsilon}_0}\right)$$

$$Y_d = \begin{cases} Y\left(1 - D\right), & p \geq 0 \\ Y\left(1 - \dfrac{D}{D_m}\right), & p < 0 \end{cases} \tag{3}$$

Table 1. Material parameters of Johnson–Cook.

Materials	$\rho_0/(g/cm^3)$	$G/(GPa)$	$A/(GPa)$	B	n	C	m
35CrMnSi	7.8	148	1.60	0	0.12	0.0	1.0
A95 ceramic	3.5	140	2.2	0	0	0.0	1.0
Steel A3	7.7	40.0	0.23	0.87	0.26	0.014	1.03
Steel 45	7.8	77.5	0.496	0.875	0.307	0.07	0.804

This equation can be regard as Johnson–Cook model which regardless of the heat effect. D is tensile damage and D_m is extremity damage, the material will lose resistance of tensile when the tensile damage reach D_m. Literature[10] has given the micro crack damage equation as follows:

$$\frac{\partial D}{\partial t} = \frac{1-v^2}{2\lambda E}\pi(\sigma^2 - \sigma_c^2)C_R D, \quad (\sigma = -P \geq \sigma_c) \tag{4}$$

In equation (4), v, E and C_R are the Poisson ratio, Young modulus and Rayleigh wave velocity, respectively; σ and σ_c are the mean tensile stress and threshold stress for tensile damage development.

Regardless of the heat effect, the following form of Gruneisen state equation is used for ceramic:

$$p = K_1\mu + K_2\mu^2 + K_3\mu^3 \tag{5}$$

Material parameters are shown in Table 1.

3 ANALYSIS OF SIMULATION RESULTS

Figure 2 shows the process of projectile penetration into composite target, corresponding to 0 μs, 25.1 μs, 64.2 μs and 120.0 μs, in which instantaneous velocity of projectile are 1054 m/s, 904 m/s, 641 m/s and 107 m/s, respectively. It can be found from the Figure 2 that the erosion of projectile, plump up of cover plate, launching of ceramic target, destruction of target materials around projectile and broaching are all getting clear description. Besides, the development of ceramic material tensile damage can be seen clearly from the Figure 2: at the initial stage of penetration, ceramic materials around and in front of projectile appear small amount of narrow crack, which expansion toward the peripheral along the radial direction and grow new at the same time. Tensile damage occurs at the interface of ceramic target and authenticate target (see Fig. 2(c)). The reason is that the impedance of A95 ceramic is greater than the impedance of steel A3, compression wave generated by the impact will reflect the tensile wave when arrive at the interface, the strength is higher than the tensile strength of A95 ceramic.

It be observed from the experimental result in Figure 3 that smashes regional in ceramic target presents turbination. There exist a crack zone in remain ceramic target and presents radioactive distribution for the center bullet hole that indicate that there existed radial crack destruction in ceramic materials due to circumference tensile during penetration. It is observed that the numerical simulation results is quite coincident with the experimental phenomena, Lemaitre yield criterion and micro crack damage equations can describe the failure characteristics of ceramic materials properly.

Table 2 is the penetration depth of composite target and reference target under penetration of projectiles, which gives the comparison of simulative depth and experimental depth. All the errors within 6% under three impact velocities no matter composite target or reference target and can be proved that the model and selected parameters in this paper are reliable.

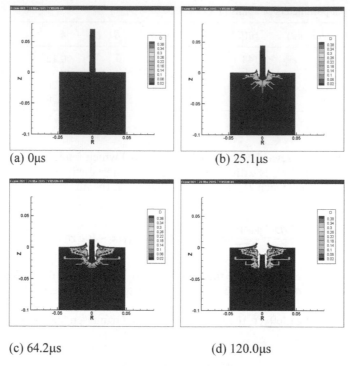

(a) 0μs   (b) 25.1μs

(c) 64.2μs   (d) 120.0μs

Figure 2.   Damage clouds of ceramic target under different time (initial velocity is 1054 m/s).

Figure 3.   Destruction of ceramic target after penetration.

Table 2.   Comparison of simulative depth and experimental depth.

Velocities (m/s)	Simulation (mm)	Experimental (mm)	Errors
Penetration depth of composite target ($\delta_1 + \delta_2 + P_{res}$)			
1054	37.8	37.5	0.8%
1177	43.8	43.7	0.23%
1265	45.4	44.4	2.3%
Penetration depth of reference target (P_{ref})			
1054	32.7	31	5.5%
1177	41.5	43.5	4.6%
1265	46.6	49.5	5.9%

Table 3. Corresponding F_s and F_m of A95 under different velocities.

Velocity (m/s)	F_s		F_m	
	Simulation	Experiment	Simulation	Experiment
1054	0.83	0.78	1.85	1.75
1177	0.92	0.99	2.06	2.21
1265	1.04	1.17	2.32	2.61

Thickness efficiency F_s and mass efficiency F_m are defined as follow equations:

$$F_s = \left(P_{ref} - P_{res} - \delta_1\right)\big/\delta_2 \tag{6}$$

$$F_m = \left(P_{ref} - P_{res} - \delta_1\right)\rho_{ref}\big/\left(\delta_2\rho_2\right) \tag{7}$$

In equations (6) and (7), ρ_{ref} and ρ_2 are the material density of reference target (Steel A3) and ceramic, respectively. By combining with the data in Table 2, corresponding F_s and F_m of A95 under different velocities can be calculated when $\delta_2 = 30$ mm and shown as in Table 3.

It be observed from Table 3 that F_s is less than 1.0 when initial velocity is 1054 m/s and 1177 m/s. F_s is more than 1.0 when initial velocity is 1265 m/s which indicate that the anti-penetration property of A95 ceramic is better than steel A3 at the same thickness. Such phenomenon is related to the mechanical properties of ceramic material: the dynamic compressive strength of ceramic increased rapidly with the increase of hydrostatic pressure and then the anti-penetration property is enhanced. Besides, we can also find that F_m maintains a high level, the anti-penetration property of A95 ceramic is equal to 1.75 times the mass of steel A3. A95 ceramic can reduce the armor weight in significant measure.

4 CONCLUSIONS

In this paper, we conducted the numerical simulation of long-rod projectile penetration into A95 ceramic composite armor and then it compared with experimental results. It is shown that Lemaitre yield criterion and micro crack damage equation introduced into EPIC-2D can reflect dynamic mechanics of ceramic material under the high speed collision and provide a good solution for numerical computation in engineering matters. Through the results of simulation and experiment, the anti-penetration property of A95 ceramic is superior to steel A3 on thickness efficiency only when initial velocity of projectile is above the critical value. Besides, A95 ceramic can reduce the armor weight in significant measure. The conclusions provide the design of armor with good ideas.

REFERENCES

[1] Donald A. & Marchand A.H., et al. 1990. Failure phenomenology of confined ceramic targets and impact rods. *International Journal Impact Engineering* 19(3): 263–275.
[2] Du Zhonghua & Zhao Guozhi, et al.. 2003. A study on the performance of a composite armor plate of ceramic/GFRP/steel. *Acta armamentaria* 24(2): 219–221.
[3] Johnson G.R. & Cook W.H. 1985. Fracture characteristics of three metals subjected to various strains, strain rates, temperatures, and pressures. *Engineering Fracture Mechanics* 21(1): 31–48.
[4] Li Yongchi & Wang Daorong, et al. 2004. A numerical simulation on anti-penetration mechanism and ceramic cone evolution of ceramic targets. *Journal of Ballistics* 16(4): 12–17.
[5] LSTC. 2007. *LS-DYNA Keyword User's Manual Version 971*. Livermore Software Technology Corporation, LSTC, Livermore.

[6] Lemaitre J. 1996. *Damage mechanics tutorial*. Beijing: Science Press.

[7] Rosenberg Z. & Dekel E., et al. 1997. Hypervelocity penetration of tungsten alloy rods into ceramic tiles: experiments and 2-D simulations. *International Journal Impact Engineering* 20: 675–683.

[8] Sun Yuxin & Li Yongchi, et al. 2005. An experimental study on the penetration confined A95 ceramic targets. *Journal of Ballistics* 17(2): 38–41.

[9] Sun Yuxin & Li Yongchi, et al. 2005. Comparison Studies on the Anti-penetration property of Toughening Ceramic and A95 Ceramic. *Journal of experimental mechanics* 20(3): 344–348.

[10] Wang Daorong. 2002. *Engineering analysis and numerical simulation research for high speed penetration*. Hefei: University of science and technology of China.

Advances in Engineering Materials and Applied Mechanics – Zhang, Gao & Xu (Eds)
© 2016 Taylor & Francis Group, London, ISBN 978-1-138-02834-0

Experimental investigation and penetration analysis of ceramic composite armors subjected to ballistic impact

W.L. Liu, Z.F. Chen, G.F. Liu & X.H. Chen
College of Material Science and Technology, Nanjing University of Aeronautics and Astronautics, Nanjing, P.R. China

X.W. Cheng & Y.W. Wang
National Key Laboratory of Science and Technology on Materials under Shock and Impact, Beijing Institute of Technology, Beijing, P.R. China

ABSTRACT: With a protected material, a standard structure consisting of alumina ceramic cylinders, Ti-6Al-4V and Ultrahigh Molecular Weight Polyethylene (UHMWPE) is put forward and designed. The areal density of the standard structure panel was 72 kg/m² and penetration of 12.7 mm Armor Piercing Incendiary (API) projectiles was investigated at the vertical velocity of m/s. In this research the penetration process and the bulletproof mechanism of different layer materials and projectile-armor interaction were investigated. Ballistic impact testing of the armor systems revealed that the standard structure was able to protect different protected materials in the fields of aircrafts, helicopters and tanks.

1   INTRODUCTION

Besides the penetration and perforation by foreign object impact, the growing need for lightweight armor structure is also of great importance in the design of protective military armors[7,9]. As Wilkins[11] pointed out, the target parameters including high hardness, compressive strength, low density and high fracture resistance should be focused on to meet the requirements for materials development. A basic armor system made of hard faced ceramic materials with composite backing is widely used on account of its comprehensive performances. Ceramic tile helps blunt as well as erode the projectile during impact and the composite backing absorbs the residual kinetic energy at the same time delays the initiation of tensile failure in the ceramic and backing plate interface[1].

Recently, an interest has been shown in the study of ballistic performance of hybrid composites. Wang et al. [10] proposed a lightweight hybrid composite armor with four layers consisting of alumina ceramics pellets, UHMWPE and two layers of Ti-6Al-4V. Evci et al. [2] developed ceramic composite armor materials with alumina ceramic frontal face and composite backing support made up of plain-weave woven and unidirectional fabrics with polyester resin. It shows that ceramic composite armors offer excellent ballistic protection solutions compared to steel.

In this paper, a standard structure was designed and constructed with three layers: ceramic as energy-absorbing layer, Ti-6Al-4V acted as resistance layer and cushion layer made of UHMWPE. By combination of the three layers as a standard structure and changing the protected material, the understanding of penetration process and projectile-armor interaction and the finding effective ways of developing light-weight solutions are to be discussed.

2 EXPERIMENTAL SECTION

2.1 *Materials design*

As Krell and Strassburger[5] found that the properties of high stiffness and high hardness in the strike-faced material are critical requirements in the penetration phase. Al 99.7 acts as the best choice owing to its higher hardness, modulus and less expensive among these materials. Figure 1a shows the shape and size of the alumina ceramic cylinder and Figure 1b shows the arrangement of ceramic layer. The other two layers were Ti-6Al-4V and UHMWPE material, respectively.

2.2 *Instruments and procedure*

A typical ballistic impact test apparatus is shown in Figure 2, and the distance between ballistic gun and target sample was about 10 m. The projectile used was a 12.7 mm API. The integral structure size of the target was 100 mm × 100 mm, as listed in Table 1. The areal density of the standard structure was 72 kg/m². In addition, Case A was composed of the standard structure layers with the protected material of Ti-6Al-4V, so were Case B and C whose protected materials were carbon fiber and aluminum alloy, respectively. The composite armors were tested at the vertical velocity of 810~825 m/s. Similar experimental device was described in Madhu's literature[6].

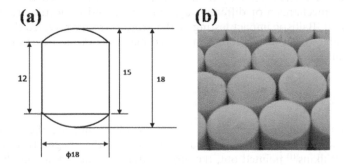

Figure 1. (a) Shape and size of the alumina ceramic cylinder (mm) and (b) arrangement of ceramic layer.

Figure 2. A typical ballistic impact test apparatus. 1-12.7 mm ballistic gun; 2-velocity measuring system; 3-target sample; 4-iron block.

Table 1. Properties of different ceramic composite armors.

	Case A	Case B	Case C
Structure	Standard structure + Ti-6Al-4V	Standard structure + carbon fiber	Standard structure + aluminum alloy
Thickness (mm)	28	36	32

The damage level of the armor plate was divided into 8 levels in which 1–4 are qualified injury and 5–8 are unqualified damage according to the evaluation criteria of GJB 59.18–88 [3].

3 RESULTS AND DISCUSSION

3.1 *Energy-absorbing mode of ceramic layer*

Ceramics, as the first layer of impacted materials, are the main energy-absorbing materials in the penetration process. In order to decelerate the projectile velocity and dissipate its impact energy, its ways of energy-absorbing include three phases, as shown in Figure 3a. In the first phase, the impact of a projectile on the surface of the ceramic material generates compressive shock waves that the ceramic cylinders erode the projectile and hinder from being penetrated. In the next phase, these stress waves reflect back as tensile waves once they reach the free surface, and then the ceramic material fractures owing to its low tensile strength performance. As a concerned point regarding the broken of the ceramic cylinders, the cracks lead to new surfaces and absorb the energy of the projectile. In the final phase, the impacted cylinders transfer the energy to the surroundings, which leads to breakdown of the neighborhood of the impacted zone thereby disintegrate and absorb energy of the projectile [4]. Figure 3b shows the original ceramic cylinders and the cracked fragments, which is compatible with the transmission analysis of stress waves. By means of different scattered ceramic cylinders, during the impact event, the fragments, formed around the impact zone and splashing towards all sides, also determine the percentage of energy absorption.

3.2 *Characterization of bulletproof mechanism*

Fracture views of layers (Case A, B, C) after ballistic testing are shown in Figure 4–6, respectively. Figure 4a shows a cross-sectional view of the integral structure, and Figure 4b-d give photographs of the impacted result of Ti-6Al-4V, UHMWPE and the protected material Ti-6Al-4V. From the three kinds of experiments, we can find that all the three protected materials are not penetrated, which indicates the standard structure can totally protect protected material of Ti-6Al-4V.

From Figure 4–6d, according to Table 1, damage levels of Case A, B and C are second level, second level and fourth level respectively, which indicates that all of the three kinds

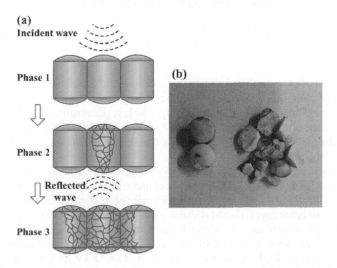

Figure 3. Ceramic layer of (a) three phases of the penetration process and (b) original ceramic cylinders and the cracked fragments.

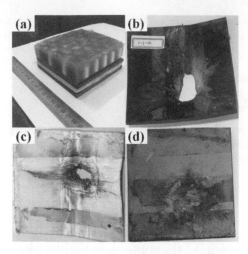

Figure 4.　Images of Case A sample (a) and fracture modes of resistance layer TI-6AL-4V (b), cushion layer UHMWPE (c) and protected material Ti-6Al-4V (d).

Figure 5.　Images of Case B sample (a) and fracture modes of resistance layer Ti-6Al-4V (b), cushion layer UHMWPE (c) and protected material carbon fiber (d).

of protected materials are qualified injury. By measuring the penetration depth we can find that the protected material of Ti-6Al-4V is 0.5 mm approximately, and the carbon fiber and aluminum alloy are 8 mm and 3 mm respectively. The less penetration depth is, the better bulletproof performance of composite armors is[8]. Adding that the thickness of Case A is the minimum and the penetration depth of protected material Ti-6Al-4V is the least, hence Case A is a better bulletproof structure.

　　The bulge height of resistance layer Ti-6Al-4V and cushion layer UHMWPE in case A, B and C is shown in Figure 7. In Case A, because the hardness of protected material of Ti-6Al-4V is the largest, as a projectile reflects back the protected material has a less effect on the bulge of resistance layer Ti-6Al-4V and cushion layer UHMWPE, therefore the bulge height of them are both 6 mm. In Case B, as a result of high elasticity of carbon fiber, as a projectile reflects back both resistance layer Ti-6Al-4V and cushion layer UHMWPE rebound. As a result of high hardness property of Ti-6Al-4V, bulge height of Ti-6Al-4V is only 2 mm. Due to high plasticity property of UHMWPE, its bulge height increases to 10 mm. In Case C, the projectile penetration depth in protected material is the most so that

Figure 6. Images of Case C sample (a) and fracture modes of resistance layer Ti-6Al-4V (b), cushion layer UHMWPE (c) and protected material aluminum alloy (d).

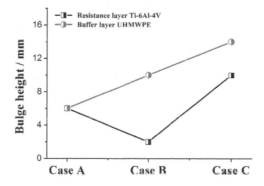

Figure 7. Bulge height varies with Case A, B and C.

when it reflects back the residual kinetic energy is not enough to provide rebound. Coupled with the high tenacity and low hardness of aluminum alloy, the bulge heights of resistance layer Ti-6Al-4V and cushion layer UHMWPE are the maximum in the three cases.

Although the standard structure can totally protect Ti-6Al-4V, carbon fiber and aluminum alloy from 12.7 mm API projectiles, Case A is the best bulletproof structure according to the comprehensive assessment of penetration depth and bulge height parameters of the three kinds of protected materials. Therefore, the standard structure is better for protecting Ti-6Al-4V material than carbon fiber in Case B or aluminum alloy in Case C.

4 CONCLUSION

In this paper, we discuss the characteristics of different layers that is used as composite material armors involving ceramic, Ti-6Al-4V and UHMWPE. Under high velocity of the projectile, the energy-absorbing mode of ceramic layer includes three phases based on the transmission analysis of stress waves. Case A structure is the best bulletproof structure according to the comprehensive assessment of penetration depth and bulge height parameters of the three kinds of protected materials. The standard structure plate taking the advantage of light-weight (areal density of only 72 kg/m²) and being taken along conveniently, can be used in vehicles, vessels, aircrafts and helicopters.

ACKNOWLEDGMENTS

The author wishes to thank the National Basic Research Program of China (973 Program) and Innovation Fund of Nanjing University of Aeronautics and Astronautics (KFJJ201440).

REFERENCES

[1] Chen W.N., Rajendran A.M., Song B., Nie X. 2007. Dynamic fracture of ceramics in armor applications. *J. Am. Ceram. Soc.* 90: 1005–1018.
[2] Evci C., Gülgeç M. 2013. Effective damage mechanisms and performance evaluation of ceramic composite armors subjected to impact loading. *J. Compos. Mater.* 48: 0021998313508594.
[3] GJB 59. 18–88 1988. Test operations procedure for armored vehicles armor plate bullet-proof test. 27 December.
[4] Gogotsi G.A. 2013. Criteria of ceramics fracture (edge chipping and fracture toughness tests). *Ceram. Int.* 39: 3293–3300.
[5] Krell A., Stranburger E. 2007. Hierarchy of key influences on the ballistic strength of opaque and transparent armor. *Ceram. Eng. Sci. Proc.* 28: 45–55.
[6] Madhu V., Ramanjaneyulu K., Bhat T.B., Gupta N.K. 2005. An experimental study of penetration resistance of ceramic armour subjected to projectile impact. *Int. J. Impact Eng.* 32: 337–350.
[7] Medvedovski E. 2006. Lightweight ceramic composite armour system. *Adv. Appl. Ceram.* 105: 241–245.
[8] Roeder B.A., Sun C.T. 2001. Dynamic penetration of alumina/aluminum laminates: experiments and modeling. *Int. J. Impact Eng.* 25: 169–185.
[9] Sarva S., Nemat-Nasser S., McGee J., Isaacs J. 2007. The effect of thin membrane restraint on the ballistic performance of armor grade ceramic tiles. *Int. J. Impact Eng.* 34: 277–302.
[10] Wang Q., Chen Z., Chen Z. 2013. Design and characteristics of hybrid composite armor subjected to projectile impact. *Mater. Design* 46: 634–639.
[11] Wilkins M.L. 1978. Mechanics of penetration and perforation. *Int. J. Eng. Sci.* 16: 793–807.

Advances in Engineering Materials and Applied Mechanics – Zhang, Gao & Xu (Eds)
© *2016 Taylor & Francis Group, London, ISBN 978-1-138-02834-0*

Phase behavior of new biodegradable polycarbonate containing diosgenyl units

Z.T. Liu, Y.N. Wang, Y.J. Liu & J.S. Hu
Center for Molecular Science and Engineering, College of Science, Northeastern University, Shenyang, China

ABSTRACT: A new biodegradable amphiphilic Liquid Crystal (LC) polycarbonate containing diosgenyl units was synthesized. The structure was characterized using Fourier transform infrared spectrometer and proton nuclear magnetic resonance spectrum. The optical texture and phase behavior were investigated with polarizing optical microscopy, differential scanning calorimetry, and X-ray diffraction. The monomer 2-diosgenylethanol showed fan-shaped texture, a melting temperature at 195°C. The polycarbonate also showed the fan-shaped texture, and a glass transition temperature at 31.5°C and a LC to isotropic phase temperature at 93.8°C. XRD showed that a sharp and intense peak at a small angle of 2.5° appeared, which indicated the polymer revealed a smectic structure.

1 INTRODUCTION

In recent year, biodegradable polymers have attracted more and more interest owing to their wide applications in biomedical areas such as drug delivery, human tissue engineering, and organ repair [1–3]. To obtain the desirable properties, many modification approaches have been used. Among them, the incorporation of the bioactive or self-assembly functional compounds to the polymer chain has currently become the most important modification strategies [4–6]. As known, Liquid Crystalline (LC) materials have been widely used in the field of electro-optical displays. In fact, LC compounds are also useful in biorelated fields because their self-organizing structures through noncovalent specific interactions are compatible with those in living systems [7,8]. At present, LC materials containing cholesteryl groups have especially attracted more and more attention in the biomaterials fields [9–12]. However, to the best of our knowledge, little research on the biodegradable aliphatic polycarbonate based on diosgenin side-functionalized chain is reported. In this study, we reported a new biodegradable LC material of diosgenin side-functionalized polycarbonate. The optical texture and phase behavior were investigated with Polarizing Optical Microscopy (POM), Differential Scanning Calorimetry (DSC), and X-ray diffraction.

2 EXPERIMENTAL

2.1 *Materials*

All chemicals were obtained from the indicated sources and used as received. Diosgenin (Wuhan, China), 1,2-ethanediol (Shenyang, China), and methoxypolyethylene glycols (mPEG$_{43}$) (Sigma-Aldrich) were used as received. 2-Diosgenylethanol was obtained from Chemistry Laboratory Center in Northeastern University of China. All other solvents and reagents used were purified by standard methods.

2.2 Measurements

FT-IR spectra were measured on a PerkinElmer spectrum One (B) spectrometer (PerkinElmer, Foster City, CA). ^{1}H NMR spectra were measured using a Bruker ARX 600 (Bruker, Germany) high resolution NMR spectrometer. The optical textures were observed with a Leica DMRX POM (Leica, Germany) equipped with a Linkam THMSE-600 (Linkam, United Kingdom) cool and hot stage. The thermal properties were determined with a Netzsch DSC 204 (Netzsch, Hanau, Germany) equipped with a cooling system at a heating and cooling rates of 10°C/min in a nitrogen atmosphere. The mesophase structure was identified using a Bruker D8 Advance (Bruker, Germany) X-Ray Diffraction (XRD) measurement with a nickel-filtered Cu-K$_\alpha$ radiation.

2.3 Synthesis of biodegradable polycarbonate

The typical synthetic route of the polycarbonate is outlined in Scheme 1. The monomer 2-diosgenylethanol (IM) and polymer mPEG$_{43}$-b-P(MCC-g-Dios)$_{12}$ were prepared as described in the literature [13].

3 RESULTS AND DISCUSSION

3.1 Optical textures

The optical textures of the monomer and polymer were observed by POM with hot stage. The visual observations under POM showed that they revealed enantiotropic smectic phase on heating and cooling cycles. When **IM** was heated to 190°C, the sample melted and the fan shaped texture gradually appeared, and the texture disappeared at 198°C. Cooling isotropic

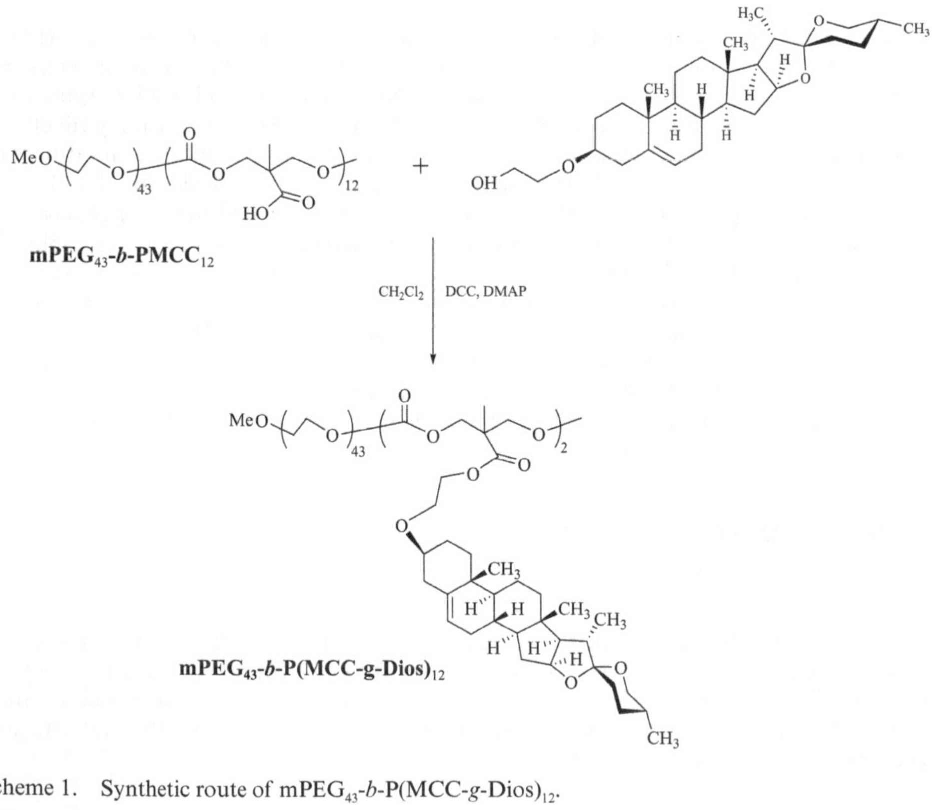

Scheme 1. Synthetic route of mPEG$_{43}$-b-P(MCC-g-Dios)$_{12}$.

584

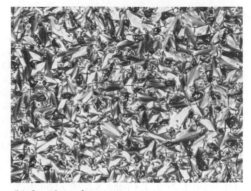

(a) fan-shaped texture   (b) fan-shaped texture

Figure 1.   Optical textures of LC-M (200×). (a) IM; (b) mPEG$_{43}$-b-P(MCC-g-Dios)$_{12}$.

state, the fan-shaped texture slowly appeared at 146°C, and crystallized at 125°C. The optical texture of **IM** is shown in Figure 1(a). POM results showed that **mPEG$_{43}$-b-P(MCC-g-Dios)$_{12}$** had strong birefringent properties. Moreover, it could exhibit a LC state below body temperature. With increasing temperature, the typical fan-shaped texture appeared at 55°C, and the birefringence texture disappeared at 98°C. The optical texture of **mPEG$_{43}$-b-P(MCC-g-Dios)$_{12}$** is shown in Figure 1(b).

3.2   *Phase behavior*

The phase behavior of the monomer and polymer was investigated with DSC and POM. All phase transitions were reversible and did not change on repeated heating and cooling cycles. On DSC heating curves, **IM** only revealed a melting transition at 195°C, and a LC to isotropic phase transition was not seen, but POM result showed **IM** had obvious mesomorphism. On cooling, an isotropic to smectic phase transition at 145°C, and the crystallization transition did not occur. The polymer **mPEG$_{43}$-b-P(MCC-g-Dios)$_{12}$** showed a glass transition temperature at 31.5°C and a LC to isotropic phase temperature at 93.8°C. To further identify mesophase structure, variable-temperature XRD studies were carried out. A sharp and intense peak at 2.5° and a weak and diffuse peak at 16.5° were observed, respectively. A sharp reflection, corresponding to the periodic distance, is characteristic of a smectic phase in which the molecules are stacked into layer with short-range, liquid-like positional order within the layers. A broad and weak peak corresponds to the distance of the local arrangement between the mesogenic side groups.

4   CONCLUSION

In this study, we reported phase behavior of a new biodegradable polycarbonate materials based on diosgenin. **IM** and **mPEG$_{43}$-b-P(MCC-g-Dios)$_{12}$** all exhibited the fan-shaped texture of a smectic phase.

ACKNOWLEDGEMENT

In this paper, the research was sponsored by Science and Technology Committee of Liaoning Province (2013020103), Science and Technology Bureau of Shenyang (F14-231-1-05), and Fundamental Research Funds for the Central Universities (N130205001) for financial support of this study.

REFERENCES

[1] Nagahama, K. Ueda, Y. Ouchi, T. & Ohya Y. 2007. Exhibition of soft and tenacious characteristics based on liquid crystal formation by introduction of cholesterol groups on biodegradable lactide copolymer. *Biomacromolecules*. 8: 3938–3943.

[2] Yu, C. Zhang, L. & Shen, Z. 2004. Ring-opening polymerization of 2,2-dimethyltrimethylene carbonate using rare earth tris(4-tert-butylphenolate)s as a single component initiator. *Journal of Molecular Catalysis A: Chemical*. 212: 365–369.

[3] Lee, A.L.Z. Venkataraman, S. Sirat, S.B.M. Gao, S.J. Hedrick, J.L. & Yang, Y.Y. 2012. The use of cholesterol-containing biodegradable block copolymers to exploit hydrophobic interactions for the delivery of anticancer drugs. *Biomaterials*. 33: 1921–1928.

[4] Klok, H.A. Hwang, J.J. & Stupp, S.I. 2002. Cholesteryl-(L-lactic acid)((n)over-bar) building blocks for self-assembling biomaterials. *Macromolecules*. 35: 746–759.

[5] Wan, T. Zou, T. Cheng, S.X. & Zhou, R.X. 2005. Synthesis and characterization of biodegradable cholesteryl end-capped polycarbonates. *Biomacromolecules*. 6: 524–529.

[6] Venkataraman, S. Lee, A.L. Maune, H.T. Hedrick, J.L. Prabhu, V.M. & Yang, Y.Y. 2013. Formation of disk- and stacked-disk-like self-assembled morphologies from cholesterol-functionalized amphiphilic polycarbonate diblock copolymers. *Macromolecules*. 46: 4839–4846.

[7] Koltover, I. Salditt, T. Rädler, J.O. & Safinya, C.R. 1998. An inverted hexagonal phase of cationic liposome-DNA complexes related to DNA release and delivery. *Science*. 281: 78.

[8] Denisov, I.G. Grinkova, Y.V. Lazarides, A.A. & Sligar, S.G. 2004. Directed self-assembly of monodisperse phospholipid bilayer nanodiscs with controlled size. *Journal of the American Chemical Society*. 126: 3477–3487.

[9] Luk, Y.Y. Tingey. M.L. Hall, D.J. Israel, B.A. Murphy, C.J. Bertics, P.J. & Abbott, NL. 2003. Using liquid crystals to amplify protein-receptor interactions: Design of surfaces with nanometer-scale topography that present histidine-tagged protein receptors. *Langmuir*. 19: 1671–1680.

[10] Zuo, T. Cheng, S.X. & Zhuo, R.X. 2005. Synthesis and enzymatic degradation of end-functionalized biodegradable polyesters. *Colloid and Polymer Science*. 283: 1091–1099.

[11] Zhang, L. Wang, Q.R. Jiang, X.S. Cheng, S.X. & Zhuo, R.X. 2005. Studies on functionalization of poly(e-caprolactone) by a cholesteryl moiety. *Journal of Biomaterials Science. Polymer Edition*. 16: 1095–1108.

[12] Hu, F.Z. Chen, S.D. Li, H. Sun, J.J. Sheng, R.L. Luo, T. & Cao, A.M. 2013. Preparation of new amphiphilic liquid-crystal diblock copolymers bearing side-on cholesteryl mesogen and their self-aggregation. *Acta Chimica Sinica*. 71: 351–359.

[13] Li, P. 2013. *Synthesis and liquid crystalline properties of carbonate monomers and polymers containing diosgenyl groups*. Mater's thesis, Northeastern University, China. 30–88.

Reaction mechanisms in formation of silicon carbide using SiO and CH_4 gases on various solid surfaces

X. Ma
SINTEF Materials and Chemistry, Blindern, Oslo, Norway

K. Tang
SINTEF Materials and Chemistry, Sluppen, Trondheim, Norway

ABSTRACT: As one of the most important non-oxide materials for abundant industrial applications, Silicon Carbide (SiC) can be produced by carbothermic reduction of silicon oxide using natural gas (methane, CH_4). In this study, Molecular Dynamics (MD) using a reactive force field ReaxFF has been employed to examine the initial stages of reactions involving a gas mixture of SiO and CH_4 on various solid surfaces, such as β-SiC, graphite sheets, iron, and amorphous SiO_2. The MD results showed that the fast formation of SiO clusters played an important role on such reactions. High SiO temperature enhanced single SiC molecule formation on SiC surface. Although SiC molecule could be formed on the graphite environment, it has not found any occurrence of such reaction on iron and SiO_2 surfaces within the simulation time so far.

1 INTRODUCTION

Carbothermic reduction of metal oxides is a classical method used for metal production. This process removes oxygen from metallic oxides by reacting with reducing agent such as carbon; thus the elementary form of metals can be produced (Ostrovski 2010). The carbon sources used in reduction processes are normally coal or coke. Recently natural gas, mostly consisting of methane (CH_4), has been used to replace coal and coke (Alizadeh 2007, Jamshidi 2008, Khoshandam 2006, Ostrovski 2006). With higher carbon activity from gaseous methane, the reduction reactions could occur at lower temperature, thus with less energy consumption and CO_2 emissions from metallurgical industries.

Oxide reduction process requires high temperature. However at high temperature, one of the non-desirable reactions that can occur is "cracking" of methane: the decomposition of methane into C and H_2. This may generate solid carbon which is similar to carbon black. To overcome the high temperature methane cracking by use of natural gas, a special technique has been developed in SINTEF recently (Monsen 2013) in which the reaction between hot SiO gas and solid carbon generates SiC which is an important intermediate product in the silicon production. In order to enable hot SiO gas to meet cold natural gas before it cracks into carbon and hydrogen, a special water-cooled injection lance is thus constructed.

However, not much is identified about the exact reaction mechanisms in the Si-C-H-O system. In the previous study (Kai 2015), the reaction mechanism for the SiO and CH_4 gaseous mixture on a β-SiC solid surface at temperature of 1600°C was examined using MD simulation with a reactive force field ReaxFF. It was found that the SiO condensation and polymerization into a ring structure played an important role. CH_4 could be decomposed into CH_3, CH_2, or CH except C atom. SiC could thus be generated by combining CH_3 with a single Silicon atom from dissociation of SiO; or a single Si atom at one end of an (SiO)n oligomer. Continuing with the previous study, the effect of the system temperature is evaluated in the current study (temperature is varied at 1200°C and 1600°C). The different solid surface materials are used, i.e., graphite sheets (simulated as carbon black), iron, and amorphous SiO_2 to examine their potentials on influencing SiC formation.

2 ADF/REAXFF SIMULATIONS

ReaxFF was developed by van Duin and co-workers (van Duin 2004) and is capable of modeling chemical reactions with atomistic potentials based on the reactive force field method. ReaxFF has been integrated into SCM/ADF and is currently available under license from SCM enabling parallel and large-scale MD simulations (SCM 2014).

2.1 *Temperature effect on formation of SiC on β-SiC surface*

A force field SiC.ff (Newsome 2012) is used to model reaction kinetics for the SiO and CH_4 gaseous mixture, which involves Si, C, H and O. In the atomic model (Fig. 1a), there are 200 SiO molecules and 300 CH_4 molecules which are randomly dispersed into a cuboid box with a dimension $92 \times 92 \times 200$ Å. The density of the gaseous mixture is 0.009 g/ml. A super-cell with a five lattice thickness of a single β-SiC crystal is inserted in the middle of the box. The temperature of CH_4 is specified as 20°C and that of SiO gas are 1200°C or 1600°C, respectively. SiC layers are spatially fixed. The velocity Verlet and Berendsen algorithms are used in the MD calculation. The time step and the damping constant are 0.25 and 500 fs respectively.

After 0.3 ns, almost every gaseous molecule is absorbed on the SiC layer (Fig. 1b). The numbers of reactant molecules decrease with time, which is shown in Figure 2. However, SiO is reduced faster than CH_4 and is exhausted after 1.5 ns. Higher temperature enhances CH_4 decomposition rate. It is also noted that temperature does not influence SiO reduction too much until 0.08 ns where there is a sharp decrease in SiO at 1600°C. The reason behind this is yet unknown and needs further study.

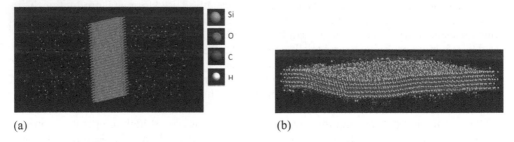

(a)   (b)

Figure 1.   Initial (a) and final (b) configurations of CH_4+SiO reaction system at 1200°C.

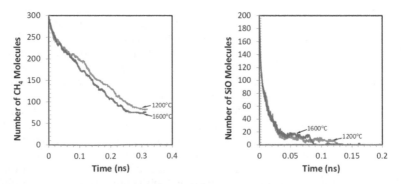

Figure 2.   Reduced number of molecules of CH_4 (left) and SiO (right) at 1200°C and 1600°C.

Modelling results in Figure 3 show there are three SiC molecules generated within the first 0.28 *ns* at 1200°C, while there might be five SiC molecules formed at high temperature 1600°C. Several snapshots are shown in Figure 4 at 1200°C in which the atoms Si and C involved are highlighted in dark green colour (Si and C atoms can be distinguished as the size of atom Si is bigger than C). The source of Si can be provided by the $(SiO)n$ cluster and confirms the findings in the previous study (Kai 2015) in which three SiC formation mechanism were suggested.

2.2 *Formation of SiC on graphite sheets*

Figure 5a shows the modeled system comprise of 100 SiO molecules, 150 CH_4 molecules, 3,4 and 5 double layers of graphite sheets with sizes of 5×4, 4×4, 3×3 lattice lengths, respectively. All these molecules are initially randomly dispersed into a cuboid box with a side length of 100 Å. The temperature of CH_4 is specified as 20°C and the remaining region of the system is set as 1600°C. The other settings are chosen as same as in Section 2.1.

After 0.32 *ns*, most of reactant molecules are attached to graphite sheets although these sheets are highly distorted (Fig. 5b). There is only one SiC molecule formed until 0.311 *ns* so far. This could be caused by reacting of CH with SiO directly (Fig. 6).

2.3 *SiO and CH_4 reaction on Fe and SiO_2 solid surfaces*

In both atomic models used for reactions on Fe and SiO_2 surfaces (Fig. 7a and Fig. 8a), there are 200 SiO molecules and 300 CH_4 molecules which are initially randomly dispersed into a cuboid box with a dimension $92 \times 92 \times 200$ Å. A super-cell with a five lattice thickness of a single α-iron crystal or a single SiO_2 structure is inserted in the middle of the boxes. The CH_4

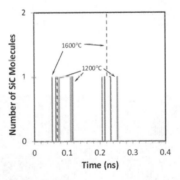

Figure 3.   Numbers of SiC molecules generated at 1200°C and 1600°C.

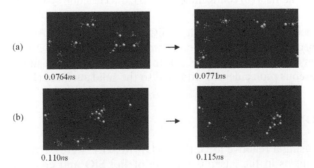

Figure 4.   Snapshots of two SiC formations at 1200°C (Si and C involved are shown in dark green colour).

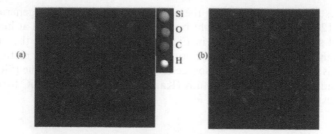

Figure 5.   Initial (a) and final (b) configurations of CH$_4$+SiO reaction system with graphite at 1600°C.

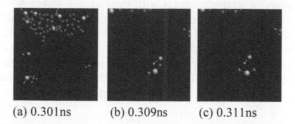

(a) 0.301ns   (b) 0.309ns   (c) 0.311ns

Figure 6.   Snapshots of SiC formation process (Si and C involved are highlighted in dark green colour).

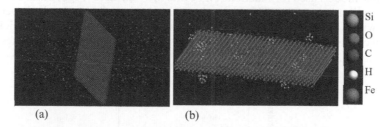

Figure 7.   Initial (a) and final (b) configurations of CH$_4$+SiO reaction system on iron at 1400°C.

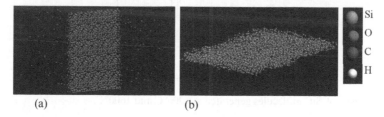

Figure 8.   Initial (a) and final (b) configurations of CH$_4$+SiO reaction system on SiO$_2$ at 1600°C.

temperature is specified as 20°C and that of SiO gas are 1400°C on Fe surface and 1600°C on SiO$_2$ surface, respectively. The other settings are same as in Section 2.1.

After a long time (longer than 0.5 ns), both simulations do not show any SiC formation although most gaseous molecules are attached to the solid surfaces (Fig. 7b and Fig. 8b).

3   CONCLUSIONS

MD simulations using a reactive force field ADF/ReaxFF were carried out to examine SiC formation mechanisms in the gaseous mixture of hot SiO and cold CH$_4$ on various solid surfaces, such as β-SiC, graphite sheets, iron, and amorphous SiO$_2$. It showed that the fast formation of SiO clusters played an important role. High SiO temperature enhanced single

SiC molecule formation on SiC surface. Although SiC molecule could be formed on the graphite environment, it has not found any occurrence of such reaction on iron and SiO_2 surface so far.

ACKNOWLEDGEMENTS

The present work is supported by the Norwegian Research Council under the project Kiselrox (Project No. 228722/O30). Special thanks to SCM/ADF for permission to use ADF/ReaxFF license and Prof. Adri van Duin for help on ReaxFF.

REFERENCES

Alizadeh, R., Jamshidi, E. & Ale-Ebrahim, H. 2007. Kinetic Study of Nickel Oxide Reduction by Methane, *Chem. Eng. Technol* 30(8): 1123–1128.

Jamshidi, E. & Ebrahim, H.A. 2008. A new clean process for barium carbonate preparation by barite reduction with methane, *Chem Eng Process* 47(9): 567–1577.

Khoshandam, B., Kumar, R.V. & Jamshidi, E. 2006. Producing Chromium Carbide Using Reduction of Chromium Oxide with Methane, *AIChE Journal* 52(3): 1094–1102.

Monsen, B., Kolbeinsen, L., Prytz, S., Myrvågnes, V. & Tang, K. 2013. Possible Use of Natural Gas For Silicon or ferrosilicon Production, in *Proceeding of the 13th International Ferroalloys Congress*, Kazakhstan, Vol. 1: 467–478.

Newsome, D., Sengupta, D., Foroutan, H., Russo, M.F. & van Duin, A.C.T. 2012. Oxidation of Silicon Carbide by O_2 and H_2O: A ReaxFF Reactive Molecular Dynamics Study: Part I, *J. Phys. Chem. C* 116(30): 16111–16121.

Ostrovski, O. & Zhang, G. 2006. Reduction and Carburization of Metal Oxides by Methane-Containing Gas, *AIChE Journal* 52(1): 300–310.

Ostrovski, O., Zhang, G., Kononov, R., Dewan, M.A.R. & Li, J. 2010. Carbothermal Solid State Reduction of Stable Metal Oxides, *Steel Research Int.* 81(10): 841–846.

SCM 2014. *ReaxFF 2014*, Vrije Universiteit, Amsterdam, The Netherlands, http://www.scm.com.

Tang, K., Ma, X., Andersson, S. & Dalaker, H. 2015. Kinetics between SiO and CH_4 at High Temperature. In Jiang, T. etc. (eds), *6th International Symposium on High-Temperature Metallurgical Processing*: 349–356, TMS (The Minerals, Metals & Materials Society).

van Duin, A.C.T. 2002. *ReaxFF User Manual.*

Advances in Engineering Materials and Applied Mechanics – Zhang, Gao & Xu (Eds)
© 2016 Taylor & Francis Group, London, ISBN 978-1-138-02834-0

A simple method for the synthesis of Mo₂N hollow nanospheres with high surface area

Y. Tang
Hubei Province Key Laboratory of Coal Conversion and New Carbon Materials,
Wuhan University of Science and Technology, Wuhan, Hubei Province, China

W. Luo
Research and Development Center of Wuhan Iron and Steel Company, Wuhan, Hubei Province, China

ABSTRACT: High-speed vibration milling method has been used to synthesize Molybdenum Nitride (Mo_2N) with hollow spherical nanostructures which has high surface area (218.5 m^2 g^{-1}) by the reaction of $MoCl_5$ and $NaNH_2$ directly. The phase structure and morphology of product are analyzed by powder X-Ray Diffraction (XRD), X-ray Photoelectron Spectroscopy (XPS), and Transmission Electron Microscopy (TEM). The thermal stability and surface area of the products are also investigated by Thermogravimetric Analysis (TGA) and BET, respectively. Such highly dispersed nitride material is expected to have applications in catalysis because of the large surface area.

1 INTRODUCTION

Transition-metal nitrides are technologically important materials with many interesting properties including excellent hardness, high melting point, abrasive resistance, good chemical stability, and high electrical conductivity, and hence have wide applications in various fields such as interconnects and diffusion barriers in microelectronics, hard wear-resistant coating in engineering, optical coatings in industry, and electrodes in semiconductor devices (Oyama 1996, Pierson 1996, Toth 1971).

Molybdenum nitride (Mo_2N), one of the transition-meta nitrides, has attracted much attention because it shows catalytic activities in a number of reactions involving hydrogen, such as hydrotreating (HDS and HDN), hydrogenolysis, hydrogenation and NH_3 synthesis (Li et al. 1998, 1999, Trawczynski 2001). Up to now, various strategies have been developed to synthesize Mo_2N. Jaggers et al. (1990) prepared Mo_2N and MoN by the reactions of ammonia with MoO_3, $(NH_4)_6Mo_7O_{24} \cdot 4H_2O$, $(NH_4)_2MoO_4$ and H_xMoO_3. Afanasiev (2002) reported the synthesis of dispersed Mo_2N with lamellar morphology and high specific surface area (158 $m^2 \cdot g^{-1}$) by the decomposition of the $(HMT)_2(NH_4)_4$-Mo_7O_{24} salt (HMT = hexamethylenetetramine) in the temperature range of 550–800°C. Cai et al. (2005) synthesized Mo_2N nanocrystals in an autoclave in the temperature range of 450–550°C, using $MoCl_5$ and NaN_3 as the reactants. In Ma's group, nanocrystalline molybdenum nitride was synthesized via a thermal reduction–nitridation route by the reaction of metallic sodium with anhydrous molybdenum pentachloride and ammonium chloride in an autoclave at 550°C (Ma & Du 2008). Lee et al. (2014) prepared single crystalline mesoporous Mo_2N nanobelts using molybdenum oxide as a template by a topotactic reaction process. Nonetheless, the surface areas of most as-prepared molybdenum nitrides were not high enough for catalytic application. Herein, we report a high-speed vibration milling method for the direct synthesis of Mo_2N hollow nanospheres with high surface area by selecting anhydrous molybdenum chlorides and sodium amide as reactants.

2 EXPERIMENTAL

All manipulations were performed in a N2 flow glove box to avoid the effect of the oxygen and moisture. Vibrating-mill reactions were carried out in a stainless steel capsule (40-mm height, 18-mm o.d., 12-mm i.d.), typically using maximum 0.5 g of total starting reagents. The capsule containing four agate balls was fixed on a vibration arm of Retsch MM200, and was vibrated vigorously at the rate of 1800 cycles per minute. Molybdenum chloride ($MoCl_5$) and sodium amide ($NaNH_2$) were purchased from Aldrich and Shanghai Chem. Co., respectively, and used without further purification. In a typical procedure, $MoCl_5$ (1 mmol) and (5 mmol) $NaNH_2$ were stirred together and then placed into the capsule to mill for 5 min. The black products were rinsed with dilute HCl solution (~1 mol·L^{-1}), absolute ethanol, and distilled water in sequence, and then dried under vacuum at 60°C for 4 h.

The composition and the structure of the as-prepared products were confirmed by the X-Ray Diffraction (XRD) pattern, using an 18 KW advanced X-ray diffractometer with Cu Kα radiation (λ = 1.54187 Å). X-ray Photoelectron Spectroscopy (XPS) was performed on ESCALAB MKII with Mg Kα (hν = 1253.6 eV) as the excitation source. Transmission Electron Microscopy (TEM) images and the Selected-Area Electron Diffraction (SAED) patterns were taken with a Hitachi H-800 transmission electron microscope with an accelerating voltage of 200 kV. Thermogravimetric Analysis (TGA) measurement was carried out on a Shimadzu TGA-50H type instrument at a heating rate of 10 K min^{-1} from room temperature to 1173 K under N_2 flow at the rate of 20.00 mL min^{-1}. A Micromeritics ASAP-2000 was used to measure the surface area of the prepared nitride by means of the adsorption of N_2 at 78 K (BET method).

3 RESULTS AND DISCUSSION

3.1 Composition of the products

The typical XRD pattern of as-obtained molybdenum nitride is shown in Figure 1. All the reflection peaks could be indexed to the (111), (200), (220) and (311) crystal planes of cubic phase Mo_2N, and no impurity peaks from element Mo or molybdenum oxides were observed in the XRD pattern. The calculated lattice constant from Figure 1 was a = 4.172 Å, well agreement with the literature value of a = 4.163 Å (JCPDS Cards No. 25-1366 for Mo_2N).

Further evidence for the composition of the product could be obtained by XPS measurements. Figure 2 shows the XPS spectra of the as-prepared Mo_2N. A doublet with binding energies at 228.5 eV for Mo $3d_{5/2}$ and 231.7 eV for Mo $3d_{3/2}$ is shown in Figure 2a.

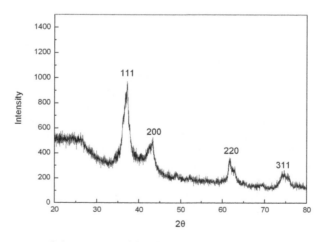

Figure 1. XRD patterns of the as-prepared Mo_2N.

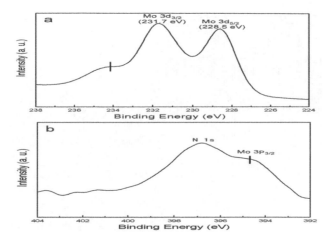

Figure 2.   XPS spectra of the as-prepared Mo$_2$N.

A shoulder around 234.1 eV observed in the curve was the characteristic of Mo^{4+} in MoO$_2$ which indicated the presence of oxidation product on the surface of the products. Since none of molybdenum oxide was detected from the XRD result, the surface oxidation degree was very small. In Figure 2b, signals shown at 394.8, and 396.8 eV corresponded to Mo (3p$_{3/2}$), and N (1s) electrons, respectively (Chiu et al. 1998). It is noteworthy that the peaks of Mo (3p) and N (1s) overlap, and the N (1s) signal is characteristic for a metal nitride material.

3.2   *Morphologies*

The morphology and structure of the as-prepared Mo$_2$N were further characterized using TEM and SAED, which are shown in Figure 3. The product was mainly made up of hollow spheres together with a small amount of peanut-like structures on the basis of a strong contrast between the dark edge and pale center. The outer diameter of the hollow spheres was in the range from about 80 nm to 300 nm, and the average thickness of the shells was around 10–20 nm. The hollow nature could be further confirmed by some broken hollow spheres shown in Figure 3c. The crystalline nature of the products was also confirmed by the SAED. The diffraction rings in the inset of Figure 3a could be indexed as (111), (200), (220), and (311) reflections, according to the cubic rock-salt structure of polycrystalline Mo$_2$N.

3.3   *TGA and BET measurements*

The thermal behavior of the as-obtained Mo$_2$N was studied by TGA. Figure 4 shows the TGA diagram of the as-prepared products with a heating rate of 10°C·min^{-1} in a stream of high purity nitrogen at a rate of 20.00 mL·min^{-1}. From the TGA curve, we can find that the weight gain of the sample has not changed significantly below 350°C. A slight weight loss indicated that this might arise from the evaporation of absorbed water on the surface of the sample. The onset of the oxidation of the Mo$_2$N sample was found to begin at about 350°C, which indicated that the sample was oxidized by oxygen to form molybdenum trioxide and nitrogen oxide. As the temperature rose, the amount of the formed molybdenum trioxide became bigger, suggesting that the oxidation rate of the sample became faster. Finally, the sample could be oxidized thoroughly at 710°C. Above 710°C, there was a continuous weight decrease corresponding to the evaporating of MoO$_3$ and the decomposition of Mo$_2$N since the calculated decomposition temperature by Gibbs Free Energy Equation was 681°C. At around 810°C, the slope of the curve changes was recorded to the melting reaction of MoO$_3$

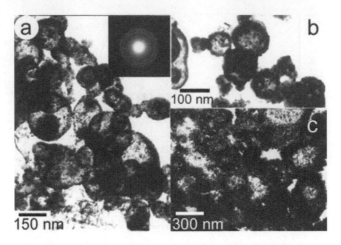

Figure 3.   TEM and SAED images of the as-prepared Mo$_2$N.

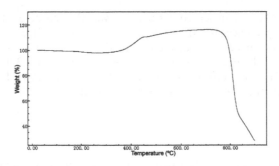

Figure 4.   Thermogravimetric analysis of the as-prepared Mo$_2$N.

(m.p. 795°C). Our study revealed that the resistance of c-Mo$_2$N to oxidation was poor because of its non-protective nature and the volatility of the oxides at moderately high temperatures.

The surface area of the prepared molybdenum nitride was determined by the adsorption of N$_2$ at 78 K. The Mo$_2$N with hollow spherical nanostructure showed a specific high BET area value (Sg = 218.5 m^2 · g^{-1}), which should present considerable interest for the catalytic applications.

3.4   *Formation mechanism*

Our high-speed vibration milling method for the direct synthesis of Mo$_2$N hollow nano-spheres is essentially based on the self-propagating reaction between MoCl$_5$ and NaNH$_2$ due to the low melting point (210°C) of NaNH$_2$. In most of the homologous solid state reactions, there is a correlation between the initiation temperature and the melting or sublimation point of the metal halides. The reaction of MoCl$_5$ and NaNH$_2$ could be tentatively indicated as follow (Parkin & Rowley 1995):

$$MoCl_5 + 5NaNH_2 \rightarrow Mo(NH_2)_5 + 5NaCl \rightarrow Mo_2N \qquad (1)$$

According to the Gibbs free energy calculations, the reactions were thermodynamically spontaneous and highly exothermic. Since the first step was self-propagating, exothermic, and rapid, the heat energy released initiated the second step almost at the same time. Thus resulted in a transient high temperature and the molten byproduct of NaCl, which was favorable to the formation and crystallization of the nitride.

4 CONCLUSIONS

In summary, we have described a high-speed vibration milling method for the direct synthesis of Mo_2N hollow nanospheres by the reaction of $MoCl_5$ and $NaNH_2$ directly. The structure of the product is confirmed to be cubic rock-salt structure of polycrystalline Mo_2N by XRD result. TEM images reveal that the as-prepared hollow nanospheres have outer diameters in the range of 80–300 nm, with the average thickness of the shells is about 10–20 nm. Their thermal stability and surface area are measured by TGA and BET, respectively, which indicate that the Mo_2N hollow nanospheres have specific high surface area ($218.5 \ m^2 \cdot g^{-1}$). Such highly dispersed nitride materials should present considerable interest for the catalytic applications.

REFERENCES

Afanasiev, P. 2002. New single source route to the Molybdenum Nitride Mo_2N. *Inorg. Chem.* 41 (21): 5317–5319.

Cai, P.J. Yang, Z.H. Wang, C.Y. Gu, Y.L. & Qian, Y.T. 2005. A simple approach to synthesize Mo_2N nanocrystals. *Chem. Lett.* 34 (10): 1360–1361.

Chiu, H.T. Chuang, S.H. Lee, G.H. & Peng, S.M. 1998. Low-temperature solution route to Molybdenum Nitride. *Adv. Mater.* 10 (17): 1475–1479.

Jaggers, C.H. Michaels, J.N. & Stacy, A.M. 1990. Preparation of high-surface-area transition-metal nitrides: molybdenum nitrides, Mo_2N and MoN. *Chem. Mater.* 2 (2): 150–157.

Lee, K.H. Lee, Y.W. Kwak, D.H. Moon, J.S. Park, A.R. Hwang, E.T. & Park, K.W. 2014. Single-crystalline mesoporous Mo_2N nanobelts with an enhanced electrocatalytic activity for oxygen reduction reaction. *Mater. Lett.* 124: 231–234.

Li, S. & Lee, J.S. 1998. Molybdenum Nitride and Carbide prepared from Heteropolyacid: II. Hydrodenitrogenation of Indole. *J. Catal.* 173 (1): 134–144.

Li, S. Lee, J.S. Hyeon, T. & Suslick, K.S. 1999. Catalytic hydrodenitrogenation of indole over molybdenum nitride and carbides with different structures. *Appl. Catal.* A–Gen. 184 (1): 1–9.

Ma, J.H. & Du, Y.H. 2008. A convenient thermal reduction–nitridation route to nanocrystalline molybdenum nitride (Mo_2N). *J. Alloys Compd.* 463 (1–2): 196–199.

Oyama, S.T. 1996. *The chemistry of transition metal carbides and nitrides.* Glasgow: Blackie Academic and Professional.

Parkin, I.P. & Rowley, A.T. 1995. Formation of transition-metal nitrides from the reactions of lithium amides and anhydrous transition-metal chlorides. *J. Mater. Chem.* (5): 909–912.

Pierson, H.O. 1996. *Handbook of refractory carbides and nitrides–properties, characteristics, processing and applications.* Westwood: Noyes Publications.

Toth, L.E. 1971. *Transition Metal carbides and nitrides.* New York: Academic Press.

Trawczynski, J. 2001. Effect of synthesis conditions on the molybdenum nitride catalytic activity. *Catal. Today* 65 (2–4): 343–348.

Advances in Engineering Materials and Applied Mechanics – Zhang, Gao & Xu (Eds)
© *2016 Taylor & Francis Group, London, ISBN 978-1-138-02834-0*

The effect of delamination on damage behavior of composite laminates subjected to three-point bending

J. Tang, S.L. Xing, J.Y. Xiao, Z.H. Wu, S.W. Wen & S. Ju
National University of Defense Technology, Changsha, Hunan, China

ABSTRACT: Delamination is an important damage pattern in composite laminates and has a great damaging influence on the mechanical properties of composite structures. In this paper, carbon fiber reinforced bismaleimide resin prepregs were used to fabricate the quasi-isotropic composite laminates with pre-inserted delamination through autoclave process. Then ultrasonic C-scanning was adopted to study the delamination propagation and the effect of pre-inserted delamination on the damage behavior of the composite laminates. The results show that (1) In the case of h > 2.034 mm, where the PTFE film was pre-inserted above layer 18, the delamination damage initiated and propagated in plane obviously at the same depth with the pre-inserted delamination. (2) In the case of 1.356 < h < 2.034 mm, where the PTFE film was pre-inserted between layer 12 and layer 18, there was merely damage propagation before ultimate failure. (3) The ultimate damage of the quasi-symmetric laminates occurred at h = 2.034 mm (1/4 thickness of the laminates). (4) The delamination was more sensitive to compressive stress than tensile stress.

1 INTRODUCTION

Advanced composites have become widely used in aircraft structures due to their high specific strength, high specific stiffness and chemical resistance. And the proportion and the position (such as wing skin, fuselage panel, hatch, etc.) of composite materials used in aircraft structures have become the main factors of evaluating whether the aircraft is advanced[1]. However, during the process of training and manufacturing of aircrafts, the composite components can be easily damaged by some impact loadings, e.g. bird striking, stone striking, tools striking[2], because these composite structures are sensitive to impact loadings[3]. Specifically, these damages include fiber fracture, matrix cracking, interlaminar delamination, fiber-matrix interface debonding and so on. The initiation, propagation and accumulation of these damages can lead to sharply decreasing of the strength and the stiffness of the composite structures, thus causing the vast degradation of structures' carrying capacity, as a result, which may result in devastated accidents[4]. The delamination is the most common damage in composite laminates[5] and also the major reason causing the deterioration of composite laminates' mechanical properties[6].

Currently, a wide range of researches have been carried out to study the delamination damages using numerical and experimental methods[7–18]. As for numerical investigations, Koloor, S.S.R. [7] et al. imitated the propagation and initiation of delamination under bending and tensile loadings through finite analysis method, also they predicted the mechanical behaviors of unsymmetric structures. Allix, O. [9] et al. established a laminate modeling with pre-existing crack by the means of a Finite Element Scheme to allow a complete simulation of delamination under the condition of relatively little experimental data. In terms of experimental studies, Amaro, A.M. [8] et al. investigated the effect of delamination on tensile strength of laminates under three-point bending. Asp, L.E. [13] et al. studied the effect of delamination growth on residual strength of laminates under impact loading. Nilsson, K.F. [14] et al. implemented an experimental investigation on delamination buckling and growth for composite panels under compressive loading. However, the experimental research of the effect of delamination on damage behavior of specific stacking sequence laminates subjected to three-point bending should be further considered.

In this paper, a series of specimens with pre-inserted delamination were fabricated by auto clave to investigate the interaction between delamination and damage behavior under three-point bending. Damage area and depth was measured with the assist of ultrasonic C-scanning nondestructive instrument. Thanks to the monitoring of damage propagation by ultrasonic C-scanning, several general interaction between delamination and damage behavior of composite laminates subjected to three-point bending can be explained to a certain extent.

2 EXPERIMENT

2.1 Materials and specimens fabrication

In this study, carbon fiber reinforced bismaleimide resin prepregs were chosen to fabricate the composite laminate with the layup sequence of [0/45/90/-45]3 s. During the process of laying up, the PTFE films were cut into circular shapes and inserted between different two adjacent layers of the laminate to produce the pre-inserted delamination. The specific parameters of the prepregs and PTFE film are shown in Table 1 and Table 2 respectively.

All composite laminates were cured in an autoclave following the curing steps shown as Figure 1. Then the composite laminates were sliced into specimens with the dimension of $350 \times 50 \times 3.0$ mm by waterjet cutting for the bending test. However, owing to the fact that the resins of prepregs were absorbed by suction fleece during the curing process, thus leading to the decrease of thickness, the average thickness of specimens was actually 2.71 mm (0.113 mm for each layer).

To investigate the effect of delamination on damage performance of composite laminates, eleven specimens with different parameters, including the diameter of the pre-inserted delamination (d) and the distance of the pre-inserted delamination from the lower surface (h), were fabricated. The schematic of the specimen is shown as Figure 2. The detailed parameters of those specimens are shown in Table 3.

Table 1.   Parameters of the prepregs.

Thickness/ mm	Resin fraction/%	Surface density/g/m²
0.125	33	132

Table 2.   Parameters of PTFE film.

Thickness/ mm	Elasticity modulus/MPa	Poisson's ratio
0.05	61.4	0.4

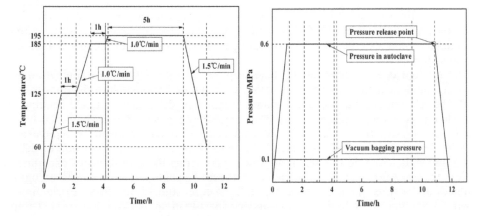

Figure 1.   The curing steps of composite laminates.

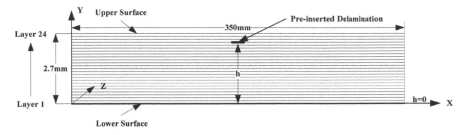

Figure 2.   Schematic of the specimen.

Table 3.   Parameters of different specimens.

Number	A1	A5	B1	B5	C1	C5	D1	D2	D3	D4	D5
h/mm	1.36	1.36	1.70	1.70	2.03	2.03	2.37	2.37	2.37	2.37	2.37
d/mm	5	25	5	25	5	25	5	10	15	20	25

h = 1.36 mm: the PTFE film was inserted between layer 12 and layer 13.
h = 1.70 mm: the PTFE film was inserted between layer 15 and layer 16.
h = 2.03 mm: the PTFE film was inserted between layer 18 and layer 19.
h = 2.37 mm: the PTFE film was inserted between layer 21 and layer 22.

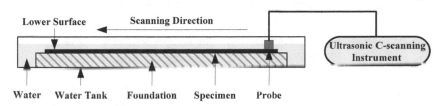

Figure 3.   Schematic of ultrasonic C-scanning experiment.

2.2   *Three-point bending test*

The nondestructive testing was performed under the condition of three-point bending test with a displacement rate of 2 mm/min and 3 specimens of each case were tested.

2.3   *Ultrasonic C-scanning analysis*

As for the ultrasonic nondestructive testing, the pulse ultrasonic wave produced by the instrument goes through the coupling agent, then the substances. Due to the differences of specific acoustic impedance between different substances, the pulse ultrasonic wave reflexes and transmits at the surface of defect[19] and the pulse echoes are detected by the probe. Then the position and area of defect can be measured resulted from the enormous differences of pulse echoes between intact substances and defects.

   In this study, the pulse width was 100.00 ns and the testing velocity was 3006.9 m/s after being revised. And water was used as coupling agent. The schematic of ultrasonic C-scanning experiment is shown as Figure 3. The depth and area of the damage were calculated by the specific software, TomoView 2.9.

3   RESULTS AND DISCUSSION

It is noticeable that the typical force-displacement curve of the specimen under three-point bending (shown as Fig. 4) indicates that the specimen experienced a progressive failure. That

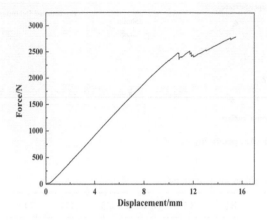

Figure 4.   Typical curve of force-displacement of specimens.

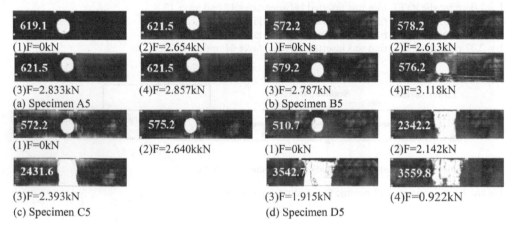

619.1	621.5
(1)F=0kN	(2)F=2.654kN
621.5	621.5
(3)F=2.833kN	(4)F=2.857kN
(a) Specimen A5	

572.2	578.2
(1)F=0kNs	(2)F=2.613kN
579.2	576.2
(3)F=2.787kN	(4)F=3.118kN
(b) Specimen B5	

572.2	575.2
(1)F=0kN	(2)F=2.640kkN
2431.6	
(3)F=2.393kN	
(c) Specimen C5	

510.7	2342.2
(1)F=0kN	(2)F=2.142kN
3542.7	3559.8
(3)F=1.915kN	(4)F=0.922kN
(d) Specimen D5	

Figure 5.   Delamination propagation of the specimen A5, B5, C5, D5.

means local damage can be produced at an early loading stage and propagate as the load increases[20].

In order to investigate how the delamination affected the damage behavior, ultrasonic C-scanning was used to monitor the propagation of delamination during the progressive failure. Specifically, when the force dropped down dramatically at first, the loading was suspended and then the ultrasonic C-scanning was used to monitor the propagation of delamination. After detecting, the force was reloaded on this specimen, when the force dropped again, the detection went on. This procedure was not finished until the specimen was entirely damaged.

In this section, the damage behavior of composite laminates under three-point bending mainly includes three aspects: (1) the damage area, (2) the damage depth and (3) the different damage behavior under tensile and compressive stress. As a result, the effect of delamination on damage behavior of composite laminates has been discussed from these three aspects respectively.

3.1   The effect of delamination on damage area

Case 1: Different depths of pre-inserted delamination with the same diameter.

Four sorts of specimens (A5, B5, C5, D5) were involved in this case. Figure 5 gives information about the propagation of delamination of A5, B5, C5, D5, respectively. The data of the damage areas (in mm²) is listed in every picture.

As for the specimen A5, the damage area kept stable at around 620 mm² before the specimen was damaged at the force 2.857 kN and the damage area of B5 was about 575 mm² whatever the force was. It is worth claiming that the fibers of lower surface were fractured occasionally at the force of 3.118 kN, which made it impossible for the scanning process to continue, as a result, this specimen had not been entirely damaged yet. The damage area of specimen C5 was just over 570 mm² before the specimen was entirely damaged and the damage area peaked at 2432 mm². It can be seen from the change of damage area of these three sorts of specimens that, in the case of $1.36 < h < 2.03$ mm, the damage area merely propagated before the specimen was entirely damaged. However, in terms of D5, it is clear that the damage area experienced a progressive expansion before the specimen was entirely damaged. Thus, in the case of $h = 2.37$ mm, the pre-inserted PTFE film could induce the propagation of delamination, while others could not.

Case 2: Different diameters of pre-inserted delamination with the same depth.

Figure 6 shows the propagation of pre-inserted delamination of four specimens (D1, D2, D3, D4).

In the case of $h = 2.37$ mm, as for $d = 5$ mm, the damage area rose from 19.1 mm² to 2644.2 mm² with the force enhancing (peaked at 2.506 kN). When it comes to $d = 10$ mm, 15 mm, 20 mm or 25 mm, the expansion of damage area shared the same tendency with the situation of D1. In addition, with d increasing from 5 mm to 15 mm, the maximum damage area increase slightly from 2644.2 mm² to 2833.6 mm², while the maximum damage areas were almost the same when $d = 20$ mm and 25 mm. It can be concluded from the discussion above that in the case of $h = 2.37$ mm, the damage area increased progressively till the specimen was entirely damaged whatever the diameter of pre-inserted delamination (d) was.

3.2 *The effect of delamination on damage depth*

Due to the fact that the upper surface of the specimen was broken during the loading process which leads to the inoperative C-scanning, the ultrasonic C-scanning was implemented on the lower surface. As a result, the larger the value of depth measured was, the closer the damage actually to the upper surface was. Figure 7 gives the details of the damage depths of A1, B1, C1 and D1 during the loading process.

As for A1, where $h = 1.36$ mm, the detected depth of delamination was about 1.3 mm, which indicates the delamination was actually the pre-inserted PTFE film by ignoring the deviation of ultrasonic C-scanning testing. After the specimen was damaged, the detected depth was around 2.0 mm. As for B1, where $h = 1.70$ mm, the detected depth of delamination was about 1.8 mm. By taking the deformation of the layers into account, the detected delamination

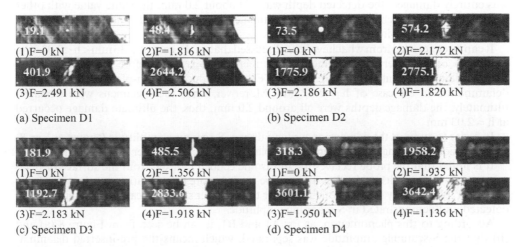

(1)F=0 kN   (2)F=1.816 kN   (1)F=0 kN   (2)F=2.172 kN

(3)F=2.491 kN   (4)F=2.506 kN   (3)F=2.186 kN   (4)F=1.820 kN

(a) Specimen D1   (b) Specimen D2

(1)F=0 kN   (2)F=1.356 kN   (1)F=0 kN   (2)F=1.935 kN

(3)F=2.183 kN   (4)F=1.918 kN   (3)F=1.950 kN   (4)F=1.136 kN

(c) Specimen D3   (d) Specimen D4

Figure 6.   Delamination propagation of the specimen D1, D2, D3, D4.

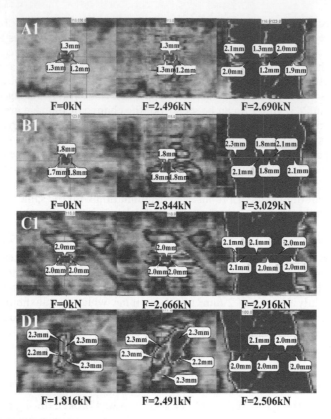

Figure 7.   Damage depths in the specimens.

could be regarded as the pre-inserted PTFE film. After the specimen beiewng damaged, the detected depth was also around 2.0 mm. When it comes to C1, where h = 2.03 mm, the detected damage depth was all about 2.0 mm before and after the specimen was damaged, which implies that the ultimate delamination damage occurred at the same plane with the pre-inserted PTFE film. When it comes to D1, where h = 2.37 mm, although the damage area increased obviously, the damage depth kept at 2.3 mm, which indicates that the pre-inserted PTFE film induced the in-plane propagation of delamination. However, when the specimen was entirely damaged, the detected depth was still about 2.0 mm, the same value with other three specimens', which illustrates that the ultimate damage occurred at about 2.0 mm whatever the depth of pre-inserted PTFE film was.

It can be concluded from what have been discussed that, in the case of 1.36 mm < h < 2.03 mm, the depth of pre-inserted delamination was merely propagated before the specimen was entirely damaged, while the pre-inserted PTFE film induced the in-plane propagation of delamination in the case of h = 2.37 mm. Moreover, when the specimens were damaged ultimately, the damage depths were all around 2.0 mm, thus, the ultimate damage occurred at h = 2.03 mm.

In order to support the conclusion we have drawn, S-scanning amplitude (shown as Fig. 8) from the side of damaged specimens was implemented to offer further information.

It is noted that the probe receives the ultrasonic echoed reflexed from the surface of delamination, which results in the fact if there are two or more delaminations at the same location (X-Y plane), only the one which is closer to the lower surface (scanning surface) can be detected and demonstrated in S-scanning amplitude.

According to this phenomenon, as for A1 and B1, it can be seen from Figure 8 (a) and (b) that the S-scanning amplitude was separated, which means the pre-inserted delamination and the delamination caused by ultimate damage were not in the same plane and the

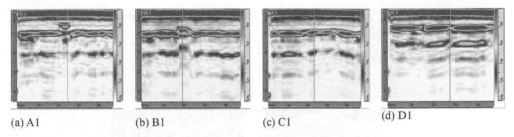

(a) A1   (b) B1   (c) C1   (d) D1

Figure 8.   S-scanning amplitude from the side of damaged specimens.

(a) Fracture section of specimen C1

Location of Pre-inserted Delamination

Location of Delamination Damage

(b) Fracture section of specimen D1

Figure 9.   Fracture section of specimen C1 and D1.

pre-inserted delamination was closer to the lower surface of the specimen. Thus, in the case of $1.36 < h < 1.70$ mm, the pre-inserted delamination did not induce the ultimate delamination damage. When it comes to C1 and D1, it is clear that the S-scanning amplitude was continent, which indicates the pre-inserted delamination was in the same plane with the ultimate damage delamination or the ultimate damage delamination was closer to the lower surface of the specimen than the pre-inserted delamination. In order to make it clear, the damaged specimens C1 and D1 were cut off from the central. As shown in Figure 9, there was only one delamination in C1, while there were two separate delamiantions in specimen D1. As a result, in the case of $h = 2.03$ mm, the pre-inserted delamination induced the ultimate damage delamination. However, in the case of $h = 2.37$ mm, although the pre-inserted delamination propagated, the ultimate damage delamination occurred in the other plane.

3.3   Damage behavior under different loadings

The structure of the laminate is quasi-isotropic, as a result, the upper half specimen mostly suffers to tensile stress, while the bottom half part mainly suffers to compressive stress when the specimen is subjected to three-point bending from the upper surface. As for the specimens, the values of tensile stress and compressive stress that the pre-inserted delamination suffers to are the same when the bending loads are added from the upper surface and lower surface of the specimen respectively, owing to the symmetric structure.

Figure 10 gives information about the different damage behaviors under three-point bending added from the upper surface and lower surface of the specimen respectively. In the case of loading from upper surface, where the pre-inserted delamination mainly suffered to compressive stress, the delamination damage propagated in plane obviously. However, when the

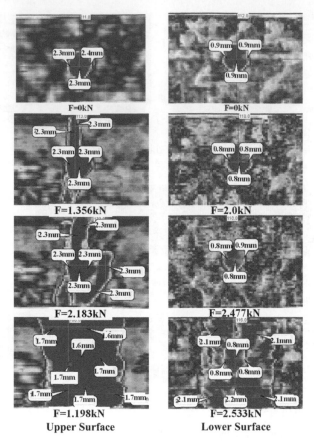

Upper Surface   **Lower Surface**

Figure 10.   Damage behavior under different loadings.

bending load was added from lower surface, thus the pre-inserted delamination mainly suffered to tensile stress, the delamination merely propagated until the ultimate damage occurred. Consequently, the delamination was more sensitive to compressive stress than tensile stress.

4   CONCLUSION

In the present paper, four types of specimens, defined by altering the location and diameter of pre-inserted PTFE film, were fabricated. To understand the effect of pre-inserted delamination on damage behavior of laminates under three-point bending, ultrasonic C-scanning was used to monitor the propagation of delamination (including damage area and damage depth). The following conclusions are made from the present investigation.

1. In the case of h > 2.03 mm, where the PTFE film was pre-inserted above layer 18, the damage initiated and propagated in plane obviously at the same depth with the pre-inserted delamination.
2. In the case of 1.36 < h < 2.03 mm, where the PTFE film was pre-inserted between layer 12 and layer 18, there was merely damage propagation before ultimate failure.
3. The ultimate damage of the quasi-symmetric laminates occurred at h = 2.03 mm (1/4 thickness of the laminates).
4. The delamination was more sensitive to compressive stress than tensile stress.

However, owing to the complexity and flexibility of the stacking sequence of composites, thus leading to the changeable stiffness matrix of each layer under certain loading, these conclusions are limited to some extent, that means they may only be suitable for the quasi-symmetry composite laminates at a stacking sequence of [0/45/90/-45]3 s or some similar structures.

FUNDING

The authors gratefully acknowledge the support of (a) the National Science Foundation of China under grant No. 51303208 and No. 11202231, (b) Major Science and Technology Projects in Hunan Province under grant No. 2011FJ1001.

REFERENCES

[1] Feng Z.M., Ou Y. 2010. Study on the drilling process without disfigurements for aero carbon fiber reinforced plastics parts. *Aeronautical Manufacturing Technology* (6): 32–34.
[2] Hawyes V.J., Curtis P.T., Soutis C. 2001. Effect of impact damage on the compressive response of composite laminates. *Composite Part A* 32: 1263–1270.
[3] Staszewsk W.J. 2002. Intelligent signal processing for damage detection in composite materials. *Compos Science and Technology* 62(9): 41–50.
[4] Liu H.X., Zhang H., Ma R.X. 2003. Nondestructive Testing Techniques for composite. *Nondestructive Testing* 25(63): 1–4.
[5] Zou, Reid, Soden. 2002. Application of a delamination model to laminated composite structures. *Composite Structure* 56: 375–389.
[6] Amaro A.M., Santos J.B. 2011. Delamination depth in composites laminates with interface elements and ultrasound analysis. *Strain* 47(2): 138–145.
[7] Koloor S.S.R., Abdul-Latif A., Tamin M.N. 2011. Mechanics of composite delamination under flexural loading. *Key Engineering Materials* 462–463: 726–731.
[8] Amaro A.M., Reis P.N.B. 2011. Delamination effect on bending behavior in carbon-epoxy composites. *Strain* 47: 203–208.
[9] Allix O., Ladeveze P. 1995. Damage analysis of interlaminar fracture specimens. *Composite Structure* 31(1): 61–74.
[10] Craven R., Lannucci L., Olsson R. 2010. Delamination buckling: A finite element study with realistic delamination shapes, multiple delaminations and fibre fracture cracks. *Composite Part A* 41(5): 684–692.
[11] Chai H., Babcock C.D., Knauss W.G. 1981. One-dimensional molding of failure in laminated plates by delamination buckling. *International Journal of Solids and Structure* 17: 1069–1083.
[12] Ernesto P.F. 2006. A computational procedure for the determination of fracture parameters at interfacial contact cracks in composite laminates. *Journal of Composite Materials* 40(8): 701–715.
[13] Asp L.E., Nilsson S., Singh S. 2001. An experimental investigation of the influence of delamination growth on the residual strength of impacted laminates. *Composite Part A* 32(9): 1229–1235.
[14] Nilsson K.F., Asp L.E., Alpman J.E., et al. 2001. Delamination buckling and growth for delaminations at different depths in a slender composite panel. *International Journal of Solids and Structure* 38(17): 3039–3071.
[15] Johnson A.F., Holzapfel M. 2006. Influence of delamination on impact damage in composite structures. *Composite Science and Technology* 66: 807–815.
[16] Kim H.J. 1996. Effect of delamination on buckling behavior of quasi-isotropic laminates. *Journal of Reinforced Plastics and Composites* 15: 1262–1277.
[17] Nilsson K.F., Asp L.E., Sjogren A. 2001. On transition of delamination growth behavior for compression loaded composite panels. *International Journal of Solids and Structure*; 38(46–47): 8407–8440.
[18] Sun X.N., Tong L.Y., Chen H.R. 2001. Progressive failure analysis of laminated plates with delamination. *Journal of Reinforced Plastics and Composites* 20(16): 1370–1408.
[19] He Y.L. 2001. Non-destructive Examination of Carbon-Carbon Composite. *Aviation Engineering and Maintenance* 4: 22–23.
[20] Chang K.Y., Liu S., 1991. Chang F.K. Damage tolerance of laminated composites containing an open hole and subjected to tensile loadings. *Journal of Composite Materials* 25: 274–301.

Advances in Engineering Materials and Applied Mechanics – Zhang, Gao & Xu (Eds)
© *2016 Taylor & Francis Group, London, ISBN 978-1-138-02834-0*

The dielectric properties of electron irradiated P(VDF-TrFE-CFE) thin film

L. Tian, L. Li, J. Sun, X.H. Zhang & Y.P. Li
Hunan Institute of Engineering, Xiangtan, Hunan, China

ABSTRACT: The dielectric properties of electron irradiated poly(vinylidene fluoride-trifluoroethylene-chlorofloroethylene) [P(VDF-TrFE-CFE)] terpolymer thin films has been investigated. Compared with the observation of the non-irradiated terpolymer film, an additional dielectric loss peak with frequency dispersion is observed at higher temperature range for irradiated terpolymer films. This loss peak is suppressed with increasing dc bias and the temperature corresponding to it is independent of dc bias. Besides, the tunability decreases and the relaxor feature is enhanced for irradiated sample.

1 INTRODUCTION

The organic polymer have drawn great attention due to the fact that they can be applied as energy density capacitor, refrigerator, transducers and artificial muscles[1–3]. Recently, intensive investigations of the electron irradiation effect on poly(vinylidene fluoride-trifluoroethylene) [P(VDF-TrFE)] copolymer have been carried out[4–6]. Most results have revealed the dielectric properties of P(VDF-TrFE) films can be greatly changed by electron irradiation. For example, the irradiated P(VDF-TrFE) exhibit relaxor characteristics[4]: a broad peak in the temperature dependence of permittivity and a frequency dispersion of permittivity near the transition temperature. It is accepted that defects introduced by the irradiation can break up the macroscopic polar domains, resulting in the conversion of P(VDF-TrFE) film from normal ferroelectric to the ferroelectric relaxor[6]. Therefore, the electron irradiation is an effective way to modify the dielectric properties of the organic polymers. The above results inspire us to study the effect of electron irradiation on P(VDF-TrFE-CFE) terpolymer in order to search useful information about the dielectric properties for irradiated terpolymer. In this work, an additional dielectric loss peak is observed for the irradiated terpolymer, which has never been found in terpolymer film. Thus, the need for studying the dielectric properties of electron irradiated terpolymer is obvious.

2 EXPERIMENT

P(VDF-TrFE-CFE) 56.2/36.3/7.5 terpolymer powders were dissolved in dimethylsulfoxide and stirred for 6–12 hours at 70 °C temperature until complete dissolution of the polymer powders. After dissolution, the terpolymer solution were spin-coated on Al coated polyimide substrate at a speed of 3000 rad/min to form films. Then the as-grown films were annealed at 118 °C for 4 hours to improve the crystallinity. Afterwards, some of these film samples were irradiated by 1.8 MeV electrons with irradiation dose of 112 Mrad. Finally, top Al electrodes were evaporated onto the films through a mask to form the Al-terpolymer-Al capacitor structure for characterizing the electric properties. The frequency dependence and dc bias dependence of dielectric properties were measured at various temperatures by an Agilent E4980A Precision LCR Meter. The temperature was varied at a rate of 1 K/min via a computer controlled cryostat (MMR Tech., Inc.).

3 RESULTS AND DISCUSSION

Figure 1(a)–(b) show the temperature dependence of permittivity (ε) for the non-irradiated and 112 Mrad electron-irradiated samples at different frequencies. As seen from the figure, both samples exhibit relaxor-like behavior: the permittivity maximum (ε_m) undergoes a decrease with increasing frequency, while the temperature of the permittivity maximum (T_m) shifts to a higher temperature with increasing frequency. The permittivity at temperature higher than T_m can be fitted with the modified Curie-Weiss law[7]:

$$1/\varepsilon - 1/\varepsilon_m = (T - T_m)^r / C \tag{1}$$

where ε_m is the permittivity maximum, and r and C are constant. The dielectric material is the normal ferroelectric when r is near 1 and is total relaxor as the r is near 2. The insets of Figure 1(a)–(b) demonstrate the $\ln(1/\varepsilon - 1/\varepsilon_m)$ vs. $\ln(T - T_m)$ plots of both samples at 10 kHz. For each sample, the experimental data fit well with the law. The r values were 1.66 and 1.90 for non-irradiated and irradiated specimen, suggesting the relaxor behavior becomes more prominent after irradiation. Similar results have also been observed in our previous study on the electron irradiated LB P(VDF-TrFE-TrFE) thin films[8].

Figure 2(a)–(b) show the temperature dependence of dielectric loss tanδ for the non-irradiated and 112 Mrad electron-irradiated samples at different frequencies. For both samples, dielectric loss peaks in the temperature region between 250 K and 270 K are observed. These dielectric loss peaks also exhibit relaxor behavior: a broad dielectric loss

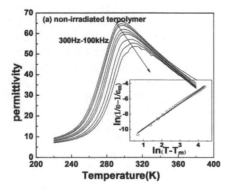
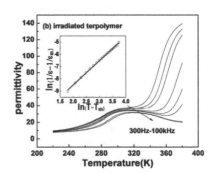

Figure 1.   (a)–(b) Temperature dependence of permittivity in the frequency range from 300 Hz to100 KHz for non-irradiated and irradiated terpolymers. The insets of Figure (a)–(b) are plot of $\ln[1/\varepsilon - 1/\varepsilon_m]$ as function of $\ln(T - T_m)$ for both samples, the solid lines are the fitting results.

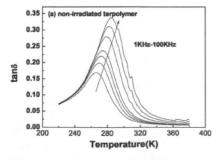
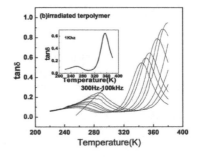

Figure 2.   (a)–(b) Temperature dependence of dielectric loss tanδ in the frequency range from 300 Hz to100 KHz for the non-irradiated and irradiated samples. Inset of Figure 2(b) is the temperature dependence of tanδ at 1 KHz.

peak, a shift of loss peak to higher temperature with increasing frequency. In the previous literature, authors attribute this broad dielectric loss peak to micro-Brownian motion of polymer chain in the amorphous region[9]. Note that compared with the observation in non-irradiated terpolymer films, some additional dielectric loss peak appear at higher temperature range especially at lower measured frequencies. As seen from inset of Figure 2(b), a noticeable dielectric loss peak is present around 350 K at measured frequency of 1 KHz for the irradiated specimen, hereafter denoted as mode A. For a more detailed insight into the relaxation process of mode A, the temperature dependence of the dielectric loss at lower frequencies were further measured and the results were shown in Figure 3. As observed from the figure, the temperature corresponding to the dielectric loss peak varies with frequency, which suggests that the peak exhibits thermal activated relaxation behavior. The temperature dependence of the relaxation frequency is analyzed by the Arrhenius law[10]:

$$f = f_0 \exp(-E / K_B T) \qquad (2)$$

where f_0 is attempt frequency, E is the activation energy for the relaxation, K_B is the Boltzmann's constant, and T is the absolute temperature corresponding to the dielectric loss peak. The peak temperature values under different frequencies are well-fitted to Arrhenius law as shown in the inset of Figure 3. The fitted value of activation energy is $E = 1.12$ eV. This value is comparable to the activation energy 0.765 eV in P(VDF-TrFE) copolymers and related to defects such as random (TG) conformation segments[11]. The present value of activation energies suggest that the associated dielectric relaxation may originate from the same source, i.e. some kinds of defects which are unavoidably introduced into terpolymer by irradiation. Therefore, we infer that mode A is a "defect mode". From previous observation in ferroelectric SrTiO$_3$ film, the "defect mode" have two important characteristics[12]: (1) the dielectric relaxation can fit with Arrhenius law; (2) temperature corresponding to the dielectric loss peak does not change with bias voltage. In our case, the first characteristic is satisfied. Therefore, to examine the second characteristic, the temperature dependence of the tanδ under different voltage bias are measured.

Figure 4 gives the temperature dependence of the tanδ under the voltage bias of 0 V, 5 V and 10 V for irradiated sample. From the Figure 4, a significant decrease in tanδ is observed by applying dc bias, which is due to the weakness of the movement of molecular chains at high dc electric field[13]. However, the temperature of loss peak remains almost the same. This is different from the "induced ferroelectric peak" in terpolymer, whose temperature increases with increasing electric field[14], but is similar with "defect mode". Therefore, the addition mode A can be assigned as "defect mode".

In our previous study, we have found after proper electron irradiation, the tunability of P(VDF-TrFE) can be improved to some extent[15]. Thus, to find out whether the tunability of the irradiated terpolymer can be improved, the permittivity under dc bias field for both

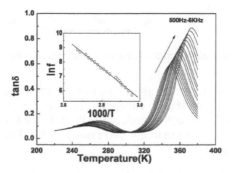

Figure 3.   Temperature dependence of tanδ for irradiated sample in the frequency range from 300 Hz to 6 KHz. Inset figure is plot of ln$f$ as function of 1000/T, the solid line is the fitting result.

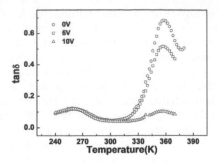

Figure 4. Temperature dependence of dielectric loss tanδ for irradiated samples at bias voltage of 0, 5, 10 V.

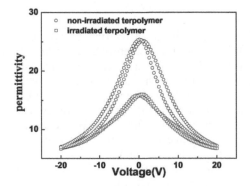

Figure 5. Dc electric voltage dependence of permittivity for non-irradiated terpolymer and irradiated terpolymer.

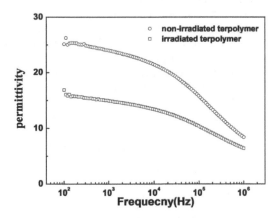

Figure 6. Frequency dependence of permittivity for non-irradiated terpolymer and irradiated terpolymer.

samples were measured and the results are shown in Figure 5. The present tunability is determined from the permittivity of the up curves of the ε–E. For non-irradiated sample, the tunability is 72%. This value is comparable with P(VDF-TrFE-CFE) fabricated with LB method[16]. After irradiation, the tunability decreases to 62%. The tunability results can be elucidated in terms of ε–E[17]:

$$\frac{\varepsilon_r(T,0)-\varepsilon_r(T,E)}{\varepsilon_r(T,E)}=1-[1+\varepsilon_0\varepsilon_r(T,0)^3 B(T)E^2]^{-1/3} \qquad (3)$$

where the left is the tunability, ε_0 is the vacuum permittivity, $\varepsilon_r(T,0)$ is the permittivity at temperature T without dc bias, $B(T)$ is the phenomenological coefficient of the fourth power of polarization that has a weak relation with the temperature, and E is the external dc bias field. The better tunability value of non-irradiated sample can be explained from the following two aspects: (1) Macroscopically speaking, according to Equation (3), the tunability is directly proportional to $\varepsilon_r(T,0)$. For the irradiated sample, the $\varepsilon_r(T,0)$ is smaller than non-irradiated sample as shown in Figure 6. Figure 6 shows the frequency dependence of the permittivity for both samples. It can be observed that the permittivity is smaller for irradiated sample in the whole frequency range. In detail, the permittivity measured at 1 kHz is 24 and 15 for non-irradiated and irradiated terpolymer. (2) Microscopically speaking, defects such as radicals, chain scission, and cross linking are easily introduced in terpolymer films after electron irradiation. Those defects results in an enhanced domain pining effect and thus reduces the tunability under a dc field. It seems irradiation had a negative impact on the tunability value of the terpolymer.

4   CONCLUSION

P(VDF-TrFE-CFE) thin film were prepared by sol-gel method and irradiated by electron with 112 Mrad dose. The permittivity and dielectric loss versus temperature were measured at different frequencies. Compared with the non-irradiated terpolymer, the relaxor feature is enhanced and an additional dielectric loss peak is observed for irradiated sample. The dielectric loss peak is suppressed with increasing dc bias and the temperature of loss peak is independent of dc bias. These results meet the characteristics of "defect mode". Besides, the permittivity versus bias electric field are studied. The tunability decreases for irradiated sample. It seems irradiation had a negative impact on the tunability value of the terpolymer.

ACKNOWLEDGMENTS

A project supported by Scientific research fund of Hunan Provincial Education Department (Grant No. 14B041) and by the Natural Science Foundation of Hunan Province (Grant No. 2015JJ6024 and No. 14JJ6040).

REFERENCES

[1] Han, Q.L.K. et al. 2014. High energy and power density capacitors from solution-processed ternary ferroelectric polymer nanocomposites. *Advanced Material* 26(36):6244–6249.
[2] Wang, Y. et al. 2007. Direct piezoelectric response of piezopolymer polyvinylidene fluoride under high mechanical strain and stress. *Applied Physics Letters* 91:222905.
[3] Neese, B. et al. 2008. Large electrocaloric effect in ferroelectric polymers near room temperature. *Science* 321(5890):821–823.
[4] Zhang, Q.M. et al. 1998. Giant electrostriction and relaxor ferroelectric behavior in electron-irradiated poly(vinylidene fluoride-trifluoroethylene) copolymer. *Science* 280(5372):2101–2104.
[5] Casalini, R. & Roland C.M. 2001. Highly electrostrictive poly(vinylidene fluoride–trifluoroethylene) networks. *Applied Physics Letters* 79:2627.
[6] Welter, C. 2003. Relaxor ferroelectric behavior of r-irradiated poly(vinylidene fluoride-trifluoroethylene) copolymers. *Physics Review B* 67:144103.
[7] Xu, Q. et al. 2013. Evolution of dielectric properties in $BaZr_xTi_{1-x}O_3$ ceramics: Effect of polar nanoregions. *Acta Materialia* 61(12):4481–4489.
[8] Tian, L. et al. 2015. Effects of Electron Irradiation on the Dielectric Behavior of Langmuir-Blodgett Terpolymer Films. *Ferroelectrics* 487(1):81–87.
[9] Meng X.J. et al. 2008. Low-temperature dielectric properties of Langmuir-Blodgett ferroelectric polymer films. *Journal of Applied Physics* 103:034110.

[10] Martirena, H.T. & Burfoot, J.C. 1974. Grain-size and pressure effects on the dielectric and piezoelectric properties of hot-pressed PZT-5. *Ferroelectrics* 7(1):151–152.

[11] Liu, P.F. et al. 2009. The Debye-like relaxation mechanism in poly (vinylidene fluoride-trifluoroethylene) ferroelectric polymers. *Journal of Applied Physics*, 106:104113.

[12] Ang C. et al. 2001. Dielectric loss and defect mode of $SrTiO_3$ thin films under direct-current bias. *Applied Physics Letters* 78:2754–2756.

[13] Meng X.J. et al. 2008. Low-temperature dielectric properties of Langmuir–Blodgett ferroelectric polymer films. *Journal of Applied physics* 103:034110.

[14] Meng XJ, et al. 2009. Threshold fields in the dc bias dependence of dielectric responses of relaxor ferroelectric terpolymer films. *Journal of Applied physics* 106:104102.

[15] Tian, L. et al. 2012. Relaxor properties and tunability of electron-irradiated thin poly(vinlidene fluoridetrifluoroethylene) copolymer film. *Advanced Materials Research* 560–561:892–898.

[16] Wang, J.L. et al. 2008. High electric tunability of relaxor ferroelectric Langmuir–Blodgett terpolymer films. *Applied Physics Letters* 93: 192905.

[17] Outzourhit, A. et al. 1995. Fabrication and characterization of $Ba_{1-x}Sr_{1-x}TiO_3$ tunable thin film capacitors. *Thin Solid Films* 259(2):218–224.

Influences of bottom heater arrangements on temperature and stress fields during initial cooling of Kyropoulos sapphire growth

S. Wang, M.Y. Wang, Z. Zhang & H.S. Fang
School of Energy and Power Engineering, Huazhong University of Science and Technology, Wuhan, P.R. China

ABSTRACT: In the paper, numerical computation is employed to study characteristics of heat transfer in a Kyropoulos furnace at the beginning of the cooling process. Comparative analyses are made to investigate effects of different heater locations on heat transfer and stress evolution in the sapphire crystal. The results show that the high von Mises stress is located in the region near the sapphire seed crystal, the crystal bottom center, and the crystal surfaces with the concave curvature. As the bottom heater gets closer to the support rod, temperature in the crystal increases obviously, in addition, the maximum von Mises stress transforms from the bottom center of the crystal to the 'bottleneck' with a downtrend. A small distance between the bottom heater and the supporter rod is helpful to reduce the stress avoiding cracks in the crystal ingot.

1 INTRODUCTION

Kyropoulos method is an important technique for the bulk-production of large-scaled and high-quality sapphire single crystals, which are widely used as substrate material in LED industry[1, 10]. In comparison with other methods, such as Heat Exchange Method (HEM)[11, 15], Czochralski (Cz)[12, 13], and edge-defined film-fed (EFG)[2, 14], sapphire grown by this method suffers low dislocation density, which is beneficial for diminishing lattice mismatch to improve the quality of GaN thin films during epitaxial growth. However, cracks induced by thermal stress, as well as high power consumption, are prominent issues during the growth.

Currently, there have a few researches on the thermal stresses during the Kyropolous process of sapphire. Demina et al.[4] developed a 2D axisymmetric model to study the Kyropoulos growth of sapphire crystal. They found that the S/L interface shape was influenced much by melt flow, which became unstable when the temperature gradient near the crystallization front was higher than a critical value. They also found that the maximal values of the von Mises stress are located in the regions near seed, along S/L interface, and bottleneck. Demina et al. further[5] study the 3D characteristics of melt convection of three certain growth stages and make a comparison with the reported results[4] of 2D modeling and experiments. The comparison results shown that maximal von Mises stress values are located in the same regions, however, the features of unsteady melt convection during the three stages were quite different from each other. Chen et al.[3] investigated the thermal and stress history of sapphire during the cooling process in a Kyropouls furnace. They concluded that the higher stress in the sapphire crystal is located in the regions near the seed, crystal centre, and the crystal surfaces with large curvature. They also found a sudden increase of the von Mises stress happened when the direction of heat flow changed, and a higher value of maximum von Mises stress was caused by the fast cooling process. Fang et al.[6] considered the effect of internal radiation at different growth stages of sapphire on S/L interface shape, melt convection, temperature and thermal stress. The result presented that internal radiation made temperature distribution more uniform, weakened natural and Marangoni convections in the melt, and diminished stress level in the crystal. They pointed out that the locally polycrystalline growth or the three-dimensionally non-uniform heating conditions are the most possible factors of crystal cracking. Fang et al. further[7] analyzed crack causes during the stable growth of sapphire crystal. Simulation results show that crystal shape has less effect on the thermal stress level. They declared that polycrystalline growth or/and other crystal defects must be another factor that made the crystal more fragile, in addition, the variation of heating condition during growth may result in high thermal stress locally that leads to the cracks at one side of the crystal. Jin et al.[9] discussed the influence of temperature-dependent thermophysical properties of sapphire on heater transfer and thermal stress in crystal during the cooling process. The results demonstrated that big deviations of temperature and stress occurred without proper applications of the thermophysical properties. They proposed that a temperature-dependent thermal expansion and thermal conductivity should be employed in the numerical computation related to thermal stress and heat transfer in sapphire crystal.

Although researches on thermal stress in sapphire single crystal by Kyropoulos method have been reported, there are few papers regarding the effect of the heater construction arrangement on thermal field and stress distribution during the Kyropoulos growth of sapphire crystal. Different arrangements of the bottom heater lead to different material consumption and energy loss, and have different influence on the stress in the crystal. The main aim of the paper is to study impacts of the locations of the bottom heater on the temperature and thermal stress distributions inner the crystal at the beginning of the cooling process, and to provide a basis for the improvement of the heater arrangement for reducing thermal stresses in crystal ingot.

2 PROCESS DESCRIPTION AND MATHEMATICAL MODELS

2.1 *System configuration*

Schematic diagram of a Kyropoulos furnace used in the present paper is shown in Figure 1. The furnace mainly includes a seed holder, a support rod, a crucible, heaters, heater shields, and ZrO_2-based thermal insulators. The heater shields are placed at the top, middle, bottom, and side of the furnace as the insulation system of the furnace with insulators to prevent heat loss into the surroundings. A 75-*Kg* sapphire ingot is grown in the crucible, 320 mm in inner diameter, 385 mm in height, and 25 mm in thickness. The sapphire crystal does not contact with the crucible during cooling process, which decreases the stress level and reduce the risk of cracks in the bulk crystal. 'L_0' stands for the distance between the bottom heater and the support rod. r_0 is the width of the bottom heater, which is constant with a value of 134 mm. Four typical locations marked by 'A', 'B', 'C' and 'D' are chosen to study the variations in

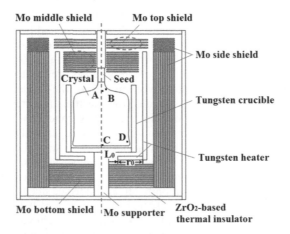

Figure 1.   Schematic diagram of a Kyropoulos furnace.

the von Mises stress and first principle stress with different locations of the bottom heater, as shown in Figure 4 and in Figure 5, respectively. The original construction of the resistance heater is shown in Figure 1. It is comprised of two parts, namely, the horizontal heater and the perpendicular heater, which are controlled independently with a constant power rate of 1:7 wherever the bottom heater locates. Heat is transferred only by conduction and radiation in the current vacuum furnace.

Thermal stresses come into being during the growth stage, release partially after the annealing, and piles up at the staring of the cooling process. In order to make a reasonable prediction of temperature and thermal stress, the temperature-dependent conductivity and nonlinear thermal expansion coefficient are adopted according to our previous research[9]. Sapphire properties can be found in the Ref.[6]. Some property parameters of other elements are listed in Table 1. In actual production, the configuration of a Kyropoulos furnace is not entirely axisymmetric, and a three-dimensional model is required for an accurate numerical analysis. However, the calculation including surface-to-surface radiation cost too much time for a transient study. Fortunately, boundary conditions for numerical computation, as well as the important components in the furnace are approximately symmetrical. Thus, a 2D axisymmetric model is adopted to save the computation load.

2.2   Mathematical models

A global steady model, considering long time annealing, is developed to calculate thermal field in the furnace and stress field in the sapphire crystal. To improve the computation efficiency, the following assumption are further employed: (1) all the components in the furnace are diffuse-gray body; (2) the sapphire single crystal is treated as opaque, and the Rosseland radiation model[9] is applied to consider the radiation effect. (3) the gravity is negligible compared to the thermal stress in the stress balance equations. (4) the thermophysical properties of the furnace elements are constant except the sapphire crystal. (5) the material properties of the sapphire crystal are taken as isotropic[3]. The governing equations can be written as follows.

Energy equation:

$$\nabla \cdot \left(k_i \nabla T \right) + \dot{q} = 0 \tag{1}$$

Table 1.   Property parameters of the furnace elements.

Properties	Value
Density, [g/cm³]	
Tungsten (heater and crucible)	19250
Molybdenum (shields and holder)	10280
ZrO$_2$-based insulator	6010
Heat capacity, [J/(Kg·K)]	
Tungsten (heater and crucible)	132
Molybdenum (shields and holder)	250
ZrO$_2$-based insulator	540
Thermal conductivity, [W/(m·K)]	
Tungsten (heater and crucible)	115
Molybdenum (shields and holder)	105
ZrO$_2$-based insulator	2.14
Emissivity	
Tungsten (heater and crucible)	0.3
Molybdenum (shields and holder)	0.28
ZrO$_2$-based insulator	0.2
Thermal expansion coefficient, [1/K]	2.137e-5 + 0.627e-8 ×
Sapphire (crystal)	T-0.134/T²

where the subscript i indicates the components of the furnace; k_i represent thermal conductivity of the material at the ith domain. $\dot{q}$ is the heat source, which is input power for the heater and is zero for other components. The diameter of the sapphire ingot in the simulation is about 0.29 mm, then, the optical thickness is nearly 5.58, which is greater than 2. To consider the radiation effect inner the sapphire ingot under high temperature, an effective thermal conductivity based on the Rosseland radiation model is employed, as shown in our previous paper[9]. To obtain the stress distribution, an axisymmetric displacement-based thermo-elastic stress model[8] is adopted, which can also be found in the Ref.[6].

To solve the above equations, the boundary conditions are described in the following. The exterior surface of the furnace, as well as the top of the seed holder, is set to room temperature, which is reasonable considering the efficient cooling by water. At the vacuum/solid interface, the temperature boundary conditions are:

$$-k_i \frac{\partial T_i}{\partial \bar{n}} = \sigma_B \varepsilon_i (T_i^4 - T_{env,i}^4) \tag{2}$$

$$T_{env,i} = \left(\frac{1}{\varepsilon_i F_i} \sum_{j=1}^{N} X_{ij} \varepsilon_j F_j T_j^4 \right)^{1/4} \tag{3}$$

where k_i is the thermal conductivity, $\bar{n}$ is the unit normal vector. ε_i, T_i, F_i, T_{envi} are the emissivity, temperature, area, and the effective environment temperatures corresponding to surface i, respectively. X_{ij} is view factor from area j to area i. Boundary conditions for stress computation in sapphire crystal are:

No traction boundary:

$$\bar{\sigma} \cdot \bar{n} = 0 \tag{4}$$

Fix constrain:

$$\bar{u} = 0 \tag{5}$$

where $\bar{\sigma}$, $\bar{n}$, $\bar{u}$, are stress tensor, the unit normal vector and displacement vector, respectively. At the crystal surface, no traction boundary condition is adopted, in addition, fix constrain is also applied on the top of the seed. At the axis, an axisymmetric condition is imposed for all variables.

3   RESULTS AND DISCUSSION

Figure 2 shows the characteristics of heat transfer in the furnace and the stress distribution inner the sapphire ingot at the beginning of the cooling process. In a Kyropoulos furnace, heat flux at the vacuum/solid interfaces is transferred only by radiation, which plays an important role especially in high temperatures. At the beginning of the cooling, the highest temperature is close to the center of the bottom heater. Heat flux in the sapphire crystal is not in a single direction. Most of the heat flux flows upward into the seed rod with lots of heat, however, only a small part of that leaves the bottom center of the crystal into the top center of the bottom crucible. The maximum von Mises stress, with a value of 29.4 MPa, appears at the bottom center of the crystal, due to higher temperature gradient caused by larger temperature difference between the crystal bottom center (point 'C', in Fig. 3) and the top center of the bottom crucible (point 'E', in Fig. 3). The isothermal becomes closer near the seed, due to a shrinking cross-section and an enhancement of heat transfer inner the sapphire ingot caused by the internal radiation effect. The intensive isothermal means the higher temperature gradients, which leads to extreme thermal stress in the regions. The higher von Mises stresses also appears in regions with concave surfaces, such as the 'bottleneck' and the sidewall of the crystal near the corner, because a concave surface is apt to cause the higher stress concentration.

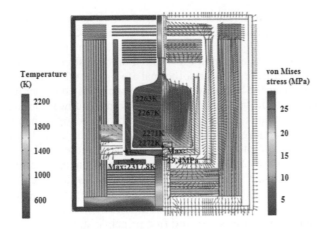

Figure 2. Distributions of temperature (left), von Mises stress (in the sapphire ingot), and heat flux vector (red arrows) at the beginning of the cooling process.

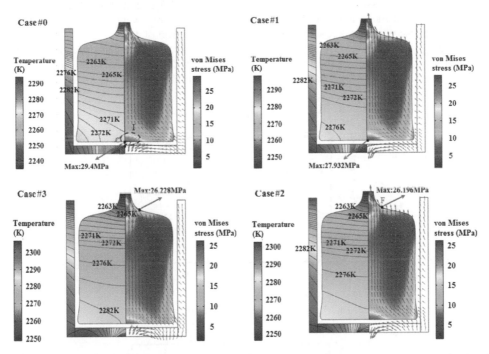

Figure 3. Distributions of temperature (left), von Mises stress (right), heat flux vector (red arrows) and isothermal (marked by numbers) at different locations of the bottom heater: case#0: L_0 = 48 mm, case#1: L_0 = 36 mm, case#2: L_0 = 24 mm, case#3: L_0 = 12 mm (temperature difference in the crystal is less than 2K).

The predicted distributions of thermal field, stress field at different L_0 are presented in Figure 3. It should be noted that data in case#0 is taken from Figure 2 by only showing the crucible and the crystal. As the bottom heater gets closer to the supporter rod, as shown in case#1, the isothermal inside the crystal moves upward entirely, which means the temperature in the crystal increases and more heat flows into the crystal. Part of the heat flux in the region I flows upwards, due to an enhancement of heat flux near the bottom of the crystal. The maximum von Mises stress still concentrates on the bottom centre of the crystal, however, it has fallen by 5%. As the bottom heater moves on, as shown in case#2, the isothermal continues moving up, and temperature in the crystal further increases. It can be clearly found that the temperature near point 'E' goes up and the position of the maximum stress in the crystal transfers from the bottom center to point 'F', with a reduced stress level. In case#3, the variation trend for the temperature and the isothermal are similar with that in case#2. Although the heat flux direction near the bottom center of the crucible is no longer horizontal, instead, pointing downward, the heat flux in region I are still pointing upward in the sapphire crystal, which indicates that the heat supplied by the bottom heater is larger than the heat extracted by the supporter rod, therefore, more heat flux enters into the crystal, which enhances the temperature gradients at point 'F', leading to a 0.12% increase of the maximum value in comparison with that in case#2. One can also find that the maximum stress level reaches up to 22.268 MPa, which is 24.26% smaller than that in Case#0, however, temperature in the sapphire crystal is higher than that in Case#0, which hints that under the same environments, power consumption in Case#3 must be lower than that in Case#0.

The variations in the von Mises stress and first principle stress with different locations of the bottom heater are traced at four positions shown in Figure 1. Figure 4 shows the variation of the von Mises stress with L_0 at different positions. It can be found that, as the bottom heater gets closer to the supporter rod, the von Mises stresses at point 'A' and 'B' almost have no changes. In contrast, the stresses at point 'C' and point 'D' decline linearly and attain the minimums with noticeable changes of 18.4% and 12.7%, respectively, since the temperature gradient at these positions keep decreasing with the reduction of L_0.

Figure 5 presents the variation of the first principle stress with L_0 at different positions. One can see that the value of the first principal stress at different positions are always positive, which implies that stress at these positions are always under tensile stress. As L_0 increases,

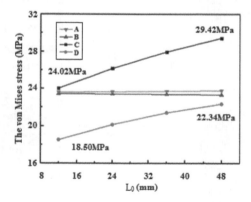

Figure 4.   Variation of the von Mises stress with L_0 at different locations.

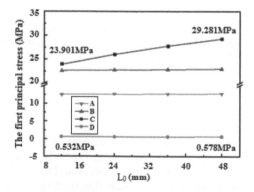

Figure 5.   Variation of the first principle stress with L_0 at different locations.

620

the tensile stress at position A, B, and D nearly keeps invariable, however, at position C, it goes up linearly to 29.281 MPa, which is far away the tensile strength limit of the sapphire $(2.73–4 \times 10^8)$[3], thus, there may be no cracking in sapphire crystal. The tensile stress at point C is always the largest, where cracks could be easier to happen if the crystal bottom comes in contact with the crucible. From the view of improving the quality of the crystal ingot, the heater arrangement in case#3 is helpful to avoid cracks in the crystal. In addition, 28% tungsten consumed for the bottom heater in Case#0 can be saved if Case#3 is adopted, which is very attractive for the actual production to lower costs.

4 CONCLUSIONS

In the paper, a steady-state model is developed to study the heater arrangements on the thermal field and stress distribution in sapphire single crystal by Kyropolous method. The following conclusion can be drawn:

1. At the beginning of the cooling, the higher von Mises stress is located in the region near the sapphire seed crystal, the crystal bottom center, and the crystal surfaces with the concave curvature.
2. In the case of the original heater construction, heat fluxes in the sapphire crystal are not in a single direction. Part of heat flux pass through the bottom center of the crystal, where the stress get a higher level, due to more heat taken away by the crucible support rod than the heat supplied by the bottom heater.
3. As the bottom heater gets close to the supporter, the direction of the heat flux near the bottom center of the crystal changes from downwards to upwards obviously, and the position of the maximum von Mises stress transfers from the bottom center to the 'bottleneck', in addition, the first principle stress level at the bottom center is always the largest and decreases linearly, which is far less than the tensile strength limit.
4. From the view of avoiding cracks in the crystal to improve the crystal quality, a closer value of L_0 is more beneficial, which is also helpful to reduce power and material consumption.
5. Low-cost and high-quality sapphire single crystal can be achieved by a reasonable arrangement of the bottom heater and appropriate power adjustment.

ACKNOWLEDGEMENTS

The work is supported by the National Natural Science Foundation of China (No. 51476068).

REFERENCES

[1] Akasaki, I. 2007. Key inventions in the history of nitride-based blue LED and LD. *Journal of Crystal Growth* 300: 2–10.
[2] Borodin, A.V. et al. 1999. Simulation of the pressure distribution in the melt for sapphire ribbon growth by the Stepanov (EFG) technique. *Journal of Crystal Growth* 198/199: 220–226.
[3] Chen, C.H. & Chen, J.C. et al. 2014. Thermal and stress distributions in larger sapphire crystals during the cooling process in a Kyropoulos furnace. *Journal of Crystal Growth* 385: 55–60.
[4] Demina, S.E. & Bystrova, E.N. et al. 2008. Use of numerical simulation for growing high quality sapphire crystals by the Kyropoulos method. *Journal of Crystal Growth* 310: 1443–1447.
[5] Demina, S.E. & Kalaev, V.V. 2011. 3D unsteady computer modeling of industrial scale Ky and Cz sapphire crystal growth. *Journal of Crystal Growth* 320: 23–27.
[6] Fang, H.S. & Jin, Z.L. et al. 2013. Role of internal radiation at the different growth stages of sapphire by Kyropoulos method. *International Journal of Heat and Mass Transfer* 67: 967–973.
[7] Fang, H.S. & Wang, S. et al. 2013. Crystal cracking analysis and three-dimensional effects during Kyropoulos sapphire growth. *Crystal Research and Technology* 48: 649–657.
[8] Fainberg, J. & Leister, H.J. 1996. Finite volume multigrid solver for thermo-elastic stress analysis in anisotropic materials. Comput. *Methods Appl. Mech. Eng.* 137: 167–174.

[9] Jin, Z.L. & Fang, H.S. et al. 2014. Influence of temperature-dependent thermophysical properties of sapphire on the modeling of Kyropoulos cooling process. *Journal of Crystal Growth* 405: 52–58.

[10] Jeong, S.M. et al. 2010. Characteristic enhancement of the blue LED chip by the growth and fabrication on patterned sapphire (0001) substrate. *Journal of Crystal Growth* 312: 258–262.

[11] Lu, C.W. & Chen, Jyh. C. 2001. Numerical computation of sapphire crystal growth using heat exchanger method. *Journal of Crystal Growth* 225: 274–281.

[12] Lu, C.W. & Chen, J.C. et al. 2010. Effects of RF coil position on the transport processes during the stages of sapphire Czochralski crystal growth. *Journal of Crystal Growth* 312: 1074–1079.

[13] Tavakoli, M.H. 2008. Numerical study of heat transport and fluid flow during different stages of sapphire Czochralski crystal growth. *Journal of Crystal Growth* 310: 3107–3112.

[14] Yu, Q.H. & Liu, L.J. et al. 2014. 3D numerical investigation and improvement to the design of the thermal field before seeding in a multi-die edge-defined film-fed growth system for sapphire ribbon crystals. *Journal of Crystal Growth* 385: 49–54.

[15] Zhang, N. & Park, H.G. et al. 2013. Simulation of heat transfer and convection during sapphire crystal growth in a modified heat exchanger method. *Journal of Crystal Growth* 367: 27–34.

The surface ring coarse grain structure and control measures of ultra-low carbon steel hot-rolled wire

K. Wang & J.M. Zhang
State Key Laboratory of Advanced Metallurgy, University of Science and Technology Beijing, Beijing, China

L.F. Wang & N.B. Lv
Shou Gang Research Institute of Technology, Beijing, China

ABSTRACT: The causes of surface ring coarse grain structure of hot-rolled low carbon steel wire rod were analyzed in this paper. The ring depth of coarse grain structure reached about 1000 μm and the grain size reaches 60–80 μm on the wire rod surface evenly, no specific orientation abnormal grain growth characteristics. Based on the annealing simulation analysis, we found that a hereditary surface ring coarse grain structure would seriously deteriorate the plastic properties. Based on the theoretical analysis and experiment, the main cause of the ring coarse grain is that the pre-finishing intensity is too large, the temperature lower than the dynamic phase transition temperature, lots of energy store in the wire rod surface grain, which promote the ferrite phase transformation and the specific orientation of the grain growth, eventually forming the coarse grain structure. By adjusting the water-cooled pre-finishing process, the finishing temperature controls at above 920°C on the ultra-low carbon steel wire rods, which can significantly reduce the occurrence of coarse grain structure.

1 INTRODUCTION

With the rapid development of high quality metal products industry, the users' production technology and equipment have made great progress and they have higher requirements on quality of steel raw materials. Ultra-low carbon steel hot-rolled wire rod is mainly used to process manufacturing applications in the communications and electronic information industry such as copper, tin and zinc galvanized clad steel wire. Its terminal products are for all types of communications cables, shielded wire and capacitance pointers. Downstream users of ultra-low carbon steel hot-rolled wire require its good drawing performance, cleanness and rod uniformity. It also needs to meet the requirements of its broken index of less than 1.5 tons of steel. For this reason the causes of hot-rolled low carbon steel wire rod coarse grain structure are studied and the measures and results are analyzed[1]-[2].

2 TYPES OF COARSE GRAIN

Microstructure analysis found the presence of coarse grain on surface of the ultra-low carbon steel hot rolled rod. The coarse grain exists on the whole rod surface except the two heads. Other parts of the rod have normal equiaxed ferrite grains with a grain size of 8. Size of coarse grain is 7 with a depth close to 1 mm and the grain size of 60–80 μm which is larger than that of the normal grain (20–30 μm). Coarse grain is uniformly distributed on the rod surface and the grain size is relatively uniform. There is no particular orientation of the crystal grains abnormal growth as shown in Figure 1.

Figure 1.   The surface ring coarse grain microstructure of the ultra-low carbon steel wire rods (a) over-all morphology (b) local morphology.

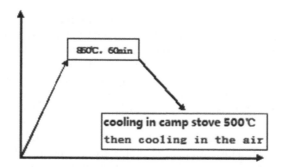

Figure 2.   The annealing process of the ultra-low carbon steel hot-rolled wire rod.

According to many samples test, this surface annular coarse grain rod is difficult to find in either of the heads but concentrated in the position of the rest of the rod surface. During the high-line production, in order to prevent the head hold, wire rod head is generally not applied water cooling technology, but the other part is under water cooling. The surface morphology and distribution of coarse grain may be caused by the fact that finishing temperature is too low and a high temperature difference between pre-cooled after finishing rod surface and center.

3   GENETIC ANALYSIS OF COARSE GRAIN STRUCTURE

Annealing experiments are carried out for the rod as to analyze whether the organization has a coarse-grained genetic character. Coarse grains' deterioration effect in its steel processing and heat treatment process is also studied. Experimental craft is as shown in Figure 2.

The ultra-low carbon hot rolled rod with a coarse grain structure is heated to 850 °C, held for 60 min, furnace cooled to 500 °C and air cooled to room temperature. After annealing the mechanical properties test and tissue analysis is conducted. Rm is found only 260 Mpa and fracture morphology of irregular shape is shown in Figure 3 (a). Micro-structure near the fracture showed rod surface appeared in a circle of coarse grains, even the rod edge has grown into a large grain as 3 (b) below. Tensile fracture showed signifi-cant linear shape and the area corresponds exactly to normal tissue organization in the center. Correspondence between the coarse grain structure after and before annealing illustrates this with a certain genetic and it would seriously deteriorate the plastic material after annealing.

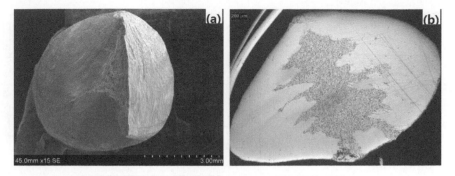

Figure 3.   Annealed low carbon steel wire rod (a) tensile fracture (b) the abnormal coarse grain microstructure near the cross-section fracture.

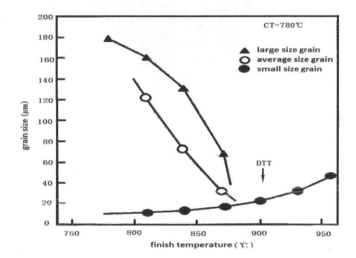

Figure 4.   The trends of surface coarse grain with finishing temperature.

4   CAUSES OF COARSE GRAIN

The classic surface coarse grain theory is that in the austenite single phase region as finishing temperature decreases, the grain is refined. When finishing temperature is below the rod dynamic phase transition temperature (two-phase region rolling), deformation will cause the surface to store a large amount of energy, which on the one hand to promote the ferrite phase transformation occurs, on the other hand, according to the principle of the lowest energy, causes a particular orientation of the grain growth and expansion. Deformation, while the two-phase region intensified surface deformation gradient, can cause an increase in deformation grain boundary migration. Because of grain boundary migration and energy consumption the grain gets coarse. In this process, the initial surface temperature is low, the heat from the wire rod center will pass to the surface, causing the surface to warm up and increase the extent of this phenomenon[3]-[8], as shown in Figure 4.

5   EFFECT OF COARSE GRAIN CONTROL

According to form hot-rolled low carbon steel wire rod surface coarse grain structure analysis combined to coarse-grained surface formation theory, the solution for ultra-low carbon hot-rolled wire rod surface coarse grain ring is temperature control after pre-finishing.

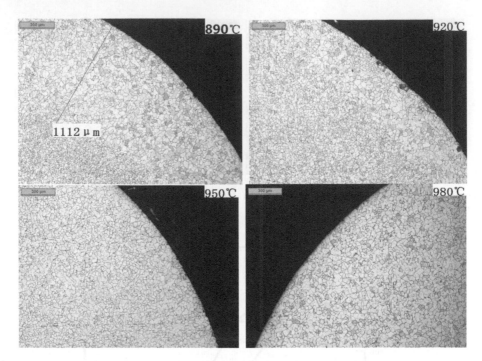

Figure 5.   The surface coarse-grained layer depth varies with finishing temperature.

According to the production characteristics of the High Line, the pre-finishing and finishing temperature water cooling technology is an important means of spinning temperature control. By adjusting the water-cooled pre-finishing process, various finishing inlet temperature control at a temperature range of 890–980 °C are carried out and the morphology of the surface characteristics of four kinds of crude crystals of the process scheme are shown in Figure 5. It is found significant coarse grain surface at a temperature range of 890–920 °C, however, when the finishing temperature rises to 950 °C, the surface of coarse grain disappeared. The lower the finishing temperature, the deeper the depth of the surface layer of coarse grain is. When the finishing temperatures is below 920 °C, the coarse grain structure depth in ultra-low carbon hot-rolled wire rod surface can reach to more than 1000μm. It can be induced that the depth of the coarse grain structure changing trend inflection point temperature is 920 °C.

6   CONCLUSIONS

1. Surface coarse grain structure is found in ultra-low carbon steel hot-rolled wire rod with depth of about 1000 μm and grain size of 60–80 μm. Coarse grain evenly distributes in the rod surface and there is no specific orientation of abnormal grain growth characteristics.
2. Annealing simulation analysis found that surface coarse grain structure has certain genetic and can cause uneven wire cross-section radial organization which would seriously deteriorate the plastic processing properties of the material after annealing.
3. Causes of coarse grains are that after finishing the pre-cooled intensity is too large, the temperature is not fully recovered, thus, the disk surface temperature is below the dynamic phase transition temperature and the rod surface has a lot of energy to promote the ferrite phase transition. The specific orientation of the grains grow and expand eventually forming a coarse grain structure.
4. To adjust the water-cooling in pre-finishing process and to fix the finishing temperature over 920 °C can significantly reduce the occurrence of coarse grain.

REFERENCES

[1] X.G. Fan, H. Yang, Z.C. Sun, D.W. Zhang. 2010. Effect of deformation inhomogeneity on the microstructure and mechanical properties of large-scale rib–web component of titanium alloy under local loading forming. *Materials Science and Engineering: A*, Volume 527, Issues 21–22, Pages 5391–5399.

[2] Z.C. Sun, H. Yang, G.J. Han, X.G. Fan. 2010. A numerical model based on internal-state-variable method for the microstructure evolution during hot-working process of TA15 titanium alloy. *Materials Science and Engineering: A*, Volume 527, Issue 15, Pages 3464–3471.

[3] Jiao Luo, Miaoquan Li, Xiaoli Li, Yanpei Shi. 2010. Constitutive model for high temperature deformation of titanium alloys using internal state variables. *Mechanics of Materials*, Volume 42, Issue 2, Pages 157–165.

[4] C. Miki, K. Homma, T. Tominaga. 2002. High strength and high performance steels and their use in bridge structures. *Journal of Constructional Steel Research*, Volume 58, Issue 1, Pages 3–20.

[5] Wang S.H., Luu W.C., Ho K.F., Wu J.K. 2003. Hydrogen permeation in a submerged arc weldment of TMCP steel. *Materials chemistry and physics*. vol. 77, no2, pp. 447–454.

[6] Jinbo Qu, Yiyin Shan, Mingchun Zhao, Ke Yang. 2001. Influence of Hot Deformation and Cooling Conditions on the Microstructures of Low Carbon Microalloyed Steels. *JOURNAL OF MATERIALS SCIENCE & TECHNOLOGY*, 17(z1).

[7] Ji Hyun Kim, Young Jin Oh, Il Soon Hwang, Dong Jin Kim, Jeong Tae Kim. 2001. Fracture behavior of heat-affected zone in low alloy steels. *Journal of Nuclear Materials*, Volume 299, Issue 2, Pages 132–139.

[8] Zhao Mingchun, Xiao Furen, Shan Yiyin, Li Yuhai, Yang Ke. 2002. Microstructural characteristic and toughening of an ultralow carbon acicular ferrite pipeline steel. *Acta Metallurgica Sinica*, vol. 38, no3, pp. 283–287.

Advances in Engineering Materials and Applied Mechanics – Zhang, Gao & Xu (Eds)
© 2016 Taylor & Francis Group, London, ISBN 978-1-138-02834-0

Effects of the temperature on characteristic parameters of the entire coal spontaneous combustion process under poor oxygen condition

H. Wen, J. Guo, Y.F. Jin, Z.F. Zhao & J.B. Fei
College of Energy Science and Engineering, Xi'an University of Science and Technology, Xi'an, China
Key Laboratory of Western Mine and Hazard Prevention, Ministry of Education of China, Xi'an, China

ABSTRACT: Temperature-programmed experiments were conducted from 20 to 450 °C in order to investigate characteristic parameters of coal spontaneous combustion at various oxidation concentrations (21%, 17%, 9%, 5%, and 3%). The experiments indicated that temperature played dominant role during the spontaneous combustion of coal. Higher temperatures would lead to larger oxidation reaction rates that represented the same trends for different oxidation concentrations. As the temperature rose, the concentrations of CO and CO_2 released from spontaneous combustion were increased at the beginning and decreased after exceeding the maximums, roughly between 370 and 430 °C. What's more, hysteresis phenomenon existed with the deduction of oxygen concentrations. So, the temperature would be higher when same CO/CO_2 outputs were reached. In addition, with the increase of temperature, the production of hydrocarbons increased exponentially. However, after exceeding 370–430 °C, temperatures of C_2H_4 and C_2H_6 were decreased but the temperature of CH_4 was increased continuously. The experiments also implied that temperature had negligible effect on the output of hydrocarbons at various oxidation concentrations.

1 INTRODUCTION

Temperature is one of the most important factors influencing coal oxidation and spontaneous combustion. Higher temperature leads to more violent spontaneous combustion. In the development process of coal fire area, the temperature of coal can increase up to several hundred degrees during the formation of spontaneous combustion in coal field. The temperature on the combustion center area can even reach to one thousand degrees[2,13]. With further extending to underground, the oxygen concentration decreased gradually from peripheral fire area to combustion center. For underground coal mines, goaf, high collapsed zone of roadway, and sealed area are believed to be dangerous. Compared with ordinary atmosphere, the oxygen concentrations in these areas are smaller. Therefore, it is of importance to investigate the effects of temperature on coal spontaneous combustion in situ where the oxygen concentration is low.

Several types of experimental systems about coal spontaneous combustion have been established and applied to simulate coal spontaneous combustion process and measure the characteristic parameters since the 1980s[4,9,10,14,15]. Oxygen consumption rate and dynamic characteristics of the high-temperature point were analyzed by Wen and Xu[6,11] based on the simulation results obtained from specially designed coal spontaneous combustion system. Zhang [16] studied the influence law of coal sample size and temperature on coal spontaneous combustion characteristic parameters through oil bath temperature programmed experiment system. Through loose coal temperature programmed experiments.

At present, although some achievements have been made on the research of coal spontaneous combustion characteristics under the condition of variable oxygen concentration, the most of which just stay at low temperature stage (lower than 200 °C), and only for determination of coal natural low temperature oxidation phase characteristic parameters, to provide theoretical support for Mine fire prevention and control. What's more, it is relatively little that research on coal spontaneous combustion characteristics under Variable oxygen concentration. Therefore, in this paper some experiments under variable oxygen concentration will be took by using XKGW-1 type coal combustion high temperature programmed device that independently researched and developed by a group. The change rule of characteristic parameters during coal spontaneous combustion process under different oxygen concentrations will have a certain guiding significance for the prevention and control of coal fire area and mine fire.

2 EXPERIMENTAL

2.1 *Experimental device*

Experiment equipment mainly includes: source gas (air pump and high pressure gas cylinder filled with different oxygen concentrations), XKGW-1 type high temperature reactor, home-made coal sample tank, gas chromatograph, etc., as shown in Figure 1.

For this experiment equipment, the vacuum atmosphere insulation structure and high temperature resistant ceramic fiber insulation materials are adopted; the maximum temperature of 1200 °C could be provided. The PID temperature control method based on artificial intelligence is adopted, to improve the stability of the heating process. In order to reduce the experimental error caused by undersize quantity of coal as far as possible, the large amount of 7.5 kg coal sample were used in experiment. In order to reduce the gas supply deviation inside the coal of coal sample tank coaxial plane, and improve the stability of body airflow inside coal tank, the annular gas path shape was adopted in the in and out of the trachea.

2.2 *Process and conditions*

The long-flame coal was used in this experiment, its industrial analysis data are shown in Table 1. The external surface of massive coal got stripped, coal core got crushed and screened into five kinds of particle size: 0~0.9 mm, 0.9~3 mm, 3~5 mm, 5~7 mm and 7~10 mm in air environment. The particle size of above-mentioned coal mixed according to the mass ratio of 1:1 into 7.5 kg coal sample, then loading it into coal sample tank. Setting the instrument's parameters, and taking experiment accord with the requirements in strictly. Supplying gas into coal sample tank for 30 min until the outlet gas composition is stable before the start of the heating, its range from 20 °C to 450 °C. When the coal heated to every integer multiples of 10 °C, extracted a gas sample, then its composition are analyzed and the data are recorded.

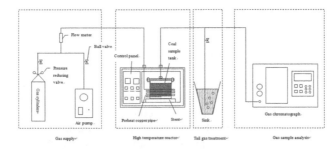

Figure 1.   Experiment device of coal fire area characteristic parameters.

Table 1.   The industrial analysis and elemental analysis of the coal sample.

Industrial analysis (%)				Elemental analysis (%.$_{daf}$)					
M_{ad}	A_d	V_{daf}	F_{cad}	C	H	O	N	S	Rank
14.90	8.38	69.67	7.06	76.32	5.02	16.25	1.01	2.4	Candle

Five kinds of oxygen concentration gas (21%, 17%, 9%, 5% and 9%) are used as gas source, of which air is used as 21% oxygen concentration gas and other four gases is made up of O_2 and N_2. The quantity of gas is controlled into 1300 ml/min by the flow meter, and the heating rate of high temperature reactor is set to 0.38 °C/min.

3 RESULTS AND ANALYSIS

3.1 *The effects of temperature on the oxygen consumption rate*

In the high temperature heating experiment, the oxygen concentration curve of coal sample tank outlet is shown in Figure 2 (a). The oxygen consumption rate of coal samples could be calculated by related formula which studied by the fire preventing and extinguishing team of XUST [12].

$$V_{O_2}^0(T) = \frac{Q \cdot C_{O_2}^0}{S \cdot L} \cdot \ln \frac{C_{O_2}^1}{C_{O_2}^2}$$

In this equation, $V_{O_2}^0(T)$ is the oxidation rate of coal samples (mol/(cm³·s)), Q is the air volume (ml), S is the cross-sectional area of coal sample tank (mm), L is the coal height (cm), $C_{O_2}^0$ is the oxygen concentration of fresh air (21%), and $C_{O_2}^1$ and $C_{O_2}^2$ are the oxygen concentration at the inlet and outlet of coal sample (%), $C_{O_2}^1 = C_{O_2}^0$.

Through the relevant experimental data are rewritten to the above formula, the curves of the oxygen consumption rate with coal temperature could be worked out, as shown in Figure 2(b).

As shown in the Figure 2, there is an obvious boundary appears near at 150 °C about the heating rate of coal samples, that's say: the oxygen consumption rate of coal samples is in a process of gradually increase below 150 °C, especially under 50 °C, the variation tendency of the oxygen concentration curve is not obvious. The oxygen consumption rate began to rapidly increase and oxygen concentration rapidly reduced above 50 °C. After 150 °C, the oxygen consumption rate reached a very high state, and showed a trend of slow increase, export oxygen concentration is at the state of very low (<1%), the lowest was 0.1%.

This is because, in the low temperature stage, coal spontaneous combustion is very inactivity, which changes from physisorption to chemical adsorption, until finally produces chemical reaction, its reaction intensity increased gradually. With the rise of coal temperature, the temperature reaches a spontaneous combustion critical temperature point, the intensity of reaction began to rapidly increase, and the trend of the oxygen consumption rate was significantly increased. When the coal temperature reached about 150 °C, the reaction of coal and oxygen began to be acute, oxygen consumption rate shown a trend that persistently slowly increase. Maybe the oxygen consumption rate curve fluctuates greatly because of drastic reaction and the model for the oxygen consumption rate calculation. Overall, the change trend of the oxygen consumption rate of coal samples, which were under different oxygen concentrations, was same. This shows that the law of temperature's influence on the oxygen consumption rate of coal samples under the condition of different oxygen concentration is same.

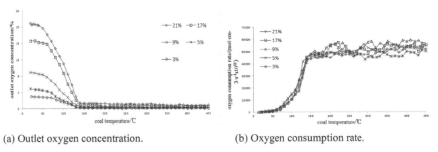

(a) Outlet oxygen concentration. (b) Oxygen consumption rate.

Figure 2. Curves of outlet oxygen concentration and oxygen consumption rate within the whole process of coal spontaneous combustion.

3.2 *The effects of temperature on the gaseous product*

The curve of the gaseous product quantity with coal temperature within coal sample oxidized spontaneous combustion whole process is shown in Figures 3~4. For convenience, Defined φ (CmHnOx) as its concentration, T (CmHnOx) as the corresponding temperature when it is being in some concentration, CO and CO_2 are known as the carbon and oxygen compounds (COx), CH_4, C_2H_4 and C_2H_6 signified with Hydrocarbons (CmHn). After processing and calculation on the experimental data, gaseous product production order under different oxygen concentration experiments are $\varphi(CO_2) > \varphi(CO) > \varphi(CH_4) > \varphi(C_2H_6) > \varphi(C_2H_4)$, temperature size order of gas generation sharp change are $T(C_2H_6) \approx T(C_2H_4) > T(CH_4) > T(CO) > T(CO_2)$. This shows that the amount of oxidation products is greater than the pyrolysis products', in the process of coal oxidation spontaneous combustion, and the former can generate at low temperature stage, while the latter is the emergence of a higher temperature. At the same time, due to the chemical stability of the product, the higher the stability of the product, the quantity is larger.

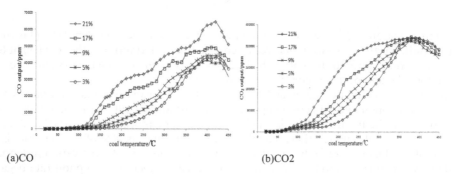

(a)CO

(b)CO2

Figure 3. The change curves of COx generation along with coal temperature within coal spontaneous combustion whole process under different oxygen concentrations.

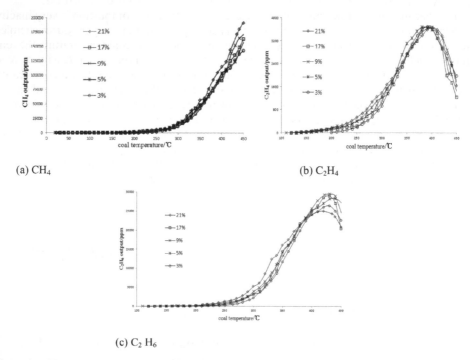

(a) CH_4

(b) C_2H_4

(c) $C_2 H_6$

Figure 4. The change curves of CmHn generation along with coal temperature within coal spontaneous combustion whole process under different oxygen concentrations.

3.2.1 *The carbon and oxygen compounds (COx)*

The change curves of carbon and oxygen compounds generation along with coal temperature within coal spontaneous combustion whole process under different oxygen concentrations are shown in Figure 3.

As shown in the Figure 3, the variation tendency of COx quantity with the increase of coal temperature under the conditions of different oxygen is basically the same, but the influence degree different. That is to say, CO production under different oxygen concentration showed a trend of rise after the fall of first, and a crest value appears near at 370~430 °C. The influence degree of the coal temperature on COx along with the increase of air supplying oxygen concentration gradually increases, which showed hysteresis on production along with the decreasing of oxygen concentration generally. That is to say, the sharp change temperature of COx amount lags, and the temperature rises when the same amount of product generated, the amount of the COx decreases obviously at the same temperature.

This is mainly because, COx is mainly generated by coal oxidation, and the oxygen concentration is a very critical factor for COx production. Moreover, with the increase of coal temperature, the chemical reaction of coal and oxygen becomes sharper, the influence degree of oxygen concentration on COx is more obvious.

3.2.2 *Hydrocarbons (CmHn)*

The change curves of Hydrocarbons (CmHn) generation along with coal temperature within coal spontaneous combustion whole process under different oxygen concentrations are shown in Figure 4.

As shown in the Figure 4, temperature's influence on CmHn generation under different oxygen concentration is slightly different, but the influence degree is basically same. That is to say, along with the rise of coal temperature, CmHn generation shown a tendency of exponential growth, but the amount of C_2H_4 and C_2H_6 decreases after it reach to the peak near in 370 ~ 430 °C while CH_4 increased all along. Along with the increase of supplying air oxygen concentration, influence effect of coal temperature on CmHn generation is not obvious. That is to say, at the same temperature the amount of CmHn generated by coal sample under different oxygen concentrations is similar, without obvious regularity.

This is mainly because [1,3,7,8], the chemical reaction of coal sample is very slow in the low temperature phase, CH_4 is mainly generated for desorption of coal. With the increase of coal temperature, chemical reaction of coal became sharply acute, CH_4 production increasing. When the coal temperature reaching a certain temperature stage gradually, coal sample began thermal decomposition under the conditions of oxygen, which gradually increased. A variety of functional groups in the molecule of coal, like ether bond, hydroxyl, carboxyl, methyl side chain, fat side chain, etc., generated kinds of hydrocarbon gas, reached the maximum in 370~430 °C temperature range. Over that temperature, hydrocarbon gas productions showed a trend of decline because of the shrinking of the corresponding functional groups. However, there is different optimum pyrolysis temperature with variety of functional groups, the corresponding temperature for peak of gaseous product's amount is different. Because CmHn is mainly produced by coal pyrolysis function, and pyrolysis reaction is very complex under the conditions of oxygen, there is a certain impact on amount of CmHn with the oxygen concentration, but it is not obvious.

4 CONCLUSIONS

The changing laws of a variety of macroscopic parameters in the whole process of coal oxidation spontaneous combustion under the conditions of five different oxygen concentrations were obtained by using the high temperature programmed heating experiment. And the influence of temperature on parameters was analyzed, its temperature range is from 20 to 450 °C, mainly including:

1. Temperature is the important factors affecting coal spontaneous combustion, the higher the temperature, the more acute on spontaneous combustion, and the oxidation rate of coal is higher. The trend of oxygen consumption rate under different oxygen concentration

along with temperature rise is same, and there is an obvious boundary temperature point near at 150 °C.

2. The influence of the temperature on carbon and oxygen compounds is obvious. The carbon and oxygen compounds yield under different oxygen concentration shown a trend of lower after the first rise with coal temperature rise, the yield peak appeared at 370~430 °C, and shown a hysteresis trend along with the oxygen concentration decreased. That is to say, the sharp change temperature of COx amount lags, and the temperature increases when the same amount of product generated, the amount of the COx decreases obviously at the same temperature.

3. The temperature's influence on CmHn generation under different oxygen concentrations is slightly different, but influence degree is basically same. Along with the rise of coal temperature, CmHn generation shown a tendency of exponential growth, but the amount of C_2H_4 and C_2H_6 decreases after it reach to the peak near in 370~430 °C while CH_4 increased all along. At the same temperature, the amount of CmHn generated by coal under different oxygen concentration is similar, without obvious regularity.

ACKNOWLEDGEMENT

This paper was supported by the Key Program of National Natural Science Foundation of China (No. 51134019), the National Natural Science Foundation of China (No. 51404195), and the Postdoctoral Science Foundation of China (No. 2014M552466).

REFERENCES

[1] Du Juan. 2008. Forming and releasing of gaseous products of coal during pyrolysis in Western China. *Journal of China University of Mining & Technology* 37(5):694–698.

[2] Jin Yong-fei. 2013. *Study on the dynamic evolution mechanism and control of coal seam outcrop fire.* Xi'an: xi'an university of science and technology.

[3] Jin Yong-fei. 2015. Experimental study on the high temperature lean oxygen oxidation combustion characteristic parameters of coal spontaneous combustion. *Journal of China coal society* 40(3):596–602.

[4] Sahay N.V. 2007. Critical temperature an approach to define proneness of coal towards spontaneous heating. *Journal of Mines, Metal sand Fuels* 55(10):510–516.

[5] Wen Hu. 2004. Experiment simulation of whole process on coal self-ignition and study of dynamical change rule in high-temperature zone. *Journal of China coal society* 29(6):689–693.

[6] Wu Shi-sheng. 2012. Product distribution and reactivity of coal pyrolysis at high temperature and in atmospheres containing O_2/steam. *Journal of Fuel Chemistry and Technology* 40(6):660–665.

[7] Xiong R. 2010. Fundamentals of coal topping gasification: characterization of pyrolysis topping in a fluidized bed reactor. *Fuel Process Technology* 91(8):810–817.

[8] Xiao Y. 2008. Research on correspondence relationship between coal spontaneous combustion index gas and feature temperature. *Coal Science and Technology* 36(6):47–51.

[9] Xu Tao. 2012. Experimental study on the temperature rising characteristic of spontaneous combustion of coal. *Journal of Mining & Safety Engineering* 29(4):575–580.

[10] Xu Yan-hui. 2005. *Research and Application on Testing System and Characteristic gas of the Coal Spontaneous Combustion.* Xi'an: xi'an university of science and technology.

[11] Xu Jing-cai. 2001. *Coal spontaneous combustion dangerous area decision theory.* Beijing: coal industry publishing house.

[12] Zhang Jian-min. 2008. *China's underground coal fire research and management.* Beijing: coal industry publishing house.

[13] Zheng Lan-fang. 2010. An experimental research on coal spontaneous combustion in different temperature stages. *Journal of Chinese people armed police force academy* 32(4):16–18.

[14] Zhong Xiao-xing. 2010. Test method of critical temperature of coal spontaneous combustion based on the temperature programmed experiment. *Journal of China Coal Society* 35(8):128–131.

[15] Zhang Yan-ni. 2010. Study on coal spontaneous combustion features of temperature rising system based on oil bathing program. *Coal Science and Technology* 38(8):85–88.

Advances in Engineering Materials and Applied Mechanics – Zhang, Gao & Xu (Eds)
© *2016 Taylor & Francis Group, London, ISBN 978-1-138-02834-0*

Synthesis and element characterization of Ciprofloxacin Biomonomer based polyurethane prodrug

Z.G. Wen

College of Chemistry, Leshan Normal University, Leshan, SiChuan, China

ABSTRACT: A polyurethane prodrug was synthesized using homemade Ciprofloxacin Biomonomer (CFB) as chain extender and was contrast to prodrug synthesized from ciprofloxacin chain extender. UV and FTIR spectrums showed that CFB was successfully conjugated into the prodrug's backbone. Contrast to ciprofloxacin, using CFB as the chain extender could availably increase the drug content ratio and drug binding ratio. In addition, using CFB as the chain extender could also narrow the distribution of molecular weight due to its' symmetric structure.

1 INTRODUCTION

Polymer prodrug is a novel drug control release material[1–3]. It shows no or little pharmacological activity until covalent-bonded drugs released from the material[4]. Ciprofloxacin (CF) is a kind of synthesized fluoroquinolones which had a broad-spectrum antimicrobial activity. It shows a good antimicrobial activity to staphylococcus, which is one of the main causes of implants' infection[5–8]. Ciprofloxacin-bonded polyurethane prodrugs (CFPU) were synthesized in preliminary study[9]. Ciprofloxacin has a secondary amine and a carboxyl in its structure. The two functional groups have different reactivity with isocyanate. This made the molecular weight and drug-load ratio of CFPU unsatisfactory. Therefore CF Biomonomer (CFB) with symmetric structure was synthesized[10]. The structures of CF and CFB are showed in Figure 1. The biomonomer CFB were supposed to solve the problems above with its two same terminal secondary amino groups.

2 EXPERIMENTS

2.1 Materials

Polycaprolactone diol (PCL), average weight 1250, and 1,6-Hexane Diisocyanate (HDI) were obtained from Aldrich Chemical Company. Ciprofloxacin Hydrochloride (CF) powder was supplied by Jingxin Pharm (Xinchang, Zhejiang, China). CFB was homemade. Dimethylsulphoxide (DMSO), Dibutyltin Dilaurate (DBTDL) and Triethylamine (TEA) were all gained from Sinopharm Chemical Reagent CO., Ltd.

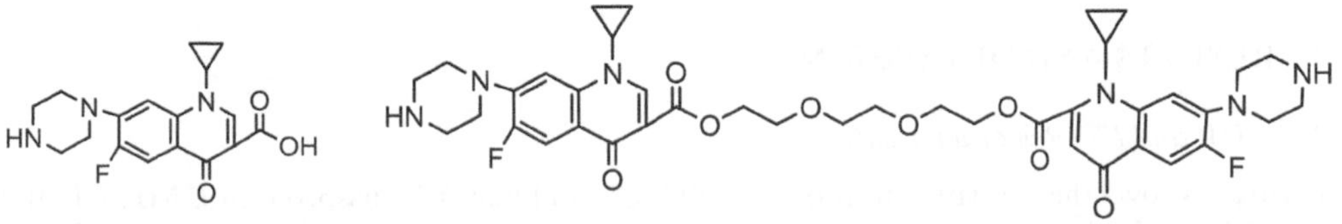

(a)CF   (b)CFB

Figure 1. Structures of CF and CFB.

Table 1.   Molar proportions of raw materials.

Serial number	HDI:PCL:CFB (or CF·HCl)
1	1:0.69:0.23
2	1:0.5:0.5
3	1:0.33:0.67

PCL and CF were dried at 60°C for one day in a vacuum oven. HDI was stored at room temperature in a desiccator over a bed of allochroic silicagel. DMSO was dehydrated by the use of molecular sieves type 4 A for 1 week.

2.2   CFBPU synthesis

Synthesis of CFPU was reported before[9]. The synthesis of CFBPU and pure PU were similar with CFPU using CFB and 1, 4-butanediol as the substitute for CF, respectively. The molar proportions of raw materials were showed in Table 1. As a contrast, CFPU with the same drug contents were synthesized. The drug content ratio, drug binding ratio and molecular weight (Mw) of CFPU and CFBPU were compared.

2.3   Calibration curve

Dissolve 5 mg CF·HCl in DMF and diluted with DMF to 100 mL as mother solution. Series of CF-DMF standard solutions were prepared by diluting 15, 10, 5, 4, 2, 1 and 0.5 mL mother solution with DMF to 25 mL, separately. Absorbance of the solutions was measured and the absorbance-concentration curve was plotted as calibration curve.

2.4   Drug binding ratio and drug content ratio

A certain quality of PU (CFPU or CFBPU) sample was dissolved in moderate volume of DMF. The concentration of CF in the sample-DMF solution was calculated according to the calibration curve. Drug content ratio of CFPU or CFBPU sample was calculated by measuring the absorbance of the CFPU-DMF/CFBPU-DMF solution at 286 nm (Cary50) using Eq. 1.

$$\text{Drug content ratio} = C \times V \times 10^{-6}/m \times 100\% \tag{1}$$

where C is the CF-concentration of drug polymer solution in DMF, μg/ml. V is the volume of the sample-DMF solution, ml. m is the weight of the drug polymer in the solution, g.

Drug binding ratio, η is calculated by Eq. 2.

$$\eta = \text{Drug content ratio/theoretical drug content ratio} \times 100\% \tag{2}$$

Theoretical drug content ratio is calculated by Eq. 3.

$$\text{Theoretical drug content ratio} = \frac{M_{CF} \times n_{CF}}{M_{CF} \times n_{CF} + M_{HDI} \times n_{HDI} + M_{PCL} \times n_{PCL}} \times 100\% \tag{3}$$

where M stands for molecular weight, and n stands for moles.

3   RESULTS AND DISCUSSION

3.1   UV and FTIR characterization

Figure 2 shows the UV spectrums of CFBPU, CF and pure PU dissolved in DMF. CF and CFBPU both show maximum absorbance at 286 nm, while pure PU has no absorbance at 286 nm. This inferred that CFBPU sample contains CF structure in it. In other words, the CFBPU was successfully synthesized.

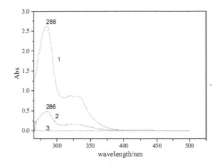

Figure 2. UV spectrums of CFBPU, CF, blank PU. 1. CFBPU, 2. CF, 3. pure PU.

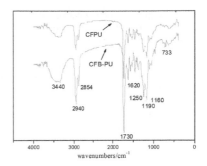

Figure 3. FTIR spectrums of CFPU and CFBPU.

Table 2. The effect of different chain extender to the drug content of polyurethane prodrug.

Sample	Material proportion	Chain extender	Theoretical drug content ratio [%]	Actual drug content ratio [%]	η
PU-1	1:0.69:0.23	CF	7.93	6.61	0.83
PU-2	1:0.69:0.23	CFB	14.79	14.43	0.98
PU-3	1:0.33:0.67	CF	27.66	21.57	0.78
PU-4	1:0.33:0.67	CFB	46.98	45.5	0.97

It could be easily inferred that, although using two kinds of chain extenders, CFBPU and CFPU have similar structures. The FTIR spectrums of CFBPU and CFPU were compared in Figure 3.

In Figure 3, CFBPU shows almost the same absorption with CFPU. The stretching vibration peak of urethane group (N-C-O) appears at 1160-1190 cm⁻¹. The strong characteristic absorption peak at 1730 cm⁻¹ corresponded to urethane group that doesn't formed hydrogen bond. Secondary amine with hydrogen bond showed an absorption peak at 3440 cm⁻¹. Peaks at 2940 and 2854 cm⁻¹ indicates that phenyl structure is existed in the sample. This makes clear that CF has been successfully bonded into the backbone of CFBPU and CFPU, because only the drug CF in the raw materials has phenyl structure in it.

3.2 Drug content ratio

The effect of the chain extender form to the drug content ratio and relative molecular weight was studied. Different structures mean different activities. CFB has a symmetrical structure with two same secondary amine end groups. Two same terminal groups mean the same activities when they reacted to isocyanate group. In contrast, CF has two different end groups, a secondary amine and a carboxyl. The former is more active than the later when reacted to the isocyanate group. Thus, CF was difficult to link into the PU backbone from two sides.

Drug content ratio is an important parameter to drug release materials. To resist the bacterial infection, in order to maintain the long-term use of implants in the body, improving the drug content of the materials is necessary. The drug content ratio and drug binding ratio η of CFBPU and CF were showed in Table 2. Results showed that, using CFB as the chain extender, CFBPU has a higher drug content ratio and a higher drug binding ratio than CFPU. CFB has two same activity secondary amine groups, so CFB is easier to be combined into polymer backbone. Otherwise, 1 mol CFB contains 2 mol CF structure. That means CFBPU could take 2 times of CF than CFPU under the same raw material proportion. That's in favor of improving the ratios of drug content and drug binding.

3.3 Molecular weight

The effect of different chain extender to the molecular weight and Polydispersity (PD) of polyurethane prodrug was showed in Table 3. Results showed that CFBPU had a narrower

Table 3. The effect of different chain extender to the molecular weight and Polydispersity (PD) of polyurethane prodrug.

Sample	Material proportion	Chain extender	Mn*	Mw**	PD
PU-5	1:0.5:0.5	CF	12370	17107	1.4
PU-6	1:0.5:0.5	CFB	12644	14608	1.2

*Mn: number-average molecular weight; **Mw: weight-average molecular weight.

polydispersity than CFPU. That could be attributed to the structural differentiation between CFB and CF mentioned above. The symmetrical structure made CFB bonded easier into the backbone of the polymer and molecular-weight distribution narrow.

4 CONCLUSION

Using CFB as the substitute of CF, a kind of polyurethane prodrug was synthesized and characterized by UV and FTIR spectrum. Results showed that, CFBPU in DMF had the same maximum absorption at 286 nm as CFPU in DMF. The maximum absorption at 286 nm is the distinctive absorption of CF in DMF. Meanwhile, pure PU has no absorption in the same case. FTIR characterization showed that CFBPU had similar spectrum as CFPU. So CFB contains CF structure in it according to UV and FTIR spectrum.

ACKNOWLEDGEMENT

It is a project supported by Scientific Research Fund of Sichuan Provincial Education Department (Grant No. 15ZB0265) and Leshan Normal University (Grant No. 0300037000773). The author also acknowledges the supportance of Science and Technology Bureau of Leshan Town (Grant No. 14SZD011).

REFERENCES

[1] Goodman, S.B., Yao, Z., Keeney, M. & Yang, F. (2013) The future of biologic coatings for orthopaedic implants. *Biomaterials*, 34, 3174–3183.
[2] Khandare, J., Minko, T. (2006) Polymer–drug conjugates: progress in polymeric prodrugs. *Progress in Polymer Science*, 31, 359–397.
[3] Rimoli, M., Ayallone, L., Caprariis, P. De. (1999) Synthesis and characterisation of poly (d, l-lactic acid)–idoxuridine conjugate. *Journal of Controlled Release*, 58, 61–68.
[4] Wang, J.L., Hao, H., Wang, Y. & Shi, M. (2011). Research Progress of Polymeric Prodrugs. *Chemistry*, 2, 131–136.
[5] Meng, Y.J., Zhu, B.D. (2006) Comparison of Antibacterial Activity of Ciprofloxacin in Vitro under Different pH Conditions. *China Pharmacy*, 17, 1300–1301.
[6] Arciola, C.R., Campoccia, D., Speziale, P., Montanaro, L. & Costerton, J.W. (2012) Biofilm formation in Staphylococcus implant infections. A review of molecular mechanisms and implications for biofilm-resistant materials. *Biomaterials*, 33, 5967–5982.
[7] Campoccia, D., Montanaro, L., Speziale, P. & Arciola, C.R. (2010) Antibiotic-loaded biomaterials and the risks for the spread of antibiotic resistance following their prophylactic and therapeutic clinical use. *Biomaterials*, 31, 6363–6377.
[8] De Giglio, E., Cometa, S., Ricci, M.A., Cafagna, D., Savino, A.M., Sabbatini, L., Orciani, M., Ceci, E., Novello, L., Tantillo, G.M. & Mattioli-Belmonte, M. (2011) Ciprofloxacin-modified electrosynthesized hydrogel coatings to prevent titanium-implant-associated infections. *Acta Biomater*, 7, 882–891.
[9] Wen, Z.G., Ye, Y.Q., Huang, J.H., Zheng, Q.Y., Dong, S.X. & Li, C.Y. (2007) Elementary study on synthesis and characterization of polymer prodrug bonded with quinolones antibiotic. *Journal of Functional Materials*, A05, 1961–1963.
[10] Wen, Z.G. (2014) Synthesis of A Derivative of Ciprofloxacin with Symmetrical Structure. *Guangdong Chemical Industry*. 16, 45–46.

Manipulating the Magnetic Anisotropy Energy of the double metallocene clusters via their oxidation state

F. Wu

School of Science, Nanjing Forestry University, Nanjing, Jiangsu, P.R. China

K. Luo

College of Information Science and Technology, Nanjing Forestry University, Nanjing, Jiangsu, P.R. China

ABSTRACT: By means of first principle calculations, we predict that the Magnetic Anisotropy Energy (MAE) of double metallocene clusters depends on the transferring of electrons in or out of the molecule. The MAE of the Pn_2Mn_2 cluster is about to be 3.74 meV. The MAE value is significantly enhanced to 9.66 meV for the $[Pn_2Mn_2]^{2-}$ while it is decreased to ~ 0.18 meV for the $[Pn_2Mn_2]^{2+}$. As for the $Pn_3(Mn_2)_2$ cluster and its ions, the ground states are all FM except for the $[Pn_3(Mn_2)_2]^{2+}$. The value of MAE is about 0.36 meV for the $Pn_3(Mn_2)_2$ cluster, which is approximate to the MAE of $PnMn_2$ nanowire. The values of MAE is increased to 2.73 meV, 3.06 meV, respectively for the cations $[Pn_3(Mn_2)_2]^-$, $[Pn_3(Mn_2)_2]^{2-}$. Basing on our analysis, the value of MAE is sensitively related to the occupation of the majority orbitals around the Fermi energy. Besides the value of MAE, the magnetization direction is also effected by changing their oxidation state.

1 INTRODUCTION

With the rapid miniaturization of the modern electronics devices, several general requirements are to be addressed, namely the ultra-high-density as well as low energy consumption and nonvolatility. It is well known that the traditional inorganic magnetic nanomaterials will reach the limit of the nanopatterning, which impels us to find some novel alternative materials. To the best of our knowledge, organometallic sandwich molecules are the comportable candidates for the future application in spintronics devices because of their intriguing electronic, magnetic, transport and optical properties. Many organometallic sandwich molecules have been well studied in the past few years, such as TM_nBz_{n+1}, TM_nCp_{n+1}, Eu_nCOT_{n+1}, $C_{60}TMCp$ etc. clusters and the corresponding one-dimensional (1D) infinite nanowires (TM = transition metal, Bz = C_6H_6, Cp = C_5H_5, COT = C_8H_8). [1–10] The electronic and magnetic properties of the organometallic sandwich molecules can be well engineered by choosing the different metal elements and the organic ligands. In recent years, the bimetallic permethylpentalene complexes Pn_2M_2 (Pn = C_8H_6, M = V, Cr, Mn, Co and Ni) have been successfully synthesised by Ashley et al. [11–12] Subsequently, the electronic and magnetic properties of 1D double metallocence nanowires, which are the periodic units of the Pn_2M_2 complexes, are theoretically studied by Wu et al[13]. They found that only the 1D $[PnMn_2]_\infty$ is Ferromagnetic (FM) and the others double metallocence nanowires are Nonmagnetic (NM) or Antiferromagnetic (AFM), which predicts that the 1D $[PnMn_2]_\infty$ nanowires and its clusters are the suitable candidates for applications in high-density information storage.

From the practical application point of view, besides a small size and switchable electronic and magnetic properties, the organometallic sandwich molecules with strong ferromagnetic coupling of local spin moments and large values of Magnetic Anisotropy Energy (MAE) are desirable as electronic and spintronic device materials. MAE is energy difference between the two magnetization directions, which gives an estimate for the energy barrier necessary to stabilize the magnetic moments against quantum tunneling and thermal fluctuations. For the progressively high storage capacity in the modern magnetic recording media, the materials with huge value of MAE are required urgently today. Usually, the directions of electron spin and the MAE can be controlled by applying External Electric Fields (EEFs).[14] However, the EEFs will become very strong in order to get a huge MAE in organometallic sandwich molecules. It is hard to realize and very difficult to control the strength of EFFs in the practical applications. A simple but effective way to control the magnetization direction in magnetic materials has proved to be feasible by Atodiresei et al through the transferring of electrons in or out of the molecules (i.e. oxidation-reduction reactions). Therefore, in the present paper we choose the $Pn_2(Mn_2)$ and $Pn_3(Mn_2)_2$ clusters as examples to investigate the magnetic properties via oxidation-reduction reactions.

2   COMPUTATIONAL METHOD

Here we performed first principles studies on the $Pn_2(Mn_2)$ and $Pn_3(Mn_2)_2$ clusters and their corresponding cations $Pn_2(Mn_2)^{+1}$, $Pn_2(Mn_2)^{+2}$, $Pn_3(Mn_2)_2^{+1}$, $Pn_3(Mn_2)_2^{+2}$ and anions $Pn_2(Mn_2)^{-1}$, $Pn_2(Mn_2)^{-2}$, $Pn_3(Mn_2)_2^{-1}$, $Pn_3(Mn_2)_2^{-2}$. All calculations are performed within the framework of a spin-polarized Density Functional Theory (DFT) as implemented in the Vienna Ab initio Simulation Package (VASP)[15-17] and the Perdew-Burke-Ernzerhof (PBE) exchange-correlation functional[18]. The interaction between valence electrons and ion cores is described by the Projected Augmented Wave (PAW) method[19-20]. To ensure a negligible interaction between the slabs, they are each separated by a vacuum gap of ~30 Å. The convergence in energy is set to 10^{-5} eV for self-consistent field calculations and 10^{-6} for the spin-orbit coupling, and that of the force in geometry optimization is set to 0.01 eV/Å. The plane-wave basis set cutoff energy is 500 eV, a k-mesh of $1 \times 1 \times 1$ with Monkhorst-Pack k-point scheme is adopt.[21] We considered two possible symmetry-determined directions of the magnetization in the molecules: axial (z)-perpendicular to the plane of the C_8H_6 ring and radial (r)-in the plane of the C_8H_6 ring. The MAE is defined as the total energy difference between these two configurations.

3   RESULTS AND DISCUSSIONS

In order to study the magnetism of the $Pn_n(Mn_2)_{n-1}$ (n = 1, 2,) clusters, we performed spin polarized DFT calculations on the clusters in FM and AFM states as shown in the Figure 1. We found that the ground states is FM, and the energies difference between the FM and AFM states (ΔE_{AFM-FM}) are about 122.88 meV/atom and 38.60 meV/atom for the $Pn_{n+1}(Mn_2)_n$ (n = 1, 2) clusters respectively. For comparison, the magnetic property of the $PnMn_2$ nanowire is also studied. The calculations predict that the $PnMn_2$ nanowire is FM metal and the ΔE_{AFM-FM} is about 62 meV/atom, which is consistent with Wu et al's results[13]. The value of MAE is 0.34 meV for the $PnMn_2$ nanowire. In Table 1 and Table 2, energies of two configurations with the spin direction along the z axis and in the xy plane, the ground states and the MAE are summarized for $Pn_{n+1}(Mn_2)_n$ (n = 1, 2) clusters and their corresponding ions. Specifically, the ground states for Pn_2Mn_2 and its corresponding ionic states are FM, and the MAE of the Pn_2Mn_2 cluster is increased to 3.74 meV (only 0.34 meV for the $PnMn_2$ nanowire). Through the way of oxidation-reduction reaction, the magnetism of the clusters is changed drastically. The MAE value is significantly enhanced to 9.66 meV for the $[Pn_2Mn_2]^{2-}$ while it is decreased to ~0.18 meV for the $[Pn_2Mn_2]^{2+}$. Although the magnitude of MAE for the $[Pn_2Mn_2]$, $[Pn_2Mn_2]^{2+}$ and $[Pn_2Mn_2]^{2-}$ is various by the way of oxidation-reduction reaction, the stable spin direction is not altered and along the z axis. However, it is found that the spin directions are transformed from the z axis to the xy plane for $[Pn_2Mn_2]^{+}$ and $[Pn_2Mn_2]^{-}$. In the case of the $Pn_3(Mn_2)_2$ and its ions, the ground states are all FM except for the $[Pn_3(Mn_2)_2]^{2+}$. The value of MAE is about 0.36 meV for the $Pn_3(Mn_2)_2$ cluster, which is approximate to the MAE of $PnMn_2$ nanowire. The values of MAE is increased to 2.73 meV,

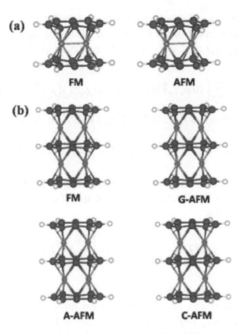

(a)

FM   AFM

(b)

FM   G-AFM

A-AFM   C-AFM

Figure 1.   The Ferromagnetic (FM) and Antiferromagnetic (AFM) sates for the (a) Pn_2Mn_2 and (b) $Pn_3(Mn_2)_2$ clusters.

Table 1.   The Ground States (GS) and energies difference of the $Pn_2(Mn_2)$ and its ionic states between the spin direction along the z axis and in the xy plane.

	$Pn_2(Mn_2)$	$[Pn_2(Mn_2)]^+$	$[Pn_2(Mn_2)]^{2+}$	$[Pn_2(Mn_2)]^-$	$[Pn_2(Mn_2)]^{2-}$
E_{xy} (eV)	−202.72034	−197.63353	−189.29423	−204.33038	−205.12929
E_z (eV)	−202.72408	−197.63057	−189.29441	−204.32561	−205.13895
MAE (meV)	3.74	−2.96	0.18	−4.77	9.66
GS	FM	FM	FM	FM	FM

Table 2.   The Ground States (GS) and energies difference of the $Pn_3(Mn_2)_2$ and its ionic states between the spin direction in the plane and out of the plane.

	$Pn_3(Mn_2)_2$	$[Pn_3(Mn_2)_2]^+$	$[Pn_3(Mn_2)_2]^{2+}$	$[Pn_3(Mn_2)_2]^-$	$[Pn_3(Mn_2)_2]^{2-}$
E_{xy} (eV)	−310.38703	−306.38295	−299.75991	−312.01068	−312.76627
E_z (eV)	−310.38739	−306.38246	−299.75633	−312.01341	−312.76933
MAE (meV)	0.36	−0.49	−3.58	2.73	3.06
GS	FM	FM	AFM	FM	FM

3.06 meV, respectively for the cations $[Pn_3(Mn_2)_2]^-$, $[Pn_3(Mn_2)_2]^{2-}$. As for the stable spin direction, only the $[Pn_3(Mn_2)_2]^+$ is in the xy plane.

The spin-orbit part of the Hamiltonian[22] in the Mn spheres can be given by

$$H_{so} = \xi \sum_{\mu_1,\mu_2,\sigma_1,\sigma_2} \langle \mu_2,\sigma_2 \mid \boldsymbol{L}\cdot\boldsymbol{S} \mid \mu_1,\sigma_1 \rangle \sum_k c^\dagger_{\mu_2,\sigma_2}(\boldsymbol{k}) c_{\mu_1,\sigma_1}(\boldsymbol{k})$$

where $\boldsymbol{k}$ is the electron wave vector, μ represents the d orbitals, σ is the spin, $c^\dagger$ and c are creation and annihilation operators. The Spin-Orbtial Coupling (SOC) interaction can be treated as a perturbation. The first order correction $\langle \mu | H_{so} | \mu \rangle$ is equal to zero due to the d orbital having $L = 0$. The second order correction δE is not equal to zero, which is equal to $\sum_{ex} \left(\langle gr | H_{so} | ex \rangle \langle ex | H_{so} | gr \rangle \right) / (E_{gr} - E_{ex})$. The δE can be divided into three parts:

$$\delta E = -\xi^2 \sum_\theta \left[A(\theta, \uparrow, \uparrow) \langle \mu_1, \uparrow | L \cdot S | \mu_2, \uparrow \rangle \langle \mu_2, \uparrow | L \cdot S | \mu_1, \uparrow \rangle \right] \tag{1}$$

$$+ A(\theta, \downarrow, \downarrow) \langle \mu_1, \downarrow | L \cdot S | \mu_2, \downarrow \rangle \langle \mu_2, \downarrow | L \cdot S | \mu_1, \downarrow \rangle \tag{2}$$

$$- A(\theta, \uparrow, \downarrow) \langle \mu_1, \uparrow | L \cdot S | \mu_2, \downarrow \rangle \langle \mu_2, \downarrow | L \cdot S | \mu_1, \uparrow \rangle \tag{3}$$

The part (1) is interaction of the majority states, the part (2) the interaction of the minority states and the part (3) is the interaction of the spin-flip transitions. The above equation can be simplified as following:

$$\delta E = -\frac{1}{4} \xi S \cdot \left[\langle L^\downarrow \rangle - \langle L^\uparrow \rangle \right] + \frac{\xi^2}{\Delta E_{ex}} \left[\frac{21}{2} S \cdot \langle T \rangle + 2 \left\langle \left(L_\zeta S_\zeta \right)^2 \right\rangle \right]$$

where ΔE_{ex} is the exchange splitting between majority and minority states.

The energy difference between two magnetization directions i.e. MAE can be written as $K = \delta E_z - \delta E_{xy}$. The K is $\sim (\xi_{so}/4) \Delta L$, if the majority states are completely full[22-23], otherwise, the K is equal to $\alpha(\xi_{so}/4) \Delta L$ and α is about 0.2–0.25.[24]

To understand the impact of the transfer of electrons in or out of the molecule on the MAE, we check the electronic and magnetic properties of Pn_2Mn_2 cluster and its corresponding ions. The ferromagnetic PDOS of Pn_2Mn_2 cluster is shown in Figure 2. We label the orbitals around the Fermi energy as numbers 1–7. As we can see, the orbital 2 is partially occupied. Specifically, the occupation of these orbitals around the Fermi energy in the molecules as a function of the oxidation state from +2 up to −2 is presented in Figure 3. It is can be seen that the value of MAE is related to the orbital occupation neighbouring the Fermi energy level. With the occupation increasing, the MAE of the molecule is correspondingly enhanced (from 0.18 meV to 9.66 meV). The MAE maximum value is ~9.66 meV when the majority state are completely full.

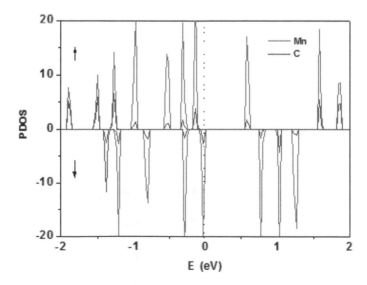

Figure 2.   The projected density of states of the Pn_2Mn_2 cluster.

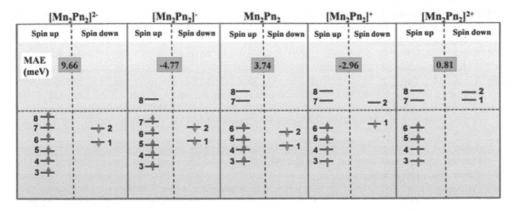

Figure 3.　Schematic view of the high occupied and the lowest unoccupied energy levels of Pn$_2$Mn$_2$ cluster and its ions.

4　CONCLUSION

In conclusion, our first principle calculations predict that the MAE of the Pn$_2$Mn$_2$ cluster is about to be 3.74 meV, and the value is significantly enhanced to 9.66 meV for the [Pn$_2$Mn$_2$]$^{2-}$ while it is decreased to ~ 0.18 meV for the [Pn$_2$Mn$_2$]$^{2+}$. In the case of Pn$_3$(Mn$_2$)$_2$ cluster and its ions, the ground states are all FM except for the [Pn$_3$(Mn$_2$)$_2$]$^{2+}$. The values of MAE is increased to 2.73 meV, 3.06 meV, respectively for the cations [Pn$_3$(Mn$_2$)$_2$]$^-$, [Pn$_3$(Mn$_2$)$_2$]$^{2-}$. According to the ferromagnetic PDOS and the occupation of these orbitals around the Fermi energy, the value of MAE is sensitively related to the occupation of the orbitals around the Fermi energy. The maximum value ~9.66 meV is occurring when the majority state are completely full.

ACKNOWLEDGMENTS

This work was supported by the NSFC (21203096, 11204137, 11474165), by NSF of Jiangsu Province (BK20130031, BK20131420, BK2012392), by the NJFU outstanding young scholars funding. We also acknowledge the support from the Shanghai Supercomputer Center.

REFERENCES

[1] Zhang, T.T.; Zhu, L.; Wu, Q.; Yang, S.W. & Wang, J. 2012. Structures and magnetism of multinuclear vanadium-pentacene sandwich clusters and their 1D molecular wires, *J. Chem. Phys.*, 137, 164309.

[2] Wang, S.J.; Li, Y.; Wu, D.; Wang, Y.F.; & Li, Z. R. 2012. *J. Phys. Chem. A.*, 116, 9189.

[3] Wang, L.; Cai, Z.; Wang, J.; Lu, J.; Luo, G.; Lai, L.; Zhou, J.; Qin, R.; Gao, Z.; Yu, D.; Li, G.; Mei, W. & Sanvito, S. 2008. Novel One-Dimensional Organometallic Half Metals: Vanadium-Cyclopentadienyl, Vanadium-Cyclopentadienyl-Benzene, and Vanadium-Anthracene Wires, *Nano Lett.*, 8, 3640.

[4] Zhu, L.; Zhang, T.; Yi, M.; & Wang, J. 2010. Ab Initio Study on Mixed Inorganic/Organic Ligand Sandwich Clusters: BzTMC$_{60}$, TM = Sc-Co, *J. Phys. Chem. A.*, 114, 9398.

[5] Zhang, X.; & Wang, J. 2010. Ab Initio Study of Bond Characteristics and Magnetic Properties of Mixed-Sandwich V$_n$Bz$_m$Cp$_k$ Clusters, *J. Phys. Chem. A.*, 114, 2319.

[6] Murahashi, T.; Inoue, R.; Usui, K. & Ogoshi, S. 2009. Square Tetrapalladium Sheet Sandwich Complexes: Cyclononatetraenyl as a Versatile Face-Capping Ligand, *J. Am. Chem. Soc.*, 131, 9888.

[7] Zhang, X.; Ng, M.F.; Wang, Y.; Wang, J. & Yang, S.W. 2009. Square Tetrapalladium Sheet Sandwich Complexes: Cyclononatetraenyl as a Versatile Face-Capping Ligand Theoretical Studies on Structural, Magnetic, and Spintronic Characteristics of Sandwiched Eu$_n$COT$_{n+1}$ ($n = 1$–4) Clusters, *ACS Nano*, 3, 2515.

[8] Zhang, X.; Wang, J.; Gao, Y.; & Zeng, X.C. 2009. Ab Initio Study of Structural and Magnetic Properties of TM_n (ferrocene)$_{n+1}$ (TM = Sc, Ti, V, Mn) Sandwich Clusters and Nanowires (n = ∞), *ACS Nano*, 3, 537.

[9] Wang, J.; Acioli; P.H. & Jellinek, J. 2005. Structure and Magnetism of V_nBz_{n+1} Sandwich Clusters, *J. Am. Chem. Soc.*, 127, 2812.

[10] Xiang, H.; Yang, J.; Hou, J.G. & Zhu, Q. 2006. One-Dimensional Transition Metal-Benzene Sandwich Polymers: Possible Ideal Conductors for Spin Transport, *J. Am. Chem. Soc.*, 128, 2310.

[11] Miyajima, K.; Yabushita, S.; Knickelbein, M.B. & Nakajima, A. 2007. Stern-Gerlach Experiments of One-Dimensional Metal-Benzene Sandwich Clusters: $Mn(C_6H_6)_m$ (M = Al, Sc, Ti, and V), *J. Am. Chem. Soc.*, 129, 8473.

[12] Ashley, A.E.; Cooper, R.T.; Wildgoose, G.G.; Green, J.C. & Dermot O.H., 2008. Homoleptic Permethylpentalene Complexes: "Double Metallocenes" of the First-Row Transition Metals, *J. Am. Chem. Soc.*, 130, 15662.

[13] Wu, X. & Zeng, X.C. 2009. Double Metallocene Nanowires, *J. Am. Chem. Soc.*, 131, 14246.

[14] Da, H.; Jin, H.M.; Lim, K.H. & Yang, S.W. 2010. Half-Metallic Spintronic Switch of Bimetallic Sandwich Molecular Wire via the Control of External Electrical Field, *J. Phys. Chem. C.*, 114, 21705.

[15] Kresse, G. & Furthmüller, J. 1996. Efficient iterative schemes for ab initio total-energy calculations using a plane-wave basis set, *Phys. Rev. B*, 54, 11169.

[16] Kresse, G. & Furthmüller, J. 1996. Efficiency of ab-initio total energy calculations for metals and semiconductors using a plane-wave basis set, *Comput. Mater.* Sci., 6, 15.

[17] Kresse, G. & Hafner, J. 1994. Ab initio molecular-dynamics simulation of the liquid-metal–amorphous-semiconductor transition in germanium, *Phys. Rev. B*, 49, 14251.

[18] Perdew, J.P.; Burke, K. & Ernzerhof, M. 1996. Generalized Gradient Approximation Made Simple, *Phys. Rev. Lett.*, 77, 3865.

[19] Blöchl, P.E., 1994. Projector augmented-wave method, *Phys. Rev. B*, 50, 17953.

[20] Joubert, D. & Kresse, G. 1999. From ultrasoft pseudopotentials to the projector augmented-wave method, *Phys. Rev. B*, 59, 1758.

[21] Pack, J.D. & Monkhorst, H.J. 1976. Special points for Brillouin-zone integrations, *Phys. Rev. B*, 13, 5188.

[22] Laan, G., 1998. Microscopic origin of magnetocrystalline anisotropy in transition metal thin films, *J. Phys. Condens. Matter* 10, 3239.

[23] P. Bruno, 1989. Tight-binding approach to the orbital magnetic moment and magnetocrystalline anisotropy of transition-metal monolayers, *Phys. Rev. B*, 39,865.

[24] Stöhr, J., 1999. Exploring the microscopic origin of magnetic anisotropies with X-ray Magnetic Circular Dichroism (XMCD) spectroscopy, *J. Magn. Magn. Mater*, 200, 470.

Influence of additives on the microwave dielectric properties of Ca[(Li$_{1/3}$Nb$_{2/3}$)$_{0.92}$Zr$_{0.08}$]O$_{3-\delta}$ ceramics

G. Xiong & H.F. Zhang
Department of Electronic and Information Engineering, Hubei University of Science and Technology, Xianning, China

ABSTRACT: The effects of B$_2$O$_3$ on the sinterability and microwave dielectric properties of Ca[(Li$_{1/3}$Nb$_{2/3}$)$_{0.92}$Zr$_{0.08}$]O$_{3-\delta}$ ceramics were investigated. The doping of B$_2$O$_3$ can effectively reduce the sintering temperature by 150~200°C. The temperature coefficient of resonator frequency τ_f increased with an increase of B$_2$O$_3$ content and sintering temperature. When the B$_2$O$_3$ content of 2 wt% was added, the optimum microwave dielectric properties: ε_r = 29.5, Qf = 16520 GHz and τ_f = −4.7 × 10^{-6}/°C were obtained at the sintering temperature of 950°C.

1 INTRODUCTION

Recently, multiplayer microwave filters were widely focused and developed in microwave circuits to meet the rapid development of advanced telecommunication. Microwave dielectric ceramics to be employed in multiplayer devices require low sintering temperature to cofired with the inner low less conductors and low melting point electrode such as Cu and Ag. Among those Low-Temperature Cofired Ceramics (LTCC) Ca(Li$_{1/3}$Nb$_{2/3}$)O$_{3-\delta}$ ceramics have been newly developed and widely investigated because of its excellent microwave dielectric properties and low sintering temperature of about 1150°C [1,2]. For the applications of LTCC, the complex perovskite should be further adjusted to lower its sintering temperature. In our preliminary work, we found the Ca[(Li$_{1/3}$Nb$_{2/3}$)$_{0.92}$Zr$_{0.08}$]O$_{3-\delta}$ ceramics having the superior dielectric properties: ε_r = 29.3, Qf = 26840 GHz and τ_f = −11.1 ppm/°C after sintering at 1170°C for 4 h. P. Liu et al. [3,4] have reported the addition of B$_2$O$_3$ were effectively in reducing the firing temperature of Ca(Li$_{1/3}$Nb$_{2/3}$)O$_{3-\delta}$-based ceramics. So, in this paper we employed B$_2$O$_3$ additive as a sintering flux to decrease the sintering temperature of the ceramics. The microwave dielectric properties were also investigated with the discussion of its relationships with the phase formation in the present system.

2 EXPERIMENTAL

The Ca[(Li$_{1/3}$Nb$_{2/3}$)$_{0.92}$Zr$_{0.08}$]O$_{3-\delta}$ powder compositions were synthesized by the conventional solid-state reaction method. High purity (≥99.9%) oxide powders of CaCO$_3$, Li$_2$CO$_3$, Zr O$_2$, Nb$_2$O$_5$, and TiO$_2$, were weighed according to the desired stoichiometry, and grounded in an agate pot with distilled water for 4 h in a planetary mill. The prepared powders calcined at 900°C for 2 h in a closed Al$_2$O$_3$ crucible. The calcined powders were milled for 3 h again with addition of B$_2$O$_3$, and then pressed into disks under a pressure of 150 Mpa. The disks were placed in a closed Al$_2$O$_3$ crucible to prevent the volatility of Li and sintered from 930°C to 1100°C for 4 h in air. The bulk densities of sintered specimens were measured by the Archimedes method. Phase formation and microstructure were examined by a X-ray diffractometer (X'Pert PRO) using the CuKα radiation. The measurement of microwave dielectric properties was performed on TE$_{011}$ mode at the resonant frequency from 4 to 7 GHz by the Hakki–Coleman's dielectric resonator method using a network analyzer (ADVANTEST R3767C). The temperature coefficient of resonator frequency (τ_f) was calculated at the range between 20°C and 80°C.

3 RESULTS AND DISCUSSION

Figure 1 shows the X-ray diffraction patterns of $Ca[(Li_{1/3}Nb_{2/3})_{0.92}Zr_{0.08}]O_{3-\delta}$ with B_2O_3 specimens sintered at 950°C for 4 h. The diffraction peaks can be indexed according to the $CaTiO_3$-type orthorhombic perovskite structure. Pure $Ca[(Li_{1/3}Nb_{2/3})_{0.92}Zr_{0.08}]O_{3-\delta}$ specimens sintered at 1170°C for 4 h was a single phase. When the B_2O_3 content increases, the peaks of superlattice diffractions of specimen 1:2 decrease until disappear, the degree of B-site 1:2 ordering will decrease, the second phase appears.

Figure 2 shows the relationship between the dielectric constant and the B_2O_3 content in $Ca[(Li_{1/3}Nb_{2/3})_{0.92}Zr_{0.08}]O_{3-\delta}$ sintered at 950°C for 4 h. The ε_r value increased by increasing the B_2O_3 content from 0.5 to 1.5 wt%, which could be contributed to the increased apparent density. However, as the B_2O_3 content became greater than 1.5 wt%, ε_r began to decrease because of the increasing of the secondary phase as confirmed in Figure 1.

Figure 3 shows the Qf value of $Ca[(Li_{1/3}Nb_{2/3})_{0.92}Zr_{0.08}]O_{3-\delta}$ specimens with B_2O_3 sintered at 950°C for 4 h. The addition of B_2O_3 greatly reduced the Qf value of $Ca[(Li_{1/3}Nb_{2/3})_{0.92}Zr_{0.08}]O_{3-\delta}$ specimens. This is expected since B_2O_3 addition inhibited the degree of 1:2 ordering in $Ca[(Li_{1/3}Nb_{2/3})_{0.92}Zr_{0.08}]O_{3-\delta}$ ceramics and thus cause the decreases of the quality factor[5,6].

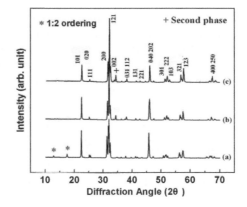

Figure 1. XRD spectra of $Ca[(Li_{1/3}Nb_{2/3})_{0.92}Zr_{0.08}]O_{3-\delta}$ specimens sintered at 950°C for 4h with the content of B_2O_3.

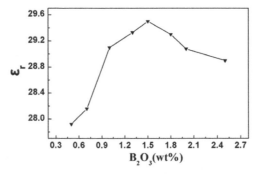

Figure 2. The ε_r values of $Ca[(Li_{1/3}Nb_{2/3})_{0.92}Zr_{0.08}]O_{3-\delta}$ specimens sintered at 950°C with the content of B_2O_3.

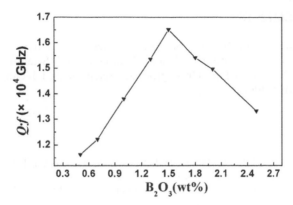

Figure 3. The Qf values of $Ca[(Li_{1/3}Nb_{2/3})_{0.92}Zr_{0.08}]O_{3-\delta}$ specimens sintered at 950°C with the content of B_2O_3.

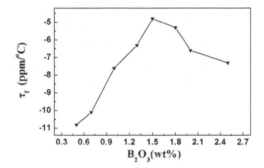

Figure 4. The τ_f values of $Ca[(Li_{1/3}Nb_{2/3})_{0.92}Zr_{0.08}]O_{3-\delta}$ specimens sintered at 950°C with the content of B_2O_3.

By increasing the B_2O_3 content, the $Ca[(Li_{1/3}Nb_{2/3})_{0.92}Zr_{0.08}]O_{3-\delta}$ specimens were gradually dense, and the Qf values of specimens first increased and then decreased. The reason for this is the appearance of the second phase in the specimen as mentioned earlier. It was reported that the Qf value relates to the relative density and second phase[7].

Figure 4 shows the temperature coefficient of resonant frequency of $Ca[(Li_{1/3}Nb_{2/3})_{0.92}Zr_{0.08}]O_{3-\delta}$ with B_2O_3 specimens sintered at the temperature of 950°C. The τ_f value ranged from negative value of –10.6 ppm/°C to negative value of –4.7 ppm/°C when the B_2O_3 content increased from 0.5 to 1.5 wt%, and then decreased. By doping 1.5 wt% B_2O_3, the $Ca[(Li_{1/3}Nb_{2/3})_{0.92}Zr_{0.08}]O_{3-\delta}$ ceramics show the optimized microwave dielectric properties: $\varepsilon_r = 29.5$, Qf = 16520 GHz and $\tau_f = -4.7 \times 10^{-6}$/°C after sintering at 950°C, indicating that the sintering temperature of $Ca[(Li_{1/3}Nb_{2/3})_{0.92}Zr_{0.08}]O_{3-\delta}$ could be reduced to 950°C without degradation of dielectric properties.

4 CONCLUSION

The doping of B_2O_3 improves the microwave dielectric properties of $Ca[(Li_{1/3}Nb_{2/3})_{0.92}Zr_{0.08}]O_{3-\delta}$ ceramics sintered at the temperature of 950°C. By increasing B_2O_3 additive content, the ε_r and the Qf values first increased then decreased from 28.6.8 and 12410 GHz to 2.5 and 10320 GHz, respectively with further doping the content of B_2O_3 from 1.8 wt% to 2.5 wt%. The τ_f value gradually moved to a positive direction with the increase of B_2O_3 content. When B_2O_3 content of 1.5 wt% was added, the optimum microwave dielectric properties: $\varepsilon_r = 29.5$, Qf = 16520 GHz and $\tau_f = -4.7 \times 10^{-6}$/°C were obtained at the sintering temperature of 950°C.

ACKNOWLEDGEMENTS

This work was supported by the Natural Science Foundation of Hubei Province (No. 2014CFC1084), the Natural Science Foundation of Scientific office of Xianning (No. XNKJ-1302) and China Undergraduate Scientific and Technological Innovation Project (No. 201410927006).

REFERENCES

[1] Choi, J.W. & Kang, C.Y.J. 1999. Microwave dielectric properties of Ca[(Li1/3Nb2/3)1-XMX]O3–δ (M = Sn, Ti) ceramics. *J. Mater. Res*. 14(9): 3567–3570.

[2] Choi, J.W. & Ha, J.Y. 2004. Microwave dielectric properties of Ca[(Li1/3Nb2/3)1-XZrX]O3–δ ceramics. *Jpn. J. APPL. Phys*. 43(1): 223–225.

[3] Liu, P. & Kim, E.S. 2001. Low-temperature sintering and microwave dielectric properties of Ca(Li1/3Nb2/3)O3–δ ceramics. *Jpn. J. APPL. Phys*. 40(9B): 5769–5773.

[4] Liu, P. & Kim,E.S. 2003. Microwave dielectric properties of Ca[(Li1/3Nb2/3)1-XTi3X]O3–δ ceramics with B2O3. *Materials Chemistry and Physics*. 79(2): 270–272.

[5] Kim, I.T. & Kim,Y.H.1995. Order-disorder transition and microwave dielectric properties of Ba(Zn1/3Ta2/3)O3 ceramics. *Jpn. J. Appl. Phys*. 34(3): 4096–4101.

[6] Davies, P.K. & Tong, J. 1997. Effect of ordering-induced domain boundaries on low-loss Ba(Zn 1/3Ta2/3)O3-BaZrO3 perovskite microwave dielectrics. *J. Am. Ceram. Soc*. 80(7): 1727–1740.

[7] Iddles, D.M. & Bell, A.J. 1992. Relationships between dopants, microstructure and the microwave dielectric properties of ZrO2-TiO2-SnO2 ceramics. *J. Mater. Sci.*, 27(23): 6303–6309.

Advances in Engineering Materials and Applied Mechanics – Zhang, Gao & Xu (Eds)
© *2016 Taylor & Francis Group, London, ISBN 978-1-138-02834-0*

Structure and physical properties of $Ho_5Co_6Sn_{18}$

W.J. Zeng, Y.Q. Chen & W. He
Key Laboratory of Nonferrous Metal Materials and New Processing Technology (Ministry of Education),
College of Materials Science and Engineering, Guangxi University, Nanning, P.R. China

C. Zeng
Guangxi International Business Vocational College, Nanning, P.R. China

ABSTRACT: The $Ho_5Co_6Sn_{18}$ alloy was prepared by arc melting and analyzed by means of the X-ray diffraction technique and Rietveld refinement. The thermal expansion of $Ho_5Co_6Sn_{18}$ was investigated from 298 K to 753 K. The magnetic property of $Ho_5Co_6Sn_{18}$ was measured from 5 K to 300 K. The electrical property of $Ho_5Co_6Sn_{18}$ was also examined in the temperature range of 2 K–300 K. The $Ho_5Co_6Sn_{18}$ compound has a tetragonal $Tb_5Rh_6Sn_{18}$-type structure (space group I4$_1$/acd (No. 142), $a = 1.3556(1)$ nm, $c = 2.7034(2)$ nm, and $Z = 8$). The average thermal expansion coefficients α_a, α_c and α_v of $Ho_5Co_6Sn_{18}$ are 1.03×10^{-5} K^{-1}, 1.39×10^{-5} K^{-1} and 3.40×10^{-5} K^{-1}, respectively. The $Ho_5Co_6Sn_{18}$ compound has a ferromagnetic transition at $T_c = 127(5)$ K and an anti-ferromagnetic transition at $T_N = 7.6$ K, respectively. The electrical property measurement indicates that the compound $Ho_5Co_6Sn_{18}$ exhibits a typical metallic behavior.

1 INTRODUCTION

The ternary RE-Co-Sn (RE = rare earth metal) alloys have been greatly attracted in recent years due to their outstanding physical properties[1-5]. A number of RE-Co-Sn ternary compounds, such as $RE_3Co_4Sn_{13}$ (RE = La, Ce, Pr, Nd, Sm, Gd, Tb, Yb), $RE_7Co_6Sn_{23}$ (RE = Sc, Ho, Y), $Tb_3Co_8Sn_4$, $RE_5Co_6Sn_{18}$ (RE = Dy, Sc, Ho) and $Pr_{117}Co_{57}Sn_{112}$[4-15] are reported in literature. Okudzeto and his co-workers[12] have grown single crystals of $Ln_5Co_6Sn_{18}$ (Ln = Er, Tm) and investigated the structural and magnetic properties of these $Ln_5Co_6Sn_{18}$ compounds. They found that $Ln_5Co_6Sn_{18}$ adopts the tetragonal $Sn_{1-x}Er_xEr_4Rh_6Sn_{18}$ structure (I4$_1$/acd, No. 142) with $a = 1.35310(2)$ nm, $c = 2.69970(4)$ nm, $V = 4.94283(13)$ nm^3 for $Er_5Co_6Sn_{18}$ and $a = 1.35190(6)$ nm, $c = 2.69760(9)$ nm, $V = 4.9302(4)$ nm^3 for $Tm_5Co_6Sn_{18}$. The effective moments of $Er_5Co_6Sn_{18}$ and $Tm_5Co_6Sn_{18}$ are 8.82 and 7.90 μ_B, respectively. The effective moment of $Er_5Co_6Sn_{18}$ with the low value of 8.82 is not close to the calculated moment of $Er^{3+} = 9.58 \mu_B$. However, this low value of 8.82 is much similar to that observed in the reentrant superconductor $Er_5Rh_6Sn_{18}$, which is isostructure with $Er_5Co_6Sn_{18}$. Lei et al.[13] have solved the crystal structure of $RE_5Co_6Sn_{18}$ (RE = Sc, Ho) by using direct methods (SHELXTL) and refined the structure by least-squares methods and their results stand well with that reported by Okudzeto et al[12]. The physical properties of $Ho_5Co_6Sn_{18}$ are not found in the open literature based on our knowledge. In order to find out the potential application of the Ho-Co-Sn alloys, it is must to investigate the physical properties of $Ho_5Co_6Sn_{18}$. The purpose of this paper is to study the physical properties, such as the thermal, electrical and magnetic properties, of the ternary compound $Ho_5Co_6Sn_{18}$.

2 EXPERIMENTAL DETAILS

The polycrystalline specimen of $Ho_5Co_6Sn_{18}$ was prepared by arc melting of initial materials of at least of 99.9 wt % under high pure argon atmosphere. The specimen was sealed in an evacuated quartz tube and annealed at 1073 K for 1000 h, and cooled down at a rate of 10 K/h to room temperature. The composition of the specimen was verified to be $Ho_{16.17}Co_{20.45}Sn_{63.37}$ by a field emission Scanning Electron Microscope (SEM, Model SU-8020) and an Energy Disperse Spectroscopy (EDS, Model X-MAX 80). The alloy button of $Ho_5Co_6Sn_{18}$ was examined by the X-Ray Diffraction (XRD) with a Rigaku D/max 2500 V powder diffractometer at room temperature. The Thermal Expansion Property (TEP) of $Ho_5Co_6Sn_{18}$ was investigated by using the High Temperature X-Ray Diffraction (HTXRD) technique in a temperature range from 298 K to 753 K. Magnetic measurements for $Ho_5Co_6Sn_{18}$ were performed in the temperature range 2–300 K employing a material property of measurement system (MPMS-5.5T, Quantum Design). The electrical property of $Ho_5Co_6Sn_{18}$ was measured from 5 K to 300 K by the standard four-probe method.

3 RESULTS AND DISCUSSION

3.1 *Crystal structure refinement*

The indexation of the XRD pattern of $Ho_5Co_6Sn_{18}$ was performed by using the Jade 6.0 program[16] and the initial lattice parameters were obtained. The initial lattice parameters and the initial structural parameters of $Ho_5Co_6Sn_{18}$ were submitted to the DBWS9807a[17] program for Rietveld refinement. The indexing results of $Ho_5Co_6Sn_{18}$ from the Jade 6.0 were used as the starting lattice parameters. The atomic positions in the unit cell of $Ho_5Co_6Sn_{18}$ were taken from the structure of Ref.[13]. During the refinement, the Pseudo-Voigt function was used as the profile fitting function. A total of 47 parameters for $Ho_5Co_6Sn_{18}$, such as scale factor, lattice constants, atomic coordinates, FWHM, preferred orientation, isotropic displacement parameters, etc. were involved in the refinement. The Rietveld refinement converges to reliability factors (R-factor) of $R_P = 5.36$, $R_F = 4.33$, $R_{exp} = 5.76$, $R_{Bragg} = 5.62$, $R_{WP} = 7.98$ and $S = 1.39$. This supports that the $Ho_5Co_6Sn_{18}$ compound crystallized in the $Tb_5Rh_6Sn_{18}$-type structure (space group $I4_1/acd$ (No. 142) and $Z = 8$). The refinement results of $Ho_5Co_6Sn_{18}$ gave the refined lattice parameters of $a = 1.3556(1)$ nm, $c = 2.7034(2)$ nm. The experimental and calculated XRD patterns of $Ho_5Co_6Sn_{18}$ from the final Rietveld refinement are shown in Figure 1.

3.2 *Lattice thermal expansion*

By analyzing the HTXRD patterns of $Ho_5Co_6Sn_{18}$ at different temperatures, it was found that there is no phase transformation but expansion in the lattice of $Ho_5Co_6Sn_{18}$ in the range of 298–753 K. The temperature dependence of lattice parameters and unit cell volume of $Ho_5Co_6Sn_{18}$, which is shown in Figure 2, can be described by the following polynomial functions:

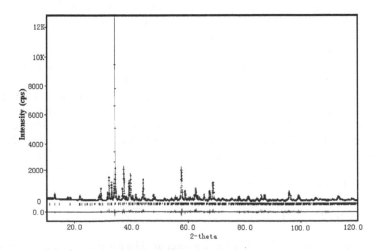

Figure 1.   The experimental and calculated XRD patterns of $Ho_5Co_6Sn_{18}$.

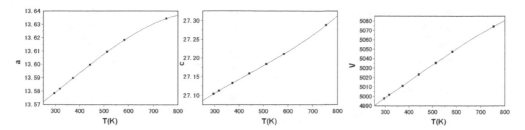

Figure 2. The temperature dependence of lattice parameters and unit cell volume of $Ho_5Co_6Sn_{18}$.

Table 1. The unit cell parameters, thermal expansion coefficients at different temperature of $Ho_5Co_6Sn_{18}$.

T(K)	a (nm)	α_a	c (nm)	α_c	V (nm³)	α_v
295	1.35784		2.71040		4.99783	
318	1.35817	1.06	2.71128	1.41	5.00168	3.37
373	1.35897	1.07	2.71331	1.38	5.01107	3.40
443	1.35998	1.06	2.71584	1.36	5.02323	3.43
513	1.36094	1.05	2.71839	1.35	5.03542	3.45
583	1.36183	1.02	2.72107	1.37	5.04738	3.44
753	1.36343	0.90	2.72875	1.48	5.07406	3.33

$$a \ (nm) = 1.3543 + 0.79297 \times 10^{-5} T + 0.18559 \times 10^{-7} T^2 - 0.17331 \times 10^{-10} T^3 \tag{1}$$

$$c \ (nm) = 2.6961 + 0.62851 \times 10^{-4} T - 0.63262 \times 10^{-7} T^2 + 0.49655 \times 10^{-10} T^3 \tag{2}$$

$$V \ (nm^3) = 4.9564 + 0.01043 T + 0.15644 \times 10^{-4} T^2 - 0.11617 \times 10^{-7} T^3 \tag{3}$$

where T is in the unit of K (Kelvin).

The measured lattice parameters of $Ho_5Co_6Sn_{18}$ listed in Table 1 were used to calculate the linear thermal expansivity (α_a) from the following equation:

$$\alpha_a = \frac{1}{a_{298}} \times \frac{a_T - a_{298}}{T - 298} \tag{4}$$

where, a_T was the lattice parameter at temperature T and a_{298} is that at 298 K. Similarly, the definition of α_a, α_c and α_v are following Eq. (4). Table 1 gives the values of α_a, α_c and α_v of $Ho_5Co_6Sn_{18}$ calculated according to Eq. (4) and these values were plotted in Figure 2. From Figure 2, it can observed that the lattice parameters a, c and volume of $Ho_5Co_6Sn_{18}$ increase slowly with temperature from 298 to 753 K. In this range of 298–753 K, the average relative Thermal Expansion Coefficient (TEC) of $Ho_5Co_6Sn_{18}$ along the a-axis is $\alpha_a = 1.03 \times 10^{-5}$ K^{-1} and that along the c-axis is $\alpha_c = 1.39 \times 10^{-5}$ K^{-1}, respectively. The average linear bulk TEC of $Ho_5Co_6Sn_{18}$, α_v, is 3.40×10^{-5} K^{-1}. This suggests that the thermal expansion along a and c axes is anisotropic with temperature. Furthermore, the TEC of $Ho_5Co_6Sn_{18}$ obeys the law of $2\alpha_a + \alpha_c = \alpha_v$ for the tetragonal structure.

3.3 Magnetic properties

Figure 3(a) displays the magnetic susceptibility vs. temperature curve of $Ho_5Co_6Sn_{18}$ measured under an applied field of 5Oe, while Figure 3(b) presents the reciprocal of magnetic susceptibility $1/\chi$ vs. temperature curve of $Ho_5Co_6Sn_{18}$. Figure 3(a) shows that $Ho_5Co_6Sn_{18}$ has a ferromagnetic transition at $T_c = 127(5)$ K and an anti-ferromagnetic transition at $T_N = 7.6$ K. From Figure 3, it can be noted that there is another phase transition occurred at T = 40(5) K.

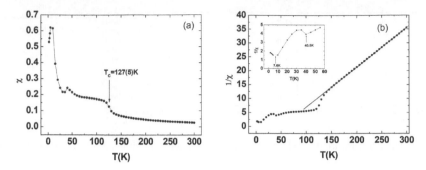

Figure 3. The magnetic susceptibility vs. temperature curve (a) and temperature dependence of reciprocal of magnetic susceptibility $1/\chi$ on temperature (b) of $Ho_5Co_6Sn_{18}$. The inset in (b) is the enlargement view at low temperatures.

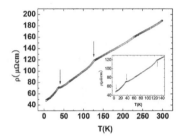

Figure 4. Temperature-dependent resistivity of $Ho_5Co_6Sn_{18}$. The inset shows the enlargement view at low temperatures.

This unusual phenomenon of $Ho_5Co_6Sn_{18}$ was attributed to the magnetic ordering types[18]. As shown in Figure 3(b), the reciprocal susceptibility of $Ho_5Co_6Sn_{18}$ above about 140 K is proportional to temperature. This points to that the Curie–Weiss behavior of $Ho_5Co_6Sn_{18}$ in the paramagnetic region. Further fitted the Curie–Weiss law to the experimental data of $Ho_5Co_6Sn_{18}$ above about 140 K and obtained the effective magnetic moment $\mu_{eff} = 8.19\ \mu_B$ per Ho atom. The paramagnetic Curie temperature was also calculated to be $\theta_p = 50.9$ K for $Ho_5Co_6Sn_{18}$. It can noted that the value of $\mu_{eff} = 8.19\ \mu_B$ is lower than the expected theoretical value of a free Ho^{3+} ion ($10.6\ \mu_B$). This further indicates that there is no contribution of the Co magnetic moment to the magnetization of $Ho_5Co_6Sn_{18}$, which is similar to those observed in the compounds of $RE_5Co_6Sn_{18}$ and $Er_5Rh_6Sn_{18}$ with similar formula [12]. This lower effective moment value is due to the anti-ferromagnetic order at lower temperature of $Ho_5Co_6Sn_{18}$. Based on the measured MT-curve of $Ho_5Co_6Sn_{18}$, $Ho_5Co_6Sn_{18}$ shows an antiferromagnetic like ordering phase transition at $T_N = 7.6$ K under a low magnetic field of 50Oe, although a large positive Curie temperature of $Ho_5Co_6Sn_{18}$ indicates a ferromagnetic ground state. The double independent crystallographic sites occupied by Ho atoms in the $Ho_5Co_6Sn_{18}$ structure may cause the competing (both antiferromagnetic and ferromagnetic) magnetic interactions between Ho atoms.

3.4 Electrical properties

The electrical resistivity of $Ho_5Co_6Sn_{18}$ was measured under a zero magnetic field condition. Figure 4 displays the dependence of the electrical resistivity of $Ho_5Co_6Sn_{18}$ on temperature ranged from 5–300 K. From Figure 4, it can be clearly observed that the resistivity of $Ho_5Co_6Sn_{18}$ decreases with the temperature decreasing. This suggests that resistivity of $Ho_5Co_6Sn_{18}$ characterizes a typical metallic behavior, which can be represented as a perfect linear relationship above 150 K. The magnitude of the resistivity of $Ho_5Co_6Sn_{18}$ at 300 K is strongly enhanced to be about 188 ($\mu\Omega cm$). With the decreasing of temperature, the resistivity of $Ho_5Co_6Sn_{18}$ levels off to a value of 48 ($\mu\Omega cm$). Thus, the Residual Resistivity Ratio

(RRR) of $Ho_5Co_6Sn_{18}$ is obtained to be 3.9. From Figure 4, it can be learned that there are three obvious transitions occurred near the T = 127 K, 40 K, 7.6 K, respectively. This is in good agreement with the magnetic measurement of $Ho_5Co_6Sn_{18}$.

4 CONCLUSION

The $Ho_5Co_6Sn_{18}$ compound was investigated by the XRD technique and its structure was refined by the Rietveld method from the XRD data. The $Ho_5Co_6Sn_{18}$ compound crystallizes in a tetragonal $Tb_5Rh_6Sn_{18}$-type structure with space group $I4_1/acd$ (No.142) and lattice parameters a = 1.3556(1) nm, c = 2.7034(2) nm. The average relative TEC of $Ho_5Co_6Sn_{18}$ along a-axis in a temperature range from 298 K to 753 K is $\alpha_a = 1.03 \times 10^{-5}$ K^{-1} and that along c-axis is $\alpha_c = 1.39 \times 10^{-5}$ K^{-1}, respectively. The TEC of $Ho_5Co_6Sn_{18}$ is anisotropic from 298 K to 753 K. The average linear bulk TEC of $Ho_5Co_6Sn_{18}$, α_V is 3.40×10^{-5} K^{-1}. The magnetic measurement for $Ho_5Co_6Sn_{18}$ shows a ferromagnetic transition at T_c = 127(5) K and an anti-ferromagnetic transition at T_N = 7.6 K. The effective magnetic moment μ_{eff} = 8.19 μ_B per Ho atom and the paramagnetic Curie temperature θ_p = 50.9 K were obtained for $Ho_5Co_6Sn_{18}$. The electrical resistivity measurement of $Ho_5Co_6Sn_{18}$ points to its typical metallic behavior and gives the Residual Resistivity Ratio (RRR) to be 3.9.

ACKNOWLEDGEMENTS

The authors would like to thank Professor C. Dong for his help in this work. The electrical resistivity measurement was performed at National Lab for Superconductivity, Institute of Physics of the Chinese Academy of Science, Beijing. This work was supported by the National Natural Science Foundation of China (Nos. 51261002 & 51461004) and the Guangxi Natural Science Foundation (No. 2012GXNSFAA053211).

Corresponding author: Prof. Wei He. College of Materials Science and Engineering. Guangxi University.100 Daxue East Road, Nanning, Guangxi, 530004. PR China.
Tel: +86-771-327-5918; Fax: +86-771-323-9406. Email: wei_he@gxu.edu.cn.

REFERENCES

[1] H. Zaigham, F.A. 2010. Characterisation of Sm-Co-Sn alloys. Khalid, *Mater. Charact.* 61, 1274.
[2] Y.H. Zhuang, J.M. Zhu, J.L. Yan, Y. Xu, J.Q. Li. 2008. Phase relationships in the Dy-Co-Sn system at 773 K. *Alloys Compd.* 459, 461.
[3] A. Gil, B. Penc, E. Wawrzyńska, J. Hernandez-Velasco, A. Szytuła, A. Zygmunt. 2004. Magnetic properties and magnetic structures of RCoxSn2 (R = Gd-Er) compounds. *Alloys Compd.* 365, 31.
[4] Y. Mudryk, P. Manfrinetti, V. Smetana, J. Liu, M.L. Fornasini, A. Provino, V.K. Pecharsky, G.J. Miller, K.A. Gschneidner Jr. 2013. Structural disorder and magnetism in rare-earth (R) R117Co54+xSn112±y. *Alloys Compd.* 557, 252.
[5] X.W. Lei, G.H. Zhong, M.J. Li, J.G. Mao. 2008. Yb3CoSn6 and Yb4Mn2Sn5: New polar intermetallics with 3D open-framework structures. *Solid State Chem.* 181, 2448.
[6] E.L. Thomas, H.O. Lee, A.N. Bankston, S. MaQuilon, P. Klavins, J.Y. Chan. 2006. Crystal growth, transport, and magnetic properties of Ln3Co4Sn13 (Ln = La, Ce) with a perovskite-like structure. *Solid State Chem.* 179, 1642.
[7] F. Canepa, P. Manfrinetti, M. Napoletano, S. Cirafici. 2001. Ferrimagnetism in Tb3Co8Sn4 intermetallic compound. *Alloys Compd.* 317–318, 556.
[8] Y. Mudryk, A. Grytsiv, P. Rogl, C. Dusek, A. Galatanu, E. Idl, H. Michor, E. Bauer, C. Godart, D. Kaczorowski, L. Romaka, O. Bodak. 2001. Physical properties and superconductivity of skutterudite-related Yb3Co4.3Sn12.7 and Yb3Co4Ge13. *Phys: Condens. Matter.* 13, 7391.
[9] R.V. Skolozdra, L.G. Aksel'rud, O.E. Koretskaya. 1985. Crystal structure of Ho7Co6Sn23, Kristallografiya. *Kristallografiya*, 30, 1003.
[10] X.W. Lei, C.Y. Yue. 2011. Crystal Structure and Band Structure of Tb3Co4Sn13. *Struct. Chem.* 30, 805.

[11] Y. Chen, B. He, J. He, W. He, L. Zeng. 2008. Crystal structure, properties and diffraction data of a new compound Dy5Co6Sn18. *Powder Diffr.* 23, 26.

[12] E.K. Okudzeto, E.L. Thomas, M. Moldovan, D.P. Young, J.Y. Chan. 2008. Physica B 403, 1628.

[13] X.W. Lei, G.H. Zhong, C.L. Hu, J.G. Mao. 2009. Syntheses and crystal structures of Y7Co6Sn23 and RE5Co6Sn18 (RE = Sc, Ho). *Alloys Compd.* 485, 124.

[14] W. He, J. Zhang, J. Yan, Y. Fu, L. Zeng. 2010. Crystal-structure and magnetic properties of the new ternary compound Pr117Co57Sn112. *Alloys Compd.* 491, 49.

[15] F. Canepa, M. Napoletano, P. Manfrinetti, S. Cirafici. 2001. Magnetism in R3Co8Sn4 compounds (R = Pr, Nd, Sm). *Alloys Compd.* 314(1–2), 29.

[16] Jade 6.0.XRD pattern processing. 2002. XRD pattern processing. Materials Data, Inc.

[17] R.A. Young, A.C. Larson, C.O. Paive-santos. 2000. User's Guide to Program DBWS9807a for Rietveld Analysis of X-ray and Neutron Powder Diffraction Patterns With a PC and Various Other Computers, School of Physics, Georgia Institute of Technology, Atlanta, GA.

[18] A.V. Morozkin. 2005. Magnetic structures of Zr6CoAs2-type Ho6FeSb2, Ho6CoBi2, Ho6FeBi2 and Ho6MnBi2 compounds. *Alloys Compd.* 395, 7.

Advances in Engineering Materials and Applied Mechanics – Zhang, Gao & Xu (Eds)
© *2016 Taylor & Francis Group, London, ISBN 978-1-138-02834-0*

Study on the storage stability of Centralizing Supply Crumb Rubber Modified Asphalt

K. Zhang & R. Luo
School of Transportation, Wuhan University of Technology, Wuhan, Hubei, China

L.L. Wang
Hanyang Municipal Construction Group Co. Ltd., China

ABSTRACT: In order to improve the restrictions of wet processing on traditional rubber asphalt, the fundamental performance test and storage stability verification of the Centralizing Supply Crumb Rubber Modified Asphalt (CS-RMA), which has a long-term storability and stability performance, were conducted through comparing with the traditional Rubber Modified Asphalt (RMA). The results show that the CS-RMA has a similar fundamental performance to RMA, without the existence of solid core rubber particles. The 22% dosage of the CS-RMA comparing with others dosage has a better workability. In addition, the CS-RMA has a more superior performance of segregation resistance as well as a better compatibility and stability performance under high temperature and a long-term static storage.

1 INTRODUCTION

Since 1990s, the crumb rubber modified asphalt mixture has been developed more than ten years and received by the industry because of the outstanding technical sense such as social significance of renewable resources, high temperature stability and noise reduction. Due to the limitation of various factors, especially the poor compatibility between rubber particles and asphalt, the unstable performance, worse resistance to long-term storage, etc., the rubber modified asphalt is difficult to be introduced in mass production[3][4][7]. The advantages of rubber asphalt mixture are obvious to all; however, how to improve the restrictions is a key issue puzzled a researcher for a long time as well as to achieve industrialization.

Crumb rubber powder is polymer elastomer which contains nature rubber or synthetic rubber, vulcanizators, vulcanization activator, carbon black and so on. Rubber particles disperse and fuse into the asphalt under the high temperature, then rubber powders swell through absorbing the light component of asphalt to keep the nature of the solid–liquid system, which causes the springback of solid rubber in the process of roller compaction[6]. Moreover, due to the rubber powders and asphalt are the inert polymer materials and most of the rubber mixes with asphalt without conducting the desulfurization process, it is difficult for rubber powders to form a stable structure with asphalt[2][5][8].

Therefore, in order to improve the restrictions of wet processing on traditional rubber asphalt, the objective of the paper is to conduct a storage stability study and verification on the Centralizing Supply Crumb Rubber Modified Asphalt (CS-RMA) that has a long-term storability and stability performance. The so-called Centralizing supply includes two aspects: on the one hand, it refers to the CS-RMA that can be stored stably in the long term and has no adverse reaction under the high temperature; on the other hand, it refers to the CS-RMA that can be manufactured like SBS modified asphalt implementing the finished production, and the specifications and quality are easy to control so as to be suitable for any scale of project. The swelling principle of the CS-RMA is introduced in the next section. Then the following section presents the preparation of the CS-RMA and fundamental performance index comparing with the traditional Rubber Modified Asphalt (RMA). The core of the paper is about the study and verification on storage stability of the CS-RMA, which is described in the third section. Finally, the findings of this paper are summarized in the last section.

2 THE SWELLING PRINCIPLE OF THE CS-RMA

In this study, the rubber powder manufactured by the way of physical desulphurization and refining in CS-RMA was designed by the compound formula, which did not contain extending oil and desulfurization catalyst due to consideration of avoiding further degradation of catalyst in the modified asphalt rubber[10]. This production of rubber powder can significantly reduce the swelling and degradation itself in the asphalt mixture in the process of storage, transportation and use, as shown in Figure 1. In addition, through the secondary granulating combined with compound formula technology, the powder particle size was reduced dramatically in the CS-RMA so as to form a thinner asphalt film in the mixture as well as greatly reduce the springback of powder particles in the process of rolling.

The specific reaction of the CS-RMA can be divided into three stages: the first stage is mainly rubber particles fuse physically with asphalt and a small amount of colloidal particles react with asphalt in the early stage of swelling. The second phase is mainly a large number of rubber particles is swelled by absorbing light component of asphalt to form a continuous and stable structure whose gel film is wrapped around the core rubber particles, while the asphalt has a high viscosity at this time. The third stage is mainly the degradation reaction of gel surrounded the core rubber particles and the solid core while the solid core disappears at this moment, and there is only small continuous gels causing the viscosity of asphalt to reduce gradually. Comparing with the RMA, the CS-RMA has the following distinctions and as shown in Figure 2.

Figure 1.   The rubber power of the CS-RMA.

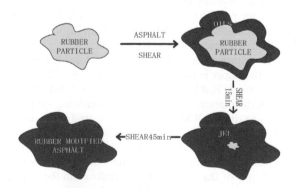

Figure 2.   The reaction process between rubber and asphalt.

1. There is not a solid core of rubber in the CS-RMA;
2. There is not a vulcanization reaction in the process of swelling;
3. There is not springback of rubber particles in the CS-RMA mixture under the rolling compaction.

3 THE FUNDAMENTAL PERFORMANCE OF THE CS-RMA

According to the preparation technology of crumb rubber modified asphalt mentioned above, the CS-RMA was prepared with different dosages (20%, 22%, and 25%). In order to compare with the performance of the CS-RMA, the traditional Rubber Modified Asphalt (RMA) was prepared as a reference with different meshes (20, 40, and 60) and dosages (15%, 20%, and 25%). A series of index tests were conducted and the result was shown in Table 1.

Comparing the indicators with CS-RMA and RMA from Table 1, it can be found as following:

1. In the three aspects of penetration, ductility and softening point, the value of the CS-RMA increases regularly with the increase of the dosage of powder particles, respectively. The nature of the CS-RMA is relatively similar to RMA.
2. In the aspect of elastic recovery, the CS-RMA is lower than the RMA. Because the elastic resilience is one of the most important indicators for asphalt to evaluate the low temperature performance and the higher the elastic recovery value is, the smaller the residual deformation of asphalt is under the load, so it is inferred that the low temperature performance of the CS-RMA is worse than RMA.
3. In the aspect of rotary viscosity, with the comparison of two kinds of rubber modified asphalt, it can be found that when the dosage of the CS-RMA is 20%, the value is lower, and when the dosage is 25%, the value is greater than 20 mesh RMA and 40 mesh RMA.
4. Combining with every index of different dosages of the CS-RMA, it is recommended to use 22% dosage of the CS-RMA as future studies.

4 THE STORAGE STABILITY OF THE CS-RMA

4.1 *The test design*

The test of storage stability of the CS-RMA consists of two parts: a. the rotary viscosity test under the different storage temperature; b. performance indicator test under the different storage time. First, conduct the test of rotary viscosity of the CS-RMA under five kinds of

Table 1. The result of index tests.

Type	Penetration (mm)	Ductility (cm)	Softening point (°C)	Recovery of elasticity (%)	Rotary viscosity (Pa·s)
CS-RMA (20%)	43.30	4.80	62.80	76.75	1.66
CS-RMA (22%)	44.10	5.00	63.80	77.80	2.28
CS-RMA (25%)	47.80	6.20	65.70	79.70	3.41
20 mesh RMA (15%)	34.30	6.85	67.70	84.20	1.20
20 mesh RMA (20%)	34.90	6.40	74.60	83.50	2.28
20 mesh RMA (25%)	32.10	7.85	77.20	84.57	3.01
40 mesh RMA (15%)	30.80	9.10	65.50	82.40	1.31
40 mesh RMA (20%)	33.70	9.80	71.70	83.60	2.56
40 mesh RMA (25%)	41.30	10.30	72.80	80.40	3.26
60 mesh RMA (15%)	32.80	9.30	64.50	83.70	1.52
60 mesh RMA (20%)	34.80	7.90	71.30	86.30	2.73
60 mesh RMA (25%)	35.60	9.80	77.00	87.95	5.38

storage temperature (160, 170, 180, 190, and 200°C). Second, prepare the CS-RMA with 22% of dosage and the RMA with 20% of dosage and 40 mesh, respectively. And place these into oven statically in the 8 h, 24 h, 48 h, 72 h, respectively, then remove the two kinds of rubber asphalt and observe the surface characteristics. It is found that the burning phenomenon is present on the RMA surface, and the surface of the CS-RMA looks better, as shown in Figure 3. The occurrence of burning phenomenon is a result of segregation of the RMA. Therefore, it is judged that the CS-RMA shows significant better performance in segregation resistance under a long-term and high temperature static storage.

4.2 Result and discussion

The Figure 4 shows the rotary viscosity of the CS-RMA under the different storage temperature and dosage. The viscosity is the core specification used to evaluate the performance of rubber modified asphalt[11]. The asphalt that has a greater viscosity causes less shear deformation under the load, and has a better recovery of elasticity and pertinence with the dynamic stability of asphalt. It can be observed from the Figure 4, no matter which dosage of the CS-RMA is, the viscosity reduces significantly with the increase of storage temperature, especially between 160°C and 170°C. The viscosity range of 25% dosage of the CS-RMA that has hyperviscosity is not in a reasonable range for construction, (i.e. it is not conductive to the construction) and the 20% dosage of the CS-RMA that has lower viscosity is not conductive to improve the pavement performance of asphalt mixtures. Finally, it is found that when the dosage is 22%, the viscosity of CS-RMA declines from 3.71 Pa·s to 1.65 Pa·s which reaches to the requirement of technical manual of the rubber asphalt. And when the temperature is 180°C, the value of viscosity is 2.28 Pa·s with good construction workability.

Then, the second step experiment was carried out (i.e. after placing two kinds of rubber modified asphalt in accordance with the provisions of test condition, the performance test was conducted again). And the result was shown in Table 2.

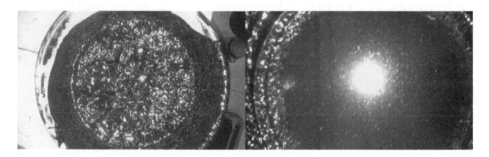

Figure 3.   The surface of rubber modified asphalt.

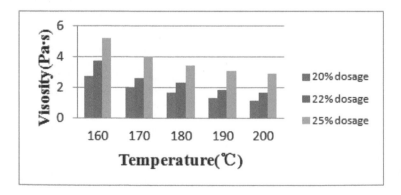

Figure 4.   The rotary viscosity of the CS-RMA.

Table 2. The results of rubber modified asphalt index under different storage time.

Type	Index	0 h	8 h	24 h	48 h	72 h
CS-RMA	Rotary viscosity (180°C) (Pa·s)	2.28	2.21	2.20	2.16	2.11
	Softening point (°C)	63.8	63.5	62.8	62.0	62.0
	Ductility (cm)	5.0	–	–	6.10	–
	Recovery of elasticity (%)	77.8	–	–	79.6	–
RMA	Rotary viscosity (180°C) (Pa·s)	2.56	2.51	2.42	2.16	1.55
	Softening point (°C)	71.7	72.8	71.5	69.5	68.0
	Ductility (cm)	9.8	–	–	10.2	–
	Recovery of elasticity (%)	83.6	–	–	82.7	–

It can be found from the data in Table 2, along with the increase of storage time, when the storage time reached to 72 h, the rotation viscosity of the CS-RMA fell by 7.46% and the RMA significantly fell by 39.45%. The softening point of two kinds of rubber asphalt was roughly presented as a slow downward trend which is not obvious, respectively are 2.82% and 5.16%. And in terms of ductility and elastic recovery, two kinds of rubber asphalt had no obvious change within 48 h, whose ductility increased by 2.31% and 2.31% respectively. In the aspect of elastic recovery, the CS-RMA increased by 2.31% and the RMA fell by 1.08%. Because the differences among the above several kinds of data are likely to be that the rubber particles continued degradation in the rubber asphalt under the continuous high temperature so as to cause the fracture of molecule chain and the decrease of molecular weight[1][9]. Nevertheless, unlike the RMA, the rubber particles in the CS-RMA can form the stable chemical bond with asphalt due to the rubber particles are relatively low molecular weight and have a sufficient swelling reaction with asphalt. Therefore, the compatibility of the CS-RMA is better than RMA.

Thus, according to the above test, the results show that the CS-RMA has a better compatibility and stability under high temperature and long-term static storage.

5 CONCLUSIONS AND FUTURE WORK

This paper aims at providing an accurate description of the storage stability of the CS-RMA based on analysis of swelling and performance test. The test of storage time and storage temperature was conducted to evaluate the storage stability of the CS-RMA by comparing with RMA. The following conclusions can be drawn from the present research:

1. Comparing with the same dosage of two kinds of rubber modified asphalt, it can be found that when the dosage is 20%, the viscosity of the CS-RMA is lower, and when the dosage is 25%, the value is greater than 20 mesh RMA and 40 mesh RMA.
2. Combining with every indictors of different dosage of the CS-RMA, it is recommended to use 22% dosage of the CS-RMA that has a better workability in this study.
3. The CS-RMA has a superior performance of segregation resistance under the long-term and high temperature static storage.
4. The CS-RMA has a better compatibility and has a better stability under the high temperature and long-term static storage.

The results of investigation show that the CS-RMA is a better kind of rubber modified asphalt used to obviously improve the restrictions of traditional rubber asphalt. It can provide a solid foundation for the subsequent research to CS-RMA mixture.

REFERENCES

[1] Haibin Li, Yanping Cheng. 2013. The research of desulfurization rubber asphalt. *The Journal of wuhan university of technology*, 35(5): 50–54.

[2] Huang Wenyuan, Zhang Yinxi. 2007. The Technical Criteria Frame of Pavement Used Asphalt Rubber in China. *Journal of Central South Highway Engineering.* 32(1): 111–114.

[3] J.G. Chehoveits. 1993. Crumb Rubber Modifier Workshop Notes, Session 9: Binder Design Procedures, Federal Highway Administration, *Office of Technology Applications.*

[4] Mull M.A., Stuart K., Yehia A. 2002. Fracture Resistance Characterization of Chemically Modified Crumb Rubber Asphalt Pavement. *Journal of Materials Science.* 37(3): 557–566.

[5] Szydlo A., Mackiewicz P. 2005. Asphalt mixes deformation sensitivity to change in rheological parameters. *Journal of materials in civil engineering.* 17(1): 1–9.

[6] Xian Liu, Fangdong Dai. 2008. Application and developing prospect of scrap tire rubber powder modified asphalt. *Journal of heilongjiang science and technology information.* (8): 71–72.

[7] Yang Yiwen, Yuan Hao. 2012. Swelling principle and pavement performance of desulfurized rubber asphalt. *Journal of highway and transportation research and development.* (29): 35–39.

[8] Y.E. Zhigang, Zhang Yuzhen, Kong Xianming. 2005. Modification of Bitumen with Desulfurized Crumb Rubber in the Present of Reactive Additives. *Journal of Wuhan University of Technology: Materials Science Edition.* 20(1): 95–97.

[9] Yiwen Yang, Hao Yuan, Tao Ma. 2012. Swelling principle and performance of desulfurizing rubber asphalt. *Highway traffic science and technology*, 29(2): 35 to 39.

[10] Zijian Chen. 2010. Low temperature performance of rubber asphalt. *Highway.* (6): 185–187.

[11] Zuwang Sun, Shunming Chen, Guangchun Zhang, Xinian Jin. 2013. *Technology application manual of rubber asphalt pavement.* China Communications Press.

Advances in Engineering Materials and Applied Mechanics – Zhang, Gao & Xu (Eds)
© 2016 Taylor & Francis Group, London, ISBN 978-1-138-02834-0

Optimal design of multi-composite activating fluxes for AZ91 magnesium alloy

G.Q. Zhang, Y.L. Ren, S.S. Zhang & Y.H. Su
School of Materials Science and Engineering, Shenyang University of Technology, Shenyang, China

ABSTRACT: The AZ91 magnesium alloy was welded by the GTAW with multi-composite activating fluxes, which contained seven components ($MgCl_2$, $CdCl_2$, $NiCl_2$, $CaCl_2$, CaO, TiO_2, and Cr_2O_3). The composite activating fluxes were designed by the uniform design method. The effects of composite activating fluxes on appearances, hardness, microstructure of welded joint and flow pattern of welding pool were analyzed after A-TIG welding. The results show that the weld penetration can be increased when effected by the all composite activating fluxes. The optimal value can be obtained when the 14th composite activating flux was used, the weld penetration of which is 3.35 times than that weld without flux, and the welded joint appearance is better. The hardness and microstructure are not affected by the 14th composite flux, and the flow pattern of weld pool is inward convection.

1 INTRODUCTION

The magnesium alloy has been considered as the lowest material within all of the structural metal materials because of its low density[1]. Due to their high specific strength/stiffness, super properties for casting, characteristics of dimensional stability, easy processing, low rejection rate and high damping coefficient, magnesium alloys have a broad application prospect in the field of automotive industry, aerospace, notebook computers, etc. With the wide application in these areas, it is particularly important to research the welding of magnesium alloys. Tungsten Inert Gas arc welding (TIG welding) is now the most commonly used method in the welding of magnesium alloys. However, due to the less weld penetration and the requirement of multi-layer welding for thick plate, the TIG welding efficiency is low. So, the emergence of A-TIG welding has provided a new direction in high efficiency and quality welding for magnesium alloy. The application prospect of A-TIG welding is broad[2-4].

The study of activating flux welding for magnesium alloys mainly focuses on the single activating flux at home and abroad. The systematic research of composite activating fluxes is less. And the research of composite activating fluxes is mainly based on AZ31 magnesium alloy[5-7], rarely based on other magnesium alloys. Therefore, studying the effect of composite activating fluxes on weld penetration, width, microstructure and properties of AZ91 magnesium alloy are of great significance for the expansion of application and product quality of magnesium alloys.

2 EXPERIMENTAL

AZ91 plates with a dimension of 50 mm × 40 mm × 5 mm were used as a base metal. Before welding, they were polished to remove grease and oxides. According to the uniform design method, seven activating fluxes $MgCl_2$, $CdCl_2$, $NiCl_2$, $CaCl_2$, CaO, TiO_2, and Cr_2O_3 were selected as components. A total of 28 kinds of composite activating flux formulations are shown in Table 1. When using U28*(28⁸) design table of which the deviation D is 0.1550 for minimum. The fluxes were scattered with acetone dispersants in a certain proportion, and coated on the surface of base metals. The welding process was implemented after the dispersants evaporated.

Table 1. Composite activating fluxes formulations/mg.

Sample no.	MgCl$_2$	CdCl$_2$	NiCl$_2$	CaCl$_2$	CaO	TiO$_2$	Cr$_2$O$_3$
1	48.9	12.9	5.25	3.74	3.02	4.21	22
2	38.6	8.34	24.1	8.09	4.86	5.45	10.6
3	33.1	4.04	6.18	35.3	8.28	6.77	6.3
4	29.3	0.25	23.6	3.95	25.7	12	5.23
5	26.3	20.5	3.4	11.8	0.34	33	4.72
6	23.8	11.4	16.7	24.1	2.98	0.38	20.7
7	21.6	5.47	2.39	4	17	9.72	39.8
8	19.7	0.88	15.8	12.5	21.3	11.2	18.6
9	18	25.1	0.26	23.7	21.3	6.44	5.19
10	16.5	13.6	10.6	1.82	1.56	40.9	15
11	15.1	6.75	40.6	6.08	4.55	24.6	2.41
12	13.8	1.6	9.38	26.7	13.6	1.87	33
13	12.6	29.7	21.2	0.22	16.3	4.64	15.4
14	11.4	15.8	5.46	8.69	41.1	7.2	10.3
15	10.4	8.04	22.9	17.8	1.87	23	16
16	9.39	2.39	3.78	62.4	3.65	14.1	4.28
17	8.44	35.1	12.3	4.36	12.2	26.2	1.48
18	7.53	18	1.02	18.9	26.3	2.52	25.7
19	6.67	9.37	14	38.7	24	1.94	5.3
20	5.85	3.24	57.7	2.34	2	12.9	16
21	5.06	42.1	6.55	10	6.8	18.5	11.1
22	4.31	20.3	30.6	20.5	8.09	13.1	3.2
23	3.58	10.8	7.41	3.41	38.8	35.4	0.64
24	2.88	4.16	28.4	11.5	45.9	0.89	6.2
25	2.2	54.1	2.33	16	2.13	7.08	16.2
26	1.55	22.8	17.9	1.05	11.9	21.6	23.2
27	0.91	12.2	2.01	12.3	26.1	30.7	15.8
28	0.3	5.13	17.3	25.3	28.9	19.3	3.7

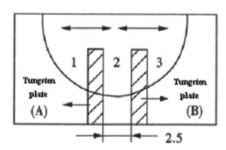

Figure 1. Schematic diagram of magnesium alloy tracer plate.

AZ91 plate was surfaced by using the AC-TIG welding device. The welding parameters are as follows: the welding current is 90A, welding speed is 300 mm/min, flow rate of argon gas is 12–15 L/min and the height of nozzle is 2–3 mm. After welding, the measurements of welded joint penetration and welded joint width were performed with metallurgical microscope, the hardness tests were carried out with a micro hardness tester (HVS-1000), the microstructures were characterized by metallurgical microscope and scanning electron microscopy.

For the analysis of molten pool flow pattern, two slots together 2.5 mm were processed on the back of magnesium alloy, the dimension of slot was 40 mm (length) × 4 mm (depth) × 1 mm (width). Two tungsten plates had the same dimensions with slot and were fixed in the slots (as shown in Fig. 1). Before welding, the activating fluxes mixed with tungsten particles (the size is about 300~500 μm) were coated on the plates. The welding process was carried out according to the parameters above after keeping still for 24 h.

3 RESULTS AND DISCUSSION

3.1 *Effect of composite activating fluxes on weld appearance*

Figures 2 and 3 show the weld penetration and depth-to-width ratio with 28 kinds of composite activating fluxes and without flux applied. It can be seen that the weld penetration and depth-to-width ratio have increased significantly and the weld width decreased with different range when coated with composite activating fluxes. By comparison, the weld penetration and depth-to-width ratio of the 14th composite formulation are maximum, reach to 2.98 mm and 0.36 respectively. The weld penetration is 3.35 times than that weld without flux. The optimal design of activating fluxes should be based on the weld penetration as the main index, but the surface forming quality should be considered meantime[8]. Figure 4 shows the weld surface appearance with 14th composite activating fluxes and without flux applied. Obviously, the weld forming quality is better (without surface defects such as porosity, crack, flash and so on) when the 14th composite activating flux applied, and worse (with surface defects such flash, porosity and so on) when no flux applied. Therefore, the effect of the 14th composite activating flux on weld not only increases the weld penetration and depth-to-width ratio to the maximum, but also makes the weld forming quality better.

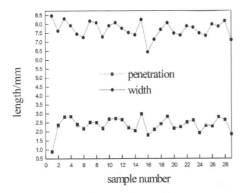

Figure 2. Weld penetration and width with different activating fluxes composite.

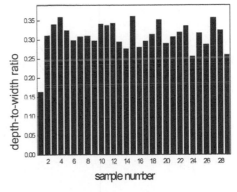

Figure 3. Depth-to-width ratio with different activating fluxes composite.

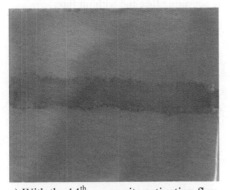

a) With the 14th composite activating flux

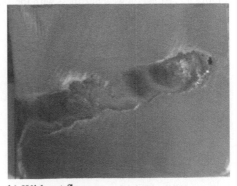

b) Without flux

Figure 4. A-TIG welding surface appearance.

3.2 *Effect of composite activating fluxes on weld microhardness*

Three test points were selected around the weld zone made with the 14th composite activating flux to analyze the hardness. The hardness curves shown in Figure 5 were obtained when compared weld hardness with and without flux. It can be seen from the Figure 5, the variation rule of hardness curve when the 14th composite activating flux applied is roughly the same with that without flux applied. The variation rule is: the hardness of weld zone is the highest, and the hardness of heat affected zone is higher than the fusion zone. The hardness of weld zone felt slightly under the 14th composite activating flux than no flux, but the hardness of HAZ and fusion zone changed little. It can be concluded from above that the coating of the optimum activating flux formulation has no obvious effect on the weld hardness.

3.3 *Effect of composite activating fluxes on weld microstructure*

Figure 6 shows the microstructure of weld made with the 14th composite activating flux and without flux. It can be seen from the figure that the weld microstructure basically consisted of matrix α-Mg and the dispersed second phase $Mg_{12}Al_{17}$ under different states. Obviously, the composition phase of the weld did not change with or without the effect of fluxes. The grain of weld made with the 14th composite activating flux was larger than the weld made without flux when comparing the size. That was because the coating of fluxes made more weld energy

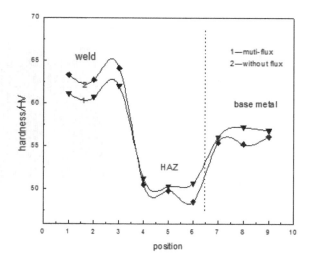

Figure 5.   Effect on weld hardness with the 14th composite activating flux and without flux.

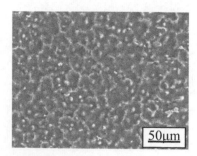
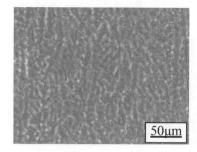

a) With the 14th composite activating flux   b) Without flux

Figure 6.   A-TIG welding microstructure.

664

a) The 14th composite activating flux   b) No activating flux

Figure 7.   Cross section of weld adopted by using tracer method.

density and higher temper, so that the crystallization time became longer and the grain size became slightly larger.

3.4   *Effect of composite activating fluxes on flow pattern of weld pool*

The 14th composite activating flux was selected as representative to analyze the change of weld pool flow pattern when coated with composite activating fluxes, because the effect of it on the weld penetration was the best. That will lay the foundations for the further study of mechanism of composite activating flux increased the penetration.

As shown in Figure 1, the weld pool is divided into three areas by the tungsten. When the molten metal in the center of the molten pool tends to flow toward the fusion line, the tungsten particles will gather in the areas 1 and 3 after welding. However, when the molten metal near the fusion line tends to flow toward the center of the molten pool, the tungsten particles will gather in the area 2. Figure 7 shows the cross section of weld adopted by using tracer method. It can be seen that the tungsten particles in areas 1 and 3 were more than area 2 when the weld made without flux, the flow pattern of weld pool was outward convection, so the weld was wide and shallow. Nevertheless, the tungsten particles in areas 1 and 3 were less than area 2 when the weld made with 14th composite activating flux, the flow pattern of weld pool was inward convection, so the weld was narrow and deep. The inward convection pattern of weld pool was due to the change of surface tension from negative to positive. So, the increase of penetration caused by the composite activating fluxes was associated with the surface tension temperature changes.

4   CONCLUSION

1. The 28 kinds of composite activating flux formulations according to the uniform design method can greatly increase the weld penetration and depth-to-width ratio, among which the 14th composite activating flux can increase these to the maximum, the value was 2.98 mm and 0.36 respectively. The weld penetration of the 14th composite activating flux was 3.35 times than that weld without flux, and the weld surface appearance is better.
2. The grain size of weld made with the optimum composite activating flux is slightly larger than the weld made without fluxes, but the weld microstructure still consists of the phase α-Mg and $Mg_{12}Al_{17}$. The microhardness of weld changes little.
3. When the weld is affected by the optimum composite activating flux, the flow pattern of weld pool is inward convection, the increase of penetration is associated with the surface tension temperature changes.

REFERENCES

[1] Zeng Rong chang, Ke Wei, Xu Yong bo, et al. 2001. Recent development and application of magnesium alloys. *Acta Metallurgica Sinica*. 37(7):673–685.
[2] Liu Jing an. 2001. Development trends and application of magnesium alloy fabrication technology. *Light Alloy Fabrication Technology*. 2(11):1–7.

[3] Nakata, Kazuhiro. 2001. Weldability of magnesium alloys. *Journal of Light Metal Welding and Construction*. 39(12):26–35.

[4] Lucas W. 2000. Activating flux improving the performance of the TIG process. *Welding and Metal Fabrication*. 68(2):6–10.

[5] Xu Jie, Liu Zi li, Shen Yi fu, et al. 2006. A-TIG welding of AZ31 magnesium alloy. *Aerospace Materials & Technology*. (6):42–45.

[6] Ma Xiang, Zhang Zhao dong, Liu Li ming, et al. 2007. Effects of single and composite component oxides activating fluxes on A-TIG welding of magnesium alloy. *Transactions of the China Welding Institution*. 28(9):39–42.

[7] Liu Zheng jun, Wu Xiao juan, Jiang Huan wen, et al. 2014. Study on activating fluxes based on oxides for magnesium alloy A-TIG welding. *Hot Working Technology*. 43(1):182–184.

[8] Du Xian Chang, Wang Yi, Guo Shu lan, et al. 2014. Uniform design and optimization of active flux for A-TIG welding of AZ31B magnesium alloy. *Transactions of the China Welding Institution*. 35(4):67–70.

Effect of glass addition on dielectric properties of Ca[(Li$_{1/3}$ Nb$_{2/3}$)$_{0.95}$Zr$_{0.05}$]O$_{3-\delta}$ –0.04TiO$_2$ ceramics

H.F. Zhang & G. Xiong

Department of Electronic and Information Engineering, Hubei University of Science and Technology, Xianning, China

ABSTRACT: The effects of ZnO-B$_2$O$_3$-Na$_2$O(ZBN) glass on the sinterability and microwave dielectric properties of Ca[(Li$_{1/3}$ Nb$_{2/3}$)$_{0.95}$Zr$_{0.05}$]O$_{3-\delta}$ –0.04TiO$_2$ ceramics were investigated. The doping of ZBN glass can effectively reduce the sintering temperature by 100~150°C. The temperature coefficient of resonator frequency τ_f is increased with an increase of ZBN glass content and sintering temperature. When the ZBN glass content of 1.8 wt% was added, the optimum microwave dielectric properties: ε_r = 31.2, Qf = 15420 GHz and τ_f = $-4.8 \times 10^{-6}/°C$ were obtained at the sintering temperature of 1000°C.

1 INTRODUCTION

Recently, multiplayer microwave filters were widely focused and developed in microwave circuits to meet the rapid development of advanced telecommunication. Microwave dielectric ceramics to be employed in multiplayer devices require low sintering temperature to cofired with the inner low less conductors and low melting point electrode such as Cu and Ag. Among those Low-Temperature-Cofired Ceramics (LTCC) Ca (Li$_{1/3}$ Nb$_{2/3}$)O$_{3-\delta}$ ceramics have been newly developed and widely investigated because of its excellent microwave dielectric properties and low sintering temperature of about 1150°C[1,2]. For the applications of LTCC, the complex perovskite should be further adjusted to lower its sintering temperature. In our preliminary work, we found that the Ca[(Li$_{1/3}$ Nb$_{2/3}$)$_{0.95}$Zr$_{0.05}$]O$_{3-\delta}$ –0.04TiO$_2$ ceramics having the superior dielectric properties: ε_r = 31.2, Qf = 21640 GHz and τ_f = -8.8 ppm/°C after sintering at 1170°C for 4 h. P. Liu et al.[3,4] have reported the addition of B$_2$O$_3$ was effectively in reducing the firing temperature of Ca(Li$_{1/3}$ Nb$_{2/3}$)O$_{3-\delta}$-based ceramics. So, in this paper we employed the ZnO-B$_2$O$_3$-Na$_2$O(ZBN) glass additive as a sintering flux to decrease the sintering temperature of the ceramics. The microwave dielectric properties were also investigated with the discussion of its relationships with the phase formation in the present system.

2 EXPERIMENTAL

The Ca[(Li$_{1/3}$ Nb$_{2/3}$)$_{0.95}$Zr$_{0.05}$]O$_{3-\delta}$ –0.04TiO$_2$ powder compositions were synthesized by the conventional solid-state reaction method. High purity ($\geq$ 99.9%) oxide powders of CaCO$_3$, Li$_2$CO$_3$, Nb$_2$O$_5$, and ZrO$_2$, were weighed according to the desired stoichiometry, and ground in an agate pot with distilled water for 4 h in a planetary mill. The prepared powders calcined at 900°C for 2 h in a closed Al$_2$O$_3$ crucible. The calcined powders were milled for 3 h again with addition of ZBN, and then pressed into disks under a pressure of 150 Mpa. The disks were placed in a closed Al$_2$O$_3$ crucible to prevent the volatility of Li and sintered from 930°C to 1100°C for 4 h in air. The bulk densities of sintered specimens were measured by the Archimedes method. Phase formation and microstructure were examined by a X-ray diffractometer (X'Pert PRO) using the CuKα radiaiton. The measurement of microwave dielectric properties was performed on TE_{011} mode at the resonant frequency from 4 to 7 GHz by the Hakki–Coleman's dielectric resonator method using a network analyzer (ADVANTEST R3767C). The temperature coefficient of resonator frequency (τ_f) was calculated at the range between 20°C and 80°C.

3 RESULTS AND DISCUSSION

Figure 1 show the X-ray diffraction patterns of $Ca[(Li_{1/3} Nb_{2/3})_{0.95}Zr_{0.05}]O_{3-\delta}$ –0.04TiO$_2$ with ZBN specimens sintered at 1000°C for 4 h. The diffraction peaks can be indexed according to the CaTiO$_3$-type orthorhombic perovskite structure, with four formula units per cell. Pure $Ca[(Li_{1/3} Nb_{2/3})_{0.95}Zr_{0.05}]O_{3-\delta}$ –0.04TiO$_2$ specimens sintered at 1150°C for 4 h were a single phase. When the ZBN content increases, the peaks of superlattice diffractions of specimen 1:2 decrease until disappear, the degree of B-site 1:2 ordering will decrease, second phase appears.

Figure 2 shows the relationship between the dielectric constant and the ZBN content in $Ca[(Li_{1/3} Nb_{2/3})_{0.95}Zr_{0.05}]O_{3-\delta}$ –0.04TiO$_2$ sintered at 1000°C for 4 h. The ε_r value increased with increasing the ZBN content from 0.7 to 2.5 wt%, which could be contributed to the increased

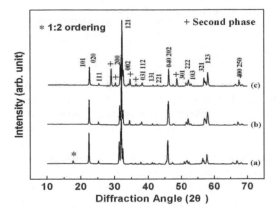

Figure 1. XRD spectra of $Ca[(Li_{1/3} Nb_{2/3})_{0.95}Zr_{0.05}]O_{3-\delta}$ –0.04TiO$_2$ specimens sintered at 1000°C for 4 h with the content of ZBN glass.

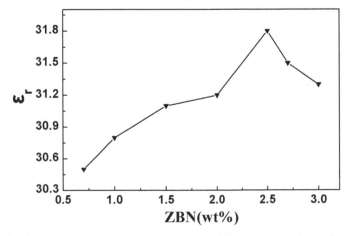

Figure 2. The ε_r values of $Ca[(Li_{1/3} Nb_{2/3})_{0.95}Zr_{0.05}]O_{3-\delta}$ –0.04TiO$_2$ specimens sintered at 1000°C with the content of ZBN glass.

apparent density. However, as ZBN content became greater than 2.5 wt%, ε_r began to decrease because of the increasing of the secondary phase as confirmed in Figure 1.

Figure 3 shows the Qf value of $Ca[(Li_{1/3}\,Nb_{2/3})_{0.95}Zr_{0.05}]O_{3-\delta}$ –0.04TiO$_2$ specimens with ZBN sintered at 1000°C for 4 h. The addition of ZBN greatly reduced the Qf value of $Ca[(Li_{1/3}\,Nb_{2/3})_{0.95}Zr_{0.05}]O_{3-\delta}$ –0.04TiO$_2$ specimens. This is expected since TiO$_2$ addition inhibited the degree of 1:2 ordering in $Ca[(Li_{1/3}\,Nb_{2/3})_{0.95}Zr_{0.05}]O_{3-\delta}$ –0.04TiO$_2$ ceramics and thus cause the decreases of the quality factor[5,6]. By increasing the ZBN content, the $Ca[(Li_{1/3}\,Nb_{2/3})_{0.95}Zr_{0.05}]O_{3-\delta}$ –0.04TiO$_2$ specimens were gradually dense, and the Qf values of specimens first increased and then decreased. The quality factor also shows the same tendency of the dielectric constant. The reason for this is the appearance of the second phase in the specimen as mentioned earlier. It was reported that the Qf value relates to relative density and second phase[7].

Figure 4 shows the temperature coefficient of resonant frequency of $Ca[(Li_{1/3}\,Nb_{2/3})_{0.95}Zr_{0.05}]O_{3-\delta}$ –0.04TiO$_2$ with ZBN specimens sintered at the temperature of 1000°C. The τ_f value ranged from negative value of –8.3 ppm/°C to negative value of –4.3 ppm/°C when the ZBN content increased from 0.7 to 2 wt%, and then decreased. The $Ca[(Li_{1/3}\,Nb_{2/3})_{0.95}Zr_{0.05}]O_{3-\delta}$ –0.04TiO$_2$ ceramics show the optimized microwave dielectric properties: ε_r = 31.3, Qf = 15410 GHz and τ_f = – 4.3 × 10^{-6}/°C after sintering at 1000°C, indicating that the sintering temperature of $Ca[(Li_{1/3}\,Nb_{2/3})_{0.95}Zr_{0.05}]O_{3-\delta}$ –0.04TiO$_2$ could be reduced to 1000°C without degradation of dielectric properties.

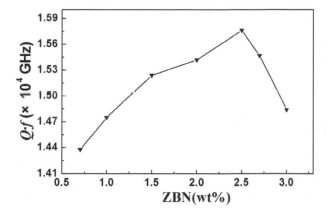

Figure 3. The Qf values of $Ca[(Li_{1/3}\,Nb_{2/3})_{0.95}Zr_{0.05}]O_{3-\delta}$ –0.04TiO$_2$ specimens sintered at 1000°C with the content of ZBN glass.

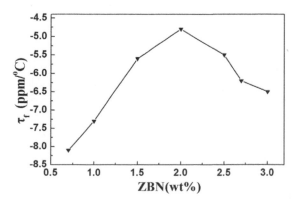

Figure 4. The τ_f values of $Ca[(Li_{1/3}\,Nb_{2/3})_{0.95}Zr_{0.05}]O_{3-\delta}$ –0.04TiO$_2$ specimens sintered at 1000°C with the content of ZBN glass.

4 CONCLUSION

The doping of the ZBN glass improves the microwave dielectric properties of $Ca[(Li_{1/3} Nb_{2/3})_{0.95}Zr_{0.05}]O_{3-\delta}$ –0.04TiO_2 ceramics sintered at the temperature 1000°C. By increasing the ZBN glass additive content, the ε_r and the Qf values first increased then decreased from 30.5 and 14380 GHz to 31.8 and 15760 GHz, respectively with further doping content of ZBN from 2.7 wt% to 3 wt%. The τ_f value gradually moved to a positive direction with the increase of ZBN content. When the ZBN content of 2 wt% was added, the optimum microwave dielectric properties: $\varepsilon_r = 31.2$, Qf $= 15420$ GHz and $\tau_f = -4.8 \times 10^{-6}/°C$ were obtained at the sintering temperature of 1000°C.

ACKNOWLEDGEMENTS

This work was supported by the Natural Science Foundation of Hubei Province (No. 2014CFC1084), the Natural Science Foundation of Scientific office of Xianning (No. XNKJ-1302), and China Undergraduate Scientific and Technological Innovation Project (No. 201410927006).

REFERENCES

[1] Choi, J.W. & Kang, C.Y.J. 1999. Microwave dielectric properties of $Ca[(Li_{1/3} Nb_{2/3})_{1-X}M_X]O_{3-\delta}$ (M = Sn, Ti) ceramics. *J. Mater. Res.* 14(9): 3567–3570.

[2] Choi, J.W. & Ha, J.Y. 2004. Microwave dielectric properties of $Ca[(Li_{1/3} Nb_{2/3})_{1-X}Zr_X]O_{3-\delta}$ceramics. *Jpn.J. APPL. Phys.* 43(1): 223–225.

[3] Liu, P. & Kim, E.S. 2001. Low-temperature sintering and microwave dielectric properties of $Ca(Li_{1/3} Nb_{2/3})O_{3-\delta}$ ceramics. *Jpn. J. APPL. Phys.* 40(9B): 5769–5773.

[4] Liu, P. & Kim, E.S. 2003. Microwave dielectric properties of $Ca[(Li_{1/3} Nb_{2/3})_{1-X}Ti_{3X}]O_{3-\delta}$ ceramics with B_2O_3. *Matericals Chemistry and Physics.* 79(2): 270–272.

[5] Kim, I.T. & Kim, Y.H. 1995. Order-disorder transition and microwave dielectric properties of $Ba(Zn_{1/3}Ta_{2/3})O_3$ ceramics. *Jpn. J. Appl. Phys.* 34(3): 4096–4101.

[6] Davies, P.K. & Tong, J. 1997. Effect of ordering-induced domain boundaries on low-loss $Ba(Zn_{1/3}Ta_{2/3})O_3$-$BaZrO_3$ perovskite microwave dielectrics. *J. Am. Ceram. Soc.* 80(7): 1727–1740.

[7] Iddles, D.M. & Bell, A.J. 1992. Relationships between dopants, microstructure and the microwave dielectric properties of ZrO2-TiO2-SnO2 ceramics. J. Mater. Sci., 27(23): 6303–6309.

Advances in Engineering Materials and Applied Mechanics – Zhang, Gao & Xu (Eds)
© 2016 Taylor & Francis Group, London, ISBN 978-1-138-02834-0

Study of removal mechanism on ZrO$_2$ ceramic using ultrasonic and ELID composite grinding

Y.Y. Zheng & Q. You
School of Mechanical and Power Engineering, Henan Polytechnic University, Henan, Jiaozuo, China

ABSTRACT: The model of undeformed chip thickness was founded that suits for the ultrasonic and ELID composite grinding, in this paper the grinding force and the surface microstructure were measured and observed through the experiment of the ultrasonic and ELID composite grinding. And then the critical condition of brittle ductile transition was founded by analyzing the change rule between grinding results and undeformed chip thickness. The results show that there is a good correlation between the grinding force and the grinding surface roughness with undeformed chip thickness. The material removal mechanism changes from plasticity to brittleness when the undeformed chip thickness arrives at 1.565 μm, it is plastic removal when the undeformed chip thickness is less than 1.565 μm, while it begans to appear brittleness removal when the undeformed chip thickness is more than 1.565 μm.

1 INTRODUCTION

The undeformed chip thickness is an important factor of grinding to characterize the grinding process. It has a great influence on the grinding process, which not only affects the size of the grinding force acting on abrasive grains, but also affects the size of specific grinding energy and the grinding zone temperature, thus producing an influence on the integrity of the machined surface. The study found that the influence of grinding wheel speed and grinding depth on the grinding effect all boils down to the change of undeformed chip thickness, and there is a good correlation between grinding force and surface roughness with the average undeformed chip thickness. The direct cause of affecting the grinding quality is the material removal mechanism, so study on the quality of ultrasonic and ELID composite grinding must analyze the material removal mechanism in the process of grinding firstly. There are mainly two removal modes that is the plasticity removal and brittleness removal for ZrO$_2$ ceramic hard brittle material, as long as the grinding chip thickness is less than critical grinding depth, ZrO$_2$ ceramic will be removed in the form of plasticity removal mode, so the quality and the integrity of grinding surface are greatly improved[1].

2 ULTRASONIC AND ELID COMPOSITE GRINDING

2.1 *Technology principle of Ultrasonic and ELID composite grinding*

The schematic diagram of ultrasonic and ELID composite grinding[2] is shown in Figure 1. Ultrasonic and ELID composite grinding technology is based on the principle of ELID grinding technology, the electric signal generated by the ultrasonic power source was converted into magnetic energy in the first coil via a wireless transmission device (in order to avoid the coil winding caused by spindle rotation), the magnetic energy was converted into electrical energy in turn in the second coil, and then the electrical signal was converted into weak mechanical vibrations by the transducer, then the amplitude of mechanical vibration is amplified to the required amplitude by horn. Finally, the enlarged vibration is passed to the

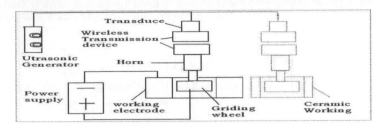

Figure 1.   Schematic diagram of ultrasonic and ELID composite grinding.

Figure 2.   Distribution of height that abrasive grains protrude the oxide film.

tool grinding wheel through a certain connection, which makes the grinding wheel to produce high frequency of the axial reciprocating vibration.

2.2   Active grains number

In the process of Ultrasonic and ELID composite grinding, ELID is actually to make the grinding wheel to keep good cutting performance, the metal binder on the surface of the grinding wheel can be electrolytic removal, at the same time a layer non-conductive oxide film is formed, due to the abrasive grains is not conductive, exposed the surface of grinding wheel gradually, as grains on the surface of grinding wheel passivation, the oxide film get thinning by shaving gradually, the electrolytic action will be strengthen, which made the protrusion height and chip space of the cutting edge of abrasive grains forms again. The thickness of oxide film ranges from 8 to 9 μm is advantageous to the dynamic grinding in ELID grinding[3], suggests that the thickness of oxide film is small in ultrasonic and ELID composite grinding, the height that abrasive grains protrude the oxide film is very similar to the height that abrasive grains protrude the surface of grinding wheel, the height of abrasive grains blade h approximately obeys the normal distribution[4]. The height that abrasive grains protrude the oxide film is shown in Figure 2, the number of the height that abrasive grains protrude in the maximum and minimum is less, most of the abrasive grains are in average height. Su Yu[5] assume that the first or semi-exposed diamond can not take participate in grinding, and can use the average value of maximum to characterize the blade height of diamond grits. So it needs to make the following assumptions before calculating dynamic active grinding grains number: 1. Ultrasonic vibration has no effect on the formation of the oxide film. 2. Abrasive grains fall off due to the grinding wheel holding force becomes smaller when protrusion height is greater than the largest protrusion height.

The maximum height that abrasive grains protrude the oxide film:

$$H_{max} = d_{max} - h$$

The minimum height that cutting grits protrude the oxide film:

$$H_{cut} = d_{max}/2$$

The average height that abrasive grains protrude the oxide film:

$$E(h) = (d_{max} - h)/2$$

672

The probability of the number of dynamic active grains:

$$P_d = \frac{1}{\sqrt{2\pi}\sigma} \int_{H_{cut}}^{H_{max}} e^{-\frac{(x-\mu)^2}{2\sigma^2}} dx = \Phi\left(\frac{H_{max} - E(h)}{\sigma}\right) - \Phi\left(\frac{H_{cut} - E(H)}{\sigma}\right) \qquad (1)$$

where d_{max} is the maximum diameter of the grains, h is the thickness of oxide film, σ is a parameter that completely defines the probability density function and it depends upon microstructure of grinding wheel.

The number of dynamic active grains per unit area is equal to the number of grains per unit area N_s multiplies the probability of the number of dynamic active grains P_d, that is:

$$C = N_s P_d = \sqrt[3]{\left(\frac{6V_m}{\pi d^3}\right)^2} \left[\Phi\left(\frac{H_{max} - E(h)}{\sigma}\right) - \Phi\left(\frac{H_{cut} - E(H)}{\sigma}\right)\right]. \qquad (2)$$

3   UNDEFORMED CHIP THICKNESS MODEL

Sanjay Agarwal[6] established the equivalent relationship between the average sectional area of undeformed chip thickness multiplied by the wheel velocity and material removal rate through analyzing the cross-section contour of grain trajectory based on the conservation of volume in plasticity. The formula of average undeformed chip thickness is deduced by using the Rayleigh's probability density function.

$$E(h) = \sqrt{\frac{a_p V_w}{2CV_s l_c}}. \qquad (3)$$

where a_p is the grinding depth, V_w is the workpiece velocity, V_s is the grinding wheel velocity, C is the number of active grains per unit area, and l_c is the wheel–workpiece contact length.

Substituting the values of C and l_c from Eq. (2) and the wheel–workpiece contact length of ultrasonic and ELID composite grinding[7] in Eq. (3), the average undeformed chip thickness of ultrasonic and ELID composite grinding can be expressed as follows:

$$h_t = \sqrt{\frac{a_p V_w}{2V_s \sqrt[3]{\left(\frac{6V_m}{\pi d_3}\right)^2} \left[\Phi\left(\frac{H_{max} - E(h)}{\sigma}\right) - \Phi\left(\frac{H_{cut} - E(h)}{\sigma}\right)\right] \sqrt{\left(1 \pm \frac{V_w}{V_s}\right)^2 + \left(\frac{2\pi f A}{V_s}\right)^2} \sqrt{\frac{d_w a_p d_s}{d_w - d_s}}}}. \qquad (4)$$

4   EXPERIMENTS

4.1   Experimental conditions

Experimental system: The specimen is ZTA nano-ceramic with the volume fraction about 15% of ZrO_2, specimen's dimensions are 60 mm in outside diameter and 35 mm in inside diameter and 45 mm in height. Three-axis machining center (VMC850E). Power supply: High-frequency pulse power of ELID (HDMD-V) with three gears voltage 60 v, 90 v, and 120 v. Scanning electron microscope (SH 4000M). White light interference surface contour graph (Talysurf CCI 6000). Three-dimensional dynamometer (KISTLER). Specific experimental conditions are shown in Table 1.

4.2   Experiment results analysis

The experimental results on grinding depth of ZrO_2 ceramics using Ultrasonic and ELID composite grinding are shown in Table 2. The change of the grinding force with undeformed

Table 1. Experimental conditions.

Experimental condition	Experiment parameter
Plastic conditions	Voltage: 120 V; Duty ratio: 5:5 μs; Plastic wheel speed: 1000 r/min; Grinding wheel speed: 1000 r/min
Sharp conditions	Duty ratio: 5:5 μs; Voltage: 120 V; Electrode gap: 1 mm; Grinding wheel speed: 1000 r/min
Grinding parameters	Feed speed: 80 mm/min; Workpiece linear velocity: 0.37 m/s; Grinding wheel linear velocity: 4.2 m/s; Grinding depth: 1, 3, 5, 7 μm
Ultrasonic parameters	Frequency: 34.835 KHz; Amplitude: 8.2 μm

Table 2. Experimental results.

Grinding depth	Grinding force		Surface roughness	Calculated value
a_p (μm)	F_n (N)	F_t (N)	R_a (μm)	h_t (μm)
1	5	2.1	0.01831	1.118
3	6.1	3.5	0.02943	1.565
5	9.3	4	0.03976	1.780
7	11.3	6.8	0.04356	1.934

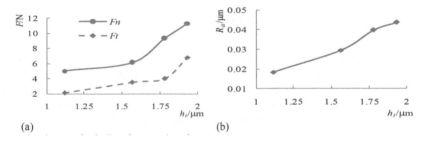

Figure 3. Change of grinding force and surface roughness with undeformed chip thickness.

chip thickness is shown in Figure 3(a), and the change of the surface roughness with undeformed chip thickness is shown in Figure 3(b). From the figure, it can be observed that with the increase of undeformed chip thickness, the normal grinding force and tangential grinding force and the surface roughness present a rising trend. This is because there is a correlation between undeformed chip thickness and grinding depth, the grinding depth increases, undeformed chip thickness will also increase accordingly. But the normal grinding force and the surface roughness value begin to sharply increase at the average undeformed chip thickness of 1.565 μm, while the tangential grinding force begins to sharply increase at the average undeformed chip thickness of 1.78 μm. It means that the material removal mechanism has changed from plasticity to brittleness, it can be explained as follows: the normal grinding force is closed to the critical load required to produce micro-crack at the average undeformed chip thickness of 1.565 μm, at this point there may be the formation of median micro cracks in the subsurface, but there is no cracks and debris on the processing surface, material removal mode begins to change from plasticity removal to brittleness removal which leads to the brittle fracture begins to appear on the surface of workpiece, and make the deterioration of the surface roughness; the median/radial cracks forms at the average undeformed chip thickness of 1.78 μm, due to the increase of the grinding depth, the normal grinding force increases sharply, at the same time, the tensile stress produced by the tangential force which consists of cutting force and friction force exceeds the ultimate stress of materials, material

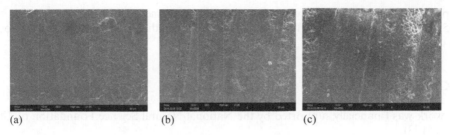

(a)   (b)   (c)

Figure 4.　SEM photos of surface topography.

removal mode performs for plasticity and brittleness mixed removal, but it is given priority to plasticity removal. While the average undeformed chip thickness reaches 1.934 μm, normal grinding force and tangential grinding force are increasing sharply, which exacerbated the median/radial crack, even make the median crack and radial crack extension become more obvious macroscopic crack.

The relationship between the material removal mode and undeformed chip thickness can be analyzed by observing the grinding surface topography at the grinding wheel linear velocity of 4.2 m/s, the workpiece linear velocity 0.37 m/s, and the grinding depth of 1 μm, 3 μm, 5 μm and 7 μm. SEM photos of grinding surface are shown in Figure 4, when the average undeformed chip thickness is 1.565 μm. It can be seen from Figure 4(a), the grinding surface is relatively smooth, and presenting the tiny grinding scratches, accumulation phenomenon formed by the material of inelastic deformation zone to flow on both sides of the grinding scratch caused by the extrusion of abrasive can be seen vaguely. Known from the analysis of the grinding force, at this point there may be the formation of median micro cracks in the subsurface, but there is no cracks or debris on the processing surface and material removal mode begins to change from plasticity remove to brittleness remove. SEM photo of grinding surface is shown in Figure 4(b) at the average undeformed chip thickness of 1.78 μm. Surface integrity is bad, and the phenomenon of brittle fracture appears, grinding scratches with clear and sharp edges is clearly visible and the bottom of the grinding scratches formed by plough of abrasive grain is smooth, and it is obvious that accumulation phenomenon formed by the material of inelastic deformation zone to flow on both sides of the grinding scratch caused by the extrusion of abrasive at this time, the brittle fracture has emerged, but the plough accumulation phenomenon caused by the plastic deformation is still dominant, material removal mode performed for plastic and brittle mixed removal. SEM photo of grinding surface as is shown in Figure 4(c) at the average undeformed chip thickness of 1.934 μm. Surface integrity is poor, there is a phenomenon of continuous mass of brittle fracture, there also appears different degrees of fracture at the bottom of the grinding scratches formed by plough of abrasive grains, the accumulation phenomena on both sides of the grinding scratches were eroded by broken, so the edges become blur, and the brittle fracture pit as half a pie formed by the extension of median cracks and the radial crack appeared. At this time, brittle fracture is more obvious, but the material removal mode is still plastic and brittle mixed, brittle fracture is dominant.

5　SUMMARY

This paper established the model of undeformed chip thickness that suits for the Ultrasonic and ELID composite grinding, it is found there is a good correlation between grinding force and grinding surface roughness with undeformed chip thickness by analyzing the change rule of grinding results with undeformed chip thickness. The surface topography was observed with the help of SEM, the results show that the material removal mechanism changes from plasticity to brittleness at the average undeformed chip thickness of 1.565 μm, it is plastic removal when the average undeformed chip thickness is less than 1.565 μm, while it begins to appear brittleness removal when the average undeformed chip thickness is more than 1.565 μm, which is consistent with the changes of the grinding force and the surface roughness taken place here.

ACKNOWLEDGEMENT

This work was financially supported by the National Natural Science Foundation of China (E51175153).

REFERENCES

[1] Han Zhenlu, Li Changhe, Wang Sheng. 2013. Experimental Investigation into Material Removal Mechanism of Nano-ZrO$_2$ Dental Ceramic Grinding, *China Mechanical Engineering*. 24 (9), 1150–1154.

[2] You Yi Zheng, Hao Pang, Qi You and Xiao Feng Jia. 2014. The Influence of Electric Parameters on the Surface Quality of Ceramic in Internal Grinding Using the Ultrasonic and ELID Composite Grinding, *Applied Mechanics and Materials*. 596, 47–51.

[3] Dai Y., Ohmori H., Lin W.M. 2006. A fundamental study on optimal oxide layer of fine diamond wheels during ELID grinding process, *Key Engineering Materials*. 304, 176–180.

[4] Nguyen T.A., Bufler D.L. 2005. Simulation of precision grinding process, part 1: generation of the grinding wheel surface, *International Journal of Machine Tools and Manufacture*. 45 (11), 1321–1328.

[5] Su Yu, Guo Hua, Wang Jin-bao, Zhang Yan-jun. 2009. Study on fashion of wearing and height of protrusion of diamond in wire saw, *Superhard Material Engineering*. 21 (2), 9–15.

[6] Sanjay Agarwal P. Venkateswara Rao. 2013. Predictive modeling of force and power based on a new analytical undeformed chip thickness model in ceramic grinding, *International Journal of Machine Tools and Manufacture*. 65, 68–w78.

[7] Zhaobo, Li Yamin, Bian Pingyan. 2013. Theoretical Research and Simulation of Grit's Motion for Internal Ultrasonic Vibration Grinding, *Key Engineering Materials*. 579–580, 133–137.

Advances in Engineering Materials and Applied Mechanics – Zhang, Gao & Xu (Eds)
© *2016 Taylor & Francis Group, London, ISBN 978-1-138-02834-0*

Effect of Low Voltage Pulsed Magnetic Field on microstructure and wear behavior of A356 alloy

W. Zhou, L. Zhang & Q. Zhou
National Defence Key Discipline Laboratory of Light Alloy Processing Science and Technology,
Nanchang Hangkong University, Nanchang, China

ABSTRACT: Effect of Low Voltage Pulsed Magnetic Field (LVPMF) on the solidified microstructure and wear behavior of A356 alloy was investigated. Experimental results show that solidified microstructure of A356 alloy was significantly refined when the LVPMF was applied during solidification. The A356 alloy with LVPMF treatment as-T6 state exhibited much higher wear resistance and damping capacity than that without LVPMF treatment. The adhesive wear was observed in A356 alloy without LVPMF treatment under higher normal load of 58.8N, whereas mainly abrasive was observed in that with LVPMF treatment.

1   INTRODUCTION

It is well recognized that the mechanical properties of aluminum alloys depend strongly on the microstructure of alloys and that a fine-grained and equiaxed microstructure is desirable in castings, because it improves mechanical properties, reduces hot tearing, facilitates feeding to eliminate shrinkage porosity, and gives a more uniform distribution of secondary phases. There are many techniques available to obtain such a microstructure[1], such as deliberate addition of master alloys containing melt inoculants, rapid solidification and physico-mechanical methods[2], which include mechanical or electromagnetic stirring, ultrasonic vibration and electromagnetic vibration. Deliberate addition of inoculants or mechanical stirring is often used in industry. However, both of them cause undesirable pollution of melts.

The application of an electromagnetic vibration in the solidification of metals and alloys during solidification is considered an effective method to improve solidification microstructure and benefit the properties of the metals and alloys[3]. Low Voltage Pulsed Magnetic Field (LVPMF) process as a new electromagnetic vibration technique has become one of the most promising new technique to vibrate the liquid metals and alloys, and it has been paid more attention in recent years. LVPMF process has several advantages such as simple equipment, high efficiency, low investment, extensive alloy application scope, as well as low pollution because of the non-contact with the melt. Moreover, compared with the conventional strong magnetic field, the process is more secure and can be used in practice. Yang et al. applied LVPMF during the solidification of Mg-Al-Zn alloy and Al-Cu alloy, and the grain refinement was achieved with proper thermal control under the action of LVPMF[4,5].

Here it should be pointed out that although the effect of LVPMF on the solidification of metals and alloys has been studied extensively, while to the best of our knowledge, the effect of LVPMF on wear behavior has not been investigated. In the present work, the effect of LVPMF on the solidified microstructure and wear behavior of A356 alloy was discussed.

2   EXPERIMENTAL PROCEDURE

The normal chemical composition of A356 alloy in this work was 7.1 wt.%Si, 0.3 wt.%Mg, 0.15 wt.%Fe and balance Al. A356 alloy was melted in a graphite crucible in an electric resistance furnace. After the melt was heated to 730°C, it was degassed for 10 min with argon gas through a graphite lance, and then cooled to 650°C. Subsequently, approximately 1000 g melt was poured into the stainless steel mold with 100 mm in diameter and 200 mm in depth that was preheated to 450°C. LVPMF was started immediately after pouring until the melt completely solidified. The LVPMF processing parameters were briefly summarized as follows, the discharging voltage was 250 V and the discharging frequency (numbers of per-second discharging) was 10Hz. The processing details were mentioned elsewhere[6].

Microanalysis samples were cut out of the ingot. They were polished and etched by Keller's reagent (2.5% HNO_3, 1.5% HCl and 1% HF in water), and the microstructure was observed using Neophot-2 optical microscopy and Hitachi S-3400 Scanning Electron Microscopy (SEM), and MIAPS (Micro-image Analysis & Process) image analyzing soft was used to determine average grain size.

Dry sliding wear tests were conducted using a pin-on-disc wear testing machine. Tests were carried out at 9.8, 19.6, 29.4, 39.2, 49 and 58.8N normal loads. The sliding velocity was kept constant at 1 ms[-1]. All tests were carried out for a sliding distance of 1800 meters. The wear testing machine was microprocessor controlled where height loss and frictional force were monitored simultaneously. The height loss data was converted to volumetric loss by multiplying it with area of cross section of the test pin. At least three tests were carried out for each set of parameter to get a representative data. The worn surface was examined under SEM, and EDS X-ray analysis was also performed.

3 RESULTS AND DISCUSSION

Figures 1 (a) and (b) show the comparison of the obtained microstructures of A356 alloys without and with LVPMF treatment at a lower magnification. When LVPMF was not imposed, the microstructure exhibited fully developed primary α-Al dendrites, and one branch of a primary dendrite was about 600 μm in length, indicating that the grain size was in the range of several millimeters since one equiaxed grain usually contained six primary dendrite arms. In contrast, the microstructure of A356 alloy treated by LVPMF which consists of majority of primary α-Al particles with rosette-like and near globular morphology, as shown in Figure 1 (b). The morphology and size of individual eutectic silicon particles were not distinguishable at such a

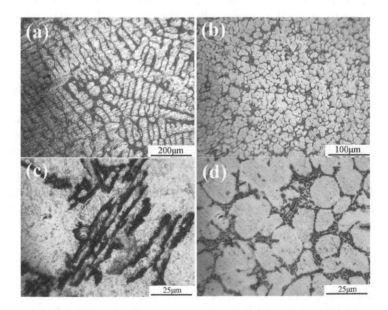

Figure 1.   Microstructures of A356 alloys without (a), (c) and with (b), (d) LVPMF treatment.

magnification. Figures 1 (c) and (d) show the micrographs of the eutectic silicon without and with LVPMF treatment at a higher magnification. The coarse acicular (plate-like) eutectic silicon observed on the A356 alloy without LVPMF treatment. However, with LVPMF treatment, fine fibrous eutectic silicon was observed. The formation of spherical primary α-Al phase and fine eutectic silicon had been reported elsewhere[6].

The specific wear rates of A356 alloys without and with LVPMF treatment as a function of normal loads are shown in Figure 2. The specific wear rate for these two materials decreased with increasing normal load as clearly shown in Figure 2. The specific wear rate was a function of volume loss, sliding distance and normal load. Generally, under the constant sliding distance the specific wear rate was inversely proportional to the normal load and remains constant if the volume loss was in multiple to the normal load[7]. However, in this study the increase in volume loss was not in multiple with increasing normal pressure and this could be due to the work hardening effect at the wear surface during sliding. It should be noted that the A356 alloy with LVPMF treatment shows less specific wear rate as compared to that without LVPMF treatment. The present experimental results of wear studies supported the microstructure studies (Fig. 1). The change in microstructure from coarse dendritic structure to fine rosette-like and near globular grain structure, and with change in coarse acicular (plate-like) eutectic silicon to fine fibrous or lamellar shape resulted in high wear resistance of A356 alloy.

Figures 3 (a) and (b) show typical worn surfaces of A356 alloys without and with LVPMF treatment and at higher normal load (58.8 N). The arrow marks show the sliding direction. One can see that the appearance of worn surface of A356 alloy without LVPMF treatment

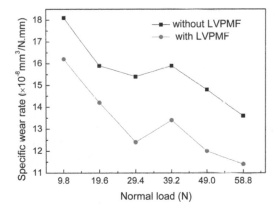

Figure 2.   Specific wear rate of A356 alloys without and with LVPMF treatment as a function of normal load.

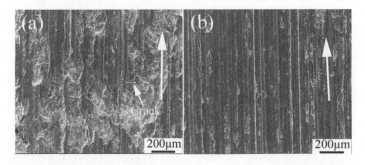

Figure 3.   SEM micrographs of worn surfaces of A356 alloys (a) without LVPMF treatment (b) with LVPMF treatment.

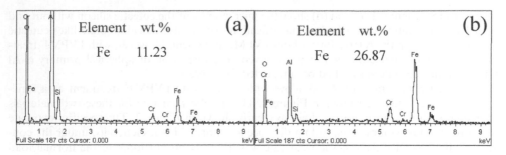

Figure 4. EDS X-ray spectrums of worn on A356 alloys (a) without LVPMF treatment, (b) with LVPMF treatment.

(Fig. 3(a)) exhibited wide grooves and irregular deep pits along the sliding direction. The presence of grooves indicated the micro-cutting effect of the counterface, while pits were the evidence of ductile fracture. Cracks also can bee found on the worn surface of A356 alloy without LVPMF treatment due to coarse acicular eutectic silicon acted as internal stress raisers. The morphologies of the worn surface suggested that the wear mechanism was adhesion in case of A356 alloy without LVPMF treatment. However, the worn surface of A356 alloy with LVPMF treatment (Fig. 3(a)) illustrated narrow grooves and a few of shallower pits along the sliding direction. The worn surface of A356 alloy with LVPMF treatment revealed the formation of less amount of adhesive wear apart from the few abrasion grooves.

Figures 4 (a) and (b) shows the EDS X-ray analysis results of the worn surfaces of A356 alloys without and with LVPMF treatment, respectively. The Al, Si, O, Cr and Fe were an indication of the mechanical mixing process accompanied by adequate oxidation reaction during the wear process. The frictional heating during the dry sliding wear facilitate easier formation of oxides. One can see that the presence of Fe on both A356 alloys without and with LVPMF treatment specimens. As for the A356 alloy with LVPMF treatment, the Fe content was much higher. The increase of the Fe level indicated better anti-friction properties of the A356 alloy with LVPMF treatment. The worn surface of the A356 alloy with LVPMF treatment was covered by layers containing Fe from the counterface material confirming the presence of Mechanical Mixed Layer (MML). The presence of MML had been reported earlier during wear studies[8]. The formation of MML on the worn surface could contribute to higher wear resistance of the A356 alloy with LVPMF treatment in comparison to that without LVPMF treatment.

4 CONCLUSION

The morphology of primary α-Al changed from coarse dendritic to fine rosette-like and near globular shape and fine lamellar or fibrous eutectic silicon particles was formed when the LVPMF was applied during solidification of A356 alloy. A356 alloy without LVPMF treatment showed much higher wear resistance than that without LVPMF treatment. The adhesive wear was observed in A356 alloy without LVPMF treatment at high normal load of 58.8N. However, mainly abrasive was observed in that with LVPMF treatment.

ACKNOWLEDGEMENTS

The research work was supported by National Natural Science Foundation of China (Grant No. 51261026 and 51401102), National Defense Key Disciplines Laboratory of Light Alloy Processing Science and Technology in Nanchang Hangkong University (GF201301005) and Nanchang Hangkong University Foundation of China (EA201403011).

REFERENCES

[1] Knuutinen, A., Nogita, K., Mcdonald, S.D. & Dahle, A.K. 2001. Modification of Al-Si alloys with Ba, Ca, Y and Yb. *J Light Metal* 1(4):229–240.

[2] Jian, X., Xu, H., Meek, T.T. & Han Q. 2005. Effect of power ultrasound on solidification of aluminum A356 alloy. *Mater Lett* 59(2–3):190–193.

[3] Vives, C. 1998. Crystallization of aluminium alloys in the presence of cavitation phenomena induced by a vibrating electromagnetic pressure. *Journal of Crystal Growth* 158:118.

[4] Fu, J.W. & Yang, Y.S. 2012. Microstructure and mechanical properties of Mg-Al-Zn alloy under a low-voltage pulsed magnetic field. *Mater Lett* 67:252.

[5] Li, Y.J., Tao, W.Z. & Yang, Y.S. 2012. Grain refinement of Al-Cu alloy in low voltage pulsed magnetic field. *J Mater Process Technol* 212:903.

[6] Zhang, L., Li, W., Yao, J.P. & Qiu, H. 2012. Effects of pulsed magnetic field on microstructures and morphology of the primary phase in semisolid A356 Al slurry. *Mater Lett* 66:190.

[7] Clarke, J. & Sarkar, A.D. 1979. Wear characteristics of as-cast binary aluminium-silicon alloys. *Wear* 54:7–16.

[8] Venkataraman, B. & Sundararajan, G. 1996. The influence of sample geometry on the friction behaviour of carbon-carbon composites. *Acta Mater* 44:461–473.

Mechanical

Advances in Engineering Materials and Applied Mechanics – Zhang, Gao & Xu (Eds)
© 2016 Taylor & Francis Group, London, ISBN 978-1-138-02834-0

The band saw blade grinding machine design

R. Zhang, Y.K. Yang & B. Tao
College of Mechanical and Electrical Engineering, Wuhan Donghu University, Wuhan, China

ABSTRACT: According to a current situation of domestic traditional wood polishing machine work, a new design of the band saw blade grinding machine has been designed, which changes the traditional working way and principle of polishing machine, and it achieves a high degree of automation, efficiency, easy operation and safety performance. All these satisfy the requirements needed in the process of production. Compared with the portable grinding machine, the grinding machine accuracy and efficiency is improved, the structure is simple, the movement is compact and coordinate, and can bear the equivalent force. More important of all, according to the morphological appearance design, it is simple and not necessary to consider the adaption of the man–machine feel. The band saw blade grinding machine has a great developing prospect.

1 INTRODUCTION

With the continuous increase of China's demand for timber, wood processing industry has been now developing rapidly; however, in the actual production of sawn timber, the efficiency of the use of saw blade is relatively low. Meanwhile, the automatic grinding device is pressing demand, and some factories are using hand-held manual polishing machine that relatively brings large workload and complicated structure. Also, it not only makes the electric tool to meet the requirements of feeling, but also consider as the aesthetic. Due to this, it is unable to be fixed and controlled, together with the operator low level of technology and improper use. In grinding machine, when the grinding starts, the saw blade cannot meet the required stiffness and sharpness. Its life shortens greatly into the bargain.

According to the above problems, we designed the grinding machine by studying the prototype of the market.

2 WORKING PRINCIPLE

The design of the grinding machine with band saw blade is mainly composed of the motor device, transmission device (including the belt drive and connecting rod drive), and polishing device. The machine works with the grinding machine frame and the band saw blade synchronous intermittent movement. As shown in Figure 1, the No. 1 motor driving device that rotates at a constant speed through the reducer and pulley drive and serrated cam drive. Cam 5 diameter, whose size is same as outer sawtooth's, installation with just a tooth bit, when it turns to the lowest point, the connecting rod tension is to be pulled back, when the grinding wheel is back to grind the blade. On the contrary, the connecting rod drives the wheel forward to the grind blade. When sawtooth counterclockwise a tooth bit in the process of grinding machine, the machine goes forward, the No. 2 motor driving machine that drives at high speed grind blade rake angle. And then back with the completion of polishing, further, the sawtooth moves a tooth, the burnish machine go forward again. The process continues until all of the saw blades are grounded.

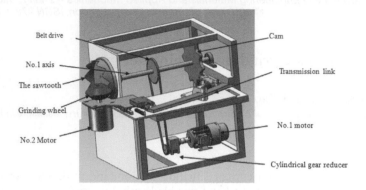

Figure 1. Sawtooth grinding machine overall structure.

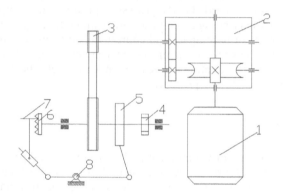

Figure 2. The overall mechanism motion sawtooth grinding machine schematic.
1—Motor; 2—Worm gear-Cylindrical gear reducer; 3—V belt transmission; 4—Bearing; 5—Cam; 6—Sawtooth; 7—Grinding wheel; 8—Fulcrum.

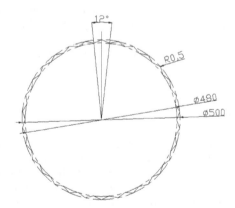

Figure 3. Cam size design.

3 CAM DESIGN

According to the requirements of sports organizations, the cam that is regarded as a main part does constant rotation and it needs to meet the cam gear travel itinerary with proportional. The cam has the same size as the outer sawtooth, installation with just a tooth bit, when it turns to the lowest point, the gear comes to the highest point, and it works in both ways.

Table 1.   Motion cycle chart.

Cam mechanism	The highest point	The lowest point	The highest point	The lowest point
Grinding wheel mechanism	Forward	Back	Forward	Back

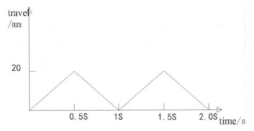

Figure 4.   Grinding wheel travel.   Figure 5.   Cam stroke.

Serrated teeth 50 mm high, with a total of 30 teeth, so the serrated outer diameter is 500 mm, the cam outer diameter is 500 mm. The distance of the highs and lows of 20 mm, the schedule of 20 mm, as shown in Figure 3.

4   THE MAIN EXECUTION DEVICE

4.1   *Transmission device*

Band saw blade grinding machine of transmission device is mainly composed of two parts the belt and the transmission. Belt drive motor is through a reduction gear and then after two belt drive, the final pass to speed shaft, the shaft is through a key connection speed passes blade chuck and cams. Connecting rod drive is through the rotation of the cam, whenever the scroll wheel on the connecting rod on the cam from a distal to the recent end finally returned to the farthest from a process that will drive the connecting rod transmission to drive the other end is fixed do a forward back the periodic motion of grinding wheel, so as to achieve the periodic cycle. The belt drive and connecting rod transmission are shown in Figure 1.

4.2   *Grinding device*

The grinding device with band serrated grinding machine uses the wheel directly connected to a motor. The high-speed operation of the motor drives the wheel, and the wheel does the forward and back and forth movement in the effect of connection in order to achieve the grinding work. The blade and grinding wheel are shown in Figure 1.

4.3   *Reduction gear*

In the process of connecting rod drive, this design between the frame and the sliding panel uses the guide rail slider components, which makes the connection on the front panel of the grinding wheel to do line reciprocating periodic movement. We chose a cylindrical gear reducer between the belt drive and No. 1 motor, make the high speed running things out of the motor to finally pass a shaft smooth transmission. The cylindrical gear reducer is shown in Figure 1.

4.4   *Three-dimensional motion simulation renderings*

Simulation of 3D motion diagram of the mechanism of this design uses the solidworks 2012 three-dimensional design. The software has the advantages of simple operation and easy to master. The main drive devices were designed, the control part, polishing device, each execution parts together, to construct the whole mechanism assembly. The 3D motion simulation diagram is shown in Figure 1.

5 CONCLUSION

This scheme design of band saw blade grinding machine, changed the traditional way and the principle of polishing machine work. It is a high degree of automation, high efficiency, and easy operation, and can be used with good safety performance. Compared with the portable grinding machine, the grinding machine accuracy is improved, the structure is simple, the movement is compact and coordinate, and can bear the equivalent force. More important of all, according to the morphological appearance design, it is simple and not necessary to consider the adaption of the man–machine feel.

REFERENCES

Cheng XianPing. 2011. *Electromechanical transmission and control.* Huazhong university of science and technology press.
2002. China Mechanical Design Canon editorial board. *China Mechanical Design Canon (All six volumes).* JiangXi Science and Technology Press.
Deng XingZhong. 2011. *Mechanical and electrical drive control, The fifth edition.* Huazhong University of Science and Technology Press.
Hao Wu, ZhenJi Zong, Lei Zhang. 2011. *Mechanical design course design handbook.* Higher Education Press.
Li ZhaoSong. 2012. *Grinding technology.* Mechanical Industry Press.
Pu LiangGui, Ji MingGang. 2013. *Mechanical design.* Higher Education Press.
Sun Huan, Chen ZuoMo, Ge WenJie. 2013. *Mechanical principles, The eighth edition.* Higher Education Press.
Wen BangChun. 2010. *Mechanical design handbook.* The fifth edition. Mechanical Industry Press.
Yang Jianmin. *Engineering materials, The second edition.* Mechanical Industry Press.
Zhong YiFang, Yang JiaJun. 2004. *Mechanical design principle and method (book 1 & book 2).* Huazhong University of Science and Technology Press.

Mechatronics

Advances in Engineering Materials and Applied Mechanics – Zhang, Gao & Xu (Eds)
© 2016 Taylor & Francis Group, London, ISBN 978-1-138-02834-0

Rapid fault diagnosis system based on the fault tree for x-type power system

D.B. Zhang, C.A. Qiao, Y. Liu & X.Y. Li
Wuhan Mechanical College, Wuhan, Hubei, China

ABSTRACT: Aiming at the uncertainty of the faults caused by the internal complexity of x-type power system, and the fact that traditional methods of fault diagnosis are unable to meet the needs of maintenance and support, the fault diagnosis of the power system is researched. This paper, beginning from its concept, expatriates the characteristics and functions of fault tree analysis, and how it is applied to solve the problems of fault location and maintenance, which provides a new and effective method for the maintenance and support of x-type power system.

1 INTRODUCTION

The performance of x-type power system is improved along with the complexity of its structure, which brings forward new problems to its fault diagnosis, among which the fault type increases so rapidly that the repair personnel feel more and more difficult to master the tens of thousands of fault knowledge. However, the status quo of the fault diagnosis toward x-type power system is that the repair technicians still diagnose the faults according to the obtained fault phenomena with their experience or by consulting the handbook on faults[3]. To x-type power system with much fault knowledge, the method of fault diagnosis depending on the men's memory or experience is inefficient and unable to meet the needs of the rapid diagnosis and maintenance in wartime. In order to locate the faults quickly, guide the maintenance, simplify the maintenance procedures and reduce unnecessary disassembly, the fault diagnosis system based on the fault tree for x-type power system need to be developed[1]. This system can solve the problems of detecting, locating and maintaining the faults, provide reliable technical support for x-type power system maintenance as well as the platform of fault diagnosis system for maintenance personnel training of the army. It is also important to improve the maintenance support ability of the power system.

2 CHARACTERISTICS AND FUNCTIONS OF FAULT TREE ANALYSIS

Fault tree analysis is a design method to determine the possible cause of the system faults and its probability, so as to compute the fault probability of the system, take the corresponding measures and improve the system reliability by analyzing the various factors (including hardware, software, environmental and man-made factors) resulting into system failure, and drawing the logic diagram (ie. fault tree). It can graphically expatiate how the system has failed, and has been applied to analyze the safety and reliability of some large complex systems.

2.1 *Characteristics of fault tree analysis*

1. Fault tree analysis is a method of graphical interpretation and logical reasoning on fault events under certain conditions. This method is applied to clearly illustrate how the system failed, and considered as a snapshot of the system under certain fault condition[2].

2. Fault tree analysis can help to link the faults organically between the system and the fault components together, and to identify all possible failure modes of the system, which are also called as all the minimal cut sets of fault tree or fault spectrum of the system.
3. Fault tree is also a kind of visualized technical data, as well as a guide of object teaching and maintenance for the equipment management and maintenance personnel who never participates in the design of the system when it is established.
4. Fault tree analysis is used to analyze the complex system, so it can't do without computer software. Nowadays has seen the rapid development and great progress of the software on fault tree analysis at the qualitative, quantitative, graphical, micro-computerized and other aspects.
5. Due to the impact of the uncertainty of statistical data, fault tree analysis is confronted with much difficulty in quantitative analysis. So in addition to the effective qualitative analysis, more people are interested in the analysis on importance and sensitivity of the fault tree, which is an important part of its qualitative analysis.

2.2 *Functions of fault tree analysis*

1. To comprehensively analyze the causes of system failure. Much flexibility exists in the fault tree analysis that is not limited to the general analysis of the system reliability, and can analyze all the conditions of system failure[6]. Not only the impact of some components faults on the system, but also the particular causes of the faults of the components are analyzed.
2. To demonstrate the internal relationship of the system, point out the logic relation between the faults of the system and its components, as well as the weak points of the system, and make clear the way and level of various potential factors' impact on the occurrence of the fault, which ensure that many problems are discovered and solved in the process of analysis and the reliability of the system is improved.
3. To calculate quantitatively the failure probability and other parameters on reliability of the complex system, and provide quantitative data for the improvement and evaluation of the system reliability.
4. To reflect clearly the relationship between the faults of the system and its units, and provide the guidance for the detection, isolation and clearance of the faults after fault tree system is established. It is equally a vivid guide about the management and maintenance for the management and maintenance personnel who never participate in the design of the system.

3 DESIGN AND REALIZATION OF THE FAULT TREE DIAGNOSIS SYSTEM

3.1 *Structure of fault diagnosis system*

When diagnosing the faults of x-type power system, one piece of fault information can induce several interconnected or overlapped results of diagnosis, from which the final diagnosis results should come out of the comprehensive analysis. Considering the complex construction of x-type power system, the diagnosis is a process of comprehensively making a decision out of ambiguous information. Fault diagnosis is based on function detection. The fault diagnosis system for x-type power system is mainly composed of the man-machine interface, the knowledge acquisition instrument, knowledge base and its management system, inference engine, database and its management system and explanation instrument. The structure of the system is shown in Figure 1.

3.2 *Man-machine interface*

The man-machine interface, as a medium between the fault diagnosis system and equipment maintenance experts, is composed of a set of programs and the corresponding hardware, and used to complete the input and output work. With this interface, the equipment maintenance

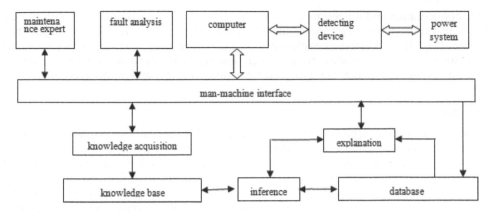

Figure 1. Diagram of diagnosis system construction.

experts can input knowledge, update and improvement of the knowledge base, and the maintenance personnel can input the fault information of the equipment into the system for the solution, and the system can output the results, answer the maintenance personnel's questions and send the order for the further fault information. In the input or output process, the man-machine interface need to complete the conversion between the internal representation and external representation[5]. When inputting, the information easily input by the equipment maintenance expert will be converted into the system's internal representation, which will be sent to the appropriate agencies to deal with. when outputting, the information to be output by the system will be converted from its internal representation to the external one which people understand easily, and then delivered to the corresponding user.

3.3 *Knowledge acquisition instrument*

This is an instrument to acquire the knowledge in fault diagnosis system. Its tasks are to input the knowledge into the knowledge base, and establish the good-performance knowledge base through maintaining consistency and integrity of the knowledge. The knowledge is acquired by the knowledge engineer from the equipment maintenance expert, and then transmitted to the knowledge base with the help of the corresponding editing software.

3.4 *The knowledge base and its management system*

Knowledge base is a memorizer for knowledge, mainly including principles about equipment maintenance, empirical knowledge of the equipment maintenance experts and relevant facts etc. The knowledge in the knowledge base is from the knowledge acquisition instrument, and at the same time is provided for the inference machine to solve problem, so it has close relationship with both of them. Knowledge base management system is in charge of the organization, retrieval, maintenance of the knowledge in the knowledge base. If any other part of fault diagnosis system need to be associated with the knowledge base, it must be realized by the management system, which ensures the unified management and use of the knowledge base.

3.5 *Inference engine*

Inference engine, as its "thinking" mechanism, is the core of the fault diagnosis system. Its tasks are to control and execute the solution to the problem by simulating the thinking process of the equipment maintenance experts. It can make use of the knowledge of knowledge base to find the solution to the fault problem and prove the correctness of the fault judgment by adopting the inference method and some control strategies according to the currently known fault phenomena. The performance and the structure of inference engine are usually connected with representation and organization of the knowledge, but have nothing to do with its content, which is conducive to guarantee the relative independence of the inference engine from the knowledge base. When the knowledge in the knowledge base is changed, it is unnecessary to modify the inference engine correspondingly, and the fault diagnosis system will be able to realize the regular reasoning, case-based fault reasoning and the fault model reasoning.

3.6 *Database*

Database is a working memory used to store the initial facts and fault description provided by the maintenance personnel, as well as the results in the process of system operation, the final results and the operation information. The results tested by the fault detection mechanism will be stored as the initial facts into the database, the content of which is constantly changing. At the beginning of finding the solution, it is stored with the initial fact provided by maintenance personnel; in the process of reasoning it is stored with reasoning results of each step. Inference engine, according to the contents of the database, choose appropriate knowledge from the knowledge base for reasoning, and then to store the results into the database.

3.7 *The explanation instrument*

The explanation instrument, used to explain its own working principle, answer such questions as why and how to make the conclusion raised by the maintenance personnel, is an important feature of the fault diagnosis system different from the common program, and also an important measure to win the trust of the maintenance personnel. In addition, its explanation for itself can also help system builders to find the errors in the knowledge base and inference engine, so as to debug and maintain the system. Therefore, the explanation instrument is indispensable for the maintenance personnel or the system itself. The Explanation instrument, composed of a set of programs, can track and record the reasoning process. When the maintenance personnel need some explanations, it can make a response respectively as required of the problem, and finally transmit the solution the user in an appointed form through the man-machine interface.

4 KNOWLEDGE REPRESENTATION

The fault tree analysis is applied for the fault diagnosis. The structure of knowledge is regarded as a fault tree, with the fault phenomenon as the top of the tree, the components or lines where the faults exist possibly as the intermediate events, and the specific fault points as the bottom events. The structure of the tree and x-type power system is consistent, as is shown in Figure 2.

Knowledge is represented by using production rules, then:

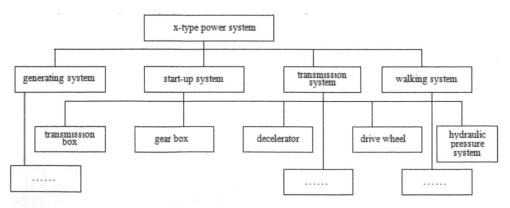

Figure 2.   Diagram about the fault tree.

IF (the condition) THEN (the conclusion)

Taking fire extinguishing system faults as an example, the rules are written as follows:

1. IF (faults exist in gear box), THEN (the transmission system does not work).
2. IF (faults exist in start-up system), THEN (power system does not work).

All the rules written according the above steps form the database of rules. We can see that the representation of rules is very intuitive and easy to be realized, suitable for the relatively simple knowledge representation, but it is unsuitable to represent the comparatively complex problems with the difficulties in knowledge acquisition. Therefore, the knowledge base is divided into shallow and deep knowledge base[4]. As for the shallow one, the knowledge and experience of fault diagnosis of the power system are learned and acquired from a large number of fault diagnosis examples of equipment maintenance experts, and stored into the knowledge base, by adopting the calculation method of reversed phase propagation learning to realize the knowledge processing.

The knowledge in the deep knowledge base is divided into two kinds, of which one is about the knowledge of positioning the faults according to the fault phenomena, which must be acquired through the learning algorithm according to the examples provided by the equipment maintenance experts. The other is the knowledge about the model of fault diagnosis principle for x-type power system, which is directly reflected to the interconnection weights of deep models and the relationship between the abstract concepts when it is used to build the deep knowledge base, unnecessary to learn and obtain from the experience.

5 CONCLUSION

The method of fault tree analysis, introduced into x-type power system maintenance and the fault diagnosis of the system, is an attempt of the new diagnosis method. With the method of fault tree analysis, the bottleneck problem of knowledge acquisition is solved, the integrity of diagnosis knowledge is ensured, and the fast and effective characteristics of fault diagnosis system are also embodied. The research and application of this system provides a new and effective method for the fault diagnosis of the power system, which will greatly improve the efficiency of battlefield maintenance of the power system.

The innovations of this article include: the method of fault tree analysis is introduced into the fault diagnosis system of equipment, which strengthens the knowledge acquisition ability of the system, and solves the bottleneck problem of knowledge acquisition of fault diagnosis system, and builds the easy-to-realize mechanism for reasoning. In this article it is researched on how to apply fault tree analysis in the power system maintenance to solve the problems of fault location and maintenance, and a new and effective method is provided for the fault diagnosis of military equipment.

REFERENCES

[1] Guan Huiling. 2000. *Principle and practice of the equipment fault diagnosing expert system*. Beijing: China Machine Press.
[2] Ni Shaoxu, Zhang Yufang, Yi Hong. 2008. Intelligent fault diagnosis method based on fault tree. *Journal of Shanghai Jiaotong University.* 42(8):1372–1375.
[3] Wang Daoping. 2000. A fault diagnosing expert system based on hybrid reasoning. *Journal of Computer Applications*. 20(7).
[4] Yang Xiaochuan, Xie Qinghua, He Jun. 2001. Fuzzy fault diagnosing method based on fault tree. *Journal of Tongji University*. 29(9):1058–1060.
[5] Yu Dejie, Zhao Dan, Zhou Anmei. 2013. Research on the diagnosis decision-making of key units based on ontology and fault tree. *Journal of Hunan University (Natural Sciences)*. 40(8):46–51.
[6] Zhang Debao. 2009. Design of portable online detector of the communication system. *Control and Automation*. (16):162.

Motor elastic suspension

Advances in Engineering Materials and Applied Mechanics – Zhang, Gao & Xu (Eds)
© 2016 Taylor & Francis Group, London, ISBN 978-1-138-02834-0

Influence of motor elastic suspension on running stability of EMUs

H. Yuan & W. Yong

State Key Laboratory of Traction Power, Southwest Jiaotong University, China

ABSTRACT: In this paper, the influence of motor elastic suspension on running stability with a simplified motor model is analyzed. Parametric studies are under taken in order to analysis the respective effects of the motor's parameters on the running stability of vehicle. It is found that the stiffness and damping properties of the motor have a great influence on the critical speed. There exists an optimum suspension frequency for motor, below which a relatively large critical speed is obtained. Finally, the effect of primary longitudinal stiffness, conicity of wheelset and series stiffness of anti-yaw damper on the optimum suspension frequency is analyzed, and it is shown that the optimum suspension frequency of motor is significantly affected by primary longitudinal stiffness and conicity of wheelset.

1 INTRODUCTION

Severe hunting of bogie emerges when EMUs operate at high speed. It has a serious effect on the safety and ride of vehicle and limits the further increasing of operating speed. Mounting motor on the bogie frame elastically through rubber node is a new technique used on high speed EMUs.

Suspending motor on bogie frame increases the mass between primary suspension and secondary suspension. It is well known that increasing the mass between primary suspension and secondary suspension is disadvantage to dynamic performance. Therefore, it is very necessary to suspend the motor on bogie frame with a careful designed parameter. Hunting stability belongs to lateral dynamics and has no relationship with longitudinal and vertical motion of vehicle. The former scholar's research shows that the yaw motion of motor has little influence on lateral running stability. Therefore, only the influence of lateral motion of motor on running stability is considered in this paper.

2 DYNAMIC MODEL

To research easily, it is assume that the suspension point of motor is only one. Its lateral stiffness and damping is K_{my} and C_{my}, so:

$$f_{my} = \frac{1}{2\pi}\sqrt{\frac{K_{my}}{M_m}}$$ (1)

$$\xi_{my} = \frac{C_{my}}{2M_m(2\pi f_{my})}$$ (2)

3 INFLUENCE OF ELASTIC SUSPENSION ON RUNNING STABILITY

In order to exclude the effect of anti-yaw damper, the influence of suspension frequency and damping of motor on critical speed without considering anti-yaw damper in whole vehicle is analyzed. The '*no motor*' and '*stiffness suspending*' case, which means decreasing or increasing the weight of bogie frame will increase or decrease the critical speed, is also analyzed. The influence of the lateral frequency f_{my} and lateral damping ratio ζ_{my} of motor on the critical speed are shown in Figure 1. The project 1 means two motors suspend on the bogie frame separately, while two motors are mounted together and suspend on the bogie frame in project 2. With the frequency fmy increasing, the critical speed first increases and then decreases and finally increases and tends to a constant value for small ζ_{my}, while it first increases and

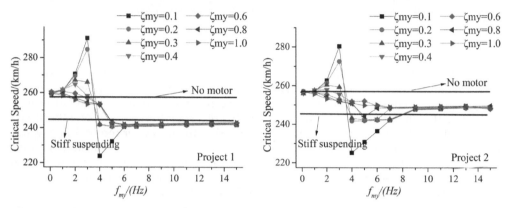

Figure 1.   The influence of f_{my} and ζ_{my} on critical speed.

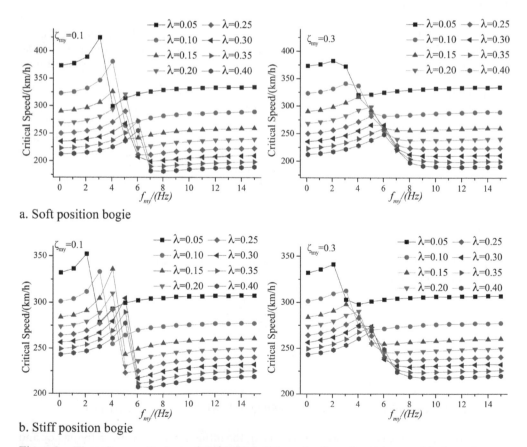

a. Soft position bogie

b. Stiff position bogie

Figure 2.   The influence of primary longitudinal stiffness, f_{my}, ζ_{my} and conicity on critical speed.

700

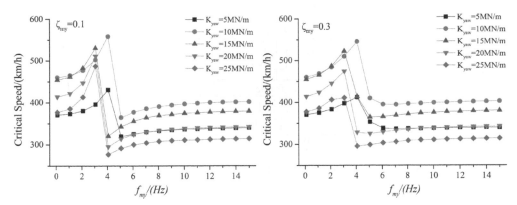

Figure 3. The influence of f_{my}, ζ_{my} and series stiffness of anti-yaw damper on critical speed.

then decreases directly to a constant value for large ζ_{my}. With an optimum motor suspension frequency, the vehicle can get the maximum critical speed which is larger than the critical speed of 'no motor' case. With an improper motor suspension frequency, the vehicle will get relatively low speed which is smaller than the critical speed of 'stiff suspending' case. It has little difference between the result of project 1 and project 2.

The influence of primary longitudinal stiffness, f_{my}, ζ_{my} and conicity on critical speed is shown in Figure 2. The 'soft position bogie' case and 'stiff position bogie' case means the primary longitudinal stiffness is 12.5 MN/m and 120 MN/m. The damping ratio of 0.1 or 0.3 is analyzed.

It can be drawn that the maximum critical speed of damping ratio of 0.1 is larger than that of 0.3 at the same condition. Increasing conicity will broaden the applicable scope of motor lateral suspension frequency. Increasing primary longitudinal stiffness will narrow the applicable scope of motor lateral suspension frequency. When conicity reach a special value, the applicable scope of motor lateral suspension frequency increases no more. This can be seen as a turning point for the influence of conicity on the applicable scope of motor lateral suspension frequency. With primary longitudinal stiffness increasing, value of the turning point increases.

For further research, the influence of anti-yaw damper is analyzed and the result is shown in Figure 3. With series stiffness of anti-yaw damper increases, the critical speed first increases then decreases and overall decreases the optimum value of lateral frequency of motor.

4 DISCUSSION ON RESULTS

The result obtained in the previous section seems to comply with certain rules and this can be explained by the theory of anti-resonance. The simplified model of whole vehicle is shown in Figure 4. The bogie frame and wheelset hunts and carbody holds still basically during vehicle hunting. So, it is assumed that carbody is stationary. The lateral coupling stiffness between motor and wheelset is negligible and the lateral degree of freedom of bogie frame, wheelset and motor is considered. External excitation input by wheelset will cause the vibration of bogie frame and motor.

According to vibration theory, dynamic magnification factor from wheelset to bogie frame (the ratio of lateral displacement amplitude of bogie and wheelset) with no damping is shown as:

$$\beta = \frac{K_{py}(K_{dy} - m_d\omega^2)}{(K_{sy} + K_{py} + K_{dy} - m_f\omega^2)(K_{dy} - m_d\omega^2) - K_{dy}^2} \tag{3}$$

$$\omega_d = \sqrt{K_{dy}/m_d} \tag{4}$$

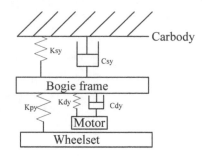

Figure 4.   Bogie hunting lateral simplified model.

where, K_{py} is primary lateral stiffness, K_{sy} is second lateral stiffness, K_{dy} is lateral position stiffness of motor, m_d is mass of motor, m_f is mass of bogie frame, and ω is excitation frequency, which is hunting frequency for vehicle lateral vibration system. ω_d is the lateral frequency of motor.

When $\omega = \omega_d$, β is equal to 0. It means that forced vibration objects keep stationary. The corresponding excitation frequency ω is called anti-resonance frequency of system. Anti-resonance frequency is different from resonance frequency. When excitation frequency of system is equal to resonance frequency, the dynamic magnification factor is equal to infinite. When excitation frequency of system is equal to anti-resonance frequency, the dynamic magnification factor is equal to 0. Since the hunting frequency is relate to operating speed, the lateral suspension stiffness of motor K_{dy} can be selected reasonably on the basis of operating speed range of vehicle. Therefore, running stability of vehicle is increased by improving suspension stiffness of motor.

5   CONCLUSIONS

This paper studies the influence of motor elastic suspension on running stability of EMUs. The following conclusions can be drawn form this study:

1. The stiffness and damping of motor have a great influence on running stability. There exists an optimum frequency of the motor suspension, below which a relatively large critical speed is obtained. Moreover, it is worth noting that a design with the frequency of the motor far away from the resonance frequency is attractive. Therefore, proper design of the motor elastic suspension is advantageous to enhance the critical speed and avoid resonance occurring at the same time.
2. The optimum suspension frequency is significantly affected by the primary suspension and wheelset conicity. The decrease of the primary longitudinal stiffness and increase of wheelset conicity will increase the optimum suspension frequency of the motor.

ACKNOWLEDGEMENTS

The authors wish to thank the European Commission for having provided financial support to the work presented here within Key project of science and technology research and development program of China Railway Corporation (2014J008-B) and Independent research project of State Key Laboratory of Traction Power of Southwest Jiaotong University (2014TPL-Z05).

REFERENCES

Alfi, S., L. Mazzola & S. Bruni (2008). Effect of motor connection on the critical speed of high-speed railway vehicles, *Vehicle System Dynamics: International Journal of Vehicle Mechanics and Mobility*, 46:S1, 201–214.

Alonso A. (2011). Yaw damper modelling and its influence on the railway dynamic stability. *Vehicle System Dynamics*. 49(9):1367–1387.

Braghin, F., S. Bruni & G. Diana (2006). Experimental and numerical investigation on the derailment of a railway heelset with solid axle, *Vehicle System Dynamics*. 44, 05–325.

Caihong Huang, Jing Zeng & Shulin Liang. (2013). Influence of system parameters on the stability limit of the undisturbed motion of a motor bogie. *Journal of Rail and Rapid Transit*. 1–13.

Jin D.C., Y. Luo & Z.H. Huang (1994). The effect of the driving motor's suspension type on the railway vehicle dynamics (in Chinese). *Journal of The China Railway Society*. 16: 43–47.

Richard JA. (1987). Stability of railway-car running at high speed. *Vehicle System Dynamics*. 16(1): 37–49.

Yao Yuan, Zhang Kailin & Zhang Hongjun (2013). Mechanism of Drive System Elastic Suspension and Its Application. *Journal of The China Railway Society*. 35(4), 23–29.

Nanotechnology

Advances in Engineering Materials and Applied Mechanics – Zhang, Gao & Xu (Eds)
© 2016 Taylor & Francis Group, London, ISBN 978-1-138-02834-0

Application of dual-modality contrast agent for medical image registration

M. Hou & M.Y. Yang
School of Computer Science and Engineering, Jiangsu Normal University, Xuzhou, China

S.Y. Qiao
School of Information and Electronic Engineering, Xuzhou Institute of Technology, Xuzhou, China

ABSTRACT: In this paper, dual-modality contrast agent can not only enhance contrast of Ultrasound Image (US), but can also be used as a contrast agent of Magnetic Resonance Image (MRI). With the help of dual-modality contrast agent, a new registration algorithm between US and MRI is introduced. To enhance the most important information of interest, the above dual-modality contrast agent was used in both ultrasound and magnetic resonance imaging process. After that, a improved affine transformation model between US and MRI was constructed, and both physical spatial transformation and intensity variation between the two modality images have been taken into consideration. The proposed algorithm combined with the dual-modality contrast agent contends with larger deformation, and performs well not only for local deformation but also for global deformation. The comparison experiments also demonstrated that US-MRI registration using the above-mentioned dual-modality contrast agent might be a promising algorithm for obtaining more precise image information.

1 INTRODUCTION

US is widely used in diagnosis and clinical applications because of its virtues of real-time, and no side effects. By using proper contrast agents, the contrast and sensitivity of US can be greatly improved (Ke, H. et al 2011, Yang F. et al 2008). However, because US is reflected very strongly when passing from gas to tissue, and vice versa, the use of US is still limited.

MRI is another non-invasive imaging tool which can provide abundant functional information (Liu Z. et al 2011). Magnetic iron oxide nanoparticles with superparamagnetic property can be used as a forceful contrast agent for MRI to further strengthen its brightness and contrast. One defect of MRI is that it can not offer real-time movement-related images.

In short, no single imaging modality holds all the merits satisfying the need of clinical applications. In many cases, US and MRI are complimentary, and both modalities are needed to discern possible pathological changes (Yang F. et al 2009). To fuse US and MRI together, US-MRI registration is required. Some researches focused on three-dimensional US-MRI registration in the field of operation navigation (Arvanitis C.D. et al 2013, Hu Y. et al 2012). The registration result depends directly on the segmentation results. It is fair to say that US image segmentation is a difficult issue at present. To solve the above problem, that only the Regions of Interest (ROI) of the US and MRI were selected was needed in the paper.

The main innovation of the paper is the introduction of the above-mentioned dual modality contrast agent to multi-modality medical image registration (Hou M. et al 2015). Using the dual modality contrast agent with the mean diameter of 3.98 μm (Yang F. et al 2008), the paper carries out the registration between ROI of US and ROI of MRI, and comes to a conclusion that with the use of dual modality contrast agent, the proposed method performs well not only for global deformation but also for local deformation. The remainder of the paper is organized as follows: the registration method based on improved affine transformation model between ROI of US and ROI of MRI is described in Section 2. Section 3 provides several groups of comparison experiments, and analyzes the experimental results, while Section 4 concludes our paper.

2 METHOD

As shown in Figure 1, dual modality contrast agent were used in both ultrasound and magnetic resonance imaging process. In order to reduce the influence of background noise on registration, ROI selection of US and MRI were carried out. And then, an improved affine transformation model between the ROIs of US (floating image) and MRI (reference image) was built, and the affine transformation vector was estimated by the Coarse-to-fine Gaussian pyramid scheme (Periaswamy S & Farid H 2003, 2006).

Let $f(x, y, t)$, $f(\hat{x}, \hat{y}, t-1)$ denote the reference and floating images, namely MRI and US. The relationship function between MRI and US is written as

$$m_7 f(x,y,t) + m_8 = f(m_1 x + m_2 y + m_5, m_3 x + m_4 y + m_6, t-1) \tag{1}$$

where $m_i, i \in (1 \ldots 6)$ represent space transformation parameters, and m_7 and m_8 respectively represent contrast and brightness change. Error function is defined as follows

$$E(\vec{m}) = \sum_{x,y \in \Omega} [m_7 f(x,y,t) + m_8 - f(m_1 x + m_2 y + m_5, m_3 x + m_4 y + m_6, t-1)]^2 \tag{2}$$

where $\vec{m}$ equals to $(m_1 \ldots m_8)^T$ and Ω represents a small neighborhood.

For (2), the first order Taylor expansion is carried out, and the corresponding vector form is as follows

$$E(\vec{m}) = \sum_{x,y \in \Omega} [k - \vec{c}^T \vec{m}]^2 \tag{3}$$

where

$$k = f_t - f + x f_x + y f_y, \vec{c} = (x f_x, y f_x, x f_y, y f_y, f_x, f_y, -f, -1)^T$$

In order to avoid degenerate solutions, smoothness constraints are added to (3) as in (4):

$$E(\vec{m}) = E_b(\vec{m}) + E_s(\vec{m}) = [k - \vec{c}^T \vec{m}]^2 + \sum_{i=1}^{8} \lambda_i \left[\left(\frac{\partial m_i}{\partial x} \right)^2 + \left(\frac{\partial m_i}{\partial y} \right)^2 \right] \tag{4}$$

For (4), derivative is carried out and the derivative result of (4) is set to 0 as in (5):

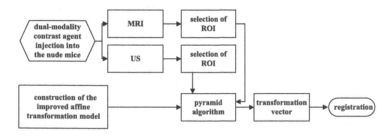

Figure 1.   Flow chart of the proposed MRI-US registration system based on dual-modality contrast agent and robust optical flow model.

$$\frac{dE(\vec{m})}{d\vec{m}} = \frac{dE_b(\vec{m})}{d\vec{m}} + \frac{dE_s(\vec{m})}{d\vec{m}} = -2\vec{c}[k - \vec{c}^T\vec{m}] + 2L(\overline{\vec{m}} - \vec{m}) = 0 \tag{5}$$

For (5), the corresponding solution is as in (6):

$$\vec{m}^{(j+1)} = (\vec{c}\vec{c}^T + L)^{-1}(\vec{c}k + L\overline{\vec{m}}^{(j)}) \tag{6}$$

where $\overline{\vec{m}}$ is mean value of $\vec{m}$ in a small neighborhood, L is a 8×8 diagonal matrix, whose diagonal entry are λ_i, and all the other elements are 0.

3   EXPERIMENTS

The algorithm is implemented based on MATLAB 2008b. With and without dual-modality contrast agent, the registration results were compared, as shown in Figure 2. According to Figure 2, the results of Root Mean Square error (RMS), Peak Signal to Noise Ratio (PSNR), correlation coefficient (COR), and Mutual Information (MI) were given in Tables 1 and 2.

When RMS is small, the similarity between the two images is greater. In contrast, the higher the PSNR, the better the quality of the registered image. If there is no relationship between the two images the COR is very low. MI approaches the maximum when the both images achieve

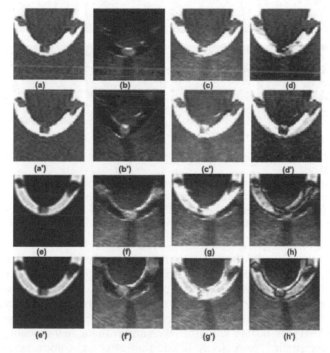

Figure 2.   Registration comparison with and without dual-modality contrast agent based on improved affine transformation.

(a) MRI without dual-modality contrast agent, (b) US without dual-modality contrast agent, (c) addition result between (a) and (b), (d) subtraction result between (a) and (b);

(a′) MRI without dual-modality contrast agent, (b′) registered US without dual-modality contrast agent, (c′) addition result between (a′) and (b′), (d′) subtraction result between (a′) and (b′);

(e) MRI with dual-modality contrast agent, (f) US with dual-modality contrast agent, (g) addition result between (e) and (f), (h) subtraction result between (e) and (f);

(e′) MRI with dual-modality contrast agent, (f′) registered US with dual-modality contrast agent, (g′) addition result between (e′) and (f′), (h′) subtraction result between (e′) and (f′).

Table 1. Quantitative comparison of performance evaluation before registration.

| | Performance evaluation | | | |
	RMS	PSNR	COR	MI
Without dual-modality contrast agent	0.2049	12.2115	0.3280	0.3883
With dual-modality contrast agent	0.1360	16.0505	0.3480	0.4956

Table 2. Quantitative comparison of registration result based on improved affine transformation.

| | Performance evaluation | | | |
	RMS	PSNR	COR	MI
Without dual-modality contrast agent	0.1870	12.9680	0.6362	0.6272
With dual-modality contrast agent	0.1219	16.9615	0.6960	0.6760

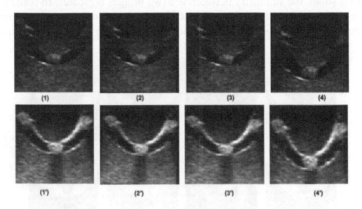

Figure 3. Registration comparison without and with dual-modality contrast agent based on the other registration algorithms.
Top (without dual-modality contrast agent): (1) registered US based on FFT, (2) registered US based on AFFINE, (3) registered US based on MMI, (4) registered US based on SIFT;
Bottom (with dual-modality contrast agent): (1′) registered US based on FFT, (2′) registered US based on AFFINE, (3′) registered US based on MMI, (4′) registered US based on SIFT.

the best registration. When RMS after registration is smaller than before registration, and three indices (PSNR, COR and MI) after registration is larger than before registration, it is called the normal variation, otherwise called abnormal variation in the following discussion.

As it can be intuitively observed, Figure 2 (f′) has obvious advantages over Figure 2 (b′) from the perspective of registration. From Tables 1 and 2, the four evaluation indexes (RMS, PSNR, COR and MI) had normal variations with and without dual-modality contrast agent, before and after registration, respectively. In addition, after registration, performance evaluation with the use of dual-modality contrast agent had better improvement than that without dual-modality contrast agent. For example, RMS decreases from 0.1870 to 0.1219, PSNR, COR and MI rises up from 12.9680 to 16.9615, from 0.6362 to 0.6960 and from 0.6272 to 0.6760, respectively. To sum up, the qualitative and quantitative analyses showed that US-MRI registration based on the proposed method is effective.

Figure 3 are registration results based on the other methods without and with dual-modality contrast agent. The other methods included Fast Fourier Transform (FFT) (Yang Z. & Penczek P.A. 2008), affine transformation (AFFINE), Mutual Information (MMI) (Sandiego C.M. et al 2013) and Scale-Invariant Feature Transform (SIFT).

Table 3. Quantitative comparison of registration result based on the other registration algorithms.

Category	Performance evaluation			
	RMS	PSNR	COR	MI
Without dual-modality contrast agent				
FFT	0.4275	8.4989	0.4195	0.3390
AFFINE	0.4251	7.3472	0.2536	0.2491
MMI	0.4541	7.3953	0.2766	0.2635
SIFT	0.2705	7.3442	0.2031	0.2564
With dual-modality contrast agent				
FFT	0.3285	9.3216	0.4007	0.4646
AFFINE	0.3268	9.6348	0.4320	0.3801
MMI	0.4588	9.6805	0.4675	0.3962
SIFT	0.4275	8.4470	0.2035	0.2600

With dual-modality contrast agent and before registration, Table 1 shows RMS, PSNR, COR and MI are 0.1360, 16.0505, 0.3480 and 0.4956, respectively. Table 3 shows Quantitative comparison of registration result based on the other registration algorithms. After registration, for FFT, only increment of COR from 0.3480 to 0.4007 is normal variation; for AFFINE, only increment of COR from 0.3480 to 0.4320 is normal variation; for MMI, only increment of COR from 0.3480 to 0.4675 is normal variation; for SIFT, the evaluation indexes are abnormal variation. In conclusion, compared with the other methods, the proposed algorithm combined with dual-modality contrast agent has the best performance.

4 CONCLUSION

The objectives of this paper were twofold. Firstly, we introduced a new type of nanomaterials, dual-modality contrast agent, into the field of medical imaging and verified that dual-modality contrast agent can indeed increase the contrast of US image and MRI, which was particularly beneficial to the subsequent processing such as medical image registration. At present, how to improve the algorithms for medical image processing had become the bottleneck, and in the short term it was difficult to have a breakthrough. By comparison, the last decade had seen advancement in every side of nanomaterials, which was estimated to be influential in the next 30–50 years, in all fields of science and technology. Unfortunately, until now nanomaterials were almost not applied in the field of medical image processing. Secondly, it was feasible that novel nanomaterials combined with good algorithm were used to solve some difficult problems in medical image field. Especially, in the cases where strong noise image was involved, nanomaterials can be considered to combine with algorithm models instead of only by improving algorithm to solve problems. Our algorithm implementation was intensity based and independent of the metric that was being used, so it can be adapted to different image modalities.

ACKNOWLEDGEMENTS

This work was supported by Foundation for the Author of Jiangsu Normal University (No. 10XLB10).

REFERENCES

Arvanitis, C.D., Livingstone, M.S. & McDannold, N. 2013. Combined ultrasound and MR imaging to guide focused ultrasound therapies in the brain. *Phys Med Biol* 58: 4749–61.

Hou, M., Chen, C., Tang, D., Luo, S., Yang, F. & Gu, N. 2015. Magnetic microbubble-mediated ultrasound-MRI registration based on robust optical flow model. *Biomedical Engineering Online* 14 (Suppl 1): S14.

Hu, Y., Ahmed, H.U., Taylor, Z., Allen, C., Emberton, M., Hawkes, D., & Barratt, D. 2012. MR to ultrasound registration for image-guided prostate interventions. *Med Image Anal* 16: 687–703.

Ke, H., Wang, J., Dai, Z., Jin, Y., Qu, E., Xing, Z., Guo, C., Yue, X. & Liu, J. 2011. Gold-nanoshelled microcapsules: a theranostic agent for ultrasound contrast imaging and photo-thermal therapy. *Angew Chem Int Ed* 123: 3073–7.

Liu, Z., Lammers, T., Ehling, J., Fokong, S., Bornemann, J., Kiessling, F. & Gätjens, J. 2011. Iron oxide nanoparticle-containing microbubble composites as contrast agents for MR and ultrasound dual-modality imaging. *Biomaterials* 32: 6155–63.

Periaswamy, S. & Farid, H. 2003. Elastic Registration in the Presence of Intensity Variations. *IEEE Transactions on Medical Imaging* 22: 865–874.

Periaswamy, S. & Farid, H. 2006. Elastic Registration with Partial Data. *Medical Image Analysis* 10: 452–464.

Sandiego, C.M., Weinzimmer, D. & Carson, R.E. 2013. Optimization of PET–MR registrations for nonhuman primates using mutual information measures: A Multi-Transform Method (MTM) *NeuroImage* 64: 571–81.

Yang, F., Li, L., Li, Y., Chen, Z., Wu, J. & Gu, N. 2008. Superparamagnetic nanoparticle-inclusion microbubbles for ultrasound contrast agents. *Phys Med Biol* 53: 6129–41.

Yang, F., Li, Y., Chen, Z., Zhang, Y., Wu, J. & Gu N. 2009. Superparamagnetic iron oxide nanoparticle-embedded encapsulated microbubbles as dual contrast agents of magnetic resonance and ultrasound imaging. *Biomaterials* 30: 3882–90.

Yang, Z. & Penczek, P.A. 2008. Cryo-EM image alignment based on nonuniform fast Fourier transform. *Ultramicroscopy* 108: 959–69.

Power system

Advances in Engineering Materials and Applied Mechanics – Zhang, Gao & Xu (Eds)
© 2016 Taylor & Francis Group, London, ISBN 978-1-138-02834-0

Research on the sewage treatment by Dielectric Barrier Discharge under the atmospheric pressure

J.P. Jia & L. Cai
Mechanical and Electrical Engineering College of Wuhan Donghu University, Wuhan, Hubei, China

L.P. Ru
Huairou Electric Power Company of State Grid Beijing Electric Power Company, Beijing, China

Y. Zhao
School of Biomedical Engineering, Hubei University of Science and Technology, Hubei Xianning, Hubei, China

ABSTRACT: The pressure of water resources shortage and water pollution makes the sewage treatment research to gradually become one of hot researches. In this paper, a sewage treatment using the atmospheric pressure plasma generated by the dielectric barrier discharge is researched. The system adopts the coaxial discharge structure and a mixture of argon and oxygen is used. And the high frequency high voltage AC power is adopted to study the characteristics of the dielectric barrier discharge. The experiments show that the discharge device is evenly distributed with blue light and kinds of active particles. And, the discharge is uniform and reliable. The common city sewage is used as sample and the treatment results by dielectric barrier discharge are analyzed. The experiments show that the low temperature plasma generated by DBD can be effectively applied to the study on the treatment of city sewage. The scheme is feasible and can be applied.

1 INTRODUCTION

Because of the rapid development of industry and agriculture all over the world and the city population increased dramatically, a large number of production and living sewage are discharged into rivers, lakes and seep into the ground. Thus, much water pollution is produced. Therefore, the waste water treatment has become the key to solve the problem of environmental pollution.

At present, the application of waste water treatment technology, especially that with high efficiency, low energy consumption and strong operability has become a research focus. Dielectric Barrier Discharge (DBD) can produce non-thermal plasma under atmospheric pressure, which contains a lot of free radicals and quasi molecules, such as OH, O, NO, etc. Its chemical property is very active, and easy to react with other atoms, molecules, or free radicals. So, it is particularly suitable for industrial application[1]. Thus, various countries in the world pay more attention to the sewage treatment technology, which using the low temperature plasma generated by dielectric barrier discharge under atmospheric pressure. In recent years, the domestic-related research department also began to pay attention to this new technology, and carry out a preliminary feasibility study.

2 DIELECTRIC BARRIER DISCHARGE

The main characteristics of dielectric barrier discharge is a layer of insulating medium placed in the discharge between electrodes, such as ceramics, enamel and glass[2].

Figure 1.   Dielectric barrier discharge with x%O2+Ar.

The reasons of the medium placed between discharge electrodes are as follows:

1. The dielectric constant of the medium is higher than air, so most of the voltage drop in on the air gap, which is to strengthen the air gap electric field intensity and the discharge is easy to form[3].
2. Restricted by technology, the surface of discharge electrodes is very difficult to make smooth, which may make the spark discharge at certain points and the discharge current over most area is very small. So, it is not conducive to make uniform discharge and may lead to deterioration or damage the surface condition of electrode. After the media layer is added between the electrodes, the breakdown voltage of the medium is much higher than that of air, so even if the electrode is local rough, nor for the breakdown of dielectric layer. When the discharge reaches a certain intensity, there may be uniform discharge phenomenon[4].

The system is not using air discharge directly, because the breakdown electric field and discharge voltage are high and the discharge is uniformity. According to the Bashin curve and considering a variety of factors, the inert gas is selected as the medium, which breakdown strength is lower.

Through the discharge space, a mixture of argon and oxygen is used. In the system, argon with low discharge voltage is as a carrier and oxygen with active discharge particle is as reactive gas, which is easy and reaction to things.

In this system, the high frequency high voltage AC power is adopted to study the characteristics of the dielectric barrier discharge[5]. The voltage is adjustable between 0 and 10 kV, and the frequency is adjustable between 20 and 40 kHz.

The experiments show that the discharge device can produce plasma under atmospheric pressure, and is easy to deal with sewage samples. The discharge is evenly distributed with blue light and kinds of active particles, as shown in Figure 1.

3 SEWAGE TREATMENT ANALYSIS

In the project, the common city sewage is used and the data are as shown in Table 1.

3.1 Treatment result about COD

COD (Chemical Oxygen Demand) is the amount of oxidant consumption with strong oxidation agent for water treatment under certain conditions[6]. It is the number of indicators of reducing substances in water. The reducing substances in water are various organic matter, nitrite, sulfide and ferrous salt, etc. But the main substance is the organic matter. Therefore, the Chemical Oxygen Demand (COD) is often used as the index of the content of organic matter in water. The measured values about Chemical Oxygen Demand (COD) determination are also different, when the reducing substances and measurement methods are different.

In Figure 2, when the treatment time is short, COD in the sample does not fall but rise. The reason is that the large molecules are decomposed into small molecules by ozone, which are difficult degradation. So, COD will increase. With the increase of treatment time, COD value decreases, and the decline speed up significantly after ten minutes.

3.2 Treatment result about BOD5

BOD5 (Biology Oxygen Demand 5) is the biochemical oxygen demand for five days, which refers to the amount of oxygen that consumed in the degradation process[7]. Because of

Table 1. Data about sewage treatment analysis.

Item	Time (min)	COD (mg/l)	BOD5 (mg/l)	Ammonia nitrogen (mg/l)	Escherichia coli (number/ml)	Color
Sample	0	150.12	65.75	17.35	1.55E+07	69
Treatment result	5	155.23	55.32	15.34	1235	58
	10	146.02	45.33	13.32	450	46
	15	115.51	31.32	10.88	25	33
	20	103.12	21.01	9.02	13	27
Removal rate	5	−3.40%	15.86%	11.59%	99.99%	15.94%
	10	2.73%	31.06%	23.23%	100.00%	33.33%
	15	23.05%	52.37%	37.29%	100.00%	52.17%
	20	31.31%	68.05%	48.01%	100.00%	60.87%

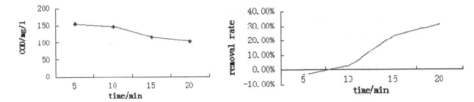

Figure 2. Treatment result about COD.

the long time required in complete degradation process, the water quality standard is the five days biochemical oxygen demand in order to standardize and improve the detection efficiency.

In Figure 3, with the increase of treatment time, the BOD5 value decreases significantly. When the treatment time was 20 min, the removal rate can reach about 70%.

3.3 Treatment result about ammonia nitrogen

The ammonia nitrogen about water treatment is the free ammonia (NH3) and ammonium ions (NH4) forms of nitrogen[8]. Nitrogen is the nutrient in the water, which can lead to eutrophication phenomenon. It is the main oxygen consuming pollutants in water and toxic to fish and aquatic organisms.

In Figure 4, the value of ammonia nitrogen is relatively slow at the beginning. With the increase of processing time, the value decreases significantly and the effect is obvious.

3.4 Treatment result about Escherichia coli

Escherichia coli is the comprehensive evaluation index of city sewage and especially an important index of sewage pollution[9].

In Figure 5, the treatment effect on Escherichia coli is the most obvious in the experiment. The main reason is that the oxygen in dielectric gas and ozone is generated by dielectric barrier discharge.

3.5 Treatment result about color

Color refers to the class of yellow and brown degree with the solubility of the substance or gelatinous substance in water[10].

In Figure 6, color processing effect is more obvious, along with the increase of processing time, the color declined.

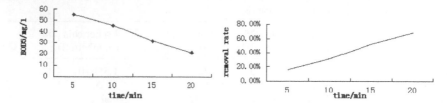

Figure 3. Treatment result about BOD5.

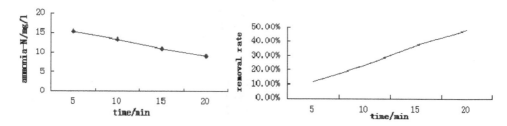

Figure 4. Treatment result about ammonia nitrogen.

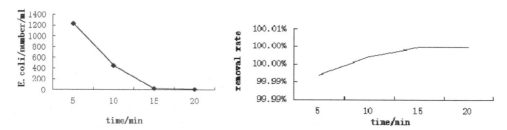

Figure 5. Treatment result about *E. coli*.

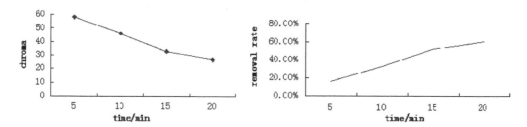

Figure 6. Treatment result about color.

4 CONCLUSION

The experiments show that the low temperature plasma generated by the DBD can be effectively applied to the study on the treatment of city sewage. The scheme is feasible and can be applied. In actual, many factors need to be considered, such as the efficiency and security of the discharge device, the optimization between power and reactor, the materials and shape optimization of discharge electrodes and the types of waste water. Moreover, how to optimize the process also need to be studied.

ACKNOWLEDGEMENTS

The source of the project: guidance project of Hubei Province education department scientific research plan, item number B2013198.

REFERENCES

[1] Yan Zuo, Guangxu Yan, Shaohui Guo. 2008. The Application of Low Temperature Plasma Oxidation Technology in Wastewater Treatment. *Water Treatment Technology*, 7:1–5.

[2] Xueji Xu, Dingchang Zhu. 1996. *Gas discharge physics*. Shanghai: Fudan University Press

[3] Zhenxin Ye, Haitao Liu. 2011. Related Factors Analysis of Dielectric Barrier Discharge Power. *Mechanical and Electrical Technology*, 1:41–43.

[4] Yuhong Han, Bing Lu, Qing Li, Zhiqiang Liu. 2007. Research progress of the water treatment technology by High voltage pulsed discharge plasma. *Journal of Hebei University* (natural science edition), 10(27):190–194.

[5] Jianshe Shao, Ping Yan. 2006. High Frequency High Voltage AC Power Applied to the Dielectric Barrier Discharge Characteristic Research. *High Voltage Engineering*, 3:78–80.

[6] Jianping Song, Ruigang Si, Linglong Meng. 2009. Gas chromatographic method for determination of organic waste water COD. *Technology Wind*, 18:189–190.

[7] Caixia Cui. 2008. Influence Factors in Determination of Wastewater BOD5. *Technology & Development of Chemical Industry*, 37(5):45–46.

[8] Huiyu Wang. 2013. Treatment of high concentration ammonia nitrogen waste water. *Theoretical study on construction of city*, 2:1–8.

[9] Long Di, Jun Ma. 2012. The fecal coliform survey of the sea outfall in Taizhou City. *Agriculture & Technology*, 32(4):160.

[10] Yuehua Huang, Tiefu Xu, liying Yang. 2013. *Water treatment technology*. Zhengzhou: The Yellow River Water Conservancy Press.

Advances in Engineering Materials and Applied Mechanics – Zhang, Gao & Xu (Eds)
© *2016 Taylor & Francis Group, London, ISBN 978-1-138-02834-0*

The study on DC bias distribution considering AC-DC hybrid transformers

Y.D. Song & W.C. Ge
Electric Power Research Institute, Liaoning Electric Power Company Limited, Shenyang, China

W.T. He
School of Electrical Engineering, Northeast Dianli University, Jilin, China

ABSTRACT: In this paper, a substation that contains AC transformers and converter transformers is treated as typical example that the substation contains multiple different transformers. In the case of the presence of the bias current, DC bias current in every transformer the neutral point distribution is calculated in detail, in which the DC bias current in AC transformers' neutral point is about 2 times of the DC bias current in converter transformers. The main DC bias suppression measures are briefly introduced, and the corresponding suppression measures are applied only to AC transformers that are used to suppress the DC bias of entire substation. Finally, this paper takes a practical DC project as basis, building a detailed electromagnetic transient model. A simulation can be obtained: the corresponding suppression measures are only applied to AC transformers can effectively inhibit the DC bias of entire substation. Considering the variety of factors, the method of connecting series capacitor in the AC transformers' neutral point has been chosen to suppress the DC bias of entire substation.

1 INTRODUCTION

UHV AC/DC power system will be developed rapidly as the backbone grid of the strong and smart grid, and the "five vertical and five horizontal" backbone grid of UHV AC and 27 project of UHV DC will be built in the future[9]. A larger bias current is detected through the substation near the DC grounding electrode. When the UHVDC is operated unipolar and full-load of a province in May 2014 (the ground current is 5000 A), the number of the substations, which has the neutral bias current nearby is greater than 200 A, between 100–200 A and between 20–100 A, is about 3, 2, 19 respectively. At present, a substation that contains AC transformers and converter transformers is adopted to the converter substations including the Shenzhen, Bao'an and Baoji in ± 500 kV HVDC project and Huidong in ± 800 kV HVDC project, in which the distribution and suppression measures are more complicated.

Under the situation of a substation containing AC transformers and converter transformers, the substation contains multiple different transformers, and it is closed to DC grounding pole relatively. The bias current through the UHV autotransformer and converter transformer of substation/converter station may be large, and the capacity of AC transformer for the DC bias current is small, which make it easier for DC bias[4][8]. So, the study of DC-biased phenomenon of AC transformers and converter transformers need to attract enough attention. In references[2–3], the DC current in Baoji 750 KV autotransformer' neutral point is calculated when BaoDe DC system is operated monopolar. In references[5][6], HVDC earth-return currents distribution in AC Power Grid is simulated and analyzed. The basic principles of the DC-biased transformer and its impact on AC and DC systems are introduced in detail in[3][10]. DC bias suppression measures and its principles have been deeply studied by some scholars, which are: reverse current method, series capacitor method, series resistance method, and series resistor-capacitor method. In references[7][10], the methods of series capacitor, series resistance, and series resistor-capacitor in transformers' neutral point are used to inhibit the DC bias of transformer. Though such approach can play a better inhibitory effect, it will affect the ground stability of power transformer and may also lead to the resonance phenomenon. The injection directional current method is used to suppress the DC bias, although its inhibitory effect is more obvious, the investment is large and the implement is more difficult.

The above researches are more focusing on the single transformers or the transformers of the same type, which cannot be directly applied to a substation with multiple different parameters and models of transformers. In this paper, the advantages and disadvantages of DC bias suppression measures are summarized, and a substation that contains the AC transformers and converter transformers is considered as typical example to derive the amount of DC bias current that flows through each transformer's neutral point. To adopt appropriate suppression measures, this paper takes a practical DC project as basis, building a detailed electromagnetic transient model, establishing the equivalent system of multiple voltage levels and detailed AC transformers and converter transformers model. Finally, the optimal suppression measures are selected based on the advantages and disadvantages of each suppression measures and the distribution of the bias current.

2 THE DC BIAS CURRENT DISTRIBUTION IN THE MULTIPLE DIFFERENT TRANSFORMERS' NEUTRAL POINT

2.1 *The typical structure and wiring of multiple different transformers*

In this paper, a substation that contains AC transformers and converter transformers is treated as the research object. In this case, two or more AC transformers is chosen in the AC substations, which can provide 70 percent capacity to high-voltage AC substation when running alone. In China, the bipolar 12 pulsations are used in HVDC transmission project, in which, the converter transformer is generally four (a spare transformer for reserve), the basic wiring is the Y0-Y and Y0-Δ. The basic structure is shown in Figure 1.

In the structure, the AC transformer is a combination of the single-phase autotransformers, the transformer is powered by its high-pressure side, the power is sent by its medium-pressure side, and its low-voltage winding is connected like Δ, which provides the reactive power compensation for substations to stabilize the AC voltage. The main wirings of transformers running in converter substation are Y0-Y and Y0-Δ, and the phase angle difference between the valve sides of these transformers is 30°. The Y0 wiring of commutation bus side and the grounding grid of high-voltage AC bus are grounded uniformly, while the valve winding is not.

2.2 *The calculation of the DC bias current in transformers' neutral point*

2.2.1 *Equivalent bias current path within the substation*
Within the substation in Figure 1, the AC transformer is operated using identical parameters of transformers in parallel, so it can be equivalent to a transformer for calculation, shown in Figure 2.

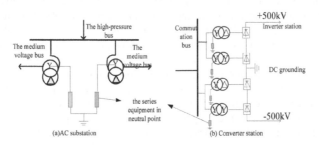

Figure 1. The grounding grid schematic of multiple different transformers.

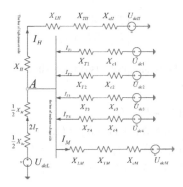

Figure 2.　The passageway schematic of bias current.

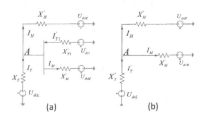

Figure 3.　The simplified passageway schematic of bias current.

In Figure 2, A is the node points of high pressure and middle pressure for high-voltage AC autotransformer, X_H and X_M are the bias current passageway impedance of high pressure side and middle pressure side for the AC transformer respectively, X_{T1}, X_{T2}, X_{T3}, X_{T4} are the bias current passageway impedance for the converter transformer, X_{C1}, X_{C2}, X_{C3}, X_{C4}, X_{HC} are the grounding impedance of the converter transformer and the AC transformer, X_{LH}, X_{TH}, X_{cH}, U_{dcH}, X_{LM}, X_{TM}, X_{cM}, U_{dcM} are the transmission line impedance, the transformer impedance, transformer grounding impedance of high pressure side and middle pressure side of the transformer, and I_H, I_M, I_T, I_{T1}, I_{T2}, I_{T3}, I_{T4} are the bias current flowing through the high-voltage bus, the medium-voltage bus, the AC transformer' neutral point and each converter transformer's neutral point.

In the identical substation or converter station, the used grounding grid is same, and the grounded potential of each transformer is consistent:

$$\begin{cases} U'_3 = U_{dc1} = U_{dc2} = U_{dc3} = U_{dc4} \\ 2I_T + I_{T1} + I_{T2} + I_{T3} + I_{T4} = I_H + I_M \end{cases} \tag{1}$$

The parameters of grounding part are consistent in converter transformer, which used the same ground network and is not directly linked to AC transformer:

$$I_{T1} = I_{T2} = I_{T3} = I_{T4} \tag{2}$$

And the Figure 2 can be simplified as Figure 3, the parameters are as follows:

$$\begin{cases} X'_H = X_{LH} + X_{TH} + X_{cH} \\ X'_M = X_{LM} + X_{TM} + X_{cM} \\ X_T = \dfrac{1}{2}(X_m + X_{hc}) \\ X'_{T1} = \dfrac{1}{2}(X_{T1} + X_{c1}) \\ X'_T = X'_{T1} // X_T \end{cases} \tag{3}$$

$$\begin{cases} I'_{T1} = 4I_{T1} \\ I'_{T} = 2I_{T} + I'_{T1} \end{cases} \quad (4)$$

The Figure 3(b) is taken as the basic analysis object, the formula can be obtained by Kirchhoff's current and voltage law:

$$\begin{cases} I'_{T} = I_{H} + I_{M} \\ I_{M} \times X'_{M} + I'_{T} \times X'_{T} = U_{dcL} - U_{dcM} \\ I_{H} \times X'_{H} + I'_{T} \times X'_{T} = U_{dcL} - U_{dcH} \\ I_{M} \times X'_{M} + U_{dcM} = I_{H} \times X'_{H} + U_{dcH} \end{cases} \quad (5)$$

The calculated result can be gotten from formula (5):

$$\begin{cases} I_{M} = \dfrac{U_{dcL} \times X'_{H} + U_{dcM} \times X'_{T} - U_{dcM}(X'_{H} + X'_{T})}{X'_{H} \times X'_{M} + X'_{T} \times X'_{M} + X'_{T} \times X'_{H}} \\ I_{H} = \dfrac{U_{dcL} \times X'_{M} + U_{dcM} \times X'_{T} - U_{dcH}(X'_{M} + X'_{T})}{X'_{H} \times X'_{M} + X'_{T} \times X'_{M} + X'_{T} \times X'_{H}} \end{cases} \quad (6)$$

The bias current of common ground point can be calculated from Figure 3 and the formulas (4)–(6):

$$I'_{T} = I_{H} + I_{M} \quad (7)$$

Therefore, the bias current flowing through each transformer's neutral point is as follows:

$$\begin{cases} I'_{T1} = \dfrac{X_{T}}{X_{T} + X'_{T1}} \times (I_{H} + I_{M}) \\ I_{T} = \dfrac{X'_{T1}}{X_{T} + X'_{T1}} \times (I_{H} + I_{M}) \\ I_{T1} = \dfrac{1}{4} \times \dfrac{X_{T}}{X_{T} + X'_{T1}} \times (I_{H} + I_{M}) \end{cases} \quad (8)$$

The bias current flowing through a single AC transformer is:

$$\dfrac{X'_{T1}}{X_{T} + X'_{T1}} \times (I_{H} + I_{M}) \times \dfrac{1}{2}$$

The neutral point bias current ratio of a single transformer with a single converter is:

$$\dfrac{I_{T}}{I_{T1}} = \dfrac{I_{T}}{0.25 I'_{T1}} = \dfrac{X_{T1} + X_{c1}}{X_{M} + X_{hc}} \quad (9)$$

where, $X_{T1} + X_{c1}$, $X_{M} + X_{hc}$ are about 10.083 Ω, 4.768 Ω respectively, so the ratio is about 2.115, that is to say, the DC bias current in AC transformers' neutral point is about 2 times of the DC bias current in converter transformers'. Meanwhile, the converter transformer is considering the factor of DC bias when design and production, so the appropriate suppression measures in the AC transformers' neutral point has been chosen to suppress the DC bias of the AC and DC systems.

3 CONCLUSION

In this paper, a substation that contains AC transformers and converter transformers is treated as typical subject that the substation contains multiple different transformers. In the case of the ground potential is determined, DC bias current in every transformer neutral point is calculated in detail, in which the DC bias current in AC transformers' neutral point is about 2 times of the DC bias current in converter transformers'.

REFERENCES

[1] Du Zhongdong, Dong Xiaohui, Wang Jianwu. 2006. Analysis on Restraining Transformer DC Bias by Changing Electric Potential of Grounding Grid. *High Voltage Engineering*. 26(8):69–73 (in Chinese).

[2] Guo Xiu-ying, Liu Hong-wei, Xu Zhong-hai. 2013. Symmetry and the Conserved Quantity of A Generalized Hamilton System. *Journal of Northeast Dianli University*. 33(1/2), pp.162–164.

[3] Kuai Di-zheng, Wan Da, Zhou Yun. 2005. Analysis and Handling of the Impact of Geomagnetically Induced Current Upon Electric Network Equipment in DC Transmission. *Automation of Electric Power Systems*. 29 (2):81–82 (in Chinese).

[4] Leng Di. 2012. *Research on Effect of ±500kV BaoDe DC Transmission on Transmission Line and 750kV Transformer near The Electrode Site*. Chongqing University. 4.

[5] Pan Zhuohong, Zhang Lu, Tan Bo. 2011. Simulation and Analysis of HVDC Earth-return Currents Distribution in AC Power Grid. *Automation of Electric Power Systems*. 35 (21):110–115 (in Chinese).

[6] Ren Zhichao, Xu Jianguo, Zhang Yikun. 2011. Study of the Simple Formula of DC Surface Potential in AC-DC Interconnected Large Power System. *Transactions of China Electrotechnical Society*. 26(7):256–264 (in Chinese).

[7] Shang Chun. 2004. Measure to Decrease the Neutral Current of the AC Transformer in HVDC Ground Return System. *High Voltage Engineering*. 30(11):52 55 (in Chinese).

[8] Xiaoping Li, Xishan Wen, Penn N. Markham. 2010. Analysis of Nonlinear Characteristics for a Three-Phase, Five-Limb Transformer Under DC Bias. *IEEE Transactions on Power Delivery*, 25(4): 2504–2511.

[9] Zhu An-ming, Wang Sen, Hu Pan-feng. 2010. Influences of the Deyang-Baoji DC Transmission Line on the 750 kV Power Transformer in Baoji Power Substation. *Power System and Clean Energy*. 26(12):67–70 (in Chinese).

[10] Zhaojie, Li Xiaolin, Lu Jinzhuang. 2006. Applying Series Resistor to Restrain Power Transformer DC Biasing. *Automation of Electric Power Systems*. 30 (12):82–89 (in Chinese).

[11] Zhu Yi-ying, Jiang Wei-ping, Zeng Zhao-hua, Yin Yong-hua. 2005. Studying on Measures of Restrsining DC Current Through Transformer Neutrals. *Proceedings of the CSEE*. 25(13):1–7 (in Chinese).

[12] Zeng Lin-jun, Lin Xiang-ning, Huang Jing-guang. 2010. Modeling and Electromagnetic Transient Simulation of UHV Autotransformer. *Proceedings of the CSEE*. 30(1):91–98 (in Chinese).

Advances in Engineering Materials and Applied Mechanics – Zhang, Gao & Xu (Eds)
© 2016 Taylor & Francis Group, London, ISBN 978-1-138-02834-0

Research on reliability assessment of the smart meter

S. Wang, J. Zuo, Q.Y. Guo & Y.W. Chu
China Electric Power Research Institute, Beijing, China

ABSTRACT: The paper studies single-phase smart meter's reliability using the reliability prediction method and accelerated life test. It adopts the component stress method and reliability prediction handbook for electronic equipment in the reliability prediction, calculates the failure rate of each unit module and mean time to failure of the smart meter. According to IEC 62059-31, it designs accelerated life test scheme through enhancing temperature and humidity, and carries out accelerated life test using same smart meters. In the end, it analyzes test results of the reliability prediction method and accelerated life test.

1 INTRODUCTION

As the user's interactive interface, the reliability of the smart meter is directly related to the vital interests of the company and consumers. Compared with the mechanical meter, it increases many new functional requirements, the conventional meter reliability evaluation method is no longer applicable[1]. Hence, it becomes extremely necessary to evaluate the reliability of smart meter[2].

The reliability of smart meter includes reliability prediction and accelerated life test[3]. According to the model and reliability data of smart meter components, the reliability prediction predicts the probability within the specified time and failure[4]. Without changing meter failure mechanism and adding new failure modes, accelerated life test accelerate the process of meter failure by increasing the stress intensity. Among of the accelerated life tests, constant accelerated life is more mature.

In this paper, it studies the reliability prediction and accelerated life test of smart meter. Through predicting the life of each functional unit of a single-phase smart meter, it can gets the reliable life of the meter. Then for this type of single-phase smart meter, According to IEC 62059-31, it is carried on accelerated life testing by high temperature and humidity test environment, then calculates its reliability life. The paper compares the smart meter reliability life calculating by the two methods, analyzes the accuracy of their results.

2 RELIABILITY PREDICTION OF SMART METER

In the Reliability prediction, it is assumed that all components of the smart meter are equal importance. The smart meter is a series system consisting of many components, failure of any one component will lead to failure of the smart meter. In conventional methods of reliability prediction, component stress method has been widely used for its strong engineering practicality. So the paper use components stress method for smart meter reliability prediction[5]. The basic principle of this method is: the reliability of components determines the reliability of the equipment; Failure of components is exponential distribution, namely the failure rate is constant. The general expression is as follows[6]:

$$\lambda_s = \sum_{i=1}^{N} N_i (\lambda_{Gi} \pi_{Qi} \pi_{Ei} \pi_{Wi}) \tag{1}$$

λ_s—total failure rate of the system;

λ_{Gi}—failure rate of the ith component under reference conditions;

π_{Qi}—the quality dependence factor of ith component under reference conditions;

π_{Ei}— the environment factor of ith component;

π_{Wi}—the stress profile factor of the of ith component;

N_i—the number of ith component;

N—the number of component species.

In the process of reliability prediction, if the manufacturer has provided the failure rate of the components, the priority of data provided by the manufacturer.

Calculation steps of component stress method:

1. Calculate failure rate of each component
   1. Analysis of the stress formula of various components, in-depth understanding of the affection of the work environment, load and other work stress
   2. Compile a list of smart meter components, including the components of the name, type, size, quantity, product standards or technical documents, performance ratings, and related design, technology, structure and work stress data and other parameters.
   3. Calculate the failure rate of each components according to the model of component failure rate.
2. Compile components working failure rate prediction table of smart meter
3. Predict the failure rate of smart meter' each module
   According to step (2) to determine the failure rate of each component, Preparation of failure rate prediction table of smart meter (data collection module, data processing module, power supply module, the carrier module). calculation module failure according to equation (2)

$$\lambda_{Pi} = \sum N\lambda_P \tag{2}$$

λ_{Pi}—the failure rate of the ith module

$\sum N\lambda_p$—the sum of all components failure rate of the ith module.

4. Predict the failure rate and average life of the smart meter. obtain the failure rate of modules according to the formula (2); Calculate the failure rate and mean time to failure of smart meter according to the formula (3) and (4)

$$\lambda_s = \sum_{i=1}^{T} \lambda_{Pi} \tag{3}$$

$$MTTF = \frac{1}{\lambda_s} \tag{4}$$

MTTF—mean time to failure.

3 ACCELERATED LIFE TEST

The accelerated life test is defined as follows: without changing the premise of meter failure mechanism, the approach obtains the reliability life by applying stress beyond the normal conditions. The accelerated life test can be divided into Constant stress accelerated life test and step stress accelerated life test[7]. constant stress accelerated life test is more mature. In this paper, constant stress accelerated life test is used to determined smart meter' reliability life.

Accelerated life test is based on two models: life distribution and life-stress model. The life stress model quantifies the manner in which the life distribution changes with different stress levels. The combination of both a life distribution and a life-stress model with time to failure data obtained at different stress levels, can provide an estimation of the characteristics at normal use conditions.

3.1 The life distribution

The Weibull distribution[8] is one of the most commonly used distribution models in reliability engineering. It can be used to model material strength, time to failure data of electronic and mechanical components, equipment or systems. its unreliability function is as follows:

$$F(t) = 1 - e^{-\left(\frac{t-\gamma}{\eta}\right)^{\beta}} \tag{5}$$

t represents time, and the remaining three parameters are constant.

3.2 The life-stress model

Accelerated life testing is carried out by changing the environmental stress on the meter, such as temperature and humidity accelerated. n this paper, it adopts the Peck's temperature-humidity model.

The Peck's acceleration factor is:

$$AF = \left(\frac{RH_u}{RH_s}\right)^{-n} e^{\frac{E_a}{k}\left(\frac{1}{T_u} - \frac{1}{T_s}\right)} \tag{6}$$

where:

RH_u is the relative percent humidity at use conditions;
RH_s is the relative percent humidity at stress conditions;
T_u is the temperature in K at use conditions;
T_s is the temperature in K at stress conditions;
K is a constant (8.617×10^{-5} eV/K);
E_a is the activation energy in electron volts (E_a is in the range of 0.3 to 1.5);
n is a constant (n is in the range of 0.3 to 1.5).
F_u and n the two coefficients of the model.

3.3 Accelerated life test program

3.3.1 Testing period

Accelerated life test is divided into three stages: inspection testing before the experience, period testing during the experience and end testing after the experience. In the current standards and technical specifications of smart meter [1] The MTTF of meter is not less than 10 years. Thus, the MTTF is defined 15 years in this paper. During the test, the equivalent time period of five-year and 10-year should be test.

3.4 Test profile

In the accelerated life test, its testing period is equivalent five years. After the test, restore to use temperature and begin testing, then enter the next cycle test.

3.5 Testing conditions

Working position: vertical
Temperature and humidity: 75 °C, 80%;
Voltage: ±20% Un;
Current: Ib;
Frequency: 50 Hz ± 5%;
Waveform: sinusoidal voltage and current, the distortion factor is less than 5%.

3.6 Test projects

a. Basic error test
b. False actuation test
c. Start test
d. Day timing error test
e. Ac voltage test
f. Communication protocol conformance test
g. Safety certification test
h. Load switch tripping and closing at 0.7Un–1.2Un
i. Power consumption test.

3.7 Failure determination

The number of samples is 43 in the test. The failure number r is calculated in the unite of sample. The same sample occurs in the reliability test session in one or more test projects failed, it is counted as a failure of the sample if the failure number r> = 3, the test suspension, and the sample is unqualified in the reliability test. Calculate the reliability life of the meter. If the failure number r < 3, the sample is qualified in the reliability test.

1. Carry out projects a–i before accelerated life test. the basic error should not exceed 60% accuracy class. If the number of failure meter is no more than three, it can be replaced by reserved meter.
2. In the process of accelerated life test, the test projects a–d should be carried out in each period, and conduct failure determination.
3. After the test, the samples should be tested projects a–i, and conduct failure determination.

4 RESULTS AND ANALYSIS

4.1 The result of reliability prediction

Single-phase smart meters with remote cost-control (carrier, switch built-in) Technical specifications: 5(60 A), 220 V, 50 Hz, 1200 imp/kWh. In the paper, component stress method is used to calculate the failure rate of each functional unit. The failure rate of each component can be get through GJB/Z 299C [9]. The prediction results of each units are as shown in Table 1.

From Table 1, we can get the smart meter' failure rate is 6748.609 FIT. The MTTF can be calculated by equation (4):

$$MTTF = \frac{1}{6478.6 \times 10^{-9} \times 8760} = 16.9(year) \tag{7}$$

It can be seen from Figure 1 that the display module, power supply module and carrier module are higher failure rate. The proportion of each module failure rate are 22%, 21%, 18%. Hence, the reliability life can be improved through increasing these three modules reliability.

Table 1. Failure rate of each module.

Modules	Measurement	Power	Cost-control	Communication	Display	Function	Carrier
Failure rate (FIT)	637.06	1422.087	911.584	428.661	1471.862	682.48	1194.375

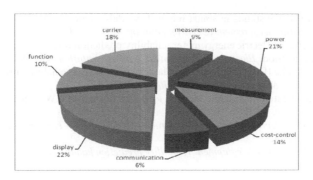

Figure 1. The proportion of each module failure rate.

Table 2. Accelerated life test.

Type of failure	Failure time	Number of failure sample	Test result
Excessive error	3rd period	2	MTTF ≥ 15

4.2 Accelerate life test results

Based on IEC62059-31, the parameter values are as follows: Ea = 0.9, n = 3. The temperature at used condition is 23°C, relative humidity is 60%. In this paper, the accelerated test conditions are chosen [75°C, 80%]. According to the Peck's temperature-humidity model, the accelerated factor is AF = 456.177.

According to accelerated factor, 288 h in accelerated life test is equivalent to 15 years at used conditions. Put the forty-three qualified single-phase smart meters into Alternating high and low temperature & humidity laboratory with 220 V, 5 A voltage and current. These smart meters are accelerated for three period, each period is 96 h.

Table 2 shows that two meters are fault because of exceed error tolerance. But the fault number of smart meter are not satisfied test termination conditions, so the sample meters' MTTF is more than 15 years.

5 CONCLUSION

Based on single-phase smart meters produced by one manufacture, the paper studies the reliability life and gets the following conclusions:

1. The MTTF of meters is 16.9 years through component stress method for smart meter reliability. the results are consist with the results of accelerated life test (15 years)
2. It can be seen by the meter prediction, the display module, power supply module and carrier module are higher failure rate. The smart meter reliability can be improved by increasing the reliability of three modules.

REFERENCES

[1] JB/T 50070-2002, *Reliability requirements and reliability compliance test for electrical energy meters.*
[2] Charles E. 2008. *An introduction to reliability and maintainability engineering.* Beijing: Tinghua university press.
[3] IEC 62059–21–2001, *Electricity metering equipment–Dependability–Collection of meter dependability data from the field.*

[4] Li X.F. 2010. IEC62059 stands in smart meter dependability prediction and verification methods application. *Electrical measurement & instrumentation.* 47(1):75–80.
[5] Ju H.J. 2013. Intelligent electric energy meter reliability prediction research and application based on component stress method. *Electrical measurement & instrumentation.* 50(11), 7–11.
[6] Michael P.A. 2002. The IEEE standards on reliability program and reliability prediction methods for electronic equipment. *Microelectronics Reliability.* 9(2), 76–82.
[7] Pu Z.Y. 2008. Accelerated life test and method for reliability test of KWH meter. *Electrical measurement & instrumentation.* 45(12):63–66.
[8] IEC 62059-31-2006, *Electricity metering equipment–Dependability–Temperature and humidity accelerated reliability testing.*
[9] GJB/Z 299C-2006, *Electronic equipment reliability prediction handbook.*

Advances in Engineering Materials and Applied Mechanics – Zhang, Gao & Xu (Eds)
© 2016 Taylor & Francis Group, London, ISBN 978-1-138-02834-0

Study application of Wide Areas Measurement technology on power system

F.S. Wu, D.W. Hao & X.J. Liao
Wuhan Mechanical College, Wuhan, Hubei, China

ABSTRACT: With the development and improvement of China's electric power system, a huge UHV AC and DC hybrid intelligent grid gradually formed. But the security and stability of this large power grid also faces enormous challenges. Wide area measurement system can monitor and analyze the operating state of the power system in dimensions of both time and space. Therefore, it is quite necessary to study WAMS technology. Begin with the structure and principles of WAMS, the paper researches the development in the past five years of WAMS in power system dynamic monitoring and fault analysis, identification and estimation, system stability control, and examines in detail the existing problems. At the end o, the paper mentions the further study issues of the wide area measurement techniques.

1 INTRODUCTION

Under the guidance of the principle "west electricity to east, north and south for mutual, national network", and with the expansion of China power grid and construction of projects like UHV power grid and west electricity to east, China is speeding up the building of inter-regional interconnected power grid adapt to large-scale power transmission. Cross-regional interconnected grids and long-distance transmission, on the one hand, can strengthen the links of grid among different regions; on the other hand, reveals such new problems threatening the safety and stability of grid as low frequency range oscillation, thus proposing new and higher requirements for stable operation and control of grid.

Based on synchronized phasor measurement technology and modern communication technology, WAMS can monitor and analyze the operation status of the power system in the vast areas, serving its real-time operation and control.[9] Information acquired through Wide-area measurement has the advantages in time synchronization and space. The results can be synchronized phasor measurements based on GPS time scale, which greatly improved the power system observability. Data obtained by WAMS has the following three features: First, synchronization in time. WAN interconnection grid will bring new grid transient and dynamic problems, which the original steady state monitoring system unable to handle. But they can be effectively improved by the synchronization of WAMS. Second, wide space. WAMS can obtain data of power grid in wide areas on the ground of synchronization, realizing real-time monitoring and treatment of wide-area information. Third, directly measured phase angle. Compared with EMS, WAMS are more accurate in measurement. The large amount of wide-area information provides data foundation for system control.

2 PMU/WAMS BASIC TECHNIQUES

2.1 *Phase angle measurement algorithm*

PMU is the foundation for realizing the application of WAMS, while phase angle measurement algorithm is the core of the PMU. Excellent PMU algorithm can improve the accuracy of measurement data; the accuracy of the algorithm directly affect the stability of the power system control, fault analysis and relay protection and other advanced applications accuracy. To modify the measurement accuracy of the algorithm to meet the requirements of engineering applications under dynamic conditions, domestic and foreign scholars have proposed a variety of phasor measurement algorithms and achieved some success. At present, the common phasor measurement algorithms are: zero-crossing detection method, a discrete Fourier transform (Discrete Fourier Transform, DFT), Kalman filtering, the instantaneous value calculation method, small phasor method, wavelet transform, minimum two multiplication. When the system is in steady state, the various methods described above are able to achieve a considerable precision, meeting the needs of general engineering applications. While as the system is in a dynamic situation, how quickly and accurately measure the corresponding phase angle turn to be the focus of each algorithm for research and improvement.[12] The structure of PUM is shown in Figure 1.

2.2 WAMS

Wide Area Measurement System (WAMS) consists of the master server in dispatch center, sub-stations distributed in key nodes of the system and high-speed digital communication network which connect the above two with each other. Its topology commonly adopts

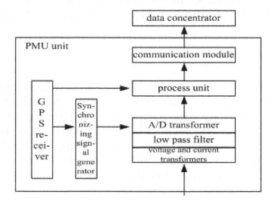

Figure 1. PUM structure diagram.

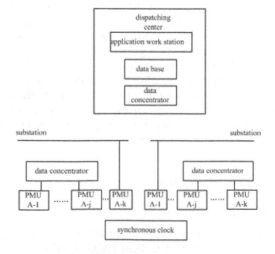

Figure 2. WANS structure schematic diagram.

master—sub-station tree structure. The sub-station means the synchronized phasor measurement device installed by power plant or transformer substation. WAMS demands for reliability, accuracy and timeliness as for synchronized phasor measurement technology. Master station refers to master control center, usually made up by basic platform and its advanced application functions. WAMS is expected to overcome the existing shortcomings on the local and distributed control system so as to achieve global optimization and coordinated control. The structure scheme of WANS is shown in Figure 2.

3 APPLICATIONS OF WAMS

3.1 *Dynamic monitoring process*

Fully based upon real PMU data and not dependent on power system simulation model, wide dynamic monitoring and fault analysis can reflect the true dynamic behavior of the whole network. At home and abroad dozens of WAMS systems have been put into practical operation. As a basic function of WAMS, dynamic monitoring provides real-time running information to dispatch center, which delivers important data for analyzing system security and stability as well as the reasons for failure. During the dynamic process of monitoring, alarming and analysis, the system detects the operation of the power grid on-line according to the real-time dynamic measurement data of key transformer substation in the grid, and gives limitation (static or dynamic) alarm when the voltage, phase angle and frequency appear abnormal. The abnormal operation status, characteristic signals and the type of faults can be observed, identified and analyzed. Observation and analysis can provide the technical support for strategy making to stably control the power system.[14]

3.2 *State estimation*

There are still some unresolved problems as for the SCADA-based state estimation. Firstly, due to time constraints of acquisition and transfer, state estimation needs relatively long period, on minute level. Besides, the same section data lack accuracy resulted from non-uniform time scale. Secondly, the nonlinear state estimation model has convergence and rapidity problem; thirdly, many a bad measurement related data indicate the lack of an unified and effective approach for detection and identification. The application of WAMS provides a guideline to solve the above problem. In recent years, due to the spreading of PMU, state estimation based on PMU and SCADA mixed measurement has become an active research direction. In addition, the introduction of PMU measurements data also push forward the research on dynamic state estimation, harmonic wave state estimation and the detection and identification of bad data. Currently, extended Kalman filter algorithm is commonly employed in the dynamic state estimation of power system, which combines the advantages of dynamic state estimation with that of PMU measurement to estimate the effect of such factors as the number of PMU, installation location and measurement weight on the accuracy of dynamic state estimation.[2–4]

3.3 *Line parameter identification*

Development and diffusion of WAMS technology offer a new angle for identify line parameters. A variety of identification algorithms have been put forward both at home and abroad, such as least-squares method, Kalman filtering, analog evolution method and maximum likelihood method. These algorithms all reflect to some extent the dynamic parameters of running transmission lines under the influence of various factors. The asymmetric incremental method can provide positive sequence parameters of the line. When the load is in bad imbalance or asymmetric failure occurs near the measured lines, it can identify zero-sequence parameters on line. Based on uniform transmission line equations and PI-shaped precise equivalent circuit, online estimation method of transmission line parameter is given after the voltage of both ends and current phasor is known. Based on sliding window total least squares, line parameter identification algorithm handles PMU data with sliding window, turning the objective function into window error minimum square hence having stronger noise immunity; at the same time, it performs respectively nuclear density estimation and point estimation to the confidence of numerous identification results so that gets the statistical results of parameter identification value.[1][5]

3.4 *Generator parameter identification*

Currently there are frequency identification method and time domain identification method for identifying synchronous generator parameter. After additional disturbance is added to running generators, dynamic PMU data records identify parameters. When the power angle can be measured, the identifiability of OLFR methods gets analyzed, and a comparison between the methods with time domain identification method is made. The paper makes an overall identification to multiple generator parameters according to the power angle trajectory of the generator collected WAMS.[10]

3.5 *Identification of excitation system and governor parameter*

Performance and parameters of the excitation system of a synchronous generator affect the static and dynamic of the system. Generator excitation system parameters usually affect the system's dynamic stability though the influence on the dynamic voltage. Slower response of governor exerts more influence on steady-state value after system disturbances. Because PMU can clearly record the waveform and the measured data of failures, constantly correction and fit made by comparing the PMU data can realize the parameter identification for excitation system and the governor.[6][13]

3.6 *Load parameter identification*

Load characteristics exert important effect on the simulating results of power system. Different characteristics have different implications to transient stability, voltage stability and low frequency oscillation. Synchronous data of power system are used in modeling, network simplification and load model identification of load areas, putting forward the concept of an integrated regional power system load model based on WAMS, and analyzing the parameter identifiability of the model. A load parameter identification method is proposed based on integrated WAN information, which taking generator relative angle as wide-area information, giving a comprehensive analysis to the trajectory sensitivity of generator power angle to load parameters under different conditions of running and failure. The analysis of sensitivity can determine the identifiability of the power angle participating in load parameter identification as well as others, so that disturbance location in favor of parameter identification can be chosen.

3.7 *Frequency stability and low frequency load shedding*

There have existed many studies on WAMS real-time information monitoring or forecasting voltage and frequency stability, but few research have conducted as for utilizing WAMS on stable control. The current study on static stability control mainly focus on load shedding. The minimum sampling time of WAMS system is 10 ms, the most suitable data source for low frequency load shedding, which can avoid slow response or too much load shedding. On the foundation of real-time wide-area data computing power disturbances provided by WAMS, we propose a two-stage adaptive load shedding strategy involving both frequency stability and voltage stability.

3.8 *Voltage stability assessment*

Online voltage stability control based on WAMS generally adopts local index method of online measurement. Researches in this area in recent years mainly concentrate on local index introduction which using wide area information. Voltage stability index is usually a variable fluctuating in the range of 0–1. The smaller is the value, the better the system voltage stability. When the index reaches 1, the system becomes instable. With local phasor information made up by the synchronized voltage phasor of the monitored load nodes and its connected equivalent voltage phasor, an equivalent two-node system is built according to simple transmission line model. Synchronized voltage phasor of the monitored load nodes as well as the half distance between the projection and amplitude of the equivalent voltage phasor are taken as the index to measure the stability of voltage.[7][8][11]

Prospect of Wide Area Measurement Technology

With the development of cross-interconnected power system, the traditional analysis tools and control strategies have been unable to meet the actual requirements. Therefore, safety monitoring stability control system based on WAMS need to be constructed. As the merits of wide area measurement techniques applying to power system stability control gradually are reflected, WAMS technology will have a bright future. It can be proved in the following areas:

1. Currently WAMS platforms are built on the provincial level. Integrated Dynamic information platform of cross-provincial or cross-regional power grid should be made if cross-regional power grid needs to analyze, monitor or real-timely and dynamically control.
2. Advanced applications based on WMAS platform shall be developed for dynamic security analysis, stability pre-warning and decision-making to the system.
3. WAMS-based real-time closed-loop control. Power systems get dynamic stability monitor and analysis on the dynamic information platform so that online security and stability control strategy can be pre-made.

4 CONCLUSION

This paper summarizes the researches and applications of WAMS technology on network dynamic monitoring of power system, state estimation, power system dynamic model identification and model calibration, voltage and frequency stability monitoring and control, etc. It forecasts the prospect of WAMS, indicating that building wide safety monitoring and control system of large-scale interconnected power systems is a right direction for research.

REFERENCES

[1] Bi Tian-shu, Ding Lan, Zhang Dao-nong. 2011, Transmission line parameters identification based on moving-window TLS. *Journal of Electric Power Science and Technology*, 5(2): 10–15.
[2] Chakrabarti S., Kyriakides E., Ledwich G. 2010, Inclusion of PMU current phasor measurements in a power system state estimator. *Generation, Transmission & Distribution, IET*, 4(10): 1104–1115.
[3] Castro Vide P.S., Maciel Barbosa F.P., Ferreira I.M. 2011, Combined use of SCADA and PMU measurements for power system state estimator performance enhancement, *Proceedings of the 2011 3rd International Youth Conference on Energetics (IYCE)*, 2011: 1–6.
[4] Defu N., Bing Y., Weimin Z. 2011, A new algorithm for power system state estimation with PMU measurements, *International Conference on Mechatronic Science, Electric Engineering and Computer (MEC)*, 2011: 114–117.
[5] Fan Qi, Mu Gang, Wang Ke-ying. 2002, Studies on lines' parameter measurement based on synchronized phasor measurements. *Journal of Northeast China Institute of Electric Power Engineering*, 26(4): 1–5.
[6] Guo Xiao-long, Xu Guo-yi, Xue An-cheng. 2008, Research of excitation system parameter identification based on PMU, *Proceedings of the 24th Conference on Power System and Automation Among Colleges and Universities of China*, 2008: 4–5.
[7] Gong Wei-zheng, Fang Xin-yan. 2011, A on-line voltage stability index based on WAMS. *Power System Technology*, 35(4): 71–75.

[8] Huang Zhi-gang, Wu Wei, Haan Ying-duo. 2002, An algorithm for voltage stability assessment based on synchronized phasor measurement. *Automation of Electric Power Systems*, 26(2): 28–33.

[9] Jin Xi, Wu Wen-hui, Wu Shi-min. 2010, Wide area measurement technology based on PMU and its applications. *Power System and Clean Energy*, 26(10): 10–13.

[10] Jin Yu-qing, Yue Chen-xi, Zhen Wei. 2007, On-line frequency response method of synchronous machine parameter identification based on rotor angle measurement. *Automation of Electric Power Systems*, 31(4): 7–11.

[11] Liao Guo-dong, Wang Xiao-ru. 2009, An on-line voltage stability monitoring method based on WAMS. *Proceedings of the CSEE*, 29(4): 8–13.

[12] Mai Rui-kun, He Zheng-you, Bo Zhi-qian. 2009, Research on synchronized phasor measurement algorithm under dynamic conditions. *Proceedings of the CSEE*, 29(10): 52–58.

[13] Shen Feng, He Ren-mu, Xie Yong-hong. 2008, Study on feasibility of excitation system parameter identification based on actual measurement of disturbance. *Power System Technology*, 32(10): 69–73.

[14] Xin Lei, Xiao Guang-ming. 2007, Wide area real-time dynamic monitoring system and its application to Northeast Power Grid. *Northeast Electric Power Technology*, 28(9): 11–14.

Author index